VTH INTERNATIONAL CONFERENCE ON TIMES OF POLYMERS (TOP) AND COMPOSITES

VTH INTERNATIONAL CONFERENCE ON TIMES OF POLYMERS (TOP) AND COMPOSITES

Hotel Continental Terme, Ischia, Italy 20-23 June 2010

EDITORS

Alberto D'Amore
The Second University of Naples-SUN

Domenico Acierno
University of Naples Federico II

Luigi Grassia
The Second University of Naples-SUN

SPONSORING ORGANIZATIONS
The Second University of Naples-SUN
University of Naples Federico II

AIP
American Institute of Physics

Melville, New York, 2010
AIP CONFERENCE PROCEEDINGS ■ 1255

Editors

Alberto D'Amore
Luigi Grassia

The Second University of Naples - SUN
Department of Aerospace and Mechanical Engineering
Via Roma 19,
81031 Aversa (CE), Italy

E-mail: alberto.damore@unina2.it
luigi.grassia@unina2.it

Domenico Acierno
University of Naples Federico II
Department of Materials and Production Engineeirng
Piazzale V. Tecchio,
80125, Naples, Italy

E-mail: acierno@unina.it

L.C. Catalog Card No. 2010906228

ISBN 978-0-7354-0804-3
ISSN 0094-243X

Printed in the United States of America

CONTENTS

Thermal Degradation and Combustion Behavior of Polypropylene/MWCNT Composites 1
G. E. Zaikov, A. D. Rakhimkulov, S. M. Lomakin, I. L. Dubnikova, A.N. Shchegolikhin, and E. Ya. Davidov

Elasticity and Inelasticity of Hard-Phase Reinforced Polyurethane Elastomers: From Sensitivity to Chemical and Physical Structure to Time Dependent Phenomena 4
C. Prisacariua, E. Scortanua, and C. P. Buckley

Investigation of Model Fuel Effects on Thermal Oxidation of Polyethylene 7
F. Djouani, E. Richaud, B. Fayolle, and J. Verdu

Radiochemical Ageing of Aromatic Polymers PEEK, PSU and Kapton® 10
E. Richaud, L. Audouin, X. Colin, C. Monchy-Leroy, and J. Verdu

Polymer-Dopant-Systems: Tailoring of Optical and Thermomechanical 13
T. Hanemann and K. Honnef

Multiblock Polyurethanes in Biomedical Applications: Fine Tuning of Degradation and Biomimetic Properties 16
G. Ciardelli, S. Sartori, A. Silvestri, P. Serafini, and F. Boccafoschi

Time Issues in Semi-Crystalline Thermoplastics Processing 19
M. Iannone

Properties of a Carbon-Fibre Composite Modified by Electrospun Poly (Vinylidene Fluoride) 22
K. Magniez, C. De Lavigne, and B. L. Fox

Strain Recovery Behavior Of Post-Yield Deformed Epoxy Resins: Effect Of The Resin/Hardener Ratio 25
S. Pandini and F. Bignotti

Synthesis and Characterization of Poly(ethylene glycol)-block-Poly(N_{α}-Boc- L-tryptophan) Copolymers 28
A. S. Voda and Q. Guo

Studying the Effect of Adsorbent Dosage on the Removal of Arsenic from Aqueous Solution by Using Polyaniline Nanocomposite 31
H. Eisazadeh, M. S. Lashkenari, and B. Davodi

Three-Dimensional Model for Whipping Motion in the Processing of Microfibers 34
Y. C. Zeng, Y. F. Sun, and X. H. Wang

Computational Protocols for Viscoelastic Material Property Characterizations Without the Use of Poisson's Ratios 37
M. Michaeli, A. Shtark, H. Grossbein, and H. H. Hilton

Design Feature Sensitivity Analysis in Numerical Modeling of Extrusion Spider Dies 40
A. G. Mamalis, G. Kouzilos, and A. K. Vortselas

In-Situ Monitoring of Polymer Crystallization by Polarization-Sensitive Optical Coherence Tomography 43
P. Hierzenberger, D. Stifter, E. Leiss-Holzinger, and G. Eder

Virus Sorbents Based on Polyaniline Interpolymer Complexes, Composites and Their Sorption Properties 46
V. F. Ivanov, V. T. Ivanova, Y. E. Kurochkina, O. L. Gribkova, M. V. Ilyina, A. A. Manykin, A. A. Isakova, and I. Yu. Sapurina

Ionic Conductivity of Membranes Based on PVdF-HFP 49
A. Garcia-Bernabé, M. Gil-Agustí, J. P. González-Gutiérrez, and A. Quijano- López

Adhesion Peptide Immobilization on Electrospun Polymers: A Photoelectron Spectroscopy Study 52
G. Iucci, G. Polzonetti, F. Ghezzo, M. Modesti, M. Roso, and M. Dettin

Evaluation of Installation Quality on Polyethelene Pipe Performance 55
N. U. Terzi, N. Dusunceli, F. Yılmazturk, and S. Yıldırım

Cyclic Behavior of High Density Polyethylene (HDPE) 58
N. Dusunceli, B. Aydemir, and N. U. Terzi

Joint Strength Control at the Fiber/Matrix Interface During the Production of Polymer Composite Materials Reinforced with High Performance Fibers 61
V. V. Kudinov and N. V. Korneeva
Assessing the Progress of Degradation in Polyurethanes by Chemiluminescence 64
M. Malíková, J. Rychlý, and K. Csomorová
Alternative Synthetic Routes for the Preparation of PLA/Montmorillonite Nanocomposites 67
S. Bronco, M. B. Coltelli, S. Castiello, F. Ciardelli, E. Taburoni, and L. Conzatti
Study of Polymer Glass Transition and Segmental Motions in Partially Crystallized Poly(ethyl acrylate)/p-xylene Mixtures 70
A. T. Stathopoulos, A. Kyritsis, F. J. R. Colomer, J. L. Gómez Ribelles, N. Shinyashiki, C. Christodoulides, and P. Pissis
Comparative Study Between Langmuir Polymer Monolayers Behavior and Their Chemical Structures 73
L. Gargallo, N. Becerra, A. Leiva, C. Sandoval, and D. Radic
Study of The Porosity Occurrence in Composite Materials by Means of Ultrasonic and In-Plane Shear Testing 76
C. Toscano, C. Vitiello, and P. Caramuta
Confinement Effects on the Structure and Dynamics in Intercalated Polymer/Layered Silicates Nanohybrids 79
K. Chrissopoulou, A. Afratisa, S. Fotiadou, B. Frick, and S. H. Anastasiadis
Implementation of Relative Humidity Control in DMA Measurements 82
M. Coletti
PP Composites with Hybrid Nanofillers: NTC Phenomenon 85
J. Sarlin and K. Immonen
Conjugated Polymer and Luminescent Nanocrystals for Ink-Jet Printing 89
E. Binetti, C. Ingrosso, M. Striccoli, P. Cosma, A. Agostiano, J. Y. Kim, J. Brugger, and M. L. Curri
Au Based Nanocomposites Towards Plasmonic Applications 92
A. Panniello, M. L. Curri, T. Placido, V. Reboud, N. Kehagias, C. M. S. Torres, D. Mecerreyes, A. Agostiano, and M. Striccoli
Direct Measurement of the Interfacial Attractions Between Functionalized Graphene and Polymers in Nanocomposites 95
M. Cai, A. J. Glover, T. J. Wallin, D. E. Kranbuehl, and H. C. Schniepp
On the Behavior of Dynamically Loaded Polymeric Materials 98
I. Emri and B. Zupancic
Consequences of Thermo- and Photo-Oxidation on End-Use Properties of Pure PE 101
J. Tireau, L. Van Schoors, K. Benzarti, and X. Colin
Inkjet Etching of Polymer Surfaces to Manufacture Microstructures for OLED Applications 104
I. A. Grimaldi, A. De Girolamo Del Mauro, G. Nenna, F. Loffredo, C. Minarini, and F. Villani
Side Chain Crystallization and Non-Equilibrium Phenomena in Nanophase Separated Poly(3-alkyl thiophenes) 107
S. Pankaj and M. Beiner
Comparing Nanofillers in Polylactide Nanocomposites 110
E. Kontou and M. Niaounakis
Intermolecular Effects in the Dynamics of Polymer Melts: Interplay of Cooperative Dynamics and Entanglements 114
M. G. Guenza
Luminescent Nanocomposites of Conducting Polymers and In-Situ Grown CdS Quantum Dots 117
C. Borriello, S. Masala, V. Bizzarr, G. Nenna, M. Re, E. Pesce, C. Minarini, and T. Di Luccio
iPP Crystallization: Micro and Nano Fillers Effects 120
E. Gioffredi, G. Cassulo, A. Frache, and P. L. Maffettone
A Comparative Study of Fiber Break-up During Compounding of Glass Fiber-Reinforced Polyamide in Batch and Continuous Mixers 123
F. Inceoglu, J. Ville, N. Ghamri, A. Durin, P. De Micheli, R. Valette, and B. Vergnes

Fabrication of a Novel Poly(3-Hydroxyoctanoate)/Nanoscale Bioactive Glass Composite Film with Potential as a Multifunctional Wound Dressing 126
R. Rai, A. R. Boccaccini, J. C. Knowles, I. C. Locke, M. P. Gordge, A. McCormick, V. Salih, N. Mordon, T. Keshavarz, and I. Roy

Plasticizer Migration from Cross-Linked Flexible PVC. 1. Effects on Tribology 129
M. Pannico, P. Persico, V. Ambrogi, and C. Carfagna

Epoxy Nanocomposites-Curing Rheokinetics,Wetting and Adhesion to Fibers 132
S. O. Ilyin, S. V. Kotomin, and V., and G. Kulichikhin

Hygrothermal Ageing of Pa 6,6 El Mazry 135
C. El Marzy, O. Correc, and X. Colin

Optimization of the Processing of HNBR/Organoclay Nanocomposite 138
R. M. Cadambi and E. Ghassemieh

Effect of the Clay Modification on the Thermal Properties of PVC 141
J. Zykova, A. Kalendova, L. Kovarova, and J. Malac

Thermal Conductivity and Heat Capacity of a Nylon-6/Multi-wall Carbon Nanotube Composite Under Pressure 145
J. Yu, B. Tonpheng, and O. Andersson

Hemp Reinforced Composites: Surface Treatment, Manufacturing Method and Fabric Type Effects . . 148
G. Cicala, G. Cristaldi, and G. Recca

Analysis of the Creep Behavior of Thin Polymer Films by Finite Elements Simulation of Micro-Bubble Inflation 152
S. Castagnet and E. Lainé

Thermal Aging of Electrical Wire Jacket 155
I. Mkacher, Y. Brument, and X. Colin

Tailoring Mater-Bi Properties by the Use of a Biowaste-Derived Additive 158
P. Cerruti, G. Santagata, G. Gomez d'Ayala,, M. Malinconico, V. Ambrogi, C. Carfagna, and P. Persico

PVC/$CaCO_3$ Nanocomposites: Influence of Nanoparticle Surface Treatment on Properties 162
I. Bonadies, M. Avella, R. Avolio, C. Carfagna, M. Emanuela Errico, and G. Gentile

Acoustic Emission Source Location in Unidirectional Carbon-Fibre-Reinforced Plastic Plates Using Virtually Trained Artificial Neural Networks 165
G. Caprino, V. Lopresto, C. Leone, and I. Papa

Rheological Behaviour, Mechanical Properties and Morphological Aspects of Thermoplastic Polyurethane Reinforced with Multiwalled Carbon Nanotubes 168
P. Russo, D. Acierno, and P. Spena

Effect of Silica Particles Concentration on the Physical Aging of PMMA/Silica Nanocomposites 172
V. M. Boucher, D. Cangialosi, A. Alegrìa, and J. Colmenero

Design of Novel Polymeric Materials by Controlled Self-Assembly 175
S. P. Piotto, S. Concilio, and P. Iannelli

Synthesis and Characterization of New Photoluminescent Oxadiazole/Carbazole Containing Polymers 178
S. Concilio, V. Bugatti, S. P. Piotto, and P. Iannelli

The Effect of processing on the PVC/Clay Nanocomposites Structure 181
A. Kalendova, J. Zykova, L. Kovarova, M. Slouf, and J. F. Gerard

Reaction in the Melt of Post-consumer Poly(ethylene terephthalate) (PET) with Ester Functionalized Polyolefins 184
M.-B. Coltelli, S. Savi, M. Aglietto, and F. Ciardelli

Effect of the Compounding Procedure on the Structure and Viscoelasticity of Polymer Nanocomposites 187
G. Capuano, G. Filippone, G. Romeo, and D. Acierno

Impact of Nanoparticles on the Microstructure and Properties of Immiscible Polymer Blends: Preliminary Investigations 190
G. Filippone and D. Acierno

Rubber Characterization with Fatigue and Crack Growth Analysis 193
L. Perier and A. Favier

A DMA Instrument with Versatile Capabilities for Rubber Testing 196
L. Perier and A. Favier

Production and Characterization of Active Transparent PET Films for Oxygen Sensitive Foods Packaging 199
M. R. Galdi and L. Incarnato
From Rheology of Nanocomposites to Rheology of Polymer Melts: Step Back or Forward? 202
V. Kulichikhin, and A. Malkin
Thermo-Mechanical and Electrical Properties of Epoxy Composites Containing Ceramic Nanofillers . . 206
A. M. Scamardella, E. Amendola, C. Petrarca, and D. Acierno
Two Timescales in Polymer Solidification: Processing vs Polymer Crystallization 209
S. Piccarolo, A. M. Poulose, and D. Carbone
Predicting the Photoageing and Photostabilization of Polymer Nanocomposites 213
S. Therias, S. Lonkar, F. Leroux, J. L. Gardette, N. T. Dintcheva, and F. P. La Mantia
Twinkling Fractal Theory of the Glass Transition and Yield Stres 215
R. P. Wool
Mass Transport in Nanocomposite Materials for Membrane Separations 218
M. Galizia, I. Puccini, M. Messori, M. G. De Angelis, and G. C. Sarti
Vibrational Properties of a Reactive Mixture Investigated During a Chemical Vitrification Process . . . 221
S. Caponi, S. Corezzi, D. Fioretto, A. Fontana, G. Monaco, and F. Rossi
Effect of Solvent Viscoelasticity on the Stress Induced Demixing 224
M. Minale
Dispersion and Characterization of Thermoplastic Polyurethane/Multiwalled Carbon Nanotubes in Co-Rotative Twin Screw Extruder 227
A. Benedito, I. Buezas, E. Giménez, and B. Galindo
Microcellular Foams Based on High Performance Thermoplastic Nanocomposites 231
L. Sorrentino, M. Gargiulo, G. Pezzullo, and S. Iannace
Nanotechnology and Food Contact Materials 234
C. D. Papaspyrides
Predicting the Ageing and the Long-Term Durability of Organic Polymer Solar Cells 237
J.-L. Gardette, A. Rivaton, S. Thérias, S. Chambon, M. Manceau, and J. Gaume
On Sliding Wear of Nanoparticle Modified Polymer Composites 240
K. Friedrich and L. Chang
Water Fluxes in Polymeric Membranes for Desalination via Membrane Distillation 243
G. A. Mannella, and V. La Carrubba
Using polarized Neutrons for Elastic and Dynamic Studies on Protein Systems 246
W. Häussler and J. Kindervater
Recycling of Chrome Tanned Leather Dust in Acrylonitrile Butadiene Rubber 249
S. H. El-Sabbagh and O. A. Mohamed
Luminescent Polymer Materials on the Basis of Aromatic Isocyanates and Open-chain Analogues of Crown Eithers 252
A. M. Gumerov, I. M. Davletbaeva, and A. I. Ahmetshina
Metalcoordinated Polyurethanes and Their Properties 255
I. M. Davletbaeva and R. S. Davletbaev
Investigation the Effect of Time Detention on the Removal of Chromuim from Aqueous Solution by Using Polyaniline/Polystyrene Nanocomposite 258
H. Eisazade, B. Davodi, and M. S. Lashkenari
Natural Fiber Reinforced PLA Composites 261
Y. S. Song, J. T. Lee, and J. R. Youn
FT-IR Investigation of Hoveyda-Grubbs 2nd Generation Catalyst in Self- Healing Epoxy Mixtures 264
L. Guadagno, P. Longo, M. Raimondo, C. Naddeo, A. Mariconda, V. Vittoria, G. Iannuzzo, and S. Russo
Self-Healing Structures in Aerospace Applications 267
L. Guadagno, M. Raimondo, G. Iannuzzo, and S. Russo
Hydrogen at Polymer Surfaces 270
J. Zemek, J. Houdkova, and B. Lesiak
On the Efect of the Oxidative Reagents on the Conductivity of Polyaniline/MMT Nanocomposites 273
A. Garcia-Bernabé, M. Gil-Agustí, G. Ortega, P. Llovera, A. Almarza, S. Vázquez, D. Amantia, and L. Aubouy
Blends of PEO in Poly(ethylene oxide)/Poly(2-vinyl pyridine)-b-poly(ethylene oxide) 276
E. Araneda, L. Gargallo, A. Leiva, N. Hadjichristidis, I. Mondragon, and D. Radić

Calorimetric Characterizacion of PLLA-MWCNTs and PLLA-MWCNTs **278**
C. Albano, G. Gonzalez, A. Karam, and M. Covis
Effects of Electron Beam Irradiation on Polyamide 12 with Fiberglass Reinforcement **281**
J.-P. Jeun, B.-S. Shin, H.-B. Kim, Y.-C. Nho, and P.-H. Kang
Preparation of Ag Nanoparticles on PAN/TiO_2 Nanofibers and Their Photocatalytic and Antibacterial Properties **284**
J.-P. Jeun, D.-K. Seo, H.-B. Kim, and P.-H. Kang
Crystallization of PCL/Clay Composites **287**
I. Janigová, D. J. Mošková, and I. Chodák
Effect of Diisocyanate Structure on Thermal Properties and Microstructure of Polyurethanes Based on Polyols Derived from Renewables Resources **290**
M. A. Corcuera, L. Rueda, B. Fernández d´Arlas, A. Saralegui, C. Marieta, A. Arbelaiz, I. Mondragon, and A. Eceiza
Experimental Design on Laminated Veneer Lumber Fiber Composite: Surface Enhancement **294**
U. Meekum and Y. Mingmongkol
Thermal Properties of PMMA and PS Filled with Organo-Silylated Montmorillonite **297**
A. De Maria, C. Borriello, and M. Schwarz
Poly(3-alkoxyphenylthiophenes) for VOCs Detection **300**
E. Massera, M. L. Miglietta, T. Polichetti, G. Di Francia, F. Borbone, A. Carella, S. Pappalardo, and A. Roviello
Infrared Thermography: Some Applications of Microencapsulated PCMs as Investigative Tool for Heat Transport Evaluation **303**
P. Persico, A. I. Renzi, and C. Carfagna
Effect of Carbon Nanotubes (CNTs) on Mechanical and Thermal Properties of Polypropylene Matrix . . **307**
J. Pascual, E. Fages, R. Balart, O. Fenollar, and T. Boronat
Mechanical Properties of Injection Moulded Binary Blends of Polyethylene with Small Additions of a Liquid Crystalline Polymer **310**
T. Ivanova, J. Zicans, I. Elksnite, M. Kalnins, and R. Maksimov
Improvement of the T-peel Strength of Polypropylene Adhesion Joints by Surface Photografting Pre-Treatment with Methyl Methacrylate **313**
R. Balart, L. Sánchez-Nácher, J. Balart, V. Fombuena, and J. M. España
Electrospinning of Drug-Loaded Polymer Systems: Preparation, Characterization and Drug Release . . **316**
G. Russo, V. Vittoria, G. Lamberti, and G. Titomanlio
Study of the Low-Pressure Plasma Effect on Polypropylene Nonwovens **319**
R. López, T. Boronat, M. Pascual, O. Calvo, and R. Balart
Improvement of Adhesion Properties of Low Density Polyethylene (LDPE) Substrate Using Atmospheric Plasma **323**
L. Sánchez-Nácher, D. García-Sanoguera, O. Fenollar, J. Balart, and V. Fombuena
Mechanical and Morphological Characterization of Novel Vinyl Plastisols with Epoxidized Linseed Oil as Natural-Based Plasticizer **326**
O. Fenollar, R. Balart, L. Sánchez-Nácher, D. García-Sanoguera, and T. Boronat
Mechanical Properties of Green Composites Based on Thermoplastic Starch **329**
F. Fornes, L. Sánchez-Nácher, O. Fenollar, T. Boronat, and D. Garcia-Sanoguera
Modeling and Stress-Strain Characteristics of Mechanical Properties of Carbon Nanotube Reinforced Polyvinylacetate Nanocomposites **333**
R. M. Meri, J. Bitenieks, M. Kalnins, and R. Maksimov
TiO_2/Polymer Nanocomposite Based Inks **336**
F. Loffredo, I. A. Grimaldi, A. De Girolamo Del Mauro, F. Villani, R. D'Amato, and C. Minarini
Investigation of PE Blend Films Through CIE L*C*h Color Scale **339**
M. Mohammadi, A. A. Yousefi, and M. Ehsani
Hole Injection Layer Containing Polyaniline-Poly (4-Styrenesulfonate) for Efficient Organic Light-Emitting Diodes **342**
A. De Girolamo Del Mauro, G. Nenna, V. Bizzarro, I. A. Grimaldi, and C. Minarini
Improvement of the Inkjet Printed VOCs Sensor Performances through the Sensing Layer Geometry . . **345**
A. De Girolamo Del Mauro, I. A. Grimaldi, F. Loffredo, E. Massera, T. Polichetti, F. Villani, and G. Di Francia

MWNT-Doped Epoxy Matrix for Detecting Impact Damages in Fiber Reinforced Composites by Electrical Resistance Measurements . . . 348
M. Monti, M. Natali, J. M. Kenny, and L. Torre
Silicone Containing Biomaterials in Cardiovascular Applications . . . 352
A. Silvestri, S. Sartori, P. Serafini, P. Ferrando, C. Mattu, S. Milione, F. Boccafoschi, and G. Ciardelli
Development of Polypropylene/Wood Flour Ecocomposites. Evaluation of Silane as Coupling Agent . . . 355
R. Bouza, L. Barral, M. J. Abad, and B. Montero
Photo-Oxidation Behaviour of EVA Antimicrobial Films . . . 358
L. Botta, R. Scaffaro, and F. P. La Mantia
Synthesis and Characterization of Linear Hyperbranched PBA Block Copolymers, as Potential Polymeric Plasticizers . . . 361
L. Ascione, V. Ambrogi, S. Battiato, C. Carfagna, P. Persico, C. Puglisi, and F. Samperi
Polylactide Nanocomposites as a New Solution for Eco-compatible Packaging Materials . . . 364
L. Incarnato, L. Di Maio, E. Garofalo, and P. Scarfato
A Study on the Gate Voltage Dependence of the Activation Energy in Meyer-Neldel Rule for Charge Mobility in Pentacene OTFTs . . . 367
M. Petrosino, R. Miscioscia, A. de Girolamo del Mauro, R. Rega, V. Cerri, C. Minarini, and A. Rubino
Porous Polyelectrolyte Hydrogels With Enhanced Swelling Properties Prepared Via Thermal Reverse Casting Technique . . . 370
A. Salerno and P. A. Netti
Weatherability Evaluation of Nanocomposite Polymeric Treatments for Surface Protection of Construction Materials . . . 373
P. Scarfato, M. Letizia Fariello, L. Di Maio, and L. Incarnato
Preparation, Characterization and Release Properties of Nanostructured Polystyrene Microparticles Containing Different Silicates . . . 376
P. Scarfato, P. Iannelli, P. Russo, and D. Acierno
Effect of clay surface silylation and dispersion method on the mechanical properties of epoxy-clay composites . . . 379
V. Romeo, F. Piscitelli, A. M. Scamardella, E. Amendola, M. Lavorgna, G. Mensitieri, and D. Acierno
Probing the Sequestering of Carbon Nanotubes in the PS Domains of SIS Block Copolymer Matrix using Electrostatic Force Microscopy . . . 382
L. Peponi, L. Valentini, and J. M. Kenny
Preparation and Characterization of Atactic Poly(vinyl alcohol)/Platinum Nanocomposites by Electrospinning . . . 386
W. S. Lyoo, Y. J. Lee, J. W. Cha, M. J. Kim, S. W. Joo, Y. S. Gal, T. H. Oh, and S. S. Han
Preparation of Syndiotactic Poly(vinyl alcohol)/Poly(vinyl pivalate/vinyl acetate) Microspheres with Radiopacity Using Suspension Copolymerization and Saponification . . . 389
W. S. Lyoo, J. W. Cha, K. Y. Kwak, Y. J. Lee, H. Y. Jeon, Y. S. Chung, and S. K. Noh
Modeling VOC Sorption and Transport in Glassy Polymeric Membranes . . . 392
M. Grazia De Angelis, L. Olivieri, and G. C. Sarti
Structural Relaxation of Amorphous PET Matrix Nanocomposites . . . 395
A. Greco, C. Esposito Corcione, A. Cavallo, and A. Maffezzoli
Torsional Stiffness Verification of an Adhesively Bonded Joint . . . 398
A. Annicchiarico, F. Caputo, G. De Angelis, F. Frascà, and G. Lamanna
Toughening the Macro Defect Free (MDF) Cements . . . 402
A. D'Amore and L. Grassia
The Anomalies of Glassy Polymers at Nanoscale from the Continuum Mechanics Perspective . . . 405
A. D'Amore and L. Grassia
Modeling the Residual Stresses in Reactive Resins-Based Materials: a Case Study of Photo-Sensitive Composites for Dental Applications . . . 408
L. Grassia and A. D'Amore

Tuning by Process of the Electrical Percolation Behavior of Multiwalled Carbon Nanotubes/Epoxy Composites . . . 411
G. Faiella, M. Zarrelli, V. Antonucci, and M. Giordano

Thermal Residual Stresses in Amorphous Thermoplastic Polymers . . . 414
L. Grassia and A. D'Amore

Isobaric PVT Behavior of Poly(Carbonate) (PC) . . . 417
L. Grassia and A. D'Amore

An Approach in Developing 3D Fiber-Deposited Magnetic Scaffolds for Tissue Engineering . . . 420
R. De Santis, A. Gloria, T. Russo, U. D'Amora, S. Zeppetelli, and L. Ambrosio

Author Index . . . 423

Thermal Degradation and Combustion Behavior of Polypropylene/MWCNT Composites

Zaikov G.E.[1], Rakhimkulov A.D.[1], Lomakin S.M.[1], I. L. Dubnikova[2], Shchegolikhin A.N.[1], Davidov E.Ya.

[1]*NM Emanuel Institute of Biochemical Physics of Russian Academy of Sciences, 4 Kosygin str., 119334 Moscow, Russia* Chembio@sky.chph.ras.ru
[2]*NN Semenov Institute of Chemical Physics of Russian Academy of Sciences, 4 Kosygin str., 119991 Moscow, Russia*

Abstract. Studies of thermal and fire-resistant properties of the polypropylene/multi-walled carbon nanotube composites (PP/MWCNT) prepared by means of melt intercalation are discussed. The sets of the data acquired with the aid of non-isothermal TG experiments have been treated by the model kinetic analysis. The thermal-oxidative degradation behavior of PP/MWCNT and stabilizing effect caused by addition of MWCNT has been investigated by means of TGA and EPR spectroscopy.
The results of cone calorimetric tests lead to the conclusion that char formation plays a key role in the mechanism of flame retardation for nanocomposites. This could be explained by the specific antioxidant properties and high thermal conductivity of MWCNT which determine high-performance carbonization during thermal degradation process.
Comparative analysis of the flammability characteristics for PP-clay/MWCNT nanocomposites was provided in order to emphasize the specific behavior of the nanocomposites under high-temperature tests.

Key-words: carbon nanotubes, composites, polypropylene, multi-walled, thermal degradation, combustion, melt intercalation, kinetic analysis, TG, TGA, ESR spectroscopy, cone calorimetric test, antioxidants, thermal conductivity.
PACS: 65.80.-g

INTRODUCTION

At present time the great attention is given to the study of properties of polymeric nanocomposites produced on the basis of well-known thermoplastics (PP, PE, PS, PMMA, polycarbonates, polyamides) and carbon nanotubes (CN). CNs are considered to have the wide set of important properties like thermal stability, reduced combustibility, electroconductivity, etc. [1-7]. Thermoplastic polymer nanocomposites are generally produced with the use of melting technique [1-12].

Development of synthetic methods and the thermal characteristics study of PP/multi-walled carbon nanotube (MWCNT) nanocomposites were taken as an objective in this paper.

A number of papers pointed at synthesis and research of thermal properties of nanocomposites (atactic polypropylene (aPP)/MWCNT) were reported [10-12]. It is remarkable that PP/MWCNT composites with minor level of nanocarbon content (1-5% by weight) were determined to obtain an increase in thermal and thermal-oxidative stability in the majority of these publications.

Thermal stability of aPP and aPP/MWCNT nanocomposites with the various concentrations of MWCNT was studied in the paper [10]. It was shown that thermal degradation processes are similar for aPP and aPP/MWCNT nanocomposites and initial degradation temperatures are the same. However, the maximum mass loss rate temperature of PP/MWCNT nanocomposites with 1 and 5% wt of MWCNT raised by 40° - 70°C as compared with pristine aPP.

RESULTS AND DISCUSSION

We believe that a higher carbonization effectiveness of MWCNTs depends on their heat conductivity. It is well known that, PP has a low thermal conductivity at standard conditions, and characterized by a minor increase with temperature up to melting point (~0.2 W/m K). On the other hand the heat conductivity of individual MWCNT is extremely high and equals to 3000 W/m K. During the high temperature pyrolysis of PP/MWCNT composition at temperatures above 300-400°C the heat conductivity can rise up to 20W/m K due to actual increase of MWCNT concentration in composition caused by volatilization of polypropylene degradation products (fig.1). The induced heating of PP/MWCNT intensifies a steady carbonization and charring of the samples and leads to decrease of RHR peak value (fig. 2).

CP1255, *V^th International Conference on Times of Polymers (TOP) and Composites*,
edited by A. D'Amore, D. Acierno, and L. Grassia

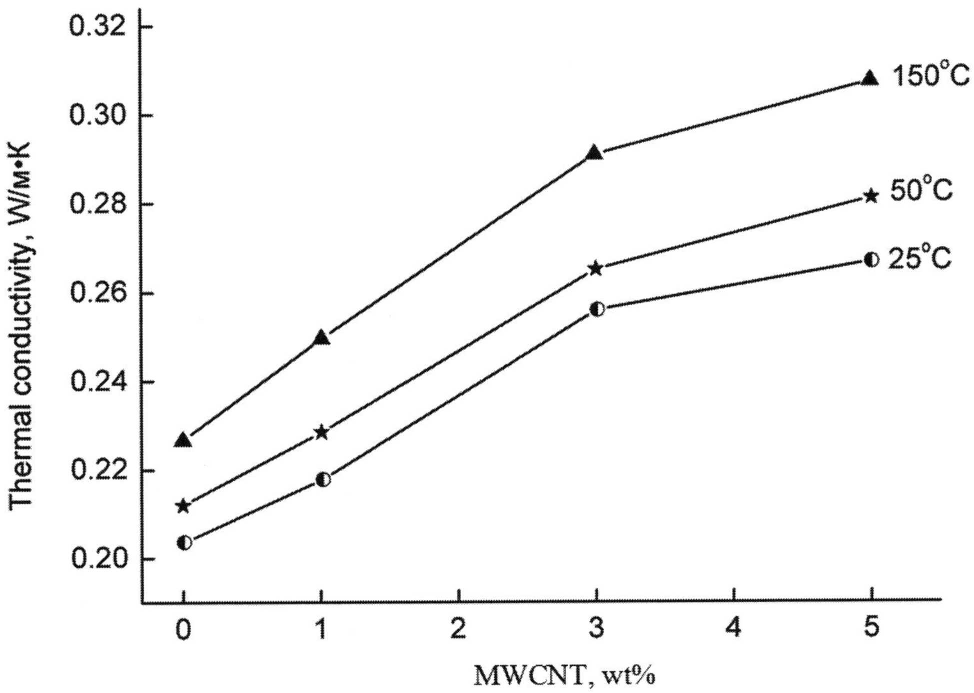

FIGURE 1. Temperature dependence of the thermal conductivity of PP/MWCNT (3) nanocomposite with different loadings of MWCNT.

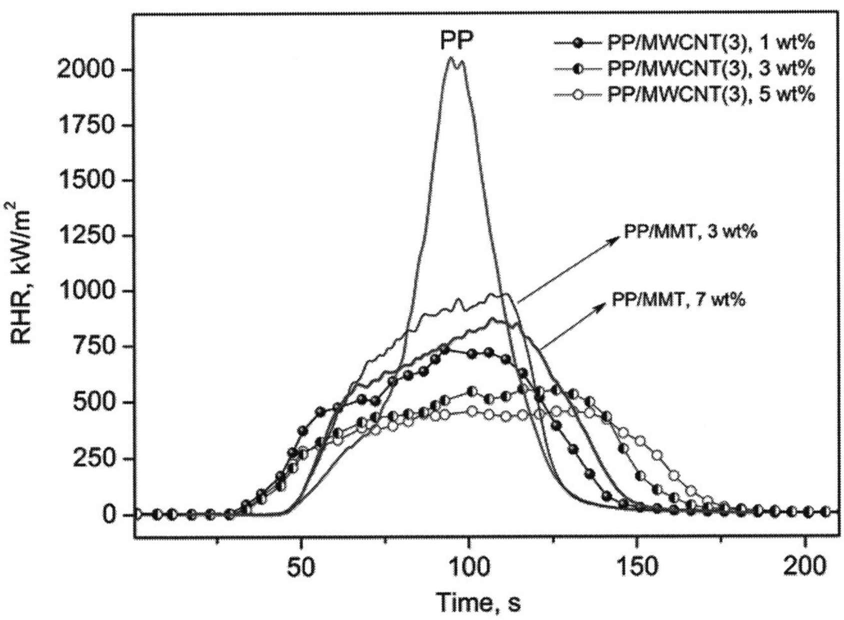

FIGURE 2. Rate of heat release vs. time for PP, PP/MWCNT and PP/MMT(Cloisite 20A) nanocomposites obtained by cone calorimeter at the incident heat flux of 35 kW m^{-2}

Fig. 3a shows EPR spectrum of the stable paramagnetic centers formed in the samples of PP/MCWNT (10% wt) heating in air at 350°C. When a heated in air PP/MCWNT specimen was placed into an EPR sample tube, a narrow singlet signal with a line width of $\Delta H_{1/2}$ = 0,69 mT and a g value of 2.003 was detected due to the stable radicals generation, analogous to those previously registered during polymers carbonization process. No EPR signal similar to that of PP/MCWNT samples were observed in the samples of pristine PP and MCWNT samples heated at 350°C in air. It should be noted that although iron impurity from MWNCT has been mentioned in other studies on pyrolysis of polymer nanocomposites as the radical traps, the EPR analyses in the current study showed the presence of paramagnetic centers relating to carbonaceous stable radicals only.

As it is seen from fig. 3b, the formation of stabilized radicals occurs with pronounced induction period which is related to antioxidant properties of MWNCT. Such a type of kinetic dependence is coincided with an oxygen uptake kinetics observed during inhibited polyolefines thermal oxidation. Moreover, no EPR signals were observed in the samples of the PP/MCWNT samples heated at 350°C in inert Ar.

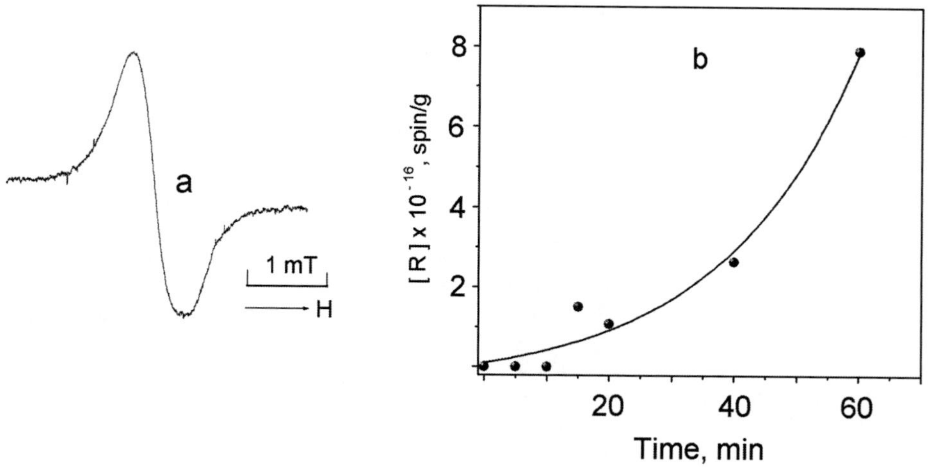

FIGURE 3. EPR spectrum of the stable paramagnetic centers formed in the samples of PP/MCWNT (10% wt) heating in air at 350°C – (a), kinetic dependence of stable radicals generation from PP/MCWNT (10% wt) under isothermal heating at 350°C in air – (b).

Thus, multi-walled carbon nanotubes are considered to be more effective filling agents than layered silicates in the terms of improvement of thermal properties and flame retardancy of PP matrix. This could be explained by the specific antioxidant properties and high thermal conductivity of MWCNT, which determine the carbonization reactions during thermal-oxidative degradation process.

REFERENCES

1. M.S.P. Shaffer and A.H. Windle, *Adv. Mater* **11**, 937-946 (1999).
2. D. Qian , E.C. Dickey, R. Andrews and T. Rantell, *Appl Phys Lett* **76**, 2868-2877 (2000).
3. Z. Jin, K.P. Pramoda, G. Xu and S.H. Goh, *Chem Phys Lett* **43**, 337-345 (2001).
4. E.T. Thostenson and T.W. Chou, *J Phys D: Appl Phys* 35, L77-84 (2002).
5. Y. Bin , M. Kitanaka, D. Zhu and M. Matsuo, *Macromolecules* **36,** 6213-6221 (2003).
6. P. Potschke and S.M. Dudkin, *Olig I. Polymer* **44**, 5023-5029 (2003).
7. B. Safadi , R. Andrews and E.A. Grulke, *J Appl Polym Sci* **84**, 2660-2671 (2002).
8. P.C.P. Watts, P.K. Fearon, W.K. Hsu, N.C. Billingham, H.W. Kroto and D.R.M. Walton, *J. Mater. Chem.* **13**, 491-505 (2003).
9. P.C.P. Watts, W.K. Hsu, D.P. Randall, H.W. Kroto and D.R.M. Walton, *Phys. Chem. Chem. Phys*. **4**, 5655-5662 (2002).
10. J. Yang, Yu. Lin and Ji. Wang, *J Appl Polym Sci* **98**, 1087-1094 (2005).
11. T. Kashiwagi, E. Grulke, J. Hilding, K. Groth, R.H. Harris, K. Butler, J.R. Shields and S. Kharchenko, *Douglas J Polymer* **45**, 4227-4239 (2004).
12. T. Kashiwagi, E. Grulke, J. Hilding, Jr.R.H. Harris and W.H. Awad WH, *Dougls J. Macromol Rapid Commun* **23**, 761-765 (2002).
13. G.E. Zaikov and S.M. Lomakin, *Combustion of Polymers*, Brill Academic Publishers, Utrecht-Boston, 2005, pp. 325.
14. G.E. Zaikov and S.M. Lomakin, *Flammability of Polymeric Materials*, Nova Science Publishers, New York, 2007, pp. 327.

Elasticity and Inelasticity of Hard-Phase Reinforced Polyurethane Elastomers: From Sensitivity to Chemical and Physical Structure to Time Dependent Phenomena

C. Prisacariu[a], E. Scortanu[a], and C.P.Buckley[b]

[a]*Petru PoniInstitute of MacromolecularChemistry of the Romanian Academy, Aleea Grigore Ghica Voda, Nr. 41 A, Iasi,700487, Romania*

[b]*Department of Engineering Science, University of Oxford, Parks Road, OxfordOX13PJ, UK*

Abstract. A study was made of how aspects of the constitutive responses of polyurethane elastomers vary with composition: the hard segment, soft segment and chain extender were varied systematically in a family of 18 segmented copolyurethane elastomers. Results were related to microstructural changes, on the basis of evidence from x-ray scattering (SAXS and WAXS). Inelastic effects were most pronounced when the hard segment crystallized, and when the phase segregation was least pronounced. Large variations were found in the responses to first loading to a given strain. Tensile modulus and work input increased significantly with degree of hard phase crystallinity, but were independent of degree of phase separation. First cycle hysteresis was found to increase with reduced phase separation. A quantitative correlation was found between the magnitude of the Mullins effect and the fractional energy dissipation by hysteresis under cyclic straining, giving a common relation that was approached by all the materials studied. The results provide new insight into the physical origin of inelastic effects in reinforced elastomers, to assist with the development of physically-based constitutive models.

Keywords: Polyurethanes; Diisocyanates; Physical-mechanical properties.
PACS: 60

INTRODUCTION

The most challenging targets remaining for the consitutive modelling of elastomers are those features of the constitutive response representing deviations from hyperelasticity: i.e. the hysteresis and time-dependence. These are small deviations in unreinforced elastomers but become prominent in the presence of filler particles, e.g. carbon black in natural rubber. In view of the hyperelastic nature of the crosslinked polymer matrix, it is clear they reflect complex interactions between the matrix and reinforcement. There has been much interest in achieving a better physical understanding of these interactions and in developing constitutive models to capture them. The present work represents a new approach to gaining insight into this problem. A study has been made of inelastic effects in the deformation of copolyurethane elastomers (PUs), where there is potential for formation of a two-phase microstructure (hard reinforcing particles embedded in an elastomeric matrix), but where the *nature* of the hard phase and the *degree* of phase separation can be controlled via the chemical structure and preparation conditions. The chemical components were: a hard segment (giving, on phase separation, hard nano-scale reinforcing particles); a soft segment (giving, on phase separation, an elastomeric matrix); and a chain extender. All three components were varied systematically and independently to create a family of 18 materials, and their constitutive responses were measured in cyclic tensile tests at room temperature, up to stretches in the range 5-6. Particular attention was paid to characterising the three inelastic features – hysteresis, Mullins effect and stress relaxation in interrupted tests – and their variations between the materials. The same materials were also studied by wide-angle and small-angle X-ray scattering (WAXS and SAXS) using synchrotron radiation, and by small-angle neutron scattering (SANS) to determine levels of crystallinity and phase segregation. The range of polymers was designed to reveal the roles of choice of macrodiol (MD) and of chain extender (CE) in determining the performance as an elastomer. The soft segment MD was either polytetrahydrofuran (PTHF) or poly(ethylene adipate) (PEA) of molar mass 2000±50. The CE used in the synthesis was ethylene glycol (EG), 1,4-butanediol (BG) or diethylene glycol (DEG).
The role of the hard segment structure was also investigated. In particular, two hard segments were compared, based on the diisocyanates: 4,4'-methylene bis(phenyl isocyanate) (MDI) and 4,4'-dibenzyl diisocyanate (DBDI) [1-4]. Rotation around the central $-CH_2$-CH_2- bridge in DBDI allows alignment of aromatic rings and hence crystallization within the hard phase, which is not available with MDI in melt-cast polyurethanes. As reported by ourselves [3] and

CP1255, *V*[th] *International Conference on Times of Polymers (TOP) and Composites,*
edited by A. D'Amore, D. Acierno, and L. Grassia

Lyman [4], the MDI molecule introduces the rigid - Ph-CH_2-Ph- moiety in the elastomeric PU hard segments. In contrast when using DBDI, the specific - Ph-CH_2-CH_2-Ph- moiety introduces a variable geometry into the hard segments due to the possibility of internal rotation of this isocyanate around the -CH_2-CH_2- ethylene bridge. This leads to the appearance of both "syn" and "anti" rotational conformations, which coexist in the DBDI based PU macromolecules, (Figure 1). As a result, in this latter case the PU macromolecules can adopt a more compact packing which enhances significantly the ability to order in crystalline structures involving predominantly the "anti" form. Thus, new polymers were achieved, with a controlled ordering of copolymer hard segment blocks on the macromolecular chain.

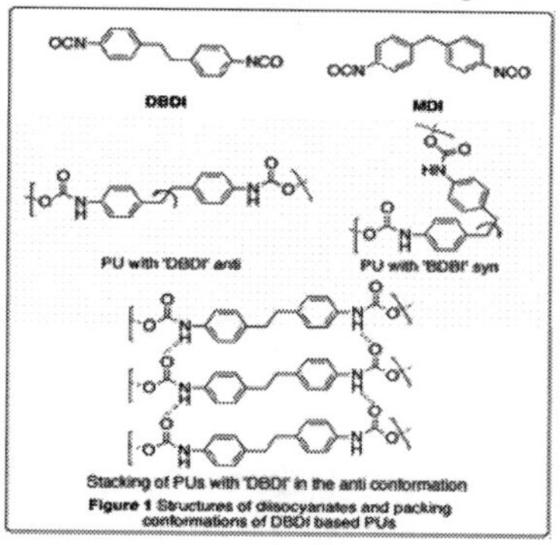

FIGURE 1. Packing conformations of MDI and DBDI.

RESULTS AND DISCUSSION

Structural Studies

WAXS and SANS measurements, and differential scanning calorimetry (DSC) of the as-moulded polymers revealed the presence of crystallinity in the DBDI-based materials (Figure 2.a.), as compared to other classical similar systems as are those derived from MDI (Figure 2.b.). There was evidence of the appearance of more or less complete phase separation associated with the formation of discrete crystalline domains.

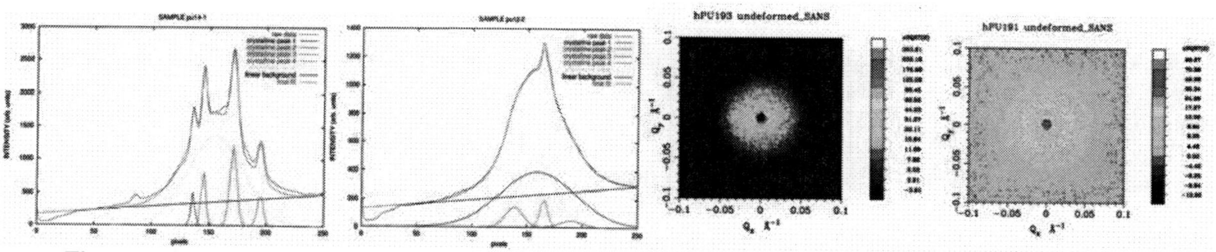

Figure 2.WAXS and SANS profiles: (a and c) Hard segments = DBDI χ=0.14; (b and d) Hard segments = MDI χ=0.

Mechanical Tests. Tensile tests results. Interpretation of mechanical response

Mechanical tests included load-unload cycles at constant rate of extension, with measurement of hysteresis and strain recovery, and stress relaxation tests. In the first phase of the study, the materials were *cycled to fixed extension*, between an extension of 3, and zero load, for 3 cycles, (Figures 3 and 4). Polymers based on DBDI hard segments, displayed higher stiffness and strength than the conventional MDI-based PUs. Lower strain recovery and strain energy recovery on cycling were observed in the case of PUs derived from dibenzyl structures. Several quantifiers of the inelasticity were calculated as shown elsewhere, [2]. These were the energy density expended on first loading to a nominal strain of 3, the relative first cycle residual strain, and the relative first cycle hysteresis energy dissipation.

The Mullins effect defined as a "Mullins factor" *M* was also studied, to have a quantitative measure of the extent to which the materials exhibit the classic response identified by Mullins. All the materials showed substantial deviations from hyperelastic response, indicating the presence of relaxation processes. The hard domains did not act simply as rigid crosslinks for the soft domains. Inelasticity measures laid in the same ranges. Similarly, all the materials showed a response on re-loading for one or more cycles that had significant Mullins-type character, with M in the range 0.8-0.95. Notable features were the pronounced hysteresis, unrecovered strain and Mullins effect. The residual strain and energy dissipation were highest for materials with hard segments of DBDI than for those of

MDI. Further mechanical tests were designed to compare inelasticity of the various materials under *cycling to increasing extension.* Each specimen was subject to a continuous pseudo-cyclic straining sequence, such that on the *n*-th cycle it was cycled between a maximum nominal strain εmax = *n* and zero load.

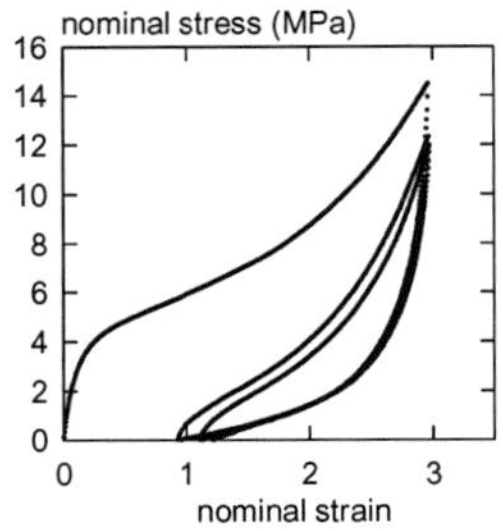

Figure 3. Example curves of nominal stress versus nominal strain when cycling to fixed extension for three cycles of a material based on MDI between zero stress and a nominal strain of 3.

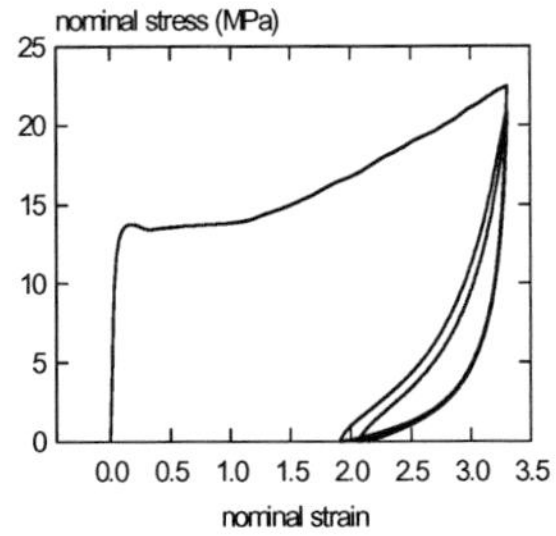

Figure 4. Same but for a material based on DBDI

Further mechanical tests were designed to compare inelasticity of the various materials under *cycling to increasing extension.* Each specimen was subject to a continuous pseudo-cyclic straining sequence, such that on the *n*-th cycle it was cycled between a maximum nominal strain εmax = *n* and zero load. Inelasticity measures, and *M* were determined for the *n-th* cycle in analogous manner to corresponding quantities defined above for the first cycle. For the MDI based PUs it observed a clear example of the Mullins effect whereby, on each cycle, the material appeared softer on reloading than during loading, until the previous maximum extension was reached, when the curve rejoined the previous curve. It was interesting to compare this result with corresponding data for DBDI based materials. Again the Mullins effect was visible, but the shape of the curves differed from the case of the MDI based PUs showing a more pronounced apparent yield phenomenon. The influence of the hard domain crystallinity was much more pronounced, leading to a doubling in the stress. These experiments revealed clearly that inelastic features in the cyclic stress-strain response of polyurethane elastomers are sensitive to microstructural detail on length scales of order nm.

Time dependent changes in the stress-strain data were followed in the case of crosslinked PUs. The evolution of mechanical properties in time, was compared with the. kinetics of the consumption of the free isocyanic (NCO) groups, kinetics of H_2O absorption and desorption and with the kinetics of CO_2 desorption resulted during the formation of urea group. CO_2 desorption appeared to be the slowest dynamic process and its rate influenced the rate of PU postcuring process.

CONCLUSIONS

The presence of DBDI hard segments instead of MDI led systematically to increases in: the input strain energy to a given elongation, hysteresis and residual strain under cyclic loading, and stress relaxation. Results showed that the Mullins effect correlated well with the relative strain energy loss by hysteresis, across the materials and, within one material, between successive cycles. Another interesting result was that hysteresis was reduced by more complete phase segregation (revealed by SAXS), but increased by increasing hard phase crystallinity (revealed by WAXS). The extent of stress relaxation in interrupted tests was found to increase with hysteresis and hence with hard segment crystallinity, reflecting the higher flow stress of the reinforcing hard domains. These and other observations have yielded insight into the physical origins of the inelastic features and provide pointers for a physically-based representation of them in constitutive models.

ACKNOWLEDGEMENTS

This work was supported by NATO Collaborative Linkage Grant CBP.EAP.CLG.981805.

REFERENCES

1. C.P. Buckley, C. Prisacariu, A. Caraculacu, *Polymer* **48**, 1388-1396 (2007)
2. C. P. Buckley, C. Prisacariu, A. Caraculacu A., Martin C.M.,"Inelasticity of hard-phase reinforced elastomers: A study of copolyurethanes with varying hard and soft segments", in *Constitutive Models for Rubber IV*, edited by P.-E. Austrell / L. Hari. ISBN 0415383463, Ed. Taylor and Francis, 2005, 465 – 470.
3. C. Prisacariu, R.H. Olley, A.Caraculacu, D.C. Bassett DC and C. Martin, *Polymer* **44**, 5407, (2003)
4. L.A. Gowerr, T.L.D. Wang and D.J. Lyman, *J. Biomater. Sci. Polymer Edn.* **6(8)**, 761, (1994).

Investigation of Model Fuel Effects on Thermal Oxidation of Polyethylene

Fatma DJOUANI, Emmanuel RICHAUD, Bruno FAYOLLE, Jacques VERDU,

Arts et Metiers ParisTech, CNRS, PIMM, 151 bd de l'Hôpital, 75013 PARIS. FRANCE.

Abstract. This work deals with a study of stabilizers extraction by ethanol-cyclohexane mixtures simulating ethanol based biofuels. It appears that the extractive power of cyclohexane is considerably stronger then ethanol one. In other words, from this point of view, ethanol based biofuels are less aggressive than a pure hydrocarbon.

Keywords: Biofuels, polyethylene, extraction, interaction, solubility, temperature,
PACS:

INTRODUCTION

Two recent trends in automotive industry, e.g. the increase of engine temperature and the increasing use of biofuels, lead to reconsider the problem of polymer-fuel interaction for all the polymer materials (polyolefins, polyamides, elastomers etc) used in motors.

The new biofuel components e.g. essentially ethanol or unsaturated esters, can affect the polymer thermooxidative stability by chemical way through cooxidation process or by physical way through stabilizer extraction. Published works on this topic are very scarce [1,2].

The aim of the present paper is to report a preliminary study focussing on the physical way only. It was chosen to study a high density polyethylene (HDPE) stabilized by a synergetic phenol-phosphite mixture, in the presence of ethanol-cyclohexane mixture having a boiling point (80°C) close to ethanol based biofuels ones.

EXPERIMENTAL

The polymer was HDPE with a density 935 kg m^{-3} at 25°C and average molar masses Mn=18 kg mol^{-1} and Mw=228 kg mol^{-1}. Films of 140µm thickness were used. Three model fuels were studied: Ethanol 100%; Ethanol-cyclohexane (50/50) and Ethanol-cyclohexane (10/90). The films were immersed in the model fuel under reflux at 80°C and periodically analyzed.

FTIR spectra were obtained on a Brucker IFS28 spectrometer in standard conditions.

DSC measurements were performed using Q10 apparatus (TA Instruments) with the following program: 1) Heating from ambient temperature to 200°C under nitrogen. 2) 5min at 200°C under nitrogen. 3) Oxygen admission at 0.1 MPa pressure and isothermal exposure at 200°C with recording of the oxidation exotherm. The end of the induction period is taken at the onset of the exotherm. The length of the induction period is directly related to the residual stabilizer concentration.

RESULTS & DISCUSSION

The effect of the model exposure on the induction time at 80°C is shown in Figure 1. In the chosen conditions, extraction is total, with both cyclohexane mixtures, in less of 30 minutes. In the investigated composition range, no influence of the composition was observed.

Extraction is considerably slower with pure ethanol. Furthermore, in this later case, the curve reaches a pseudo-asymptotic level corresponding to about one third of the initial OIT value.

CP1255, *Vth International Conference on Times of Polymers (TOP) and Composites,*
edited by A. D'Amore, D. Acierno, and L. Grassia

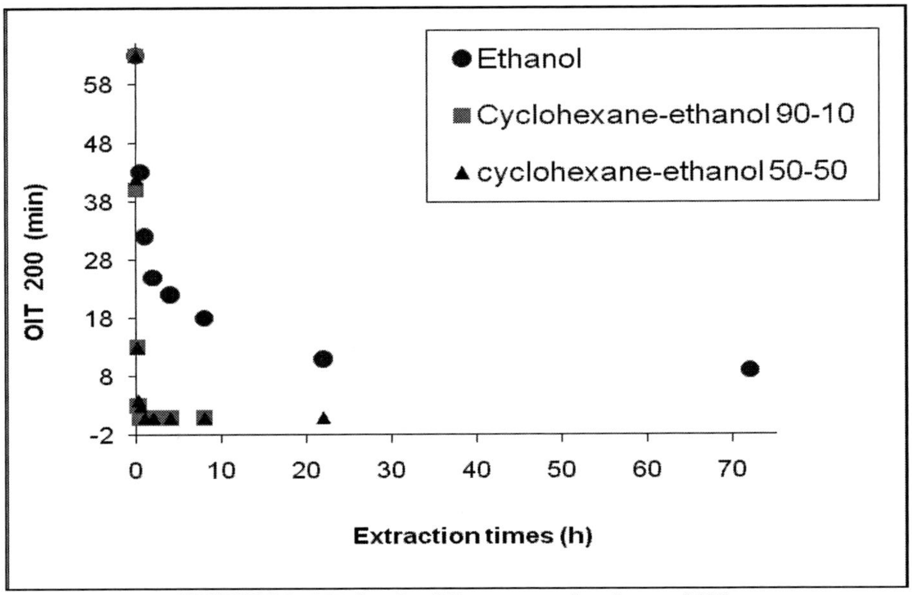

FIGURE 1. OIT data for different extraction times at 80°C.

Complementary information can be obtained from IR spectra in the 1170-1220 cm^{-1} region (Figure2). Two phosphite bands at 1191 and 1211 cm^{-1} can be observed in the initial spectrum. They disappear totally after 10 min in ethanol-cyclohexane mixtures. They disappear more gradually in pure ethanol but they have almost totally disappeared after 8h. These results can be interpreted as follows:

1. The extractive power of cyclohexane is considerably higher than the ethanol one that is not surprising since HDPE and cyclohexane solubility parameters are very close (δ=16.8$MPa^{1/2}$). Cyclohexane is expected to swell HDPE and thus to favor antioxidant diffusion.
2. In thin films, stabilizer transport is expected to depend of two terms:
   The diffusivity into the polymer matrix and the exchange at the polymer-solvent interface (extraction). The above result suggest that extraction at the polymer/ethanol interface is easier for the phosphite then for the second (phenolic) component of the stabilizing system. This difference could explain the existence of a non negligible induction time after 8h in ethanol while the phosphite has been almost entirely extracted.

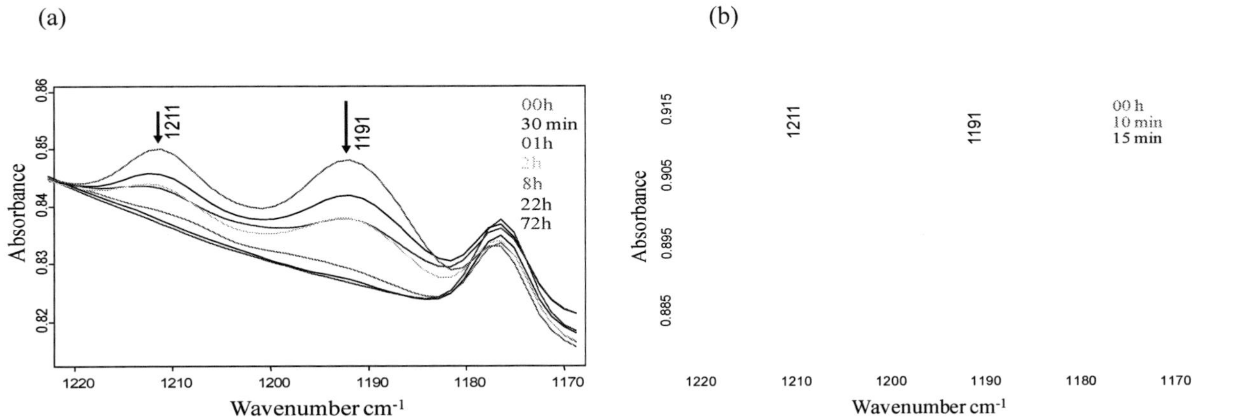

FIGURE 2. Phosphite extraction evolution of the infrared spectra FTIR spectra of a PE films upon exposure to (a) pure ethanol, (b) cyclohexane-ethanol mixture (50-50) at 80°C in the 1170-1220 cm^{-1} domain.

CONCLUSIONS

This study puts clearly in evidence the fact that ethanol displays a lower extractive power towards HDPE stabilizers than hydrocarbons. Furthermore, extraction could be partially selective, phosphite loss being faster than phenol one for the system under study. From the punctual point of view, it appears that, if the presence of alcohol in fuels alters the polymer thermooxidative stability, this will be not due to stabilizer extraction.

REFERENCES

1. M.M. Maru *et al.*, Biodiesel compatibility with carbon steel and HDPE parts, Fuel Processing Technology 90 (2009) 1175–1182
2. Biodiesel handling and use guidelines. Report DOE/GO-102006-2358 Third Edition September 2006. US department of energy, p. 16. http://www.nrel.gov/ vehiclesandfuels/npbf/pdfs/40555.pdf

Radiochemical Ageing of Aromatic Polymers PEEK, PSU and Kapton®

E. Richaud[a,b], L. Audouin[a], X. Colin[a], C. Monchy-Leroy[b] and J. Verdu[a]

[a] *Arts et Metiers ParisTech, CNRS, PIMM, 151 bd de l'Hôpitâl, 75013 PARIS. FRANCE.*
[b] *EDF R&D. MMC. EDF R&D, MMC, Site des Renardières, 77818 MORET SUR LOING. FRANCE.*

Abstract. This article deals with degradation mechanism of three aromatic polymers submitted to γ-rays in air at 60°C. T_g measurements and GPC results indicated that thin samples (thickness lower than 200 µm) undergo mostly chain scission whereas sol gel analysis and rheometric measurements showed that thicker ones undergo mainly crosslinking. Both results are explained by oxygen diffusion control of oxidation resulting in the formation of a superficial oxidized layer experimentally observed by µATR InfraRed in which oxidative processes orientate rather to chain scission than crosslinking. Experimental results also allowed discussing relative oxidative stability, sensitivity of T_g to chain scission concentration, and crosslinking mechanism (Y or H).

Keywords: Aromatic polymers, chain scission, crosslinking, oxidized layer, radiochemical yields, glass transition.
PACS: 81.05.Lg, 81.70Pg, 82.30.Cf, 82.35.Lr, 82.80.Bg

Nuclear plants were first designed for 40 years in the early 70's. A current issue is to know if the initial 40 years duration for which they were designed can be exceeded. It tackles the question of irradiation induced changes of mechanical properties of polymers. Here, the study will focus on the case of poly(ether ether ketone) (PEEK), bisphenol A polysulfone (PSU), and poly(pyromellitimide) (Kapton®). Previous studies [1,2,3] were dedicated to their degradation under conditions at which influence of oxidative processes is underestimated (high temperature, high irradiation, vacuum) whereas these latter probably play a major role in failure in device conditions. Here, we were interested in studying crosslinking, chain scissions and the way they are influenced by oxygen diffusion. Classical means (molar mass measurements by GPC for detecting chain scission, and sol-gel analysis for crosslinking) and less classical ones (T_g measurements by DSC and rheometry at molten state) were used.

EXPERIMENTAL

Materials under study were bisphenol A polysulfone (Solvay Chemicals), poly(ether ether ketone) (Victrex), and poly(pyromellitimide) (Dupont). Two types of samples were investigated:
- Thin films (200 µm for PSU, 60 µm for PEEK and 125 µm for Kapton)
- Thick samples: injection molded 2 mm thick PSU and 250 µm thick PEEK sheath.

They were submitted to ^{60}Co γ rays (24 kGy h^{-1}) radiations under 1.5 bar air at 60°C up to 30.7 MGy doses. Degradation was monitored by measurements of:
- T_g using DSC (TA Q10, nitrogen flow, aluminum pan covered with lid).
- Molar mass using Gel Permeation Chromatography (Styragel SE 05 column, THF flow) for PSU.
- Insoluble fraction using sol-gel analysis (in trichloroacetic acid for PEEK, in chloroform for PSU).
- Zero shear viscosity by rheometry at molten state (Aeres rheometer, 220°C, 1 mm gap).
- Build-up of oxidized species using FTIR spectroscopy in transmission mode for films (Brücker IFS 28) and Thickness of Oxidized Layer (TOL) using µATR for thick samples (Spotlight 300 apparatus by Perkin Elmer).

CP1255, *Vth International Conference on Times of Polymers (TOP) and Composites,*
edited by A. D'Amore, D. Acierno, and L. Grassia

RESULTS AND DISCUSSION

Radiodegradation of Thin Films

Irradiated PSU thin films were first analyzed by GPC. Their molar mass was decreased highlighting thus a major chain scission process. Equations proposed by Saito [4] $1/M_N-1/M_{N0} = 10^7.[G(s)-G(x)].\delta$ and $1/M_W-1/M_{W0} = 10^7.[G(s)/2-2G(x)].\delta$ allowed an estimation of radiochemical yields in chain scission G(s) = 0.17 and in crosslinking G(x) = 0 (Fig. 1a). DSC analysis of PSU films revealed a decrease of T_g (Fig. 1b).

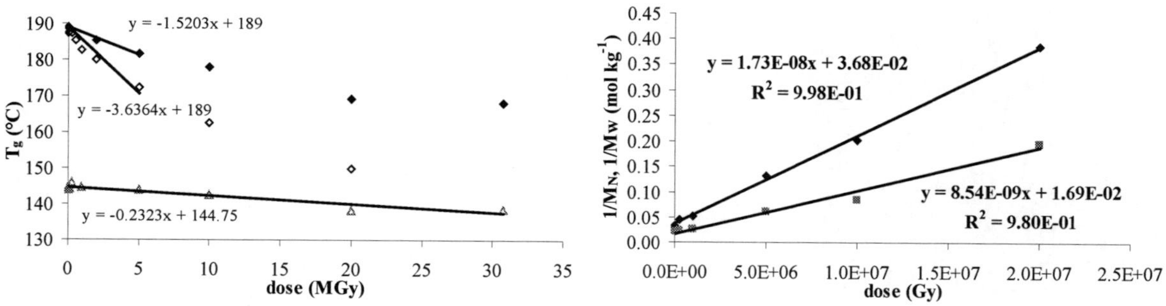

(a) (b)

FIGURE 1. Glass transition temperature change for PEEK thin films (△), PSU thin films (◇) and PSU thick samples (◆) (a) and reciprocal number (◆) and weight (■) molar mass changes with dose from GPC results.

Since this latter is linked to average molar mass, the Fox-Flory relationship [5] $T_g = T_{g\infty} - k_{FF}/M_N$ was derived to assess G(s) using Fox-Flory parameter calculated from theoretical considerations. Using k_{FF}(PSU) = 220 K mol kg^{-1}, G(s) was found close to 0.15 (G(x) being equal to 0) which validates the use of T_g measurement for estimating the chain scission build-up in radiodegradated polymers.

DSC analyses proved that Kapton® films were unaffected by irradiation while PEEK samples also display a T_g decrease which is characteristic of a chain scission process. Using k_{FF}(PEEK) = 185 K mol kg^{-1}, G(s) was found on the order of 0.012 i.e. ten times lower than PSU value.

Radiodegradation of Thick Samples

Thick samples of PEEK and PSU became insoluble after irradiation which demonstrates the existence of a major crosslinking process. This was confirmed by rheometric analysis of PSU sample showing the disappearance of Newtonian plateau and an increase of infinite viscosity. Charlesby-Pinner approach [6] was implemented:

$$w_S + w_S^{1/2} = \frac{G_H(s)}{2G_H(x)} + \frac{10^7}{\overline{M}_{W0}.G_H(x).\delta} \qquad (1)$$

$$1 + 3w_S^{1/2} = \frac{2G_Y(s)}{G_Y(x)} + \frac{4.10^7}{\overline{M}_{W0}.G_Y(x).\delta} \qquad (2)$$

where $w_S = 1 - w_I$ is the soluble fraction, $G_H(s)$, $G_H(x)$, $G_Y(s)$ and $G_Y(x)$ are the radiochemical yields respectively for chain scission and crosslinking for an H- or Y- crosslinking mechanism, δ being the dose (Gy).

Confrontation of yields values derived from rheometry with the ones derived Charlesby-Pinner approach in H- or Y- crosslink mechanism suggested that PSU H-crosslinking predominates in our experimental conditions whereas Y-crosslink was reported for higher temperatures [7]. Following values were found: G(s) ~ 0.12 and G(x) ~ 0.15.

DSC results showed that thick samples T_g decrease was half thins sample ones. This last result was in good agreement with the respective values of Fox-Flory and DiMarzio's [8] parameters expressing respectively T_g sensitivity to chain scissions and crosslinking.

Radiochemical yields for PEEK G(s) and G(x) were found respectively close to 0 and 0.02 indicating that crosslinking predominates in thick samples whereas it is negligible compared to chain scission in thin samples.

Thin Oxidized Layers Measurements

Best explanation is to reconcile the fact that G(x) tends towards 0 in thin samples whereas it is of the same order than G(s) in thick samples was the existence of a superficial oxidized layer in which oxygen reaction with irradiation generated radicals prevents these radicals to react together by coupling to give an insoluble network. Oxidation of thin PSU or PEEK films was indeed observed by the build-up of carbonyl and hydroxyl compounds in the 1700-1800 cm^{-1} and 3000-3800 cm^{-1} range (Kapton ® spectra remained unchanged). Using ATR microscopy, the changes of these absorptions with depth were observed for the thick samples. Thickness of Oxidized Layer was estimated close to 100 µm for both PEEK and PSU.

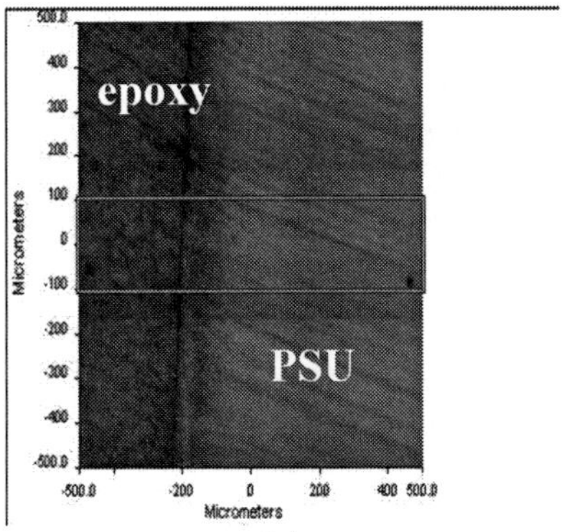

(a)

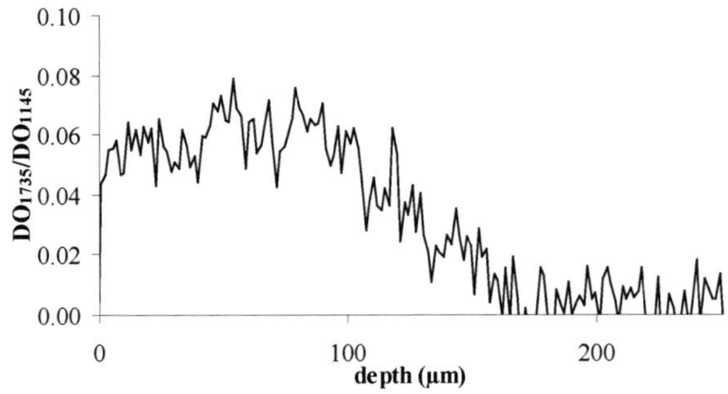

(b)

FIGURE 2. Microscopy picture of 30.7 MGy irradiated thick sample PSU with epoxy embedding (a) and profile of 1735 absorbance normalized by 1145 cm^{-1} one (antisymetric SO_2 stretching) (b).

TOL measurement allows the calculation of oxidation rate r_{OX} [9] which is linked to $k_p[PH](r_i/k_t)^{1/2}$, [PH] being the substrate concentration, r_i the initiation rate estimated from previous study [10], k_p the kinetic constant for POO° + PH propagation reaction and k_t for POO° + POO° termination one. k_p/k_t expresses the oxidative stability of a given polymer, and will be discussed considering structure-properties relationships.

CONCLUSIONS

This article was dedicated to the irradiation induced degradation of three aromatic skeleton polymers. It was shown that Kapton® was more radiation resistant than PEEK which was itself more stable than PSU. It is in good agreement with the aromaticity degree of each polymer and their content in unstable moieties such as isopropylidene and sulfone. T_g measurements were successfully employed as a simple method for estimating chain scission build-up of PEEK and PSU. Comparison of rheometry and sol-gel analysis suggested that PSU H-crosslinks under the investigated exposure conditions, whereas previous studies shown a Y-crosslinking at more elevated temperatures. Oxidation profiles recorded by micro-ATR demonstrated that aromatic polymer undergo a radio-oxidative process leading to heterogeneous degradation.

REFERENCES

1. D. J. T. Hill, H. Kudoh, and T. Seguchi, *Radiat. Phys. Chem.* **48**, 569-576 (1996).
2. A. S. Vaughan and G. C. Stevens, *Polymer* **36,** 1531-1540 (1995).
3. E. S. A. Hegazy, T. Sasuga, T. Seguchi, *Polymer* **33**, 2911-2914 (1992).
4. O. Saito, *J. Phys. Soc. Jap.* **13**, 198-206 (1958).
5. T. G. Fox, P. J. Flory, *J. Appl. Phys* **21**, 581-591 (1950).
6. J. Olejniczak, J. Rosiak, A. Charlesby, *Rad. Phys. Chem.* **37**, 499-504 (1991).
7. D. J. T. Hill, D. A. Lewis, J. H. O'Donnell, A. K. Whittaker, *Polym. Adv. Tech.* **9,** 45-51, (1998).
8. E.A. Di Marzio, *J. Res. Nat. Bur. Standards-A. Phys. Chem.* **68**, 611-617 (1964).
9. L. Audouin, V. Langlois, J. Verdu, J. C. M. de Bruijn, J. Mat. Sci **29,** 569-583 (1994).

10 K. Heiland, D. J. T. Hill, J. L. Hopewell, D. A. Lewis, J. H. O'Donnell, P. J. Pomery, A. K. Whittaker, *Adv. Chem. Ser* **249,** 637-650 (1996).

Polymer-Dopant-Systems: Tailoring of Optical and Thermomechanical Properties

Thomas Hanemann[a,b] and Kirsten Honnef[b]

[a]*Karlsruhe Institute of Technology (KIT Campus North), Institute for Materials Research III, Herman-von Helmholtz-Platz 1, D-76344 Eggenstein-Leopoldshafen, Germany)*
email: thomas.hanemann@kit.edu
[b]*University of Freiburg, Department of Microsystems Engineering, Georges-Koehler-Allee 102 D-79110 Freiburg, Germany*

Abstract. For modern application in microoptics the refractive index of polymers can be adjusted in a certain range by the addition of electron-rich organic dopants. As an unwanted side-effect a pronounced plasticizing occurs. In this work the addition of a crosslinker (divinylbenzene) to the unsaturated polyester matrix, doped with phenanthrene for refractive index adjustment, allows for a successful suppression of the plasticizing effect measured after polymerization. Even at high phenanthrene concentrations the glass transition temperature of the initial polymer could be almost retained. An increase of the polymers refractive index from 1.5684 up to a value of 1.6425 could be achieved.

Keywords: polymer guest-host systems, organic dopants, refractive index, glass transition temperature, viscosity.
PACS: Replace this text with PACS numbers; choose from this list: http://www.aip.org/pacs/index.html

INTRODUCTION

In non-silicon Microsystem Technologies polymers are the most important material class for the realization of devices in the application fields of microfluidics and microoptics. In the latter case polymer based microoptical free-space components like lenses, prisms, a.o., or polymer waveguide devices like beam splitters (3dB-coupler) are favourable e.g. due their low weight in comparison to glass based devices. Established molding techniques like UV-molding, hot embossing or injection molding enable low cost replication [1,2]. Only a few highly transparent commercial thermoplastic polymers are available, the most prominent ones are polymethylmethacrylate (PMMA) and polycarbonate (PC). In case of curable reactive resins many different (meth)acrylates, epoxies and unsaturated polyesters are available yielding thermoplastic polymers or thermosets after polymerization. The optical transmittance in the visible range is around 90%, the refractive index spans mainly a range from 1.49 (PMMA) to 1.56 (PC, unsaturated polyesters). With respect to application as lenses or waveguides a tailoring of the refractive index is desirable. A refractive index increase is possible by chemical synthesis using suitable monomers carrying electron-rich molecular moieties. The addition of nanosized inorganic or organic particles or dopants enables a refractive index rise also if the largest particle sizes are below 30 nm avoiding Raleigh-scattering retaining the good optical transmittance. Unfortunately it was found, that polymer-based composites containing nanosized ceramics like Al_2O_3 or TiO_2 show poor optical transmittance values due to a pronounced agglomerate formation of the nanosized primary particles (10-30 nm) even at a small solid content of 1% [3,4]. The addition of electron-rich molecules like phenanthrene or benzochinoline to a polymer matrix forming polymer guest-host-systems allow a pronounced increase of the refractive index due to the good solubility of phenanthrene in the polymer, but as side-effect a significant drop of the polymer's glass transition temperature (TG) due to plasticizing occurs [5].

In this contribution the influence of a crosslinker on the composites′ optical and thermomechanical properties targeting a suppression of the plasticizing effect originated by phenanthrene as dopant will be discussed.

CP1255, *V[th] International Conference on Times of Polymers (TOP) and Composites,*
edited by A. D'Amore, D. Acierno, and L. Grassia

EXPERIMENTAL

A commercially available unsaturated polyester/styrene resin was used as curable polymer matrix. Phenanthrene was added (0-40 wt%) with respect to refractive index increase, divinylbenzene (DVB) was added (0-32 wt%) as crosslinking agent for a suppression of the phenanthrene's plasticizing effect. All components were mixed together using a high speed stirrer (IKA, Ultraturrax T8) up the solubility limit of phenanthrene retaining a clear and highly transparent mixture. Methylethylketonperoxide (MEKP) was used as cold hardener for resin solidification under ambient conditions followed by a postcure at 80°C for finishing polymerization. The viscosity of the uncured mixtures were measured using a Bohlin CVO50 Rheometer at 20°C, the refractive index (589 nm, 20°C; Krüss Abbe Refractometer AR2008), the glass transition temperature (Mettler-Toledo FP900, equipped with a FP85 DTA cell) and the Vickers hardness (Paar-Physica MHT10) were measured after polymerization to suitable test specimen.

RESULTS AND DISCUSSION

The addition of divinylbenzene and phenanthrene cause a reduction of the mixtures viscosity which is favorable for device fabrication using the different reactive resin molding techniques. As shown earlier increasing phenanthrene concentrations cause an almost linear increase of the refractive index n [5]. The addition of the divinylbenzene should cause a further rise of n due to the presence of easy polarisable electrons in the molecule. Figure 1 shows the change of n with both additive amounts (balls: refractive index of all systems; triangles: projection refractive index as function of DVB-content; diamond: projection refractive index as function of phenanthrene-amount; squares: projection overview about all DVB and phenanthrene concentrations). While increasing phenanthrene amounts cause a significant refractive index increase, the addition of DVB leads only to a slight n-increase. As positive side-effect the solubility of phenanthrene in the polymer matrix increases with higher DVB-concentrations (0 wt% DVB: 20 wt% phenanthrene; 32 wt% DVB: 35 wt% phenanthrene). Starting with a refractive index of 1.5684 for the additive-free unsaturated polyester resin a refractive index of 1.6425 for a mixture containing 32 wt% DVB and 35 wt% phenanthrene could be realized. Increasing phenanthrene contents cause a yellowing of the mixture reducing the applicability in the visible range.

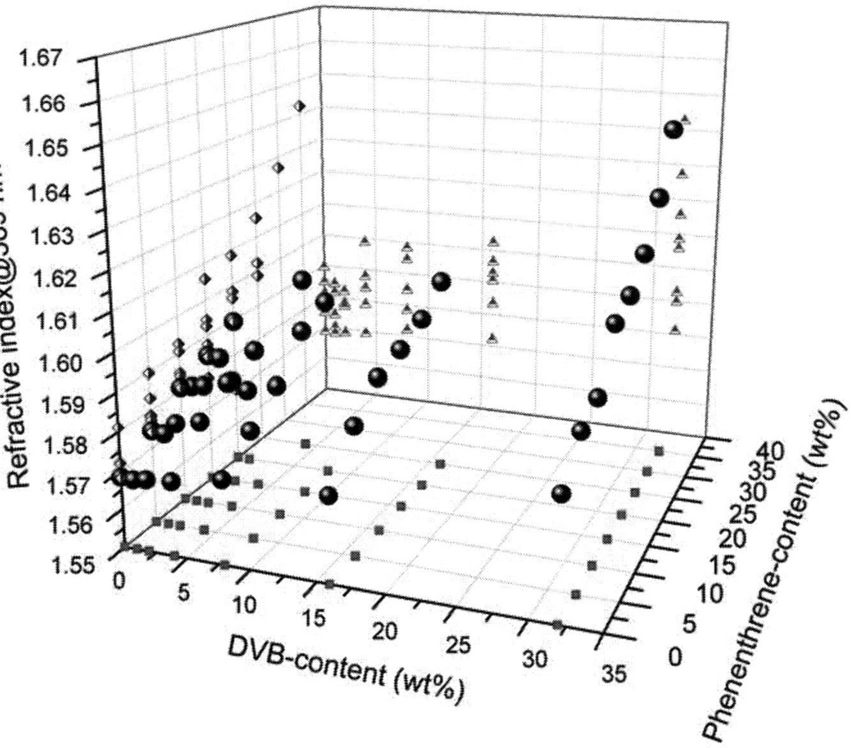

FIGURE 1. Change of the refractive index with phenanthrene and DVB content.

It was shown earlier that increasing amounts of inorganic nanoparticles or organic dopants cause a pronounced depression of TG [4,5]. Figure 2 shows the influence of phenanthrene, and more important, of DVB on the mixtures

TG after curing (balls: TG of all systems; triangles: projection TG as function of DVB-content; diamond: projection TG as function of phenanthrene-amount; squares: projection overview about all DVB and phenanthrene concentrations). At constant DVB-concentration increasing phenanthrene amounts cause a pronounced TG decay. At constant phenanthrene load ascending DVB-concentrations lead to a TG-gain. Without any additive the polymer possesses a TG of 89°C, 32 wt% DVB-addition enables a TG-increase up to 110°C. A mixture containing 25 wt% phenanthrene shows a TG of 36°C, the further addition of 32 wt% DVB yields an increase up to 82°C. The results show, that the plasticizing effect of phenanthrene can almost be compensated by the copolymerization with DVB.

The measurement of the Vickers hardness of all systems confirms the TG-results. Increasing dopant amount causes a pronounced softening, increasing DVB-contents an increase of the Vickers hardness. But combining both trends (32 wt% DVB, 40 wt% phenanthrene) cannot recover the initial hardness value of the polymer.

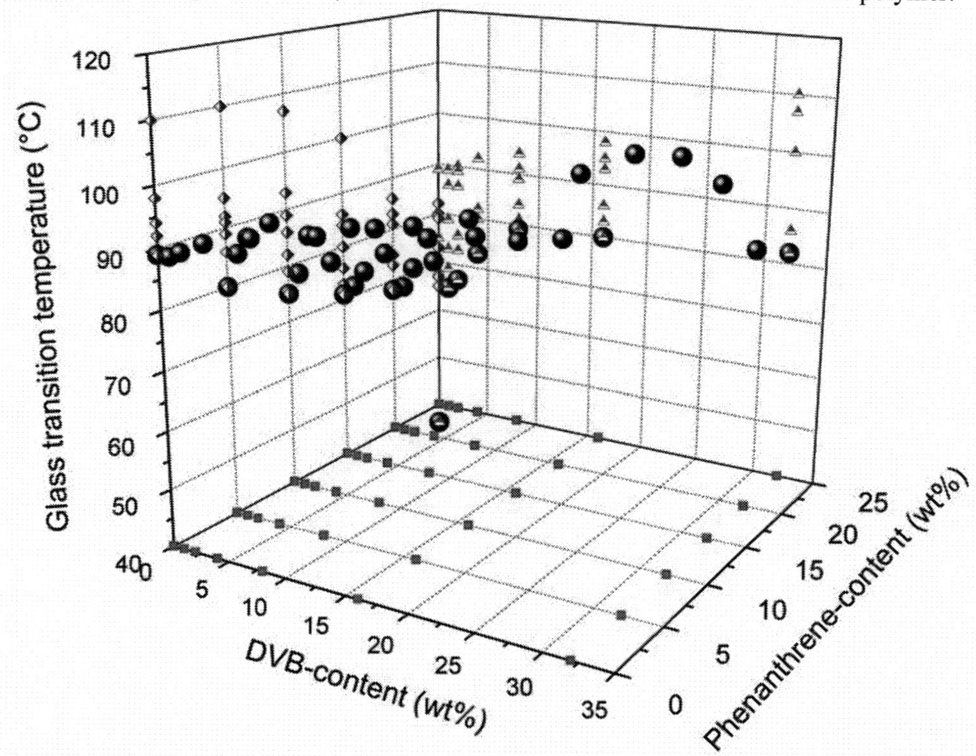

FIGURE 2. Change of the glass transition temperature with phenanthrene and DVB content.

CONCLUSION AND FUTURE WORK

The refractive index of a polymer matrix, here unsaturated polyester/styrene, can be significantly raised by the addition of organic dopants. The coeval glass transition temperature decay can be partially compensated by the use of a difunctional monomer like divinylbenzene generating a network like structure.

With respect to thermoplastic device fabrication with adjusted refractive index and good thermomechanical properties the applied approach will be extended to meth(acrylates) allowing the use of established molding techniques.

REFERENCES

1. S. Bäumer (Editor), *Handbook of Plastic Optics*, Weinheim, Wiley-VCH, 2005.
2. S. Sinzinger, J. Jahns, *Microoptics*, Weinheim, Wiley-VCH, 1999.
3. E. Ritzhaupt-Kleissl, J. Böhm, J. Haußelt, T. Hanemann, *Adv. Eng. Mater.* **7**, 540-545 (2005).
4. T. Hanemann, J. Haußelt, E. Ritzhaupt-Kleissl, *Microsystem Technologies* **15**, 421-427 (2009).
5. T. Hanemann, J. Böhm, K. Honnef, E. Ritzhaupt-Kleissl, J. Haußelt, *Macromol. Mater. Eng.* **292**, 285-294 (2007).

Multiblock polyurethanes in biomedical applications: fine tuning of degradation and biomimetic properties

Gianluca Ciardelli[a], Susanna Sartori[a,b], Antonella Silvestri[a], Piero Serafini[a], Andrea Caporale[c], Francesca Boccafoschi[d]

[a]*Department of Mechanics, Politecnico di Torino, Corso Duca degli Abruzzi24, 10129 Torino, Italy*
[b]*Proplast srl Strada Savonesa, 9 - 15057 Rivalta Scrivia (AL), Italy*
[c]*Department of Chemical Sciences, University of Padua, Via Marzolo, 1-35131 Padova, Italy*
[d]*Department of Clin. and Exp. Medicine, University of Eastern Piedmont, Via Solaroli 17, 28100 Novara, Italy,*

Abstract. Biodegradable and biostable soft segment based polyurethanes (PURs) were synthesized as new polymers for medical applications. Poly(ε-caprolactone) (PCL) was chosen as macrodiol for biodegradable PURs for its biodegradability and biocompatibility properties. Polydimethylsiloxane (PDMS) and polytetramethylenoxide (PTMO) were selected as macrodiols for biostable PURs because of their high hydrolysis resistance[1]. 1,6-Diisocyanatohexane (HDI) was used as diisocyanate component producing non toxic metabolytes. 1,4-Cyclohexane dimethanol (CDM), a peptide-based diammine and a lysine ethyl ester with grafted RGD peptide sequence are the chain extenders. The latter shows biomimetic properties able to promote material cellular adhesion. Chemical and mechanical characterizations of the obtained polymers highlight their suitability for applications in the cardiovascular field.

Keywords: Polyurethanes, cellular adhesion, myocardial applications.
PACS: 87.85.J, 87.85.Qr, 87.85.Rs, 87.85.Lf

INTRODUCTION

PURs are considered to be excellent biomedical materials having good mechanical properties and showing a satisfactory blood compatibility and have been used for several examples both in regenerative medicine and medical devices. PURs biopolymers requirements are becoming more and more specialized to meet increasingly demand corresponding to specific environmental stresses and biomedical applications. In this work a series of PURs containing aliphatic diisocyanate are proposed. Both biodegradable and biostable soft segment were investigated. Moreover a library of chain extender were used, ranging from an aliphatic cyclic diol, to peptide-based diamines. Peptide units were selected from cell adhesive molecules and enzymatic degradable sequences in order to achieve a synthetic Extracellular Matrix analogue. PURs synthesized were characterised by Size Exclusion Chromatography (SEC), Fourier Transform Infrared Spectroscopy (FT-IR), ^{1}H-Nuclear Magnetic Resonance (^{1}H-NMR), Differential Scanning Calorimetry (DSC), Dinamico-Mechanical Analysis (DMA) and stress-strain tests.

MATERIALS AND METHODS

Polyurethanes Synthesis

The synthesis of the peptide-based diammine Ala-Ala-($NH(CH_2)_4NH_2$), used as biodegradable chain extender, was carried out in solution. $NH_2(CH_2)4NH_2$ was selectively protected and the coupling of aminoacids was carried out by 1-hydroxy-benzotriazole (HOBt)/2-(1H-benzotriazol-1-yl)-1,1,3,3-tetramethyluronium hexafluorophosphate (HBTU). The synthesis of the other kind of chain extender, grafted RGD lysine ethyl ester , is described elsewhere[2].

CP1255, *Vth International Conference on Times of Polymers (TOP) and Composites,*
edited by A. D'Amore, D. Acierno, and L. Grassia

PURs were synthesized by using a two steps procedure[2]. A biodegradable polyurethane's synthesis is described. Poly(ε-caprolactone) diol (PCL, M_n=1250 or 2000, Aldrich, USA) was dissolved in 1,2-dichloroethane (Carlo Erba, Italy) and the solution azeotropically dried. The diisocyanate (1,6-diisocyanatehexane, HDI, Sigma-Aldrich) was reacted (2:1molar ratio) with the soft segment at 85 °C in anhydrous DCE, to form the prepolymer. At the end of prepolymer formation, (150 min, dibutyltin dilaurate as catalyst) the solution was concentrated. The chain extender (1,4-cyclohexane dimethanol, CDM, Aldrich, mixture cis and trans; lysine ethyl ester, Sigma-Aldrich, USA, custom made peptides) previously dried, was added at 1:1 molar ratio with respect to the prepolymer. Chain extension reaction proceeded for 18 h at 25° C and then it was stopped by the addition of methanol. The polymer was obtained by precipitation in petroleum ether. In order to remove low molecular weight impurities, polymers were redissolved in dimethylformamide (DMF, Aldrich) and precipitated in methanol. Mixed polydimethylsiloxane (PDMS) and polytetramethylenoxide (PTMO) based PURs were synthesized by the same procedure of biodegradable ones, using PDMS (Tegomer H-Si-2311, Evonik, M_n=2500g/mol) and PTMO (Terethane, Aldrich, M_n = 2000 g/mol) as soft segment macrodiols in different molar ratio, tetrahydrofuran (THF, Aldrich) as solvent, 70 °C as first step temperature and distilled water as precipitation solvent.

The code name used for all PURs synthesized is X-Y-Z-N-type, where X, Y and Z indicate chain extender, diisocyanate and macrodiol respectively. In the case of biostable PURs, macrodiols are two and each one is indicated by two letters (PDMS=PD, PTMO=PT). N indicates the macrodiol molecular weight for biodegradable PURs, the soft segment molar composition respect to PTMO for biostable ones. For example A-H-C-2000 is the name code of a PU obtained by Ala-Ala-($NH(CH_2)_4NH_2$) as chain extender, HDI as diisocyanate, PCL as macrodiol with M_n=2000 g/mol. When X=C and X=L the chain extenders are CDM and grafted RGD lysine ethyl ester respectively.

Polymers Characterization

Molecular weights (M_n) and molecular weight distributions (M_w/M_n) of PURs were measured by SEC using a Water 1525 binary system equipped with a Water 2414 RI detector. The measurements were performed by using four Styragel columns (range 103-106 Å) THF as eluent (1.0 mL/min), narrow polystyrene standards were used for calibration. FT-IR and DSC characterizations were performed by using a Perkin Elmer Spectrum GX FTIR System and a TA INSTRUMENTS DSC Q20, respectively. Biostable polyurethanes were chemically characterized also by 1H-NMR. NMR spectra were recorded on a Bruke Bruker AVANCE 400 operating at 400 MHz for ^1H. The ^1H chemical shifts are referenced to $SiMe_4$ using the residual protio impurities of the deuterated solvents as external reference. Tensile tests were performed on PU films to study their mechanical properties. Stress-strain tests and DMA were performed by using a MTS QTest/10 Elite Controller and a TA ISTRUMENTS DMA Q800, respectively.

RESULTS AND DISCUSSION

Biostable and biodegradable PURs were successfully synthesized as shown by chemical-physical analysis. The observed FT-IR spectra PU peaks are collected in table 1. As regards PDMS-PTMO based PURs, C-O-C stretching band is not visible because it is probably covered by the large and strong Si-O-Si stretching band: both vibrations occur at the same region. As regards PCL based PURs, C=O related to Urethanes and ester group are overlapped.

TABLE (1). Assignments of FTIR spectrum of obtained polyurethanes.

Group	Wavelenght(/cm)	note
ν (N-H)	3420	PCL and PDMS-PTMO PURs
ν (C=O)	1720	PCL based PURs
ν (C=O) urethan	1700-1690	PCL and PDMS-PTMO PURs
δ (N-H) + ν (C-N)	1530	PCL and PDMS-PTMO PURs
δ (N-H) + ν (C-N), β(C-H)	1290	PCL and PDMS-PTMO PURs
w (CH_2)), ν (C-O-C) ester	1250	PCL based PURs
δ (N-H) + ν (C-N)	1225	PCL and PDMS-PTMO PURs
ν (C-O-C) urethan	1180-1080	PCL and PDMS-PTMO PURs
ν (Si-O- Si)	1100	PDMS based PURs

SEC analysis show that biostable PURs exhibit molecular weights encompassing from 20.000 to 26.000 g/mol with M_w/M_n in the range 1,41-2,03. Biodegradable PURs show higher molecular weights: 22.000 g/mol for A-H-C-

2000, 100.000g/mol for L-H-C-2000, 77.000 g/mol for C-H-C-1250 with M_w/M_n in the range 1,40-2,40. ^{1}H-NMR was used to confirm the PTMO presence in the polymer chain. In the ^{1}H-NMR spectra of PDMS-PTMO based PURs (not shown) the signal at 3,4 ppm corresponds to the $-O-CH_2-$ group and is not present in the ^{1}H-NMR spectrum of polyurethane containing 100% of PDMS in the soft segment (C-H-PDPT-0).

DSC analysis show the almost complete amorphous nature of PDMS-PTMO based PURs and a melting point above physiological temperature for biodegradable PURs. In table (2) experimental data about thermal and mechanical characterizations are shown. From stress-strain analysis turns out that an higher percentage of PTMO in the soft segment improves biostable PURs mechanical properties, especially Young Modulus and Break Load. Mechanical properties of C-H-PDPT-60 and C-H-PDPT-80 make these two polymers suitable for cardiac surgery devices. In particular, they can be utilized for annuloplastic rings, that have to be capable of mimicking the physiological 3D motion of the native mitral annulus and accommodating its anatomical saddle shape. As regards biodegradable polyurethanes, the presence of degradable peptide decreases the molecular weight and subsequently the modules. Furthermore the decreasing of the molecular weight of PCL diol involves a decrease in modules, probably due to a lower crystallinity. High elongations and low values of Young Modulus make this PURs attractive for myocardial tissue engineering.

TABLE (2). Data concerning thermal and mechanical properties of obtained polyurethanes.

Sample	T_g (°C)	T_m(°C)	Young Modulus (MPa)	Average Max % Elongation	Break Load (MPa)
C-H-PDPT-20	-80	-	1,5	260	0,15
C-H-PDPT-40	-67	-	1,2	220	0,22
C-H-PDPT-60	-60	-	9	520	1,2
C-H-PDPT-80	-45	-	7,6	780	1,1
A-H-C-2000	n.d.	50	8,2	192	0,4
L-H-C-2000	n.d.	41	51	750	4,4
C-H-C-2000	n.d.	45	66	400	4,3
C-H-C-1250	-53	24.7/60	40	250	3,8

Preliminary biocompatibility tests on the obtained PURs were performed with fibroblasts and mesenchymal stem cells. In particular, promising results were achieved by RGD functionalized PU, that shows high biomimetic properties and consequently a better cellular adhesion.

CONCLUSIONS

PCL and PDMS-PTMO based PURs have been successfully synthesized, as revealed from spectroscopic analysis. Depending on their chemical and thermo-mechanical properties they seems to be promising materials for biodegradable mimetic myocardial tissue engineering and biostable mitralic valve applications respectively.

ACKNOWLEDGMENTS

Co-financing by the Regione Piemonte through the Bando Regionale - 2006 per la Ricerca e lo Sviluppo Precompetitivo project nr.10-BIADS, and the POR: Piattaforme Innovative (F.E:S.R. 2007-2013) "Active-Advanced Cardiovascular Therapies"; is acknowledged. Susanna Sartori is grateful to Progetto Lagrange Fondazione CRT for the Research Grant. The authors thank Dr. S. Milione of "Università degli Studi di Salerno" for his technical support in SEC analysis.

REFERENCES

1. P.A. Gunatillake, D.J.Martin, G.F. Meijs, S.J. McCarthy, R. Adhikari, Aust. J. Chem., 56, 545-557 (2003)
2. G. Ciardelli, A. Rechichi, S. Sartori, M. D'Acunto, A. Caporale, E. Peggion, G. Vozzi, P. Giusti, Polym. Adv. Technol. 17, 786–789 (2006)

Time Issues in Semi-Crystalline Thermoplastics Processing

Michele Iannone

Alenia Aeronautica, Pomigliano D'Arco (Naples, Italy)

Abstract. Cooling rate of semicrystalline thermoplastic (PEEK) based composite parts has been assessed by means of a thermal flow model. Cold tool thermoforming was found suitable for thin parts, but the use of a preheated tool is suggested. Cooling rate is critical for automated lay-up; consequently the optimization of this technique requires the use of a complex apparatus, able to provide both heating of the part during lay-up and a control of the cooling rate. Alternatively a modified cooling rate vs. crystallinity behavior of the material must be achieved.

Keywords: Composite, Thermoplastics, Processing, Cooling rate.
PACS: 81.05.Qk

INTRODUCTION

Semi-Crystalline Thermoplastics show some very interesting properties, among which a very high service temperature and high resistance to the environment and to the organic solvents. These properties allow to considering their utilization as matrix in fiber reinforced composites for aeronautical applications.

In fact some thermoplastics show very challenging mechanical properties, mainly toughness. Unfortunately for amorphous thermoplastics T_g is generally unsatisfactory for aeronautical structural applications, as well as environmental and solvent resistance. Some amorphous thermoplastic, with very high T_g, like PEI, has been considered for aeronautical structural applications, and was sometimes successful, but only for secondary structures, due to solvent resistance concerns. The real solution for the very demanding requirements is given by the semi-crystalline thermoplastics, like PEEK, PEKK, PPS. In fact the presence of a crystalline phase improves the solvent resistance, and reduces the properties decrease at T_g; consequently PEEK, PEKK and (in a lower extension) PPS are strong candidates also for highly loaded aircraft structures, when used as matrix in carbon reinforced composites.

These materials must be heated above T_f in order to be processed, and this is a very severe technological challenge; in fact Tf for PPS is about 280 °C, for PEEK and PEKK about 340 °C. Whatever the processing technique is, the material must be heated at very high temperature, for a time short enough to prevent the thermal degradation, formed and then cooled to the solid state. For semi-crystalline thermoplastic materials also cooling rate must be controlled, in fact the crystallinity content depends on cooling rate.

For PEEK generally a slow cooling rate (about 1°C/min) produces the maximum crystallinity content (about 40%). A relatively constant crystallinity of about 30% is obtained at a cooling rate ranging from 10 to 300 °C/min. Finally, for very high cooling rate (from 1000 °C/min up) an amorphous polymer is obtained (ref. 1).

For certified applications, like the aeronautical ones, the "standard" crystallinity is required. In fact the crystallinity increase could improve some property like the elastic modulus, but can decrease the material toughness. On the other side crystallinity decrease can affect the sensitivity of the material to the environment (e.g. humidity) and to aggressive agents (e.g. organic solvents).

In this work two processes for thermoplastics are evaluated, thermoforming and automated lay up, and the cooling rate effects on both are discussed for processing of carbon fiber reinforced thermoplastic composites, specifically with PEEK matrix. The cooling rate is assessed for the two different processes, considering different process parameters, by means of a thermal model, based on the Fourier Law, to evaluate temperature vs time in different points of the material thickness.

CP1255, *V[th] International Conference on Times of Polymers (TOP) and Composites,*
edited by A. D'Amore, D. Acierno, and L. Grassia

THERMOFORMING

There are different thermoforming techniques. Differences can involve the kind of tools (metal-metal, metal-rubber, diaphragm, etc.). Generally a more compliant tool is required for more complicate shapes.

Another major difference is given by the thermal history, and thermoforming processes can be grouped in two families:

- Melt thermoforming, in which the forming process is developed above the melting temperature. Both tools and formed part stay above the melting temperature during forming, and then are cooled together.
- Cold tool thermoforming, in which the part is preheated above the melting temperature, and then is formed by a tool staying at a lower temperature. In that case the tool operates as forming tool as well as cooling tool. A preheated tool can be used to reduce the cooling rate.

Melt thermoforming is potentially more adequate for very complex shapes. Double diaphragm forming is a very interesting version of this technique. For thermoplastics also forming by autoclave can be included in this process family. In the meantime heating, forming and cooling times can be very long, and consequently the process is very expensive.

Cold tool thermoforming is potentially cheaper, mainly if adequate automation is provided for heating and forming plus cooling procedure. Motion issues must be investigated, as well as size of the parts that can be obtained with this technique.

A schematic description of cold tool thermoforming is shown in fig. 1.

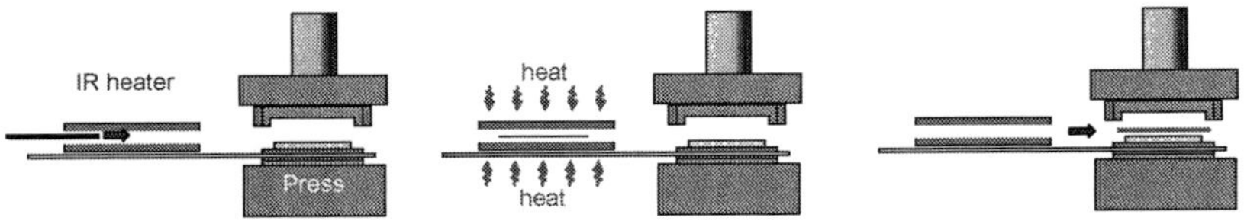

FIGURE 1. Schematic view of the cold tool thermoforming process

Some examples of different cooling rate history for cold tool thermoforming are reported in the following. In fig. 2.a temperature vs. time is shown for a 1 mm thick carbon fiber reinforced PEEK sheet, melt at 360 °C and then cooled with a 20°C cold tool, recorded in points placed at different depth (the last one is in the middle, other points are equally spaced up to the surface). In fig. 2.b cooling history is described with the same approach for the same sheet, cooled with a tool at 200°C. In fig, 2.c cooling is outlined for a 10 mm thick sheet, cooled with a 20°C tool.

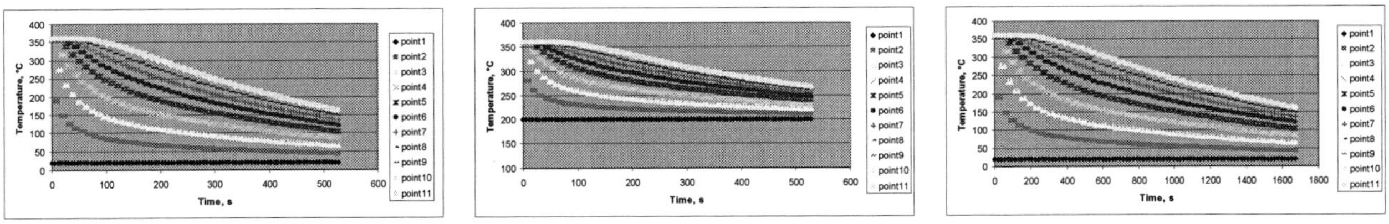

(2.a) (2.b) (2.c)

FIGURE 2. Temperature vs. time in different points of a schematic view of the cold tool thermoforming process of a 1 mm thick carbon fiber reinforced PEEK sheet, melt at 360 °C and then cooled with a 20°C cold tool (a); : 1 mm thick carbon fiber reinforced PEEK sheet, melt at 360 °C and then cooled with a 200°C cold tool (b); 10 mm thick carbon fiber reinforced PEEK sheet, melt at 360 °C and then cooled with a 20°C cold tool (c)

In fig. 3.a, 3.b and 3.c the cooling rate of the different points from 360 to 340 °C is respectively reported for the cooling process described respectively in fig. 2.a, 2.b and 2.c. Considering a processing window with a cooling rate ranging from 0,16 to 5 °C, for the 1 mm thick plate cooled with a cold tool 20 % of the thickness is cooled too fast, 10% slightly above the limit. In this case the poorly crystallized area is large. Using a pre-heated tool only 10% of the points are out from the window, and 10% is close to the boundary limit. The poorly crystallized area is in this case very limited. For the thicker sheet (fig 6) cooling rate is (relatively) above the limit only for 10 % of the points, also with a cold tool, but for many of the thicker points (30%) the cooling rate is too slow, with potential excess of crystallization. Of course the problem can be more severe for thicker parts.

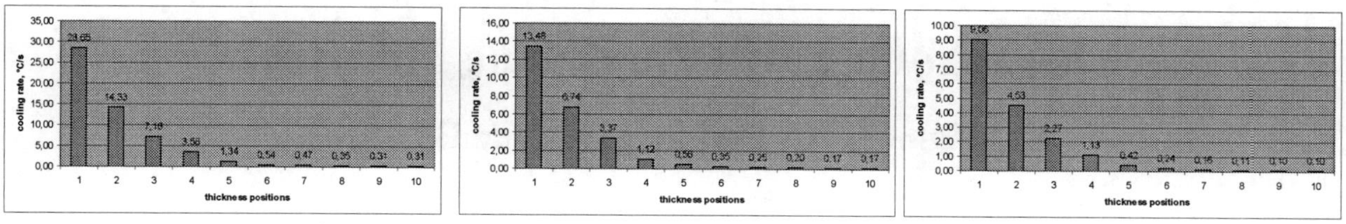

3.a 3.b 3.c

FIGURE 3. Cooling rate in points at different thicknesses for a 1 mm thick carbon fiber reinforces PEEK sheet, melt at 360 °C and then cooled with a 20°C cold tool (3.a); 1 mm thick carbon fiber reinforces PEEK sheet, melt at 360 °C and then cooled with a 200°C cold tool (3.b); 10 mm thick carbon fiber reinforces PEEK sheet, melt at 360 °C and then cooled with a 20°C cold tool (3.c)

The conclusion is that for PEEK, when a very strict control is required for crystallinity content (e.g. for aeronautical structural applications) some concerns exist for the application of the thermoforming technique to thick parts. For thin part preheated tools are suggested.

AUTOMATED TAPE LAY-UP (ATL) AND FIBER PLACEMENT (AFP)

These techniques give several potential advantages. The lay-up can be optimised, quality can be more carefully standardised, and manpower costs are reduced. Both ATL and AFP are currently utilised in performing lay-ups of thermoset matrix composites, which are cured in autoclave after aying-up. Thermoplastic based composites aren't tacky, so they can be laid-up only if a sufficient heat is provided to melt them. Thus ATL and AFP of thermoplastics can be performed with a fixture using a heated lay-up shoe; if the technique is successful, it allows to performing the final part without further curing. Also in this case one of the critical process issues is cooling rate; in fact the melt prepreg is laid-up on a cold lay-up, giving a fast cooling.

In fig. 4.a the thermal history is shown of the points at different thickness of a 0,125 mm thick prepreg tape laid up on a cold composite lay-up. In fig. 4.b the cooling rate evaluated in the same points from 360 °C to 340 °C is reported

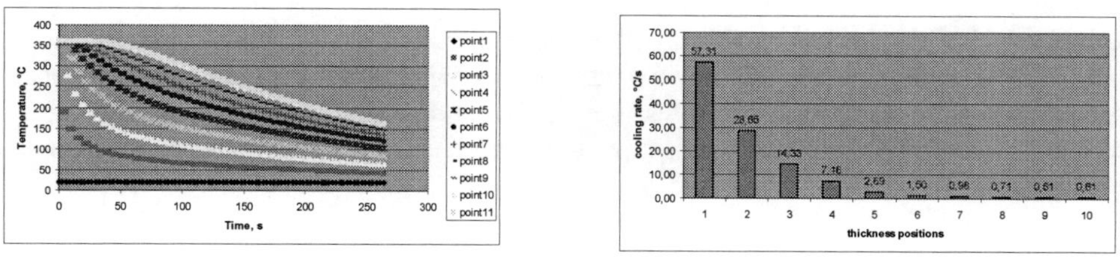

4.a 4.b

FIGURE 4. Temperature vs. time in points in different positions of a .125 mm thick carbon fiber reinforced PEEK prepreg, automatically laid up on a cold lay-up (4.a); cooling rate in the same points in fig. 4.b.

The analysis of the thermal history stresses that a large part of the prepreg (40%) cools down with a rate higher than the upper limit of the processing window (1,6-5 °C/min), in some case much faster. That indicates that heating is required also in the cooling phase, in order to prevent amorphization. It means that the processing apparatus must be very complex, and consequently very expensive. A different approach requires to modifying the material chemical structure, in order to obtain a lower sensitivity of the crystallinity content to the cooling rate.

REFERENCES

1) M. J.Kenny, A.D'Amore, L.Nicolais, M.Iannone and B.Scatteia – Processing of Amorphous PEEK and Amorphous PEEK Based Composites – SAMPE Journal Vol.25, N°4, July/August 1989

Properties of a Carbon-Fibre Composite Modified by Electrospun Poly (Vinylidene Fluoride).

K. Magniez [a], C. De Lavigne [ab] and B.L. Fox [a]

[a] *Institute of Technology Research and Innovation, Deakin University, Victoria, Waurn Ponds, 3217, Australia*
. [b] *École nationale supérieure des arts et industries textiles (ENSAIT), Roubaix, 59056, France.*

Abstract. *The interlaminar toughening of a carbon-fibre reinforced composite by incorporation of electrospun polyvinylidene fluoride (PVDF) nanofibrous membranes was explored in this work. The nanofibres were electrospun directly onto commercial pre-impregnated carbon fibre materials under optimised conditions and PVDF was found to primarily crystallise in its β phase polymorphic form. There is strong evidence from DMTA analysis to suggest that a partial miscibility between the amorphous phases of the PVDF nanofibres and the epoxy exists. The improved plastic deformation at the crack tip after inclusion of the nanofibres was directly translated to a 57% increase in the mode II interlaminar fracture toughness (in-plane shear failure). Conversely, the fracture toughness in mode I (opening failure) was slightly lower than the reference by approximately 20%, and the results were interpreted from the complex micromechanisms of failure arising from the changes in polymorphism of the PVDF.*

Keywords: polyvinylidene fluoride; electrospinning; composites.
PACS: N/A

INTRODUCTION

For the past decade, the improvement in the fracture toughness of composite has been realised through the development of novel resin systems involving the use of rubbers and thermoplastics as toughening agents. Interlayer toughening described as the inclusion of discrete layers of a secondary material, in film, fibrous or particulate form, in the inter-plies region has also been successfully applied [1]. Very recently, the interlayer toughening of carbon fibre reinforced composites using a nanofibrous substrate produced by electrospinning has been reported [2,3] but thus far this approach remains novel as very limited literature has been published. Polyvinylidene fluoride (PVDF) is a well-known thermoplastic that has shown excellent compatibility with epoxy resin [4]. For this reason, the interlaminar toughening of a carbon-fibre reinforced composite by incorporation of electrospun PVDF nanofibrous membranes was explored in this paper. The work aimed to determine the complex interfacial interactions between the nanofibres and the epoxy matrix, and understand how a secondary interlayer phase influences on the resistance to crack propagation in the resulting composites. The mode I and mode II fracture toughness results were correlated to the micromechanisms of failure and changes in polymorphic crystal forms in order to establish the structure-property relationships.

RESULTS

The electrospinning process a high molecular weight PVDF was found to be optimum at (20kV, 25wt%) where nanofibres free of defects with average diameter of 340nm ± 150nm and orientation in random directions could be observed (Fig. 1). PVDF nanofibrous membranes were electrospun directly onto pre-impregnated woven material (Toray G83C).

Double Cantilever Beam (DCB) (mode I) and End Notch Flexure (ENF) (mode II) fracture toughness tests were performed on the resulting composites. The mode I and mode II fracture toughness represent the resistance of a material to delaminate in an opening and in-plane shear modes of failure, respectively upon crack propagation. It can be noted that the unmodified reference sample out-performed the other modified samples and displayed the best resistance to crack propagation with an average value G_{IC} of 1.02 kJ/m². The resistance to crack propagation for the samples containing the PVDF films was found to be very poor with average G_{IC} values as low as 0.2 kJ/m², and poor interfacial strength between the film and the epoxy was accounted for this result. The resistance to crack propagation for the samples containing the PVDF nanofibrous membranes was found to be slightly lower than the one of the

CP1255, *V[th] International Conference on Times of Polymers (TOP) and Composites,*
edited by A. D'Amore, D. Acierno, and L. Grassia

reference sample with an average value G_{IC} of 0.8 kJ/m². This trend is somehow consistent with the recently reported literature showing no significant effects on the mode I results after incorporation of electrospun membranes in composite [2]. A cross section analysis of the DCB samples (not shown here for brevity) showed that the crack path transited from the mid-section of the nanofibrous interlayer region to the interfacial interlayer/resin region.

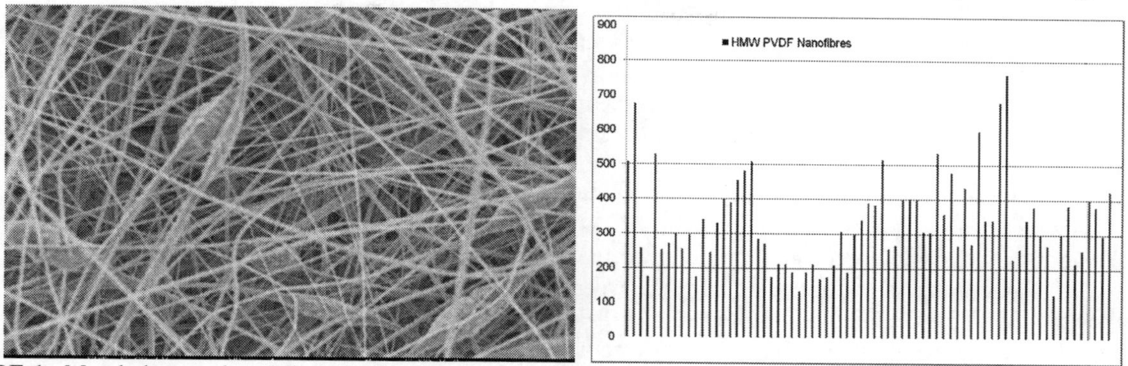

FIGURE 1. Morphology and statistical distribution of the high molecular weight (HMW) PVDF nanofibres (polymer flow rate 1.5 ml/min, tip to collector distance of 15 cm, DMF /acetone).

Displayed in Fig. 2 are the the plots of the mode I fracture toughness (referred to as G_{IC}) as a function of the crack growth and the mode II results.

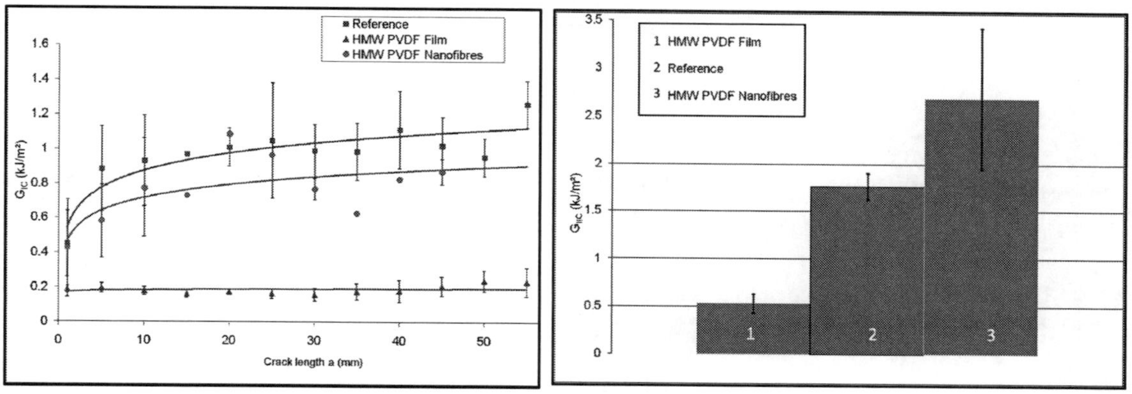

FIGURE 2. Mode I and mode II fracture toughness results (left and right images, respectively) for the reference composite and composite modified with high molecular weight (HMW) PVDF film and nanofibres.

This type of behaviour has previously been described in interlayer toughened systems and it was attributed to complex changes in micromechanisms of failure during the loading [5]. SEM surface fracography analysis showed a combination of brittle and ductile regions which evidenced the complexity of the micromechanisms of failure. Plastic deformation of the matrix and the nanofibres was visible in the semi-ductile regions and typical brittle interlaminar fracture surfaces accompanied with the presence of nanofibres embedded within the epoxy matrix were also visible (Fig. 3).

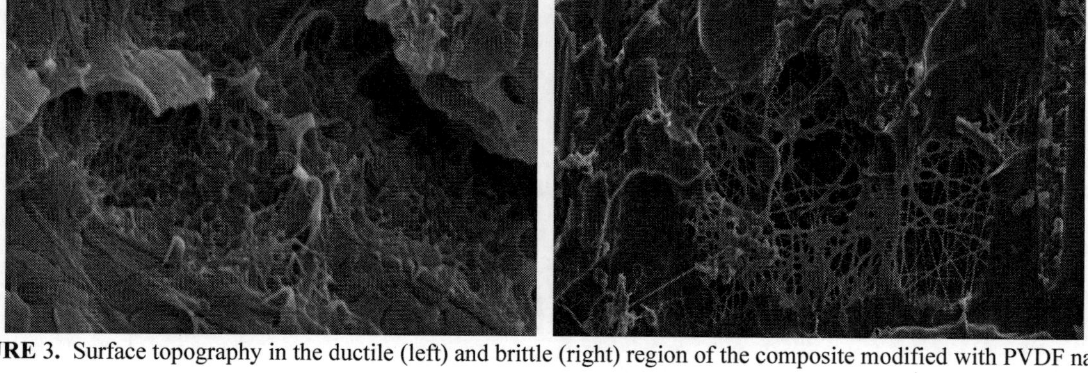

FIGURE 3. Surface topography in the ductile (left) and brittle (right) region of the composite modified with PVDF nanofibres.

The polymorphic behaviour exhibited by the PVDF was detected by FT-IR spectra where the predominance of the less ductile β crystal phase was visible (not shown here for brevity). In light of these results, it is speculated that the polymorphic behaviour displayed by PVDF in the nanofibrous membranes strongly influenced the fracture toughness of the resulting composites.

Mode II results (Fig. 2) showed that the insertion of a PVDF film into the composite underperformed which highlights the inability for the epoxy matrix to plastically deform ahead of the crack tip. On the other hand, the introduction of electrospun PVDF membranes improved the G_{IIC} by approximately 57%. The complex surface morphology after fracture with the presence of mixed brittle and ductile regions was also visible from the SEM images. It was found that the ductility of the epoxy matrix in the presence of the nanofibres was enhanced at the crack tip. The complexity in the crack path and micromechanisms of failure which have been aforementioned might have been in themselves contributing to the observed improvements.

DMTA analysis also showed no improvement in the dynamic response of the composite after inclusion of PVDF film or nanofibres. On the other hand, and despite the occurrence of interfacial phase separation, there is strong evidence to suggest that a partial miscibility between the amorphous phases of the epoxy and PVDF exists. Annealing at 185°C for 45 minutes resulted in a shift in glass transition temperature (T_g) which implicates partial miscibility between the amorphous phases and high compatibility (Fig. 4).

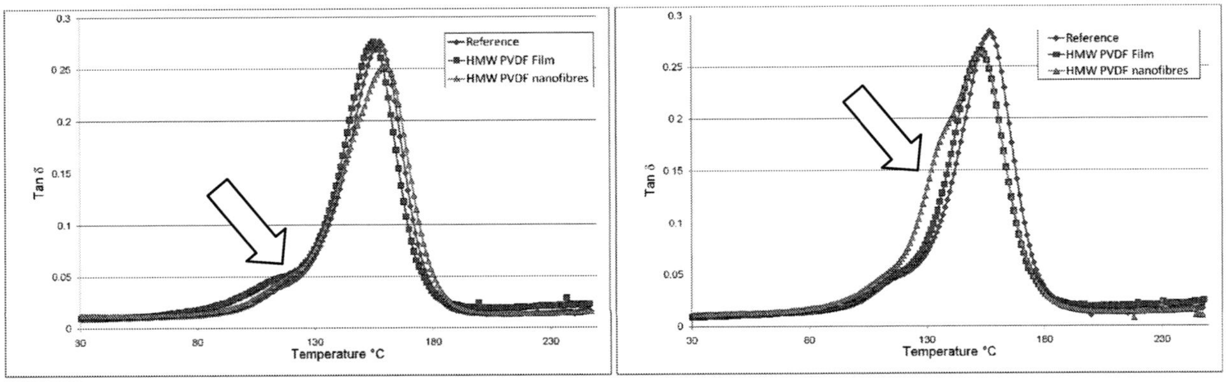

FIGURE 4. Tan (δ) of the reference composite and composites modified with high molecular weight (HMW) PVDF film and nanofibres as such (left) and annealed (right).

ACKNOWLEDGMENTS

The authors thank Alfred Deakin Prof. Xungai Wang from the textile group at Deakin University for financial support of this research project. The authors gratefully acknowledge the helpful assistance of Associate Prof. Tong Lin and Dr. Jian Fang during this work and Glenn Bridgford from Arkema providing the Kynar® films.

REFERENCES

1. J.C. Hedrick, N.M. Patel and J.E. McGrath, 'Toughening of Epoxy Resin Networks with Functionalized Engineering Thermoplastics' in Toughened Plastics I, edited by C.K. Riew and A.J. Kinloch, American Chemical Society, 1993, pp. 293–304.
2. Y. Dzenis Y and D. Reneker, US Patent No. 626,333 (2001). .
3. G. Li, P. Li, C. Zhang, Y. Yu, H. Liu, S. Zhang, X. Jia, X. Yang, Z. Xue, S. Ryu, Composites Science and Technology **68,** 987-994 (2008).
4. A. Beehag, K. Horton, B. Cartwright, US Patent No. EP20060291496 (2006).
5. M. Hojo., S. Matsuda, M. Tanaka, S. Ochiai, A. Murakami, Composites Science and Technology **66**, 665-675 (2006).

Strain Recovery Behavior Of Post-Yield Deformed Epoxy Resins: Effect Of The Resin/Hardener Ratio

Stefano Pandini and Fabio Bignotti

Department of Mechanical and Industrial Engineering, University of Brescia, Via Branze 38, Brescia, Italy - stefano.pandini@ing.unibs.it, fabio.bignotti@ing.unibs.it

Abstract. The effect of the resin/hardener ratio on the strain recovery kinetics of various post-yield deformed epoxy resins is investigated in isothermal tests and in thermally stimulated recovery tests. Recovery is seen to take place following a two-stage process, in which the different stages can be distinguished on the time and temperature scales. These two recovery contributions are differently affected by the resin/hardener ratio and by the applied deformation.

Keywords: epoxy, resin/hardener ratio, post-yield, strain recovery.
PACS: 46.35.+z; 81.05.Lg; 06.60.Mr

INTRODUCTION

The post-yield deformational behavior of cross-linked systems is known to be strongly dependent on the material molecular architecture, and in a previous study on epoxy resins we have shown that large deformation phenomena, such as yielding, strain softening and strain hardening, are affected in different manners by chain mobility and by crosslink density [1]. As suggested by literature works [2-4], the deformational mechanisms of largely deformed polymers can be explored in strain recovery experiments, which depict a complex dependence of the recovering strain on time, temperature and deformation conditions. In this paper we attempt at a description of the recovery behavior of epoxy resins obtained with various resin/hardener ratios and deformed at various post-yield strain levels.

EXPERIMENTAL

The systems investigated are obtained by curing a DGEBA epoxy resin (DER 330) with a diamine hardener (Jeffamine D230) for resin/hardener equivalent ratios, q, ranging between 0.7 and 1.25. By changing the value of q different cross-link densities, ν, are achieved. A tighter cross-linked structure is shown for q close to 1 (ν = 1814 mol/m^3 for q = 0.96), while as the excess of resin or hardener increases lower ν values were found (ν = 1154 mol/m^3 for q = 0.70, and ν = 616 mol/m^3 for q = 1.25).

Cylindrical specimens (height ~10mm; diameter ~ 7.5mm) are realized by casting the resin in silicon moulds and applying an optimized curing cycle. The obtained specimens are compressed by means of an electro-mechanical dynamometer at a nominal deformation of 50%, and then rapidly unloaded. After unloading, the strain recovery behavior is measured in two sets of tests: i) isothermal recovery tests, in which the specimen dimensional changes are measured as a function of time at various temperatures; ii) thermally stimulated recovery (TSR) tests, in which recovery is measured on specimens subjected to a heating ramp (1°C/min) in a DMTA instrument. The strain recovery is evaluated from the change in specimens height, and the residual deformation is calculated as ε_{res} = $(h/h_0)\cdot 100$, h_0 and h representing the initial (undeformed) and actual height of the specimen, respectively. Additional TSR tests are performed on specimens deformed at strain levels between 5% (i.e. right above yielding) and 70%.

RESULTS AND DISCUSSION

Isothermal recovery tests, carried out on deformed specimens after unloading, allow the description of strain recovery as a time-delayed and thermally activated process. In these experiments the residual deformation is

CP1255, *Vth International Conference on Times of Polymers (TOP) and Composites,* edited by A. D'Amore, D. Acierno, and L. Grassia measured as a function of time at various temperatures, by monitoring the specimen height via a PC controlled photo-camera; Figure 1 provides a schematic representation of the specimen dimensional changes during recovery.

The results of isothermal experiments are represented in Figure 1 for the resin with q = 1.07, whose glass transition temperature, T_g, was evaluated around 74°C by DSC measurements at a heating rate of 10°C/min; the first 20 minutes of the recovery process are not represented here, since in this time interval the specimens, although recovering the strain, have not reached yet thermal equilibrium condition. The representation suggests that, after a fast recovery during unloading, evaluated of about 15% for all the resins, the residual deformation decreases with time following a process that become faster as the test temperature increases. The recovery process is seen to achieve significant recovery rates when heated close or above T_g, and for temperatures slightly higher than T_g full recovery can be achieved in less than an hour, finally restoring the specimen original dimensions. Also the other resins investigated in isothermal experiments (q = 0.70; 0.78; 0.96) have shown a similar behavior, although it has to be remarked that for these systems the recovery rate becomes significant already for temperature 10°C below T_g.

Starting from the isothermal recovery curves, the application of a time-temperature reduction scheme leads to tentative representation of a strain recovery master curve, here represented for a reference temperature T_0 = 23°C. As the graph suggests, the residual deformation decreases with a regular trend for times up to 10^9 s, while a steeper decrease occurs at longer times. According to literature [2], the different decreasing rates could be associated to two distinct recovery mechanisms differently distributed on the reduced time-scale. Also the other resins exhibited these two recovery stages, the second stage typically occurring at shorter reduced times for resins with lower T_g.

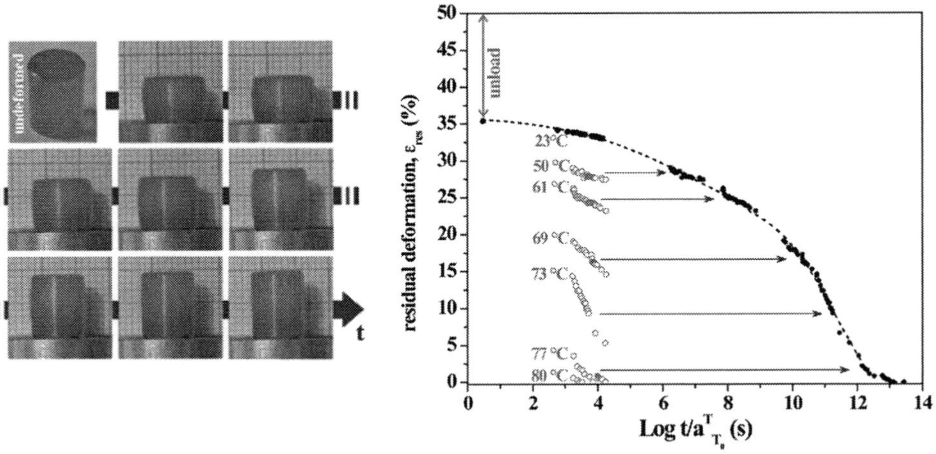

FIGURE 1. Recovering strain during isothermal recovery tests, and representation of isothermal recovery curves (open symbols) and of the recovery master curve (full symbols; reference temperature T_0 = 23°C) for the resin with q = 1.07 (T_g = 74°C).

The typical result of a TSR experiment is represented in Fig. 2a, for the resins with q = 1.07. In the graph, which shows evident analogies with the strain recovery master curve representation, the residual deformation reduction is plotted as a function of the increasing temperature. It is possible to distinguish the presence of two different decreasing processes on the temperature scale, that can become more evident when representing the results in terms of so-called "recovery rate", $d\varepsilon_{res}/dT$. Similarly to what reported in literature for cold-worked epoxy systems [3,4], the recovery rate trace exhibits a narrow peak for temperatures well below T_g, followed by a sharp peak distributed on the glass transition region.

TSR experiments carried out on epoxy resins with various q are represented in Figure 2b. The study of the recovery rate seems to suggest that all the resins, independently from the value of q, exhibit a sub-T_g recovery process with similar features and average rates, whereas the high-temperature process shows that the peak maximum and its distribution on the temperature scale are affected by the value of q. In fact, as q becomes closer to 1 the peak shifts toward higher temperatures, with a decrease in maximum rate and a broader distribution. Similar results are presented in a literature work [5], in which the recovery behavior of epoxy systems deformed at small strains was investigated for different resin/hardener ratio and in excess of resin, and the effect is interpreted as due to the higher mobility related to a less tight network structure. It is anyway interesting to observe that in our case such dependence on the cross-link density can be considered true only when describing separately the case of resin or hardener excess. This becomes particularly evident when comparing the traces of the resins with q = 0.70 and q = 1.25, which show similar recovery rates, although possessing different crosslink densities (the value of ν of the former being twice of that of the latter), suggesting that recovery behavior can not be regarded as only governed by the network density.

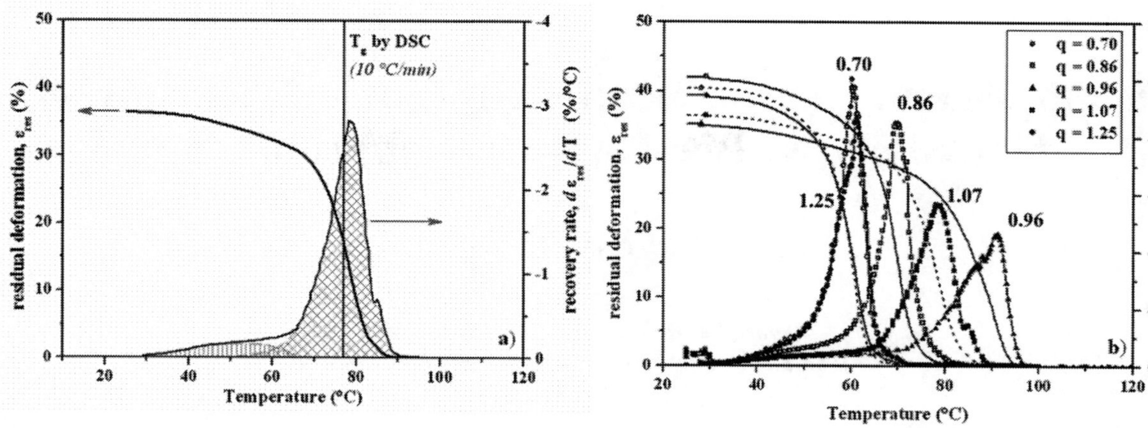

FIGURE 2. a) Typical TSR result (epoxy resin with q=1.07) and b) TSR curves for epoxies with various resin/hardener ratios, q.

TSR tests were performed on some of the resins (q = 0.70; 0.96; 1.07) deformed at various strain levels above yielding. The results of these tests are reported in terms of recovery rate in Figure 3a for the resin with q = 0.96, and similar data were obtained also for the other resins. It can be seen that as the applied deformation increases the intensity of both recovery signals becomes more pronounced, and the height of the high temperature recovery peak increases with a faster rate than that related to the sub-T_g recovery process, becoming predominant for deformation levels above 20%. The amount of deformation recovered in the two processes, i.e. below and above the onset temperature of the high temperature peak, are represented in Figure 3b as a function of the applied deformation for the various epoxies; the true stress *vs.* nominal strain curve of the resin with q = 0.96 is also reported on the graph, as indication of the post-yield deformation phenomena occurring at the various applied strain levels. It can be seen for all the resins that the low and high temperature components exhibit similar deformation values for strains between 5 and 10% (i.e. right above yielding). The difference between the two components becomes important for higher deformation levels, in particular for nominal deformations above 20% (i.e. at the onset of the strain hardening region). At such strain levels, in fact, the sub-T_g recovered strain exhibit a very slow increase, suggesting in some cases a steady state condition, while the high temperature recovery component follows a regular increasing trend.

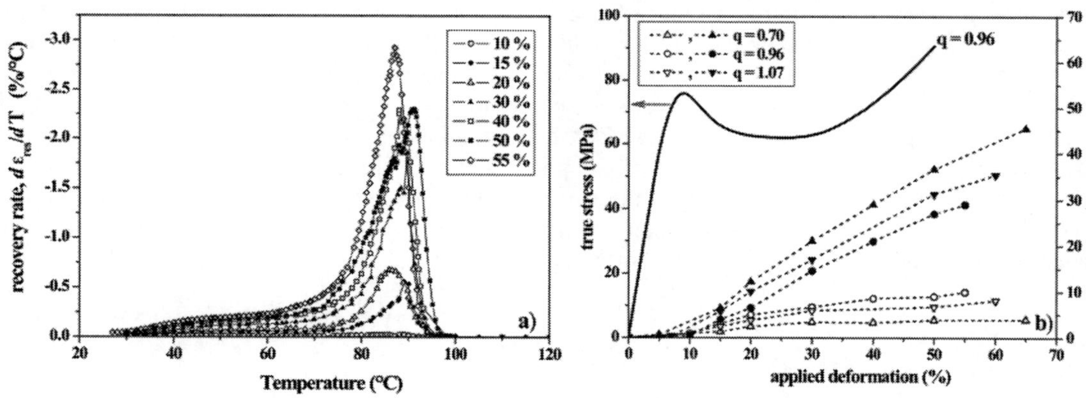

FIGURE 3. a) TSR recovery rate curves for the epoxy resin with q = 0.96 at various strain levels; b) low temperature (open symbols) and high temperature (full symbols) recovery components as a function of the applied deformation, and true stress *vs.* nominal deformation curve for the resin with q = 0.96.

REFERENCES

1. S. Pandini, F. Baldi, R. De Santis, F. Bignotti, "Epoxy/layered-silicate nanocomposites: effect of the matrix composition on large deformation and fracture behaviour" in ECCM 13 Conference Proceedings (June 2-5, 2008, Stockholm).
2. R. Quinson, J. Perez, M. Rink, A. Pavan, *Journal of Material Science* **31 (16)**, 4387-4394 (1996).
3. E. F. Oleynik, "Distortional Plasticity of Organic Glassy Polymers" in *High Performance Polymers: structure, properties, composites, fibers,* edited by E. Baer, A. Moet, Munich: Hanser Publishers, 1991, pp. 79-102.
4. H. Kawakami, H. Yamanaka, Y. Nanzai, *Polymer* **46 (25)**, 11806-11813 (2005).
5. J. Boye, J.J. Martinez, C. Lacabanne, *Journal of Thermal Analysis* **37 (8)**, 1775-1782 (1991).

Synthesis and Characterization of Poly(ethylene glycol)-block-Poly(N_α-Boc-L-tryptophan) copolymers

Andreea S. Voda and Qipeng Guo

Centre for Material and Fibre Innovation
Institute for Technology Research and Innovation
Deakin University, Geelong, Victoria 3217, Australia

Abstract. Amphiphilic diblock and triblock copolymers were synthesized from poly(ethylene glycol) (PEG) and N_α-Boc-L-tryptophan. A number of reaction conditions were investigated in order to achieve optimum yield and maximum chain extension for the newly formed block copolymers. Characterization of the novel copolymers was evaluated by means of 1D and 2D solution state nuclear magnetic resonance (NMR), infrared (IR) spectroscopy and size exclusion chromatography (SEC). The 2D NMR investigation of the block copolymers provided evidence to suggest that protonation of the nitrogen atoms present along the newly formed peptide back bone was occurring when the solvent environment was under acidic conditions.

Keywords: Amphiphilic block copolymers, poly(ethylene glycol), tryptophan, NMR.
PACS: N/A

INTRODUCTION

Block copolymers such as poly (ethylene-gycol)-b-poly (lactide) (PEG-PLA) have long been employed for use in drug delivery systems as well as tissue engineering[1, 2], however over the past few decades much research attention has been dedicated to the synthesis of a new class of amphiphilic block copolymers incorporating poly (amino acids) (PAAs)[3]. Reasons for the surge in interest include their biocompatibility as well as their ability to be enzymaticaly degraded [4] and harmlessly excreted from the body. Amphiphilic copolymers are of particular importance due to their ability to self assemble and incorporate hydrophobic drugs into their inner core whilst at the same time travelling unrecognized through the blood stream[5]. The copolymers mentioned incorporate the well known biocompatible polymer PEG, as the hydrophilic block, and the PAAs of hydrophobic amino acids such as L-leucine [6] and phenylalanine [7]. Also investigated as hydrophilic components are negatively charged amino acids such as L-aspartate [8] and glutamate [9]. This study includes the synthesis and characterization of novel diblock and triblock copolymers of PEG and poly N_α-Boc-L-tryptophan using nuclear magnetic resonance (NMR), fourier transform infrared spectroscopy (FTIR) and size exclusion chromatography (SEC) as means of analysis.

RESULTS

The novel diblock and the triblock copolymers (Fig. 1) were synthesised through the ring opening polymerization of N-carboxyanhydride (NCA) and N_α-Boc-L-tryptophan. The well documented ring opening polymerisation relies on a primary amine initiated nucleophillic attack at the C5 position of the NCA [10]. In order to optimize the synthesis several solvent environments were investigated; these solvents include dichloromethane (DCM), tetrahydrofuran (THF), dimethylformamaide (DMF) dioxane as well as a 3 : 2 solvent mix of dioxane and chloroform. Other parameters explored include reaction temperature, time allowed for synthesis, reagent concentrations and the reagent to initiator ratio.

The solvent environment which proved most desirable for the polymerization reaction was a dioxane : chloroform mix (3:2). This solvent mixture provided the greatest peptide chain extension whilst also being relatively

CP1255, *Vth International Conference on Times of Polymers (TOP) and Composites,*
edited by A. D'Amore, D. Acierno, and L. Grassia

easy to remove during the work up process. Additionally, a reaction time of 72 hours at 40 °C was found to be the optimum conditions for the ring opening polymerization reaction.

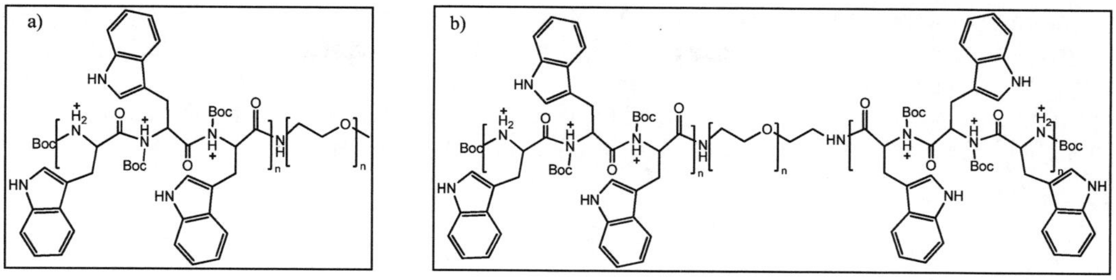

FIGURE 1. Proposed structures of PEG-poly(N_α-Boc-L-trp) (a) diblock and (b triblock.

The ^{1}H NMR spectra (Fig.2 b) for both the diblock and the triblock copolymers present new signals at 4.8 and 3.3 ppm which are characteristic of the α proton and the β protons of the polymerised tryptophan, indicated by the red square. In the tryptophan monomer the α proton is observed as a quartet at 4.94 ppm and the β protons resonate as two doublets of doublets centred at 3.72 and 3.50 ppm respectively. The considerable change in chemical shift for the three protons has been attributed to the electronegative effect gained from the coordination to the newly formed peptide bond; the appearance of a new proton signal at 5.1 ppm was also observed which correlates to the newly formed peptide link. The new proton was observed to correlate to the carbon of the α proton as well as to the carbonyl of the Boc protecting group in an HMBC experiment. A COSY experiment also found the new proton correlated to the α proton. The unknown proton in the spectrum was assigned to the hydrogen of the amide bond where the Boc group remains attached. This was believed to occur due to the acidity of the chloroform solvent during the analysis. The theory aforementioned was supported by the loss of the extra proton signal following the complete evaporation of chloroform and the measurement of the proton spectra in deuterated dimethyl sulfoxide (d6-DMSO) (Fig. 2 c)

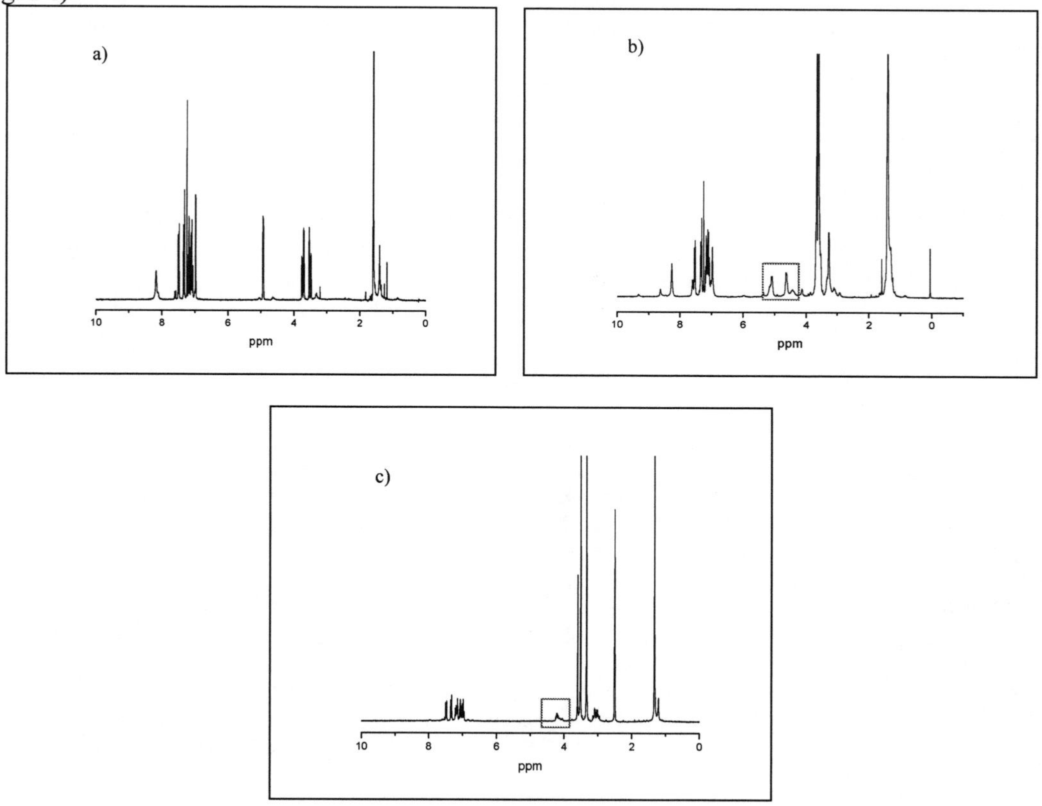

FIGURE 2. ^{1}H NMR data of (a) N-carboxyanhydride (NCA) N_α-Boc-L-tryptophan starting material in $CDCl_3$, (b) triblock copolymer in $CDCl_3$ and (c) triblock copolymer in d6-DMSO

The ^{13}C NMR spectrum of the tryptophan starting material (Fig. 2 a) exhibits three carbonyl signals at 166.5, 148.0 and 146.1 ppm respectively. Upon polymerization, there was an apparent change in chemical shift of the signals. From the spectrum it was noted that one signal completely disappeared from the spectrum leaving only two carbonyl signals; one at 172.9 ppm and the other present at 155.39 ppm. From 2D NMR studies it can be concluded that the carbonyl signal which resonates at 172.9 ppm was that of the Boc protecting group and the second signal belongs to the carbonyl now part of the newly formed peptide bond. The displacement of the third carbonyl occurred as expected (along with the oxygen) in the form of CO_2 as the peptide bond forms.

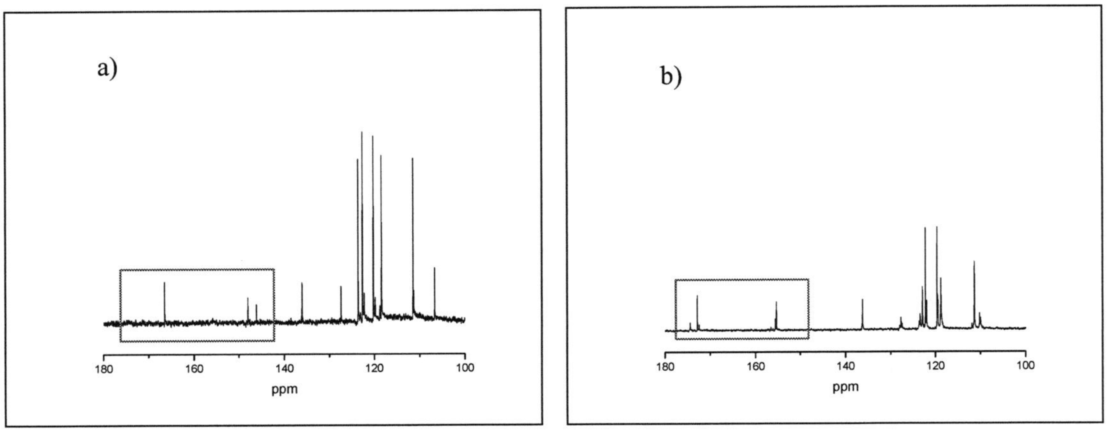

FIGURE 3. ^{13}C NMR spectra in $CDCl_3$ of (a) NCA N_α-Boc-L-tryptophan (starting material) and (b) synthesised triblock copolymer (only region from 180 ppm to 100 ppm shown for clarity).

The ^{13}C NMR data was supported by the IR spectra and peptide chain extension was further supported by successful SEC molecular weight measurements. IR and SEC data were not included due to space constraints.

ACKNOWLEDGMENTS

The authors would like to acknowledge Alfred Deakin Professor Xungai Wang and Professor Peter Hodgson for providing the financial support required to attend this conference and Deakin University for awarding A. S. Voda the Deakin University Postgraduate Research Scholarship.

REFERENCES

1. Q. Liu, C. Cai and C.-M. Dong, *Journal of Biomedical Materials Research Part A*, 2009, **88A**, 990-999.
2. C. Hiemstra, Z. Zhong, L. Li, P. J. Dijkstra and J. Feijen, *Biomacromolecules*, 2006, **7**, 2790-2795.
3. K. K. Younsoo Bae, *Advanced Drug Delivery Reviews*, 2009, **61**, 768-784.
4. C.-H. Wang, K.-R. Fan and G.-H. Hsiue, *Biomaterials*, 2005, **26**, 2803-2811.
5. S. Stolnik, L. Illum and S. S. Davis, *Advanced Drug Delivery Reviews*, 1995, **16**, 195-214.
6. K.-S. Kim, S. Chung, I.-J. Chin, M.-N. Kim and J.-S. Yoon, *Journal of Applied Polymer Science*, 1999, **72**, 341-348.
7. X. D. M. Yuan, *European Polymer Journal*, 2001, **37**, 1907-1912.
8. M. Prabaharan, J. J. Grailer, S. Pilla, D. A. Steeber and S. Gong, *Biomaterials*, 2009, **30**, 5757-5766.
9. L. Wang, S. Wang and J. z. Bei, *Polymers for Advanced Technologies*, 2004, **15**, 617-621.
10. D. L. Pickel, N. Politakos, A. Avgeropoulos and J. M. Messman, *Macromolecules*, 2009, **42**, 7781-7788.

Studying the effect of adsorbent dosage on the removal of Arsenic from aqueous solution by using polyaniline nanocomposite

Hossein Eisazadeh[a], Mohammad Soleimani Lashkenari[b] and Behzad Davodi[c]

[a] *Prof in chemical engineering department of Babol University of Technology, P.O.Box 484, Babol, Iran*

[b,c] *Ms students of Babol University of Technology, P.O.Box 484, Babol, Iran*

Abstract: In this study, the effect of dose of polymer on the removal of arsenic from aqueous solution by using polyaniline nanocomposite was studied. The conductive polyaniline was prepared using ammonium peroxy disulfate as an oxidant at room temperature. The results indicate that the polyaniline are able to remove arsenic from aqueous media. The chemical structure of obtained product was determined by FTIR spectroscopy.

Keywords: Polyaniline, chemical structure, FTIR, removal, arsenic.

INTRODUCTION

Arsenic is one of the most toxic and carcinogenic chemical elements and is regarded by the World Health Organization as the first priority issue among the toxic substances [1]. Chronic exposure to arsenic contaminated drinking water is the major cause of arsenic poisoning especially in developing countries, e.g. Bangladesh, Vietnam and China, where millions of villagers are using groundwater with arsenic concentration above the WHO guideline of 10 g/L. Some countries such as the United States and China have issued new standard for arsenic in drinking water at 10 g/L, which means that many public water systems have to adopt arsenic removal processes in their water treatment systems in order to meet the stringent drinking water standards. Various technologies are currently available to remove arsenic from water, such as ion exchange [2], coagulation (coprecipitation) [3], reverse osmosis [4], bioremediation [5], and adsorption [6,7]. Adsorption is a common practice for arsenic removal from drinking water due to technological and cost advantages. Conductive electroactive polymers such as polyaniline can be used for the removal of heavy metals from water and waste water [8, 9]. In this research the ability of polyanilin nanocomposite to remove arsenic from water and also the effect of dose of polymer have been investigated.

EXPERIMENTAL

Instrumentation

Magnetic mixer model MK20, digital scale model GF600, PH meter model pH211, scanning electron microscope (SEM) model XL30, fourier transform infrared (FTIR) spectrometer model shimadzu 4100 and inductively coupled plasma (ICP) model vista-pro were employed.

Reagents and standard solutions

All reagents were used as received without further purification, unless stated otherwise. Distilled deionized water was used throughout this work. Aniline monomer was purified by simple distillation. Materials used in this work were ammonium peroxy disulfate, arsenic chloride, aniline monomer, hydroxypropylcellulose (HPC, $Mw=10^6$) and sulphuric acid from merck.

Polyaniline preparation

For preparation of polyaniline nanocomposite (PAn), 1g of $(NH_4)_2S_2O_8$ and 0.5g of HPC were added to 100 mL of sulphuric acid (1M) and then uniform solution was resulted by using magnetic mixer.

CP1255, *Vth International Conference on Times of Polymers (TOP) and Composites,*
edited by A. D'Amore, D. Acierno, and L. Grassia

Then, 1 mL fresh distilled aniline monomer was added to stirred aqueous solution. The reaction was carried out for 5 h at room temperature. Consequently, the resulted polymer was filtered on filter paper and to separate the oligomers and impurities, product was washed several times with deionized water and dried at room temperature.

Method of removal

Completely mixed batch reactor (CMBR) technique was used to remove arsenic from water. The average dose of polymer in these experiments was approximately 0.25, 0.5, 0.75 and 1 g. A 50 mL of solution were added to the beaker containing one of the adsorbents with concentration of 0.125, 0.5, 0.75 and 1 g mixed for 30 min by magnetic mixer with its rotating speed of 500 rpm. Then the adsorbent was separated from the solution using filter paper. The analysis of As(III) was carried out using atomic adsorption method.

RESULTS AND DISCUSSION

The chemical method can be a general and useful procedure to prepare conductive polymer. The results indicate that the removal of arsenic from the solution is dependent on the ability of surface absorption. The effect of adsorbent dosage in the removal of arsenic are shown in Table 1. As can be seen in Table, the removal of arsenic was related to amount of adsorbent. Removal percentage increases by increasing the dose of adsorbent. In Figure1, SEM micrograph of polyaniline nanocomposite is shown.

TABLE 1. Effect of adsorbent dosage on the adsorption percentage of As

Type of adsorbent	Adsorbent dosage (g)	Initial conc. (ppm)	Final conc. (ppm)	Removal percentage (w/w %)
	0.125	100	13.2	
PAn		100		
	0.5	100	9.08	90.92
	1	100		

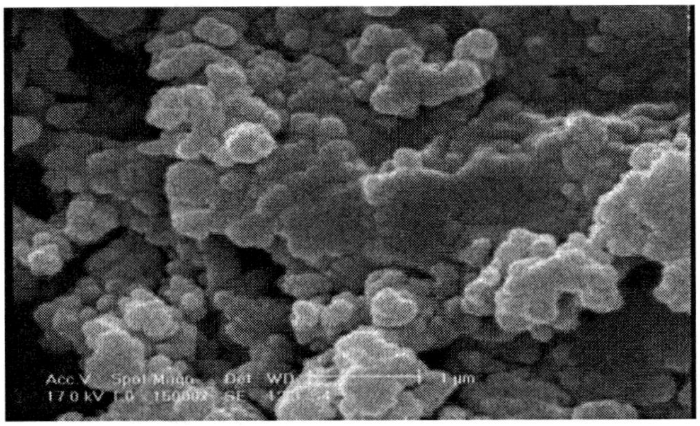

FIGURE 1. Scanning electron micrograph of PAn in aqueous media. Reaction conditions: (ammonium peroxy disulfate =10 g/L, aniline monomer 10.75×10^{-2} mol/L, hydroxypropylcellulose = 5 g/L, volume of solution 100 mL (H_2S0_4 1M), reaction time 5 hours at room temperature).

The structure of obtained product was determined by FTIR spectrum. The FTIR spectroscopy has provided valuable information regarding the formation of polyaniline nanocomposite. FTIR analysis has been done to identify the characteristic peaks of product.

FTIR spectra in the 2500-400 cm^{-1} region, for PAn is shown in Figure 2. As can be seen, the peak at about 1568 cm^{-1} is assigned to the C=C stretching vibration of the quinoid ring, 1460 cm^{-1} is assigned to the stretching vibration of C=C of the benzenoid ring, 1298 cm^{-1} correspondence to C-N stretching vibration, 1168 cm^{-1} C-H in-plane deformation and 812 cm^{-1} C-H out-of-plane deformation.

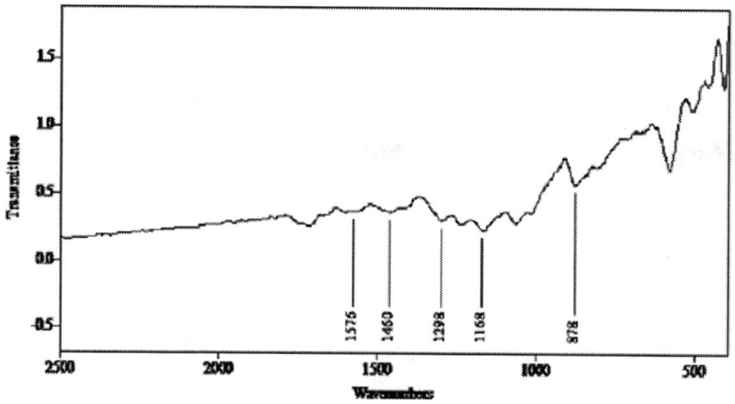

FIGURE 2. FTIR spectra of polyaniline/HPC generated in aqueous media

CONCLUSIONS

In this study, the effect of polyaniline nanocomposite and other adsorbents on the removal of arsenic from aqueous media were investigated. The results indicated that the removal percentage related to the dose of adsorbent. Removal percentage increased by increasing the dose of adsorbent. SEM micrographs show that surface morphology plays a major role in the removal percentage, because the total surface area increases as the particle size decreases.

REFERENCES

1. S. Lo, H. T. Jeng and C. H. Lai, *Water Sci. Technol.* **35,** 63–70 (1997.).
2. J. Kim and M. M. Benjamin, *Water Res.* **38**, 2053–2062 (2004).
3. A. Zouboulis and I. Katsoyiannis, *Sep. Sci. Technol.* **37**, 2859–2873 (2002).
4. R. Y. Ning, *Desalination*, **143**, 237–241 (2002).
5. T. M. Gihring, G. K. Druschel, R. B. Mccleskey, R. J. Hamers and J. F. Banfield, *Environ. Sci. Technol.* **35**, 3857–3862 (2001).
6. Q. L. Zhang, Y. C. Lin, X. Chen and N. Y. Gao, *J. Hazard. Mater.* **148**, 671–678 (2007).
7. A. Sperlich, A. Werner, A. Genz, G. Amy, E. Worch and M. Jekel, *Water Res.* **39,** 1190–1198 (2007).
8. H. Eisazadeh, *Chinese Journal of Polymer Science.* **25**, 393-397 (2007).
9. H. Eisazadeh, *Journal of Applied Polymer Science*. **104**, 1964-1967 (2007).

Three-Dimensional Model for Whipping Motion in the Processing of Microfibers

Y. C. Zeng, Y. F. Sun and X. H. Wang

College of Textiles, Donghua University, Songjiang, Shanghai 201620, People's Republic of China

Abstract. Electrospinning and Melt blowing are the most commonly used processes to produce microfibers from extruded polymer solution or melt. The present study deals with the dynamic modeling of the instabilities and related processes during producing microfibers. A bead-viscoelastic element fiber model is employed in modeling three-dimensional paths of the fiber whipping in the processes of electrosping and melt blowing. The mathematical model provides a reasonable representation of the previous published predicted and experimental data.

Keywords: Microfiber, processing, modeling, extrude
PACS: 89.20.Kk

One of the most significant developments in recent years has been the technology to extrude extremely fine fibers while maintaining all the characteristics expected by textile manufacturers and consumers. Electrospinning and melt blowing are the most commonly used processes to produce microfibers. Electrospinning can make fibers with the diameters in the range of 100 nm and even less; while the fibers produced via melt blowing are usually not smaller than 1-2 μm. Polymer microfibers find use in a range of life sciences, medical, and filtration application as well as sensors and military protective clothing.

In the process of electrosping, a high voltage is applied to create an electrically charged jet of polymer solution out from the spinneret. Before reaching the collector, the jet evaporates or solidifies, and is collected as a web of electrospun fibers. Melt-blown fibers are produced by extruding a polymer melt through the spinneret and drawing down the extrudate with high-velocity hot air stream. During fiber formation in the processes of electrospinning or melt blowing, the fiber undergoes bending instability, which is called whipping. Whipping process plays a key role in fiber elongation. During whipping motion, the jet bends into a complex path and is stretched by very large ratios. This is the key physical element responsible for enormously strong stretching and formation of microfibers.

Reneker and coworkers [1, 2] explained the observed whipping motion by a bending instability, and the physical mechanism of the electrospinning was explained and described following them. Compared to the electrospinning research, much fewer has been done on the whipping dynamics in the melt blowing process. Shambaugh's group [2, 3, 4] did a series work on modeling and measuring fiber motion. The model developed by Shambaugh's group accounts for the effect of whipping on attenuation. However, for the experimental conditions that they used, these effects (both measured and predicted effects) were small. This present work deals with the mechanism and (electro) hydrodynamic modeling of the whipping processes resulting in electrospinning and melt blowing of microfibers. We represent the fiber by a bead-viscoelastic elements model, and the Lagrange approach is used to describe the three-dimensional paths of the fiber motion.

PHYSICAL MODEL AND GOVERNING EQUATIONS

The polymeric jet in the spinning process behaves like a non-Newton fluid. We model the jet as a series of beads connected by viscoelastic elements, as shown in Fig. 1. Each bead possesses a mass m and a diameter d. A pair of adjacent beads, i-1 and i, form the jet element (i-1, i) with a length l. The forces acting on the bead are: (1) external force $\mathbf{F}_{ex}$ such as gravity, aerodynamic force, and electric force; (2) internal force $\mathbf{F}_{in}$ such as viscoelastic force and Coulomb force; and (3) bending restoring force $\mathbf{F}_b$ to restore the rectilinear shape of the bending part of the jet.

CP1255, *Vth International Conference on Times of Polymers (TOP) and Composites,*
edited by A. D'Amore, D. Acierno, and L. Grassia

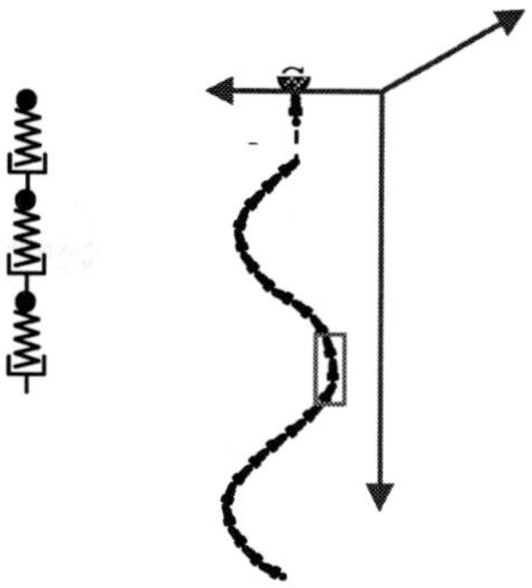

FIGURE 1. Schematic of the bead-viscoelastic element fiber model

The total number of beads, N, increases over time as new beads are inserted at the top to represent the flow of solution or melt into the jet. Therefore, Lagrange equation for fiber motion is:

$$m_i \frac{d^2\mathbf{r}_i}{dt^2} = \mathbf{F}_{\mathrm{ex},i} + \mathbf{F}_{\mathrm{in},i} + \mathbf{F}_{\mathrm{b},i} \tag{1}$$

here, the subscript i indicates bead i.

As regards the electrospinng process, a high voltage is applied so that every bead possesses a charge e. $\mathbf{F}_{\mathrm{ex}}$ is the electric force created by the applied high voltage. $\mathbf{F}_{\mathrm{in}}$ includes two forces, Coulomb force from all the other beads, and viscoelastic force from the adjacent jet elements. $\mathbf{F}_{\mathrm{b}}$ is the surface tension force tending to keep a liquid stream straight. In the melt blowing process, the aerodynamic force, instead of the electric force, pays the key role in producing the fiber. Hence, $\mathbf{F}_{\mathrm{ex}}$ is the air drag on the bead. To calculate the air drag, the airflow field under the melt blowing spinneret is obtained by solving the Navier-Stokes equations in Euler scheme through the commercial software FLUENT 6.2. There is no Coulomb force imposed on the bead, and $\mathbf{F}_{\mathrm{in}}$ is the viscoelastic force. We use Maxwell model to describe rheological behavior of the polymer. As is the case for electrospinning, the surface tension is the force to restore bending during melt blowing.

Perturbations of the lateral position and lateral velocity of the fiber lead to the development of the bending instability, and become the observed whipping motion. To model the 3-D whipping motion, an initial perturbation should be added. In our model, the initial perturbation is added by inserting small displacements in x and y coordinates of bead i

$$x_i = a\sin(\omega t) \tag{2a}$$

$$y_i = a\cos(\omega t) \tag{2b}$$

here a is the initial perturbation amplitude, and ω is the initial perturbation frequency.

RESULTS AND DISCUSSION

Fig. 2 shows the simulation results of a whipping develops with the time evolution in the processes of electrospinning and melt blowing. It is shown that with the time development, the jet follows a bending, spirling and looping path in three dimensions. Whipping in the electrospinning process is an electrically driven bending instability; while in the melt blowing process, it is an aerodynamic-driven bending instability. This mechnism leads to a suppression of the whipping during melt blowing. The shape of the whipping in electrospinning has been shown based on high speed photography [1]. The images illustrate that the expanding spiral is a typical shape of a whipping. As the perimeter of spiraling loops increase, the jet is stretched, resluting in the formation of microfibers. By resolving the governing equations, the final fiber diameter can be predicted. The mathematical model involves variables related to processing parmaters and polymer properties. By changing the values of the variables, the effects of parameters on the formed fiber diameter can be studied.

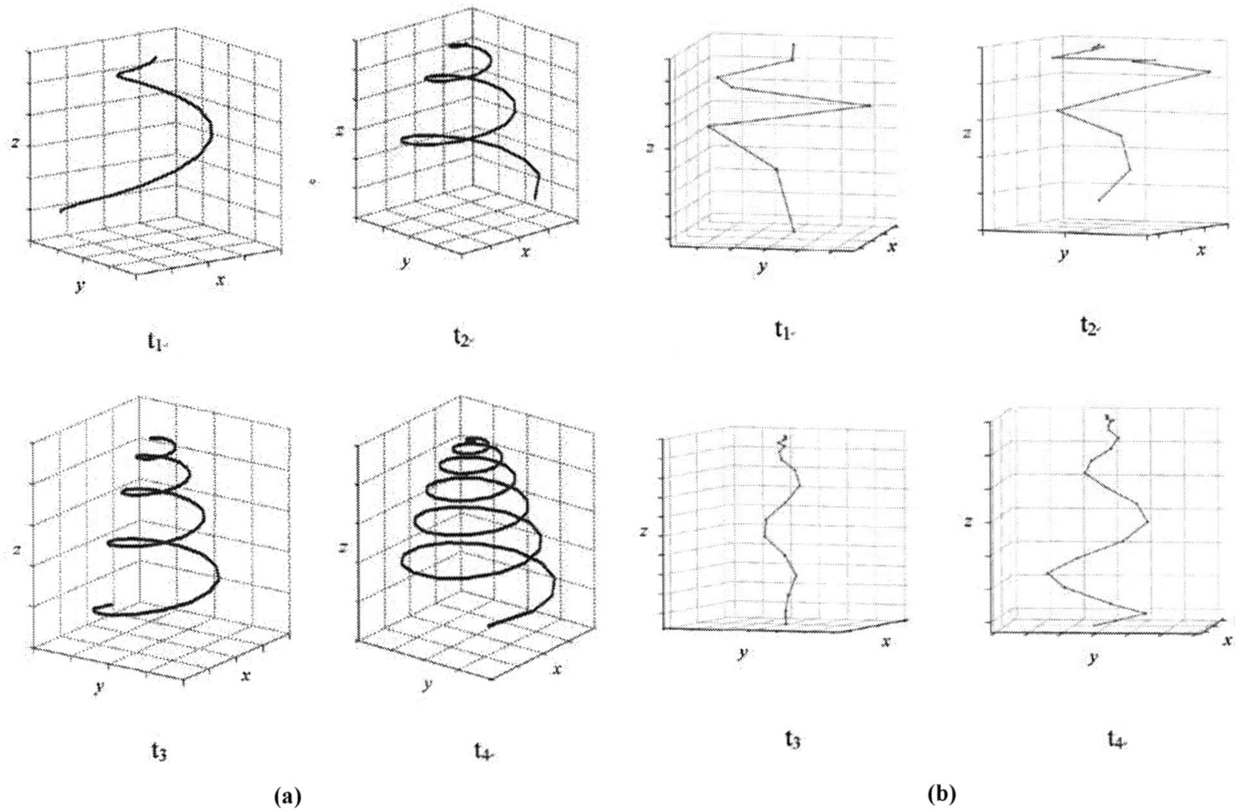

FIGURE 2. The time evolution of a whipping development. (a) electrosping process, (b) belt blowing process

CONCLUSIONS

We have proposed a method for modeling the whipping motion in electrospinning and melt blowing, which are the most commonly used processes to produce microfibers. We represent the fiber by a bead-viscoelastic elements model, and the Lagrange approach is used to describe the three-dimensional fiber paths. The whipping develops with the time evolution in the processing of microfibers is simulated. The model can also predict the formed fiber diameter, as well as the effects of parameters on fiber diameter. The theoretical results provide a useful way to optimize processing parameters.

ACKNOWLEDGMENTS

This work was supported by The National Natural Science Foundation of China (no. 10972052), and Foundation for the Author of National Excellent Doctoral Dissertation of the People's Republic of China (no. 2007B54).

REFERENCES

1. D. H. Reneker, A. L. Yarin, H. Fong and S. Koombhongse, *J. App. Phys.* **87**, 4531-4547 (2000).
2 A. L. Yarin, S. Koombhongse and D. H. Reneker, *J. App. Phys.* **89**, 3018-3026 (2001).
3. R. S. Rao and R. L. Shambaugh, *Ind. Eng. Chem. Res.* **32**, 3100-3111 (1993).
4. V. T. Marla and R. L. Shambaugh, *Ind. Eng. Chem. Res.* **42**, 6993-7055 (2003).
5. J. H. Beard, R. L. Shambaugh, B. R. Shambaugh and D. w. Schmidtke, *Ind. Eng. Chem. Res.* **46**, 7340-7352 (2007).

Computational Protocols for Viscoelastic Material Property Characterizations without the Use of Poisson's Ratios

Michael Michaeli,[a] Abraham Shtark,[a] Hagay Grossbein[a] and Harry H. Hilton[b]

[a]IMI, POB 1044, Ramat Hasharon 47100 Israel
Email: michael.michaeli@gmail.com, shtarka@gmail.com, g.hagay@gmail.com
[b]Aerospace Engineering Department in the College of Engineering and Private Sector Program Division of The National Center for Supercomputing Applications, 104 South Wright Street, Urbana, IL, USA.
Email: h-hilton@illinois.edu

Abstract. The objective of this work is to present the numerical implementation for the alternative determination of viscoelastic material properties without using Poisson's ratios as presented in [1] – [3]. The presented method is based on the 3–D generalized constitutive relations viscoelastic materials with hereditary integrals. The numerical procedures are based on experiments using photogrammetric and tensile testing instrumentation, which provide stress data in the 1–D loaded direction and strains in both longitudinal (loaded) and transverse directions. Measurements and data analyses include both starting transient and steady-state loading conditions. The paper presents the implementation of solutions for the linear case, where the relaxation time values are prescribed according to to the scheme presented in [1] and [3]. Convergence of the Prony series representations is evaluated.

Keywords: constitutive relations, experimental & analytical material characterizations, Poisson's ratio, viscoelasticity.
PACS: 02, 06, 61.

INTRODUCTION

There are four major problem areas associated with viscoelastic material property characterization [1] – [13]:

Strain gages – Electrical resistance strain gages, which have seen active and valuable service for three quarters of a century, are primarily intended for testing in conjunction with metals. Their matching of moduli with these materials creates a difference of three to six orders of magnitude if they were installed on high polymers. Consequently, instead of measuring true strains in polymer specimen, strain gages provide substantial local reinforcements and provide otherwise absent strains [1] – [3]. These shortcomings can be overcome by using optical strain measurement techniques instead of strain gages as has been done in the present paper.

Poisson's ratios – Isotropic elastic material properties can be fully characterized by any two of the following: Young's, shear, bulk moduli, Lamé constants, corresponding compliances and Poisson's ratios. While these elastic constants have readily definable viscoelastic counterparts in the form of time dependent functions, only the PRs stand out as unacceptable descriptors of viscoelastic material behavior. This is due to the fact that linear viscoelastic PRs are not only time dependent but also functions of stresses and their time histories. Thus, these PRs cannot be defined as material behavior parameters in the class of moduli, compliances, and creep and relaxation functions [4] – [8]. Furthermore, PRs do not contribute to the elastic – viscoelastic correspondence principle and their use in stress–strain analysis as well as in material characterization is counter indicated.

Starting transients – Elastic media, being conservative, have no capacity for memory effects and/or energy dissipation. Consequently, the manner of initiating loadings or strains, unless severe dynamic effects are involved, does not affect subsequent responses. Viscoelastic materials, on the other hand, being hypersensitive to both phenomena, are influenced by loading cycles [12], [13]. Therefore, care must be exercised to include starting transient effects in the analyses of experimental data when characterizing responses.

The instantaneous elastic modulus – Independently of starting transients, the determination of the correct and proper values of the various moduli from quasi-static experiments remains a high threshold to pass though [12], [13]. The only foolproof approach to correct determinations of initial/instantaneous viscoelastic moduli remains by means

CP1255, *V[th] International Conference on Times of Polymers (TOP) and Composites,*
edited by A. D'Amore, D. Acierno, and L. Grassia

of wave propagation experiments [14] – [16]. In the absence of dynamical experiments, several quasi-static protocols for the determination of instantaneous moduli are formulated and evaluated. These are most suitable for analyzing the high fidelity data harvested by experimental 1–D quasi-static photogrammetric procedures.

THEORETICAL, COMPUTATIONAL AND EXPERIMENTAL OVERVIEW

The analysis of viscoelastic PRs has been extensively covered in the literature [4] – [11]. Unfortunately, their definitions have been wrongly interpreted in a sizable number of publications as is shown in detail in [5] – [11]. The original classical PR definition in terms of two orthogonal normal strains [17] applies to all deformable media. Upon substituting the viscoelastic constitutive relations, one notes that that even in the 1 – D loading case the stresses are trapped inside both time integrals and that the PRs are not only time functions but dependent on the stress components and their time histories as well [5] – [7]. Consequently, the viscoelastic PRs are not material properties to be used for material characterizations as they are non-universal functions that depend on the specific ad hoc local processes which are used preferentially and/or randomly to define the experiments.

Different experiments will produce highly distinct PR time functions, whereas moduli, compliances, creep and relaxation functions are universal material properties. The use of viscoelastic PRs for material characterizations, as well as in stress–strain analyses, is counter indicated for all types of stress fields including 1–D ones and must be avoided [4] – [7].

As a matter of convenience, 1–D characterization experiments are conducted under one of the following conditions: constant stress, constant stress rate, constant strain or constant strain rate. Physically, none of these loading conditions can be achieved instantaneously and they must undergo a brief loading cycle from zero to their constant value. The important influence of starting transients on accurate viscoelastic material property determinations is analyzed and evaluated in [1] – [3], [12] and [13]. Data must be collected during these loading cycles and used in the viscoelastic characterization curve fitting procedures.

For the 1 – D stress state, two constitutive relations are available to determine the two unknown creep relaxation functions. This can be readily accomplished with measurements of one stress and two strains in perpendicular directions at selected times. The integrals contain the Prony series representing the viscoelastic material properties and C^n_{ii11} are the unknown parameters to be determined from the experimental data, while the τ_n values are pre-selected

$$\underbrace{\epsilon_{ii}(t)}_{i=1,2} - \underbrace{C^0_{ii11}\sigma_{11}(t)}_{\substack{=\text{ instantaneous}\\ \text{elastic}\\ \text{response}}} = \underbrace{\sum_{n=1}^{N_{pr}} \frac{C^n_{ii11}}{\tau_n}\int_0^{t_1} \exp\left(-\frac{t-t'}{\tau_n}\right)\sigma_{11}(t')\,dt'}_{=\text{ start up transient response}} + \underbrace{\sum_{n=1}^{N_{pr}} \frac{C^n_{ii11}}{\tau_n}\int_{t_1}^{t} \exp\left(-\frac{t-t'}{\tau_n}\right)\sigma_{11}(t)\,dt'}_{=\text{ response to stabilized loadings}} \quad t > t_1$$

In summary, the numerical protocol is as follows:

Numerical differentiation: In order to determine initial instantaneous compliances, the strain – time derivatives are calculated numerically, using the forward difference formula. Sufficient quantities of very short time data must be collected.

Numerical integration: The convolution integrals in the constitutive relations are calculated numerically with Simpson's rule for odd and even number of intervals.

Linear least square (LSQ): In order to display the numerical implementation, one considers the linear case where the relaxation times are pre-selected according to procedures described in [1] – [3]. A least square formulation is used where one solves for the Prony series coefficients using the sampled stress, strain and time.

ACKNOWLEDGMENTS

Support by grants from IMI at Ramat Hasharon, Israel and from the Private Sector Division of the National Center for Supercomputing Applications (NCSA) at the University of Illinois at Urbana-Champaign (UIUC) is gratefully acknowledged.

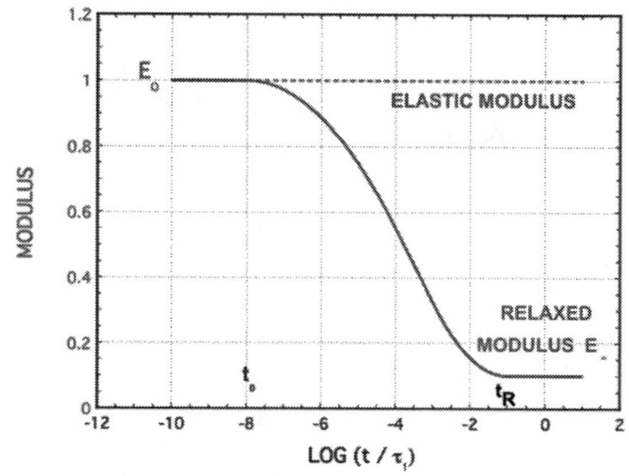

FIGURE 1. Relaxation modulus

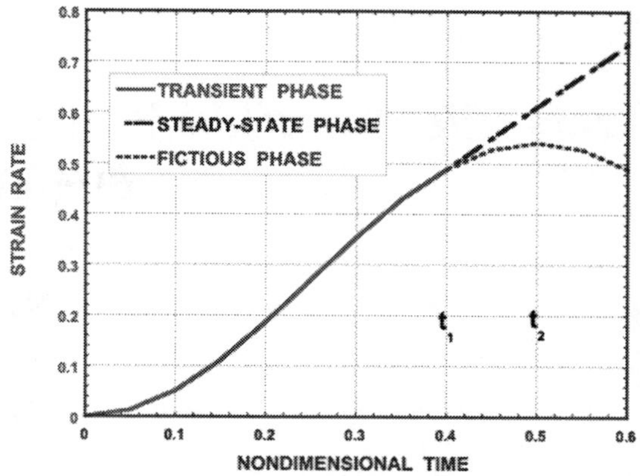

FIGURE 2. Initial loading phase

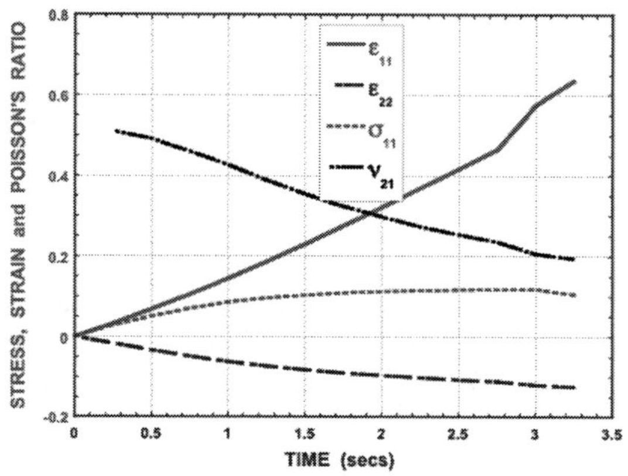

FIGURE 3. Typical experimental data

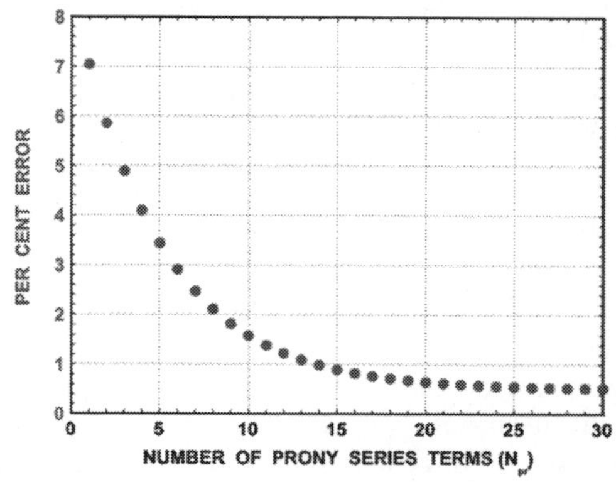

FIGURE 4. Convergence of LSQ fit

REFERENCES

1. A. Shtark, H. Grosbein, G. Sameach and H. H. Hilton, in *Proc. Int. Mech. Eng. Cong,. ASME IMECE2007-41068,* 2007.
2. A. Shtark, H. Grosbein and H. H. Hilton, in *Proc. Int. Mech. Eng. Cong., IMECE2009-10332*, 2009.
3. A. Shtark, H, Grosbein, G. Sameach and H. H. Hilton, In press in *ASME J. App. Mech., JAM08- 1361*, 2010.
4. H. H. Hilton, *Int. J. Mech. Comp. Mat. Struct.* **3**, 97–100 (1996).
5. H. H. Hilton and S. Yi, *Int. J. Sol. and Struct.* **35**, 3081–3095 (1998)
6 H. H. Hilton, *J. Elast.* **63**, 221–251 (2001)
7. H. H. Hilton, *J. Mech. Mat. Struct.* **4**, 1341–1364 (2009)
8. K. Khan and H. H. Hilton, *J. Therm. Str.* **33**, 29–36 (2010)
9. N. W. Tschoegl, *Mech. of Time–Dep. Mat.* **1**, 3–31 (1997)
10. N. W. Tschoegl, W. G. Knauss and I. Emri, *Mech. Time-Dep. Mat.* **6**, 3–51 (2002)
11. R. S. Lakes and A. Wineman, *J. Elast.* **85**, 45–63 (2006)
12. C. E. Beldica and H. H. Hilton, in *Proc. IUTAM Symp. on Mech. Waves for Comp/ Str. Char.,* 2002, Chania, Crete,, 27-32.
13. C. E. Beldica and H. H. Hilton, in *press J. Sand. Str. Mat.* (2010)
14. D. R. Bland, *The Theory of Linear Viscoelasticity.* Pergamon Press, New York, 1960
15. H. Kolsky, *Stress Waves in Solids.* Dover Publications, New York, 1963
16. J. F. Doyle, *Wave Propagation in Structures: Spectral Analysis Using Fast Discrete Fourier Transforms.* Springer, New York, 1997
17. S. D. Poisson, *Mémoires de l'Académie Royal des Sciences de l'Institut de France* **8**, 357–570, 623–627 (1829)
18. H. H. Hilton, "An introduction to viscoelastic analysis," *Engineering Design for Plastics*, edited by E. Baer, 1964, pp. 199–276. Reinhold Publishing Corp., New York.
19. R. M. Christensen, *Theory of Viscoelasticity - An Introduction,* Academic Press, New York, 1982

Design Feature Sensitivity Analysis in Numerical Modeling of Extrusion Spider Dies

A.G. Mamalis[a], G. Kouzilos[a] and A.K. Vortselas[a]

[a]*Laboratory of Manufacturing Technology, National Technical University of Athens, 9 Iroon Polytechniou str., 15870, Athens, Greece*

Abstract. The work described here, is aiming to the development of an optimum design for a spider die used for the extrusion of high density polyethylene (HDPE) tubes. For this purpose, a computational fluid dynamics (CFD) based model using the generalized Newtonian approach was employed, to investigate pressure drop, flow and temperature uniformity in the die.

Keywords: HDPE, extrusion, spider die, Carreau model, non-Newtonian fluid, design optimisation.
PACS: 47.11.Hj; 47.50.Cd; 81.20.Hy

INTRODUCTION

The production of extruded polyethylene film, rods, tubes and pipes is a common industrial process that has been the subject of investigations by the authors and others over many years [1, 2, 3, 4]. The majority of plastic tubes and components for fluid transport and other industrial applications are mainly manufactured by polymer extrusion using an annular 'in-line' or 'straight-through' die. What is most important when designing such a die is to secure the inner mandrel of the die to its external casing by means of ties (spider legs).

The work described here, is aiming to the development of an optimum design for a spider die used for the extrusion of high density polyethylene (HDPE) tubes. For this purpose, a computational fluid dynamics (CFD) based model using the generalized Newtonian approach was employed, to investigate pressure drop, flow and temperature uniformity in the die. CFD-based approaches have been developed previously by other researchers to study various aspects of flow in extrusion dies [5, 6].

EQUIPMENT AND PROCEDURE

The spider die was manufactured using IMPAX tool steel and consists of 5 zones (Fig. 1a):

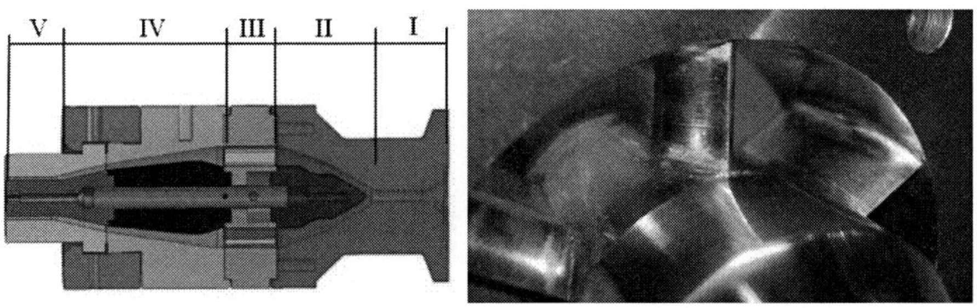

(a) (b)

FIGURE 1. a) Spider die zones, b) detail of reference geometry spider legs.

- **Zone I** constitutes the fluid entrance zone in the die.
- **Zone II** is the area, where the fluid diverts from the extrusion axis is distributed through the mandrel cone into a ring-shaped cross-section.
- **Zone III** is the spider zone, in which one end of the die (male) is constrained by the spider legs (Fig.1b).

CP1255, *V[th] International Conference on Times of Polymers (TOP) and Composites,*
edited by A. D'Amore, D. Acierno, and L. Grassia

- **Zone IV** is the relaxation zone, where the flow marks generated by the spider legs, can be extinguished, leading in this way to a uniform flow.
- **Zone V** is the final stage before the extrusion product exits the die; in this zone the pipe takes its' final dimensions and shape.

Since the spider die is intended to be used for the production of HDPE tubes, it was mounted on a single screw Johnson Plastics Extruder with characteristics: length/diameter ratio 24:1, screw diameter 38 mm and compression ratio 2.75. High density polyethylene tube with external diameter of 32 mm and wall thickness 2.4 mm was produced using the above mentioned equipment. The polymeric material used was the SABIC B 5823. [1]

NUMERICAL MODELLING

A three-dimensional conjugate heat transfer model has been developed for non-Newtonian materials being processed in the extrusion die. The numerical solution assumes that, a homogeneous and isotropic HDPE melt with a uniform temperature of $T=469K$ is flowing into the spider die. The temperature of the die wall is kept at a constant value, $T_w=469K$ and the volumetric flow rate of the melt is fixed at $\dot{V}_{max} = 7,9 \cdot 10^{-6} m^3/\sec$.

Polymer melts are non-Newtonian fluids and various models have been developed to describe the dependence of viscosity on shear rate and temperature. Flexibility is provided by the Carreau–Yasuda model [7], described as

$$n = a_T n_0(T_R)[1 + (a_T \dot{\lambda}(T_R)\dot{\gamma})^{\alpha}]^{(n-1)/a} \tag{1}$$

where $n_0=5.43 \cdot 10^3\ Pa \cdot s$, $\dot{\lambda}=0.063s$, $\alpha=2$ and $n=0.2723$ are the fitting parameters of the model and a_T is the shift factor given as

$$a_T = \exp[Eo/R(1/T_R - 1/T)] \tag{2}$$

The temperature dependence is introduced in Eq. (2) through the shift factor, a_T, while $E_0=6.5kcal/mol$ is the activation energy and R is the universal gas constant.

Considering the characteristics of the polymer melts flow in the die channel when the steady-state of the extrusion is achieved, the following assumptions are made:

- Incompressible steady laminar flow prevails, i.e. this is a steady-state simulation.
- No slip conditions on the wall surface pertain, i.e. the melt flow velocity on the fluid-die/mold interface is equal to the moving velocity of the die wall.
- Inertial and gravitational forces are neglected, since the Reynolds number of polymer melts is relatively low.

This numerical model incorporates the effect of viscous dissipation, which causes an increase in the fluid temperature. Fig.2a shows the pressure distributions in the entire domain of the die, as calculated by the FE simulation for the reference geometry. The pressure decreases continuously from die inlet to outlet, for a total pressure drop of $8.9MPa$, which is validated by experimental results.

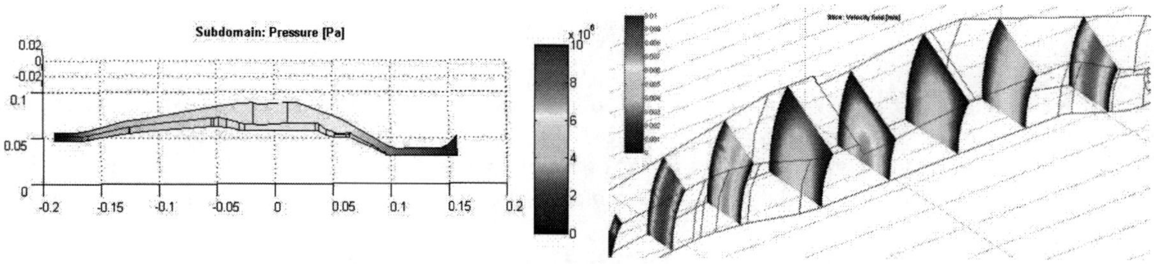

(a) (b)

FIGURE 2. a) Pressure distribution. b)Velocity distribution at various cross-sections around the spider leg.

SPIDER DIE GEOMETRY ANALYSIS

In polymer extrusion processes, product quality is affected by the homogeneity of flow speed and temperature at the die exit. Inhomogeneity causes shape distortions as well as material weak spots (weld lines). [8] The major cause for such inhomogeneity in spider dies is the discontinuity and distortion caused to the flow by the spider legs (Fig.2b), whereas the die sections aft of the spider work to re-homogenise the flow. Five geometric characteristics of the spider leg cross-section, plus the length of the relaxation zone, were selected for a sensitivity analysis, to help us determine the areas where design optimisation would me most effective.

Taking design limitations into consideration, three levels were selected for each parameter and a Taguchi design of experiments was drafted, for a total of 27 experiments. For each one of them, the respective spline curve was created (Fig.3a) and used to modify the geometry of the spider part in a CAD environment. The fluid volume was extracted from the die assembly and $1/12^{th}$ of it was meshed (taking advantage of symmetry planes) and imported in the FEM environment (Fig.3b). The signal-to-noise ratio for flow velocity at the nodes of the melt exit cross-section was chosen as the objective function for homogeneity (less is better):

$$\mathrm{NPM} = 10 \cdot \log_{10}\left(\frac{\overline{X}^{b}}{s}\right)^{2} \tag{3}$$

$\overline{X}$, s are the mean and standard deviation of velocities and b is derived from the least-squares regression: $\log_{10}(s_i) = a + b \cdot \log_{10}(\overline{X}_i)$. All the above steps were scripted in a programming environment, enabling us to perform the number of simulations required with time restrictions imposed only by the run time of the FE solver (up to 10 min each on a PC).

(a) 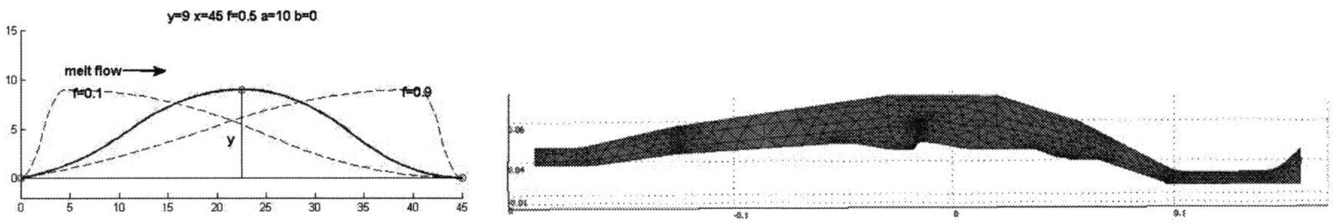 (b)

FIGURE 3. a) Spider leg cross-section (spline curve) to be optimised b) Adaptive meshing of the melt volume.

NPM values range between 7.49 and 7.18 for the optimum point (Table 1). ANOVA results indicate that the effect of exit length is overwhelming, thus confirming the empirical choice of maximising the length of zone V. Of the other parameters, only the spider leg length X has a significant effect.

TABLE (1). Optimisation parameters, DOE levels, optimum point and ANOVA sensitivity results.

	Parameters	Level 1	Level 2	Level 3	Optimum	ANOVA
L	Exit length (mm)	24	34.5	45	**45**	82%
Y	Leg width (mm)	9	10.5	12	**9**	1.6%
X	Leg length (mm)	30	55	80	**80**	12,6%
f	Spline form factor	0.1	0.5	0.9	**0.5**	1.9%
a	Spline front angle (deg)	0	5	10	**0**	0.5%
b	Spline back angle (deg)	0	5	10	**10**	0.1%

CONCLUSIONS

A non Newtonian CFD model of the spider die polymer extrusion process, correlating well with experimental results, has been created. A streamlined design process, enabling the optimisation of die design features has been implemented. A sensitivity analysis performed on the effect of six design parameters on flow homogeneity is used to validate design choices and pinpoint the features for which further optimisation would be feasible.

REFERENCES

1. A.G. Mamalis, K.N. Spentzas, G. Kouzilos, I. Theodorakopoulos and N.G. Pantelelis, *J. Advances Polym Techn* (in-Press).
2. H. Yan and J. Xia, *Sci. and Tech. of Advanced Mat.* **7**, 127-131 (2006).
3. F. Ding and A. J. Giacomin, *Polymer Eng. and Sci.* **44**, 1811-1827 (2004).
4. R. K. Krishnaswamy, D. C. Rohlfing, A. M. Sukhadi and K. R. Slusarz, *Polymer Eng. and Sci.* **44**, 2266-2273 (2004).
5. M. Avcı and O. Aydın, *C. R. Mecanique* **334**, 164-169 (2006).
6. B. Yang and L. J. Lee, *Polymer Eng. and Sci.* **27**, 1079 – 1087 (1987).
7. V. Nassehi: *Practical Aspects of Finite Element Modelling of Polymer Processing*, London: John Wiley & Sons, 2002.
8. W. Michaeli, *Design and Engineering Computation, 2nd ed.*, Munich: Hanser Publishers, 1992.

In-situ Monitoring of Polymer Crystallization by Polarization-Sensitive Optical Coherence Tomography

Peter Hierzenberger[a], David Stifter[b], Elisabeth Leiss-Holzinger[c], Gerhard Eder[a]

[a]*Institute of Polymer Science, Johannes Kepler University Linz, Altenbergerstr. 69, 4040 Linz / Austria*
[b]*Christian Doppler Laboratory for Microscopic and Spectroscopic Material Characterization, Center for Surface- and Nanoanalytics, Johannes Kepler University Linz, Altenbergerstr. 69, 4040 Linz / Austria*
[c]*Research Center for Non Destructive Testing GmbH, Hafenstrasse 47-51, 4020 Linz / Austria*

Abstract. The use of polarization-sensitive optical coherence tomography for monitoring the progress of a classical shear experiment in a slit die leads to new insights in the mechanisms of polymer crystallization. For the first time it is possible to get both depth and time resolved information on the development of different layers as a result of the processing conditions. By additionally measuring the transmitted light the method is at the same time fully compatible to conventional birefringence measurements. Microscopic analysis of the extruded stripes completes the description of the crystallization process.

Keywords: shear-induced crystallization, optical coherence tomography, birefringence measurements
PACS: 42.30.Wb, 42.25.Hz, 81.10.Fq, 81.10.Aj, *47.85.md*

INTRODUCTION

For the study of the influence of flow on the crystallization kinetics, duct flow experiments, where the undercooled polymer melt undergoes a shear treatment in a slit die, have been proven to be a useful tool for the investigation of polymer solidification. Especially so-called short term shearing experiments, as introduced by *Liedauer et al.* [1], provide insights into the mechanisms of flow-induced crystallization, since the effects of nucleation and growth are effectively separated. *Liedauer et al.* used a slit die of large aspect ratio connected to a single screw extrusion line including a relaxation part, an accumulator for hot polymer melt and two valves to apply box-like pressure profiles on the die. Close to the exit of the die, in the large side walls, were two glass windows through which polarized laser light was directed. By following the change of polarization as a consequence of the formation of birefringent structures during crystallization the kinetics of crystallization could be investigated. Additional information was obtained by characterizing the morphology of the extracted sample after cooling to room temperature. As is well known, three zones can be discerned: in the middle of the sample, where only low shear gradients exist, a spherulitic core is formed whose morphology is basically identical to that of a sample cooled under quiescent conditions. This core passes into a so-called fine grained layer, where the spherulites are too small to be recognized. Finally, close to the interface between sample and die the high shear gradients provoke the formation of highly-oriented and therefore highly birefringent layers, in which the morphology changes to shish kebabs.

For monitoring of the progress of crystallization *Liedauer et al.* used the increasing optical retardation: the time to reach half the wavelength of the laser light was used to compare results obtained under variation of temperature, pressure and shearing time. However, these observations can only be performed in the transparent regime when crystallization has not yet proceeded too far. Besides, the pure birefringence signal only gives integral information over the whole sample thickness, but not spatially resolved information on the individual layers. A solution to these problems is provided by a rather new technique called optical coherence tomography (OCT). Originally used for biomedical diagnostics, *Stifter* [2] showed its applicability also for non-destructive material testing. In particular, OCT provides cross-sectional images of non-transparent, scattering media in a contact-free way. In this paper, we show, to the best of our knowledge, for the first time the application of an extended OCT technique, namely polarization-sensitive OCT (PS-OCT), for the monitoring of shearing experiments and first qualitative results.

CP1255, *V^th International Conference on Times of Polymers (TOP) and Composites,*
edited by A. D'Amore, D. Acierno, and L. Grassia

APPARATUS

Figure 1 shows the experimental set-up of our apparatus, consisting of three main parts: the PS-OCT system, the extrusion line, and the transmission detection unit. The extruder part is similar to the one used by *Liedauer et al.* with a slit die (width 1 mm, height 10 mm, length 110 mm) with two windows close to the die exit.

The PS-OCT system operates in reflection geometry and basically consists of a Michelson interferometer, with one mirror replaced by the sample, i. e. the die, which is irradiated by broadband light with a center wavelength of 1560 nm. Structures in the sample partially reflect the light which is interfered with the light of the reference arm of the interferometer. By spectral analysis of the obtained interference spectrum spatial information over the sample depth is obtained (i.e. spectral-domain OCT [2]). Lateral scanning with a galvano scanner leads to cross-sectional images of the sample, leading to so-called reflectivity images. For the analysis of the polarization state of the light polarizers and quarter wave plates are added, giving additional information on the birefringence properties of the sample, encoded in so-called retardation images, assigning different colors to different retardation values.

The last part is the transmission detection unit which corresponds to the measurement set-up of *Liedauer et al.* and is used for additionally obtaining conventional transmission data. All the experimental raw data, together with temperature and pressure data of the extruder, is digitally recorded and evaluated.

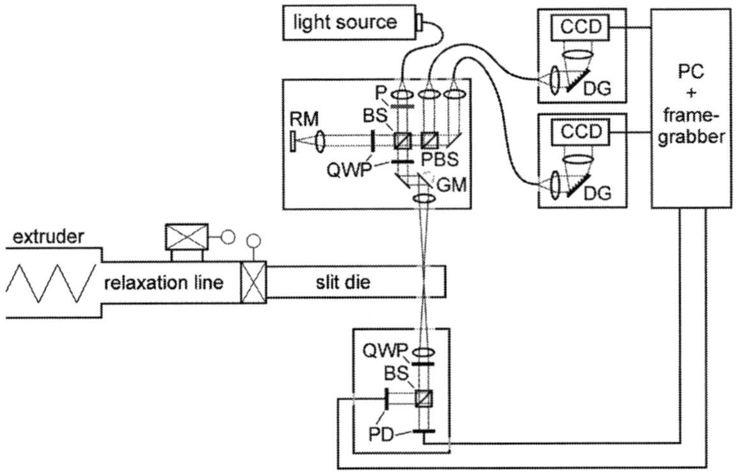

FIGURE 1. Sketch of the experimental set-up, containing the PS-OCT system, the extrusion line and the transmission detection unit. Abbrevations: reference mirror (RM), polarizer (P), beamsplitter (BS), polarizing beamsplitter (PBS), quarter wave plate (QWP), galvano scanner mirror (GM), diffraction grating (DG), line camera (CCD), photodiodes (PD)

RESULTS AND DISCUSSION

The material investigated was an isotactic poly(propylene) DM55pharm from Borealis. It was first extruded through the die heated to 200 °C, then held 15 min at that temperature and finally quenched to 140 °C within another 15 min. As soon as the temperature was equilibrated, image acquisition was started and pressure applied as described in [1] and [3]. Fig. 2 shows the resulting reflectivity images at different stages of crystallization. The horizontal white lines at the top and bottom of the images correspond to the polymer / glass interfaces, with the flow perpendicular to the image plane. With 150 s, a quite long shearing time was used in this experiment and images (a) and (b) were taken still during the shearing period. The growth of a scattering layer is clearly recognised, starting immediately or at least shortly after the onset of shear and with decreasing speed over time. On the last image (f), the formation of vacuoles can be seen caused by the volume reduction in the solidified polymer (area surrounded by bright lines in the middle of the sample).

The corresponding retardation images give complementary information on the birefringence, as shown in Fig. 3. In this figure the heavily sheared sample of Fig. 2 is compared to a sample that was sheared for only 25 s; each sample is presented in a reflectivity image of the cooled sample, the corresponding retardation image, and a picture of a microtome section of the sample taken with a standard optical microscope under crossed polarizers. For the retardation images, a change from blue to red signifies an increase of the optical retardation by 90°.

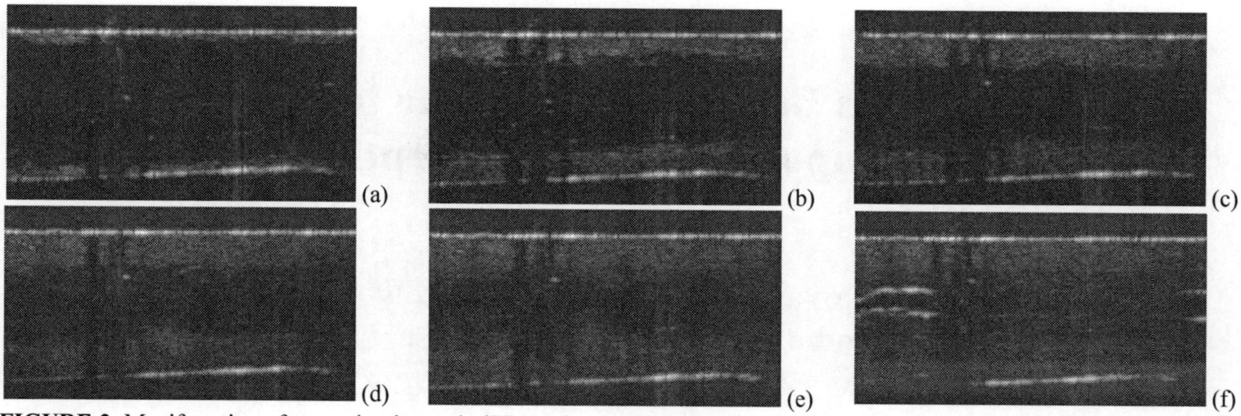

FIGURE 2. Manifestation of scattering layers in iPP as observed in reflectivity images after different time intervals: a = 50, b = 148, c = 250, d = 600, e = 1460 and f = 4520 s after the onset of shear; temperature 140 °C, pressure 200 bar, shearing time 150 s.

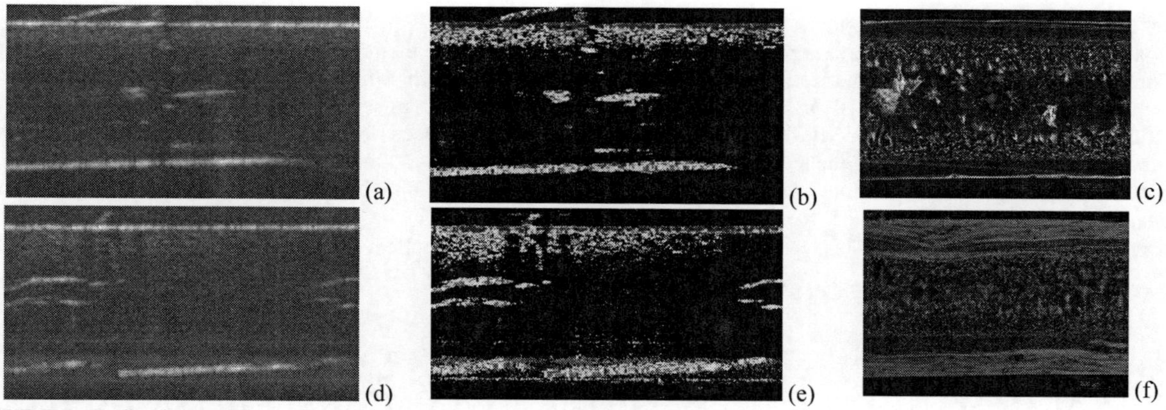

FIGURE 3. Reflectivity and retardation images from OCT compared with ex-situ optical microscopy images of the cooled samples. a-c: shearing time 25 s, d-f: shearing time 150 s

As can be seen, for the mild shearing conditions there is only a small change from blue to green (~45°). At the same time the OCT-reflectivity hints to a thin uniform scattering layer corresponding to a rather uniform layer of similar thickness in the microscopy image. In contrast, retardation in the heavily sheared sample reaches in average higher values (yellow to red, ~60°-90°), corresponding to a thicker scattering layer extending over the outermost third of the sample cross-section.

Further experiments are currently underway to increase our knowledge and interpretation of PS-OCT images and time sequences, also by comparison with traditional transmission measurements, with the goal to obtain quantitave information on flow induced crystallization. The fact that we are now able to derive data from the inside of turbid samples enables for the first time a proper analysis for longer shearing times, in which – as in processing – solidification and flow occur simultaneously.

ACKNOWLEDGEMENTS

The financial support by the Austrian Science Fund (project P19751-N20), the Federal Ministry of Economy, Family and Youth and the National Foundation for Research, Technology and Development is gratefully acknowledged.

REFERENCES

1. S. Liedauer et al., *Int. Polym. Proc.* **3**, 236-244 (1993)
2. D. Stifter, *Appl. Phys. B* **88**, 337-357 (2007)
3. G. Eder and H. Janeschitz-Kriegl, in *Processing of Polymers*, edited by H.E.H. Meijer, *Mat. Sci. Techn.* **18**, 269-342 (1997)

Virus Sorbents Based on Polyaniline Interpolymer Complexes, Composites and their Sorption Properties

Victor F. Ivanov[a], Valeria T. Ivanova[b], Yanina E. Kurochkina[b], Oksana L. Gribkova[a], Maria V. Ilyina[b], Anatoly A. Manykin[b], Aleksandra A. Isakova[a], Irina Yu. Sapurina[c]

[a]*A.N. Frumkin Institute of Physical Chemistry and Electrochemistry RAS, 119991 Moscow, Leninsky Pr., Russia,*
[b]*D.I. Ivanovsky Institute of Virology RAMS, 123098, Moscow, Gamaleya St.16, Russia,*
[c] *Institute of Macromolecular Compounds RAS, 199004, St. Petersburgr. Bolshoy Pr.31,Russia*

Complexes between influenza viruses proteins and antibodies to them create the precipitation lines in polyaniline (PAn) films which were detected by polarization optical microscopy after coating the film from the top the eutectic mixture of nematic liquid crystals MBBA:EBBA. Polyaniline and it's composites interact as sorbents with influenza human and bird viruses apart from antigenic properties of surface proteins as well as with bacteriophage T4-D. The sorption of virions occurred in different solutions in the temperature range (8-25 °C) during 15 - 60 min. After contact with sorbent the HA titers of influenza virus in solutions were decreased in 4 -2048 times. The immobilized influenza viruses were able to interact with homological antibodies from immune sera in solution.

Keywords: Polyaniline, composites, virus, bacteriophage T4-D, sorption, optical detection.

PACS: 81.16.Fg, 81.70Fy, 82.35.Cd, 83.80.Xz

OPTICAL DETECTION OF VIRUS – ANTIBODY COMPLEXES IN POLYANILINE FILMS

In this report we present the results of the investigation of interaction between the biological objects and nanocomposites of polyaniline (PAn). PAn relates to typical conducting polymers stable in various environmental conditions. It can be easily modified and changes reversibly the oxidation degree thus influencing on their sorption properties. We have studied the interaction of influenza viruses proteins and antibodies to them on the surface of PAn films using the eutectic mixture of nematic liquid crystals (NLC) MBBA:EBBA(n-methoxybenzyliden-n-butylaniline, ethoxybenzyliden-n-butylaniline) [1]. It was found that the influenza viruses with homological antibodies (homological system) formed the crystal like complexes on the surface of PAn films (Fig.1) detected by polarization microscopy. This phenomenon was not observed in the in the case of using of heterogeneous antibodies from immune serum. This result remembers the precipitation lines in reaction of immune diffusion in agar or agarose for detection the homological system (antigen + antibody) in biology.

VIRUS INTERACTIONS WITH DIFFERENT SORBENTS

The supramolecular complexes of polyaniline in acidic forms (PAn – HCl and PAn- PAMPSA) were synthesized in the presence of the low molecular hydrochloric acid (HCl) and polymeric acid PAMPSA (poly-2-acrylamido-2-methyl-1-propanosulfonic acid). The sorption properties of PAn base and salt forms with respect to some influenza human viruses have been investigated [2]. Now we present the results of investigation of pandemic, epidemic human influenza strains, influenza bird viruses and bacteriophage with the same and other sorbents - the carbon nanotubes and nanocomposites - carbon nanotubes covered by PAn [3].

CP1255, *Vth International Conference on Times of Polymers (TOP) and Composites,*
edited by A. D'Amore, D. Acierno, and L. Grassia

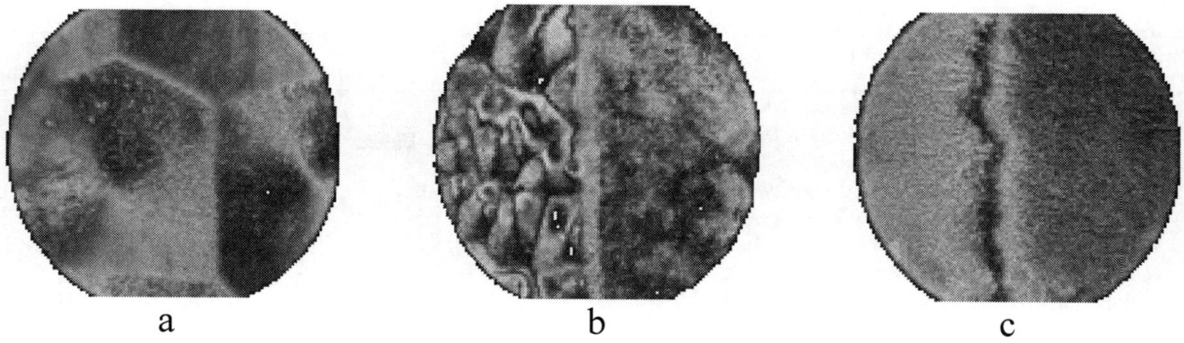

FIGURE 1. Detection of influenza viruses type B by the eutectic mixture of nematic liquid crystals MBBA:EBBA, x 500: a) polyaniline:HCl layer / layer of complex of antibody to influenza virus type B with influenza virus type B (homological structure) / NLC eutectic mixture of MBBA:EBBA on the whole surface; b) test object: polyaniline : HCl layer on the whole surface/ layer of antibody type A on the left/ antigen type B on the left/ NLC eutectic mixture of MBBA:EBBA on the whole surface; c) polyaniline layer on the whole surface/ polyaniline layer treated by influenza virus type A on the right.

In all experiments the mass of sorbent was 5 mg and the water solution volume was 0.2 ml. Recently many cases of human diseases caused by influenza bird viruses A(H5N1) and A(H7N7) were observed in the world. Influenza viruses multiply in the bird intestines and then emit to the environment that is why the new pandemia caused by these viruses can appeared. The pandemic influenza strain A(H1N1)v, related to swine viruses circulated in the world since 2009, caused gastro-intestinal problems - diarrhea, so it can spread by water path. Bacteriophage T4 -D is the classical model for biochemical studies. It is the also the indicator of the solution contained bacterium. The method of investigation virus-sorbent interaction was developed by Ivanova et al [4]. It includes some stages: intensive contact virus + sorbent on shaker during different time diapasons (15min -1 h) in temperature range (8-25°C), low speed centrifugation (2000 circle /min), the determination virus in solutions before contact and after sorption in supernatant solutions. In the case of influenza viruses hemagglutination test with human erythrocytes was used, for bacteriophage - the agar layer plaque assay. The sediment (virus + sorbent) after centrifugation represents the immunosorbent. The reaction of inhibition of hemagglutination (HI-test) was used to established the virus activity in immunosorbent with antibodies. The experiments were carried out with the follow solutions: H_20 after previously special treatment by heating to destroy other pathogens, 0.15 M NaCl in H_20, 0.1 M Tris-HCl buffer (pH 7.2), allantoic fluid of chicken embryos and nutrient medium for tissue cell cultures MDCK. There were investigated both purified influenza viruses as without purifications, just viruses after their cultivations in the chicken embryos and in cell cultures MDCK. The purified viruses had high start titers $\geq$ 512 HAE. The results of investigations are presented in Tables 1, 2.

TABLE 1. Sorption of influenza viruses by different forms of polyaniline

Sorbent	Virus	Type	Virus Titer in Solutions	
			Before Sorption	After Sorption
Pan-0	A/Sichuan/2/87	A(H3N2)	4096	2
	A/IIV- Moscow/01/09	A(H1N1)v	64	8
	B/Malaysia/2506/04	B	2048	8
	A/Mallard/Pennsylvania/1024/84	A(H5N2)	128	4
	A/FPV/Weybridge/34	A(H7N7)	512	4
	Bacteriophage T4 -D		3.4×10^7	0
Pan-HCl	A/New Caledonia/20/99	A(H1N1)	512	8
	B/Malaysia/2506/04	B	1024	8
	B/Tomsk/58/04	B	256	16
Pan- PAMPCK 1:1	A/Tomsk/110/08	A(H3N2)	64	4
	A/IIV- Moscow/01/09	A(H1N1)v	64	2
	A/Kaleningrad/03/06	A(H1N1)	64	4
	B/Malaysia/2506/04	B	2048	8
	A/FPV/Weybridge/34	A(H7N7)	2048	4
	Bacteriophage T4-D		3.4×10^5	0

TABLE 2. Sorption of viruses with carbon nanotubes covered and without polyaniline

Sorbents	Virus	Type	Virus Titer in Solutions	
			Before Sorption	After Sorption
Carbon nanotubes, covered with PAn	A/Moscow/72/07	A(H3N2)	64 *	<2
	A/IIV- Moscow/01/09	A(H1N1)v	64	8
	B/Malaysia/2506/04	B	2048	256
	A/Mallard/Pennsylvania/1024/84	A(H5N2)	128	2
	A/FPV/Weybridge/34	A/H7N7	512	32
	Bacteriophage T4-D		5.0×10^{7} **	0
Carbon nanotubes	A/ Moscow/72/07	A(H3N2)	64	16
	A/IIV- Moscow/01/09	A(H1N1)v	64	16
	B/Malaysia/2506/04	B	2048	512
	A/Mallard/Pennsylvania/1024/84	A(H5N2	128	16
	A/FPV/Weybridge/34	A/H7N7	512	128
	Bacteriophage T4-D		2.4×10^{7}	620

Note: T=25°C, contact t= 30 min; *Hemagglutination titer /0.2ml (HAE),** bacteriophage titer /ml.

We investigated the influenza strains, isolated in period 1934 to 2009 in different countries. It was shown that these influenza viruses were immobilized at the sorbents regardless of antigenic structure of surface proteins. In the case of supramolecular complexes of polyaniline the HA titers were reduced in 2048-16 times depending on the initial titer. In the case of carbon nanotubes covered with PAn and without PAn the influenza viruses titers in the sorption were decreased in 64-4 times. It depends of initial viral titer and did not depend of antigenic structure of surface proteins. So for influenza viruses the supramolecular complexes of PAn were best sorbents then nanotubes. Bacteriophages actively interacted with supramolecular complexes of polyaniline. The initial titers of bacteriophage in different solutions after interaction with sorbents were droped. When the aqueous 0.15 M NaCl solution was used the titer was decreased from 5.0×10^{7} to 0 in the case of the carbon nanotubes covered by PAn. In experiments with nanotubes withought PAn the titer 2.4×10^{7} was diminished to 620. The interaction of bacteriophages with this sorbent was more weakly and depended of solution's nature. It was shown that influenza viruses A and B, bacteriophages were immobilized at the sorbents, the active sorption of them were intensive with Pan sorbent, less with carbon nanotubes covered Pan and the least with nanotubes without Pan. All this data were carried out at T=25°C. The same results were obtained for supramolecular complexes of PAn at T=8°C. The activities of all nanotubes were less at T=8°C. The immunosorbents (sorbents + viruses) were analyzed with homological antibodies in HI-test. After contact the immunosorbents with serum during 18 h, T=4°C serum titer was deminished since 1/640 to 1/40. So it was established that viruses connected with sorbent could also interact with homological antibodies from solutions.

The high affinity of PAn based sorbents to viruses may be attributed to high affinity of PAn toward proteins that are constituent external part of viruses [5]. Partially this affinity may be attributed to acid-base and donor-acceptor interactions between viral proteins and semi-oxidized PAn and it's interpolymer complexes used in our work. It is well known that semi-oxidized PAN contains fully oxidized, semi-oxidized and fully reduced polymer fragments distributed along Pan chain [6]. They may take part in donor-acceptor interactions with proteins components of viruses.

ACKNOWLEDGMENTS

The research was partially support by ISTC grants #3070, 3718.

REFERENCES

1. V.F.Ivanov et al.,*Journal of Optical Technology,* **73,** No 8, 90-92 (2006).
2. V.F. Ivanov et al.,*Rus. J. Electrochemistry,* **40,** No 3, 299-304 (2004).
3. Yu. Sapurina et al., Rus J. Electrochemistry, **43,** No 5, 528-536 (2007).
4. V.T. Ivanova et al., *Problems of Virology* ,**54,** No 3, 21-26 (2007).
5. S. Piletsky, E. Piletska, A. Bossi, et al., Biotechnology and Biochemistry, **82,** 86-92 (2003).
6. V.F. Ivanov, O.L. Gribkova., S.V. Novikov et al., Met., **152,** 153-156 (2005).

Ionic Conductivity of Membranes Based on PVdF-HFP

A. Garcia-Bernabé, M. Gil-Agustí, J.P. González-Gutiérrez, A. Quijano-López

Energy Technological Institute (ITE), Av. Juan de la Cierva 24, Parque Tecnológico de Valencia, 46980 Paterna-Valencia (Spain)

Abstract. Membranes based on PVdF-HFP have been prepared by solution-casting methode. Lithium perchlorate ($LiClO_4$) and lithium trifluoromethane sulfonate ($LiCF_3SO_3$) were used as lithium salts, and ethylene carbonate and a mixture of ethylene and propylene carbonate (1:1 wt. %) as plasticizers. The ionic conductivity was determined by means of impedance spectroscopy. The Nyquist plot was fitted with a general equivalent circuit. The ionic conductivity increases to plastificizer content. The membrane consisting of 50 w% PVdF-HFP, 35w% PC+EC and 15w% $LiCF_3SO_3$ has the highest conductivity with a value of 6.892×10^{-3} S/cm at room temperature.

Keywords: PVdF-HFP, GPE, ionic conductivity, Nyquist Plot
PACS: 68.35 bm, 66.70. HK, 82.47 Aa

INTRODUCTION

Poly(vinylidene fluoride-*co*-hexafluoropropylene) (PVdF-HFP) has been extensively used in the preparation of Gel Polymer Electrolyte (GPE) for Polymer-Lithium batteries.[1] PVdF-HFP is a fluorinated polymer with low crystallinity, and high dielectric permittivity ($\varepsilon = 8.4$) that helps for greater ionization of lithium salt,. In spite of this greater ionization, the ionic conductivity of lithium salt solution in PVdF-HFP is lower for use in Polymer-Lithium Batteries applications. The addition of some plasticizers as such ethylene carbonate (EC), propylene carbonate (PC) or γ-butyrolactone (BL) increases the ionic conductivity of electrolyte. These plasticizers are organic solvents with high dielectric permittivity, high boiling point and low vapor pressure.

GPEs based on PVdF-HFP with a combination of plasticizers and different lithium salts have been studied. The conductivity decreases with the increase of polymer content in the gel. PVdF–HFP gel electrolytes with LiBF4 as salt exhibited maximum conductivity. These polyelectrolytes are thermally stable up to 70ºC[2]. GPE consisting of 25 wt% PVdF-HFP, 65wt% EC+PC and 10 wt% $LiN(CF3SO_2)_2$has an ionic conductivity of 1.2×10^{-4} S/cm.[3] However, GPE whit $LiN(C_2F_5SO_2)_2$ shows diffusion coefficients lower than the GPE with $LiN(CF3SO_2)_2$.[4]

In this study, the ionic conductivity of GPE based on PVdF-HFP is determined.

EXPERIMENTAL PART

All reagents were used without any further purification, and they were obtained from Sigma-Aldrich. PVdF-HFP, lithium salt and plasticizers were dissolved in 15 ml of tetrahydrofuran (THF). The polymer solution was cast as film and THF was allowed to evaporate slowly at room temperature. The obtained films are regular and with a thickness of 0.1 mm. the ionic conductivity was determined by sandwiching of the sample between two gold electrodes. The measurements were realized using a Potentiostat from AUTOLAB in the frequency range of 10^{-3} to 10^{6} Hz at room temperature.

RESULTS AND DISCUSSION

Two lithium salts and two plasticizers have been tested in GPE based on PVdF-HFP. The lithium salts were lithium perchlorate ($LiClO_4$) and lithium trifluoromethane sulfonate ($LiCF_3SO_3$), and plasticizers were ethylene

CP1255, *V^th International Conference on Times of Polymers (TOP) and Composites,*
edited by A. D'Amore, D. Acierno, and L. Grassia

carbonate (EC) and a mixture propylene carbonate (PC) + ethylene carbonate (EC) (1:1 wt. %). The conductivity was determined by fitting of the Nyquist plots considering general equivalent circuit as shown figure 1. The general equivalent circuit is a combination of Resistances (R) and Constant Phase Elements (CPE) in series or parallel, which can be simplified according to PVdF-HFP membrane.

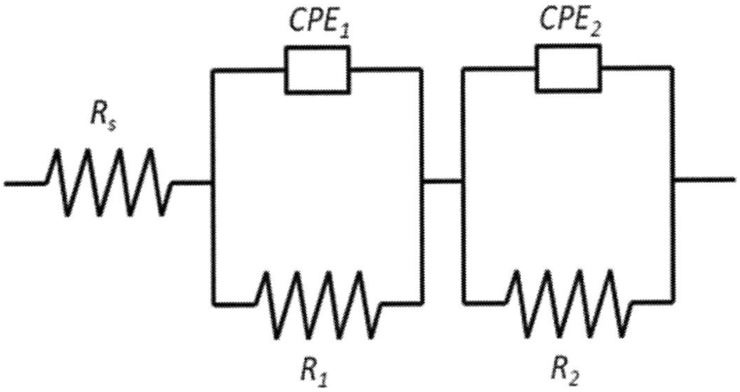

FIGURE 1. General Equivalent Circuit.

Nyquist plot of GPE with PVdF-HFP (70% wt), lithium trifluoromethane sulfonate (5% wt), PC+EC (25% wt) is represented in the figure 2. A semicircle can be observed at higher frequency. The impedance value in the intersection of the semicircle with the linear behavior is the resistance contribution of the membrane. The DC conductivity (σ_{DC}) can be calculated by means of the follow equation.

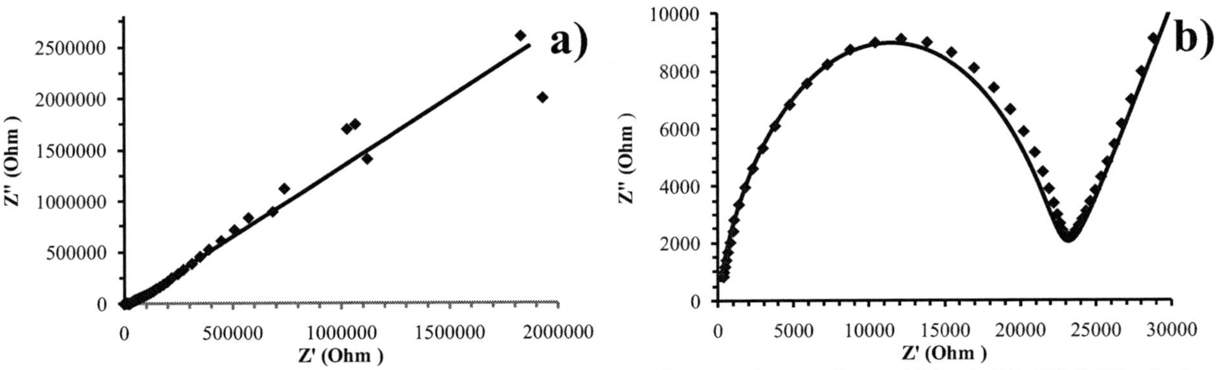

FIGURE 2. Nyquist plot of GPE with PVdF-HFP (70% wt), lithium trifluoromethane sulfonate (5% wt), PC+EC (25% wt). a) Nyquist plot for all range frequency, b) Nyquist plot at higher frequency.

$$\sigma_{DC} = \frac{l}{A\,Z} \qquad (1)$$

where l is thickness of the sample, A is the area of the electrodes and Z is the resistance contribution.

Ionic conductivity of GPE based on PVdF-HFP with 5w% of lithium salts is shown in figure 3, where we study the variation of conductivity regard to plasticizer content, type of salt or plasticizer. We can observed that the conductivity increases with plasticizer content, or decreases with polymer content. The values of ionic conductivity of the membrane at the lowest plasticizer content are very low, smaller than 10^{-8} S/cm. These values increase until value of 4.4×10^{-3} S/cm for GPE with PC+EC as plasticizer (45 %) and lithium perchlorate (5%). The conductivity of the membrane with lithium perchlorate is better using EC as plasticizer. However, if we use lithium trifluoromethane sulfonate, the best plasticizer is PC+EC at low plasticizer content, or EC at high plasticizer content.

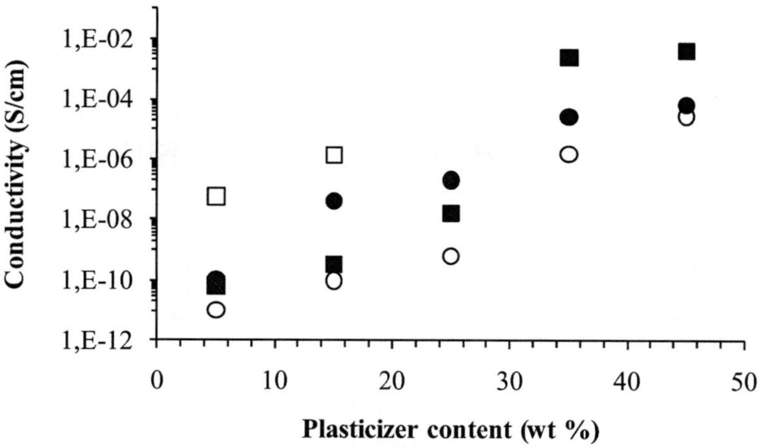

FIGURE 3. Conductivity of PVdF-HFP membranes with 5%w lithium salts. Lithium perchlorate (open symbol) and lithium trifluoromethane sulfonate (solid symbol). Plasticizers: EC (square) and PC+EC (circle).

In the figure 4, we study the variation of the conductivity regard to salt content, which its tendency is not very obvious. The conductivity of membranes with 50%w PVdF-HFP is very high, with order of 10^{-4} or 10^{-3} S/cm.

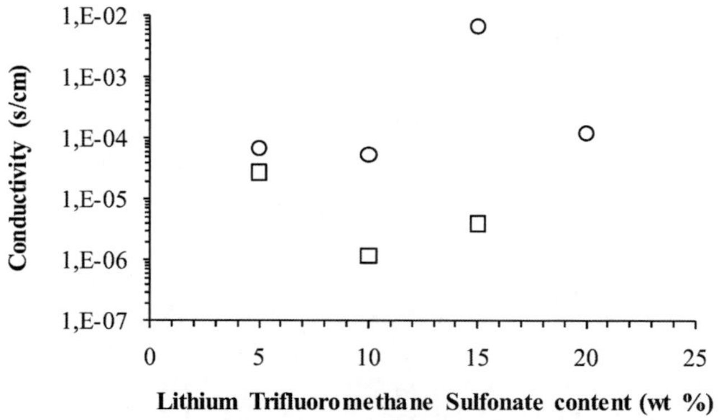

FIGURE 4. Conductivity of PVdF-HFP membranes with 50%w PVdF-HFP (circle) and 60%w PVdF-HFP (square), and EC+PC (1:1 wt. %) as plasticizer.

CONCLUSION

GPEs based on PVdF-HFP have been prepared using solution-casting method with different lithium salts and plasticizer. The main factor in ionic conductivity is the plasticizer content. The highest ionic conductivity was 6.89×10^{-3} S/cm.

ACKNOWLEDGMENTS

This work was supported by the European Regional Development Fund (ERDF) and the Ministry of Science and Innovation (Spain) in the context of the National Plan for Scientific research, Development and Innovation 2009-2011 (ETF: project ref. PID-590000-2009-4).

REFERENCES

1. A. M. Stephan, *Eur. Polym. J.* **42**, 21-42 (2006).
2. A. M. Stephan, S. G. Kumar, N.G. Renganathan and M. A. Kulandainathan. *Eur. Polym. J.* **41**, 15-21 (2005).
3. H.-S. Kim, P. Periasamy and S.-I. Moon. *J. Power Sources* **141**, 293–297 (2005).
4. C. Capiglia, Y. Saito, H. Kataoka, T. Kodama, E. Quartarone and P. Mustarelli *Solid State Ionics* **131**, 291–299 (2000).

Adhesion Peptide Immobilization on Electrospun Polymers: a Photoelectron Spectroscopy Study

G.Iucci[a], G. Polzonetti[a], F. Ghezzo[b], M. Modesti[b], M. Roso[b], M. Dettin[b]

[a] *Department of Physics, INSTM and CISDiC, University "Roma Tre", Via della Vasca Navale, 84 00146 Rome*
[b] *Department of Chemical Process Engineering, University of Padova, Via Marzolo, 9 - 35131 Padova*

Abstract. The immobilization of an oligopeptide, mimicking the adhesion sequence of fibronectin, on the surface of polymer films prepared by electrospinning was investigated by XPS spectroscopy. Films of polycaprolactone (PCL) and poly(l-lactide-caprolactone) [P(LLA-CL)] were prepared by electrospinning onto aluminium substrates. A modified adhesion peptide containing a photoreactive group was immobilized on the surface of the polymer nanofibers by incubation in peptide solution and subsequent UV irradiation. XPS analysis yield evidence of successful peptide immobilization on the polymer surface; the amount of immobilized peptide increases with the concentration of the mother solution.

Keywords: elecrospinning, nanofibers, polymers, peptides, XPS
PACS: 73.61

POLYMER NANOFIBERS

Electrospinning is one of the most promising methods for the preparation of micro- and nanoporous polymeric materials. Nanofibers of biocompatible polymers such as polycaprolactone and poly(l-lactide) are currently object of increasing interest as possible candidates for the preparation of biomedical devices for tissue engineering, such as small calibre blood vessels [1-3]. The biomimetic properties of the polymer nanofibers can be improved by surface modifications [2] or by protein incorporation in the polymer blend [4]. Extracellular matrix (ECM) proteins, such as fibronectin and vitronectin, play a major role in the adhesion of endothelial cells to materials surface; cell adhesion can be promoted by immobilization of oligopeptides mimicking the adhesion sequence of ECM proteins [5,6].

In this paper we present a XPS study on the immobilization of a peptide containing the RGD (Arg-Gly-Asp) adhesion sequence on the surface of biomimetic polymeric scaffolds consisting of nanofibers of polycaprolactone (PCL) and poly(l-lactide-caprolactone) copolymer [P(LLA-CL)] prepared by electrospinning.

SAMPLE PREPARATION

Polymer films were prepared by electrospinning under the following conditions: Solution, 10% wt P(LLA-CL) (70-30) (PURAC Biochem–Gorinchen, Holland) or PCL (M_n=60000 Da Sigma.Aldrich, St. Louis, USA) in a DCM:DMF = 70:30 wt solution; 0.4 mm diameter needle; supply at 1 ml/h; tension, 8kV [7]. The aluminum foil (15x15 cm) used as the collector was cut in circular samples of 1.2 cm diameter.

The peptide synthesized for the crosslinking of the polymeric surface has the following sequence: H-Phe(p-N_3)-Gly-Arg-Gly-Asp-Ser-Pro-Gly-Arg-Gly-Asp-Ser-Pro-Gly-Arg-Gly-Asp-Ser-Pro-Gly-Arg-Gly-Asp-Ser-Pro-Phe(p-N_3)-NH_2. It was synthesized as a p-amino analogue with Fmoc chemistry using a Syro-I synthesizer (Multisyntech Gmbh – Germany). After the conversion of amino groups into azido groups by reaction in solution with sodium nitrite and sodium azide, the crude product was purified by RP-HPLC and characterized by mass spectrometry. The peptide pre-conditioning was obtained through the incubation of samples with a solution of 1 mg peptide in 1 ml MilliQ water, 0.1 mg/ml or 0.01 mg/ml for 1 h at room temperature. The samples were dried under vacuum in the presence of P_2O_5. Each sample was irradiated for 1 h at 254 nm and at 366 nm (distance from the lamp, 1 cm).

CP1255, *V[th] International Conference on Times of Polymers (TOP) and Composites,*
edited by A. D'Amore, D. Acierno, and L. Grassia

XPS RESULTS

XPS measurements were performed in an instrument of our own construction and design consisting of a preparation and an analysis chamber, equipped with a 150 mm mean radius hemispherical electron analyzer followed by a 16-channel detector. Mg Kα non-monochromatized X-ray radiation (hν = 1253.6 eV) was used for recording core level spectra of the polymer samples, pristine and after peptide immobilization (C1s, N1s, O1s). The spectra were energy referenced to the binding energy (BE) of the C1s signal of aliphatic carbons, located at BE=285.0 eV [8]. Atomic ratios were estimated from peak intensities by using Scofield's cross section values. The C1s and O1s spectra were analyzed by curve-fitting analysis, using gaussian curves as fitting functions.

The C1s and O1s spectra recorded for pristine PCL and P(LLA-CL) electrospun films, and related curve fitting analysis, are shown in fig.1. The Al2p signal from the underlying aluminium substrate could not be detected, probably due to the thickness and homogeneity of the polymer overlayer. The chemical structure of PCL and PLLA [poly(l-lactide)] is also shown in the figure; carbon and oxygen atoms having similar chemical environment in the two polymers have been labelled with the same number.

From literature data [9], the C1s spectra of PCL and PLLA consist of three signals, related to aliphatic carbons, labelled 1 (C_1) in the polymer structures shown in fig 1, to C-O carbons (C_2) and to O-C=O carbons (C_3), in increasing BE order. The expected numeric ratio between the three different types of carbon is $C_1/C_2/C_3$ =4/1/1 for PCL, 1/1/1 for PLLA. O1s spectra of PCL and PLLA show two peaks, related to C=O (O_1) and C-O (O_2) oxygens, respectively; the expected atomic ratio is O_1/O_2=1/1 for both polymers.

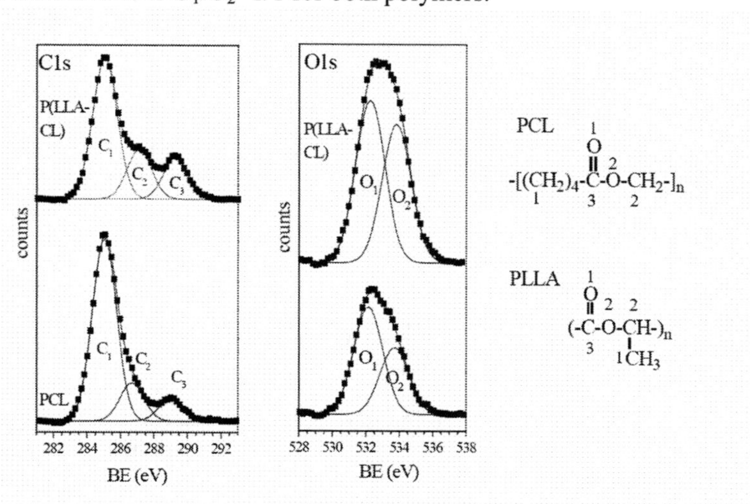

FIGURE 1. C1s and O1s photoemission spectra and related curve fitting analysis for PCL an P(LLA-CL); markers represent experimental points, lines fitting components and calculated spectra. Chemical structure of PCL and PLLA is also shown.

The BEs measured from curve-fitting analysis for the two polymer films, shown in table 1, are in good agreement with literature data [9] (PLLA is used for comparison for the P(LLA-CL) film). The BE value of C_2 is slightly higher for P(LLA-CL) than for PCL, since C_2 in PLLA is also a to α carbonyl group.

TABLE (1). Binding energies

signal	assignment	measured BE (eV)		Reference BE (eV) [9]	
		PCL	P(LLA-CL)	PCL	PLLA
C_1	C-C	285.0	285.0	285.0	285.0
C_2	C-O	286.7	287.1	286.54	286.98
C_3	O-C=O	289.0	289.2	289.08	289.06
O_1	O=C	532.2	532.2	532.24	532.24
O_2	O-C	533.6	533.7	533.54	533.54

Table 2 shows the measured atomic ratios for the two polymer films, both pristine and after peptide immobilization. For pristine PCL , the measured C_1/C_3 and C/O ratios are well above the expected values, showing the presence of surface contamination from aliphatic carbons; upon peptide immobilization the surface contamination form aliphatic carbons apparently decreases. A better agreement between measured and calculated values is found for pristine P(LLA-CL).

Table 2 measured atomic ratios

material	Peptide conc. (mg/ml)	C_1/C_3	C_2/C_3	O_1/O_2	C/O	N/C	N/O
PCL	pristine	8.6	1.5	1.6	5.2		
	1.0	6.3	1.7	1.5	4.4	0.023	0.1
	0.10	6	1.6	1.5	4.5	0.008	0.033
	0.010	7	1.6	2.4	4	0.005	0.02
P(LLA-CL)	pristine	1.9	1.06	1.1	1.85		
	1.0	2.38	1.16	1.2	2.36	0.022	0.057
	0.10	2.16	1.1	1.1	2.0	0.009	0.019
	0.010	2.3	1.1	1.1	2.28	0.006	0.014

Peptide immobilization on the polymer surface took place for both materials, as evidenced by the presence of a N1s peak at a BE = 400.0 eV, in agreement with the expected value for the peptide amide nitrogens [8]. The N1s signal can be detected even for the lowest peptide concentration (0.01 mg/ml). The amount of immobilized peptide, measured by the N/C and N/O ratios, increases with peptide concentration of the mother solution. A plot of the measured N/C and N/O ratios as a function of peptide concentration is shown in fig. 3; the trend is very similar for the two polymer films.

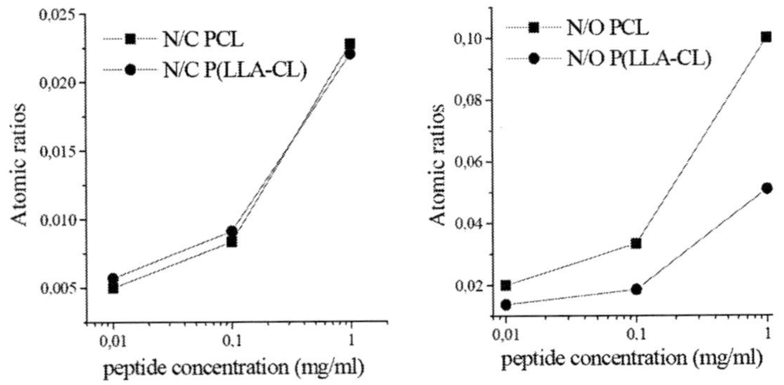

FIGURE 2. Plot of the measured atomic ratios (N/C and N/O) as a function of peptide concentration in the mother solution

The reported data yield evidence of successful immobilization of the RGD adhesion peptide onto the surface of both electrospun polymer films. Preliminary assays show that RGD immobilization increases cell adhesion on the polymer nanofibers.

ACKNOWLEDGMENTS

Financial support from MIUR is gratefully acknowledged.

REFERENCES

1. M. Spasova, N. Malonova, D. Paneva, R. Mincheva, P. Dubois, I. Rashkov, V. Maximova and D. Danchev *Biomacromol.* **11**, 151-159 (2010)
2. D. Paneva, F. Bougard, N. Malonova, P. Dubois and I. Rashkov, *European Polymer Journal* **44**. 566-578 (2008)
3. E. Luong-Van, L. Grøndahl, K. Ngiap Chua, K. W. Leong, V. Nurcombe and S.M. Cool *Biomaterials* **27**, 2042-2050 (2006)
4. H. I. Kim, R. Matsuno, J-H. Seo, T. Konno, M. Takai and K. Ishihara *Current Applied Physics* **9**, e249-51 (2009)
5. M. Dettin, A. Bagno, R. Gambaretto, G. Iucci, M. T. Conconi, N. Tuccitto, A. M. Menti, C. Grandi, C. Di Bello, A. Licciardello and G. Polzonetti *J. Biomed.Materials Res.* A **90a**, 35-45 (2009)
6. M. Dettin, T. Herath, R. Gambaretto, G. Iucci, C. Battocchio, A. Bagno, F. Ghezzo, C. Di Bello, G. Polzonetti and L. Di Silvio *J. Biomed.Materials Res. A* **91A** 463-79 (2009)
7. K.H.Lee, H.Y. Kim, M.S.Khil, Y.M. Ra, D.R.Lee *Polymer* **44**, 1287-1294 (2003)
8. J.F. Moulder, WF Stickle, Sobol PE, Bomben KD. *Handbook of X-ray Photoelectron Spectroscopy* Physical Electronics Inc. eds.
9. G. Beamson, D. Briggs *High resolution XPS of organic polymers-The Scienta ESCA300 Database* John Wiley & Sons 1992

Evaluation of Installation Quality on Polyethelene Pipe Performance under Surcharge Loads

Niyazi U. Terzi[b] Necmi Dusunceli[a],Ferruh Yılmazturk[c],S. Yıldırım[d]

[a]Civil Engineering Department, Aksaray University 68100 Aksaray/TURKEY
[b]Mechanical Engineering Department, Aksaray University 68100 Aksaray/TURKEY
[c]Geodesy Photogrammetry Department, Aksaray University 68100 Aksaray/TURKEY
[d]Civil Engineering Department, Yıldız Technical University Istanbul/TURKEY

Abstract. In this study, a comprehensive laboratory test on 100 mm diameter HDPE 100 polyethylene flexible pipe buried in quartz sand is described. The laboratory test was performed in a 40 mm thick plexiglass-fronted test tank which replicated a classical trench section in field conditions. The polyethylene flexible pipe was positioned against the glass with its longitudinal axis perpendicular to the glass. This allowed direct observation of the backfill-pipe interactions and pipe performance under surcharge loads. Three high definition photogrammetric cameras were taken photogrammetric images through the glass allowing the discrete measurement and image processing of the deformation patterns of the polyethelene pipe conduit during the pipe installation and incremental surcharge loading. Vertical loads were applied in increments of 10 kPa to 150 kPa using air pressure membranes. According to the test results, it is understood that the installation technique and backfill properties have an important effect on circumferential strains, performance and deformation characteristics of polyethelene HDPE pipe. It was also observed that close-range image processing is a very simple and appropriate method for measuring three dimensional pipe deformations under various conditions.

Keywords: HDPE flexible pipe, Soil-Pipe Interaction
PACS: 46.35.+z, 62.20.-x, 81.70.Bt, 62.20. D-.

INTRODUCTION

Medium and high density polyethylene has become the accepted material for the renewal of existing water mains distribution systems, and over the past two decades, there has been a continuing increase in the utilization of high density polyethylene (HDPE) based thermoplastic flexible pipes. This kind of polyethylene based pipes has a crucial role in the collection, transportation and removal of fluids and contaminants in a variety of municipal and environmental applications. Typical municipal applications of plastic pipes consist of storm sewer, water recharge, potable water and other gravity flow systems. Environmental applications include landfill sewer collection systems, industrial liquid collection and treatment systems. Due to the diversity of applications, HDPE flexible pipes can be subjected to shallow or deep burial loading conditions with or without extra live loads. Although there has been a steady increase in the use of flexible pipes and developments in their design; there are still many issues that are not fully understood about three dimensional soil-pipe interactions, behavior of its thermoplastic characteristics and pipe behavior, especially in terms of the deflection type and strain limits of the pipe wall sections. Full-scale field tests can provide valuable and realistic information to resolve engineering problems. However, field tests are expensive, time consuming, and difficult to modify for varying test conditions. A possible alternative is to use laboratory tests to represent a reasonable practical and realistic simulation of field conditions. With the help of advanced instrumentations and efficient modeling techniques, field assumptions can be replicated in laboratory conditions. The prime advantages of laboratory tests are that the variables can be measured precisely and controlled much more easily than in field tests.

In this experimental investigation, the behavior of soil-pipe interaction and polyethelene deformation pattern under surcharge loads were studied. A series of tests on a small single pipe of 100 mm diameter, buried in sand backfill material subjected to surcharge loads, were performed in accordance with ASTM D 2321. The testing program was planned to evaluate the role of backfill installation quality on the pipe behavior. In each test, the percent vertical diameter change (DD) was found by dividing the vertical and lateral diameter change (DD_v), by the initial external diameter (D), It should be noted that 5% maximum vertical diameter change (DD) is considered as the criteria for the safety of the pipe. The 5% limit is based on a safety factor of four applied to the historically accepted limit of 20% deformation to avoid snap-through buckling and damage to the pipe, Rogers et al (1995).

CP1255, *V[th] International Conference on Times of Polymers (TOP) and Composites,*
edited by A. D'Amore, D. Acierno, and L. Grassia

TESTING APARATUS

The test equipment consisted of four main parts, the test tank, pluviation system, data acquisition system and Instrumentation equipments. (Figure 1.a,b,c).

a) Test equipments b) Camera positions c) Specific targets

Figure 1. General View of Laboratory Facilities

Brief information of each part is as follows:

Testing Tank

Based on the ASTM D 2321 standards, minimum trench width is not to be less than either 1.5 times the pipe outside. Therefore, based on recommendation, the width of the trench was selected as 700 mm, hence, and the testing tank given in Figure 1b was designed as a rigid steel box, 500 mm in length, 1000 mm in height, and 700 mm in width, encompassing the reinforced soil and model pipe, which stood vertically along its square face while testing was in progress. The rigidity of the test tank was guaranteed using three stiff steel sections of U-220 on the sides and bottom of the tank. Extra stiffeners were used to restrict the excessive deflections of the sidewalls. Thereby the K_o condition was successfully achieved.

Pluviation system

The backfill installation to the test tank was made by two methods; (1), air pluviation (rain-throughair) and (2), scooping. The air pluviation system was used in shown in Fig 1a. The method used in the installation of the testing tank at known and uniform density was based on the raining technique. The intensity of the backfill deposition is governed by four parameters (1) the spacing and the size of the openings in the perforated base, (2) the sizes of the openings of the screens, (3) the escalation rate of the container and finally (4) spacing between backfill surface and container.

Instrumentation and Data Acquisition System

In this study, the basic objective of the photogrammetric work was to determine the magnitude and direction of the motion of a number of discrete points on the object's surface, at each loading increment. The entire system was mounted on a tripod to facilitate vertical movement and rotate around the vertical axis. The multi-image acquisition system was based on progressive scan CCD cameras provided with a sensor of 780x582 pixels and IEEE-1394 port. Through the IEEE-1394 interface, the computer communicates with the camera, allowing the digital transmission of images and the software control of some of the camera parameters.

BACKFILL PROPERTIES

In the experimental studies, Sile sand was selected as the backfill material. The sand was fine, washed, dried and sorted by a size that was of rounded to sub-rounded particles. The specific gravity of the backfill particles was determined by the gas jar method. Three tests were carried out to obtain an average value of 2.657. The maximum and minimum void ratios of Sile Sand were obtained as 0.52 and 0.87 respectively.

PIPE PROPERTIES

The PE 100 type HDPE plastic pipe , supplied by the Dizayn Pipe Corporation with a production standard of DIN 8074 was used in the studies. The Elastic modulus and Poisson's ratio of the pipe material were 900 MPa and 0.40 respectively. Other engineering parameters such as the mass per unit area and ultimate tensile strength of the pipe material were 0.802 kg/m^2 and 6.1 kN/m respectively.

EXPERIMENTAL WORKS AND RESULTS

In classic trench installations, the vertical dead and live loads are the governing parameters and the flexible pipes must be designed to have sufficient strength, stiffness and the durability to resist the load under both loading conditions. Buried flexible pipes are not only affected by the engineering properties of the flexible pipe itself, including flexural-circumferential rigidity and durability and the ductility of polyethylene raw material, but also by the shear strength and stiffness parameters of the backfill material. In this study the experimental research program was carried out in uniform and ununiform backfill conditions. For uniform conditions, before installing the hopper for the backfill in the test box the raining device was adjusted using different heights of pouring and perforated plates to ensure the desired density could be selected for each test. After the installation of a 150 mm thick bedding layer, HDPE flexible pipe was located. The soil in the bedding zone was compacted with a plate of 200x200 mm^2 and operating mass of 4.5 kg ASTM D 1557. The soil compaction was checked to ensure that a maximum dry density of 0.92 % was achieved. Cable connections between the data acquisition system and instruments were carried out and primer calibrations were made. Pluviation was continued until the soil box was fuelled. The membrane was laid out over the whole surface of the embedment and the box lid was closed and locked. The testing of the flexible buried pipe began by applying a load to the pressurized air membrane until the first loading step (10 kPa) was achieved. For ununiform conditions embankment was installed without using pluviation system. As seen in Figure 1c, the targets that were used in the image processing were mounted on the front side of the pipe section. Deformations of crown and invert under uniform and ununiform backfill conditions were shown in Fig 2

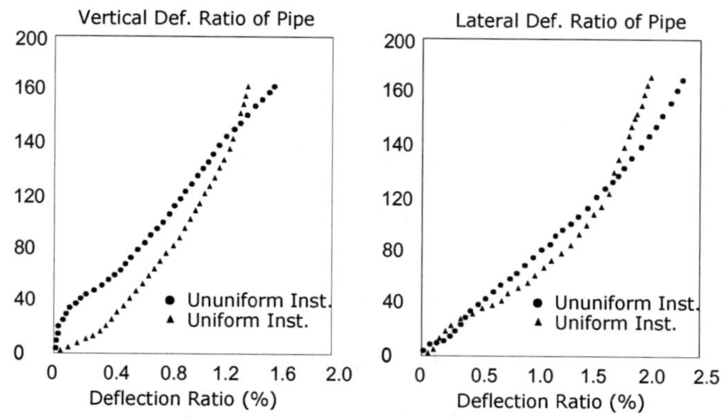

Figure 2 Evaluation of installation quality on pipe deflection behavior.

Many flexible buried pipes, have been known to deflect up to 20% from their original diameter without structural failure, however 5% initial deflection design limit is recommended Rogers (1995), for the following reasons; 1.) If the pipe is poorly restrained laterally (e.g poor compaction or weak embedment material used), failure may occur due to excessive deformation.2) Flexible plastic pipes will continue to deflect with time due to the creep characteristics of pipe 3)Excessive deflection could cause infiltration and exfiltration to occur as joints become unsealed.

CONCLUSION

As seen in experimental studies, deflections at the vertical and lateral axes of the pipe wall in all backfill conditions were less than 5%. In the case of incremental vertical surcharge loading the pipe wall shrank vertically and elongated horizontally. This phenomenon continued as the vertical load increases and the maximum deflection of the pipe occurred at the vertical axis (pipe invert). Even the final deflection percent of the pipe conduit in different backfill conditions approximately equal, deflection pattern of the ployethelene pipe is different. For uniform backfill condition deflection curve of pipe seems to be more lineer than ununiform backfill condition. This phenomenon is basically related to the resistance quality of the homogenous medium.

REFERENCES

ASTM D 2321 "Standard Practice for Underground Installation of Thermoplastic Pipe for Sewers and Other Gravity-Flow Applications" ASTM International Publication (2008)

Rogers, C. D. F., Fleming, P. R., Loeppky, M. W. J. and Faragher, E. (1995). "Structural Performance of Profile-Wall Drainage Pipe - Stiffness Requirements Contrasted with Results of Laboratory and Field Tests", Transportation Research Record, 1514, pp 83-92.

ASTM D 1557 "Standard Test Methods for Laboratory Compaction Characteristics of Soil Using Modified Effort (56,000 ft-lbf/ft3 (2,700 kN-m/m3), (2005)

Cyclic Behavior of High Density Polyethylene (HDPE)

Necmi Dusunceli[a], Bulent Aydemir[b] and Niyazi U. Terzi[c]

[a]*Mechanical Engineering Department, Aksaray University 68100 Aksaray/TURKEY*
[b]*TUBITAK National Metrology Institute, 41489 Kocaeli/TURKEY*
[c]*Civil Engineering Department, Aksaray University 68100 Aksaray/TURKEY*

Abstract. This article presents the mechanical behavior of high density polyethylene (HDPE). Samples were prepared by extracting extruded HDPE pipe. Cyclic and strain rate jump behavior of HDPE were studied under uniaxial tensile loading conditions. The strain jump tests indicated that mechanical behavior of HDPE has deformation memory. Further, it was found that increasing cycle number on cyclic loading test increased strain accumulation amount and HDPE exhibited ratcheting behavior at on loading-unloading-reloading at constant stress level.

Keywords: Viscoelasticity, Mechanical properties of solids, Mechanical Testing, Stress-Strain relation.
PACS: 46.35.+z, 62.20.-x, 81.70.Bt, 62.20. D-.

INTRODUCTION

Polymers are intensively used in mechanical applications such as pipe manufacturing, load bearing, biomedical joint etc. where stress-strain and lifetime analysis are important task for the material selection and prediction. Numerous studies have been carried out mechanical behavior determination and its modeling of polymeric materials in the last two decades. These investigation included loading-unloading, creep relaxation and cyclic loading. Polymeric materials may be subjected to cyclic loading conditions in many applications such as load bearing and biomedical joint. For this reason, the cyclic behavior of polymeric materials was studied different aspects and different loading conditions.

High performance thermoplastics such as high density polyethylene (HDPE) are a semi crystalline polymeric material having a microstructure of that has both crystalline and amorphous regions. High density polyethylene (HDPE) has a ratio of up to 90% crystallinity however; low density polyethylene (LDPE) has a ratio of up to 40% crystallinity. Thus, it can be seen that the greater the crystalline regions the higher the density of polymeric materials and this affects a number of physical and mechanical properties of HDPE. Generally, increasing degree of crystallinity results in greater yield strength and stiffness. The mechanical behavior of HDPE is viscoelastic and because of its semi crystalline composition is very complex being directly related to time and temperature. The crystalline region that account for the elastic response to forces is stiff; the amorphous region accounting for the viscous fluid-like response is soft [1-2].

Even though there are large numbers of experimental and analytical investigations on HDPE, few have examined the effects of manufacturing techniques on the small and finite deformation behaviors of HDPE. Understanding the deformation behavior under different loading can offer the designer of HDPE products reliable data relevant to practical applications. The study of the loading history is important in engineering design because the mechanical properties and deformation mechanism may be heavily dependent on the applied loading types [3-5].

In the literature, much investigation has been conducted on the effect of strain rate, loading types and temperature on the mechanical behavior of polymeric materials. A few studies have investigated the effect of cyclic deformation loading. The objective of this experimental work was to explore the mechanical behavior of HDPE associated with cyclic loading unloading and strain jumping loading history. Cyclic loading-unloading tests were conducted for constant stress level at constant strain rate. Strain jumping test were studied at two strain rate levels in tension.

CP1255, *V[th] International Conference on Times of Polymers (TOP) and Composites,*
edited by A. D'Amore, D. Acierno, and L. Grassia

EXPERIMENTAL WORKS AND RESULTS

Experimental work was conducted at room temperature by using a testing machine Zwick 250. The load-cell capacity of the tensile machine can be manipulated by adjusting with different load-cells in the capacity of 5, 10, 20, 50, and 100 kN. A load-cell in capacity of 10 kN is used in tensile tests. The test parameters can be adjusted with software called TestXpert and the test results are calculated with this software. Displacement controlled quasistatic tension experiments were performed on extruded HDPE.

First the extruded samples for the tensile tests were obtained from PE100 (HDPE) pipe with a diameter of 160 mm and a 5 mm wall thickness of pipe. The extruded samples were created according to the ISO 6259-1 and ISO 6259-3 standards for the determining of the properties of tensile deformation in polymeric pipes [6-7]. The density of this material is 0.954 g/m^3 and trade name is Total XS10B. Fig. 1 shows the dimensions of the extruded samples.

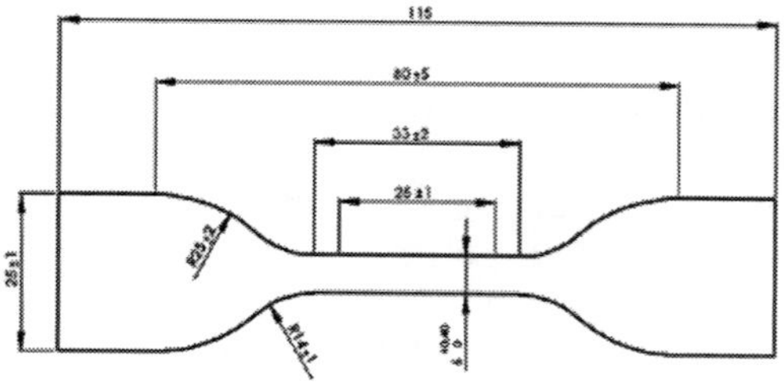

FIGURE 1. Sample dimensions.

The first test was loading-unloading tensile test up to 0.15 strain level at 1.e-4 1/s strain rate and is depicted in Fig. 2. Stress-strain curves of loading and unloading demonstrated elastic-viscoelastic-viscoplastic mechanical behavior. The shape of the unloading curves is nonlinear and concave curvature curve. Viscoelastic recovery during the unloading was observed and the magnitude of the recovered strain magnitude is 10.34%.

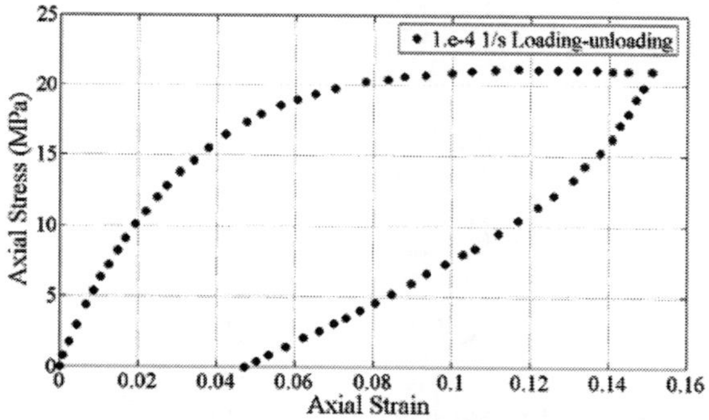

FIGURE 2. Stress-strain curves of HDPE in uniaxial loading-unloading test at 1.e-4 /s..

The second test was cyclic loading. This test included loading with constant strain rate (1.e-4 1/s) up to maximum stress level (17 MPa.), unloading down to the zero stress with the same strain rate and reloading up to maximum stress level (17 MPa.). The stress level was determined in the sub-yield region, in order to decaying of maximum stress. Numbers of reloading and unloading are 10 cycles. Stress-strain curve was shown in Fig. 3. The stress-strain curves of loading-unloading-reloading are nonlinear. Strain accumulation in each cycle of deformation was observed. This accumulation increased with strain rate up to the maximum stress level. Accumulation strain amount on the loading at maximum stress level and on unloading at minimum stress level gradually evolved with increasing number of cycles. These strain amounts at each of cycle was given in Table 1.

TABLE 1. Strain accumulation on cyclic loading-unloading.

Number of cycle	Strain at maximum stress level (%)	Strain at minimum stress level (%)	Number of cycle	Strain at maximum stress level (%)	Strain at minimum stress level (%)
1	3.66	0.81	6	4.12	1.32
2	3.82	1.01	7	4.17	1.37
3	3.92	1.12	8	4.22	1.4
4	4.02	1.22	9	4.27	1.45
5	4.07	1.27	10	4.3	1.5

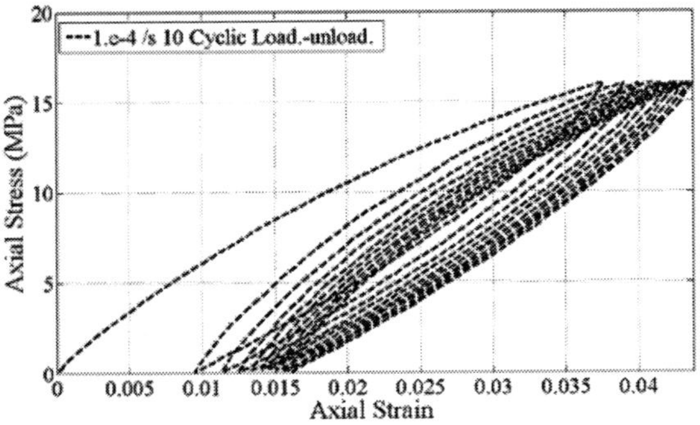

FIGURE 3. Stress-strain curves of HDPE in uniaxial cyclic loading-unloading test at 1.e-4 /s. Number of cycle is 10, $\sigma_{max.}$=10 MPa..

The last test was strain rate jump loading. To investigate memory effects associated with strain rate changes, the loading-unloading test at 15% strain level was performed. This test included that the initial strain rate (1.e-3 1/s.) was sharply changed to 1.e-4 1/s. for strain interval of 3% before a jump back to the initial strain rate. Stres-strain curves were given in Fig. 4. These results indicated that mechanical behavior of HDPE has deformation memory.

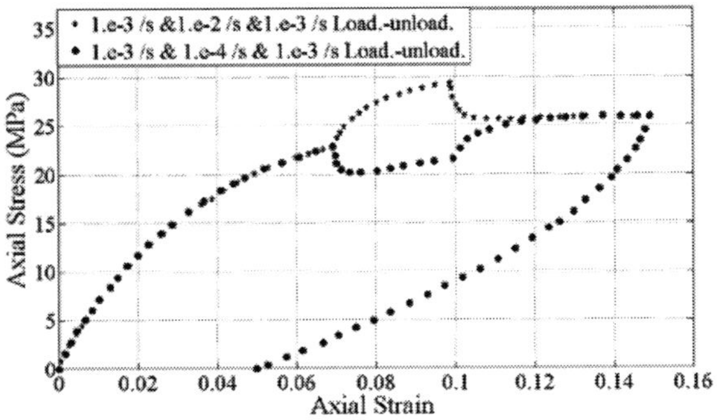

FIGURE 4. Stress-strain curves of HDPE under uniaxial tension loading-unloading. • symbols: stress-strain curves on loading at 1.e-3 /s at 7% strain level-on loading at 1.e-4 /s at 10% strain level-on loading at 1.e-3 /s at 15% and on unloading at 1.e-3 1/s. * symbols: stress-strain curves on loading at 1.e-3 /s at 7% strain level-on loading at 1.e-2 /s at 10% strain level-on loading at 1.e-3 /s at 15% and on unloading at 1.e-3 1/s

CONCLUSIONS

A series of experimental work were performed to explore on mechanical response of HDPE at uniaxial cyclic and loading-unloading tensile deformation. Mechanical behavior of HDPE was found to be sensitivite to cyclic loading. It was observed that strain accumulation on cyclic loading increases with increasing number of cycle. The loading-unloading curve is highly nonlinear and viscoelastic. Deformation behavior of HDPE has a memory effects. The analysis of inelastic deformation behavior of engineering materials has become an important element in design, life prediction and fatigue performance of components subjected to severe mechanical cycling. It is important that designers of HDPE products are fully aware of these impacts and accessible accurate data must be made available.

REFERENCES

1. N.Dusunceli and O.U. Colak, *Mater. Design* 29, 1117-1124 (2008).
2. O.U. Colak anf N.Dusunceli, *J Eng. Mater.-T. ASME* 128, 572-578 (2006).
3. A.D. Drozdov and J de C. Christiansen, *Polymer* 48, 3003-3012 (2007).
4. A.D. Drozdov and J de C. Christiansen, *Comp. Mater. Sci.* 39, 465-480 (2007).
5. C. Dreistadt, A.S.Bonnet, P. Chevrier and P. Lipinski, *Mater. Design* 30, 3126-3140 (2009)
6. International Standard ISO 6259-1, 1997.
7. International Standard ISO 6259-3, 1997.

Joint Strength Control at the Fiber/Matrix Interface during the Production of Polymer Composite Materials Reinforced with High Performance Fibers

Vladimir V. Kudinov[a] and Natalia V. Korneeva[b]

[a]*Laboratory of Plasma Coatings and Composite materials, A.A. Baikov Institute of Metallurgy and Materials Science, Russian Academy of Sciences, 49, Leninskii Prospect, Moscow, 119991 Russian Federation*
[b]*Laboratory of Reinforced Plastics, N.N. Semenov Institute of Chemical Physics, Russian Academy of Sciences, 4, Kosygin St., Moscow, 119991 Russian Federation*

Abstract. The paper presents the results obtained in the study of the joint strength between polymer matrix and high performance polyethylene fiber. The fiber/matrix joints simulate the unit cell of the fiber-reinforced composite materials. Effect of heat treatment on the composite properties at the interface was estimated by a multifilament wet-pull-out method. It was found that the joint strength may be increased with the help of extra heart treatment. Both the energy to peak load and the energy to failure for CM joints at various stages of loading were determined.

Keywords: Interface, Bonding, Multifilament Wet-Pull-Out Method, High Performance Polyethylene Fiber, Fiber Capillary Rise, Pull-Out Force, Wettability, Fiber Impregnation with the Matrix, Interfacial Shear Strength, Heat Treatment, Energy to Peak Load, Energy to Failure
PACS: 81.05.Ni; 81.05.Qk; 81.40 Jj; 83.10.-y

INTRODUCTION

At present time multifilament fiber (a single bundle of filaments) is used for reinforcing of the CM. Such fiber can consist of several thousands of the filaments. Therefore, the interface effects on the properties of composites reinforced with multifilament high performance fibers.

High performance fibers are usually considered those that exhibit high specific strength (tensile strength to weight ratio) combined with high elongation before break, such as *p*-aramids (Kevlar®, Twaron®, Rusar® etc.) and high performance polyethylenes (Dyneema® and Spectra®).

Composites reinforced with multifilament fiber possess the interface developed. The thin is filament diameter the larger is the contact area between the filaments and the matrix. That is why the interaction at the interface greatly influences on the CM properties and their keeping under load. The strength of such composites is mainly determined by two parameters. First, this is the degree of fiber impregnation with the matrix. The better is fiber impregnation the larger is the contact area between the filaments and the matrix. Secondly, this is the strength at the interface.

The objectives of the work were first to study with the help of wet-pull-out (W-P-O) method the wettability, fiber impregnation with the matrix, and interfacial shear strength (IFSS) for the simulated CM joints of epoxy matrix with high performance polyethylene (HPPE) fibers, and secondly to investigate the effect of heat treatment on the CM properties and energy to failure for CM.

EXPERIMENTAL

A multifilament W-P-O method allows one on the same sample to estimate the wettability and impregnation of the fiber with the liquid matrix and measure the fiber capillary rise *h* and interfacial shear strength after curing process finished [1]. The method considers a total interaction between the matrix and filaments. It is the advantage of this method in comparison with the tests of individual monofilament [2]. Such total interaction between the

CP1255, *V[th] International Conference on Times of Polymers (TOP) and Composites,*
edited by A. D'Amore, D. Acierno, and L. Grassia

filaments in the bundle and the matrix always takes place both at production and loading of CM. The samples tested (Fig. 1) simulate the unit cell of the fiber-reinforced composite materials.

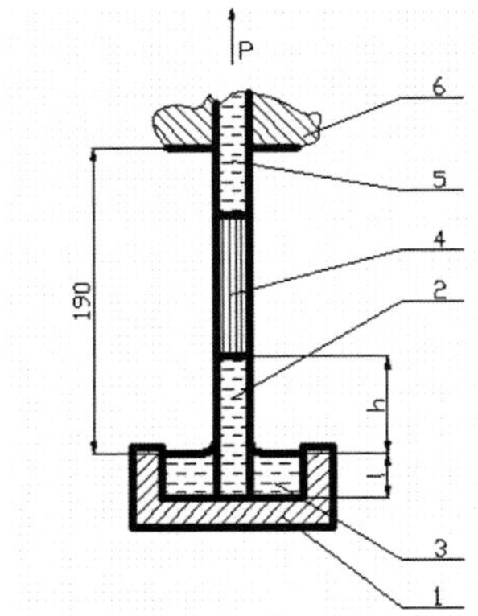

FIGURE 1. Scheme of the Sample: 1 – Mould, 2 – Fiber Impregnated with the Matrix, 3 – Length *l* of the Matrix, 4 – Part of the Initial Fiber, 5 – Part of the Fiber Impregnated with the Matrix, 6 – Fiber Fixation, *h* – Fiber Capillary Rise, *P* – Pull-out Force

As reinforcement, we used Dyneema® SK-75 multifilament high performance polyethylene fiber from DSM. The SK-75 fiber had a tensile strength of 3.4 GPa, an elastic modulus of 110 GPa, a density of 0.97 g/cm^3, elongation at break of 3.8% and a filament titre of 1760 dtex. There are 1150 filaments in a single bundle of filaments. The matrix was an epoxy resin Epicot-828 from Shell Chemical Co. cured with polyethylene polyamine.

The samples of fiber/matrix joints were prepared in tephlon moulds (1) at room temperature. The bundle of filaments (2) was located in the middle of mould. The liquid polymer matrix (3) was poured into the moulds. Fiber capillary rise *h* was the height the liquid matrix was lifted. The samples were cured into a hard block during a day. To finish curing process heart treatment of the hard samples was used. Four curing modes were used: mode I – at 45°C for 4 h, mode II – at 90°C for 4 h, mode III – at 90°C for 13 h, mode IV – at 125°C for 5 hours. All curing modes are given in Table 1. Modes II, III, and IV were used in estimating the effect of heart treatment (temperature and its duration) on the properties of CM interface. All samples were heart treated by the curing mode I. Further to such heart treatment the samples of № 2, № 5 and the samples of № 3, № 4 were treated with the help of curing mode II and III, respectively. The sample of № 6 was treated according to the curing mode IV.

All measurements were carried out by the multifilament W-P-O method [1]. Controlled parameters were fiber capillary rise *h* and pull-out force *P*. The joint strength was estimated with the help of the value of the normalized pull-out force **(***P/l***)**. We measured the value of *P* (N) required to pull the fiber out from the cured matrix. We changed the force changing the layer length *l* (mm). The value of *l* was changed by the way of cutting off the lower butt of the sample. To estimate wetting and quality of fiber impregnation we measured the value of *h* (mm). To delete skews the part (4) of initial fiber was used. The W-P-O test was carried out on the Instron 3382 universal testing machine, at a cross-head speed of 20 mm/min. The energy to failure ΣA (J) for the samples was calculated from the force - displacement curve (Fig. 2). It was the area under the whole curve. The energy to peak load *A* (J) was the area under the force - displacement curve from the zero loading to the peak.

TABLE 1. Curing Modes Used

Curing mode number	Sample number	Curing mode	
		Temperature, °C	Time, h
I	1, 7	45	4
II	2, 5	90	4
III	3, 4	90	13
IV	6	125	5

RESULTS AND DISCUSSION

We believe that the diameter *D* of multifilament fiber pulled out from the matrix is constant. Then the value of *P*/*l* determines the joint strength between the fiber and the matrix.

It is shown (Fig. 2), that the destruction of the unit cell of the CM reinforced with multifilament fiber occurred due to fiber shear at the interface and crack formation. After all, the fiber was retained in the matrix only by friction.

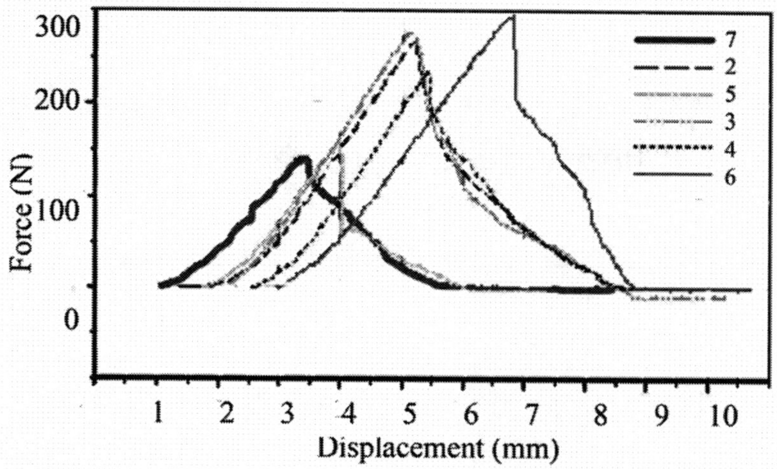

FIGURE 2. Force - Displacement Curves for Samples under Investigation

The value of *h* defines both the wettability and the fiber impregnation, and the joint strength at the interface. Higher is the value of *h*, larger is the value of *P/l*. At the same curing mode I for the samples of № 1 and № 7, the values of *P/l* were 25 N/mm and 31.3 N/mm for the values of *h* equal 12 and 17 mm, respectively. Higher is the value of *P/l*, larger is the value of normalized energy to peak load (*A/l*) for CM joints. The lowest strength, and the lowest values of *P/l* and *A/l*, and the highest value of normalized energy to failure *ΣA /l* for CM equal 0.205 J/mm were obtained for № 1sample which remained without extra heat treatment.

Extra heat treatment of the hard samples allows one to increase the joint strength and energy to peak load. After heat treatment of № 2 and № 3 samples according to curing mode II and III, the values of *P/l* increased by a factor of 1.94 (from 25 to 48.4 N/mm) and 1.84 (from 25 to 46 N/mm), respectively. For the same samples the values of *A/l* increased by a factor of 1.57 (from 0.044 to 0.069 J/mm) and 1.47 (from 0.044 to 0.065 J/mm), respectively. Highest values of *P/l* and *A/l* were obtained for the sample of № 6 due to extra heat treatment by mode IV.

TABLE 2. Effect of Wettability and Fiber Impregnation and Heat Treatment on the Joint Strength of Dyneema® SK-75 fiber with Epoxy Matrix. Base length was 190 mm.

Sample number	*h*, mm	*P*, N	*l*, mm	*P/l*, N/mm	*ΔL*, mm	*A*, J	*A/l*, J/mm	*ΣA*, J	*ΣA /l*, J/mm	*A/l*, %
1	12	299	12	25	4.05	0.53	0.044	2.46	0.205	21
7	17	141	4.5	31.3	2.37	0.14	0.031	0.26	0.058	53
2	20	266	5.5	48.4	3.85	0.38	0.069	0.65	0.118	58
5	12	161.7	4.0	40.4	2.25	0.15	0.038	0.22	0.055	68
3	18	276	6.0	46	3.03	0.39	0.065	0.67	0.112	58
4	15	234.8	5.3	44.3	2.97	0.29	0.055	0.52	0.098	56
6	15	294.8	5.0	59	3.93	0.51	0.102	0.74	0.148	69

CONCLUSIONS

Results obtained show that the interface defines the composite performance characteristics such as strength and energy to failure for CM. Composite heat treatment effects on strength properties of CM. It was found that the joint strength grows with increasing temperature and duration of heat treatment according to curing modes studied.

REFERENCES

1. V.V. Kudinov and N.V. Korneeva, *Macromol. Symp.*, **286**, 187-194 (2009).
2. L.J. Broutman, in *25th Annual Technical Conference Reinforced Plastics/Composites Divisions*, Soc. of the Plast. Ind., Washington, DC, Section 13-D, 1970, p.1.

Assessing the Progress of Degradation in Polyurethanes by Chemiluminescence

Marta Malíková [a], Jozef Rychlý [a], Katarína Csomorová [a]

[a]*Polymer Institute, Slovak Academy of Sciences, Dubravska cesta 11, 842 36 Bratislava, Slovakia*

Abstract. The thermooxidative and thermal stabilities of polyurethanes synthesized from polyisocyanate and polyol were assessed by chemiluminescence and thermogravimetry. The effect of varying heating rates, synthesis procedures (i.e., with or without a crosslinking catalyst), atmospheres and long-term ageing under laboratory conditions were examined and a mechanism of oxidative degradation was proposed based on comparisons of the - separately determined - oxidizability of polyisocyanate and polyol to that of polyurethane.

Keywords: Chemiluminescence, Thermogravimetry, Polyurethanes
PACS: 81.05.Lg

INTRODUCTION

In order to fill the gap in the understanding of both photooxidation and thermal oxidation mechanisms, the present study involves the chemiluminescence from unstabilized polyester urethanes in order to assess their thermal stability. Polyurethanes were prepared from the polyol Desmophene and the polyisocyanate Desmodur with and/or without the presence of a crosslinking catalyst. The effect of varying heating rates and atmospheres were described. The chemiluminescence runs after 18 months of sample storage in the laboratory were compared to their original counterparts, and the possible mechanism of light emission from polyurethane as well as its link with the material's degradation was outlined.

RESULTS AND DISCUSSION

If we measure the chemiluminescence from polyol and polyisocyanate in oxygen and compare it with that of polyurethane itself, we see that both components of the polyurethane network are strongly luminescent. However, the start of the light emission from the polyisocyanate precedes that from the polyol. The chemiluminescence from polyurethane then represents the combination of both (Fig. 1). According to this figure, the oxidation process should initially be induced in the polyisocyanate segments and only later spread to the polyol moieties. This has already been seen in thermogravimetry experiments in nitrogen, and such a behavior is quite the opposite of the suggestion expressed in the study of photoinitiated oxidation of polyether urethanes [1] where two stages of the process were proposed.

CP1255, *Vth International Conference on Times of Polymers (TOP) and Composites,*
edited by A. D'Amore, D. Acierno, and L. Grassia

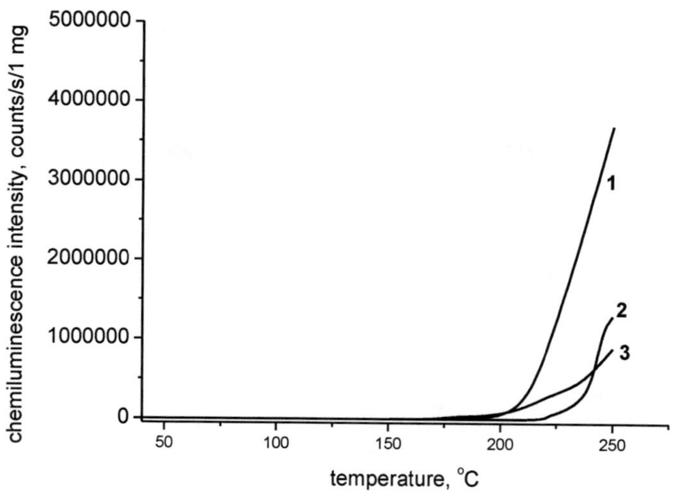

FIGURE. 1. Non-isothermal chemiluminescence runs for the two polymers composing polyurethane, namely the polyisocyanate Desmodur (3) and the polyol Desmophene (2), as well as that for polyurethane itself (1).

Isothermal chemiluminescence experiments (Fig. 2) provided the lines with a maximum which could be adequately fitted by Eq. 1 [2].

$$I = \frac{A\exp(-k_1 t)}{[1+Y\exp(-k_2 t)]^2} \tag{1}$$

The values of rate constants for fast-decomposing peroxides (k_2) and slow-decomposing peroxides (k_1) for a given set of temperatures are given in Table 1. The rate constants k_1 at 150 and 200 °C are quite close to those for the heating rate of 1 °C/min from nonisothermal experiments. This may be evidence of the slow-decomposing isolated hydroperoxides being the main species leading the oxidation process. At higher rates of heating, one may observe a significant discrepancy between rate constants from isothermal and nonisothermal experiments, probably due to higher values of activation energy in the latter case with a significant contribution from polymer chain scission to the formation of free peroxyl radicals. The relatively low values of parameter A, related to the relative increase in the hydroperoxide concentration to above the initial concentration, indicates that the length of the kinetic oxidative chains was rather short for polyurethanes.

TABLE 1. Parameters of Eq. 1 for isothermal CL runs in oxygen shown in Fig. 2.

Temperature, °C	A	k1, s^{-1}	Y	k2, s^{-1}
150	1 040	2.0e-6		1.4e-4
170	152 201	-6	1.47	-4
180	256 342	1.0e-5	1.17	5.3e-4
190		6.0e-5	1.72	1.32e-3
200	700 379	6.0e-5	0.589	8.9e-4
220		-4	1.12	-3

It is interesting to compare the chemiluminescence vs. temperature runs on the original polyurethane prepared without the catalyst to that kept 18 months under laboratory conditions in air. The difference is striking, indicating that significant changes occurred in the sample during its long-term storage. These changes may be attributed to the effect of light on the sample during the storage.

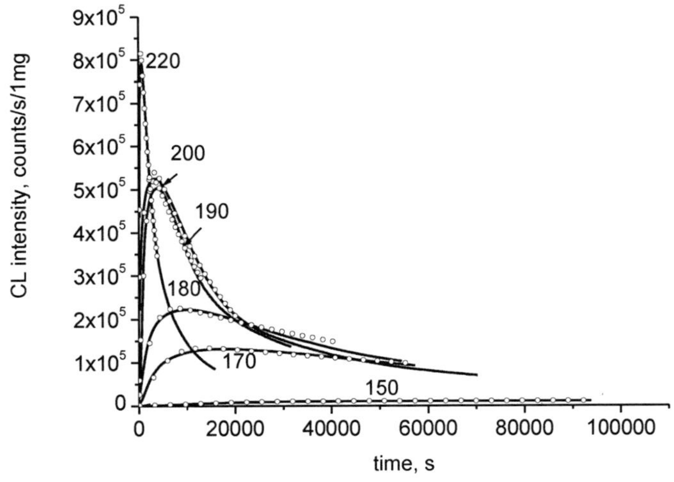

FIGURE. 2. Isotherms of chemiluminescence runs in oxygen for polyurethanes prepared with a catalyst. The symbols represent the theoretical fit by Eq. 1.

According to Zhang [3] and others [4], there are three main pathways for the initial degradation of the urethane linkage, in a nitrogen atmosphere. In the presence of oxygen, the following sequence of reactions may be proposed: Thermal oxidation of polyether urethanes starts in polyisocyanate segments, probably on methylene units adjacent to NH groups. This was confirmed by chemiluminescence measurements in oxygen of either polyisocyanate or polyol in comparison to polyurethane prepared from the two components. The increase of the light emission from polyisocyanate precedes that from polyol. A similar pattern was revealed by nonisothermal thermogravimetry in nitrogen.

ACKNOWLEDGMENTS

The present research has received funding from the European Community's Seventh Framework Programme FP7/2007-2013 under the grant agreement no. 212218 - Popart: Strategy for the preservation of plastic artifacts in museum collections.

The authors also gratefully acknowledge the support from the Grant Agency VEGA, Project No. 2/0115/09. This publication is the result of the project implementation: Centre for materials, layers and systems for applications and chemical processes under extreme conditions supported by the Research & Development Operational Programme funded by the ERDF.

REFERENCES

1. D.F. Mielewski et al., *Polym. Degrad. Stab*. **33**, 93-104 (1991).
2. J. Rychlý et al., *Polym. Degrad. Stab*. **78**, 357 (2002).
3. Y. Zhang et. al., *Polymer Testing* **28**, 264-269 (2009).
4. E. Dyer and G.E.J. Newborn, *J. Am. Chem. Soc*. **80** 5495-5498 (1958).

Alternative synthetic routes for the preparation of PLA/montmorillonite nanocomposites

Simona Bronco[a], Maria Beatrice Coltelli[b] and Stefania Castiello[b], Francesco Ciardelli[b], Elisa Taburoni[b], Lucia Conzatti[c]

[a] *CNR-IPCF Istituto per i Processi Chimco-Fisici, via Risorgimento n. 35, 56126 Pisa (Italy), simona@dcci.unipi.it*
[b] *Dipartimento di Chimica e Chimica Industriale via Risorgimento n. 35, 56126 Pisa (Italy)*
[c] *CNR-ISMAC Genova, via dei Marini 6, 16149 Genova (Italy)*

Abstract. Although the improvement of tensile Modulus and thermal stability due to the addition of innovative organoclays to PLA were negligible, our investigation allowed to evidence that the preliminary preparation of an inorganic rich composite through in situ polymerization of lactyde gave a final morphology improved with respect to that achieved by simply melt dispersing organoclay powder. Anyway the former preparation method should be further investigated in order to control PLA structural features resulting from the ring opening polymerization synthesis.

Keywords: Bioplastics, nanocomposites, montmorillonites, biocompatible cations, melt blending, ROP.
PACS: 61.46.-w: http://www.aip.org/pacs/index.html

INTRODUCTION

A large part of the scientific community is focusing the attention in the last year in the development of renewable polymeric materials with enhanced barrier properties [1]. High performance environment friendly biodegradable plastics, named as bioplastics, can be produced by using aliphatic polyesters [2] and in particular poly(lactic acid) (PLA). Anyway, it is well-known that its thermo-mechanical and gas barrier properties cannot allow the replacement of traditional polymers in all the application fields and for these reasons innovative methods to modulate its properties are investigated [3-5].

Take the advantage from the knowledge already available on the polyolefinic materials, the preparation of layered silicate nanocomposites was extensively investigated by intercalation in the melt and in situ polymerization. In the former case commercial or properly modified organoclays were used in order to favour the dispersion of the polar inorganic filler in the apolar polymeric matrix [6], in the latter case the *in situ* ring opening polymerization of lactide onto clay or organoclay was performed.

The present paper is focused onto the study of nanostructure, thermal and mechanical properties of nanocomposites composed of PLA and a constant amount of organoclay (3 wt%). Properly modified organoclays were prepared in order to allow the homogeneous dispersion of the hydrophilic clays in the apolar polymer matrix; in particular, 2-hydroxyethyl-trimethylammonium (choline), polyethyleneoxide(15)-(hydrogenated tallow) were used as surfactants as their structure can match the requirements of a biocompatible material. These OMMTs were used for preparing composites by melt blending or by *in situ* ring opening polymerization (using the clay surfactant as initiator for the polymerization) followed by melt dispersion into a PLA matrix. Structural, morphological and thermo-mechanical properties of the products are compared in order to assess advantages and disadvantages of the two different preparation routes.

EXPERIMENTAL

Materials

CP1255, *Vth International Conference on Times of Polymers (TOP) and Composites,*
edited by A. D'Amore, D. Acierno, and L. Grassia

PLA 2002D was a NatureWorks product. Sodium montmorillonite aqueous suspension and Dellite D43B were kindly supplied by Laviosa Chimica Mineraria.

Two different types of alkylammonium salts were used to increase the surface hydrophobicity of pristine MMT, 2-hydroxyethyl-trimethylammonium chloride (choline chloride, named Colcl) and polyethyleneoxide(15)-(hydrogenated tallow)-ammonium chloride (Ethoquad, named Etho). Modification of sodium montmorillonite was carried out in water. The preparation of PLA/organoclay composites with an inorganic content of 3% by weight, was performed in a Brabender discontinuous mixer, at 180°C, 100 rpm for 20 minutes.

For PLA synthesis Rac-3,6-dimethyl-1,4-dioxane-2,5-dione (lactide) and Tin(II) octanoate (assay > 96% and 98% respectively) were used. The reaction was carried out in bulk in a continuously stirred glass equipment at 95°C under nitrogen atmosphere. The product was purified through reprecipitation in methanol.

Characterization

Phase morphology was investigated through Transmission Electron Microscopy (TEM) and collecting X-ray diffraction data by WAXD.

Thermal stability of reagents and products was determined using a Mettler Toledo TGA/SDTA 851e apparatus connected to a dry N_2 purge. The analysis was run from 25 °C to 600 °C with a scanning rate of 10 °C/min under nitrogen or oxidative atmosphere.

Mechanical properties of the polymer were studied at 10 mm/min with a Tinius Olsen H10KT dynamometer, according to ASTM D638 procedure.

RESULTS AND DISCUSSION

For each organically modified montmorillonite through spectrophotometric titration with Methylene Blu, the degree of substitution was determined. For all the samples the substitution degree was close to 100%. The ion exchange experiments allowed to easily replace Na+ with potentially biocompatible cations, not yet employed for such application. The ion exchange study showed a relationship between the degree of exchange and the basal spacing d_{001} of montmorillonite. The latter increased when an increased degree of exchange was achieved, steric hindrance of intercalated ions being equal.

The prepared organophilic montmorillonites were dispersed in the melt in a discontinuous mixer. The PLA composite containing montmorillonite modified with Ethoquad (nEtho) showed mainly microaggregates and a partially intercalated structure. The d_{001} of D43B was not much enlarged thanks to the dispersion in the PLA matrix.

In the case of the PLA composite containing montmorillonite modified with choline chloride (nColcl) only microaggregates could be observed by TEM.

Table 1: results of TEM, WAXD, TGA and tensile tests characterization

Sample	**Main Morphology**	**d_{001}**	**T^a_{ip}**	**E^b**
PLA	-	-	361	3.0±0.2
D43B	-	1.80	-	-
nD43B	Mainly exfoliated	3.24	368	3.5±0.2
nEtho	Partial intercalation / Micro aggregates	1.84	365	2.7±0.2
nEthoMs	Microaggregates / Intercalated stacks	1.84	361	2.8±0.2
nColcl	Micro aggregates	1.48	363	2.5±0.1
nColclMs	Micro aggregates	1.46	362	2.4±0.2

[a] T_{ip}= temperature of inflection point= temperature of maximum rate of degradation (°C)
[b] Tensile Modulus (**GPa**)

The reaction, initiated by Tin(II) octanoate, proceeded through the insertion of the monomer into the metal-oxygen bond of the initiator. The presence of the clay in the bulk seemed to affect the final molecular weight of the polymer. From our preliminary studies it was verified that the presence of layered silicates resulted in a decrease of molecular weight with respect to the synthesis of pure polyester.

This did not allow to prepare useful nanocomposites by this process; therefore montmorillonite rich composites (masterbatches) were prepared by ring opening polymerization starting from Colcl and Etho. Successively they were diluted in pure PLA in the laboratory mixer in order to prepare 3% by weight clay composites, nColclMs and nEthoMs respectively.

The TEM (Fig. 1) and WAXD characterization showed the presence of microaggregates and intercalated stacks in samples obtained by the preliminary preparation of the montmorillonite rich composites (Tab.1), but on the other hand, the tensile Modulus and thermal stability resulted not improved. The latter properties were probably affected by the presence of the PLA fraction at low molecular weight of the masterbatch prepared by in situ polymerization.

The use for comparison of a commercial montmorillonite (D43B) (modified by dimethyl benzyl hydrogenated tallow quaternary ammonium salt) dispersed in the melt resulted in mainly exfoliated compounds, with an increase of the tensile Modulus from 3000 up to 3500 MPa and a 7 degrees increase in inflection point degradation temperature (Tab. 1).

Fig. 1: TEM micrographs obtained from nEtho (i) and nEthoMs (ii)

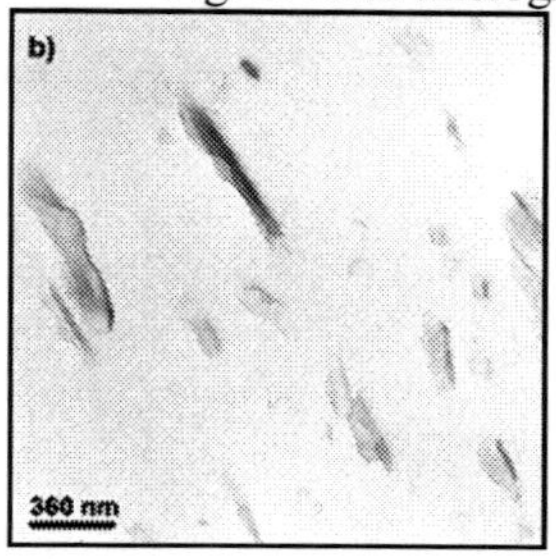

(i)

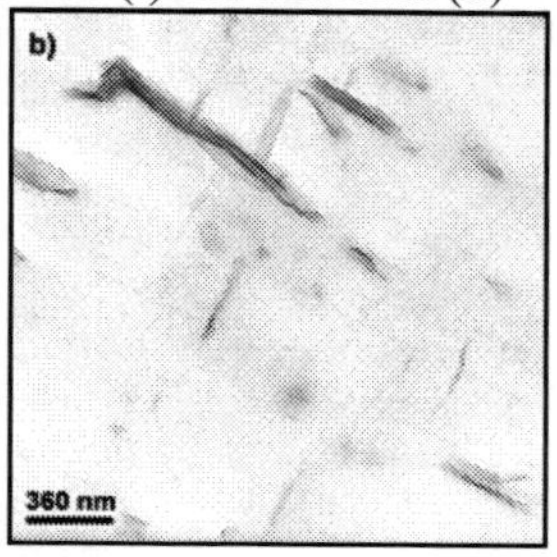

(ii)

CONCLUSIONS

Although the improvement of tensile Modulus and thermal stability due to the addition of innovative organoclays to PLA were negligible, our investigation allowed to evidence that the preliminary preparation of an inorganic rich composite through in situ polymerization of lactide gave a final morphology improved with respect to that achieved by simply melt dispersing organoclay powder. Anyway the former preparation method should be further investigated in order to control PLA structural features resulting from the ring opening polymerization synthesis

ACKNOWLEDGMENTS

Financial support by MIUR (NANOPACK FIRB 2003 D.D.2186 grant RBNE03R78E) is also kindly acknowledged.

REFERENCES

1. J. Lunt, *Polym. Degrad. Stab.* **59(1-3)**, 145-152 (1998).
2. Yu, K. Dean, L. Li, *Prog. Polym. Sci.* **31**, 576-602 (2006).
3. R. Auras, B. Harte, S. Selke, *Macromol. Biosci.* **4(9)**, 835-864 (2004).
4. A. Sorrentino, G. Gorrasi, V. Vittoria, *Trends in Food Science & Technology* **18(2),** 84-95 (2007).
5. M-B. Coltelli, I. Della Maggiore, M. Bertoldo, F. Signori, S. Bronco, F. Ciardelli, *J. Appl.Polym. Sci.* **110(2),** 1250-1262 (2008).
6. R. Masirek, Z. Kulinski, D. Chionna, E. Piorkowska, M. Pracella, *J. Appl. Polym. Sci.* **105(1),** 255-268 (2007).

Study of Polymer Glass Transition and Segmental Motions in Partially Crystallized Poly(ethyl acrylate)/p-xylene Mixtures

A. T. Stathopoulos[a], A. Kyritsis[a], F. J. Romero Colomer[b], J. L. Gómez Ribelles[b,c,d], N. Shinyashiki[e], C. Christodoulides[a], P. Pissis[a]

[a]*Department of Physics, National Technical University of Athens, Zografou Campus, 15780 Athens, Greece*
[b]*Centro de Biomateriales e Ingenería Tisular, Universidad Politechnica de Valencia, P.O. Box 22012, E-46071 Valencia, Spain*
[c]*Regenerative Medicine Unit, Centro de Investigación Príncipe Felipe, Autopista del Saler 16, 46013 Valencia, Spain*
[d]*CIBER en Bioingeniería,Biomateriales y Nanomedicina, Valencia, Spain*
[e]*Department of Physics, Tokai University, Hiratsuka, Kanagawa 259-1292, Japan*

Abstract. A poly(ethyl acrylate network), PEA, was swollen with several contents of p-xylene, c_{px}, from xerogel until saturation ($0 \leq c_{px} \leq 0.86$). Differential Scanning Calorimetry (DSC), Thermally Stimulated Depolarization Currents (TSDC) and Dielectric Relaxation Spectroscopy (DRS) were performed to probe the thermal transitions and the polymer molecular dynamics in these mixtures. The results show, that the mixtures are homogeneous in the whole temperature range studied for p-xylene contents $c_{px} < 0.20$, whereas are partially crystallized for $c_{px} \geq 0.20$ (co-existence of homogeneous PEA/p-xylene phase and p-xylene crystallites). For $0.20 \leq c_{px} \leq 0.30$, the mixtures are partially crystallized only on heating (above the glass transition temperature, cold crystallization), while for $c_{px} > 0.30$ during cooling. The partially crystallized mixtures exhibit a critical p-xylene content estimated at 0.12-0.15 that remains un-crystallized regardless of the p-xylene composition. The respective polymer glass transition temperature remains constant at -47 ±2°C due to the presence of this specific amount of un-crystallized p-xylene for $c_{px} > 0.30$, whereas it decreases significantly with increasing c_{px} ($0 < c_{px} < 0.30$). The α relaxation polymer segmental dynamics, related with the polymer glass transition, sustains its cooperative nature in the partially crystallized mixtures, whereas the distribution of α process relaxation times exhibit a slight broadening compared to the xerogel and homogeneous mixtures, ascribed to the spatial concentration fluctuations induced by the presence of p-xylene crystallites. Finally, the α process dielectric strength as well as the heat capacity increment of glass transition undergo significant depression compared to the homogeneous mixtures once the p-xylene crystallites are present.

Keywords: poly(ethyl acrylate), p-xylene, binary mixtures, plasticization effect, glass transition, segmental molecular dynamics
PACS: 64.70.Pj, 64.75.Va, 77.22.Gm

INTRODUCTION

Polymer/solvent binary mixtures have attracted both technological and scientific interest. The investigation, on a molecular level, of the organization of these small molecules within the polymer matrix and their intermolecular interactions with polymer chains as well as their transport properties and mobility is necessary in order to establish a general predictive relationship between the physical and mechanical properties of the mixtures and the nature and content of the small molecules. The polymer dynamics regarding primary processes related with the glass transition under the presence of solvent and vice versa has been studied extensively[1-4], mainly in homogeneous mixtures. The behavior of the polymer and solvent dynamics in partially crystallized mixtures are not still wide studied, though quite intriguing[5]. Interesting enough, in partially crystallized mixtures, a specific amount of solvent remains un-crystallized regardless of the solvent content. According to Rault et al., the presence of an amount of un-crystallized

CP1255, *Vth International Conference on Times of Polymers (TOP) and Composites,*
edited by A. D'Amore, D. Acierno, and L. Grassia

solvent is not explained by invoking any specific interactions like hydrogen bonding, but it is governed by kinetic hindrances as the temperature lowers approaching the glass-transition[6,7].

The present work deals with the investigation of the phase transitions and the polymer dynamics of poly(ethyl acrylate)/p-xylene swollen polymer network spanning a wide range of non polar p-xylene compositions (from the xerogel to saturation, c_{px} = 0.86). Differential Scanning Calorimetry (DSC), Thermally Stimulated Depolarization Currents (TSDC) and Dielectric Relaxation Spectroscopy (DRS) were performed to this direction. These methods allow the construction of the phase diagram of this binary mixture and the study of the polymer segmental dynamics in both homogeneous and partially crystallized mixtures, in terms of the chain mobility, shape and strength of the α relaxation process. Correlations between the results of DSC, TSDC and DRS on PEA/p-xylene mixture composition are of great interest.

EXPERIMENTAL

Poly(ethyl acrylate) polymer network with 2 wt.% of EGDMA as crosslinker agent was exposed to p-xylene atmosphere for variant times or immersed in pure p-xylene (99% pure). The p-xylene content c_{px} was calculated gravimetrically by dividing the solvent mass in the swollen sample to the total mass. Differential Scanning Calorimetry (DSC) was performed in a Mettler Toledo DSC 823E apparatus. The samples were subjected to cooling and heating scans between -90 and - 25°C, both scans with a rate of 10 deg/min. Nitrogen gas was let through the DSC cell with a flow rate of 30ml/min. Thermally Stimulated Depolarization Currents (TSDC) measurements were performed by means of a KEITHLEY 617 electrometer in the temperature interval between -150 and 20°C under nitrogen flow, controlled by a Novocontrol Quatro Cryosystem. Typical conditions were E_p=1-5 kV/cm for the polarizing field, t_p=5 min for the polarization time at polarization temperature T_p=-10°C, 10 grad/min for the cooling rate to T_o=-150°C, and b=3 grad/min for the heating rate.Dielectric Relaxation Spectroscopy (DRS) measurements were performed with an Alpha Analyzer combined with a Quatro Temperature Control system, both by Novocontrol. Complex dielectric permittivity $\varepsilon^*(f)=\varepsilon'(f)-i\varepsilon''(f)$ was measured isothermally in steps of 5°C in the temperature interval -150 to 20°C and in the frequency range from 10^{-1} to 10^6 Hz under nitrogen flow.

RESULTS AND DISCUSSION

The DSC thermograms reveal that the PEA/p-xylene mixtures are classified into three qualitatively different groups in terms of the respective phase transitions in the recorded thermograms with changing p-xylene content. For p-xylene contents $0 < c_{px} < 0.20$ the DSC thermograms exhibit no p-xylene first order phase transitions indicating homogeneity in PEA/p-xylene mixtures. For intermediate p-xylene contents $0.20 < c_{px} \leq 0.30$, the first order phase transitions of p-xylene are emerged mainly on heating and after the glass transition, whereas the crystallization of p-xylene on cooling is absent. This behavior is ascribed to supercooling effects that favors the crystal nuclei formation on cooling, however, the increased viscosity by lowering temperature hinders their growth, which occurs only on heating and after the glass transition. For higher p-xylene contents $c_{px} > 0.30$ the crystallization of excess p-xylene occurs during cooling rendering the mixtures partially crystallized at a wide measured temperature interval.

The phase transitions of p-xylene (crystallization and melting) as well as the polymer glass transition temperature are presented in Fig.1. It is obvious the plasticization of polymer glass transition with adding p-xylene which is well conformed to the well known Fox expression, even in the intermediate p-xylene contents $0.20 < c_{px} \leq 0.30$ (since cold crystallization occurs after the recorded glass transition). However, for the mixtures with $c_{px} > 0.30$ for which the mixture are partially crystallized from cooling, the polymer glass transition remains constant regardless of the p-xylene content at a specific temperature -47 ± 2 °C. This is explained through the fact that once the p-xylene molecules form crystals the polymer concentration in the polymer / un-crystallized p-xylene phase increases and subsequently the polymer glass transition temperature corresponds to a mixture with lower p-xylene content. By extrapolating the values of constant glass transition towards the Fox curve (Fig. 1), the amount of un-crystallized p-xylene in the liquid can be estimated, which lies between 0.12 and 0.15.

The polymer segmental dynamics (α relaxation process) related to polymer glass transition is studied by means of TSDC and DRS in a wide p-xylene content interval. Special interest attract the partially crystallized mixtures ($c_{px} > 0.30$) for which the polymer dynamics remains active despite the presence of p-xylene crystallites. The segmental polymer motions maintain their cooperativity, though variances were observed regarding the shape and the respective dielectric strength compared to the xerogel and the homogeneous mixtures. Both DRS and TSDC results shows the broadening of the α relaxation process in partially crystallized mixtures, which is ascribed to the concentration fluctuations in the liquid phase caused by the presence of p-xylene crystallites. Interesting enough is also the depression of dielectric strength (measured by TSDC and DRS) as well as of the heat capacity increment of glass transition for $c_{px} > 0.30$, implying the presence of restricted mobility due to reduced number of available configurations for the polymer chains to rearrange (Fig. 3).

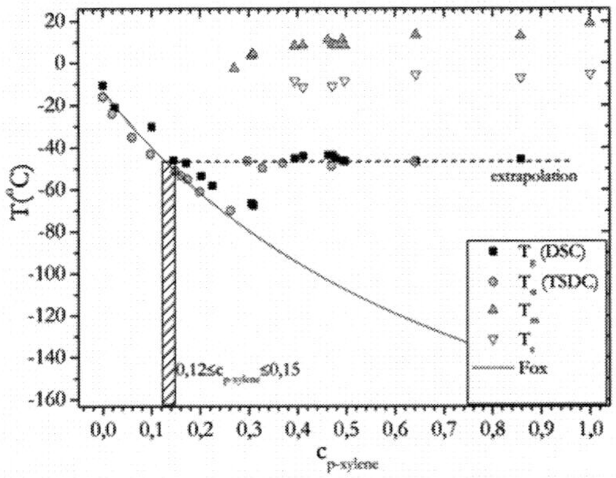

FIGURE 1. The phase diagram. The horizontal axis corresponds to p-xylene content. The plot shows the first order phase transitions of p-xylene (T_m, T_c), as well as the polymer glass transition temperatures extracted from DSC (T_g) and TSDC (T_α) measurements. The solid line corresponds to the respective Fox curve. The horizontal dot line corresponds to the extrapolation towards Fox curve.

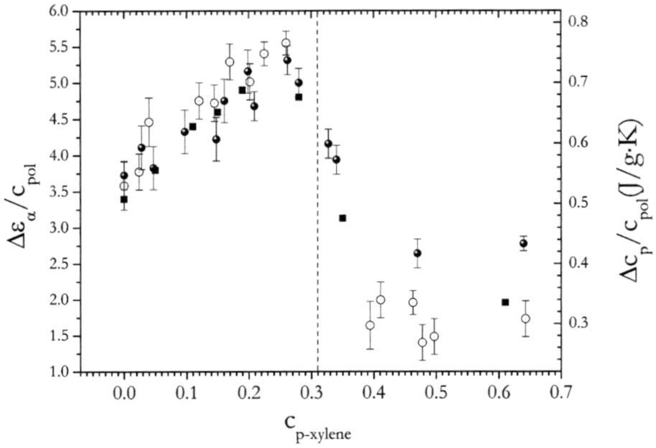

FIGURE 2. The α relaxation process dielectric strength (solid circles –extracted from TSDC, solid squares -from DRS) compared with the heat capacity increment at glass transition (open circles-right axis) all divided by the polymer content in mixtures against p-xylene content.

CONCLUSIONS

The PEA/p-xylene mixtures were studied with DSC and dielectric techniques (TSDC and DRS) for a wide range of p-xylene contents. The results have shown that the PEA/p-xylene mixtures are partially crystallized over a wide measured temperature range for c_{px}>0.30, while are homogeneous mixtures in the whole measured temperature range for c_{px}< 0.20. For intermediate p-xylene contents the mixtures are homogeneous on cooling, while they present cold crystallization on heating and after the glass transition. In partially crystallized mixtures (c_{px}>0.30) a specific amount of p-xylene was found to remain un-crystallized. The content of un-crystallized solvent has been estimated about 0.12-10.15. This constant amount of p-xylene maintains the polymer glass transition constant at -47 ± 2°C regardless the amount of p-xylene crystallites, while it yields broadening of α relaxation times distribution and significant depression both for the α process dielectric strength and heat capacity increment at glass transition compared to the homogeneous mixtures.

ACKNOWLEDGMENTS

A.T.S. would like to thank the Department of Physics, National Technical University of Athens for the financial support of this work.

REFERENCES

1. S. Cerveny, G. A. Schwartz, A. Alegría, R. Bergmann, J. Swenson, *J. Chem. Phys.* **124**, 194501 (2006)
2. N. Shinyashiki, S. Sudo, S. Yagihara, A. Spanoudaki, A. Kyritsis, P.Pissis, *J.Phys:Condens. Matter* **19**, 205113 (2007)
3. M. Yada, M. Nakazawa, O. Urakawa, M. Morishima, K. Adachi, *Macromolecules* **33**, 3368 (2000)
4. D. Cangialosi, A. Alegría, J. Colmenero, *J. Chem. Phys.* **128,** 224508 (2008)
5. N. Shinyashiki, M. Shimomura, T. Ushiyama, T. Miyagawa, S. Yagihara, *Journal of Physical Chemistry B* **111**, 10079 (2007)
6. J. Rault, A. Lucas, R. Neffati, M. Monleón Pradas, *Macromolecules* **30**, 7866 (1997)
7. M. Salmerón Sánchez, M. Monleón Pradas, J.L. Gómez Ribelles, *Journal of Non-Crystalline Solids* **307-310**, 750 (2002)

Comparative study between Langmuir polymer monolayers behavior and their chemical structures

L.Gargallo, N.Becerra, A.Leiva, C.Sandoval , D. Radic

Laboratorio de Química Fisica de Macromoleculas, Departamento de Química Física. Facultad de Quimica, Pontificia Universidad Católica de Chile. Casilla 306, correo 22.Santiago, Chile. E-mail: lgargall@puc.cl

Abstract

The monolayer of a diblock copolymer was studied by Brewster Angle Microscopy (BAM) during compression, which allowed the estimation of the collapse pressure π_c of the polymeric system. Molecular Dynamic Simulation (MDS) was performed in order to explain the experimental behavior of the functionalized poly(methacrylate)s studied at the air-water interface.

Keywords: Surface pressure,air-water interface,Diblock copolymer, Poly(phtalimidoalkyl methacrylate)s

INTRODUCTION

In the last decades extensive studies have been performed to elucidate the chemical-structural relationship surface behavior and intermolecular forces of 2D arrays of macromolecules at the air-water interface.Particularly, there has been growing interest in the study of basic properties of functionalized polymers and block copolymers at interfaces and the surface free energy of the polymers.[1-2] The molecular orientation of polymers with different hydrophobicity degree at interfaces is of fundamental significance for biological and tecnological processes.

The aim of this contribution is to investigate the changes in the molecular organization of the monolayers of functionalized poly(methacrylate)s and one ampliplilic diblock copolymer respectively.

RESULTS AND DISCUSSION

For all the studied systems,the Langmuir monolayers were obtained by spreading solutions onto MILLI-Q water as subphase. After spreading, the monolayers were compressed at a velocity of 5 $cm^2 min^{-1}$ at 298 K[3-4] The chemical structures of the functinalized poly(methacrylate)s are shown in Scheme 1 and the chemical structure of the diblock copolymer studied is shown in Scheme 2

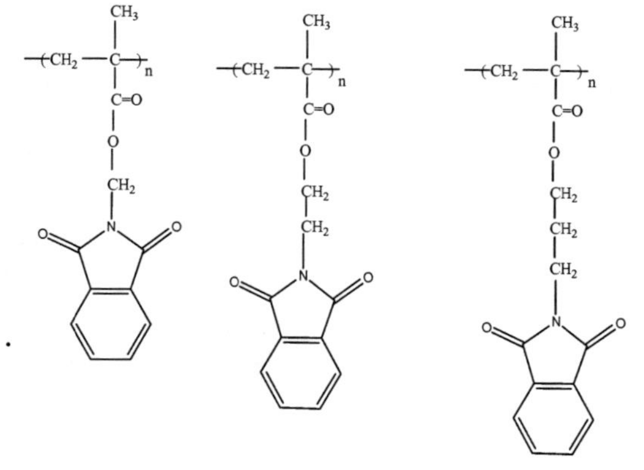

1 2 3

Scheme 1: 1-)Poly(N-phthalimidomethyl methacrylate) (PNPMMA) 2).-Poly(N-phthalimidoethyl methacrylate) (PNPEMA); 3-) Poly(N-phthalimidopropyl methacrylate) (PNPPMA)

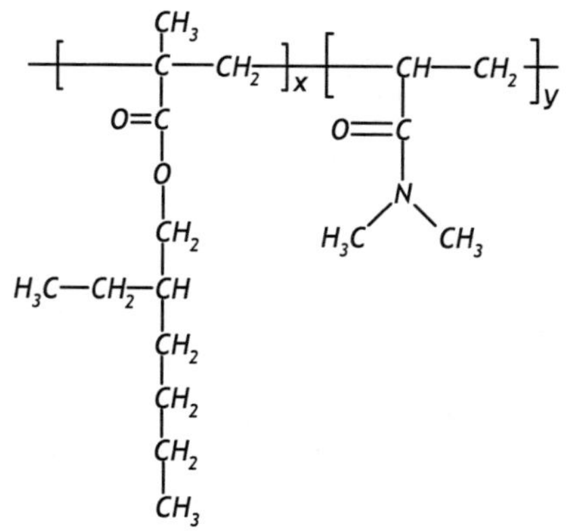

Scheme 2. poly(N,N-dimethylacrylamide)-b-poly (2-ethyl hexyl methacrylate) (PDMA-b-PHEMA)

CP1255, *V*[th] *International Conference on Times of Polymers (TOP) and Composites,*
edited by A. D'Amore, D. Acierno, and L. Grassia

The poly(methacrylate) derivatives form stable film at the air-water interface. The π-A curves are the condensed type and tend clearly towards isotherms more condensed when the number of the methylene groups in the lateral chain increases. As the compression continues, the surface pressure sharply increases, giving compacted films for the three systems. (See Figure 1) . Significant differences in the overall shape of the isotherm were observed for the diblock copolymer PDMA-b-PHEMA (see Figure 2) The monolayer of the copolymer presents Newtonian behavior and is highly elastic. The isotherm presents a particularly steep slope and are very similar to those obtained for molecular aggregates. This behavior, together with the linearity of the isotherms predicts a little variation of the elasticity at the different surface pressures considered.

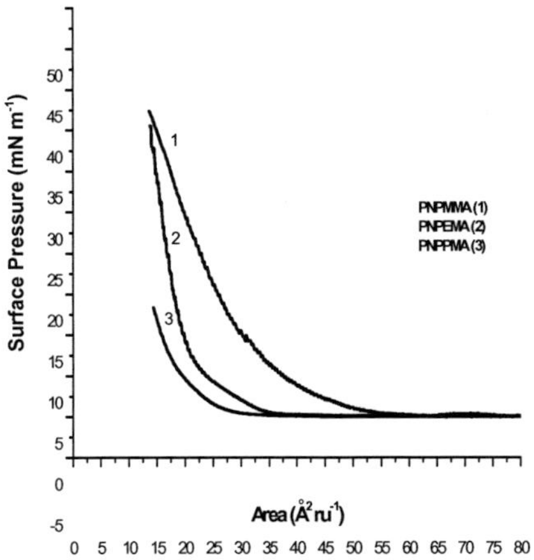

Figure 1.Surface pressure isotherms (π-A) of the polymers studied.

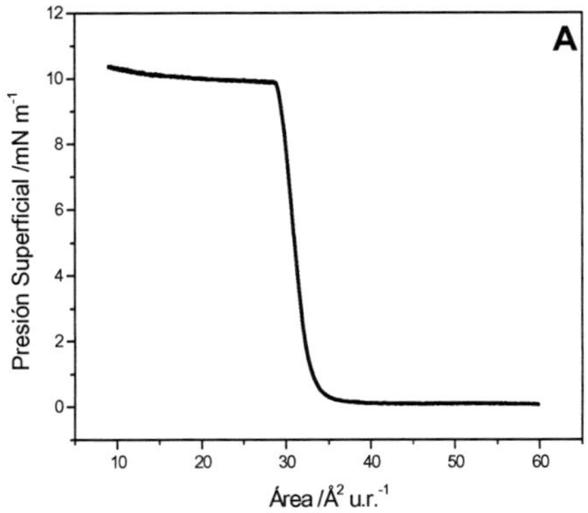

Figure 2. Surface pressure isotherms (π-A) of the Diblock copolymer.

Tables 1, 2 and 3 compile the results of the main surface properties of all these polymeric materials.

TABLE 1. A_0 and critical exponent υ values obtained from isotherms π–A for PNPMMA, PNPEMA and PNPPMA.

Polymer	A_0 (Å^2/ru)	υ (± 0.01)
PNPMMA	27	0.69
PNPEMA	20	0.57
PNPPMA	17	0.56

TABLE 2. Surface free energy SE, dispersión energy contribution γ_{disp} and polar energy contribution γ_{polar} values for ,PNPMMA, PNPEMA and PNPPMA.

Polymer	CA * Degrees	CA ** Degrees	SE (mN/m)	γ_{disp} (mN/m)	γ_{polar} (mN/m)
PNPMMA	51.9 ± 0.9	28.0 ± 0.9	55.7	41.0	14.7
PNPEMA	47.2 ± 0.9	43.3 ± 0.9	54.8	33.4	21.4
PNPPMA	67.1 ± 0.9	34.9 ± 0.9	46.9	39.6	7.3

* water
** diiodomethane

The diblock copolymer monolayer was studied by BAM during compression, which allowed the determination of the collapse pressure π_c of the system. *Table 3 compile some surface properties for the diblock copolymer and their homopolymers.*

Table 3. Surface properties of the diblock copolymer and υ exponent of PEHMA-b-PDMA diblock copolymer and PDMA,PEHMA homopolymers respectively.

Diblock Copolyme and homopolymersr	A_0 $(Å^2\, r.u^{-1})$	π_C $(mN\, m^{-1})$	A_C $(Å^2\, r.u^{-1})$	$\upsilon\,(\pm 0.02)$
PEHMA-b-PDMA	32.6	10.4	28.7	0.52
PDMA	32.4	-	-	0.85
PEHMA	39.2	10.0	31.8	0.50

Order parameters S(r)[5] calculated from MDS for the three functionalized poly(methacrylate)s studied are shown in Figure 3 from lying orientation.

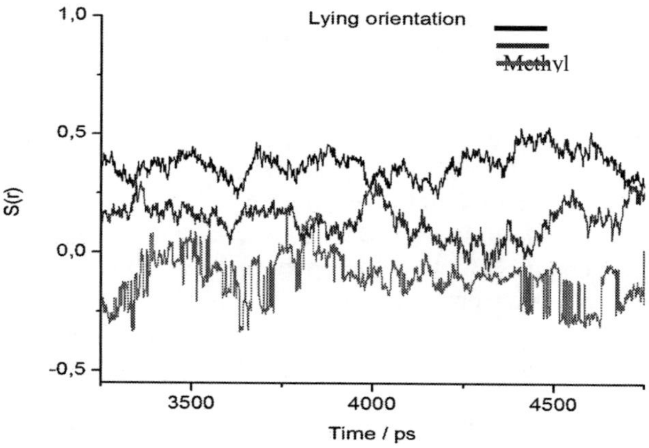

Figure 3. Order parameter, S(r), for PNPMMA, PNPEMA, PNPPMA between methyl groups of different chains from lying orientation.

ACKNOWLEDGEMENTS

Acknowledgememts is made to DiPUC and Fondecyt projects 1080026, 1080007 and 1080238. N.B. thanks to Fondecyt for Doctoral Fellowship

REFERENCES

1. Miranda, B.; Hilles, H.M.; Rubio, R.G.; Ritacco, H.; Radic, D.; Gargallo,L.; Sferrazza M.; and Ortega F. Langmuir, 25(21), 12561-12568 (2009)
2. Leiva, A.; Farias, A.; Gargallo, L.; Radic´.D. Eur.Polymer J. 44, 2589–2598 , (2008)
3. Leiva,A;Gargallo,L;Gonzalez,A,Radic,D, Eur.Polym.J.,40, 2349 (2004)
4.- Gargallo,L.;Vargas,D;Leiva,A;Radic,D.,J.Colloid Inter.Sci.,301,607 (2006)
5. Gargallo, L.; Vargas, D.; Becerra, N.; Sandoval, C.; Saldias, C.; Leiva, A.; Radic´,D. Macromol. Symp, 278 , 80 – 88 (2009)

Study of The Porosity Occurrence in Composite Materials by Means of Ultrasonic and In-Plane Shear Testing

Cinzia Toscano, Carlo Vitiello and Pietro Caramuta

C.I.R.A. Scpa – Italian Aerospace Research Centre
Via Maiorise s/n – 81043 - Capua (CE), Italy
mailto: c.toscano@cira.it

Abstract. The induced porosity in CFRP specimens, with different staking sequence, has been inspected by means the Pulse-Echo Ultrasonic testing. The specimens with [45°/-45°] staking sequence have also been processed by in-plane shear testing, to study the influence of the porosity on the degradation of interlaminar shear strength.

Keywords: CFRP, porosity, ultrasonic testing, in-plane shear test
PACS: 62.20.de, 77.84.Lf, 81.70.Cv, *43.35.

INTRODUCTION

The presence of defects in Carbon Fiber Reinforced Plastics (CFRP) may emerge since their production, because of the several different parameters, such as pressure and temperature, that are involved in the manufacturing processes [1-2].

The detection of voids and porosity is particularly crucial because they are responsible of the delamination occurrence (inter lamina debonding). The local failure can induce severe degradation of the stiffness of the structures, causing damages during the in-service life of the components [3].

A careful porosity monitoring by means of proper Non Destructive Testing (NDT) can prevent the above cited problems. The Ultrasonic Pulse-Echo testing [4] allows to get informations not only concerning the evaluation of the amount of porosity, but also to investigate how the particular staking sequence play in the porosity distribution.

The aim of this work is to prepare a data base in order to create a relationship among types of porosity, ultrasonic testing response and porosity dependent-variation of the composite mechanical properties.

SPECIMEN MANUFACTURING AND TESTING

Pre-preg M21/IM7 laminas provided by HEXCEL has been used for the specimen manufacturing. By hand laying up, specimens with four different staking sequences have been prepared, and, in order to induce porosity, some of them have been cured by applying only a percentage of the prescribed pressure (7 bar as indicated in the material datasheet), during the autoclave process. In the table 1, the used staking sequences and the pressures applied are resumed.

TABLE (1). Specimens staking sequences and pressures used to induce porosity during the autoclave process (as percentage of the HEXCEL datasheet prescribed pressure).

Staking sequence	No. plies	100%	50%	0%
P1: [0°]	24	3	3	3
P2: [90°]	24	3	3	3
P3: $[(45°/-45°)_6]_S$	24	3	3	3
P4: $[45/0/-45/90/45/0/-45/90//90/-45/0/45/90/-45/0/45]_S$	32	3	3	3

CP1255, *V^th International Conference on Times of Polymers (TOP) and Composites,*
edited by A. D'Amore, D. Acierno, and L. Grassia

All the specimens have been inspected by Immersion Ultrasonic Pulse-Echo testing. Specimens type P3 (see the table 1) have been processed by in-plane shear testing in accordance with the standard [5].

Results of the Tests

The presence of voids or porosity in composite material has a strong influence on the attenuation of a ultrasonic signal passing through it: the voids can scatter[4] the wave increasing the total attenuation in the material. Of course this effect depends on the frequency of the ultrasonic wave, the dimension and the density of the air bubbles. Moreover, the voids cause a variation in the material density, and then, a variation of the acoustic impedance of the material. As a consequence, an increase in the total attenuation of the ultrasonic signal is related with the porosity "intensity" in the specimens. Generally speaking, in Pulse-Echo testing, some part of the signal energy entering the material is lost: that depends on the double reflections[6] at the interface between water and the specimen, as well as on the bulk attenuation over two passes in the material. In order to measure the total attenuation of the signal the attenuation on the Back Wall Echo, BWE, (the echo coming from the last interface of the specimen) has been measured.

In-plane shear tests has been performed to measure the interlaminar mechanical properties of the composite. According with the ASTM D3518 standard, the P3 type specimens have been tested using a set of three sample for each curing pressure. The obtained Shear Modulus and Shear Strength have been related to the attenuation data.

Ultrasonic Pulse-Echo Test

All the inspections have been carried out using a narrow band 5 MHz probe. The attenuations have been measured by averaging over hundreds of points on the obtained BWE attenuation C-scan. The obtained attenuations and the applied pressure (as percentage of the 7 bar prescribed), have been reported for each staking sequence in table 2.

TABLE (2). Measured attenuations at different pressure percentage.

Specimen type	Atten. at 100% (dB)	Atten. at 50% (dB)	Atten. at 0% (dB)
P1	15,5	18,7	19
P2	15,3	18,0	20,1
P3	15,4	17,3	21
P4	16,4	18,9	22,1

P4 is the thicker specimen and its attenuation is stronger at 0%, as expected. P4 and P1 doesn't seem to behave linearly with respect to the cure pressure (figure 1). This behavior will be matter of further investigations.

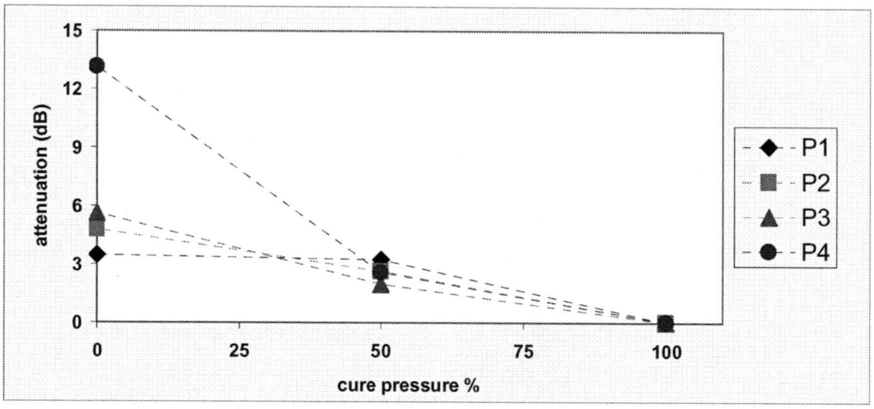

FIGURE 1. Normalized attenuations with respect the (minimum) attenuation, obtained at 100% of the pressure.

In-Plane Shear Test

In the following table 3, the results of the Shear test are reported in terms of Shear Modulus and Shear Strength.

TABLE 3. Results of the In-plane Shear Test.

Specimen P3	Shear Modulus (GPa)	Shear Strenght (GPa)
100%	4,6	95
50%	4,39	83,97
0%	3,96	78,29

Comparing the results with the correspondent measured attenuations (figure 2) it can be seen that in-plane shear modulus and strength are influenced by the presence of the porosity into the resin. The overall reduction of the composite mechanical properties is anyway smoothed by the presence of the fibers that are, obviously, not affected by matrix defects. The reduction of the in-plane shear properties, due to the porosity, is more evident at high strain values. In fact the shear strength is reduced more than the stiffness, that is typically measured at very low deformations.

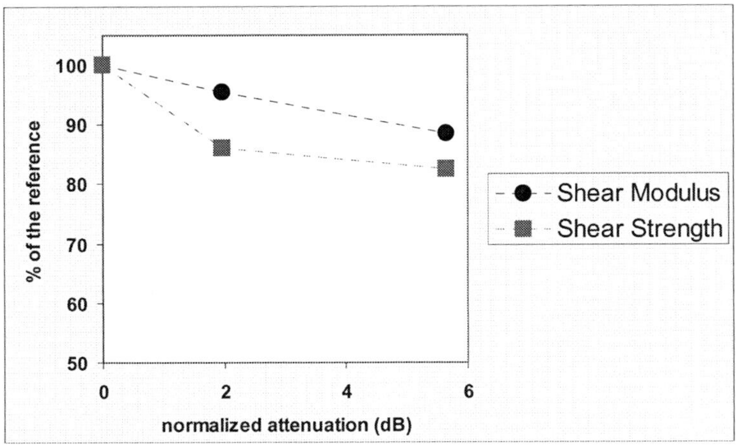

FIGURE 2. Shear modulus and Shear strength normalized with respect to the reference data (at 100% pressure) versus the related attenuation.

CONCLUSION

CFRP composites have been prepared by hand lay-up, in order to study the effect of the induced a porosity on the Ultrasonic testing and In-plane Shear testing. The first obtained results show that the porosity distribution is dependent on the staking sequence, and that the Shear Strength is influenced by the porosity amount. This results represents just a first phase of a porosity characterization testing campaign. The characterization will involve several different staking sequences and different amount of induced porosity. Both the influence of the porosity on physical and mechanical properties of the CFRPs will be deeply studied. More accurate and effective non destructive and destructive testing will be performed in order to increase the repeatability of the results.

ACKNOWLEDGMENTS

This work has been funded by the Italian PRORA project.

REFERENCES

1. F. C. Campbell et alt., *Journal of Advanced Materials*, **26**, 18-33 (1995)
2. R. D. Adams and P. Cawley, *NDT International*, **21**, 208-222 (1988)
3. H.T. Yoshida, T. Ogasa, R. Hayashi, *Comp. Sc. and Tech.* **25**, 3-18 (1986)
4. J. Ktrautkrämer and H. Ktrautkrämer, Ultrasonic Testing of Materials, Heidelberg DE, Springer-Verlag, 1990, 4th edition
5. ASTM Standard D3518: Standard Test Method for In-Plane Shear Response of Polymer Matrix Composite Materials by Tensile Test of a ±45° Laminate, ASTM International 94 (2007)
6. H. Jeong and D. K. Hsu, *Ultrasonics*, **33**, 195-203 (1995)

Confinement Effects on the Structure and Dynamics in Intercalated Polymer / Layered Silicates Nanohybrids

K. Chrissopoulou[a], A. Afratis[a], S. Fotiadou[a,b], B. Frick[c] and S. H. Anastasiadis[a,d]

[a]*Foundation for Research and Technology-Hellas, Instutute of Electronic Structure and Laser, P.O. Box 1527, 711 10, Heraklion Crete, Greece*
[b]*Aristotle University of Thessaloniki, Department of Chemical Engineering, 541 24 Thessaloniki, Greece*
[c]*ILL-Institut Laue-Langevin, F-38042 Grenoble Cedex, France*
[d]*University of Crete, Department of Chemistry, P.O. Box 2208, 710 03, Heraklion Crete, Greece*

Abstract. The structure and dynamics of PEO/Na^+ MMT nanocomposites are investigated by XRD, DSC, and quasielastic neutron scattering (QENS). For concentrations up to 20 wt% the PEO chains within the galleries form either a single- or a double-layered structure of intercalated chains; at higher PEO content only double layers of intercalated PEO chains are formed within the 0.9nm galleries. Moreover, it is only for polymer content above 70 wt% that peaks corresponding to crystalline PEO and DSC melting transition are observed, indicating that the confined chains remain liquid-like and only the excess polymer outside the completely full galleries can crystallize. QENS investigated the dynamics of PEO in bulk and in confinement. A jump of the bulk PEO dynamics at T_m is observed whereas the dynamics of confined PEO shows only weak wavevector and temperature dependence and goes smoothly through the bulk T_m.

Keywords: Layered Silicates, Intercalation, Quasielastic neutron scattering, Confinement, Dynamics.
PACS: 81.07.Pr, 81.05.Qk

INTRODUCTION

Polymers in restricted space can behave very differently than in the bulk especially when the chains are confined to dimensions of the order of their size. The dynamics of macro-molecules in thin films or in porous media has attracted the scientific interest for more than a decade [1]. There has been experimental evidence for significant reductions in the glass transition temperature, T_g, of thin polymer films with decreasing film thickness, indicative of enhanced segmental motion with the free surface and/or a non-wetting hard interface playing the key role; a strongly attractive surface may alter the behavior.

Besides planar polymer films, other experimental geometries have been utilized to study the effect of confinement on the segmental motion where the equivalence in the behavior between polymer nanocomposites and thin polymer films has been verified [2]. Utilizing polymer/layered-silicate nanohybrids has been proven particularly advantageous in this respect. Mixing polymers with layered silicates may result in three different micro- or nano-structures depending on the interactions between the chains and the inorganic surfaces: *phase separated*, where the polymer and the inorganic are mutually immiscible, *intercalated*, where the polymer chains intercalate between the layers of the inorganic material, and *exfoliated*, where the inorganic platelets are dispersed within the polymeric matrix [3]. The intercalated nanocomposites, in which a 0.8-2.5 nm polymer film resides between parallel inorganic layers in a well-ordered multilayer with a repeat distance of a few nm, offer a unique avenue to study the behavior of macromolecules in nanoscopic confinement by utilizing conventional analytical techniques on macroscopic specimens.

A number of investigations probed the variety of macromolecular dynamics in polymer/layered silicate nanocomposites. We report herein our investigation of the local dynamics of poly(ethylene oxide) confined within hydrophilic sodium-montmorillonite, Na^+-MMT [4,5,6]. The segmental dynamics is significantly accelerated in confinement relative to that in the bulk and it now exhibits much weaker *T*-dependence [4]. At the same time, the very local processes like the methyl side-group motion or the so-called β-relaxation remain unaffected by the

CP1255, *Vth International Conference on Times of Polymers (TOP) and Composites,*
edited by A. D'Amore, D. Acierno, and L. Grassia

confinement [4]. The semicrystalline PEO remains amorphous within the galleries; this influences significantly both the segmental dynamics and the ionic motion [4,5,6].

EXPERIMENTAL

Hydrophilic poly(ethylene oxide)/Na^+-MMT, PEO/Na^+-MMT, nano-composites were synthesized by direct melt mixing of PEO (M_w= 100,000, T_g=-67°C and T_m=65°C, Aldrich) with Na^+-MMT (Southern Clay) in compositions that covered the range from pure polymer to pure clay. The hybrids as well as the silicates were characterized by X-ray diffraction (XRD) in order to determine the interlayer spacing; the increase of the d_{001} spacing of the nanohybrid with respect to that of the parent silicate by approximately 10Å verifies intercalation.

Quasielastic high-resolution neutron backscattering experiments were performed at the IN16 spectrometer of ILL. Energy variation is performed by moving the monochromator and exploring the Doppler effect; the range of energy transfer is $-14.7<\Delta E=\omega h<14.7\mu eV$. The incident wavelength is λ=0.6271 nm, resulting in a range of accessible scattering vectors $0.19\leq q\leq 1.9Å^{-1}$. The experimental resolution function is obtained by measuring the silicate at 1.5K whereas the energy resolution is 0.49μeV. The spectrometer provides the dynamic structure factor $S_{exp}(q,\omega)$. In addition to the measurements of the quasielastic spectra, the energy resolved elastic intensity scattered from the samples is studied as a function of temperature for $1.5\leq T\leq 380$K with a heating rate of 0.3K/min. With an identical setup of the analyzers, the intensity $I_{el}(q,\omega\approx 0)$ is recorded with the Doppler driver at rest. At 1.5K, the samples represent elastic scatterers; at higher T, the molecular dynamics is enhanced, which leads to a broadening of the scattering function and the corresponding loss of the elastically scattered intensity.

RESULTS AND DISCUSSION

Fig. 1 shows the XRD patterns for pure PEO, Na^+-MMT and the nanohybrids [4]. The pure Na^+-MMT shows a main diffraction peak at 2θ=8.8° (1.0 nm interlayer spacing). As the PEO content increases, the Na^+-MMT peak disappears and two other peaks emerge at 6.8° and 4.8° that correspond to interlayer distances of 1.30 and 1.85 nm with relative intensities that depend on the PEO content. At concentrations up to 20 wt% PEO, either a single- or a double-layer structure of PEO chains forms within the galleries. Further increase of the PEO concentration reveals only double-layers of intercalated PEO chains. Moreover, for PEO content below 70 wt% no peaks are observed that can be assigned to the crystalline structure of PEO signifying that the PEO chains within the galleries are liquid-like; this is verified by DSC and Raman scattering as well. It is only for PEO concentrations higher than 70 wt% that two peaks at 18.9° and 23.2° emerge, which agree with the diffraction peaks of bulk crystalline PEO due to its monoclinic crystal structure. This indicates crystallization only of the excess polymer outside the completely full galleries and it shows that the presence of the MMT affects the morphology of the intercalated polymer.

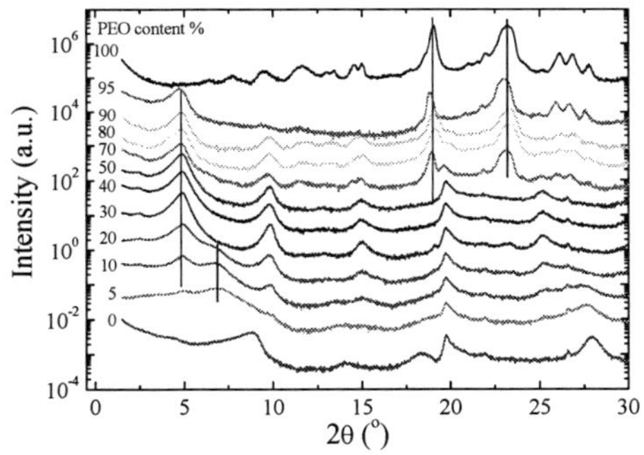

FIGURE 1: XRD pattern for the PEO/Na^+-MMT nanocomposites [4]

Fig. 2 shows the energy resolved elastic scans for the PEO, the Na^+-MMT, and the 30wt% PEO hybrid for various q's. The scans for PEO exhibit two distinct relaxation steps. The first begins above 200K and is related to the segmental motion of the amorphous portion of the semi-crystalline polymer above its T_g~220K (it may also include motion of traces of water). The second is an abrupt drop of the intensity at the T_m of PEO, which corresponds to the melting of the polymer crystallites and indicates that the polymer motion has become very fast. The data for the Na^+-MMT exhibit a continuous small decrease with increasing T around the T_m of water and it reaches a plateau at higher T's. The intensity of the nanocomposite exhibits a very broad decrease with a pronounced q-dependence.

There is no evidence of a sharp drop due to melting similar to the one observed in the homopolymer, which also signifies that there is no crystalline PEO. The behavior is also evident in the T dependence of the mean square displacement, calculated from the q-dependence of the elastically scattered intensity at each T, which shows a dramatic acceleration of the msd of the bulk polymer at PEO T_m and the absence of any transition in the composite.

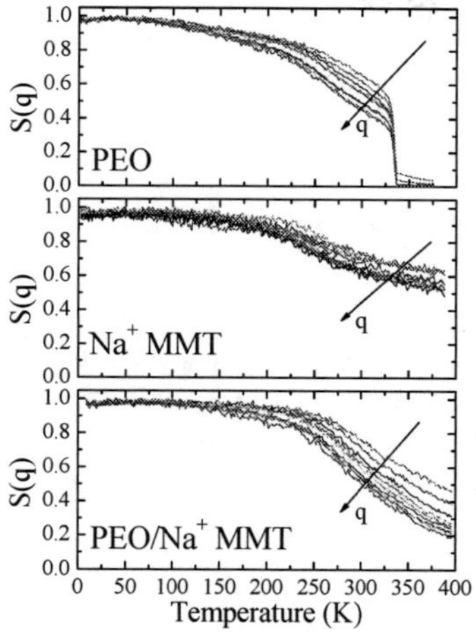

FIGURE 2: Elastically scattered intensity of the PEO homopolymer, the Na^+-MMT and the hybrid with 30wt% PEO.

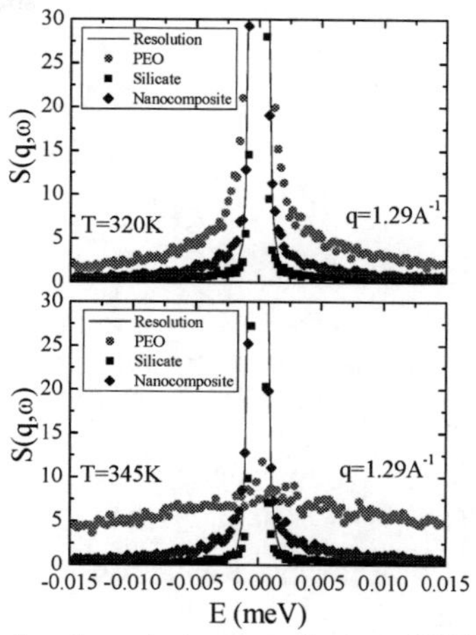

FIGURE 3: Incoherent structure factor of PEO, Na^+-MMT and the 30% hybrid at q=1.29$Å^{-1}$ for (a) T=320K and (b) T=345K. The line is the instrumental resolution [6].

Fig. 3 shows the incoherent structure factor of PEO, Na^+-MMT, and the composite with 30wt% PEO at two temperatures below and above the PEO T_m. The Figure shows the inelastic wings of the spectra whereas the instrumental resolution is included for comparison. The spectrum for Na^+-MMT is very similar to the resolution function. The PEO spectra are very different at the two temperatures. At 320K there is an elastic and a quasielastic contribution to the scattered intensity, whereas at 345K there is almost no scattered intensity and there is a very broad (almost background-like) quasielastic contribution because all PEO motions are already too fast to be measured. On the contrary, the existence of dynamics is evident at both temperatures for the confined system. The absence of dynamics in the two parent samples proves that the observed motion in the nanohybrid is due to the polymer, which is strongly affected by the confinement.

ACKNOWLEDGMENTS

Part of this research was sponsored by the Greek General Secretariat for Research and Technology (ΠΕΝΕΔ and ΠΑΒΕΤ Programme), by the European Union (G5RD-CT-2002-00834 and NMP3-CT-2005-506621), and by NATO's Scientific Affairs Division.

REFERENCES

1. M. Alcoutlabi and G. B. McKenna, *J. Phys.: Condens. Matter* **17,** R461 (2005). C. B. Roth and J. R. Dutcher, *J. Electroanal. Chem.* **584**, 13 (2005).
2. A. Bansal et al. *Nature Mater.* **4**, 693 (2005).
3. E. P. Giannelis, *Adv. Mater.* **8**, 29 (1996). R. K. Krisnamoorti, R. A. Vaia and E. P. Giannelis, *Chem. Mater.* **8,** 1728 (1996).
4. M. M. Elmahdy, K. Chrissopoulou, A. Afratis, G. Floudas and S. H. Anastasiadis, *Macromolecules* **39**, 5170 (2006).
5. K. Chrissopoulou, A. Afratis, S. Fotiadou and S. H. Anastasiadis, in preparation (2010).
6. K. Chrissopoulou, A. Afratis, S. H. Anastasiadis, M. M. Elmahdy, G. Floudas and B. Frick, *Eur. Phys. J.* **141**, 267 (2007).

Implementation of Relative Humidity Control in DMA Measurements

Marco Coletti[a]

[a]*TA Instruments – Via A. Grandi, Vimodrone (MI), Italy*

Abstract. A novel apparatus to perform Dynamic Mechanical Analysis tests under controlled conditions of Relative Humidity and Temperature is presented. It is possible to assess material properties which are dependent on RH (such as the Coefficient of Hygroscopic Expansion) or classical DMA parameters under more real-life conditions (Storage and Loss Moduli in a wet environment). Three case studies are presented reporting about mechanical properties of Nylon 6, the Coefficient of Hygroscopic Expansion of Nylon 6 and the Stress Relaxation behavior of Nafion 112, a material which is used in applications such as fuel cells.

Keywords: DMA, Relative Humidity, Coefficient of Hygroscopic Expansion
PACS: 46.35.+z

IMPLEMENTATION OF THE RELATIVE HUMIDITY ACCESSORY

Dynamic Mechanical Analysis (DMA) is a well known technique to measure mechanical properties of material such as Storage Modulus, Loss Modulus, Tan Delta and others. These are extremely important information in many application fields. The possibility to control temperature and applied frequency allows the user to simulate most of the real-life operating conditions under which the material will be actually used.

However, materials are often employed in conditions where a certain Relative Humidity (RH) is present, but commercially available DMA's usually cannot control this parameter (or this control is limited to relatively low temperatures): the sample is either exposed to a dry nitrogen atmosphere or regular air (either dried or not).

In this work, I present an implementation that is intended to allow a DMA to actively control both temperature and RH inside the chamber, thus measuring the effects of humidity on the mechanical properties of the material.

This accessory has been mounted directly over the measuring clamp of a TA Instruments Q800 DMA and is implemented in the software, so that the user can control both temperature and RH via the regular Thermal Advantage software that controls the instrument. The RH Accessory consists in a chamber with Peltier elements to control the temperature and an external humidifier that brings a flow of gas at precisely controlled RH into the chamber. The Peltier chamber replaces the usual furnace of the DMA and is user – exchangeable.

The final apparatus is capable of controlling the temperature of the furnace between 5°C and 120°C, and the RH between 0% up to 95% (depending on temperature). Relative Humidity can be either maintained at a constant value, and temperature ramped, or increased (decreased) in steps.

All standard clamps available for DMA testing can fit into the accessory chamber, so measurements can be performed on all the samples the DMA can regularly analyse.

Three case studies are presented here: RH dependence of Mechanical Properties of Nylon 6; measurement of the Coefficient of Hygroscopic Expansion of Nylon 6; RH and temperature dependence of Stress Relaxation behavior in Nafion 112.

Effects of RH on Dynamic – Mechanical properties of Nylon 6

Nylon 6 is widely used in many applications due to its exceptional physical, thermal and chemical properties, as well as good mechanical properties. However, Nylon is often brought in contact with water and it is well known[1] that it is strongly plasticized by humidity.

CP1255, *Vth International Conference on Times of Polymers (TOP) and Composites,*
edited by A. D'Amore, D. Acierno, and L. Grassia

In order to study the dependence of Nylon 6 mechanical properties on the environment RH, a thin sample was tested in a standard Single Cantilever clamp under the Humidity Accessory. It is very important for the sample to be thin as this will affect the water transportation inside the sample.

Two kinds of test were performed: several temperature ramps, each at a constant value of RH, and a RH ramp at a constant temperature.

Results of the first test (conditions: 0.5°C/min, 1Hz, 0.03% strain) are shown in FIGURE 1. It is clear that both the absolute value of the Storage Modulus is affected (mainly at ambient temperature) and the glass transition of the polymer (identified by the temperature at which the maximum of tandelta occurs) is shifted to lower values as the RH increases. This clearly shows the plasticizing effect that water has on Nylon 6, expecially at RH levels above 50%.

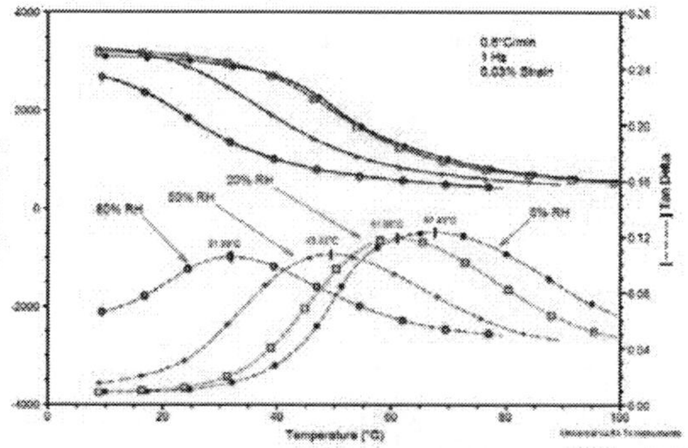

FIGURE 1. Different temperature ramps performed on Nylon 6 at several RH values. The Glass Transition temperature is identified as the maximum of tandelta

On the other hand, if we keep the temperature constant at 50°C and we increase the RH (steps of 1%, held for 5 minutes) over the range 5% - 95% under the same dynamic conditions, it is possible to measure the decay of the mechanical properties of the Nylon 6 and in particular the humidity induced Glass Transition (that occurs when RH reaches about 50%). Results are presented in FIGURE 2.

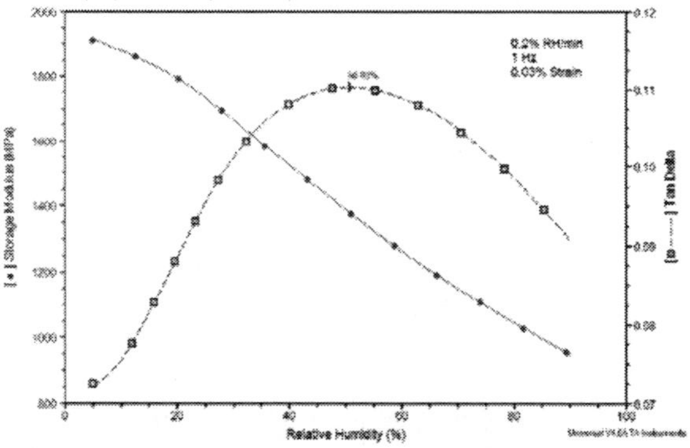

FIGURE 2. Nylon 6 under a stepped ramp of RH values. The humidity induced Glass Transition can be measured as the peak of tandelta.

Measurement of the Coefficient of Hygroscopic Expansion of Nylon 6

The Coefficient of Hygroscopic Expansion (CHE) is defined analogously to the Coefficient of Thermal Expansion as the ratio between the expansion due to humidity and the RH level. This is a parameter of particular importance as if different CHE materials are coupled together, they might detach if exposed to high RH levels (like paperback book covers). This parameter can be easily measured with a DMA and the RH accessory by exposing a thin Nylon 6 sample to an increasing level of RH. Results are shown in FIGURE 3. The CHE is measured at different values of RH and it is clear than Nylon 6 exhibits an increasing CHE with RH, going from around 12μm/m% at 35% up to 26μm/m% at 85%. This shows again the hygroscopic properties of Nylon 6 and the plasticizing effect of water on it.

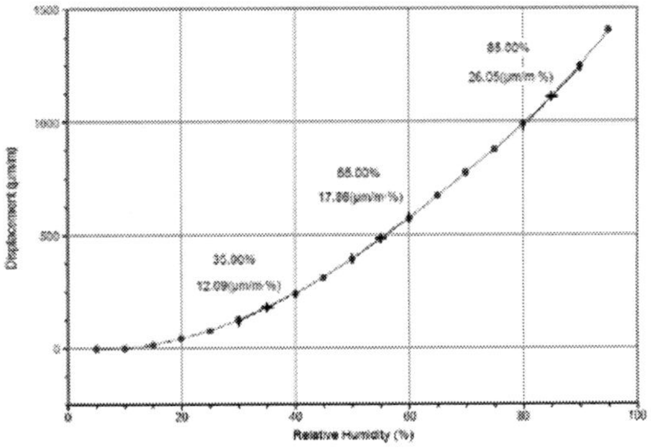

FIGURE 3. Measurement of the CHE of Nylon 6. Note that the actual value strongly depends on the RH level.

Effects of RH on Stress Relaxation Behaviour of Nafion 112

Nafion 112 is a polymer widely used to manufacture membranes inside Proton Exchange Membrane Fuel Cells[2]. As water is the primary by-product of this kind of Cells and their operating temperatures can range between 80°C and 110°C, it is very important to assess the mechanical properties of polymers like Nafion 112 when they are used in such applications. Stress Relaxation tests has been performed at different temperatures and RH levels in order to assess the mechanical resistance of Nafion 112. Results are shown in FIGURE 4. It is evident that upon increasing temperature and RH, the Stress Relaxation behavior of Nafion 112 changes dramatically: the Relaxation Modulus (relatively stable at ambient conditions of 25°C and 50% RH) is strongly reduced when temperature and RH are increased to 85°C and 85%, respectively. In addition to that, the time dependence of the modulus is more profound.

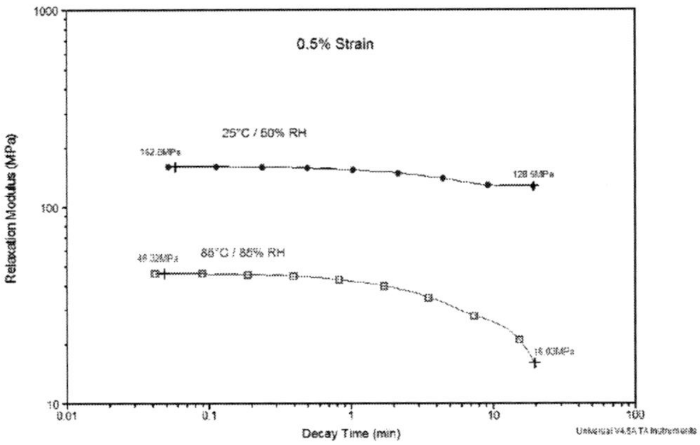

FIGURE 4. Effects of temperature and RH on the Stress Relaxation behavior of Nafion 112

ACKNOWLEDGMENTS

I would like to thank S. Aubuchon, from TA Instruments, New Castle, for providing the DMA data presented here.

REFERENCES

1. Jin, X.; Ellis, T.S.; Karasz, F.E. *J. Poly. Sci.:Poly. Phy. Ed.* **1984**, *22*(10), 1701–17.
2. a) Verbrugge, M.W.; Hill, R.F. *J. Electrochem.Soc.* **1990**, *137*, 3370; b) Springer, T.E.; Zawodzinski, T.A.; Gottesfeld, S. *J. Electrochem. Soc.* **1991**, *138*, 2334; c) Wnek, G.E.; Ehrenberg, S.G.; Serpico, J.M.; Sheikh-Ali, B.M.; Tangredi, T.N.; Zador, E. *Second International Symposium on New Materials for Fuel Cell and Modern Battery Systems*, Canada, 1997; d) Uan- Zo-li, J.T. M.S. Thesis, Virginia Polytechnic Institute and State University, 2001.

PP composites with Hybrid Nanofillers: NTC phenomenon

Juha Sarlin and Kirsi Immonen

VTT Technical Research Centre of Finland, Sinitaival 6, P.O.Box 1300, FI-33101 Tampere,

Abstract. Electric conductive plastic composites have a wide potential for commercial applications, some examples are EMI shielding housings and components in automotive industry and in consumer electronics, equipments in health care sector and fuel cell components. A phenomenon in conductive composites, especially in composites with carbon based fillers, is change of thermal induced change in conductivity as a result of morphological transitions. Usually the observed changes are practically irreversible. The phenomenon may cause increasing resistivity, usually called as "positive temperature coefficient" (PTC) or decreasing resistivity, called "negative temperature coefficient (NTC), where the new morphology created by heat treatment is more favorable for electric conductivity compared to the original state. The existence of NTC is a sing of the lost potential in material design and processing. Therefore detailed information about the phenomenon gives us tools to develop high performance conductive materials. It this paper we discuss about NTC phenomenon observed in PP composites with CNT or in-situ synthesized CNT-PANi hybrid nanofiller with an amphiphilic dispersing agent. The goal of the paper is not to present a comprehensive model of this phenomenon; we present some experimental results which may be related to polymer-filler interactions. These details are a part of this complicated phenomenon.

Keywords: Reinforced polymers and polymer-based composites; Nanotubes; Organic-inorganic hybrid nanostructures; Electrical and magnetic properties related to treatment conditions; Conducting polymers; Nanoparticles in polymers; Solids.
PACS: 81.05.Qk; 81.07.De; 81.07.Pr; 81.40.Rs; 82.35.Cd; 82.35.Np; 82.35.Jk.

INTRODUCTION

Carbon is common filler in polymeric composites. Addition of carbon in plastics and rubbers increases stability of composites. Electric conductivity is a property which may be created in plastics using carbon fillers. Today there exist a wide range of commercial carbon black (CB) filler grades for plastics and rubbers. Carbon fibers (CF) added a new perspective to the exploitation of carbon filled plastic composites. Industrial production of nanostructured carbon, fullerenes and especially carbon nanotubes (CNT), has opened new potential for carbon filled composites. Today CNT is a key component to develop plastic composites characterized by high strength and modulus, high electric conductivity and heat conductivity. The present progress in graphene technology increases the commercial potential of modern carbon based fillers in polymeric matrices.

An anomaly observed in many carbon based composites is systematic change of electric conductivity in composites as a result of high temperature treatment. The observed change may increase resistivity of the materials, and the phenomenon is called "positive temperature coefficient" (PTC), and the inverse phenomenon is "negative temperature coefficient" (NCT).[1-3] The phenomena may be reversible or usually (practically) irreversible. The phenomenon may be desired, a number of functional applications can exploit this change of conductivity. Smart materials characterized by PTC or NTC phenomenon offer self regulating functionality are a starting point to construct cost effective applications. PTC effect is frequently studied to realize smart material applications with current limiting functionality. But these phenomena may be harmful. An example is high conductive injection molded parts; where a post treatment of molded parts to achieve desired conductivity by NTC effect may be technically and economically unrealistic. Sometimes PTC and NTC phenomena are used only for cases where the mentioned phenomena are more or less reversible, but in this paper these terms are used for both reversible and irreversible cases.

Conductive plastic composites have wide potential in present industry. Only for an example, low cost plastic based composites for EMI shielding materials for injection molded housings and other components have commercial

CP1255, *Vth International Conference on Times of Polymers (TOP) and Composites,*
edited by A. D'Amore, D. Acierno, and L. Grassia

interest in automotive industry and in consumer electronics, especially in applications where light weight and low cost are favorable. To produce high conductive plastic composites the possible existing NCT effect highlights the lost potential in processing and in material design concept. The phenomenon gives evidence of a high conductive state which is not produced during the manufacturing process. The origin of NCT (or PTC) is closely related to the percolation, but this is a straightforward expression, to understand thermal-induced percolation properties of matrices, fillers and filler-matrix interactions should be considered. Understanding of NTC and PTC phenomena gives tools to design new high performance conductive composites.

Today organic-inorganic materials are a key concept to create new material properties, and these kinds of materials are promising candidates for fillers in plastic composites. CNT is a common subject for this purpose. Interactions between CNT and polymers are crucial challenges which may overcome by organic modification of CNT. Principally organic groups on CNT can compatibles tubes in polymeric matrix. Organic groups on CNT may have a number of other functional properties, the idea is widely used in many sophisticated concept to produce advanced functional nanomaterials.

In this paper we report a CNT based hybrid nanomaterial, properties of polypropylene (PP) composites with this hybrid filler and some selected properties of this composite, especially focused on the measured thermal-induced decrease of resistivity. Our objective is not to give a comprehensive concept about the NCT phenomenon; the idea is to highlight some aspects which may have effects on this observed phenomenon.

CNT-PANi HYBRID FILLER IN PP

For conductive composites CNT offer an attractive potential. Depending on structure and purity carbon nanotubes are characterized by high conductivity and by high aspect ratio which together may realize low percolation threshold. Low density and excellent stability are both favorable properties for high performance composites. Inherently conductive polymers are another group which can produce conductive plastics by blending non-conductive matrix polymer and conductive polymer. Polyaniline (PANi) is an inherently conductive polymer characterized by high stability and low cost, which both favor industrial applications, although the conductivity is moderate, typically 10^{-3} ... 10^{1} S/cm.

We have *in-situ* synthesized CNT-PANi hybrid nanoparticles without covalent bonds between CNT and PANi. Covalent bonds between these two materials require breakage of the ring architecture of nanotubes which decreases conductivity of tubes, and strong modification of CNT may destroy totally observed conductivity. In our work typical CNT/PANi ratios were 1/1 – 3/1. According our experience higher PANi amounts in hybrids does not favor conductivity or other desired properties. The polymer matrix is polypropylene (PP), in this work a high impact injection molding grade with mold flow index (MFI) 7.5. PP was a commercial copolymer with high impact properties and the material shows β crystallinity which is typical for this kind of PP. Compounding and molding were carried out by a laboratory scale equipment (Haake) whit a batch size about 6 grams. A part of molded samples included an amphiphilic dispersing agent, ethyl gallate (EG), in amount of 20 w% compared to the matrix polymer. The family of gallates includes low molecular weight molecules characterized by an aromatic ring and an alkyl tail. The length of the alkyl tail gives a tool to control interactions between dispersant and matrix. The aromatic structure may favor compatibility with PANi and maybe with CNT; in both materials we can find the same ring structure. In our previous work we have observed highest conductivity (26 S/cm) in composites with EG. PP composites including CNT-PANi hybrid filler and gallates with longer alkyl tails realized lower conductivity values. [4]

PROPERTIES OF PP-HYBRID FILLED COMPOSITES

Morphology

SEM analysis showed a tendency of EG to separate from PP. This may be obvious, the aromatic end of EG may dominate interactions between dispersing agent and PP before the alkyl tail is a short one. Increasing length of the alkyl tail increases compatibility in PP. In systems without conductive fillers EG forms in PP more or less elongate drop like inclusions. Melt flow in molding process produces this elongation, and the shape of EG phase may show some fibril like form. Addition of conductive fillers, neat CNT or CNT-PANi hybrid filler changes essentially the morphology of composites. The EG phase forms long fibril like structure, and especially samples with the hybrid filler this fibril like minority phase was characteristic. In composites with neat CNT or the hybrid filler without the dispersing agent showed quite homogenous dispersion of these fillers in matrix with some amount of agglomerates of fillers. Fracture surface micrographs of composites with EG showed only a small amount of separate fillers, probably used fillers were included in the EG phase, especially in the case of the hybrid fillers the tendency was well observable. The ability of EG to incorporate these fillers was obvious. [4]

Thermal-Induced Conductivity

In order to measure possible thermal induced changes in conductivity of PP composites with (a) neat CNT, (b) CNT-PANi hybrid and (c) CNT-PANi hybrid and EG, samples were heated slowly up to 160 °C (melting temperature of PP) and surface resistances were measured. Strong NTC effect was observed in the case of PP-CNT composite, the resistance decreased from 10^8 Ω to 10^3 Ω when temperature reached the maximum value. The observed decrease of resistance started when temperature was increased over 60 °C. Slowly cooling to the room temperature caused only a minor increase of resistance. Thermal induced change in the surface resistance was 5 decades. A composite with the hybrid filler owned resistance about $5*10^5$ Ω before treatment, and the treatment decreased resistance to the value of 10^3 Ω. Again the measured change was stable. The decrease of resistance started about a temperature of 80 °C. Composite with the hybrid filler and the dispersing agent showed lowest surface resistance, after molding the resistance was about $8*10^3$ Ω and after treatment the measured value was $7*10^2$ Ω, so the change was only one decade. So, although there were large differences in measured resistances after molding of composites, the used thermal treatment produces almost similar surface resistances.

Relaxation Properties

Dynamic mechanical thermal analysis (DMTA) in tensile mode showed only a weak development in storage modulus when neat CNT of CNT-PANi hybrid filler were added in PP. Addition of the dispersing agent in PP decreased storage modulus approximately by an amount corresponding the volume fraction of EG. This may indicate the biphasic structure of the PP/EG system. Mechanical load does not transfer over phase boundaries and the measured storage modulus is only the mechanical response of PP phase. In system where both the hybrid filler and EG were added in PP storage modulus increased, and similar but weaker effect was observed in PP/CNT/EG system. An explanation for the phenomenon in the hybrid filled system is molecular recognition, which is predicted. [5] Addition of fillers in PP modified loss modulus. In this work we are interested in properties of PP when temperature is above Tg. In semicrystalline PP we can observe α relaxation which is assigned on motions in crystalline phase. Addition of the both type of fillers, neat CNT and CNT-PANi hybrid caused an observable α relaxation peak, but in the hybrid filled system the intensity of this relaxations was essentially stronger compared to the composites without the dispersing agent.[6]

Crystallinity

Wide angle X-ray diffraction (WAXS) analysis showed a strong effect of CNT on polymer matrix. The crystalline structure of the used PP forms of α crystallinity, but included a portion of β crystalline form. This is an expected result before β crystallinity favor high impact strength, and b nucleating is today an industrial technology for PP. Addition of CNT-PANi hybrid filler with or without the dispersing agent decreased the amount of β crystallinity from 15% to 9 % when Turner-Jones equation was applied. Addition of neat CNT in PP hinder β form crystallization, the observed amount of β crystallinity in PP/CNT composite was less than 2%. CNT is an effective nucleating agent for α structure, but the hybrid filler where CNT is coated by PANi this kind of phenomenon does not take place.[6]

Crystallisation

DSC melting thermograms of filled composites showed almost similar peak temperatures, about 165 oC. Calculated crystallinity (X_c) in all materials was practically identical. The detailed analysis of melting thermograms showed a high temperature shoulder in PP-CNT composite which was not observable in PP or in the hybrid filled PP composite. The ln(-ln(1-Xc) vs. ln(t) plot, which is used for modified Avrami method to analyze crystallization kinetics, showed the phenomenon to be almost similar at cooling rates from 2.5 up to 20 °C/min. An explanation for this high temperature shoulder is transcrystallisation. CNT may nucleate crystal growth where growth follows normal of the CNT surface. [6]

DISCUSSION

Heat induced changes in plastic composites are usually results of morphological transitions. NTC indicates unsatisfactory morphology for conductive pathways in composite. A part of potential conductivity is lost due to the processing and material design.

The phenomenon can be related to the percolation; small morphological enhancement may realize conductive pathways and measured conductivity may increase by decades. The structure of percolation paths is more or less stable. Therefore interactions between conductive fillers are a relevant topic for research. Recently there are reports describing the phenomenon by formation of conductive agglomerates.[7] Thermal induced self assembly where separate particles organize and form high conductive volumes in percolation paths is proposed to be a process causing NTC effect. A proof for this phenomenon is shear sensitive conductivity of blends.[8] The formation of these conductive agglomerates allows discussion about charge transfer mechanism; if hopping controls longer range transfer rates may tunneling effects dominate in these conductive agglomerates.[9] The formation of agglomerates is usually proofed by transmission electron microscopy (TEM) analysis, but our experience to apply TEM indicated the sensitivity of the method on the sample preparation. We can speculate the observed effects of filler concentration on the NCT phenomenon depends on the filler-filler interactions. Low filler concentrations do not favor formation of stable high conductive volumes in percolation paths.

We are focused our work to study filler-matrix interactions. If filler-filler interactions are essential for NTC effect, filler-matrix interactions and morphology transitions in matrix forms preconditions for filler structure in conductive pathways. According SEN analysis in our composites low molecular weight dispersing agent with low viscosity in processing conditions gives a phase forming continuous structure which is ore attractive for fillers compared to PP melt. After treatment observed surface resistivities were almost similar in all cases, but low viscosity minority phase allows rapid formation of conductivity pathways during processing. This process can be supported by dispersant – filler interactions which was observed in CNT-PANi filler system with EG. This highlights the role of filler-matrix interactions.

Typical nucleating effects of nanoparticles can lock conductive fillers into polymer phase. Transcrystallisation creates stable crystalline polymer structure covering nanoparticles which separate individual fillers and freeze thermal induced motions of these particles. Both phenomena restrict filler–filler interactions and limit thermal induced formation of high conductive percolation paths.

To design high conductive plastic composites with low filler loadings we should support rapid formation of conductive pathways. This means control both filler-filler and filler-matrix interactions.

ACKNOWLEDGMENTS

This work has been carried out in European IP project POLYCOND (IP 515835-2), partially funded by EC. The financial support is gratefully acknowledged.

REFERENCES

1. Q.Cao, Y.Song, Y.Tan and Q.Zheng, Polymer **50.** 6350-6356 (2009).
2. V.Panwar, V.K.Sachdev, R:M:mehra, Eur.Polym.J. **43.** 573-585 (2007).
3. F.Liu, X.Zhang, W.Li, J.Cheng, X.Tao, Y.Li and L.Sheng, Cpmposites Part A **40.** 1717-1721 (2009).
4. K.Immonen, K.Nättinen, J.Hartikainen and J.Sarlin, J.Appl.Polym.Sci **114** 1494-1502 (2009).
5. O.Ikkala, Lo.O.Pietilä, P.Passiniemi, Y.Cao and A.Andreatta, U.S.Patent No. 5,783,111 (21 July 1998).
6. J.Sarlin and K.Immonen, J.Polym.Sci. B. Submited
7. I.Alig, T.Skipa, M.Engel, D.Lellinger, S.Pegel, S.Pegel, P.Postchke, Physica Status Solidi B **244.** 4223-4226 (2007).
8. H.Deng, T.Skipa, R.Zhang, D.Lellinger, E.Bilotti, I.Alig and T.Peijs, Polymer **50.** 3747-3754 (2009).
9. P.Murugaraj, N.Mora-Huertas, D.E.Mainwairing, Y.Ding and S.Agrawal, Composites Part A **39.** 308-313 (2008).

Conjugated Polymer and Luminescent Nanocrystals for Ink-Jet Printing

E. Binetti[ab], C. Ingrosso[ab], M. Striccoli[b], P. Cosma[ab], A. Agostiano[ab], J. Y. Kim[c], J. Brugger[c], and M. L. Curri[b]

[a] *Dept. of Chemistry, University of Bari, Via Orabona 4, I-70126 Bari, Italy*
[b] *CNR-IPCF Div. Bari, c/o Dept. of Chemistry, University of Bari, Via Orabona 4, I-70126 Bari, Italy*
[c] *Microsystems Laboratory, Ecole Polytechnique Fédérale de Lausanne (EPFL), CH-1015 Lausanne, Switzerland*

Abstract. Nanocomposite solutions formed of Poly [(9,9-diesilfluorene)-(2,7 diyl)-alt-(2,5-dimetil-1,4-phenilene)] and highly luminescent nanocrystals of CdSe coated with a shell of ZnS (CdSe@ZnS) with different size have been obtained by using a common solvent and characterized by means of absorption and emission spectroscopy. Viscosity has been investigated, in order to prove the suitability of the inks for the printing processing. The homogeneous nanocomposites have been precisely dispensed by ink-jet printing, using drop-on-demand mode onto glass substrates. The deposited micro-scale pixels have been morphologically characterized by means of fluorescence microscopy, 3D profilometry and Atomic Force Microscopy (AFM). The results demonstrate that the spectroscopical properties of the components have been effectively conveyed to the final nanocomposites, retaining the size dependent feature of the inorganic moiety, and providing a suitable ink with for fabricating reproducible microstructures.

Keywords: Nanocrystals, Nanocomposite, Hybryd Organic-Inorganic –Materials, Ink-Jet Printing.
PACS: 81.16.Be, 81.07.Ta, 81.05.Fb, 78.67.Sc

INTRODUCTION

Organic-inorganic hybrid materials are interesting systems that can present novel synergic properties due to the combination of organic polymers and inorganic nanosized components, respectively, in original composite materials. Nanocomposites will thus combine the peculiar characteristics of the organic moiety, namely specific mechanical and electronic properties of the polymer component, with the original size- and shape dependent properties of inorganic nanocrystals (NCs) [1]. Therefore the potential of the resulting functional materials can burst enormously, offering unique opportunities to exploit their fundamental properties in a wide range of applications. In addition, the possibility of incorporating NCs into an organic host polymers provides a convenient way to bridge the meso- to the nano-scale, resulting in a highly processable functional materials which can be suitably patterned in a variety of microstructures, ready to be integrated in advanced systems and devices.

Among the available patterning methods, ink jet printing (IJP) is a dispensing technique that allows accurate and local deposition of minute quantities of liquid materials and melts with a reproducible precision and low reagent consumption. Such technique allows the fabrication of 2D and 3D micro-scale structures on planar and curved surfaces or on pre-made patterns [2] and of unique multifunctional structures with high added value properties, promising for advanced MEMS device and miniaturized system applications. The technological applicability of inkjet printing has been recently extended by employing colloidal NCs as components for manufacturing inks. In particular, the organic compatible properties of the NC surface have provided the manufacturing of nanocomposites, in which the excellent processability of polymers can be joint with the outstanding size- and shape- dependent properties of the nano-objects. Recently, luminescent single-and multi-color pixel arrays of nanocomposite solutions based on polystyrene functionalized with highly luminescent colloidal cadmium sulfide (CdS) and differently sized CdSe@ZnS NCs, dispersed in chloroform carrier medium have been ink-jet printed by the author's group [3-4].

In this work, nanocomposites formed of a conductive conjugated polymer, namely Poly [(9,9- diesilfluorene)-(2,7 diyl)-alt-(2,5-dimetil-1,4phenilene)] (PF-DMB), modified by organic capped differently sized highly

CP1255, *V[th] International Conference on Times of Polymers (TOP) and Composites,*
edited by A. D'Amore, D. Acierno, and L. Grassia

luminescent colloidal core shell type NCs (CdSe@ZnS), have been investigated in view of their applicability in IJP. The prepared nanocomposites have been characterized by means of absorption and emission spectroscopy, both steady state and time resolved. Electrochemical characterization has been also performed to complete the evaluation of the opto-electronic characteristics of the prepared functional nanocomposites. Finally, a viscosimetric investigation has been carried out, in order to prove the suitability of the inks for the printing processing. The homogeneous nanocomposites have been then deposited by means of a piezoelectric inkjet printer in drop-on-demand (DOD) dispensing mode. The deposited micro-scale pixels have been morphologically characterized by means of fluorescence microscopy, 3D profilometry and Atomic Force Microscopy (AFM) and the effect of the NC size on the final printed nanocomposite microstructures has been evaluated. The results demonstrate that the investigated inks can be efficiently and reliably printed without any specific control on the environmental conditions or substrate pre-treatment and that luminescent pixels can be dispensed in order to obtain reproducible microstructures.

MATERIALS PREPARATION

Organic capped CdSe@ZnS NCs (2.8, 3.5 and 5.0 nm in diameter) have been synthesized by thermal decomposition of precursors in hot coordinating solvents, as reported in [5]. PF-DMB have been purchased by Sigma-Aldrich. NCs and polymer have been dissolved in chloroform solutions, with concentration of 2.0 x 10-5 M and 4.0 mg/ml respectively. Then nanocomposites have been obtained by mixing equal volumes of chloroform solutions of the two components for 1 h by under stirring. The list of the prepared samples of nanocomposites is reported in table i). The nanocomposites have been deposited by spin-coating at 1500 rpm, and by IJP onto glass and ITO coated glass substrates, respectively, for microscopical, spectroscopical and electrocehemical experiments.

Inkjet printing has been performed by using a piezo-actuated inkjet printer from Microdrop Technologies. The nozzle (diameter 70 μm) has been actuated in squeeze mode. The pressure in the inkjet nozzle was controlled to optimize the liquid flow, the wetting of the nozzle, and the ejection properties. The substrate has been positioned relatively to the printing nozzle with a motorized high-precision x-y-θ stage.

TABLE i). List of prepared nanocomposites.

NC diameter (nm)	NC Concentration (mol/l)	Polymer Concentration (mg/ml)	Molar ratio (pol/NC)
2.8	1.0 x 10-5	2.0	27
3.2	1.0 x 10-5	2.0	27
5.0	1.0×10^{-5}	2.0	27

RESULTS AND DISCUSSION

In Figure 1 the absorbance and PL spectra of the three differently sized NC based nanocomposite reported in Tab. 1 and deposited in thin films are reported. The figure evidences both the typical absorbance and PL signals in the near-UV and visible spectral regions ascribed to PF-DMB and to CdSe@ZnS NCs, respectively. Indeed the polymer presents a strong emission peak in the blue region of the spectrum that maintains shape and position after embedding the NCs. The CdSe@ZnS NCs show both broad absorption bands and narrow band-edge emission peaks in the orange-red spectral region, whose spectral positions are directly correlated to the NC size [5]. As confirmed by Figure 1, the characteristic size-dependent properties of the NCs and the typical optical features of PF-DMB are combined in the nanocomposites.

Viscosimetry of the prepared nanocomposites has been investigated, revealing that the viscosity is mainly dependent on the polymer concentration. The estimated nanocomposite mean viscosities are about 0.7 cP for all the investigated NC sizes, thus confirming the suitability of the prepared nanocomposites for IJP deposition.

Cyclic voltammetry on the nanocomposites allowed the determination of their HOMO and LUMO levels. Nanocomposites have been deposited in arrays of pixels onto glass substrates by means of a piezoelectric inkjet printer with a 70μm nozzle diameter. The deposited pixels have been characterized by optical and fluorescence and atomic force microscopy (AFM). Fluorescence micrographs of the pixel arrays deposited on glass by IJP from nanocomposite solutions of CdSe@ZnS NCs based on 2.8nm (A), 3.5nm (B) and 5.0 nm (C) in diameter respectively, are shown in Figure 2. The occurrence of regularly shaped pixels, with an average diameter of about 500 μm can be observed.

FIGURE 1. Absorbance and PL spectra of thin films of the prepared nanocomposites deposited onto glass substrates.

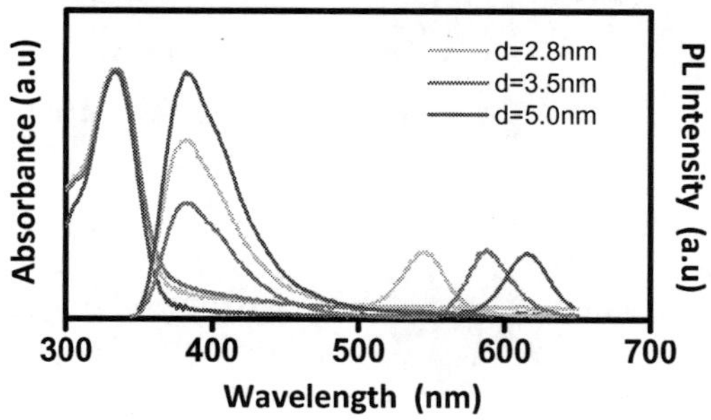

The luminescence of nanocomposite pixels can be accounted to the combination of the emissive contribution of polymer, in the near-UV region, and of the band edge emission peak of the differently sized NCs in the yellow-red region. For all the investigated pixel arrays, the micrographs of Figure 2 point out the occurrence of an accumulation of ink at the pixel edges, known as coffee staining effect, which is typically due to a fast solvent evaporation rate[6]. In addition, the pixels reported in panel A of Figure 2 show an asymmetric material accumulation, known as pinning-effect, which is caused by the low polymer concentration [7]. Finally, AFM investigations of the printed pixels show a topography mainly characterized by grain-type features whose density and dimension increase with the size of the embedded NCs. Such features can be ascribed to NC aggregation phenomena, typically due to interdigitation of the ligand alkyl chains coordinating the NC surface [8].

FIGURE 2. Fluorescence microscopy images of single color pixel arrays formed of nanocomposites based on CdSe@ZnS NCs of 2.8 nm (A), 3.5nm (B) and 5.0nm (C) in diameter. Panel (D) display an AFM micrography of a pixel of nanocomposite containing 2.8nm diameter CdSe@ZnS NCs.

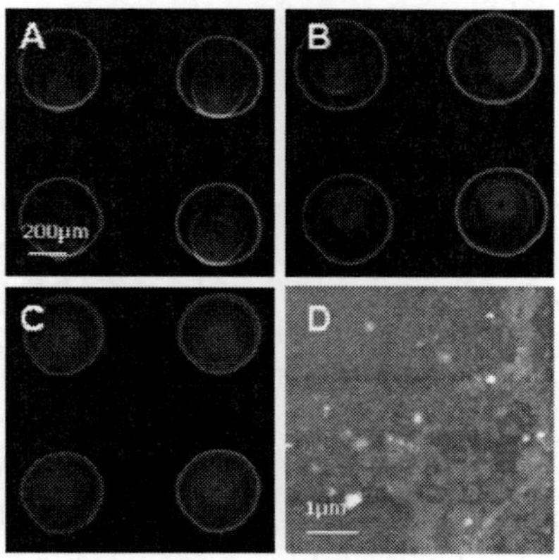

REFERENCES

1. A.P. Alivisatos, *Science* **271**(5251), 933-937 (1996).
2. D. Jang, D. Kim, and J. Moon, *Langmuir* **25**(5), 2629-2635 (2009).
3. C. Ingrosso, J.Y. Kim, E. Binetti, V. Fakhfouri, M. Striccoli, A. Agostiano, M.L. Curri and J. Brugger, *Microelectronic Engineering* **86**(4-6), 1124-1126 (2009).
4. J.Y. Kim, C. Ingrosso, V. Fakhfouri, M. Striccoli, A. Agostiano, M.L. Curri and J. Brugger, *Small* **5**(9), 1051-1057 (2009).
5. X. Peng, L. Manna, W. Yang, J. Wickham, E. Scher, A. Kadavanich and A.P. Alivisatos, *Nature* **404**(6773), 59-61 (2000).
6. R.D. Deegan, O. Bakajin, T.F. Dupont, G. Huber, S.R. Nagel and T.A. Witten, *Nature* **389**(6653), 827-829 (1997).
7. R.D. Deegan, O. Bakajin, T.F. Dupont, G. Huber, S.R. Nagel and T.A. Witten, *Physical Review E* **62**, 756-765 (2000).
8. M. Tamborra, M. Striccoli, M.L. Curri, J.A. Alducin, D. Mecerreyes, J.A. Pomposo, N. Kehagias, V. Reboud, C.M. Sotomayor Torres and A. Agostiano, *Small* **3**(5), 822-828 (2007).

Au Based Nanocomposites Towards Plasmonic Applications

A. Panniello[a*], M.L. Curri[a], T. Placido[b], V. Reboud[c], N. Kehagias[c], C. M. Sotomayor Torres[c,d], D. Mecerreyes[e], A. Agostiano[a,b] M. Striccoli[a]

[a]*IPCF-CNR Bari, c/o Department of Chemistry, Via Orabona, 4, I-70126 Bari, Italy.*
[b]*Department of Chemistry University of Bari, Via Orabona, 4, I-70126 Bari, Italy.*
[c]*Catalan Institute of Nanotechnology, Bellaterra, 08193 Barcelona, Spain*
[d]*Catalan Institute of Research and Advanced Studies ICREA, 08010 Barcelona, Spain*
[e]*CIDETEC – Center for Electrochemical Technologies, Paseo Miramó n 196, 20009 San Sebastian, Spain.*

Abstract. Incorporation of nano-sized metals in polymers can transfer their unique features to the host matrix, providing nanocomposite materials with improved optical, electric, magnetic and mechanical properties. In this work, colloidal Au nanorods have been incorporated into PMMA based random co-polymer, properly functionalized with amino groups and the optical and morphological properties of the resulting nanocomposite have been investigated by spectroscopic and AFM measurements. Au nanorods have demonstrated to preserve the plasmon absorption and to retain morphological features upon the incorporation, thus making the final metal modified polymer composite exploitable for the fabrication of plasmonic devices. The prepared nanocomposites have been then patterned by Nano Imprint Lithography technique in order to demonstrate the viability of the materials towards optical applications.

Keywords: Au nanorods, PMMA copolymer, nanocomposites, Nano-Imprint Lithography
PACS: 78.67.Qa, 78.67.Sc, 82.35.Np, 81.16.Nd

INTRODUCTION

Metallic nanoparticles (NPs) have been extensively studied during the past decades and exploited in a variety of applications, such as catalysis, electronic, photonics and for fabrication of biomedical, biosensing and optical devices.[1] Due to the low dimensionality, metal nano-clusters are characterized by a typical plasmon absorption due to the collective oscillation of the electrons that are free to move on the metal surface.[1] The plasmonic band is specific for each metal and it is responsible for the distinctive color of noble metal colloid solution (i.e. Cu, Au or Ag NP). Chemical environment and surroundings, cluster organization, temperature, and especially NP's morphology affect the plasmon absorption band. Accordingly, while spherical NPs possess a single plasmonic resonance, depending on their size, rod-shaped NPs are characterized by two distinct plasmons, ascribable to the transverse and longitudinal dimension, respectively. Despite the original size-dependent properties of such promising materials, their practical application is limited by the poor processability and handiness of colloidal NPs. Conversely, incorporation of metal NPs in a host matrix, such as polymers, allows the combination of the physical-chemical properties of inorganic nanostructures with the mechanical and structural properties and, above all, the high processability of the organic matrix. Moreover, the presence of the host matrix can enhance material processability, structurability and conductivity making the resulting nanocomposites appealing candidates for lithographic processes towards the fabrication of optical devices, plane waveguides and optical fibers.[2,3] In particular, recent research works have been focused on the preparation of Au NP based nanocomposite materials, due to the peculiar size-dependent properties exhibited by Au nanoclusters. Au NPs have been incorporated into polyethylene to investigate the magnetic properties [4,5], into electroconductive polymers for exploiting the improved electrical features of the nanocomposites in chemical sensing applications [6,7], and into photocurable acrylate to obtain holographic materials exhibiting nonlinear optical properties [8,9]. In this work, colloidal Au nanorods (NRs) have been incorporated into amino functionalized polymethyl methacrylate (PMMA) by following

* a.panniello@ba.ipcf.cnr.it

CP1255, *V[th] International Conference on Times of Polymers (TOP) and Composites,*
edited by A. D'Amore, D. Acierno, and L. Grassia

an *ex situ* procedure, consisting in embedding Au NPs in pre-made polymers, using a common solvent. The addition of suitable functionalities to the polymeric chain represents an effective enhancement of the affinity between NP surface, which is covered by an organic capping layer, and the host matrix, allowing the homogeneous dispersion of the inorganic nanostructures in the polymer, thus preventing their aggregation and thus preserving their characteristic optical features.

MATERIALS AND METHODS

Au NRs have been prepared by following a reliable procedure from the literature [10], by using $HAuCl_4$ as precursors, ascorbic acid and $NaBH_4$ as reducing agents, cetyltrimethylammonium bromide (CTAB) as surfactant and in the presence of Ag^+ ions ($AgNO_3$) for allowing the anisotropic growth of the nanoclusters. Water soluble CTAB capped Au NRs have been transferred in toluene by using a ligand exchange performed with mercaptosuccinic acid (MSA) and tetraoctylammonium bromide (TOAB) [11]. TEM measurements have been performed by using a Jeol Jem 1011 microscope operating at an accelerating voltage of 100 kV, in order to investigate the morphology of the prepared NPs, which results in a rod shape, with a mean size of 20 nm in diameter and 40 nm long (See the inset of Figure 1). Amino functionalized PMMA random copolymer (PMMA-co-DMAEMA) has been prepared starting from methyl methacrylate and dimethylamino-methyl methacrylate, as the monomers, in accord to the procedure reported in literature [12]. Nanocomposite solutions have been prepared by adding the proper amount of Au NR toluene solution, with a fixed concentration of 2.5 mM, to the amino functionalized PMMA co-polymer, to obtain the final polymer concentration of 50 mg/ml.

RESULTS AND DISCUSSION

The effect of the incorporation of Au NRs in PMMA-DMAEMA functionalized co-polymer has been investigated by UV-Vis absorption spectroscopy, since possible aggregation, ripening and/or dissolution phenomena can be monitored by a variation of the absorption maximum and, in general, by changing in the spectral features. Moreover, the stability of the Au NR PMMA-co-DMAEMA based nanocomposite solutions has been studied by monitoring absorption spectra during time, to verify the preservation of the optical properties and the time stability of the prepared materials, as shown in Figure 1(a). The absorption properties of the nanocomposite solutions are retained upon the incorporation in PMMA-co-DMAEMA polymer, since both transversal and longitudinal plasmon resonances of anisotropic shaped Au NPs do not evidence significant variation in their form and intensity after the incorporation in the organic matrix. In addition, the absorption spectra of Au NR based nanocomposite solutions during time shows that the optical features of the inorganic nanostructures are preserved even two months after the nanocomposite preparation, thus highlighting the stability of the prepared materials. The optical properties of the Au NR based composites as thin films have been also studied. For this purpose a small amount of the nanocomposite solution has been deposited onto a proper cleaned glass substrate by drop casting. The comparison between absorption spectra of the NP based composite solution and cast film, reported in Figure 1(b), evidences that the optical features of the metal NP in the nanocomposite films are retained.

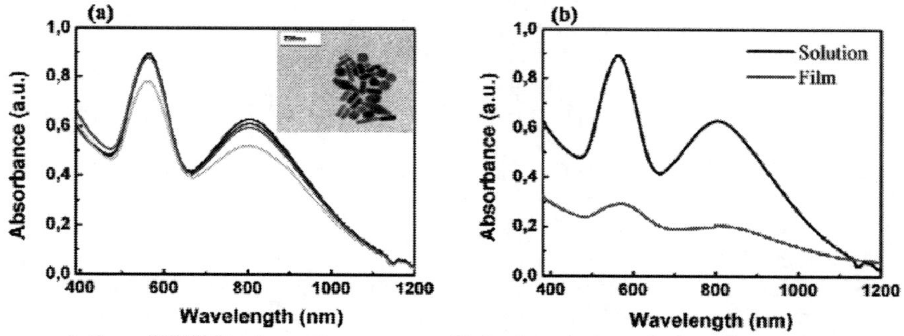

FIGURE 1. **(a)** Time evolution of UV-Vis absorption spectra of 2.5 mM toluene solution of Au NRs incorporated into PMMA-co-DMAEMA copolymer (50 mg/ml), as prepared (black line), after 24 hours (blue line), 72 hours (red line), 2 months (green line). In the inset: TEM image of the Au NRs. **(b)** UV-Vis absorption spectra of Au NR-amino functionalized PMMA based co-polymer nanocomposite in toluene solution and as film prepared by drop casting

In order to investigate the distribution of the Au NRs into the polymeric matrix and the topographic features of the nanocomposite thin films, non-contact mode atomic force microscopy (AFM) measurements have been carried out by a PSIA Xe-100 Scanning Probe Microscope, in non-contact mode. NRs-PMMA samples for AFM measurements have been prepared by spin-coating at 3000 rpm a dilute composite solution onto suitable cleaned glass substrates. AFM measurements highlight the occurrence of regular features 20 nm high, which are compatible with the transversal dimension of the rod shaped particles (Figure 2a), as estimated by the TEM analysis. The AFM investigation does not show any large structure, thus suggesting the absence of any relevant coalesced aggregates or ripening phenomena in the Au NR-amino functionalized PMMA nanocomposite, evidencing a homogeneous and uniform dispersion of the metal nanostructures in the polymer. The uniform dispersion of the NRs in the host matrix can reasonably ascribed to the effect of the PMMA functionalized with tertiary amino groups which can improve the interaction of the NP surface with the polymeric chains[12]. Here the proposed approach of dispersing NPs in a polymer host matrix, by using a common solvent has been demonstrated to preserve the physical chemical features of both the components. In addition the proper functionalization of the polymeric chain, by chemical groups possessing high affinity with the colloidal NP surfaces, limits the formation of aggregates and preserving the optical properties of the NRs. The reported nanocomposite preparation procedure provides an homogeneously dispersed Au NR- PMMA based co-polymer nanocomposite, which have been exploited in further processing and patterning techniques. Finally, Nano Imprint Lithography (NIL) has been successfully applied to fabricate patterned structures with micro and submicron resolution (Figure 2b), thus demonstrating the viability of the functional nanocomposites in the fabrication of optoelectronic and photonic devices.

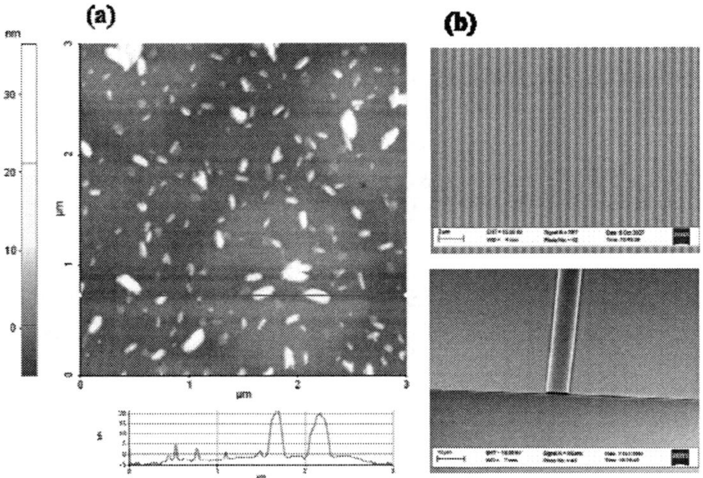

FIGURE 2. In **(a)** 2D view topography AFM image of the Au NR-PMMA-co-DMAEMA nanocomposite thin film. with cross-section line profiles along the red line. In (b) SEM images of waveguides fabricated in the prepared nanocomposite by NIL

REFERENCES

1. S. Eustis and M. A. El-Sayed, *Chem. Soc. Rev.* **35**, 209-217 (2006).
2. K. E. Gonsalves, L. Merhari, H. Wu and Y. Hu, *Adv. Mater.* **13**, 703-714 (2001).
3. V. Reboud, N. Kehagias, M. Striccoli, T. Placido, A. Panniello, M. L. Curri, M. Zelsmann, F. Reuther, G. Gruetzner and C. M. S. Torres, *J. Vacuum Sci. Technol. B* **25**, 2642-2644 (2007).
4. A. Pucci, N. Tirelli, E. A. Willneff, S. L. M. Schroeder, F. Galembeck and G. Ruggeri, *J. Mater. Chem.* **14**, 3495-3502 (2004).
5. J. de la Venta, A. Pucci, E. Fernández Pinel, M. A. García, C. de J. Fernandez, P. Crespo, P. Mazzoldi, G. Ruggeri and A. Hernando, *Adv. Mater.* **19**, 875-877 (2007).
6. T. Del Castillo-Castro, E. Larios-Rodriguez, Z. Molina-Arenas, M. M. Castillo-Ortega and J. Tanori, *Compos. Pt. A-Appl. Sci. Manuf.* **38**, 107-113 (2007).
7. J. Mathiyarasu, S. Senthilkumar, K. L. N. Phani and V. Yegnaraman, *Mater. Lett.* **62**, 571-573 (2008).
8. R. A. Vaia, C. L. Dennis, L. V. Natarajan, V. P. Tondiglia, D. W. Tomlin and T. J. Bunning, *Adv. Mater.* **13**, 1570-1574 (2001).
9. L. M. Goldenberg, O. V. Sakhno, T. N. Smirnova, P. Helliwell, V. Chechik and J. Stumpe, *Chem. Mater.* **20**, 4619-4627 (2008).
10. N. R. Jana, *Small* **1**, 875-882 (2005).
11. J. Yang, J.-C. Wu, Y.-C. Wu, J.-K. Wang and C.-C. Chen, *Chem. Phys. Lett.* **416**, 215-219 (2005).
12. M. Tamborra, M. Striccoli, M.L. Curri, J.A. Alducin, D. Mecerreyes, J.A. Pomposo, N. Kehagias, V. Reboud, C.M. Sotomayor Torres and A. Agostiano, *Small* **3**, 822-828 (2007)

Direct Measurement of the Interfacial Attractions between Functionalized Graphene and Polymers in Nanocomposites

M. Cai,[1] A. J. Glover,[1,2] T. J. Wallin[2], D. E. Kranbuehl,[1,2] H. C. Schniepp[1,*]

[1] *Department of Applied Science, The College of William and Mary, Williamsburg, VA 23187.*
[2] *Department of Chemistry, The College of William and Mary, Williamsburg, VA 23187.*
[*] *Author to whom correspondence should be addressed (e-mail: schniepp@wm.edu).*

Abstract. The objective of this work is to characterize and thereby achieve a fundamental scientific understanding at the molecular level of the relationship of the inter molecular forces measured at the nano scale between an individual nano particle and a polymer system. In this paper, we developed a method to directly characterize and measure the relative strength of the interfacial attractive forces between graphene nano particles and the polymer on the nano length scale of an individual particle using AFM techniques. The method was successfully applied to study the interfacial attractive strength in graphene polymer nano-composites. Two polymers poly(methyl methacrylate) (PMMA) and polyvinyl alcohol (PVA) were studied. The results support that this method is capable of characterizing the interfacial attractive forces directly at the nano-scale of graphene with a macromolecule material. The method can be applied to other nano particle macromolecular systems allowing determination of the interaction strength between nano-particles and the macromolecular material. This information paves the way to characterize, then to predict and finally optimize a nano particle, macromolecular system's performance properties.

Keywords: AFM, graphene, nanocomposite, interfacial force

PACS: 68.37.Ps

INTRODUCTION

Polymer nanocomposite materials, incorporating nanoparticles (reinforcing) into polymers, have advanced properties and great potential to be multi-functional, high-performance, lightweight materials. Graphene, the single atomic plane of graphite, is predicted to possess outstanding mechanical, electrical, and barrier properties. The monolayer graphene has a Young's modulus as high as 1.0 TPa, which make graphene a

CP1255, *V[th] International Conference on Times of Polymers (TOP) and Composites,*
edited by A. D'Amore, D. Acierno, and L. Grassia

promising reinforcing phase for ultra strong structural composite materials. However, they are highly hydrophobic, and the stable dispersion of the pure graphene sheets can only be achieved in a small amount of solvents. This limits the application of pure graphene sheets in nanocomposites. Functionalized graphene sheets, which have functional side groups, such as epoxy (C-O-C), and hydroxyl (C-OH) groups, attached to the carbon backbone surface, are compatible with a wider variety of solvents. Moreover, large-scale production of single graphene sheets is possible. This new nanocomposite has the potential for widespread applications as structural materials in aerospace, civil engineering, sports equipments and biological implants, due to its significant enhancement in the performance properties.

The mechanical properties of composites depend not only on the dispersion and mechanical properties of the nanofiller in the polymer but also very strongly on the properties of the polymer particle interface. Thus of major fundamental importance is developing experimental methods to measure the inter molecular forces between the particle and the polymer. Then the next fundamental scientific challenge is to understand the relationship between the particle polymer interfacial forces measured at nano dimensions to the resulting macroscopic mechanical properties. Here we are particularly interested in functional graphene sheets with a C:O ratio of 2(FGS_2,) and the interfacial interaction with it and with polyvinyl alcohol (PVA) – and poly(methyl methacrylate) (PMMA) as well as the resulting effect on these nano composite's mechanical performance.

In this paper, we have developed a method to directly measure the attractive strength between two materials using AFM, which can be applied to measure the interfacial strength between nanoparticles (especially 2-dimensional fillers) and the polymer material in the nanocomposite. In this method, the nano particle graphene sheets are sandwiched between two different materials possessing atomically smooth surfaces. After separation of these two materials where we peel off one material from the other one, the sheets are prone to stick on one of the surfaces. Thereby depending on the interfacial attraction strength between the sheets and these two materials one can assess the relative strength of the interfacial forces between the nano particle and the top surface of the polymer relative to the interfacial forces of the nano particle and the bottom substrate. In this manner one can assess the relative strength of the interfacial focus between one type of particle compared to another with a particular polymer, thereby choosing an optimum particle surface chemistry to obtain strong interfacial attraction. Or one can use this method to select a polymer chemical structure which optimizes the interfacial attractive forces for a particular nano particle.

Here this method is used to characterize and understand the relative strength of the interfacial attractive forces between FGS_2–and poly(methyl methacrylate) (PMMA), and polyvinyl alcohol (PVA).

Experimental

The FGS filler sheets are made by first completely oxidizing graphite powder and then ultrasonicating to fully exfoliate. Unlike montmorillonite and graphite nanotubes, graphene sheets can be dispersed as individual nanoparticles in a fluid and in a polymer. This is a very large advantage to these two widely studied systems. The oxidization of graphite powder (Asbury Carbons, Asbury, NJ, grade 3243, 99.5 %) was completed according to the Hummers method [1].

Results and Discussion

As described in Figure 1, the FGS_2 nano sheets are spin-coated on the base material. Then the polymer is spin coated on top of the base material. In this way, the nano particle sheets of graphene are sandwiched between two different materials. By sandwiching the nano sheets, we have constructed two interfaces: the base material- nano sheet interface and the polymer material-nano sheet interface. Next, the polymer material is peeled off from the base material breaking one of the above two interfaces. Then, we used AFM to examine both surfaces for the presence of nano sheets.

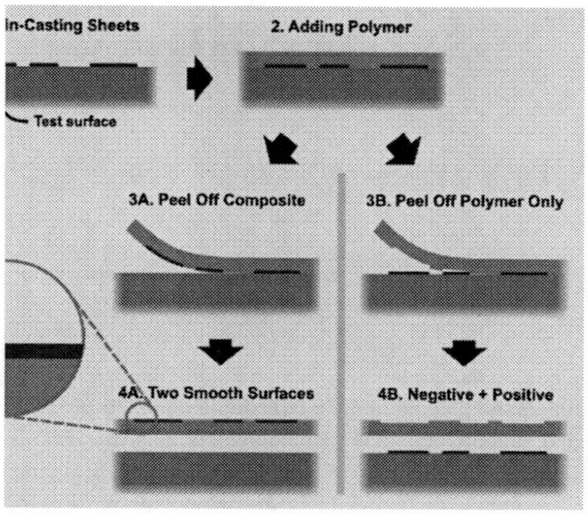

FIGURE 1 *Illustration of method to make nanocomposites located at the composite surface. Two situations can be observed depending on the interfacial attractions between the* FGS_2 *and polymer, and the* FGS_2 *and the substrate. For some polymer/substrate combinations it is found that after the peeling step (3.), the sheets attach to the polymer (3A.–4A.) or stay on the substrate (3B.–4B.)*

Our results show that all the sheets stick only to one of the two surfaces. Consequently, two scenarios happen due to differences in the interfacial attractive forces. If all the sheets stick to the attached material (see 3A-4A), a featureless surface would be found for the base material surface. At the same time, the surface of the attached material would be found to be flat and contain graphene nano sheets embedded inside. This situation implies that the attached polymer-nano particle interface has stronger attractive forces than the base material- nano particle interface. In the second scenario, all the sheets are still on the base material after peeling off the polymer material. At the same time, this leaves the imprints on the peeled-off polymer (see 3B-4B). When examining both surfaces, we should find that the base material surface contains topographical features of single-layer graphene, while the peeled-off attached material shows negative topography, which are the nano scale imprints of the graphene. Therefore, this scenario implies that the interfacial attractive forces between the polymer material and the nano particles is weaker than the base material.

We applied this method to examine the interfacial attraction between graphene nano particles and the polymers PVA and PMMA relative to the attractive forces between graphene nano particles and mica or HOPG. HOPG consists of graphite layers which have a similar surface as FGS_2: the planar sp^2 carbon network. The AFM experiments show the following interfacial attraction strength order: HOPG < PMMA < mica < PVA.

Conclusion

We developed methods to directly characterize and measure the relative strength of the interfacial attractive forces between nano particles and the polymer matrix. The method was successfully applied to study the interfacial attractive strength in FGS_2–polymer nanocomposites using two polymers (PMMA and PVA).

References

1. W.S. Hummers, Jr., R.E. Offeman, "Preparation of graphitic oxide", *Journal of the American Chemical Society,* 1958, 8, 1339.

On the Behavior of Dynamically Loaded Polymeric Materials

Igor Emri[a] and Barbara Zupancic[a]

[a]*Center for Experimental Mechanics, Faculty of Mechanical Engineering, University of Ljubljana, Askerceva 6, 1000 Ljubljana, Slovenia*

Abstract. This paper presents the analysis of time-dependent behavior of polymeric material under cyclic loading, modeled with the tooth-like loading conditions in terms of the shear stress that are determined by the operating frequency and the ratio between unloading and loading phase within one loading cycle. Assuming viscoelastic behavior of the material, the intensity of strain accumulation process is analyzed regarding the effects of the operating frequency, and number of loading cycles for series of time-dependent material properties expressed with discrete retardation spectrum. Results show that by knowing spectrum distribution of the time-dependent material property we may predict the critical operating frequencies that may lead to the highest level of accumulated strain.

Keywords: Polymers, Viscoelasticity, Retardation Spectrum, Cyclic Loading, Strain Accumulation, Durability.
PACS: 81.05-Lg

INTRODUCTION

Fatigue response of polymers strongly depends on polymers' viscoelastic nature. There are two basic competitive mechanisms that can cause the failure of cyclically loaded polymer; for higher frequencies the problem of local heating occurs due to the production of hysteresis energy during fatigue, while at lower frequencies fatigue failure is essentially caused by localized strain accumulation process that leads to the crack initiation, followed by its propagation.

In our study we focused on the strain accumulation process in viscoelastic material starting from certain type of loading conditions in terms of the shear stress that the cyclically loaded polymeric product may be exposed to during its operation [1]. In Fig.1a shear stress loading conditions with the magnitude τ_0 are presented for one loading cycle. a presents lasting of loading phase, b lasting of unloading phase within one loading cycle, from which we determined the ratio between the unloading and loading phase, $\kappa = \frac{b}{a}$. $\xi = a + b$ in Fig.1a presents the time period of one loading cycle from which frequency of cyclic loading was determined, $\nu = \frac{1}{\xi} = \frac{1}{+}$. Fig.1b shows the time-dependent response of viscoelastic material in one loading cycle under specified loading conditions.

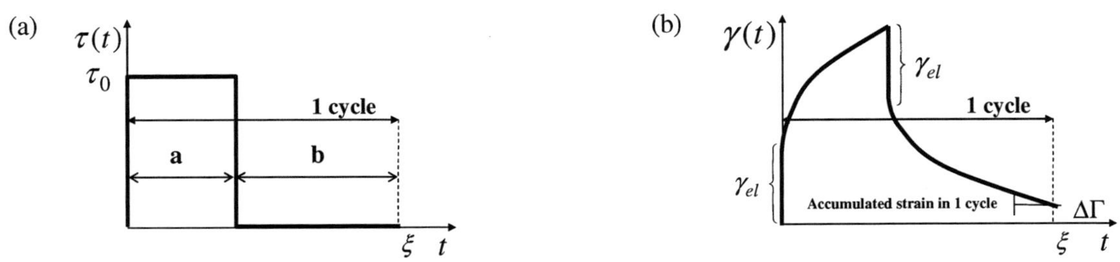

FIGURE 1. (a) Loading conditions in terms of the shear stress within one loading cycle, (b) shear strain response of viscoelastic material within one loading cycle under this loading conditions.

CP1255, *V[th] International Conference on Times of Polymers (TOP) and Composites,*
edited by A. D'Amore, D. Acierno, and L. Grassia

Under loading phase within the cycle time-dependent material starts to creep, and when it is unloaded the retardation process begins. Under certain operating conditions the retardation process between two consecutive cycles is not finished to the strain-free state, which means that material enters next loading cycle with a strain residual, what causes the strain accumulation process.

PROCESS OF STRAIN ACCUMULATION

Loading conditions described above were for N loading cycles and $t \le N(a+b)$ modeled with the following expression,

$$\tau(t) = \tau(0)h(t) + \tau_0 \sum_{j=1}^{N} \left\{ h\left[t-(j-1)(a+b)\right] - h\left[t-a-(j-1)(a+b)\right] \right\}, \tag{1}$$

where $h(t)$ represents Heaviside step function.
The shear strain response of the time-dependent material is described with the integral equation [2],

$$\gamma(t) = \tau(0)J(t) + \int_{0^+}^{t} J(t-s)\frac{\partial \tau(s)}{\partial s} ds, \tag{2}$$

where $J(t)$ is shear creep compliance that we expressed with discrete retardation spectrum as

$$J(t) = J_g + \sum_{i=1}^{K} L_i \left(1-e^{-\frac{t}{\lambda_i}}\right). \tag{3}$$

J_g denotes glassy creep compliance, K the number of retardation times, λ_i, where K corresponding spectrum lines with intensities $L_i = L(\lambda_i)$ are located. Introducing Eq.1 and 3 into the Eq.2 we obtained the following shear strain expression for the time $t \le N(a+b)$,

$$\gamma(t, t \le N(a+b)) = \tau(0)J(t) + \tau_0 \sum_{j=1}^{N}\sum_{i=1}^{K} L_i \left[e^{-\frac{t-\frac{1}{\nu(1+\kappa)}-\frac{(j-1)}{\nu}}{\lambda_i}} - e^{-\frac{t-\frac{(j-1)}{\nu}}{\lambda_i}} \right], \tag{4}$$

where $\kappa = \frac{b}{a}$, and $\nu = \frac{1}{\xi} = \frac{1}{a+b}$. Second term in this expression presents total accumulated strain that is for N loading cycles rewritten as

$$\Gamma_N(t) = \tau_0 \sum_{j=1}^{N}\sum_{i=1}^{K} L_i \left[e^{-\frac{(N-j+1)(1+\kappa)-1}{\nu\lambda_i}} - e^{-\frac{N-j+1}{\nu\lambda_i}} \right]. \tag{5}$$

To analyze the process of strain accumulation as a function of operating frequencies in the range of logarithmic values from -3 to 3 [1/sec], and the number of loading cycles, $N = 10000$, $\kappa = \frac{b}{a} = 1$ was selected, $\tau_0 = 1MPa$, and material retardation spectrum with the Gauss distribution ($\sigma = 1, \mu = 0$) of spectrum lines' intensity in logarithmic time scale of retardation times from -10 to 10 for $J_g = 0.001 \frac{1}{MPa}$. Introducing values of input parameters into Eq.5 we obtained the following evolution of accumulated strain as a function of operating frequency and number of loading cycles (see Fig.2a).

(a) (b)

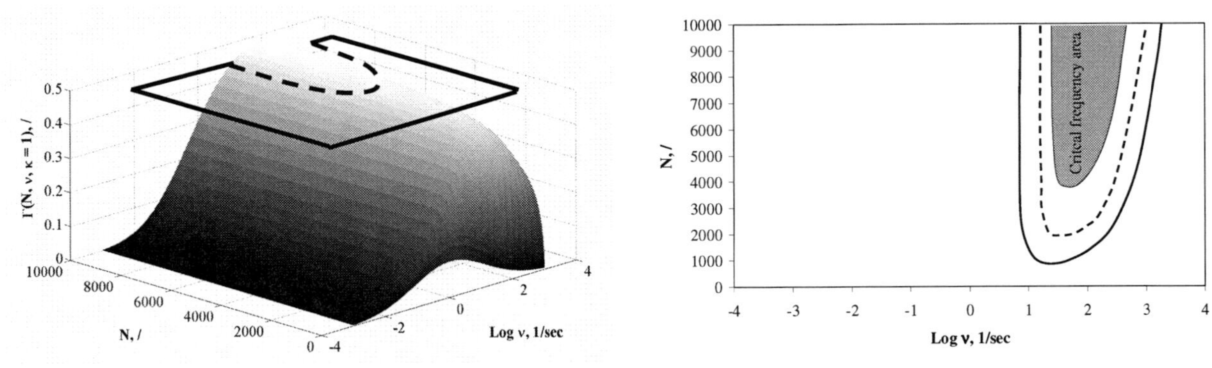

FIGURE 2. (a) 3D diagram of accumulated strain, $\Gamma(N, \nu, \kappa = 1)$, as a function loading cycles, N, and frequency, ν, and (b) frequency area corresponding to three different values of the cumulative accumulated strain.

If we define certain upper limits of the accumulated strain, as shown with the cross-section plane (solid black contour) in Fig.2a, and project the cross-section curve (dashed black line) onto $N - \nu$ plane, we obtain different levels of critical frequency areas as shown in Fig.2b. Within the "critical frequency area" (Fig.2b) we may expect the most intensive strain accumulation. Such diagrams may be used for predicting the expected life span of dynamically loaded polymeric products.

CONCLUSIONS

We presented the theoretical approach for modeling the strain accumulation process in dynamically loaded polymeric products. The strain accumulation process was analyzed for materials with the »Gauss« retardation spectrum distribution, which mimics well the behavior of real materials. The strain accumulation process was analyzed as function of the loading frequency and the number of loading cycles. Based on this analysis we have proposed the methodology for predicting the durability (the life span) of dynamically loaded polymeric products.

REFERENCES

1. I. Emri, J. Kramar, A. Hribar, A. Nikonov, U. Florjančič, U. "Time-dependent constitutive modeling of drive belts – I.The effect of geometry and number of loading cycles", *Mechanics of Time-Dependent Materials*, 2006, pp. 245-262.
2. N. W. Tschoegl, "The Phenomenological Theory of Linear Viscoelastic Behavior: An Introduction", 1989.

Consequences of thermo- and photo-oxidation on end-use properties of pure PE

Jonathan Tireau[a,b], Laetitia Van Schoors[a], Karim Benzarti[a], Xavier Colin[b]

[a]*Université Paris-Est, LCPC, Division PCM, 58, Boulevard Lefebvre, 75732 Paris cedex 15, France*
[b]*PIMM (UMR CNRS 8006), Arts et Métiers ParisTech, 151, Boulevard de l'hôpital, 75013 Paris, France*

Abstract. Thin films of polyethylene have been exposed to thermal and photochemical accelerated agings at different temperatures and different UV intensities. In both cases, mechanical tests performed on aged samples revealed a severe embrittlement of the PE material. However, a multi-scale analysis showed that i) under thermal oxidative conditions, such embrittlement mainly resulted from a reduction of molecular mobility of the amorphous phase due to annealing phenomena, while ii) it was rather associated to a competition between chain scission and cross-linking processes under photo-oxidative conditions. As regards the oxidation mechanisms, it was found that thermo-oxidation is clearly initiated by a bimolecular decomposition of hydroperoxides, while photo-oxidation seems rather initiated by the photolysis of ketones and hydroperoxides. Over a critical threshold of the hydroperoxide concentration, thermolysis of hydroperoxides seems to predominate and even to control the overall photo-oxidation kinetics.

Keywords: Polyethylene, thermo-oxidation, photo-oxidation, embrittlement.
PACS: 82.35.-x

INTRODUCTION

Polyethylene (PE) is commonly used in many civil engineering applications (geomembranes, protective sheaths of cable stayed bridges, gas and water pipes...) because of its low cost, easy processability, good impermeability and chemical resistance. However, in external environment, several factors such as ultraviolet radiations, seasonal temperature variations or washing by rain water can lead ultimately to a deterioration of the PE material[1]. It is usually admitted that photo-oxidation is the main cause of degradation for PE[2]. Nevertheless, other physical and chemical phenomena, such as physical loss and chemical consumption of antioxidants[3,4], thermo-oxidation[5] or annealing phenomena[6] can also contribute to an overall degradation of the material properties.

The objective of this study was to investigate the respective effects of these various phenomena on the durability of pure PE. With this intention, thin PE films (200-300 µm thick) were exposed to two types of ageing conditions: thermal ageing in a ventilated oven at different temperatures (40, 60, 90, 100 and 110°C) and photochemical ageing in a UV chamber at different intensities ($0.51 W.m^{-2}$; $0.65 W.m^{-2}$ and $0.75 W.m^{-2}$) at 38°C. Exposed samples were then characterized using a multi-scale approach: i) infrared spectrometry (IR) at the molecular level, ii) rheology in the melt state and dynamic mechanical analysis (DMA) at the macromolecular level, iii) differential scanning calorimetry (DSC) , X-ray diffraction (XRD) and density measurements at the microstructural level and iv) tensile tests at the macroscopic scale.

CP1255, *V[th] International Conference on Times of Polymers (TOP) and Composites,*
edited by A. D'Amore, D. Acierno, and L. Grassia

RESULTS AND DISCUSSION

The main results obtained for exposed specimens are presented and discussed in this section.

At the molecular scale, IR spectroscopic analysis coupled with chemical derivatization (NH_3 treatment) made it possible to identify and quantify the different oxidation products (Figure 1). As regards thermo-oxidation, the initiation stage consisted mainly in the bimolecular decomposition of hydroperoxides. Differently, for photo-oxidation, photolysis of hydroperoxides and ketones (by Norrish II) were identified as the main initiation reactions. Beyond a critical threshold of the hydroperoxide concentration, thermolysis of hydroperoxides seemed to predominate over the two other initiation reactions and even to control the overall kinetics of oxidation.

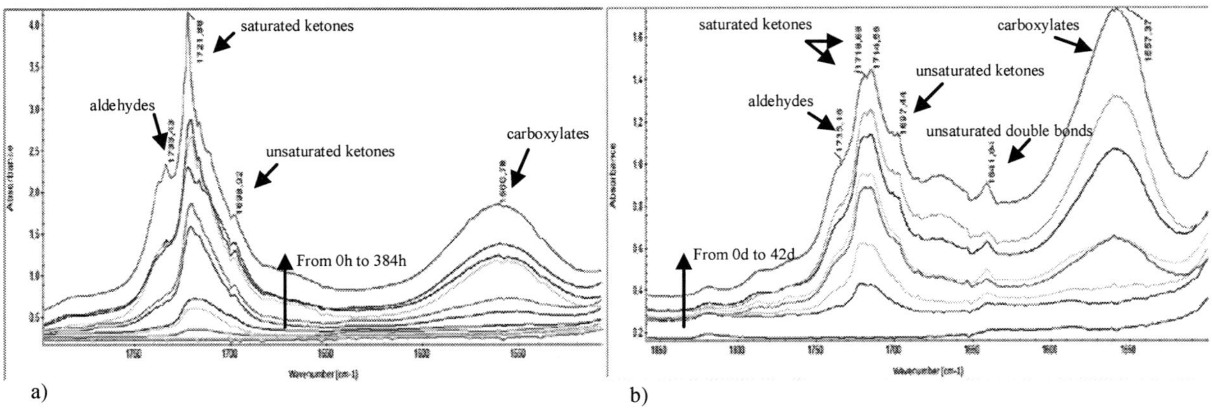

FIGURE 1. Changes in the absorbance spectra after NH_3 treatment on films of pure PE (200 μm) exposed a) in a ventilated oven at 100°C for periods up to 384h, b) in a Weather-o-meter chamber at 0.75W.m^{-2} and 38°C in atmospheric air for periods up to 42 days.

At the macromolecular scale, in the case of thermo-oxidation, a large part of the carbonyl species resulted from the β-scission of alkoxyl radicals. Chain scissions (s) lead to a sharp decrease in the weight average molecular mass (Figure 2a). As regards photo-oxidation, rheology in the melt state revealed, on the one hand, the occurrence of chains scissions (s) probably due to the photolysis of hydroperoxides and ketones, and on the other hand, the appearance of a cross-linking phenomenon (x) from the early period of exposure (Figure 2b).

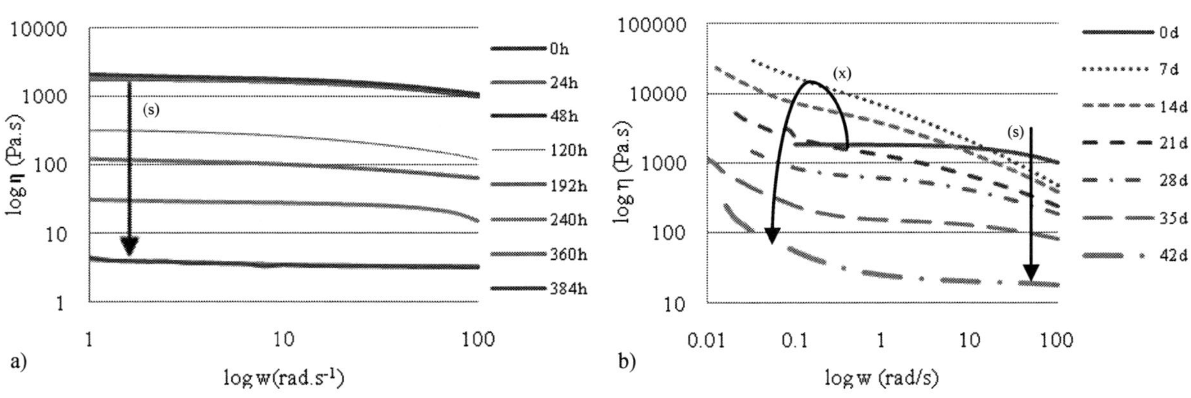

FIGURE 2. Evolutions of the rheological behaviours for pure PE films (200μm) aged a) in a ventilated oven at 100°C, b) in a Weather-o-meter chamber at 0.75W.m^{-2} and 38°C in atmospheric air.

At the microstructural scale, DSC analysis revealed microstructural changes due to annealing phenomena during the first moments of the thermo-oxidative treatment. Due to an increase in the molecular mobility, the smaller crystalline lamellae (lc) seemed to coalesce, hence forming thicker lamellae exhibiting a melting temperature higher than the ageing temperature (Figure 3a). Thus, annealing could lead to a more uniform crystalline structure in PE. After the oxidation induction time, chain scissions destroyed the entanglement network located in the rubbery amorphous phase and thus liberated small macromolecular segments. These segments were able to diffuse toward interfaces between amorphous and crystalline regions, leading to a global increase in the lamellae thickness (Lc) and crystallinity ratio[7]. Such effects were even more pronounced in the case of photo-oxidation due to chimicrystallisation phenomena (Figure 3b).

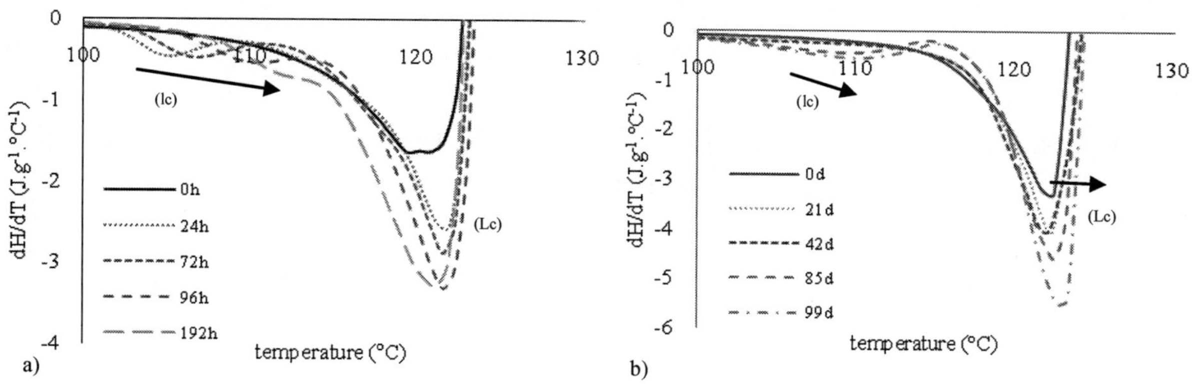

FIGURE 3. DSC derivative-thermograms for samples aged a) in a ventilated oven at 100°C, b) in Weather-o-meter chamber at 0.75W.m^{-2} and 38°C in atmospheric air (heating rate of 10°C/min).

At a macroscopic scale, ultimate tensile elongation of PE decreased sharply during thermal ageing (Figure 4a). Embrittlement appeared before the oxidation induction time, hence before the onset of chain scission. Thus, embrittlement might result mainly from a reduction of molecular mobility of the amorphous phase, which is itself a consequence of the increasing crystalline ratio. Differently, for photochemical ageing, mechanical properties of PE seemed directly affected by changes in the weight average molecular mass. A competition between chain scission and cross-linking processes led rapidly to the embrittlement of the material (Figure 4b).

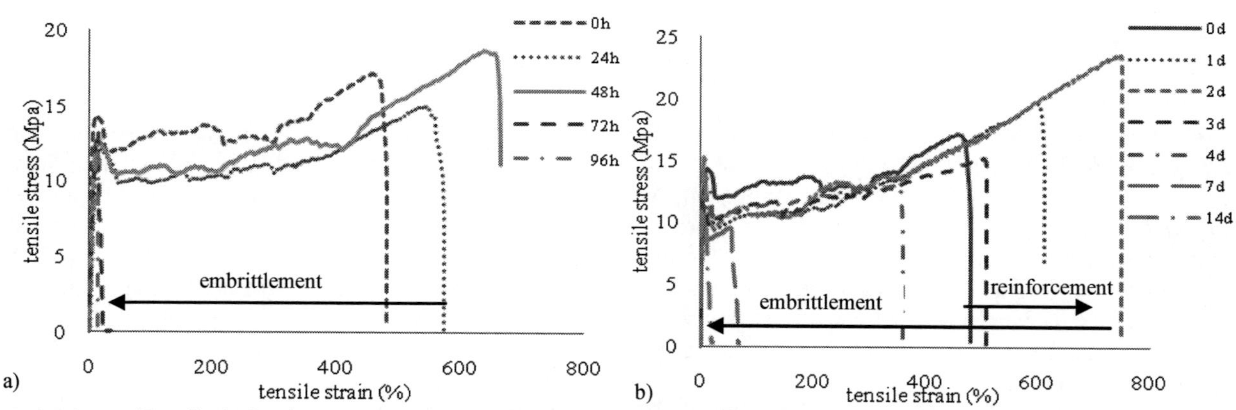

FIGURE 4. Tensile behaviours of pure PE films (200μm thick) exposed for various periods of time a) in a ventilated oven at 100°C, b) in Weather-o-meter chamber at 0.75W.m^{-2} and 38°C in atmospheric air.

REFERENCES

1. S. Halim Hamid and I. Hussain, "*Lifetime Prediction of Plastics*", Handbook of Polymer Degradation, Second Edition, Revised and Expanded, New York, Marcel Dekker, 2000, pp. 669-726
2. P. Gijsman, "*The long term stability of polyolefins*", wibro dissertatiedrukkerij, Helmond, 1994
3. X. Colin, L. Audouin, J. Verdu, M. Rozental-Evesque, B. Rabaud, F. Martin, F. Bourgine, Polym Eng Sci **49**, 1642-1652 (2009)
4. X. Colin, L. Audouin, J. Verdu, M. Rozental-Evesque, B. Rabaud, F. Martin, F. Bourgine, Polym Eng Sci **49**, 1429-1437 (2009)
5. F. Delor, G. Teissedre, M. Baba, J. Lacoste, Polym Degrad Stab **60**, 321-331 (1998)
6. S. Castagnet and Y. Deburck, Mater Sci Eng A **448**, 56-66, (2007)
7. Gedde U.W. et Ifwarson M., Polym Eng Sci **30**, 202-210 (1990)

Inkjet Etching of Polymer Surfaces to Manufacture Microstructures for OLED Applications

I. A. Grimaldi, A. De Girolamo Del Mauro, G. Nenna, F. Loffredo, C. Minarini and F. Villani

Portici ENEA Research Center, Piazzale E. Fermi I, 80055 Portici (NA), Italy

Abstract. A soluble polymer layer can be structured by means of solvent drops etching . The inkjet etching (IJE), innovative technique that consists of ejecting liquid material in controlled manner in terms of both volume amount (10-100 pL) and positioning on the substrate (μm resolution), allows to manufacture suitable shaped microstructures on top of polymer surfaces. In the present work we fabricated IJ etched polymer microstructures whose shape varies from concave to convex by employing a pure solvent or solvent mixtures with suitable mixing ratio of the solvents. The manufactured structure profile has been analyzed and the related geometrical parameters, i.e. depth, diameter and height, have been valuated as function of the drops number. Preliminary tests have been performed in order to employ the IJ etched polymer microstructures with the aim to improve the out-coupling efficiency of organic light-emitting diodes.

Keywords: organic light-emitting diode; inkjet etching; polymer microstructures; out-coupling efficiency.
PACS: 82.35.Gh Polymers on surfaces

INTRODUCTION

Inkjet printing (IJP) is a tool for positioning small volumes of liquid material precisely on the substrate. Inkjet etching (IJE) is an application of IJP technique and consists in depositing solvent drops on a soluble polymer layer, inducing the structuring of the polymer film itself. When the solvent drop hits the soluble polymer layer, the polymer is locally dissolved, carried from the centre to the edge and deposited there at the end of drying process, creating micro-cavity shaped structures [1-3]. This flow inside the drop and directed towards its rim, where the evaporation rate is higher, is known as the 'coffee-stain effect' [1,4]. The shape of the polymer microstructures can be varied from concave to convex by employing solvent mixtures with suitable mixing ratio of the solvents.

By changing the number of drops of the pure solvent or solvent mixtures, the geometrical parameters of the microstructures also change, i.e. the depth, diameter and height can be modified.

The structuring of a polymer surface by means of IJE technology is employed in several electronic applications [5,6]. In particular, the microstructures can be also employed in order to improve the out-coupling efficiency of organic light-emitting diodes (OLEDs) [7,8]. The portion of the light generated in the emissive layer of OLED is trapped inside the device by the total internal reflection (TIR) and wave-guided along the device until it is lost by absorption and/or emission at the edge of the substrate.

In this work we fabricated ordered structures by inkjet etching a previously spin-coated poly(styrene) layer. By inkjet printing drops of toluene or mixture of toluene and n-methyl pyrrolidone onto the polymer layer, spherical symmetrically microstructures have been manufactured. We analyzed the microstructure shape by varying the etching solvent and the volume ratios of the solvents. Moreover, as the employed soluble polymer has got suitable optical properties, such as high transparency and very low degree of scattering in the visible region, we investigated the application of IJ etched microstructures to OLED devices in order to improve their out-coupling efficiency. A microstructures array has been manufactured on the opposite side of an OLED device. A preliminary analysis of the electro-optical properties of structured electroluminescent devices has been performed.

CP1255, *Vth International Conference on Times of Polymers (TOP) and Composites,*
edited by A. D'Amore, D. Acierno, and L. Grassia

EXPERIMENTAL

The microstructures array has been fabricated on a layer of poly(styrene) (PS) (Aldrich, average Mw ~ 192,000), 9 μm thick. PS film has been spun on glass substrate by employing a 15% (w/w) chlorobenzene solution.

Successively, drops of toluene (TOL) or mixture of toluene and n-methyl pyrrolidone (TOL:NMP) have been inkjet printed on the polymer layer. The inkjet equipment (Aurel S.p.A.) uses a piezoelectric Drop on Demand (DoD) technology to jet liquid droplets through Microdrop print-heads with 30 and 50 μm opening nozzles. Sequences of droplets have been printed at 10 Hz drop emission frequency and 1 mm/s print-head speed to create a square array of concave- or convex-shaped microstructures. By changing the number of drops, the geometrical parameters of the structure have been modified.

A surface profilometer (KLA Tencor P-10 Surface Profiler) has been used in order to analyze the profile of the structures manufactured on PS film and to evaluate geometrical parameters.

As application, we employed the IJ etched microstructures in OLED device. The electroluminescent device has been fabricated on glass substrate on the opposite side respect to the polymer layer and its structure was ITO/TPD/Alq_3/Al.

RESULTS AND DISCUSSION

The structuring of the polymer layer has been performed by inkjet etching pure solvent (TOL) or solvent mixtures (TOL:NMP) drops. In both cases, the etching materials are known to be good solvents for PS polymer. The polymer/glass substrate was heated during drops printing in order to improve the structuring uniformity. The typical profile of the structure performed upon the TOL drop print onto PS layer is shown in figure 1. Y1 and Y2 represent the cavity depth and the edges height, respectively. Both these geometrical parameters that characterize the micro-cavity profile increase as the solvent drop number increases. The cavity depth and edges height as functions of drops number are shown in figure 2.

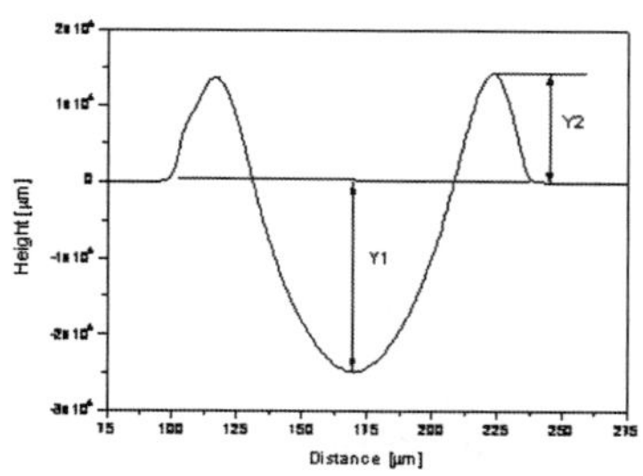

FIGURE 1. PS cavity profile performer by TOL drop etching. Y1 and Y2 are cavity depth and edges height, respectively.

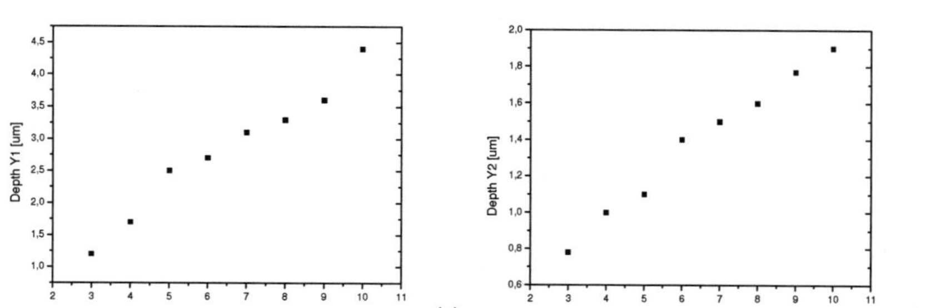

FIGURE 2. Cavity depth [Y1] (a) and edges height [Y2] (b) as functions of drops number.

The employing of solvent mixtures allows of modifying the microstructure profile shape by varying the solvents volume ratios. The profiles of the structures formed by sessile droplet deposition of the mixture of toluene and n-methyl pyrrolidone at volume ratios from 1:1 to 2.3:1 onto PS substrates are shown in figure 3 (a). The geometrical parameters of these microstructures vary as the drops number increases as shown in figure 3 (b)-(d) for each solvents volume ratio.

As application, we employed in OLEDs the fabricated polymer microstructures in order to improve their out-coupling efficiency. Indeed, the PS polymer with suitable optical properties and the fabricated microstructures geometry could lower the EL radiation losses due to the TIR. A preliminary analysis of electro-optical properties of structured devices confirms this hypothesis and indicates that the emission efficiency increases of 3-5% compared to an unstructured OLED.

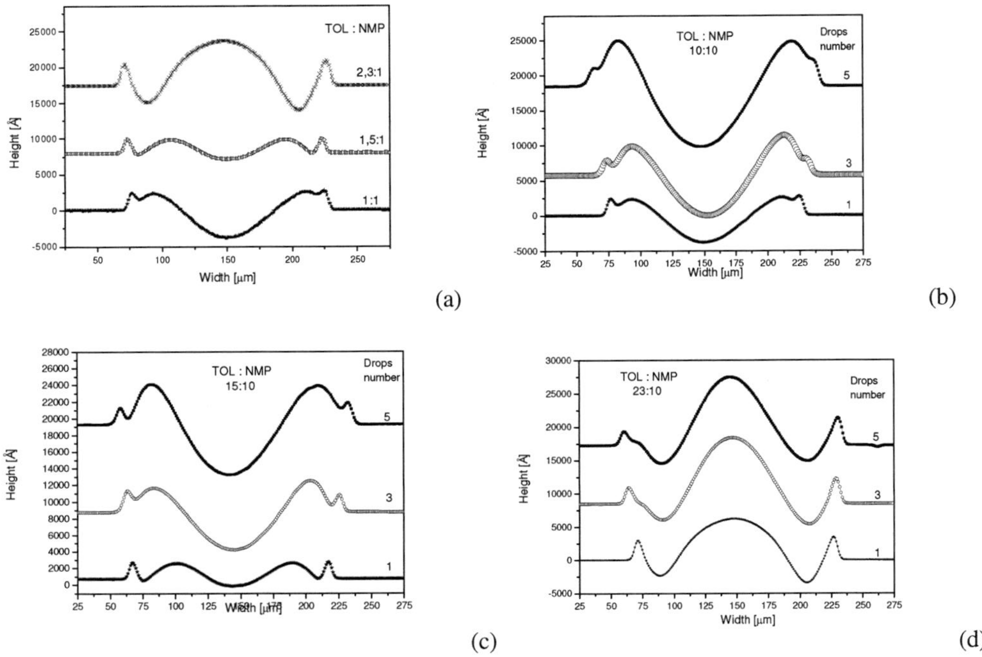

FIGURE 3. Profiles of the structures formed by TOL:NMP droplet deposition at volume ratios 1:1, 1.5:1 and 2.3:1 onto PS layer (a). Microstructures geometry as the drops number changes ((b)-(d)) for each solvents volume ratio.

CONCLUSIONS

The feasibility to employ the IJE technique to perform polymer microstructures has been investigated. The microstructures profile can be controlled by employing suitable etching materials. We demonstrated that the microstructures shape can be varied from concave to convex using pure solvent or solvent mixtures with suitable mixing ratios of the solvents. The structures geometrical parameters can be also modified by changing the drops number. Finally, we performed a preliminary test on the application of these polymer microstructures to OLEDs in order to better guide the generated EL radiation externally.

ACKNOWLEDGMENTS

This work was supported by the TRIPODE project (DM 20160) financed by the Ministero dell'Università e della Ricerca (MUR). The authors acknowledge Tommaso Fasolino for technical support in electro-optical measurements.

REFERENCES

1. R. D. Deegan, O. Bakajin, T. F. Dupont, G. Huber, S. R. Nagel, T. A. Witten, *Nature* **389**, 827-829 (1997).
2. R. D. Deegan, *Phys. Rev. E* **61**, 475-485 (2000).
3. R. D. Deegan, O. Bakajin, T. F. Dupont, G. Huber, S. R. Nagel, T. A. Witten, *Phys. Rev. E* **62**, 756-765 (2000).
4. E. Tekin, E. Holder, D. Kozodaev, U. S. Schubert, *Adv. Funct. Mater.* **17**, 277-284 (2007).
5. Y. Xia and R. H. Friend, *Appl. Phys. Lett.* **90**, 253513 (2007).
6. R. Pericet-Camara, A. Best, S. K. Nett, J. S. Gutmann, E. Bonaccurso, *Opt. Exp.* **15**, 9877-9882 (2007).
7. W. Li, R. A. Jones, S. C. Allen, J. C. Heikenfeld, A. J. Steckl, *J. Dispay Technol.* **2**, 143-152 (2006).
8. W.-K. Huang, W.-S. Wang, H.-C. Kan, F.-C. Chen, *Jpn. J. Appl. Phys.* **45**, L1100-L1102 (2006).

Side chain crystallization and non-equilibrium phenomena in nanophase separated poly(3-alkyl thiophenes)

Shireesh Pankaj and Mario Beiner

Institut für Physik, Martin-Luther-Universität Halle-Wittenberg, D-6099 Halle, Germany

Abstract. A series of regio-random poly(3-alkyl thiophenes) containing $C = 6$-12 carbon atoms per side chain is studied using shear, calorimetry and x-ray scattering methods. Dynamic and structural data indicate a nanophase separation of thiophene main chains and alkyl groups. Alkyl nanodomains of size of 1-3 nm are formed as indicated by pre-peaks in scattering data at $0.2\ \text{Å}^{-1} \leq q \leq 0.6\ \text{Å}^{-1}$. Apart from the conventional α relaxation, an additional α_{PE} process appears at low temperatures reflecting an independent dynamics within the alkyl nanodomains. The influence of slow side chain crystallization and densification processes during physical aging below T_g on the relaxation behavior is studied. Long term experiments show that short alkyl groups do not crystallize, whereas for longer alkyl groups, side chain crystallization is occurring but hindered by immobile and rigid main chains. Slow side chain crystallization processes are indicated in poly(3-decyl thiophene) and poly(3-dodecyl thiophene) by melting peaks in DSC scans, a decreasing α_{PE} intensity and an increase in the plateau modulus. Isothermal crystallization experiments on these samples show that the alkyl groups can exist in different polymorphic states.

Keywords: Side chain polymers, glass transition, physical ageing, polymorphism, nanophase separation
PACS: 64.70.pj, 83.80.-k, 61.50.Ks, 81.40.Cd

INTRODUCTION

Poly(3-alkyl thiophenes) (P3ATs) belong to the class of conjugated comb-like polymers having favourable electronic and optoelectronic properties [1, 2, 3]. Based on the head-tail arrangement along the main chain, they are classified as regio-regular (100% head-tail-head-tail, HT-HT) or regio-random samples (random distribution of HT-HT, TT-HT, HT-HH, TT-HH). Efficient charge transport is realised in regio-regular P3ATs due to packing of thiophene rings on a well defined crystal lattice supporting planarity and π-π orbital overlap. Alkylation of regio-regular P3ATs improves processibility, however also introduces additional driving forces affecting the structure formation process. Studying the dynamics and the packing of the side chains appears to be difficult in these systems as the thiophene rings crystallize. Regio-random systems appear to be advantageous for that purpose as their thiophene backbones remain amorphous.

A typical feature of comb-like polymers with similar microstructure is that alkyl side chains separate from the main chain to form alkyl nanodomains of the size of 1-3 nm. This phenomenon has been called nanophase separation and it is found for poly(n-alkyl methacrylates), poly(n-alkyl acrylates) [4] and also reported for regio-random P3ATs recently [5]. Nanophase separated side chain polymers are characterized by (i) a pre-peak in the scattering data at intermediate q values ($0.2\ \text{Å}^{-1} \leq q \leq 0.6\ \text{Å}^{-1}$) as well as (ii) unique low temperature dynamics within the alkyl nanodomains (α_{PE}) which is depending on the side chain length and is independent on the main chain properties. The α_{PE} process appears usually at temperatures much below the conventional α relaxation incorporating the main chains.

Primary aim of this work is to study how the dynamics of regio-random P3ATs change on long and application relevant time scales due to the tendency of alkyl groups to densify and crystallize. Of special interest are changes in side chain packing during annealing near and below T_g and the consequences of these changes on the dynamics.

MATERIALS AND METHODS

A series of regio-random (1:1:1:1) (HT-HT:HT-HH:TT-HH:TT-HT) P3ATs with C=6-12 carbon atoms per side chain is studied. Samples were purchased from Rieke Metals Inc. Their synthesis is described elsewhere [6].

X-ray scattering measurements in the range of $0.2\ \text{Å}^{-1} \leq q \leq 0.6\ \text{Å}^{-1}$ were performed on an instrument assembled by JJ X-rays based on 2D detector Hi-Star from Bruker and a Rigaku rotating anode. A Rheometrics RDAII is used for dynamic shear measurements in the frequency range 0.01-100 rad/s in strip geometry (18 x 4 x 1 mm^3) with a control strain of 0.1%. DSC heating scans with a rate of 20 K/min are performed on a Perkin-Elmer DSC7.

CP1255, *Vth International Conference on Times of Polymers (TOP) and Composites,*
edited by A. D'Amore, D. Acierno, and L. Grassia

RESULTS AND DISCUSSION

The results of dynamic shear measurements on the series of regio-random P3ATs are presented in Fig. 1(a). Shear loss modulus data clearly show that two independent relaxation processes occur. The α_{PE} process at low temperature corresponds to cooperative motions within the alkyl domains (often designated as "polyethylene like glass transition") and appears at a comparable frequency-temperature position like in other side chain polymers containing alkyl groups with identical length. The cooperative α process at high temperature is accompanied by softening of the entire system and incorporates the thiophene rings. The α process shifts to lower temperature if the side chain length increases due internal plasticization as discussed by Heijboer [7], while the α_{PE} process moves to higher temperature. The latter trend has been interpreted in the hindered glass transition picture [8] as confinement effect resulting from a conflict of alkyl nanodomain size with the size of cooperative rearranging regions (CRRs) [4]. Nanophase separation of main and side chains in P3ATs is reflected by a pre-peak in scattering data in the intermediate q range indicating main chain to main chain distances between 16 Å for poly(3-hexyl thiophene) (P3HT, $C = 6$) and 24 Å for poly(3-dodecyl thiophene) (P3DDT, $C = 12$) [5]. On comparing the peak positions for the α_{PE} process with those for poly(n-alkyl methacrylates), the trends are found to be similar as shown in Fig. 1(b). A glass transition is resolvable in DSC heating scans preformed after cooling the samples from 80 to -90°C with a rate of -20 K/min (Fig. 1(c)). The shift in T_g with side chain length is similar to that of the α process in shear relaxation data. For P3DDT, a wavy structure is seen above T_g indicating side chain crystallization. Three polymorphic forms are found during isothermal crystallization at different temperatures [5]. Interestingly all polymorphic forms can coexist. P3HT ($C = 6$) and poly(3-octyl thiophene) (P3OT, $C = 8$) seem to be fully amorphous as their heat capacity curves show only a glass transition. For poly(3-decyl thiophene) (P3DT, $C = 10$), a tiny peak is seen above T_g indicating either structural relaxation or a very small melting endotherm.

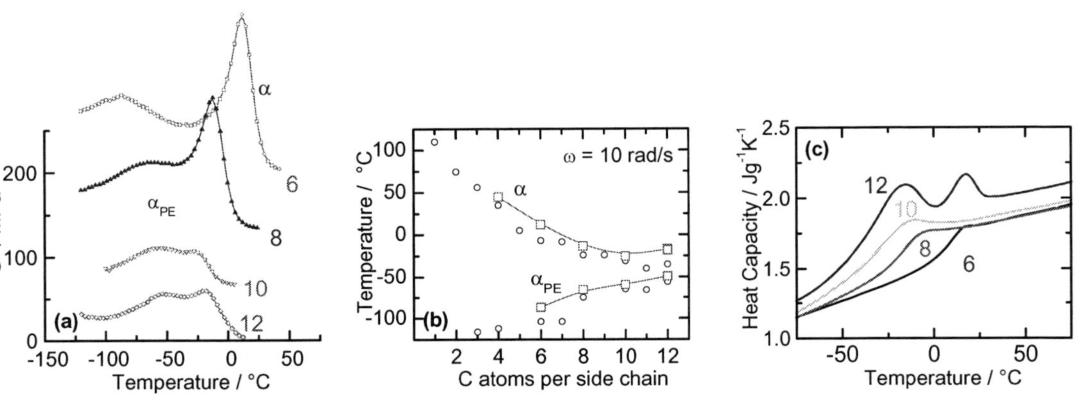

FIGURE 1. (a) Shear loss modulus G'' vs temperature at angular frequency of 10 rad/s. Curves are vertically shifted for the sake of clarity. (b) Relaxation temperature vs number of carbon atoms for P3ATs (squares) and poly(n-alkyl methacrylates) (circles) [4]. Data is derived from fitting isochronal data with gaussian functions. (c) DSC heating scan with heating rate of $dT/dt = +20$ K/min for the series of regio-random poly(3-alkyl thiophenes) with different alkyl side chain lengths.

Isothermal crystallization experiments on P3DT performed at -25°C and 5°C show that side chain crystallization occurs (Fig. 2(a)) like reported for P3DDT recently [5]. Thick line represents a scan for an as received sample. A glass transition around -32°C is seen followed by an endotherm at 48°C designated as melting of form I. The estimated degree of crystallinity of the alkyl groups is around 5%. On varying the crystallization times from 5 min to 7 h at T_c=-25°C, heating scans subsequently show an increase in the intensity of the melting peak for form III. Heat of fusion and melting temperature (around 10°C) do increase systematically. Similar trends occur when we anneal the sample at T_c=5°C leading to a tiny melting peak centered around 28°C (form II). In contrast to the findings for P3DDT, form I does not grow after annealing at 26°C on time scales of hours. Possibly longer times or slightly different temperatures are required. Note that polymorphism has not been reported for other side chain polymers containing alkyl groups with similar length. Eventually the occurence of different polymorphs is related to main chain sequences with different head-tail arrangement.

Dynamic shear measurements were performed to quantify the effects of slow side chain crystallization processes on the dynamics. Obviously, both relaxation processes, α_{PE} and α, are affected by crystallization. The conventional α process slows down and it becomes broader (Fig. 2(b)). More interestingly there is a significant decrease in the intensity of the α_{PE} process during crystallization as seen by comparing the situation before and after crystallization

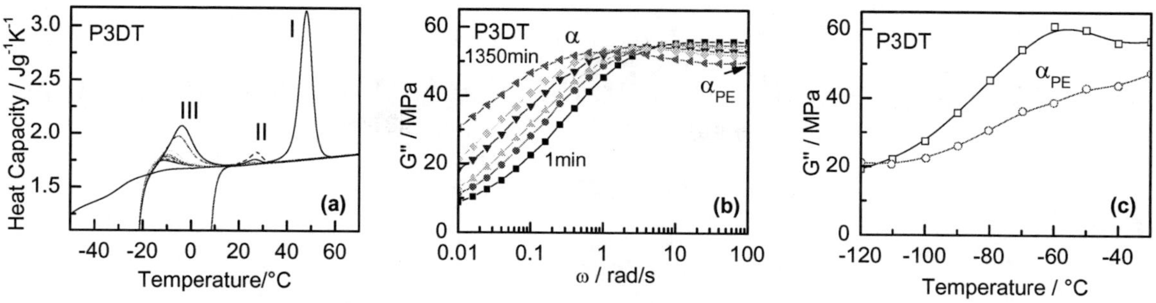

FIGURE 2. (a) DSC heating scans (dT/dt=20K/min) for P3DT performed after crystallization for different times t_c (5, 10, 15, 20, 25, 30, 60, 240 and 420 min) at T_c= -25°C and 5°C. Thick line represents a scan for an as received sample. The melting peaks of different forms are labled with I, II and III. (b) Shear loss modulus $G''(\omega)$ for P3DT measured after isothermal crystallization at -25°C for 1, 50, 100, 250, 400 and 1350 minutes. (c) Shear loss modulus $G''(\omega)$ for P3DT at angular frequency of 10 rad/s. Heating (squares) and cooling (circles) performed before and after isothermal crystallization at -25°C for 24 hours.

at -25°C for 24 h (Fig. 2(c)). The intensity of the α_{PE} process in the semi-crystalline state is significantly reduced. This finding indicates side chain crystallization since crystalline methylene units cannot contribute to the α_{PE} process. The observed changes in the conventional α process might be an indication for an indirect influence of the side chain packing on the main chain dynamics [9]. In general, there seems to be a strong coupling between main and side chains causing long term effects in relaxation and crystallization behavior of nanophase separated side chain polymers. The rigid main chains do hinder side chain crystallization while the packing state of the side chains is strongly affecting the dynamics of the thiophene rings in the main chain.

CONCLUSIONS AND OUTLOOK

Nanophase separation appears to be a common finding for comb-like polymers. Features in dynamic and structural data for regio-random P3ATs are similar to that in other polymer series containing long alkyl groups. An independent dynamics of the methylene units within the alkyl nanodomains is indicated by an α_{PE} process at low temperature. It is shown that the side chains crystallize partly in P3DT and P3DDT if samples are annealed for several hours at temperatures between -25 and 6°C. The results of this study indicate that long term effects in these systems are a consequence of a conflict between crystallization tendency of the alkyl groups and kinetic hindrance by immobile main chains in the disordered state. The main chains prevent fast side chain crystallization while the packing state of the alkyl groups influences the main chain dynamics and the mechanical properties. A peculiar finding for regio-random P3DT and P3DDT is that the alkyl groups exist in different polymorphic states. This is unusual and not observed for other side chain polymers containing alkyl groups with similar length. The reported findings seem to be interesting for understanding the properties of regio-regular P3ATs showing changes in side chain packing at particular temperatures. Experiments on frustrated alkyl groups in small, self-assembled alkyl nanodomains might be also a tool in studying early stages of polyethylene crystallization [10].

REFERENCES

1. M. Winokur, "Structural Studies of Conducting Polymers," in *Handbook of Conducting Polymers*, edited by T. Skotheim, R. Elsenbaumer, and J. Reynolds, Marcel Dekker, New York, 1998, vol. 2, pp. 707–726.
2. M. J. Winokur, and W. Chunwachirasiri, *Journal of Polymer Science Part B-Polymer Physics* **41**, 2630–2648 (2003).
3. R. McCullough, *Advanced Materials* **10**, 93–116 (1998).
4. M. Beiner, and H. Huth, *Nature Materials* **2**, 595–599 (2003).
5. S. Pankaj, E. Hempel, and M. Beiner, *Macromolecules* **42**, 716–724 (2009).
6. T. Chen, X. Wu, and R. Rieke, *Journal of the American Chemical Society* **117**, 233–244 (1995).
7. J. Heijboer, *Physics of Non-Crystalline Solids; Prins, J. A., Ed.; North Holland: Amsterdam* p. 231 (1965).
8. E. Donth, *Glasübergang*, Akademie, Berlin, 1981.
9. S. Pankaj, and M. Beiner, Long-term behavior and side chain crystalliztaion of poly(3-alkyl thiophenes), submitted.
10. M. Beiner, *Journal of Polymer Science Part B-Polymer Physics* **46**, 1556–1561 (2008).

Comparing Nanofillers in Polylactide Nanocomposites

Evangelia Kontou[a] and Michael Niaounakis

[a]Department of Applied Mathematical and Physical Sciences
Section of Mechanics, National Technical University of Athens
5 Heroes of Polytechnion, GR-15773, Athens, Greece

Abstract. The present study compares the effects of two different types of nanosized fillers (silica and montmorillonite) at three different weight fractions as well their mixtures on the thermomechanical properties of polylactide (PLA). The role of aggregation and interphase was investigated with several experimental techniques including DSC, WAXS, SEM and tensile measurements. The experimental results clearly suggest that silica and montmorilonite have different reinforcing and toughnening effects on PLA, while the combination of the two different nanofillers seems to have a detrimental effect on the total interphase volume of the material. Four micromechanics models (Mori-Tanaka, Chen, Taya-Chu and Odegard), describing the Young's modulus of the nanocomposites, were used to study the different matrix-nanofiller interactions.

Keywords: nanocomposites, polylactide, interphase, aggregation, silica, montmorillonite, models
PACS: 82.35.Np, 82.35.Pq, 83.80, 81.05.Lg, 81.05.Qk

INTRODUCTION

The aim of this work is to study the effect of two different types of nanosized fillers at three different weight fractions as well their mixtures on the thermomechanical properties of polylactide (PLA). Silica nanoparticles, a commercial modified montmorillonite and their mixture were used for the preparation of PLA/Si, PLA/MMT and PLA/MMT/Si nanocomposites, respectively. The majority of previous works has been dealt with layered silicates based nanocomposites, while in the present work a comparative study of the thermomechanical properties between the two nanofiller types was made, to detect the different matrix-filler interactions, occurring between PLA and either silica or montmorillonite nanoparticles. The experimental study of the thermomechanical properties was supplemented with a theoretical study involving the application of four micromechanics models, describing the Young's modulus of the three series of nanocomposites for the interpretation of the matrix-nanofiller interactions.

MATERIALS-EXPERIMENTAL

Two different types of nano-fillers and their mixtures were introduced in pristine PLA: Silica Aerosil R972 (supplied by Degussa Chemicals) is hydrophobic fumed silica after treated with dimethyldichlorosilane based on hydrophilic fumed silica, with a specific surface area of 130 m^2/g. The average primary size is 16 nm. Montmorillonite clay is of type Nanomer I30 E by Nanocor modified by octadecylamine. Based on PLA and the two nanofiller-types, three series of PLA nanocomposites were prepared by melt mixing. The first two series consisted of nanocomposites with a filler content of 2, 3 and 5 wt.% of either silica or MMT; the third series contained nanocomposites with a filler content of 4 wt.% and a MMT/Si ratio of 60/40 and 40/60. The experimental techniques have been described in a previous work[1].

CP1255, *V[th] International Conference on Times of Polymers (TOP) and Composites,*
edited by A. D'Amore, D. Acierno, and L. Grassia

RESULTS AND DISCUSSION

From the DSC experimental results of Table 1, it is observed that the T_g of the PLA/Si materials is reduced slightly, while the T_g of the PLA/MMT ones is increased substantially with respect to the T_g of the pristine PLA material. In both series of nanocomposites the nanofiller content variation does not affect the T_g values of the nanocomposites examined. In the PLA/MMT/Si nanocomposites the T_g increases by increasing the MMT content.

TABLE 1. DSC results.

Sample type	**Tg (°C)**	**Heat of fusion, ΔH_m (J/g)**	**Melting point T_m(°C)**	**Percent crystallinity**	**Cold crystn (n heating) T_c (°C)**	**Crystn. (on cooling) T_{cc} (°C)**
PLA	52.0	19.5	153.4	20.9	110	94
PLA/Si/2	50.9	30.7	151.8	33.0	-	110 (broad)
PLA/Si/3	50.9	30.3	152.3	32.5	-	slightly broad
PLA/Si/5	50.7	29.2	150.2	31.3	-	94
PLA/MMT/2	57.9	20.7	152.0	22.2	97	93, 121 (bimodal)
PLA/MMT/3	57.9	29.1	151.7	31.3	96.5	119.3
PLA/MMT/5	57.9	25.7	152.9	27.6	97.6	90
PLA/MMT40/Si60	58.0	17.7	152.8	19.0	-	99.3
PLA/MMT60/Si40	60.0	19.6	153.4	21.1	-	100-140 (broad)

The T_g increases of the PLA/MMT nanocomposites have been related to the restricted segmental motions of the matrix within an interaction region around the montmorillonite platelets that hinders cooperative motion by raising the energy barrier for intermolecular chain movement[2,3]. On the other hand, the T_g reduction of the PLA/Si nanocomposites is attributed to the increment of polymer mobility[4,5]. In such a case, the interphase between matrix and nanofillers must be poor, due to the lack of connecting polymer from the nanoparticles to the surrounding polymer matrix.

The experimental fact that the T_g of the PLA/Si and PLA/MMT nanocomposites is not affected by the nanofiller content is in contradiction to the established prior art opinion[11] that an increase of the nanofiller content is expected to diminish the interparticle distances, and shift the T_g of the material to higher temperatures. A possible explanation to this discrepancy is the creation of aggregates in the matrix. The SEM images of all nanocomposites –see for example Fig. 3- display large aggregates the size of which increases with the nanofiller content. However, the increase of the nanofiller content does not lead necessarily to an extra increase of the total nanofiller surface area due to the creation of aggregates. The MMT nanoparticles have greater tendency to create aggregates than the silica nanospheres due to their larger surface are.

In Fig. 1 the PLA sample shows a broad cold crystallization (exotherm) area around 110 °C. When cold crystallization occurs, less perfect crystallites are formed, which melt during the DSC heating run. The PLA/Si nanocomposites do not display a cold crystallization area. The PLA/MMT nanocomposites exhibit a cold crystallization area shifted to lower temperatures, around 97 °C, while the total cold crystallization area increases with the MMT content -see also Table 1. The lowering of the cold crystallization temperature in respect to pristine PLA indicates that the incorporation of clay considerably promotes kinetics and extend of PLA crystallization on heating[12]. Apparently, the MMT nanofiller acts as an effective nucleating agent speeding up PLA crystallization. On the other hand the PLA/MMT/Si nanocomposites do not display cold crystallization and their thermal behavior is similar to pristine PLA –see Fig. 2.

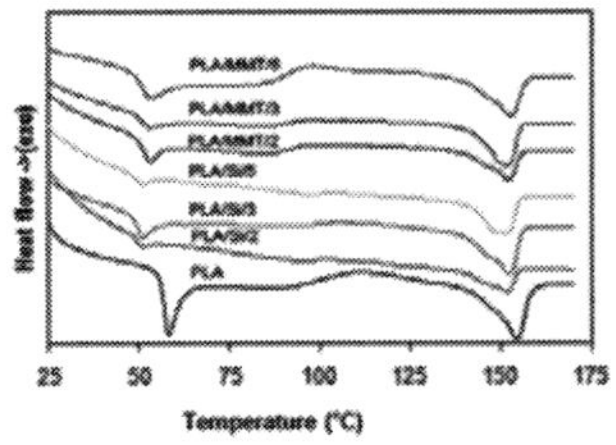

FIGUE 1. DSC thermograms of PLA nanocomposites (Setaram DSC 141 instrument; heating rate 10 ˚C).

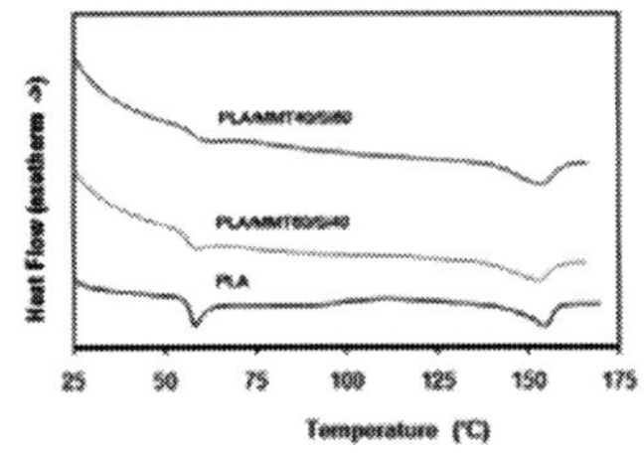

FIGURE 2. DSC thermograms of PLA/MMT/Si nano-composites (Setaram DSC 141 instrument; heating rate 10 ˚C).

Regarding the melting behavior, the PLA sample shows a melting point at about 153 ˚C and a heat of fusion of 19.50 J/g. The incorporation of silica nanoparticles or MMT clays affects slightly the melting point of the materials, but leads to a substantial increment of heat of fusion, as shown in Table 1. That means that both nanofiller types act as nucleating agents, with the silica nanoparticles leading to a relatively higher crystallinity content.

Tensile results

The tensile results of pristine PLA and PLA nanocomposites are presented in Fig. 4, 5 and 6 and summarized in Table 2. Regarding the PLA/Si nancomposites, shown in Fig. 4, the highest modulus is obtained at a silica content of 2 wt.% (37.4% increment), while the ductility is retained close to that of the pure matrix. The PLA/MMT nanocomposites, shown in Fig. 5, exhibit a somehow different behavior. The highest modulus reinforcement occurs at 3 wt.% (32.6% increment), while the PLA/MMT nanocomposites seem the be more brittle than the PLA matrix and the PLA/Si/3. These results are in consistency with the DSC data where the two specific nanocomposites with the highest modulus appear to a have a maximum crystallinity content of 30.79% and 29.1%, respectively, and given that the Young's modulus in a semicrystalline material depends on the crystallinity content.

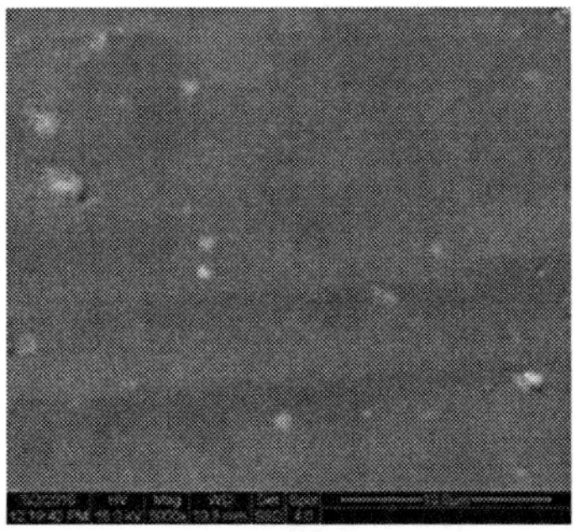

Figure 3. SEM micrograph of PLA/MMT/3 nanocomposite

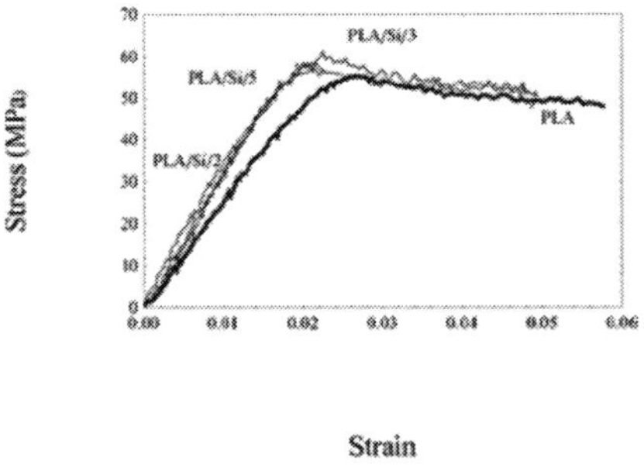

Figure 4. Tensile results of PLA/Si nanocomposites (Instron 1121 tester).

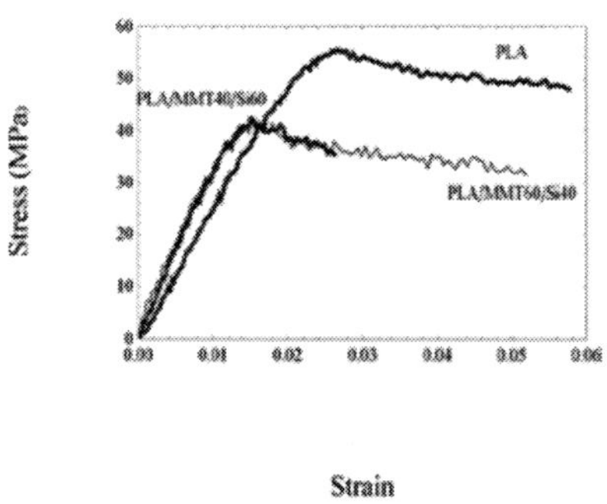

Figure 5. Tensile results of PLA/MMT nanocomposites (Instron 1121 tester).

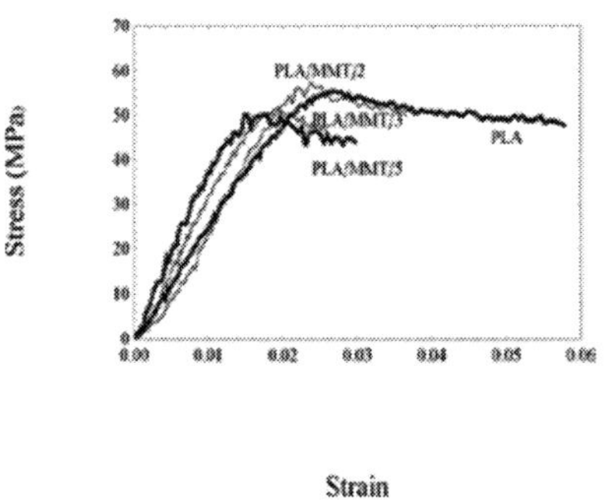

Figure 6. Tensile results of PLA/MMT/Si nanocomposites (Instron 1121 tester).

Table 2. Tensile results and micromechanics models.

Sample type	E_c/E_m (experimental)	Mori-Tanaka model	Chen model	Taya-Chou model	Odegard model	Young's modulus (MPa)	Yield stress (MPa)	Tensile strength (MPa)	Strain at break
PLA	1.00	-	-	-	-	2800	55.2	47.8	0.056
PLA/Si/ 2	1.37	1.02	1.02	1.02	1.36	3846	57.7	51.8	0.050
PLA/Si/ 3	1.18	1.03	1.03	1.03	1.18	3300	61.0	53.0	0.048
PLA/Si /5	1.23	1.06	1.06	1.05	1.20	3436	57.0	55.0	0.023
PLA/MMT/2	1.14	1.02	1.02	1.02	1.17	3200	56.0	51.0	0.037
PLA/MMT/3	1.33	1.03	1.03	1.03	1.33	3713	52.0	48.0	0.027
PLA/MMT/5	1.30	1.06	1.06	1.05	1.31	3650	51.0	48.0	0.030
PLA/MMT60/Si40	1.18	1.04	1.04	1.04	1.18	3300	41.5	32.0	0.052
PLA/MMT40/Si60	1.18	1.04	1.04	1.04	1.18	3323	41.6	39.0	0.026

The modulus of the PLA/Si nanocomposites changes slightly with the silica content, while the modulus of the PLA/MMT nanocomposite increases with the MMT content. WAXS and SEM revealed the presence of intercalated, aggregated and partially exfoliated structures. At a MMT content of 5 wt.% the nanocomposite becomes more brittle and its tensile strength drops, apparently due to the formation of agglomerates. Both PLA/MMT/Si nancomposites exhibit the same modulus which is higher than the modulus of pristine PLA. Both PLA/MMT/Si nancomposites exhibit the same yield stress, which is lower than the yield stress of pristine PLA. Apparently, the mixture of nanofillers disrupts the interphase region and promotes debonding. In a system where the interfacial adhesion is not high, debonding could occur at a lower tensile stress than the yield stress of pristine PLA[6,7]. The PLA/MMT60/Si40 is more ductile than the PLA/MMT40/Si60 nanocomposite. On the basis of the experimental and theoretical results, it is postulated that the aggregation effect acts antagonistically to the interphase effect. The increase of the nanofiller content above a certain amount is not accompanied by a corresponding increase of the interphase because the total nanofiller surface area is countered balanced by the creation of aggregates which compromise the expected properties improvement.

Modeling

In Table 2 the experimental values of relative Young's modulus i.e. the composite's modulus over the modulus of PLA matrix (E_c/E_m), for the various studied material types were compared with the corresponding theoretical values predicted by four well known micromechanics models (Mori-Tanaka, Chen, Taya-Chu and Odegard[8]). From these results it becomes obvious that the first three models introduced for conventional composite materials are not able to describe adequately the reinforcement induced by the nanofillers, given that the filler content is very low (up to 5 wt.%). On other hand, when the specific region developed around nanoparticles, the so-called "interphase", is taken into account according to the Odegard model, the reinforcement can be successfully modeled. In all material types the interphase modulus E_i was fitted to be equal to 7 GPa, while the interphase volume fraction of the PLA/Si and PLA/MMT nanocomposites was varied from 0.2 up to 0.55. The interphase volume fraction of the PLA/MMT/Si nanocompsites was 0.18. The reduced total interphase volume is in agreement with the corresponding tensile results - see Fig. 6- meaning that the combination of the two different nanofillers has a detrimental effect on the properties of the material.

REFERENCES

1. E. Kontou and G. Anthoulis, *J. Appl. Polym. Sci.*, **105**, 1723-1731 ((2007).
2. Q. Zhou and M. Xanthos, *Polym. Degrad. Stab.* **93**, 1450-1459 (2008).
3. Ray S. Sinna and M. Makhatha, *Polymer* **50**, 4635-4643 (2009).
4. B. Ash, R. Siegel and L. Schadler, *J. Polym.Sc. Part B, Polym. Phys.* **42**, 4371-4383 (2004).
5. D. Zax, D. Yang, R. Santos, H. Hegermann, E. Giannelis and E. Manias, *J. Chem. Phys.* **112**, 2945-2951 (2000).
6. L. Jiang, J. Zhang and M. Wolcott, *Polymer* **48**, 7632-7644 (2007).
7. W. Zuiderduin, C. Westzaan, J. Huetink and E. Gaymans, *Polymer* **44**, 261-275 (2003).
8. G.M. Odegard, T. Clancy T.C. and T.S. Gates *Polymer* **46**, 553-562 (2005).

Intermolecular Effects in the Dynamics of Polymer Melts: Interplay of Cooperative Dynamics and Entanglements

Marina G. Guenza

Department of Chemistry and Institute of Theoretical Science, University of Oregon, Eugene, OR 97403 USA

Abstract. The generalized Langevin equation for cooperative dynamics is a microscopic, system-specific approach that describes the motion of macromolecules in a melt. The approach is formally derived from the first principle Liouville equation by applying projection operator techniques. Unlike other approaches for polymer melt dynamics, it explicitly includes the forces generated by the presence of intermolecular correlations. Comparison with simulations and experiments shows a quantitative agreement for the mean-square displacement and the dynamic structure factor across the transition from unentangled to entangled dynamics.

Keywords: polymer dynamics, subdiffusive behavior, entanglements
PACS: 61.25.Em; 61.25.H-; 61.25.hk

A MANY-CHAIN APPROACH TO POLYMER MELT DYNAMICS

The dynamics of polymer melts is characterized by the complex interplay between chain connectivity and intermolecular interactions. Entanglements, which constrain chain dynamics, are the manifestation of the uncrossability of chains due to the hard-core repulsions between monomers belonging to different macromolecules. The physics of entanglements is ultimately a many-body problem, where intermolecular interactions play a dominant role. Because of the difficulty of solving many-body problems, one has to resort to mean-field descriptions. For example, the "reptation" model is a single chain mean-field theory where an entangled polymer moves inside an effective tube formed by the surrounding chains. The reptation model is fully phenomenological, and represents correctly the scaling behavior of long chains moving in a field of fixed obstacles. Its description is, however, not quantitative, nor microscopic. Chain length fluctuations and tube relaxation have been included a posteriori in the model to improve agreement with experiments. However, the theory is still unable to reproduce quantitatively and consistently all the available experimental results.[1]

Our approach, a generalized Langevin equation (GLE) for cooperative dynamics, is a mean-field description where a group of chains move in the field of the surrounding molecules.[2,3] Macromolecules interact through the intermolecular pair potential, and are represented at the atomistic level through a freely rotating chain type of structural model. Because the intermolecular interactions are explicitly included in the theory, it is straightforward to account for chain uncrossability, which is the origin of entanglements.

Our many-chain approach is motivated by an analysis of simulations, which show that polymer melts display interconverting regions of fast and slow dynamics. These dynamical heterogeneities vanish at a characteristic time, τ_{decorr}, which corresponds to the time necessary for a molecule to diffuse outside the range of the intermolecular potential. The observed separation of timescales between slow and fast dynamics allows one to use projection operator techniques and isolate from the full dynamical equation of the liquid, i.e. the Liouville equation, the dynamics of the slow molecules in the field of the fast ones. Each slow molecule, either entangled or unentangled, is dynamically correlated to any molecule that happens to be at a distance shorter than the range of the intermolecular potential, and uncorrelated with any molecule situated at a distance larger than the same. The correlation vanishes when two molecules interdiffuse a distance larger than the range of the potential, which is of the order of the molecular dimension, i.e. the radius-of-gyration R_g.

An analytical expression for the potential acting between the centers-of-mass of a pair of polymers was derived recently.[4,5] Because of its Gaussian form, it allows for a formal solution of the set of coupled Langevin equations

CP1255, *Vth International Conference on Times of Polymers (TOP) and Composites,*
edited by A. D'Amore, D. Acierno, and L. Grassia

through normal modes transformation. Time correlation functions are expressed as a function of eigenvalues and eigenvectors of the two matrices describing intra- and inter-molecular potentials. Full dynamical behavior is predicted as a function of the thermodynamic parameters (temperature, T, and liquid density, ρ) and the molecular structure (degree of polymerization, N, polymer semiflexibility, i.e. the freely-rotating-chain parameter g, and monomer friction coefficient, ζ). Any dynamical quantity of interest, including the dynamic structure factor, is easily calculated once the experimental parameters are known.

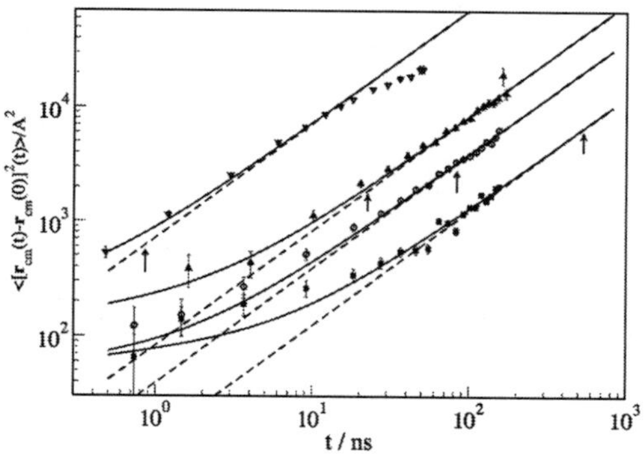

FIGURE 1. Center-of-mass mean square displacement, extracted from the data at q=0.3 nm^{-1}. Symbols: experimental data for N=36 (down triangles), N=106 (up triangles), N=192 (open circles), N=377 (filled squares). Vertical arrows: decorrelation time, τ_{decorr}. Full lines: cooperative dynamics generalized Langevin equation. Dashed lines: free diffusion. Reprinted with permission from Ref.6. Copyright 2008.

In recent papers, the GLE for cooperative dynamics has been tested against simulation data of polymer melts, obtaining quantitative agreement for physical quantities such as the mean-square displacement of monomer and center-of-mass.[3] The theory also predicts for the static correlation of the normal modes, $<x_p(0)^2>$, at fixed N, a change in the scaling with mode index p. For flexible systems, $g=0$, the theory recovers the Rouse scaling of p^{-2}, while for stiff polymers, $g\rightarrow 1$, the high-index modes appear to follow a p^{-3} scaling in agreement with simulations and experiments.

Recently, we compared theoretical predictions for polyethylene chain dynamics with data of Neutron Spin Echo measurements, performed by Richter and coworkers, and Farago and coworkers.[6] Samples have degree-of-polymerizations (N=36, 102, 192, 377) that cross the transition from unentangled to entangled dynamics, as N_e=130. Parameters are reported in Table 1. For polyethylene the degree-of-flexibility is $g=0.785$. Thermodynamic parameters include the temperature $T= 509\ K$, and density $\rho=0.733\ g/cm^3$. The comparison of our theory with Neutron Spin Echo measurements covers an extended time range, and fully samples the short-time window where subdiffusive behavior appears as well as the long time diffusive regime. Figure 1 displays the agreement between theoretical predictions (lines) and data extracted from experiments (symbols) for the different molecular weights. Dashed lines represents purely Brownian diffusion. We also report as vertical arrows the values of the calculated decorrelation time, τ_{decorr}. All samples display anomalous subdiffusive dynamics at short time and the transition to normal (Brownian) diffusion at long times. The measured long-time diffusion coefficients are found to be in quantitative agreement with previous measurements obtained with different experimental methods.[6] For each sample we extract from the long-time diffusive regime the single-chain diffusion coefficient, ζ_{eff}., reported in Table 1. Because the samples cover the crossover from unentagled to entangled dynamics, the friction coefficients are molecular weight dependent.

We also report direct comparison with NSE data of dynamic structure factors, at different lengthscales. Figure 2 shows that theoretical predictions are in good agreement with the experiments in the whole range of timescales investigated and for two different molecular weights. The only adjustable parameter in the theory is the number of chains undergoing cooperative dynamics. The theory is quite sensitive to this value, as is shown in the left panel of Fig.2 where curves for two different values of n are displayed. The optimized number of chains undergoing cooperative dynamics is reported in Table 1 and roughly scales as the square root of the degree of polymerization, as predicted by the theory where $n\propto\rho N^{1/2}l$.

In conclusion, we find that the observed quantitative agreement between our theory and experiments is encouraging. The experimental data presented here clearly displays a subdiffusive center-of-mass regime at time scales shorter than the characteristic decorrelation time, at which the single chain dynamics becomes subdiffusive. Conventional approaches to polymer melt dynamics, including Rouse theory and the reptation model, are mean-field theories of a single chain and predict free diffusion in the short time regime. Our equation relates the observed subdiffusive behavior to the presence of dynamical heterogeneities in the liquid. Subdiffusive dynamics is due to the presence of dynamical correlation between polymers, when they are at a distance smaller than the range of their intermolecular potentials.

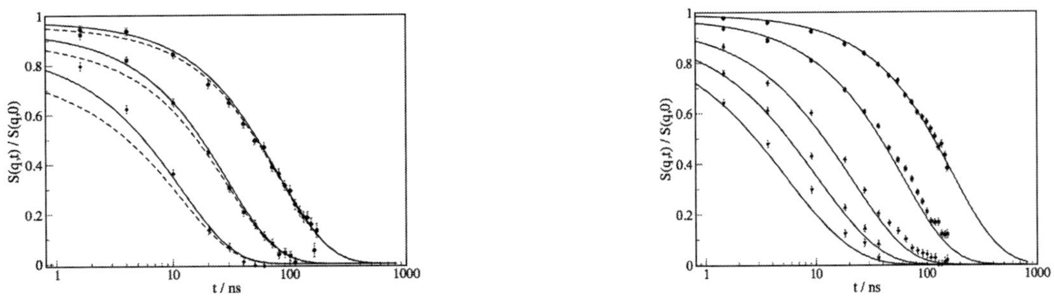

FIGURE 2. Dynamic structure factor for N=106 (left panel) and N=192 (right panel). Symbols: NSE data for q=0.3 nm^{-1} (circles), q=0.5 nm^{-1} (squares), q=0.77 nm^{-1} (diamonds) and q=0.96 nm^{-1} (triangles). Full lines: theoretical prediction of the many-chain cooperative dynamics approach (CDGLE). Left panel: N=106, CDGLE with n=4 (full lines) and n=3 (dashed lines). Right panel: N=192 with n=9 (full lines). Reprinted with permission from Ref. 6. Copyright 2008.

We notice that the microscopic nature of our dynamic equation allows us to represent the specific monomeric structure of the polymer investigated. Data in this paper are for polyethylene chains; however, our approach is general and has been successfully applied to the dynamics of chains with different monomeric structures.[7]

Finally we observe that the explicit presence of theintermolecular potential in our equation of motion allows us to directly evaluate the effect of chain uncrossability and entanglements on the single chain dynamics, as well as to evaluate the relative importance of intermolecular vs intramolecular contributions to the polymer dynamics.

TABLE 1. Number of segments *N*, diffusion coefficient *D*, effective monomer friction coefficient ζ_{eff}, square radius-of-gyration R_g^2, fitted number of correlated chains *n*, and decorrelation time $\tau_{decorr}=R_g^2/D$

N	*D* [A^2/ns]	ζ_{eff} *[dyne ns/A]*	R_g^2 [A^2]	*n*	$\tau_{decorr}=R_g^2/D$ [ns]
36	119.7	$0.0163 \cdot 10^{-7}$	102	2	0.85
106	13.77	$0.048 \cdot 10^{-7}$	300	4	21.79
192	6.5	$0.056 \cdot 10^{-7}$	544	9	83.69

ACKNOWLEDGMENTS

We acknowledge the support of the National Science Foundation.

REFERENCES

1. R. G. Larson, *Rheology of Complex Fluids*, Oxford University Press, New York, 1999.
2. M. Guenza, *Phys. Rev. Lett.* **88**, 25901-25904 (2002).
3. M. Guenza, *J. Chem. Phys.* **119**, 7568 (2003).
4. G. Yatsenko, E. J. Sambriski, M. A. Nemirovskaya and M. G. Guenza, *Phys. Rev. Lett.* **93**, 257803-257806 (2004).
5. E. J. Sambriski, G. Yatsenko, M. A. Nemirovskaya and M. G. Guenza, *J. Chem. Phys.* **125**, 2349021-23490212 (2006).
6. M. Zamponi, A. Wischnewski, M. Monkenbusch, L. Willner, D. Richter, P. Falus, B. Farago, and M. G. Guenza *J. Phys. Chem. B* **112**, 16220-162229 (2008).
7. M. G. Guenza *J. Phys.: Cond. Matt.* **20**, 033101-0331024 (2008).

Luminescent nanocomposites of conducting polymers and *in-situ* grown CdS quantum dots

***C. Borriello,*[1] *S. Masala,*[1] *V. Bizzarro,*[2] *G. Nenna,*[1] *M. Re,*[3] *E. Pesce,*[3] *C. Minarini,*[1] *and T. Di Luccio*[1]**

[1]ENEA, UTTP NANO C. R. Portici, piazzale E. Fermi, 80055 Portici (NA), Italy
[2]IMAST , piazzale E.Fermi, 80055 Portici (NA), Italy
[3] ENEA MATCOMP, C. R. Brindisi SS7 Appia Km 713, 72100 Brindisi, Italy
e-mail: carmela.borriello@ enea.it

Abstract. Luminescent PVK:CdS and P3HT:CdS nanocomposites with enhanced electrooptical properties have been synthesized. The nucleation and growth of CdS nanoparticles have been obtained by the thermolysis of a single Cd and S precursor dispersed in the polymers. The size distribution and morphology of the nanoparticles have been studied by TEM analyses. Monodispersive and very small nanoparticles of diameter below 3 nm in PVK and 2 nm in P3HT, have been obtained. The application of such nanocomposites as emitting layers in OLED devices is discussed.

Keywords: Nanocomposites, CdS nanoparticles, conducting polymers, electroluminescence
PACS: 82.35Np, 81.07b, 78.67Sc

INTRODUCTION

Nanocomposites obtained by combining conducting polymers and semiconductor nanoparticles have so far attracted much interest for manifold applications such as tuneable luminous sources for field emission displays and electroluminescent devices [1]. Nanocomposites have several advantages with respect to fully organic or fully inorganic systems [2]. As an example, polymers embedding semiconducting nanoparticles combine their good processability, high transparency, digital printing flexibility also on a large scale with the unique optical properties of the nanoparticles. In particular the incorporation of II–VI nanocrystals (or quantum dots) in conjugated polymers has been found to enhance the electroluminescence properties of these materials [3] and to improve the long term stability, colour tunability and quantum efficiencies of electroluminescent devices based on such nanocomposites.

Conventionally nanocomposites of nanocrystals and conjugated polymers are obtained by mixing a given polymer with nanocrystals in presence of surfactants to prevent the growth of larger crystals. However the presence of surface ligands may reduce the charge transfer between the polymer matrix and the nanocrystals [4].

This work presents the realization of conducting and luminescent hybrid nanocomposite materials by nucleation and growth of CdS nanoparticles directly in two organic polymers, Polyvinylcarbazole (PVK) and Poly (3-Hexylthiophene) (P3HT). The purpose of our research is to improve the optoelectronic properties of the polymers, by enhancing the luminescence intensity or tuning the emission wavelength. CdS nanoparticles have been synthesised in-situ within the polymers by thermolysis of a cadmium bis(thiolate) precursor avoiding the use of any surfactant or other stabilizing agents.

The size of CdS nanoparticles was controlled by varying the annealing conditions. The nanoparticle morphology and dimensions were investigated by TEM analyses.

The P3HT based nanocomposites have been used for the realization of a single layer OLED device with structure ITO/P3HT:CdS/LiF/Al which transport and emission properties have been compared to the analogous pure polymer device.

CP1255, *Vth International Conference on Times of Polymers (TOP) and Composites,*
edited by A. D'Amore, D. Acierno, and L. Grassia

RESULTS AND DISCUSSION

$Cd(SC_{12}H_{25})_2$ cadmium dodecylthiolate has been used as Cd and S precursor. The polymers P3HT and PVK have been dissolved in chlorobenzene at a concentration of 25mg/mL. The precursor has been successively added to this solution at 20% wt. A bulk foil has been obtained by casting and annealed at different temperatures under vacuum.

Previous studies on the nucleation and growth of CdS nanoparticles by thermolysis have shown that the formation of the nanoparticles starts at temperatures of about 170°C in a thermoplastic copolymer [5]. However the choice of the annealing temperatures depends also on the glass transition of the polymers. Therefore the annealing processes have been conducted at temperatures higher than their Tg to allow a better diffusion and distribution of the Cd dodecylthiolate in the polymeric matrix. The annealing in PVK (Tg=226°C) have been carried on at 250, 265 and 275 °C while the annealing in P3HT (Tg<-3°C) have been carried on at 200 and 240 °C.

TEM analyses of the samples, realised by a TEM TECNAI G2 F30, have confirmed the formation of CdS nanoparticles. The samples for TEM have been prepared by dropping few μL of the precursor polymer solution on Cu grids. Spherical and very small nanoparticles are present in both polymers, as shown in Fig. 1a for P3HT:CdS. The HRTEM image in the inset shows regular lattice fringes at a distance of about 0.31 nm. Statistical analysis over about 100 nanoparticles (see istogram in Fig. 1b) has given a narrow nanoparticles size distribution peaked at about 2 nm for the nanocomposite annealed at 240 °C. The results obtained in the case of PVK:CdS indicate nanoparticles of about 3 and 5 nm upon annealing at 265°C and at 275°C, respectively.

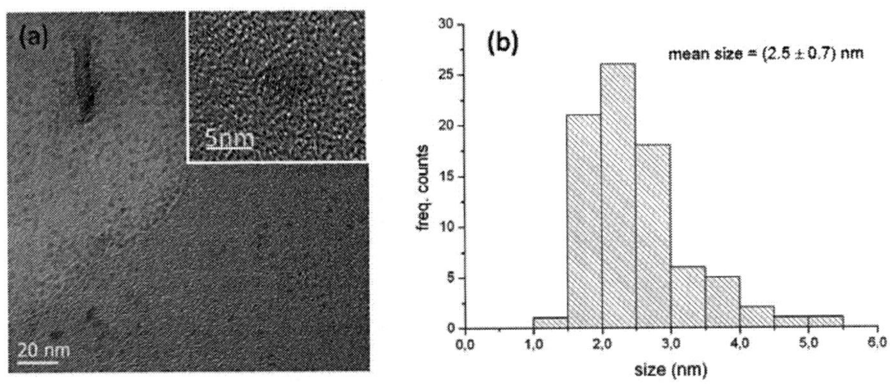

FIGURE 1. (a) Bright Field TEM image of CdS nanocrystals in P3HT matrix after annealing at 240 °C; (b) Statistical distribution of the nanoparticle size evaluated from sampling about 100 nanoparticles.

The P3HT:CdS prepared at 240°C of Fig. 1 have been employed to prepare simple OLED devices ITO/P3HT:CdS/LiF/Al. The nanocomposite solution was spin coated (2000 rpm) on a patterned ITO/glass substrate used as anode electrode and covered by a LiF/Al electrode (100 nm thick) obtained by evaporation. Current-voltage characteristics (Fig. 2a) and electroluminescence at V ~ 8.5Volts (Fig. 2b) are reported for the device including the nanoparticles and compared with the device of P3HT only. The current-voltage characteristics show a decrease of the turn-on voltage from the pure polymer (5.8V) to the P3HT:CdS (5.4V). The electroluminescence of P3HT is characterized by a double peak between 550 and 800 nm due to π-π* transitions [6]. The introduction of the CdS nanoparticles induces a strong enhancement of the electroluminescence signal with loss of the double structure of the P3HT electroluminescence signal. The study of the nanoparticles influence on the electroluminescence mechanism of P3HT:CdS in currently under study together with possible effects of the annealing process on the morphology and transport properties of P3HT [7]. To this scope, additional devices will be realised using P3HT:CdS at 200°C and the PVK related nanocomposites.

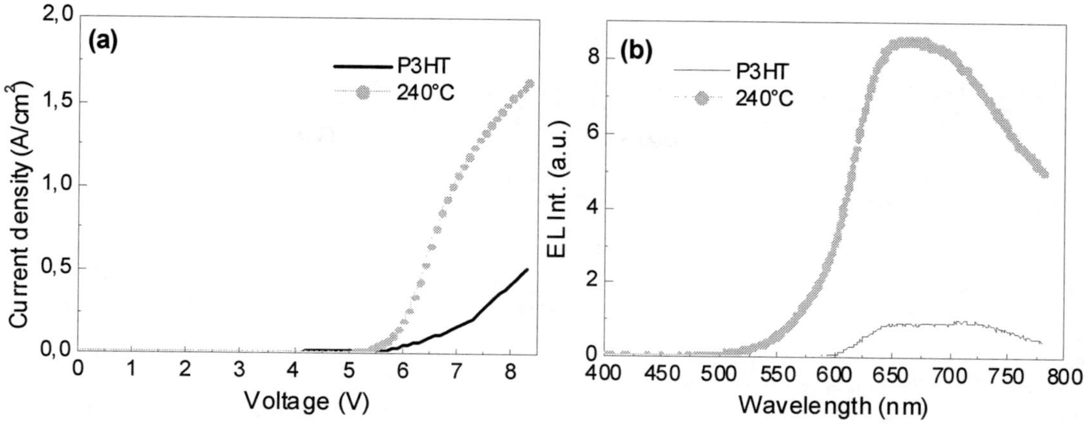

FIGURE 2. Current-voltage characteristics (a) and electroluminescence at V~8.5Volts (b) on ITO/P3HT/LiF/Al and ITO/P3HT:CdS/LiF/Al where P3HT:CdS has been obtained after annealing at 240°C.

CONCLUSIONS

Luminescent PVK and P3HT nanocomposites containing CdS nanoparticles have been synthesized by in-situ thermolisis of a cadmium bis(tiolate). This synthetic approach allows the use of a single, air stable and easy to prepare precursor for both Cd and S. Moreover the nucleation and growth of CdS nanoparticles directly in the polymer avoids the use of surfactants. The CdS nanoparticles obtained are very small (few nm) and present a narrow size distribution in both polymers as shown by TEM analysis of the samples.

P3HT nanocomposites have been used to prepare simple OLED devices ITO/P3HT:CdS/LiF/Al whose electric and electrooptical properties have been measured. A strong increase of electrooptical signal has been obtained in the case of the nanocomposite at 240°C with respect to pure P3HT.

ACKNOWLEDGMENTS

A..M. Laera from ENEA C.R. Brindisi is aknowledged for providing the precursor.

REFERENCES

1. N. Shipway, E. Katz, and I. Willner, *Chemphyschem* Wiley-VCH-Verlag, Weinheim 2000, **1**, pp 18.
2. B.R. Saunders, M.L. Turner Advances in Colloid and Interface Science, **138**, pp 1–23 (2008)
3. S. Coe-Sullivan, W. K. Woo, J. S. Steckel,Moungi Bawendi , V. Bulovic, *Organic Electronics* **4** (2003)
4. C. B. Murray; D. J. Norris. M. G. Bawendi, *J. Am. Chem. Soc.* **115**, pp 8706, (1993); J. Ouyang, J. Kuijper, S. Brot, D. Kingston, X. D.M. Wu, Leek, M.Z. Hu, J.A. Ripmeester, K. Yu *J. Phys. Chem. C* **113**, pp. 7579–7593.
5. T. Di Luccio, A.M. Laera; L. Tapfer, S. Kempter, R. Kraus, B. Nickel, *J. Phys. Chem. B*, **110** pp 12603-12609 (2006).
6. A. Ruseckas, E. B. Namdas, T. Ganguly, M. Theander, M. Svensson, M. R. Andersson, O. Inganas and V. Sundstrom, *J. Phys. Chem. B*, **105**, pp 7624–7631 (2001).
7. D. Liu, M. Fina, L. Ren, S. S. Mao, *Applied Physics A: Materials Science & Processing*, **96**, pp. 7 (2009).

iPP Crystallization: Micro and Nano Fillers Effects

Emilia Gioffredi[a], Gabriele Cassulo[a], Alberto Frache[a] and Pier Luca Maffettone[b]

[a]*Politecnico di Torino – Sede di Alessandria, V.le T. Michel, 5, 15121 Alessandria, Italy*
[b]*Università degli Studi di Napoli 'Federico II', P.le V. Tecchio 80, 80125 Napoli, Italy*

Abstract. Since the properties of semicrystalline polymers depend on the morphology, studies on effect of fillers on the composites crystallization are of great interest. In this work, a micrometric talc and a nanoclay are dispersed in a polypropylene matrix and the influence of two different fillers and temperature on the polymer crystallization behavior in quiescent conditions is assessed by means of differential scanning calorimetry (DSC) and rheological characterization through linear viscoelasticity (SAOS) [1]. The DSC tests lead to the half crystallization time, depending on overall crystallization rate (i.e. nucleation and growth) whereas by rheological measurement one can deduce also the induction time, depending only on nucleation [2].

Keywords: Nanocomposites, Crystallization, Linear Viscoelasticity, Polypropylene.
PACS: 81; 83

INTRODUCTION

Nanocomposites are a new class of composites, for which at least one dimension of the dispersed particles is in the nanometer range. Fillers are incorporated into polymeric materials in order to improve their qualities or to produce new chemical and physical properties. Polymer-layered silicate nanocomposites have been widely investigated because the starting clay materials are easily available and because their intercalation chemistry has been studied for a long time [3].

MATERIALS AND METHODS

A commercial isotactic polypropylene (Moplen HP400R, Basell, iPP) was melt mixed with maleic anhydride grafted polypropylene as compatibilizer (Polybond 3200, Crompton Corp., PPgMA) and organoclay (Cloisite20A, Southern Clay Products, Clo) using a twin screw microextruder. Three systems were prepared by increasing the organoclay loading at fixed iPP/PPgMA ratio (13.3): 2wt%Clo (B2), 3.5wt%Clo (B3.5) and 5wt%Clo (B5). The microcomposite is prepared in the same conditions of nanocomposites using Talc (HTP4, Imifabi, talc) at 2wt%talc (Bt2). The reference matrix is the blend iPP + 7wt%PPgMA (matrix).

Structural characterization was carried out by means of X-ray diffraction (XRD) and scanning electron microscopy (SEM) analyses.

Material behavior in the molten state was studied conducting rheological measurements in small-amplitude oscillatory shear with rotational rheometer.

DSC and SAOS were employed to investigate the crystallization behavior in quiescent conditions of the prepared micro and nano composites.

RESULTS AND DISCUSSION

XRD patterns of blends displayed a peak position small shift with respect to neat Cloisite20A revealing a low level of intercalation of iPP into the interlayer space. The peak position of Bt2 is unchanged compared with that one of neat talc demonstrating no intercalation of iPP into the interlayer space.

CP1255, *Vth International Conference on Times of Polymers (TOP) and Composites,*
edited by A. D'Amore, D. Acierno, and L. Grassia

SEM images show the presence of some clay aggregates of micrometric dimensions suggesting a poor clay dispersion within the polymer matrix. In the case of blend with talc one can see only microaggregates.

During the SAOS tests, the linear viscoelastic range of the sample must not be exceeded to avoid structure modification. Thus, it is critical to determine the linear viscoelastic range by applying a strain sweep before testing the viscoelasticity of the polymer and the composites under a frequency sweep (in order to obtain the corresponding curves for the storage modulus G', loss modulus G" and complex viscosity η^*). All data shown in this paper are got in the linear viscoelastic range.

The effect of clay addition is a general increase of moduli at all frequencies. As the particle loading increases, elastic and also loss moduli increase. The change is more evident for G': a filler network is generated by the strong interactions (steric or Van der Waals interactions) among silica layers or silica particles [4]. The network appears already in the blend at smaller filler percentage (B2).

The effect of talc addition is a general rise of moduli and viscosity. The filler network doesn't appear.

Polymer Crystallization with DSC and SAOS Measurements

DSC results allow to evaluate the half crystallization time ($t_{1/2}$), defined as the time required to reach 50% of the complete crystallization (depending on overall crystallization rate, i.e. nucleation and growth), the melting temperature and the crystallinity percentage of matrix and blends.

$t_{1/2}$ increases with increasing the crystallization temperature (T_C) for all samples, but, as shows in Table 1, the presence of clay causes an improvement of crystallization rate at the same T_C. Increasing the filler concentration seems to produce a small increase of crystallization rate, especially at lower temperatures. The addition of talc determines a significant rise of the crystallization rate, an effect much stronger than that due to clay. So Cloisite20A can act as an heterogeneous nucleation agent for iPP crystallization, but traditional nucleation agent (i.e. talc) appears to be more efficient in this respect.

TABLE (1). Half Crystallization Time Proportional Decrease of Blend Compared with Matrix.

Materials	$t_{1/2}$ @ T_C=130°C	$t_{1/2}$ @ T_C=132°C	$t_{1/2}$ @ T_C=134°C	$t_{1/2}$ @ T_C=136°C	$t_{1/2}$ @ T_C=138°C
B2	-29.7%	-18.4%	-31.2%	-20.1%	-4.4%
B3.5	-35.2%	-8.3%	-30.8%	-12.7%	-12.7%
B5	-27.4%	-24.3%	-33.0%	-29.4%	-33.8%
Bt2	-87.3%	-86.3%	-89.4%	-88.2%	-89.0%

No significant differences are detected in the melting temperatures of micro and nano composites. The crystallinity percentage for all the specimens crystallized at the same T_C is nearly constant. This observation suggests that the addition of filler has no obvious effect on the crystallinity of matrix.

In the recent years, melt rheometry has been proved to be a reliable technique to study the crystallization kinetics of polymer materials [2, 5] by means of dynamic time sweep tests.

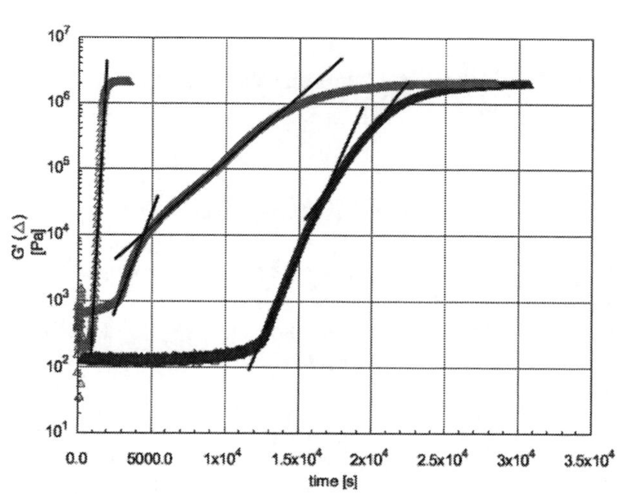

FIGURE 1. Isothermal Crystallization of Matrix (Dark Blue Line), B2 (Green Line) and Bt2 (Red Line) @ T_C=146°C.

SAOS results obtained during isothermal crystallization of samples in the rheometer are shown in Figure 1 for the Matrix, the micro-composite (Bt2) and one of the nano-composites (B2).

In this plot one can clearly observe a different evolution of storage modulus versus time during crystallization. A plateau of G' at very short times is followed by a monotonic increase after a few minutes indicating the induction stage of nucleation. Then, the storage modulus abruptly increases with time during the crystallites growth, and the homogeneous melt system changes into the heterogeneous system. Finally, G' shows a second plateau at long time when the system reaches an equilibrium and the crystallization process is over [6].

For the matrix and the micro-composite the trend is sigmoidal. For the nano-composites the trend is again sigmoidal, but the shape is rather irregular. This behavior is probably due to the coexistence in the nano-composites of some filler microaggregates and a global network. During early stages, the microaggregates act like nucleation agents (initial steep slope of the curve), while afterwards, in the phase of crystallites growth, this action is contrasted by the network.

Both half crystallization time (calculated with a relation proposed in [2]) and induction time (t_{ind}, defined as the time required to induce nucleation and depending only on nucleation, located at the end of first plateau) are obtained. Both parameters increase with increasing the crystallization temperature (T_C) for all samples, but the presence of clay causes an improvement of crystallization rate.

The times proportional decrease is shown in Tables 2 ad 3. An addition of filler seems to produce an increase of crystallization rate. Again, the effect on the crystallization rate of talc addition is more pronounced with respect to that due to clays.

TABLE (2). Half Crystallization Time Proportional Decrease of Blend Compared with Matrix.

Materials	$t_{1/2}$ **@ T_C=140°C**	$t_{1/2}$ **@ T_C=144°C**	$t_{1/2}$ **@ T_C=146°C**
B2	-47.5%	-35.2%	-30.7%
B3.5	-53.7%	-61.4%	-53.7%
B5	-75.7%	-70.5%	-67.4%
Bt2	-98.9%	-93.8%	-93.5%

It is possible to observe that t_{ind} decreases with the filler presence, and this fall is stronger than that observed for $t_{1/2}$.

TABLE (3). Induction Time Proportional Decrease of Blend Compared with Matrix.

Materials	t_{ind} **@ T_C=140°C**	t_{ind} **@ T_C=144°C**	t_{ind} **@ T_C=146°C**
B2	-85.8%	-93.6%	-95.3%
B3.5	-96.0%	-96.3%	-97.1%
B5	-97.3%	-97.5%	-97.9%
Bt2	-98.3%	-98.6%	-97.9%

The above findings suggest that the addition of filler has a large effect on the induction time while the effect on the half crystallization time is less pronounced.

Therefore, rheological characterization of iPP crystallization confirms DSC data (Cloisite20A can act as a weak heterogeneous nucleation agent), and, furthermore, suggests that clay addition mainly promotes nucleation but has scarce influence on the growth of iPP crystallites.

REFERENCES

1. D. Nichetti and N. Grizzuti, *Polymer Engineering and Science* **44**, 1514-1521 (2004).
2. Y.P. Khanna, *Macromolecules* **26**, 3639-3643 (1993).
3. S. Pavlidou and C.D. Papaspyrides, *Progress in Polymer Science* **33**, 1119-1198 (2008).
4. M.J. Solomon et al., *Macromolecules* **34**, 1864-1872 (2001).
5. S. Acierno et al., *Macromolecules* **35**, 5043-5048 (2002).
6. D.W. Chae et al., *Polymer* **47**, 3609-3615 (2006).

A Comparative Study of Fiber Break-up During Compounding of Glass Fiber-Reinforced Polyamide in Batch and Continuous Mixers

Funda Inceoglu [a], Julien Ville [b], Nessim Ghamri [a], Audrey Durin [c], Pascal De Micheli [c], Rudy Valette [c], and Bruno Vergnes [c]

[a] *Centre de Recherche, ARKEMA, CERDATO, 27 470 Serquigny (France)*
[b] *POLYTECHS, B.P. 14 - 76 450 Cany-Barville (France)*
[c] *MINES ParisTech, CEMEF, UMR CNRS 7635 BP 207, 06904 Sophia Antipolice, France*

Abstract. Glass fiber breakage during compounding was determined for both batch and continuous mixers including an internal mixer and two different co-rotating twin screw extruders; of laboratory and industrial scales. The effect of processing conditions such as screw speed, feed rate and screw profiles on fiber damage was investigated. Finally, the results of experimental studies were discussed, compared and interpreted using a model for fiber breakage.

Keywords: Glass fiber, polyamide, internal mixer, twin screw extruder
PACS: 81.05.Qk, 81.20.Hy

INTRODUCTION

The average fiber length and fiber length distribution in glass-fiber reinforced thermoplastics significantly affect the final product quality, mainly the mechanical properties. However, due to the strong flow conditions generated during the compounding process, severe fiber breakage occurs and the fiber length can be reduced to 10% of its initial length, resulting in a weaker reinforcement of the part formed in subsequent processes such as injection molding. Several researchers have investigated the extent of fiber breakage during processes as well as the effect of process parameters [1-6]. Reduction in fiber length during compounding mainly depends on both processing conditions (shear-rate and residence time) and melt viscosity.

The aim of the present study is to make a comparison study of fiber degradation for different mixers including (i) an internal mixer, (ii) a laboratory-scale co-rotating twin screw extruder, and (iii) an industrial scale co-rotating twin screw extruder. The effect of processing conditions as well as the different screw configurations on fiber damage was investigated. Experimental results were finally interpreted by a model of fiber breakage, derived from the one initially proposed by Shon et al. [7].

EXPERIMENTAL

The mixing equipments used were (i) a Haake Rheomix 600P internal mixer, (ii) a ThermoFisher PTW 24 co-rotating twin screw extruder, and (iii) a Werner Pfleiderer WP40 co-rotating twin screw extruder. The experiments conducted on an internal mixer include the incorporation of 50 g of dried PA12 (Rilsan A, supplied by Arkema, France) into the chamber and mixing for 2 min. at 240°C. The glass fiber chops with an initial length of 3380 μm were then added, in an amount of 30 wt.% of PA12, to the chamber and mixed with molten polymer at rotor speeds of 20, 40, 80 and 120 rpm, and mixing times of 30,60,90 and 120 s. For the trials made with co-rotating twin screw extruders, dried polymer granules were fed through the main feeder and mixed with 30 wt.% of glass fibers which were added through the side feeder. The compounding was performed at screw speeds of 200, 300, 400, 500 and 600 rpm and feed rates of 5, 7 and 9 kg/h with ThermoFisher PTW 24 twin screw extruder, and at screw speeds of

CP1255, *V[th] International Conference on Times of Polymers (TOP) and Composites,*
edited by A. D'Amore, D. Acierno, and L. Grassia

100, 200 and 300 rpm and feed rates of 10, 20, 50, 80 and 100 kg/h with Werner Pfleiderer WP40 twin screw extruder. Similar screw profiles were used on both machines. Fiber length measurement was done by optical microscopy and image analysis, after burning the samples in a furnace at a temperature of 800°C for 1 hour and dispersing the recovered fibers in glycerol.

RESULTS AND DISCUSSION

Figure 1(a) illustrates the number average fiber length values for different rotor speed and mixing time combinations in the internal mixer. It is clear that fiber lengths decreased drastically with increase in rotor speed. The effect of mixing time on fiber degradation leads to a rapid decrease until reaching a minimal value, depending on rotor speed. By considering that the strain is related to the number of turns, a good agreement for the averaged fiber length L_n with the buckling-induced fiber break-up kinetic model proposed by Shon et al. [7] (equation (1)), can be obtained (see Fig.1(b)).

$$L(strain) = [L_0 - L_\infty]\exp(-k.strain) + L_\infty \tag{1}$$

where k is the fiber break-up rate constant, L_0 is the initial fiber length and L_∞ is the fiber length below which there is no more fiber break-up by buckling.

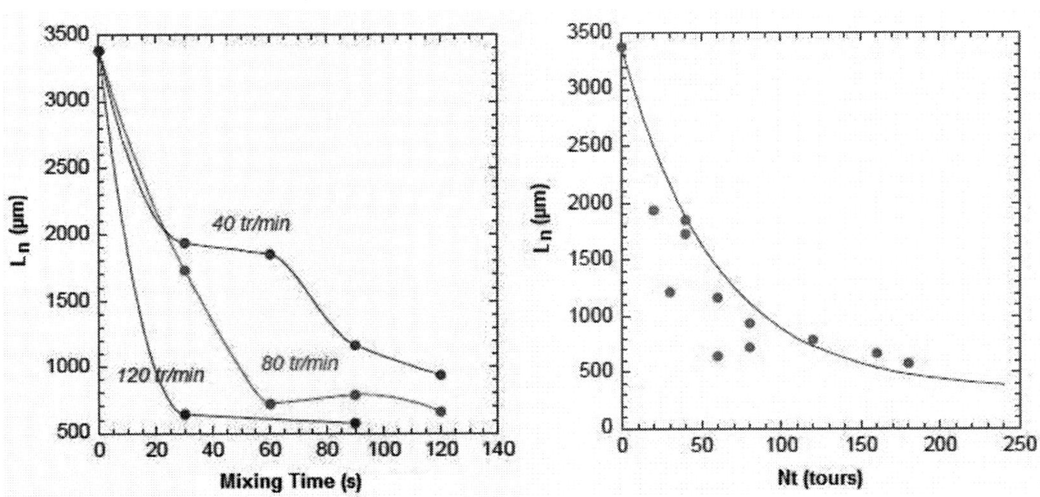

(a) (b)

FIGURE 1. **(a)** Effect of rotor speed and mixing time, and **(b)** effect of number of rotor turns, on length average fiber length (L_n) for internal mixer.

Fig. 2 shows the effect of screw speed/feed rate (N/Q) ratio on mean fiber length distribution for two different twin screw extruders. It is observed that fiber length decreases with increasing the screw speed and decreasing the feed rate for both types of extruders. However, the effect of these parameters on fiber breakage is quite small as average fiber length remains almost the same for different N/Q ratios. It also appears that the extent of fiber length breakage is more pronounced in twin screw extruder than that in an internal mixer. The reason for this behavior is that larger strain was generated during twin-screw extrusion. When considering specific mechanical energy (SME), instead of total strain in the proposed model, experimental results show a very good agreement, as shown in Fig 3. Moreover, it appears that breakage rate is large up to a critical SME level of 100 kWh/t. According to the results obtained in the present study, we can conclude that the evaluation of fiber length can be described by using fiber break-up kinetic model taking into account SME.

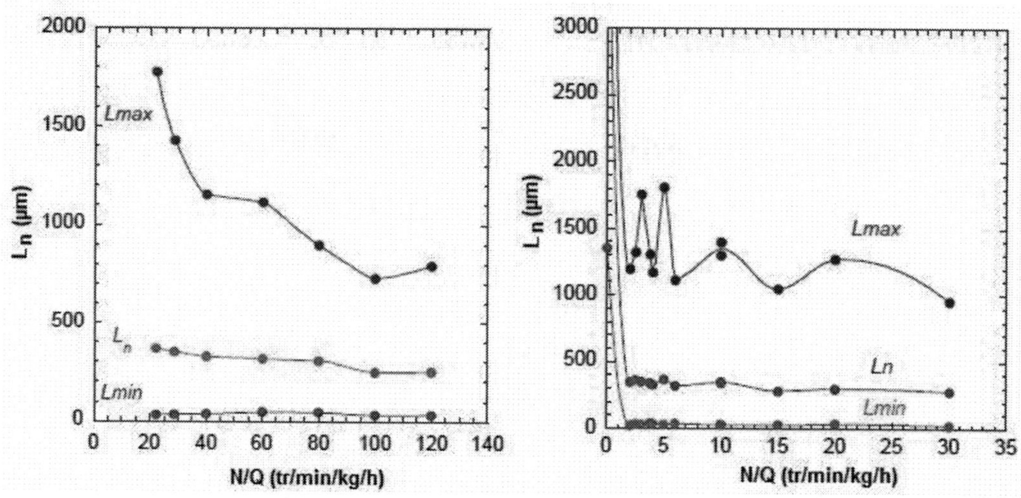

(a) **(b)**

FIGURE 2. Change in minimum (L_{min}), average (L_n) and maximum (L_{max}) fiber length with different *N/Q* ratios for **(a)** ThermoFisher PTW 24 twin screw extruder, and **(b)** Werner Pfleiderer WP40 twin screw extruder.

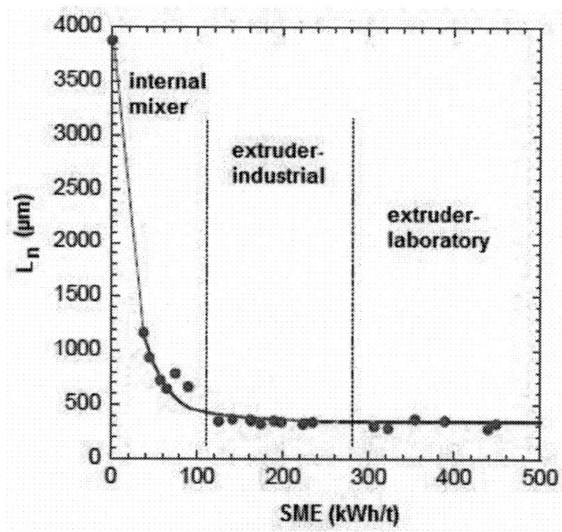

FIGURE 3. The effect of SME values for different mixers on average fiber length.

REFERENCES

1. K. Stade, *Polym. Eng.Sci.* **17**, 50-57 (1977).
2. L. Czarnecki and J .L. White, *J. Appl. Polym. Sci.* **25**, 1217-1244 (1980).
3. D. Wall, *Polym. Compos.* **10**, 98-102 (1989).
4. V. B. Gupta, R. K. Mittal and P. K: Sharma, *Polym. Compos.* **10**, 16-27 (1989).
5. K. Ramani, D. Bank and N. Kraemer, *Polym. Compos.* **16**, 258-266 (1995).
6. K. Shon and J.L. White, *Polym. Eng. Sci.* **39**, 1757-1768 (1999).
7. K. Shon, D. Liu and J.L. White, *Int. Polym. Process.* **3,** 322-331 (2005).

Fabrication of a novel poly(3-hydroxyoctanoate) / nanoscale bioactive glass composite film with potential as a multifunctional wound dressing

Ranjana Rai[a], Aldo R. Boccaccini[b,c], Jonathan C. Knowles[d], Ian C. Locke[e], Michael P. Gordge[e], Aine. Mc Cormick[f], Vehid Salih[d], Nicola Mordon[d] ,Tajalli, Keshavarz[a] and Ipsita Roy[a]*

[a]*Department of Molecular and Applied Biosciences, University of Westminster, London W1W 6UW, UK*
[b]*Department of Materials, Imperial College London, London SW7 2AZ, UK*
[c]*Department of Materials Science & Engineering, University of Erlangen, Nuremberg Cauestr. 6. 91058, Germany*
[d]*Division of Biomterials and Tissue Engineering, UCL Eastman Dental Institute, London WCIX 8LD, UK*
[e]*Department of Biomedical Sciences, School of Lifesciences, University of Westminster, London W1W 6UW, UK*
[f]*Haemophilia Reference Centre, St. Thomas' Hospital , Lambeth Palace Road, London SE1 7EH, UK*

Abstract. Fabrication of a composite scaffold of nanobioglass (n-BG) 45S5 and poly(3-hydroxyocatnoate), P(3HO) was studied for the first time with the aim of developing a novel, multifunctional wound dressing. The incorporation of n-BG accelerated blood clotting time and its incorporation in the polymer matrix enhanced the wettability, surface roughness and bio-compatibility of the scaffold.

Keywords: polyhydroxyalkanoates; poly(3-hydroxyoctanoate); nanobioglass 45S5; Keratinocyte; wound dressing

INTRODUCTION

Medium chain length polyhydroxyalkanoates, mcl-PHAs, are polyesters of 3-hydroxyacids produced mainly by *Pseudomonas* sp. as intracellular carbon reserves using renewable resources. Mechanically these polymers exhibit flexible, elastomeric properties; also their biodegradability and biocompatibility are well established, hence these are attracting interest as biomaterials for soft tissue engineering and drug delivery.[1] We have successfully produced a homopolymer poly(3-hydroxyoctanoate), P(3HO), from *P. mendocina*. Detailed structural, mechanical, thermal and molecular weight analysis of this polymer has been carried out.[2] In this work we describe for the very first time the fabrication of a composite P(3HO) film by incorporating bioactive glass n-BG to produce a material with the potential to be used as a multifunctional wound dressing, which will act as both a scaffold for skin regeneration and also accelerate wound haemostasis, thus potentially reducing blood loss following tissue injury.

MATERIALS AND METHODS

Fabrication and characterization of scaffolds

Composite 2D scaffolds (films) were prepared of 5 and 10 wt % polymer solutions containing 1 wt % bioactive n-BG of the nominal 45S5® composition (46 wt % SiO2, 23 wt % Na2O, 27 wt % CaO and 4 wt % P2O5) with spherical shape and a mean particle size of 29 nm. 5 and 10 wt % polymer only scaffolds were used as controls. The films were made using the solvent casting method described elsewhere.[3]

Microstructural studies for surface topography and roughness of the polymer only and composite P(3HO) films were carried out using white light interferometry on a Zygo® New View 200 OMP0407C analyzer. The wettability of the films was determined according to the method described in Misra *et al.* (2008) using a KSV Cam 200 optical contact angle meter (KSV Instruments Ltd, Helsinki, Finland).

CP1255, *V[th] International Conference on Times of Polymers (TOP) and Composites,*
edited by A. D'Amore, D. Acierno, and L. Grassia

In vitro haemostatic study using Thromboelastography®,(TEG®)

The haemostatic effect of the n-BG was studied by Thromboelastography using a TEG® 5000 series Hemostasis Analyzer (Medicell Ltd, London, UK). Briefly, 20 μl of 0.2M CaCl2 was added to 360 μl of citrated whole blood, followed by the n-BG. The following clotting parameters were then monitored – R time (reflecting the time delay before the clotting process begins to be detectable), alpha angle (reflecting the rate at which clotting proceeds) and maximum amplitude (reflecting the strength of the formed clot). Three different amounts of n-BG, i.e. 1, 2 and 4 mg were used for the study, in order to look for dose-related changes.

In vitro cell culture study

The *in vitro* cell culture studies were carried out on the P(3HO) neat and P(3HO)/n-BG composite films using the human keratinocyte cell line HaCaT which was cultured in Dulbecco's modified Eagles Medium (DMEM) supplemented with 10% foetal calf serum, 1% Glutamine and 1% penicillin and streptomycin solution. The cells were incubated at 37°C in a 5% CO_2 humidified atmosphere with the culture medium changed every 2 days.

For cell seeding on the scaffolds, 1 cm^2 polymer / composite samples were UV sterilised for 30 minutes on each side and passivated for 12 hrs. A cell density of, 20,000 cells were used to seed the scaffolds kept in 24 well plates. The cells were analysed after 1, 4 and 7 days for cell adhesion, proliferation and morphology. Cell culture studies were carried out on triplicate samples per experiment.

Cell adhesion and proliferation studies were carried out using the Neutral Red (NR) assay as described elsewhere.[4] The total NR uptake was a measure of the cells proliferation and viability (% NR uptake is directly proportional to the number of live and uninjured cells and was compared to a control population grown on tissue culture plates. The control is normalised to 100%. The samples for the SEM analysis of cell morphology were prepared as described in Misra et al. (2008)

RESULTS AND DISCUSSION

The addition of bioactive n-BG altered the surface morphology of the P(3HO) neat films by introducing surface roughness as evident in composite films. White light interferometry studies showed a root mean square (RMS) value of 0.440 μm for the composite films as opposed to 0.238 μm for polymer only films. (Figures 1)

FIGURE 1. Surface study of (A) neat P(3HO) and (B) P(3HO)/nanobioglass composite film.

The incorporation of n-BG also enhanced the wettability of the polymer as confirmed by static water contact angle, (θ_{H2O}) measurements. The θ_{H2O} values observed were; $\theta_{H2O} = 76.63\pm2^o$ for 5 wt% P(3HO)/1 wt% nBG; $\theta_{H2O} = 77.3\pm3^o$ for 5 wt % P(3HO) neat film; $\theta_{H2O} = 75.43\pm5^o$ for 10 wt% P(3HO)/1 wt% nBG and $\theta_{H2O} = 78.5\pm3^o$ for 10% neat P(3HO) film

The haemostatic effect of silicate bioactive glasses was reported by Ostomel et al. (2006) although the glass used by them was of different composition to the material used here.[5] The n-BG was found to be haemostaticaly active and increasing amounts of n-BG showed corresponding increase in the initiation of clotting (R time) (Figure 2; Table I), although rate of clotting (alpha angle) was not altered in a clearly dose-related manner and clot strength (maximum amplitude) was slightly reduced.

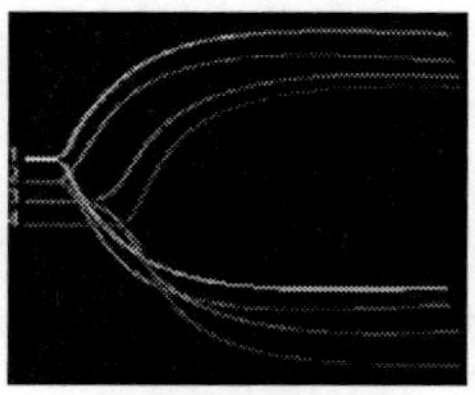

FIGURE 2. Combined thromboelastograph plots.

Sample	R (min)	Alpha (degree)	Maximum Amplitude (mm)
4 mg nB (Profile 1)	3.8	61.4	57.3
2 mg nBG (Profile 2)	5.6	50.7	57.95
1 mg nBG (Profile 3)	6.25	52.5	60.9
Control (Profile 4)	9.7	60.2	66.1

TABLE 1. *In vitro* TEG® clotting parameters.

In vitro cell culture studies demonstrated that, seeded HaCaT cells were able to attach and proliferate on these scaffolds. However, cell proliferation on the composite films was higher than that on the comparable polymer only film with differences in cell morphology also evident (Figures 3). This may be a result of the increased surface roughness and wettability of the composite film because of n-BG incorporation which provided improved anchorage for attachment of the seeded cells. This result was in accordance to that observed by Paschel et al. (2007) where proliferation rate of HaCaT cells increased with decreasing hydrophobicity of a PHA scaffold. [6]

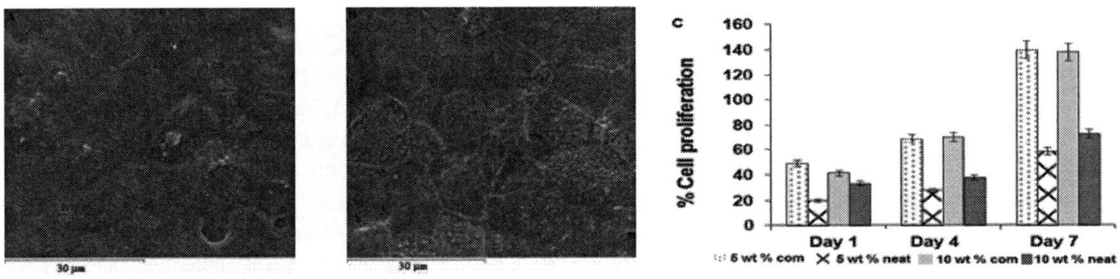

FIGURE 3. SEM images of the HaCaT cells on (A) 10 wt % polymer only film and (B) 10 wt % composite film; (C) Cell proliferation measurements for Day 1, 4 and 7 using neutral red assay. (n = 3, error bars = ± s.d)

CONCLUSIONS

Novel composites of P(3HO) and bioactive n-BG have been developed and studied here for the first time. The main conclusions from these studies are that n-BG 45S5 has a haemostatic effect and that the incorporation of n-BG into P(3HO) polymer films increases the surface roughness of the films and also enhances its wettability; both of which may be factors in the increased attachment and proliferation of HaCaT cells observed in these composite scaffolds.

ACKNOWLEDGEMENTS

The authors gratefully acknowledge the financial support received from the Quintin Hogg Foundation of the University of Westminster. We also acknowledge D. Mohn and Prof. W. Stark (ETH Zurich) for providing the n-Bg.

REFERENCES

1. G.Q. Chen and W. Qiong (2005). "The application of polyhydroxyalkanoates as tissue engineering materials." Biomaterials **26**: 6565–6578.
2. R. Rai, A.R. Boccaccini, J.C. Knowles, T. Keshavarz, and I. Roy (2010). "Poly-3-hydroxyoctanoate P(3HO), a unique medium chain length polyhydroxyalkanoate homopolymer: production using *Psuedomonas mendocina* and its detailed characterisation." In preparation for submission to Biomacromolecules.
3. S.K. Misra, S.E. Philip, W. Chrzanowski, S.N. Nazhat, I. Roy, J.C. KNowles, V. Salih, and A.R. Boccaccini (2008). "Incorporation of vitamin E in poly(3hydroxybutyrate)/Bioglass composite films: effect on surface properties and cell attachment." Journal of the Royal Society Interface.
4. A. Prashar, I.C. Locke, and C.S. Evans (2004). "Cytotoxicity of lavender oil and its major components to human skin cells." Cell Proliferation **37**: 221-229.
5. T.A. Ostomel, Q. Shi, C.K. Tsung, H. Liang, and G.D. Stucky (2006). "Spherical Bioactive Glass with Enhanced Rates of Hydroxyapatite Deposition and Hemostatic activity." Small **2**(11): 1261-1265.
6. G. Peschel, H.M. Dahse, A. Konrad, G.H. Wieland, P.J. Mueller, D.P. Martin, and M. Roth (2007). "Growth of keratinocytes on porous films of poly(3-hydroxybutyrate) and poly(4-hydroxybutyrate) blended with hyaluronic acid and chitosan." Journal of biomedical materials research: 1073-1081.

Plasticizer migration from cross-linked flexible PVC. 1. Effects on tribology

M. Pannico [a], P. Persico [a,b], V. Ambrogi [a], C. Carfagna [a,b]

[a] *Department of Materials and Production Engineering, University of Naples "Federico II", Piazzale Tecchio, 80125 Napoli, Italy*

[b] *Institute of Polymer Chemistry and Technology – CNR, Via Campi Flegrei 34, 80078 Pozzuoli (Na), Italy*

Abstract. Utilization of soft PVC is restricted by plasticizer migration that can affect material properties, as well as its toxicity. Modifying the chemical structure of PVC is one of the most effective tool to reduce the diffusion of plasticizer. In this work, a soft cross-linked PVC was obtained using a difunctional amine, namely isophoron diamine (IPDA) as the cross-linking agent. The gel content (wt %) was evaluated by weighting the insoluble portion obtained through solvent extraction technique. Thermogravimetric analysis (TGA) revealed that cross-linking reactions promote thermal degradation phenomena in the polymer matrix. Tribological properties of soft uncross-linked, cross-linked and rigid PVC were determined. Soft formulations were held in contact for 32 days with rigid PVC sheets. Plasticizer migration towards the interface causes an increase of dynamic friction compared to that of the reference rigid PVC.

Keywords: Poly(vinyl chloride), cross-linking, tribology, plasticizer migration.
PACS: 82.35.Gh; 81.40.Pq; 81.05.Lg

INTRODUCTION

Polyvinyl Chloride (PVC) is one of the oldest and most common plastics in use today, because it is easily processed and compounded. However, in order to optimize its properties plasticizers are frequently used. In general, PVC plasticizers are low-molecular weight (M) compounds: esters, phosphates, and epoxides as well as citrates and adipates [1]. Because of their low M, they have a high tendency to leach out from the polymer [2, 3]. The loss of plasticizers is one of the most dominant adverse factor contributing to the aging of flexible PVC, [4, 5] making it useless for many applications [6, 7] because of a general loss of properties. In addition, the plasticizers themselves may be harmful to health or the environment.

A distinct option aimed at reduction of plasticizer migration is cross-linking of PVC [7, 8, 9]. Formation of more chemical bonds reduces the free volume and segmental mobility, thus affecting plasticizer diffusion through the matrix. Thus, Romero Tendero et al. used a difunctional amine, namely isophoron diamine (IPDA) as a cross-linking agent for PVC plastisols [8]. In this work, IPDA was used as cross-linking agent for flexible PVC dry blends. Three different percentages of IPDA were applied. Since the cross-linking process was expected to enhance the degradation of the polymer, heat stabilizers were also added in the formulations [8, 10, 11].The extent and the efficiency of cross-linking was estimated. The effect of the addition of IPDA on PVC thermal stability was studied. Tribological tests were also performed on rigid PVC after being held in contact with soft PVCs.

RESULTS

PVC of K value 70 (sPVC), the plasticizer dioctyl phthalate (DOP), heat stabilizer Ca/Zn, co-thermal stabilizer epoxy soybean oil (ESBO) and the cross-linker isophoron diamine (IPDA) have been used. Flexible samples prepared are reported in detail in Table 1. All the components were dry blended in a mixer. The blends were recovered and then roll-milled in a plasticorder at 40 rpm, T = 90°C, 15 min. The flexible sheets were obtained using a hot press at 150°C and 200 bar for 7 min. Rigid PVC sheets were obtained at 190°C, 200 bar for 7 min.

CP1255, *V[th] International Conference on Times of Polymers (TOP) and Composites,*
edited by A. D'Amore, D. Acierno, and L. Grassia

TABLE 1. Flexible PVC formulations.

Components	A	B1	B2	B3
	Samples			
sPVC	100phr	100phr	100phr	100phr
DOP	70 phr	70 phr	70 phr	70 phr
ESBO	2 phr	2 phr	2 phr	2 phr
Ca/Zn	1 phr	1 phr	1 phr	1 phr
IPDA	-	1.2 %	2.2 %	3.2 %-

The efficiency of IPDA as a chemical cross-linker for PVC was evaluated by calculating the weight percent of the insoluble fraction recovered after solvent extraction procedure. In the case of sample A, although no cross-linking agent was used in the formulation, a residual gel amount≈ 13 wt. % was found. This finding is probably due to cross-linking of polyene sequences formed upon dehydrochlorination (DHCl) caused by fairly high processing temperatures [12, 13]. In the case of B1, B2, B3 the gel content increased up to 24, 28, and 48 wt% respectively. The effect of IPDA on the gel quantity formation can be described in terms of two distinct simultaneous phenomena: during amine cross-linking the chlorine atom is extracted from the main polymer chain and replaced by a cross-linker unit leading to the formation of N-C bonds [8]. At the same time, due to the cross-linking reaction, concentration of HCl molecules increases. This enhances the catalytic action of PVC dehydrochlorination and consequently the cross-linking.

In order to evaluate the effect of cross-linking on the plasticizer migration extent, weight changes of rigid PVC in contact with soft sPVC-based samples were monitored. After 32 days the weight increase of rigid PVC sheets kept in contact with soft cross-linked PVCs was about 18 % lower than that of sheets in contact with sample A. This finding proves that chemical cross-linking of PVC is an attractive tool to overcome the disadvantage related to the plasticizer migration.

As demonstrated by TGA, the cross-liking mechanism enhances degradation of the polymer. Results revealed that the thermal degradation process proceeds in two basic stages. In the case of sample A, within the range of 200–350°C, mainly loss of DOP (T = 280°C) and PVC DHCl (T = 291°C) take place. On the other hand, samples B1, B2 and B3 exhibit one broad degradation peak, centered at lower temperatures (T≈265°C), indicating simultaneous occurrence of the two phenomena. It is likely that the increased concentration of HCl due to the use of IPDA is responsible for faster degradation of PVC [14, 15]. In the second degradative stage at T > 400 °C which corresponds to scission of covalent bonds all the samples exhibit the same behavior.

Curves showing dynamic friction results for rigid and flexible sPVC-based samples as a function of the number of revolutions are shown in Fig. 1. We find that rigid PVC is characterized by an average value of dynamic friction≈ 0.15. As expected, the A sample stands apart. The cross-linked samples B1, B2, and B3 exhibit comparable friction values. In Fig. 2 dynamic friction values for rigid PVC sheets held in contact for 32 days with flexible sPVC-based systems are presented. Rigid PVC in contact with sample A exhibits the highest friction values; a large amount of DOP moved towards the interface, increasing the adhesion between the relative sliding parts.

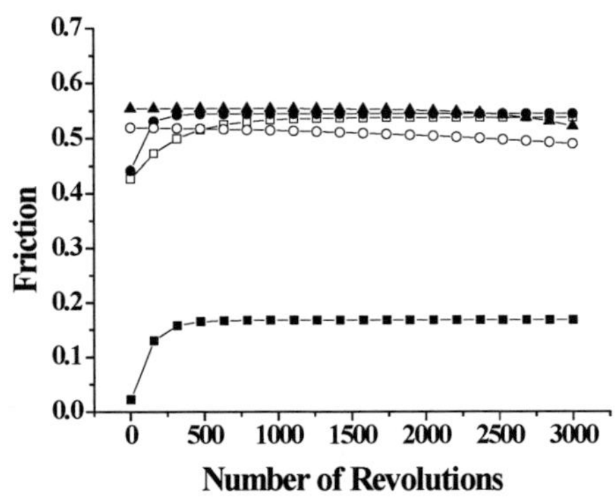

FIGURE 1. Friction of unaged samples: rigid PVC (-■-), A (-□-), B1 (-●-), B2 (-▲-), B3 (-○-).

Rigid PVC in contact with B1, B2 and B3 shows comparable friction behavior, relatively close to neat rigid PVC. This finding validated the effectiveness of cross-linking in preventing plasticizer migration.

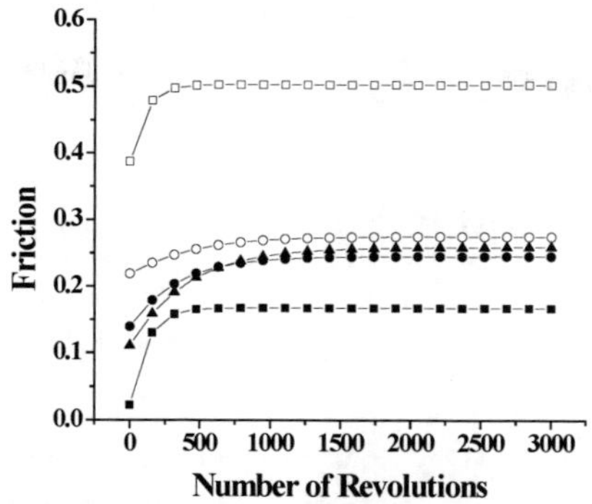

FIGURE 2. Friction of unaged rigid PVC (-■-), aged rigid PVC after being in contact for 32 days with A (-□-), B1 (-●-), B2 (-▲-), B3 (-○-).

CONCLUDING REMARKS

Isophoron diamine was shown to be an efficient cross-linking agent for PVC dry blends. This was confirmed by solvent extraction tests indicating that insoluble fraction increases with increasing cross-linker concentration. Plasticizer migration was reduced, however the cross-linking process enhanced the degradation of the polymer. After plasticizer diffusion, friction behaviour of rigid PVC changed. This was due to a change on its surface and thus to an increased friction.

ACKNOWLEDGMENTS

The authors wish to thank for the financial support provided by Faraplan S.p.A., Vicenza, Italy.

REFERENCES

1. P. Persico, V. Ambrogi, D. Acierno, C. Carfagna, Jnl Vinyl. Addit. Techn. 15 (3), 139-146 (2009).
2. T. D. Stark, H. Choi, P. W. Diebel, 57th Canadian Geotechnical Conference, Geo Quebec, Canada (2004).
3. Marcilla, S. Garcia, J. C. Garcia-Quesada, Polymer Test. 27, 221–233 (2008).
4. K. Z. Gumargalieva, V. B. Ivanov, G. E. Zaikov, Ju. V. Moiseev, T. V. Pokholok, Polymer Degrad. Stab. 52, 73-9, (1996).
5. I. Jakubowicz, N. Yarahmadi, T. Gevert, Polymer Degrad. Stab. 66, 415-21, (1999).
6. M. L. Marin, J. López, A. Sánchez, J. Vilaplana, A. Jiménez, Bull Environ Contam Toxicol. 60, 68–73, (1998).
7. S. Lakshmi, S. Jayakrishnan, Polymer, 39, 151–7, (1998).
8. P. M. R. Tendero, A. Jimenez, A. Greco, A. Maffezzoli, Eur. Polymer Jnl. 42, 961-969, (2006).
9. A. Marcilla, J. C. Garcia, Eur. Polymer Jnl. 33, 357–63, (1997).
10. M. I. Beltrán, J. C. García, A. Marcilla, M. Hidalgo, C. Mijangos, Polymer Degrad Stab. 65, 65-73, (1999).
11. G. Arias, R. Benavides, E. L. Castillo Garza, M. M. Téllez Rosas, Jnl Vinyl Addit. Techn. 12 (2), 49-54, (2006).
12. W. H. Starnes Jr., Ge. Xianlong, Macromolecules 37, 352-359, (2004).
13. X-G Zheng, L-H Tang, Na. Zhang, Q-H Gao, C-F Zhang and Z-B Zhu. Energy & Fuels 17, 896-900, (2003).
14. B. Li, Polymer Degrad. Stab. 68, 197-204, (2000).
15. D. Braun. Deutsches Kunststoff-Institut, Darmstadt, Germany, (1981).

Epoxy Nanocomposites – Curing Rheokinetics, Wetting and Adhesion to Fibers

Ilyin S.O., Kotomin S.V. and Kulichikhin V.G.

A.V.Topchiev Institute of Petrochemical Synthesis, 29, Leninskii Prospect, Moscow, Russia 119991

Abstract. Epoxy nanocomposites considered as challenging polymeric matrix for advanced reinforced plastics. Nanofillers change rheokinetics of epoxy resin curing, affect wetting and adhesion to aramid and carbon fibers. In all cases extreme dependence of adhesive strength vs filler content in the binder was observed. New experimental techniques were developed to study wettability and fiber-matrix adhesion interaction, using yarn penetration path length, aramid fiber knot pull-up test and electrical admittance of the fracture surface of CFRP.

Keywords: epoxy resins, nanocomposites, rheology, wetting, adhesion, aramid fibers, carbon fibers
PACS: 83.80.-k, 81.07.Wx, 82.35.Gh, 68.08.Bc, 47.57.Qk

OBJECTS

Oligomer binder - epoxy-amine composition containing diglycidyl ether of bisphenol-A (DGEBA) and 4,4'-diaminodiphenyl sulfone (DDS) in stoichiometric ratio. **Reinforcing fibers** –polyamidbenzamidazol fiber (SVM grade, 29 tex) and carbon fiber yarn (UKN grade, 5000 tex). **Fillers** -modified montmorillonite (MMT) Cloisite 30B, carbon nanotubes (NT) two types (I, II) with different specific surface (> 120 m2/g and 700-900 m2/g), nanodiamonds of detonation synthesis (ND). Nanofillers were dispersed in liquid oligomer with a high speed mechanical stirrer for 30 minutes at 30000 rpm.

EXPERIMENTAL

Curing rheokinetics of filled epoxy oligomers studied on Rheostress ThermoHaake RS600 rotational viscometer and DSC TA Instruments 2920

Wetting studied in terms of wetting rate during capillary impregnation while curing

Adhesion of aramid fibers. New test procedure using yarn (thread) knot pull-up was developed. One end of the aramid fiber yarn is placed inside a loop forming knot from the same yarn, and then after adding of a few binder drops in the contact area between two yarns the knot is tighten (Fig.1 a). After curing of epoxy binder the two ends of the sample were pulling on INSTRON testing machine. Maximal pulling force F (Fig.1 b) characterizes the adhesion strength of the joint fiber – matrix, and normalized value is calculated as F/T where T is linear density of the yarn expressed in tex (weight in grams of the yarn with length 1 km). Although directly the contact adhesion area is not calculated, it should be proportional to the tex value of the yarn. That procedure proved to be much easier to carry out for experiments as compared with traditional pull-out test with a single fiber (filament), and relevant for evaluation of adhesion for various binders to one-type aramid fiber.

Adhesion of carbon fibres. The simple experimental technique developed earlier for aramid and glass fibers cannot be used for CF because they are too thin and brittle. Conventional tests based on study of single fiber fragmentation including acoustic emission analysis are complicated and rather time consuming. For evaluation of adhesion of carbon fibers (CF) there is well-known approach on analysis of length for ends of broken CF on the fracture surface of CFRP. That value directly corresponds to critical fiber length which characterize the adhesion strength, but need to use electron microscope puts some restriction on the method in case of many samples. We proposed to evaluate the length of CF ends on the fracture surface by measurement of electrical resistance of the

CP1255, *V*[th] *International Conference on Times of Polymers (TOP) and Composites,*
edited by A. D'Amore, D. Acierno, and L. Grassia

fracture surface CFRP imbedded in electrolyte solution (NaCl in water) . The longer are the fiber ends on the fracture surface, the less is conductivity. After proper calibration procedure (Fig.2 b) one may easy determine critical fiber length and on the base of it – the adhesion strength.

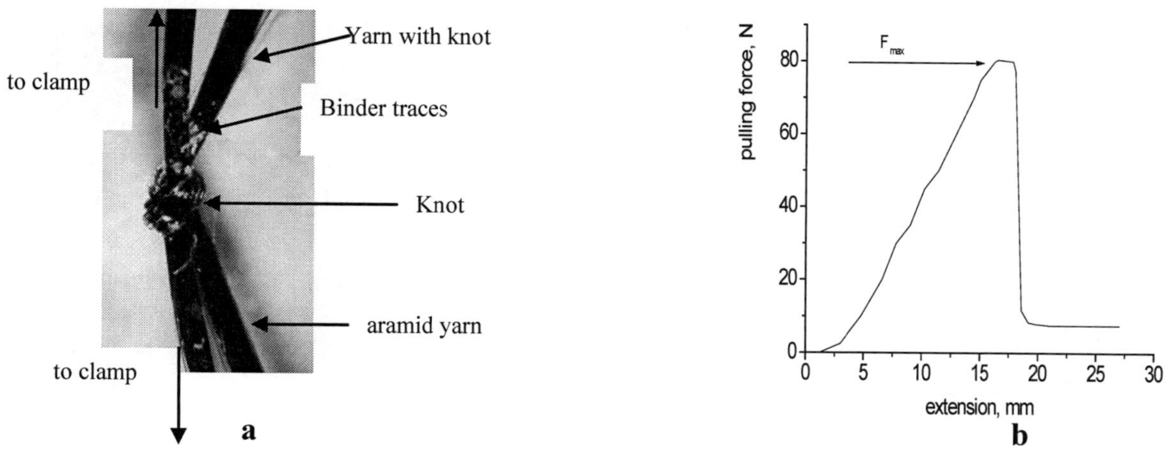

FIGURE 1. Sample with knot after pull-up test (a) and typical test curve for extension (b)

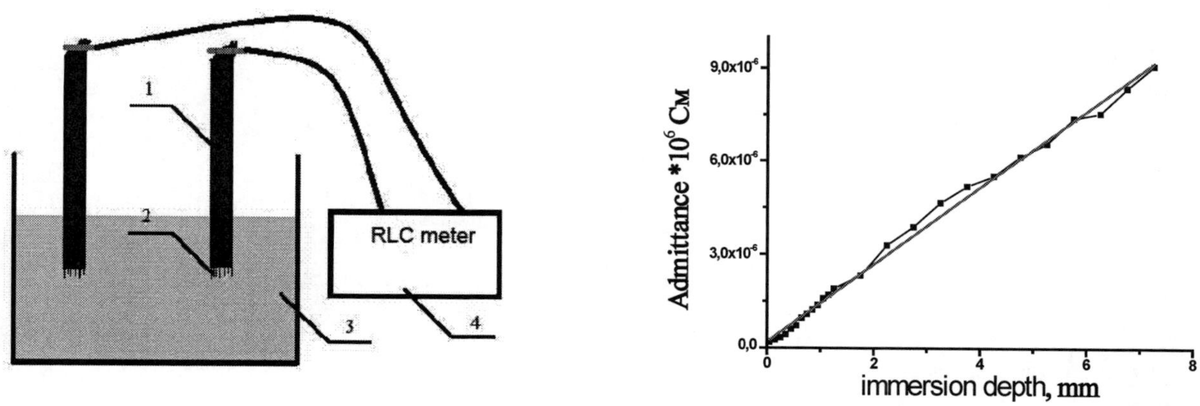

$$\tau_s = \frac{d_f \sigma_f}{4l}$$

τ $_f$-adhesion strength, d_f- diameter, σ_f- tensile strength, l_f – length of the fiber

FIGURE 2. Test scheme of resistance measurement for two parts of broken microplastics immersed in electrolyte (a) and typical calibration curve (b) 1 –microplastic, 2 – fiber ends on fracture surface, 3 – electrolyte, 4 –RLC meter

RESULTS AND DISCUSSION

Introduction of nanoparticles in the epoxy binder influences the kinetics of the curing process. The results of rheokinetic studies show that the introduction of filler delays the onset of gelation. According to DSC data, the maximum peak of reaction heat shifted toward earlier time, which indicates the acceleration of curing. The total thermal effect of the resin curing with presence of nanoparticles increases that characterizes conversion level of reaction.

Possible reason for acceleration - the presence of nanoparticles on the surface functional groups, which catalyze the process of disclosure of epoxy cycle due to the protonation of oxygen atoms with hydroxyl groups located on the surface of nanoparticles. In addition, the acceleration of curing may be related to the orientation of oligomer molecules in the adsorption layer. It is possible that displacement of a point of gelation is due to selective adsorption of different oligomer fractions on the surface of nanoparticles, which is more effective for low-molecular fractions.

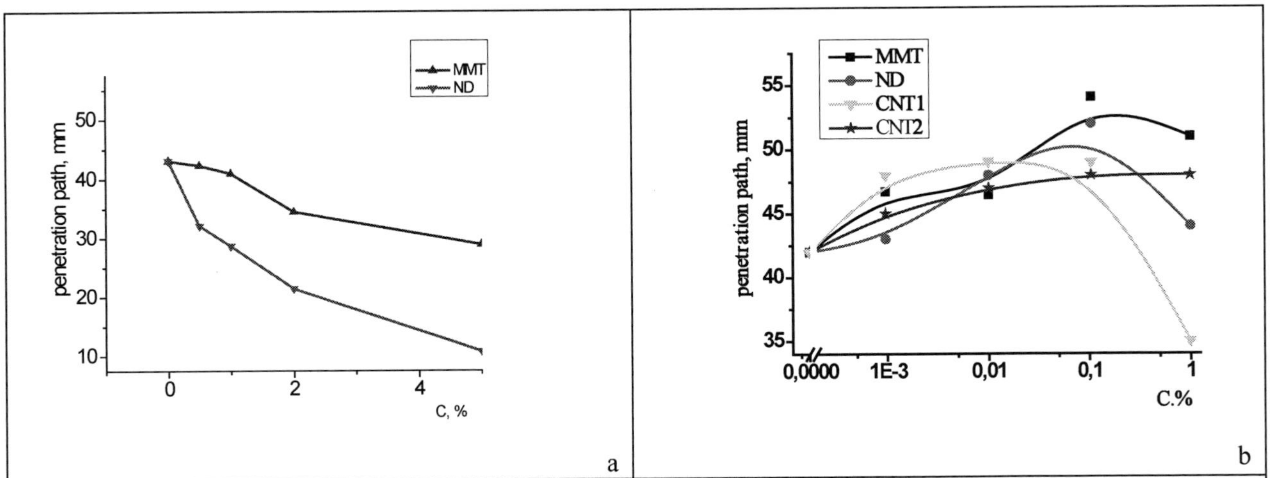

FIGURE. 3 Penetration path vs filler content in resin for aramid (a) and carbon (b) fibers

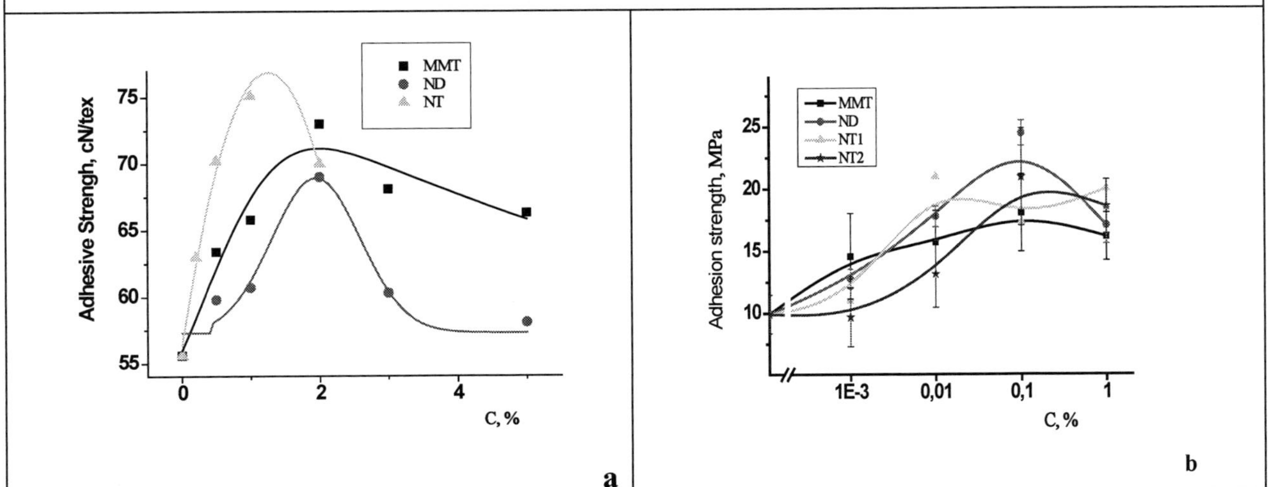

FIGURE. 4 Adhesive strength vs filler content in resin for aramid (a) and carbon (b) fibers

An important step in interaction binder with reinforcing fiber is wetting, studied during capillary impregnating of yarn by resin droplets at a temperature of curing. Low viscosity of binder provides good wettability of the fiber that contributes to strong adhesion between fiber and matrix. With the introduction of nanoparticles into epoxy resin the increase in impregnation of carbon fiber on its length to a certain concentration of particles was observed up to 0.1% content (Figure 3, b)

It was found the extreme dependence of matrix-fiber adhesion strength on the concentration of nanoparticles in resin (Figure 4 a,b). In the case of the introduction of the modified clay binder adhesion strength of connection to the SVM fiber increases by 33%, with the introduction of Nd - 25%, while for NT - increased to 40% of the values for the unfilled binder.

CONCLUSIONS

Presence of nanofillers in epoxy resins slows up the growth of the binder viscosity during the curing, which improves processing conditions of reinforced plastics. At the same time, speed and depth of the reaction increased, thereby reducing the time of curing cycle of binder. Introduction of nanosized particles to epoxy resin provides improving the adhesive interaction of fibers with a binder. New experimental technique for measurement of adhesive strength of aramid and carbon fibers proved to be useful for study of fiber-matrix interaction.

HYGROTHERMAL AGEING OF PA 6,6

Chaker EL MAZRY[1,2], Olivier CORREC[1], Xavier COLIN[2]

[1)] CSTB, Centre Scientifique et Technique du Bâtiment, Université Paris Est, 11 Rue Henri Picherit, 44300 Nantes, France

[2)] ARTS ET METIERS ParisTech, Laboratoire PIMM (UMR 8006), 151 boulevard de l'Hôpital, 75013 Paris, France

Abstract: Polyamides are moderately polar polymers absorbing significant amounts of moisture in wet environments. As an example, PA 6,6 uptakes more than 8% by weight of moisture at ambient temperature in 100% RH. In addition, amide is a well-known hydrolysable group. It is thus suspected that such a polymer perishes by hygrothermal degradation. To check this assumption, sorption/desorption experiments were performed between 40 and 70°C in various wet atmospheres (typically between 0 and 90% RH), to determine the water transport properties of PA 6,6. In parallel, hydrolysis experiments were made between 60 and 90°C in distilled water (100% RH) to determine the different structural changes of PA 6,6 at different pertinent scales by conventional laboratory techniques. Main results are presented and discussed.

Keywords: Polyamide 6,6, water sorption/desorption, hydrolysis, embrittlement.

PACS: 82.35.-x

1. INTRODUCTION

The use of safety parts made of polymer materials in drinking water distribution networks, in particular inside building, constitutes an economic and scientific challenge in the field of valves and fittings.

The aim of the present study is to elucidate the degradation mechanisms of polyamide 6,6 disconnectors in real use conditions, i.e. in permanent contact with water, chlorinated disinfectants and subjected to moderate hydrostatic pressures. It is admitted that such a polymer can perish by two distinct chemical processes: thermal oxidation and hydrolysis. This latter constitutes the object of the present communication.

2. EXPERIMENTAL

1.1. MATERIAL

Pure PA 6,6 pellets were supplied by Rhodia. Their main characteristics are: melting temperature of 266,9 ± 1,2°C, crystallinity ratio of 42 ± 2 %, density of 1,14 $kg.m^{-3}$ and weight average molar mass of 37 $kg.mol^{-1}$. Before processing, pellets were carefully dried at 80°C during 72h in a primary vacuum oven in order to prevent hydrolysis at high temperature in molten state. Then, thin PA 6,6 films, of about 30 or 100 µm thickness, were compression molded at 270°C under 17 MPa pressure. They were kept in a desiccator containing silica gel in order to prevent any moisture uptake before hygrothermal ageing experiments.

1.2. SORPTION VAPOR TEST

Water sorption/desorption experiments were carried out at 40, 50, 60 and 70°C in various wet atmospheres (between 0 and 90% RH) in a DVS (Dynamic vapor sorption) apparatus provided by SMS (Surface Measurement System, London, England) company. This latter is equipped by a microbalance of Cahn with an electromagnetic compensation, allowing to detect mass variations of 0,1 µg.

1.3. HYDROLYTIC AGING

PA 6,6 films were immersed in distilled water in a glass jar maintained at a constant temperature of 60, 70, 80 or 90°C during several hundreds of hours. At precise times of exposure, films were removed from the glass jar and dried at room temperature in a desiccator containing silica gel. After complete drying, films were characterized with conventional laboratory techniques to determine structural modifications at different pertinent scales (molecular, macromolecular, microscopic and macroscopic scales): IRTF spectrophotometry (Brüker, IFS 28 spectrometer, minimal resolution of 4 cm^{-1}), rheometry in melt state (Rheometrics, ARES rheometer, parallel plate geometry, plate diameter of 25 mm, gap of 1 mm, deformation of 5%, temperature of 265°C), Differential Scanning Calorimetry (TA Instruments Q10 apparatus, heating rate of 10°$C.min^{-1}$), and tensile testing (Instron 4310 machine, strain rate of 7.5×10^{-3} s^{-1}, 23°C and 50% RH).

CP1255, *V[th] International Conference on Times of Polymers (TOP) and Composites,*
edited by A. D'Amore, D. Acierno, and L. Grassia

3. RESULTS AND DISCUSSION

3.1. WATER TRANSPORT PROPERTY

As an example, the water sorption/desorption isotherms of PA 6,6 films of 30 µm thickness exposed at 50, 60 and 70°C between 0 and 90% RH are presented in Figure 1. They call for the following comments:

- Water sorption/desorption isotherms superimpose perfectly (Figure 1a). It can be thus concluded that, in the temperature range under study, water absorption is totally reversible for PA 6,6.

- Water sorption isotherms are linear at low RH, but display clearly a positive concavity at high RH. This deviation to the classical Henry's law (for which the sorption isotherm is linear in the whole RH range) has been already reported by Hernandez **[1]** and Hernandez and Gavara **[2]** in the case of PA 6. It is attributed to cluster formation and the mathematical equation of the sorption isotherm can be derived from Zimm-Lundberg theory **[3]** to determine the cluster characteristics.

- These curves are almost temperature independent (Figure 1b), i.e. the equilibrium water concentration is almost temperature independent. It is a general behavior of moderately polar polymers. An explanation, valid for the Henry's regime, but which can be extended to other sorption modes, was given by Merdas et al. **[4]**. In these polymers, water-polymer interactions (through hydrogen bonds) are highly exothermic, so that the heat of water dissolution is negative and its absolute value is close to the heat of water vaporization. As a result, the resulting activation energy of equilibrium water concentration, which is the sum of these contributions, is close to zero.

- On the contrary, the coefficient of water diffusion into PA 6,6 is temperature dependent. In the temperature range under study, it obeys an Arrhenius law.

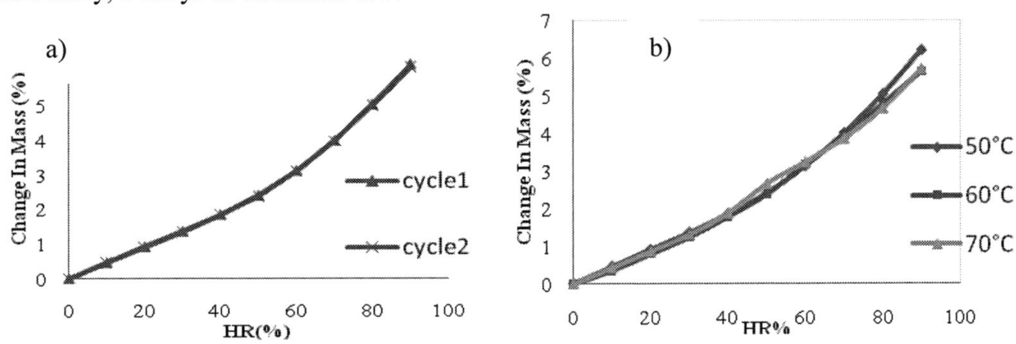

Figure 1 : a) Example of sorption/desorption isotherms isotherm of PA 6,6 at 50°C; b) Sorption isotherms of PA 6,6 at 50, 60, and 70°C.

3.2. HYDROLYTIC AGEING

As an example, the structural changes and their consequences on the mechanical properties of PA 6,6 films of 100 µm thickness exposed at 70 and 80°C in distilled water (100% RH) are presented in Figures 2, 3, 4 and 5.

3.2.1. Morphological changes

Changes in crystallinity ratio, against time of exposure, were monitored by two distinct techniques: IRTF spectrophotometry and DSC, giving quite similar results.
As an example, the kinetic curves determined from IR peaks located at 906 cm^{-1} (for crystalline phase) and 1140 cm^{-1} (for amorphous phase) **[5]** are presented in Figure 2. A significant increase in crystallinity occurs in the first 300 hours of exposure at 70 and 80°C: about 11% at 70°C and 14% at 80°C. It was attributed, in a first approach, to cold crystallization phenomena.

3.2.2. Molecular changes

It is generally reported that hydrolysis destroys the C-N bonds of PA 6,6 monomer unit, transforming them into amine NH_2 and carboxylic acid COOH chain-ends groups. Changes in concentration of C-N and C=O groups were monitored by IRTF spectrometry respectively at 3080 cm^{-1} and 1735 cm^{-1}.
As an example, the kinetic curves of C-N bonds concentration are presented in Figure 3. They displays a peculiar shape constituted of three distinct domains: i) an initial period during which the C-N bonds concentration decreases pseudo-exponentially with time of exposure; ii) a pseudo-steady-state period during the concentration remains almost constant; iii) and a final period during which the concentration decreases slowly with time.
In the two first domains, changes in C-N bonds concentration are closely correlated with changes in crystallinity ratio. It can be thus concluded that cold crystallisation phenomena are the only causes of modification of IR

spectra in this timescale, hydrolysis being too slow to induce measurable effects at the molecular scale. Such an assumption was easily checked from the changes in average molar mass (see next section).
At the opposite, in the final domain, the changes in C-N bonds concentration result probably of a chemical degradation, perhaps thermal oxidation. Hygrothermal experiments in deoxygenated water (100% RH) are in progress at the laboratory in order to check this assumption.

3.2.3. Macromolecular changes

Each hydrolysis event leads to a chain scission. As a result, it is generally found that average molar masses decrease dramatically in wet environments. Changes in weight average molar mass M_W were monitored by rheometry in melt state, using the classical Bueche's law **[6]** for determining the values of M_W from the Newtonian plateau: $\eta = K \times M_W^{3.4}$.
The resulting kinetic curves (Figure 3) reveal that chain scissions occur essentially in the early periods of exposure, reducing the chain length of about 25% after only 5 hours of exposure. Then, M_W seems to tend towards an equilibrium value depending on the exposure temperature, which suggests that PA 6,6 hydrolysis would be reversible (i.e. hydrolysis would be equilibrated by condensation).

3.2.4. Fracture properties

Cold crystallization phenomena (leading to an increase in crystallinity) and hydrolysis (generating chain scissions) are responsible for a catastrophic decrease in the ultimate elongation of PA 6,6 films in the first 7 hours of exposure. If ones defines lifetime using the usual industrial endlife criterion (i.e. a reduction of 50% of the initial ultimate elongation), ones finds that lifetime is lower than 5 h at 70 and 80°C (Figure 5).

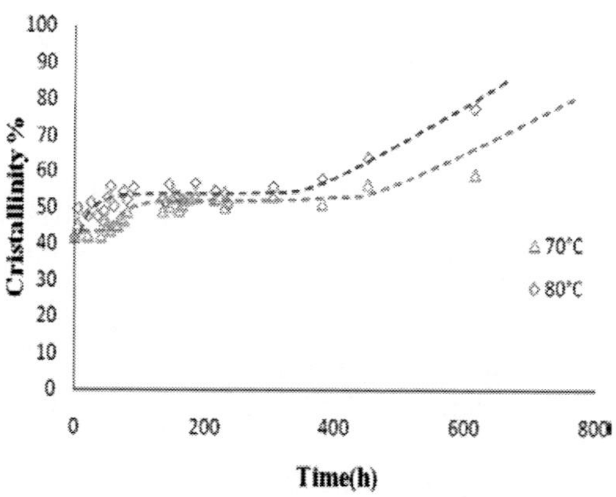

Figure 2: Changes in crystallinity ratio at 70 and 80°C at 100% RH.

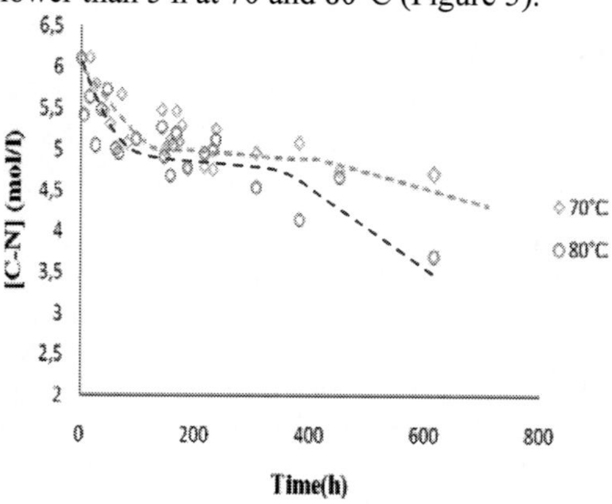

Figure 3: Change in C-N bonds concentration at 70 and 80°C at 100% RH.

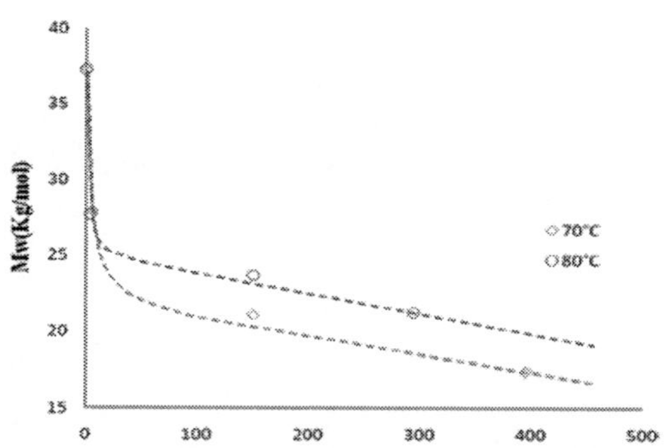

Figure 4: Changes in weight average molar mass at 70 and 80°C at 100% RH.

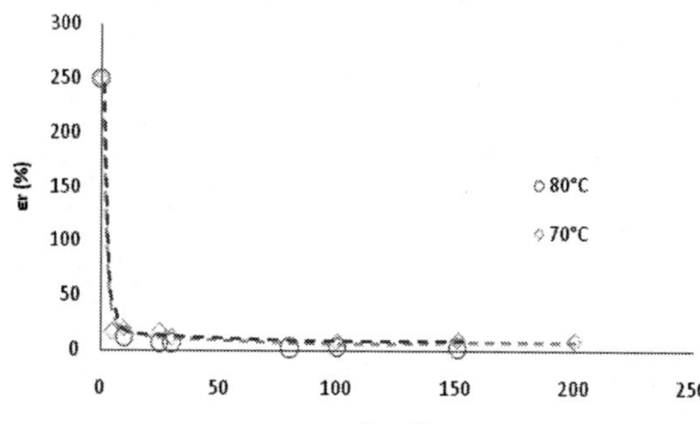

Figure 5: Changes in tensile ultimate elongation at 70 and 80°C at 100% RH.

REFERENCES

[1] Hernandez, R. J. J Food Eng 1994, 22, 495.
[2] Hernandez, R. J.; Gavara, R. J Polym Sci B 1994, 32, 2367.
[3] Zimm B.H.; Lundberg J.L., J. Phys. Chem. 1956, 60, 425.
[4] Merdas I.; Thominette F.; Verdu J. Polym. Degrad. Stab. 2003, 79, 419.
[5] Zimmerman, J. Encyc. Polym. Sci. Eng., 2d ed., ed. by H.F. Mark, et al. John Wiley and Sons, New York, 1985-1989.
[6] Bueche, F. J. Chem. Phys. 1954, 22(4), 603.

Optimization of the Processing of HNBR/Organoclay Nanocomposite

Rahul M. Cadambi[a] and Elaheh Ghassemieh

Department of Mechanical Engineering, The University of Sheffield, S1 3JD, United Kingdom
[a]R.Cadambi@Sheffield.ac.uk

Abstract. In the current investigation, rubber/clay nanocomposites were prepared by two different methods using hydrogenated nitrile butadiene rubber (HNBR) and the organoclay namely Cloisite 15A (C15A). A new novel approach involving swelling of C15A by ulltrasonication in HNBR solution has been carried out for improving the exfoliation and compatibilization of organoclays with HNBR matrix. With the addition of 5phr of clay, the elongation at break and tear strength improved by 16% and 24% respectively. The effect of coupling agents namely amino functional silane and tetrasulfido silane on the nanocomposites have been investigated. The elongation at break and tear strength improved by 46% and 77% respectively with the use of silanes. The improvement in the mechanical properties attributes to improved interaction between the organoclays and HNBR matrix. This interaction has been studied by X-ray diffraction and transmission electron microscope. Pre-dispersion technique clearly suggests very good improvement in the dispersion and properties due to better filler-rubber compatibility.

Keywords: Hydrogenated nitrile butadiene rubber, organoclays, nanocomposites, silanes.
PACS: 78.67.Sc

INTRODUCTION

Hydrogenated nitrile butadiene rubber (HNBR) are widely used for high demanding critical applications due to their unmatched resistance to heat, oils, ozone and very good resistance to swelling by aliphatic hydrocarbons [1]. Conventionally, large amount of fillers are added into elastomeric matrices to achieve the required mechanical properties. Recently Toyota researchers [2] showed that small addition (<10 wt%) of nanoclays lead to very good improvement in physical, mechanical, chemical and thermal properties. Due to the high aspect ratio and reinforcement at nanoscale level, the nanofillers may be used as a substitute for traditional fillers in elastomeric matrices.

The aim of this investigation is to optimise the best suitable processing method and filler adhesion agent for synthesising HNBR/clay nanocomposites. Two different methods have been employed for the preparation of nanocomposites. A new novel approach has been employed which involves sonication of organoclays with HNBR solution. The effect of different silane coupling agents to the rubber solution has also been investigated. The literatures reports more about sulphur cured vulcanized HNBR nanocomposites. Our focus is mainly on preparing peroxide cured HNBR/clay nanocomposites for developing a rubber with improved life at high temperature and pressure. The optimum percentage of the organoclay added to the HNBR matrix has been found [3] to be 5phr for our formulation. The synthesised HNBR nanocomposites have been analysed by morphological studies and mechanical testing.

EXPERIMENTAL

Materials

A fully saturated HNBR containing 36% acrylonitrile and all the other rubber compounding materials were kindly supplied by James Walker Co, UK. The compounding formulation was followed as per James Walker's recipe. Commercially available organoclay (C15A) containing sodium montmorillonite was procured from Southern Clay Products Inc. The basic recipe was kept common for all the sample preparations. Xiameter OFS-6011 an amino functional silane (AS) and Xiameter Z-6940 a tetrasulfidosilane (SS) kindly supplied by Dow Corning were used as coupling agents.

Preparation of HNBR/clay nanocomposites

CP1255, *Vth International Conference on Times of Polymers (TOP) and Composites,*
edited by A. D'Amore, D. Acierno, and L. Grassia

HNBR/clay nanocomposites were prepared by conventional mechanical shear mixing technique. Addition of organoclays to HNBR was evaluated by two different methods, namely direct addition of clays and use of ultrasonically pre-dispersed rubber/clay master-batch. The pre-dispersion of rubber/clay was carried out as follows. Initially, HNBR was swollen in methyl ethyl ketone (MEK). The organoclay was dissolved in MEK separately by continuously stirring it to obtain a uniform mixture. Both the clay solvent and swollen HNBR was mixed together and subjected to ultrasonication for a period of 30mins by using Branson Sonifier 450 composed of solid titanium stepped horn. The HNBR/clay master-batch after sonication was evaporated under vacuum at room temperature to obtain a pre-dispersed rubber/clay master-batch. In order to analyse the influence of different silane coupling agents on organoclays and HNBR, the separate rubber/clay master-batch was prepared by adding 5wt% of both AS and SS coupling agents in the similar way as described above. HNBR nanocomposites were prepared using Francis Shaw K0 intermix under 25rpm at 90°C for a period of 5mins. In the first method, C15A subjected to freeze drying was added directly to the mixing chamber along with other compounding materials. In the second method, the sonicated HNBR/C15A master-batch was added to obtain the nanocomposite without any change in the basic recipe. The amount of clay loadings for all the prepared samples was kept at 5phr. The material dropped out from the mixer was passed through a two-roll mill. The rubber test sheet samples were cured for 6mins at 185°C.

In order to simplify reference to different formulations, abbreviations are used in the rest of this report. For example H15AMBAS5 indicates a formulation of HNBR (H) with 5phr (5) of C15A added as organoclay produced by the method of master batch (MB) using the coupling agent amino silane (AS).

Characterisation

XRD analysis was performed on Siemens 500-D diffractometer with Ni filter and Cu Kα (λ=1.5406Å) radiation. The diffraction angle was scanned from 3° to 15° in steps of 0.02 at a rate of 0.5°/min for rubber nanocomposites. High resolution transmission electron microscopic (TEM) analysis was performed on Philips CM200 FEG TEM operated at an accelerating voltage of 80kV. The samples for TEM analysis were taken from the direction of milling. Ultrathin sections of approximately 50nm in thickness were cut with a diamond knife at -100°C using Lecia ultracut cryo microtome. Sections were carefully placed on the 200 mesh formvar-coated copper grids before performing TEM.

Tensile testing was carried out as per BS 903 Part A2 / ISO 37. Tear test was carried out using a crescent shaped samples according to BS 903 Part A3 / ISO 34-1. Both the tensile and tear tests were carried out at room temperature on Instron 3366 instrument. Hardness test was performed on IRHD instrument as per BS 903 Part A26N / ISO 48. An average of atleast five measurements were considered for the analysis. Compression set test was carried out to as per BS 903 Part A6 / ISO 815. The specimens were compressed by 25% at 150°C for duration of 24hrs. The compression set readings on three samples were taken after 30mins of recovery.

RESULTS AND DISCUSSION

Structural Characterisation

Figure 1 shows the XRD patterns of various prepared HNBR nanocomposites. C15A shows an average basal spacing of d = 23.7Å and 13.8Å (2θ = 3.72° and 2θ = 7.26°). The XRD patterns for various samples synthesized by different methods shows the disappearance of the diffraction peaks to the corresponding C15A basal spacing. This clearly shows that there is exfoliation of the silicate nanolayers. XRD patterns of pre-dispersed HNBR/C15A samples by ultrasonication shows more exfoliation compared to the direct addition method. Pre-dispersed master-batch seems to be more effective for dispersing organoclays into the HNBR matrix for obtaining an exfoliated nanocomposite structure. Also, it is known that the XRD measurements cannot always discriminate between intercalates and a partially or fully exfoliated morphology [4]. TEM (figure 2 (a) and (b)) shows uniform dispersion of silicate layers in the HNBR matrix at two different magnifications.

Mechanical Properties

Tensile strength, tear strength, elongation at break and modulus at different elongations (50%, 100% and 200%) of the prepared nanocomposites are compiled in table 1. It can be observed that with small amount (5phr) of C15A, there is noticeable improvement in the modulus by 40%. This indicates the presence of good interaction of the highly anisotropic silicate layers with the rubber matrix obtained by both the methods. Also, an improvement in the tensile strength without any loss in the elongation at break by the use of pre-dispersed HNBR/clay master-batch could be observed. This shows that the pre-dispersion technique has resulted in improved reinforcement of organoclays in the rubber matrix. Silane samples H15AMBAS5 and H15AMBSS5 shows enhanced mechanical properties.

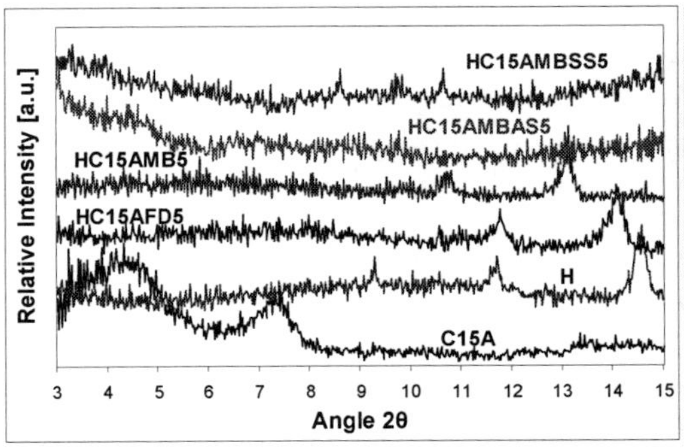

Figure 1. X-ray diffraction patterns of various prepared HNBR/C15A nanocomposites

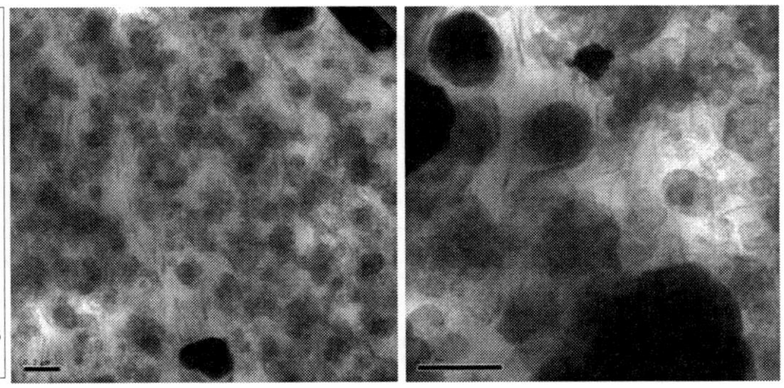

(a) (b)

Figure 2. TEM micrographs of HNBR/C15A nanocomposites

With the use of SS silane, there is very good improvement in tensile strength and elongation by around 10% and 46% respectively. The results indicate that silanes improve the adhesion of the nanoclays onto the HNBR matrix. With the small amounts of incorporation of organoclays, there is considerable improvement in tear strength for all the prepared samples. SS coupling agent has improved the tear strength by 77%. This clearly suggests that SS is an effective coupling agent for the preparation of nanocomposites. The improved hardness relates to the higher strength of the samples with the use of small amounts of organoclays. It is worth to pointing out that there is no loss of compression set with the addition of the organoclays indicating that there is no loss in the elastic properties. From the above results, it can be proposed that the silicate layers are uniformly dispersed and well exfoliated. Synthesis of HNBR/clay nanocomposites by using pre-dispersed rubber/clay master-batch is an optimum method for improving the compatibility of nanofiller in the rubber matrix.

TABLE 1. Mechanical properties of different HNBR/clay nanocomposites

	H	**H15AFD5**	**H15AMB5**	**H15AMBAS5**	**H15AMBSS5**
Tensile strength (MPa)	28.4 ± 0.88	28.0 ± 0.68	28.85 ± 0.74	30.7 ± 0.65	31.3 ± 0.75
Elongation at break (%)	215 ± 7	232 ± 11	250 ± 11	240± 8	313 ± 12
Modulus, 50% (MPa)	3.37 ± 0.15	6.65 ± 0.08	6.73 ± 0.17	4.3 ± 0.15	3. 7 ± 0.14
Modulus, 100% (MPa)	9.7 ± 0.38	13.66 ± 0.20	13.65 ± 0.19	9.98 ± 0.22	7.41 ± 0.33
Modulus, 200% (MPa)	23.56 ± 0.73	25.32 ± 0.5	26.69 ± 0.4	25.73 ± 0.5	20.7 ± 0.50
Tear strength (MPa)	39.6 ± 0.88	47.7 ± 0.19	48.9 ± 0.07	49.6 ± 0.25	70.2 ± 0.18
Hardness, IRHD	79 ± 0.8	80 ± 0.7	79 ± 0.3	84 ± 0.5	82 ± 0.6
Compression set (%)	24.1 ± 1.2	15.6 ± 1.1	15.5 ± 0.9	16.3 ± 1.2	16.4 ± 1.0

CONCLUSIONS

The structural characterization and mechanical properties of HNBR/clay nanocomposites synthesized by two different methods i.e. direct addition and pre-dispersed master-batch method have been studied. For 5phr level of clay loadings, both the methods give rise to well dispersed and exfoliated structure as verified by XRD and TEM studies. The pre-dispersion technique gives rise to improved compatibility with the filler and matrix compared to the other method. With the use of silanes, there is very good improvement in the mechanical properties and SS seems to be an effective coupling agent for improving the adhesion between the nanofiller and rubber matrix.

REFERENCES

1. Michael E. Wood, *Rubber Technology-Compounding and Testing for Performance*, Ed. by John S. Dick, Hanser Publishers, Munich, 2001, pp. 201-203.
2. Y. Kojima, A. Usuki, M. Kawasumi, A. Okada, Y. Fukushima, T. Kurauchi, and O. Kamigaito, *Journal of Materials Research,* **8**, 1185-1189 (1993)
3. R. M. Cadambi and E. Ghassemieh, *unpublished results.*
4. B. M. Alexander and W. G. Jeffrey, *Journal of Applied Polymer Science,* **87**, 1329-1338 (2003).

ACKNOWLEDGEMENT

The authors are grateful to James Walker Co, United Kingdom for helping us in the formulation of the compound and valuable suggestions in the preparation of the samples.

EFFECT OF THE CLAY MODIFICATION ON THE THERMAL PROPERTIES OF PVC

J. Zykova[a], A. Kalendova[a], L. Kovarova[a], J. Malac[a]

[a]*Department of Polymer Engineering, Tomas Bata University in Zlin, Faculty of Technology, TGM 275, 762 72 Zlin, Czech Republic*

Abstract. The intercalation of organic compound is necessary to functionalize the montmorillonite surface. So, the intercalation of diethylene glycol and polyethylene glycol and the influence of plasticizer, namely tricresyl phosphate and isodecyldiphenyl phosphate as co-intercalating agents were studied. For the PVC/clay nanocomposites the suspension type of PVC was used and the compound was prepared by the melt intercalation method. Two kinds of MMT (Cloisite®Na$^+$, Cloisite®30B) and laboratory modified MMT by intercalation (diethylene glycol, polyethylene glycol) and co-intercalation (tricresyl phosphate and isodecildifenyl phosphate) agents were used for the set of nanocomposite samples. The thermal stability of nanocomposite specimens was tested using of thermo gravimetric analysis, DMA and pH method.

Keywords: PVC, montmorillonite, thermal stability
PACS: 81.07.Pr Organic-inorganic hybrid nanostructures

INTRODUCTION

Polymer composites belong to the most dynamically development materials in the 21st century. Because, these materials are able to offered many advantages and due to the optimal ration of polymer and filler could bring materials whose have exactly corresponding properties of the specific PVC product.

Polyvinyl chloride (PVC) is one of the most widespread thermoplastic materials in the world owing to its valuable properties, wide applications, high chemical resistance, barrier properties and low cost [1,2]. On the other hand, thermal stability and processibility of PVC is inferior compared with common polymers, for example polyethylene, polypropylene, polyamide. The poorer thermal stability is close-knit with the manufacturing properties and reduction of usage time of the products. Therefore, it is necessary to search PVC with a good thermal stability [3,4].

Preparing PVC/clay nanocomposites can be one of the effective ways to improve PVC final properties. For the preparation of polymer/clay nanocomposites, modification of layered silicate is very important factor due to achievement of polymer/clay exfoliated structure. During modification of the clay the metal ions on basal surfaces of the layered clay must be exchanged to organic cations that change polarity of the gallery environment and the forces between platelets are decreased and hence the compatibility of polymer-clay is improved [5,6,7]. Melt intercalation is the most widely used method for preparation of nanocomposite materials, but this process requires high temperatures. If this temperature is higher than the thermal stability of the applied organic agent, the decomposition of this agent occurs [7-10].

In this paper, the effect of modification of MMT on the thermal stability of PVC/MMT nanocomposites is investigated. Two kinds of MMT were used and for the modification of MMT, various kinds of intercalation and co-intercalation agents were applied. Then, thermal stability was measured.

EXPERIMENTAL

Materials

In this work, suspension polyvinyl chloride (Neralit 652) was suplied by Spolana a.s., Czech Republic. Two different types of montmorillonite (MMT) were obtained from Southem Clay Products, Inc. Texas USA: natural unmodified MMT Cloisite Na$^+$, which has cation exchange capacity of 92.6 mequiv./100g, and MMT organically modified by alkyl quaternary ammonium salt Cloisite 30B, which has cation exchange capacity of 90 mequiv./100g. Finally, intercalation agents (diethylen glycol (DEG) and polyethylene glycol (PEG)), and co-

CP1255, *Vth International Conference on Times of Polymers (TOP) and Composites,*
edited by A. D'Amore, D. Acierno, and L. Grassia

intercalation agents (tricresyl phosphate (TCP) and isodecyl diphenyl phosphate (IDDP)) were utilized as modifiers of montmorillonite.

The first aim of this research was to prepare modified fillers from two types of MMT: MMT Cloisite Na^+ and Cloisite 30B. The ion–dipole method was employed for the modification of nanofillers by intercalation and co-intercalation agents at 85 °C in laboratory mixer.

Nanocomposite materials were prepared by mixing montmorillonite with suspension type of PVC on Buss KO-kneader MKS 30. The operating temperature ranged from 130 to 150 °C and the screw speed was 70 rpm.

Methods

pH method was used for determination of the tendency of PVC compounds to evolving hydrogen chloride and any other acidic products. A sample in a test tube was heated in nitrogen atmosphere at 200°C until a predetermined change in the pH of absorbing solution was obtained, required for a certain amount of hydrogen chloride. Times required for pH change to 3.8 were measured and named the stability times.

Dynamic mechanical analysis (DMA) of unfilled PVC and PVC/clay nanocomposite samples was carried out in tensile mode at 30 °C, a constant frequency of 1 Hz and force amplitude of 1N.

Thermo gravimetric analysis (TGA) was carried out using TGAQ500 analyser. The mass loss vs. temperature was measured at a heating rate of 20 °C/min in inert atmosphere from laboratory temperature to 600°C.

Result and discussion

For the determination of tendency of unfilled PVC and nanocomposite samples to evolve hydrogen chloride and any other acid product at 200 °C, pH method was used. The results of the stability tests of the nanocomposite materials are specified in Table 1. As can be seen, the PVC without clay possesses the longest stability time from all PVC samples (65 min). On the other hand, all of the laboratory intercalated nanocomposite samples had longer stability times (37 – 49 min) than PVC filled by the commercially intercalated MMT (Cloisite 30B) (23 min). When PVC samples filled by MMT modified by DEG or PEG were compared, the PVC/MMT/DEG had about 59 % higher stability time than PVC/MMT/PEG. If the half amount of intercalation agent was replaced by co-intercalation agents (TCP, IDDP), extremely rise of the stability occurred. PVC/MMT intercalated by PEG and co-intercalated by TCP or IDDP had about 57 % higher stability time (49 and 48 min) in comparison with PVC/MMT intercalated only by PEG (26 min), in particular. It could be said that the phosphate co-intercalation agents could be suitable for improving thermal stability of PVC/clay nanocomposites.

TABLE 1. Stability times of PVC measured by pH method

Samples	Time to degradation (min)	Temperature(°C)	Ratio*
PVC	65	200	-
PVC+Na^+	37	200	1 : 0.5
PVC+Na^++DEG	44	200	1 : 0.25 : 0.25
PVC+Na^++DEG+TCP	46	200	1 : 0.25 : 0.25
PVC+Na^++DEG+IDDP	47	200	1 : 0.25 : 0.25
PVC+Na^++PEG	26	200	1 : 0.25 : 0.25
PVC+Na^++PEG+TCP	49	200	1 : 0.25 : 0.25
PVC+Na^++PEG+IDDP	48	200	1 : 0.25 : 0.25
PVC+30B	23	200	1 : 0.5
PVC+30B+TCP	31	200	1 : 0.25 : 0.25

* It means ratio among nanofiller, intercalation and co-intercalation agents, respectively.

The next part of the study investigated the polymer-filler interaction using DMA. Table 2 illustrates the modulus obtained by DMA analysis of un-filled PVC and PVC/MMT nanocomposite samples at 30 °C. It was found that the storage and loss moduli increased about 22 % when a small amount of MMT was introduced to PVC. Furthermore, it could be concluded that PVC samples filled by MMT modified by intercalation and co-intercalation agents exhibit improvement of moduli. This increase may be attributed to the better dispersion of MMT in PVC matrix and storage interaction between MMT and PVC.

TABLE 2. The results of DMA analysis at 30 °C

Samples	Storage modulus (MPa)	Loss modulus (MPa)
PVC	169	68
PVC+Na^+	229	87
PVC+Na^++DEG	279	94
PVC+Na^++DEG+TCP	309	104
PVC+Na^++DEG+IDDP	320	108
PVC+Na^++PEG	311	103
PVC+Na^++PEG+TCP	353	111
PVC+Na^++PEG+IDDP	379	114
PVC+30B	451	132

Figure 1a and 1b represent thermo gravimetric (TG) and differencial thermo gravimetric (DTG) curves of un-modified PVC and PVC/clay nanocomposites, and the corresponding data are summarized in Table 3. The curves reveal two degradation steps of un-filled PVC and PVC/MMT nanocomposite samples. The first step can be attributed to dehydrochlorination of PVC, and the second step corresponds with condensation and fragmentation phases. For each degradation steps, the weight lost onset temperature (T_{onset}), which indicated 20 wt. % mass loss, and the temperature of the maximum mass loss (T_{max}) were characterised. Generally, T_{onset} of the first thermal degradation step decreased for all nanocomposites. However, T_{max} of the first thermal degradation step slightly increased when PVC was filled by unmodified MMT. PVC/MMT modified by PEG and TCP or IDDP had comparable T_{max} with un-filled PVC. Futhermore, T_{onset} of the second degradation step was also reduced for almost all nanocomposite specimens compared to unfilled PVC. Though, PVC filled by MMT modified by DEG and IDDP had the highest T_{onset} of the second degradation step (454 °C) and the lowest T_{onset} had nanocomposites PVC/MMT and PVC/30B (439 °C) when compared it to unfilled PVC (445 °C). It means that phosphate intercalation agent, especially TCP, could improve thermal stability of PVC. T_{max} of the second degradation step was increased for almost all nanocomposite samples.

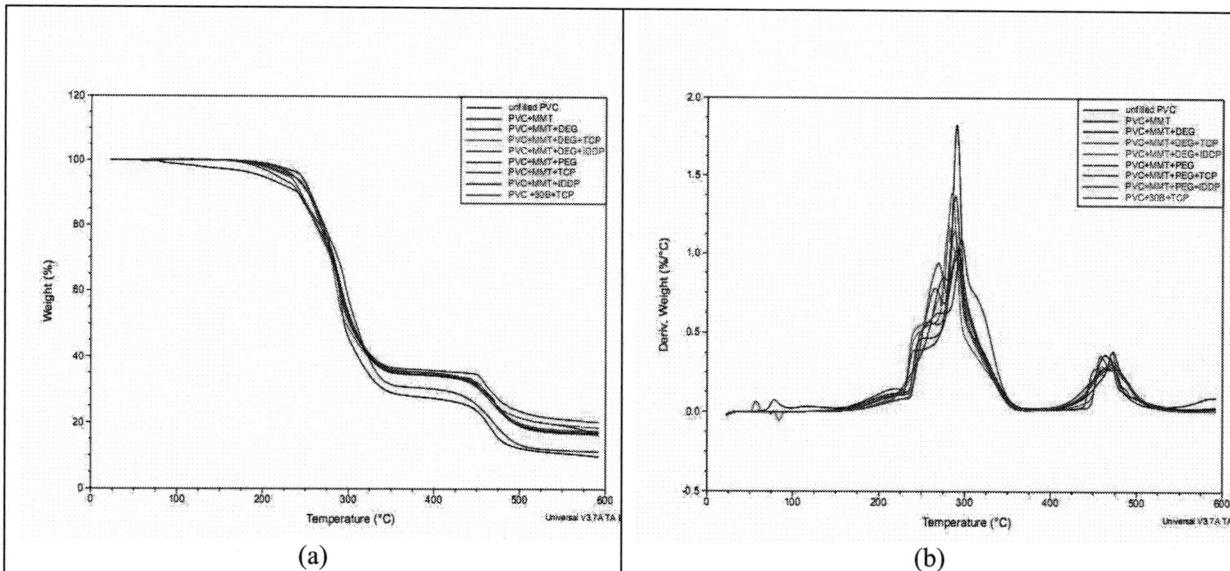

FIGURE 1. a) TGA and b) DTGA curves for unfilled PVC and PVC nanocomposite samples

TABLE 3. The results of TGA analysis

Samples	First step		Second step	
	T_{onset} (°C)	T_{max} (°C)	T_{onset} (°C)	T_{max} (°C)
PVC	268	290	445	464
PVC+Na^+	257	295	439	471
PVC+Na^++DEG	259	288	449	469
PVC+Na^++DEG+TCP	252	287	445	473
PVC+Na^++DEG+IDDP	256	285	454	471
PVC+Na^++PEG	251	291	452	472
PVC+Na^++PEG+TCP	248	290	450	458
PVC+Na^++PEG+IDDP	245	290	442	472
PVC+30B	243	280	439	463
PVC+30B+TCP	255	282	441	467

CONCLUSION

PVC is one of the most useful plastics in the word due to its wide application and low cost. However, during processing thermal degradation of PVC is usually evolved. Therefore, there is one way for the reduction of this problem, which is formation of PVC/clay nanocomposites. The aims of this research were to prepare organoclay using ion-dipole method with intercalation agents (diethylene glycol, polyethylene glycol), and co-intercalation agents (trycresil phosphate, isodecyldiphenyl phosphate) and also the preparation of PVC/clay nanocomposites using melt intercalation method. Finally, the influence of laboratory modified MMT on thermal properties of PVC were studied and compared to PVC/clay modified by quaternary amoniumsalt and un-filled PVC.

From the results of DMA it was shown that the moduli of PVC nanocomposites increase with using intercalation and co-intercalation agents. A possible explaination could be that clay modified by intercalation and co-intercalation agents is more polar than the clay modified only by the intercalation surfactant, and hence it is more compatible with the polymer matrix. However, the moduli of PVC/MMT laboratory modified are still lower than PVC/MMT commercially modified. It means that PVC/MMT commercially modified had the best exfoliation structure.

The thermal stability of PVC/MMT nanocomposites prepared by melt intercalation method was lower than unfilled PVC. The low thermal stability is supposedly because of the presence of metals, free and interlayer water in MMT. On the other hand, almost all PVC/MMT laboratory modified nanocomposites had enhanced thermal stability compared to PVC/clay commercially intercalated by quaternary ammonium salts.

ACKNOWLEDGMENTS

This work was supported by the internal grant of TBU in Zlín No. IGA/16/FT/10/D funded from the resources of specific university research.

REFERENCES

[1] D. Y. Yang, Q. X. Liu, X. L Xie, F. D. Zeng, *Journal of Therrmal Analysis and Technology* **84**(2), 355-359 (2006).
[2] F. Gong, M. Feng, Ch. Zhao, S. Zhang, M. Yang, *Polym Deg Stab* **84**, 289-294 (2004).
[3] M. Pan, X. Li. X. Shi, H. Hu, L. Zhang, *J Appl Pol Sci* **94**, 277-286 (2004).
[4] J. Ren, Y. Huang,Y. Liu, X. Tang, *Polymer Testing* **24**, 316-323 (2005).
[5] Z. M. Lisng, Ch. Y. Wan, Y. Zhang, P. Wei, J. Yin, *J Appl Pol Sci* **92**, 567-575 (2004).
[6] W. H. Awad, G. Beyer, D. Benderly, et col., *Polymer* **50**, 1857-1867 (2009).
[7] B. Yalcin, M. Cakmak, *Polymer* **45**, 6623–6638, 2004.
[8] A. Marongiu, T. Faravelli, *Dipartimento di Chimica*, 2003, 519-193.
[9] T. Ren, J. Yang, Y. Huang, J. Ren, Y. Liu, *Polymer composites*, 2006, DOI 10.102/pc.
[10] A. Leszczyńska, J. Njuguna, K. Pielichowski, J. R., Banerjee, *Thermochimica Acta* **453**, 75-96 (2007).

Thermal Conductivity and Heat Capacity of a Nylon-6/Multi-wall Carbon Nanotube Composite Under Pressure

Junchun Yu[a], Bounphanh Tonpheng[a] and Ove Andersson[a]

[a]*Department of Physics, Umeå University, Umeå, Sweden*

Abstract. The thermal conductivity, κ, of nylon-6 increased 22% whereas the heat capacity per unit volume, ρc_p, decreased 10% by adding 2.1 wt% Multi-Wall Carbon Nanotubes MWCNTs. Simultaneously, the glass transition temperature, T_g, which was detected as a weak sigmoidal increase in ρc_p and a decrease in $d\kappa/dT$, increased 11 K. These results show that the MWCNTs-nylon-6 interaction restricts the segmental mobility of nylon-6 and decreases c_p of nylon-6.

Keywords: Thermal conductivity, heat capacity, nylon-6, multi-wall carbon nanotube, high pressure, glass transition.
PACS: 62.50.-p, 65.60.+a, 81.05.Lg, 81.07.De

INTRODUCTION

Carbon nanotubes (CNTs) attract intensive attention because of their extraordinary thermal and mechanical properties, and high electrical conductivity. This new form of carbon material has great potential to be utilized in thermal management, high performance composites, electronics etc. [1]. The thermal conductivity κ of single-wall carbon nanotubes (SWCNTs) and multi-wall carbon nanotubes (MWCNTs) has been studied experimentally. Fujii *et al.*[2] measured κ of individual SWCNT and found κ>2000 W m^{-1} K^{-1} for a CNT diameter of 9.8 nm, which further increased with decreasing diameter. Kim *et al.*[3] reported κ>3000 W m^{-1} K^{-1} for MWCNT at room temperature while that of SWCNTs and MWCNTs in bundles was ~250 and ~20 W m^{-1} K^{-1} respectively [4, 5]. Thus, adding relatively small amount of CNTs may dramatically improve κ of poor thermal conductors such as polymers. Tonpheng *et al.*[6] reported an increase of κ by 120% for polyisoprene with 5wt% SWCNTs. The improvement was roughly proportional to the percentage of SWCNTs in the polymer composite, but the increase was lower than that predicted by simple mixture rules, which is likely due to high interfacial thermal resistance. In this work, we study κ, heat capacity and glass transition behavior for nylon-6 and its MWCNT composite and also explore the possibility of improved interfacial contact and percolation through high-pressure densification.

EXPERIMENTAL SECTION

MWCNTs (Nanocyl-3150) were purchased from Nanocyl, Belgium and purified before use. A typical purification procedure included oxidation in air, washing in 37% HCl acid (Sigma-Aldrich, ACS reagent) and, subsequently, rinsing by water, filtration and drying [7]. The synthesizes of the nylon-6/MWCNTs composite was done by in-situ polymerization [8], using sonication to disperse MWCNTs (0.24 g or 2.1 wt%) in melted ε-caprolactam (10 g, Aldrich, 99%) at 80 °C, adding 6-aminocaproic acid (1 g, Sigma, ≥99%) as initiator afterwards and heating to 250 °C with simultaneous mechanical stirring under nitrogen atmosphere for 6 h. A pure nylon-6 control sample was synthesized using the same procedure.

The nylon-6/MWCNT and nylon-6 samples were hot pressed into plates (diameter: ~37 mm, thickness: ~ 3mm) at 250°C in an argon atmosphere, and inserted in a custom made Teflon® cell. The Teflon® cell, contained a type K thermocouple and a hot-wire probe (0.1 mm Ni-wire) for measurements of κ and ρc_p using the hot-wire method [9]. The cell was mounted in a piston cylinder apparatus and later loaded in a hydraulic press. The sample could be

CP1255, *Vth International Conference on Times of Polymers (TOP) and Composites,*
edited by A. D'Amore, D. Acierno, and L. Grassia

heated up to ca. 530 K by an electric heater surrounding the cylinder and cooled down to ca. 100 K by liquid nitrogen. The estimated inaccuracies are ±2% in κ and ±5% in ρc_p.

Transmission electron microscope (TEM, JEM-1230) and tapping mode Atomic Force Microscopy (AFM, MultiMode, Veeco Metrology) were performed to exam the MWCNTs purity and the dispersion of MWCNTs in nylon-6. Differential scanning calorimetry (DSC, Pyris Diamond) was used to analyze the degree of crystallinity.

RESULTS AND DISCUSSION

As shown in Figure 1, height and amplitude images of the 2.1 wt% nylon-6/MWCNT composite suggest a fairly good dispersion of MWCNTs in the nylon-6 matrix. The long, tubular and twisted structures, which are the polymer wrapped MWCNTs, have a typical diameter of ~50 nm. The TEM image (insert) shows that highly purified MWCNTs, with no visible amounts of amorphous carbon or metallic catalyst, are obtained after purification.

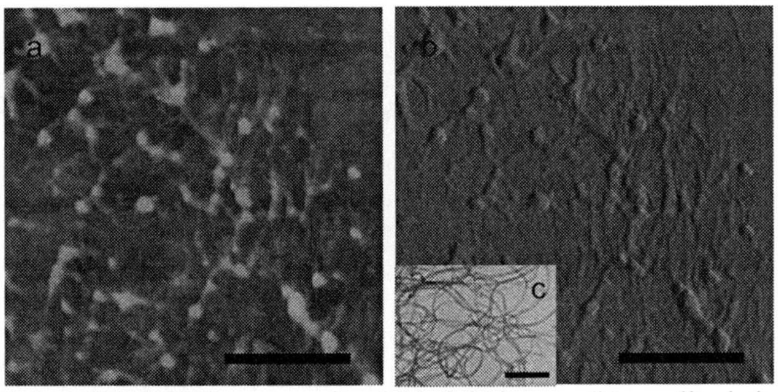

FIGURE 1. AFM image of Nylon-6/MWCNTs composite: (a) height image, (b) amplitude image (scale bar is 1 μm). The insert (c) is a TEM image of the purified MWCNTs (scale bar is 200 nm)

A pressure scan of both nylon-6 and its composites was performed at 298 K from low pressure to 1.0 GPa, as indicated in Figure 2a. The thermal conductivity of both samples increases due to densification, which reduces the inter-particle distances and, thus, improves the heat transfer between polymer chains. However, there is no indication of a stronger increase for the composite, which shows that the densification is insufficient to form a percolation network of the MWCNTs and that the thermal contact between the MWCNTs and nylon-6 does not improve sufficiently to make an impact on $\kappa(p)$. The behavior of κ below 1 GPa is therefore mainly governed by the changes in the nylon-6 matrix.

Figure 2b shows $\kappa(T)$ from 290 to 440 K at 0.8 GPa for both nylon-6 and the 2.1 wt% nylon-6/MWCNT composite. The thermal conductivity of nylon-6 shows a distinct change in $d\kappa/dT$ from positive to negative near 368 K, which is a typical behavior at the glass transitions temperature T_g of polymers [10]. This result is also confirmed by the simultaneously measured data for ρc_p (figure 2c). These show the (stretched weak) sigmoidal increase, which is characteristic of a glass transition. Thus, T_g of nylon-6, which is difficult to observe in several other characteristics that normally show clear T_g features, gives a distinct change in the temperature dependence of κ. The corresponding results for the composite are similar but with a change in $d\kappa/dT$ at 379 K, i.e. the 2.1 wt% MWCNTs raised T_g about 11 K. The shift of T_g is due to restricted segmental mobility, which is apparently caused by the MWCNT-nylon-6 interaction or, in terms of the free volume theory, the MWCNTs reduces the free volume or voids in the polymer.

The κ of the composite is 0.549 W m^{-1} K^{-1}, which is ~22% higher than that of nylon-6 (0.452 W m^{-1} K^{-1}) at 0.8 GPa, 298 K. As shown in Table 1, the degree of crystallinity of nylon-6 and the composite was the same, ~20% (melting enthalpy, ΔH, of 100% α crystalline nylon-6 is 241 J g^{-1}) [11], which shows that the enhanced κ of the composite was caused by the high κ of the MWCNTs and not by a change of crystallinity. Another interesting observation is the significant MWCNT induced decrease of ρc_p from 1.19 MJ m^{-3} K^{-1} for nylon-6 to 1.07 MJ m^{-3} K^{-1} for the composite at atmosphere pressure and 298 K (not shown in figure), i.e. a decrease of 10%. Since this is significantly larger than 0.2% calculated by the mixture rule [6, 12], it means that the MWCNTs decreases the heat capacity of nylon-6. Moreover, since the relative decrease is about the same above T_g as that below (figure 2c), the MWCNT- nylon-6 interaction apparently reduces the vibrational and configurational heat capacities of nylon-6 to about the same extent, and this effect should make the nylon-6 matrix slightly more rigid.

TABLE 1. Melting enthalpy (ΔH) and crystallinity (C_{DSC})determined by DSC

Sample	ΔH **(J/g)**	C_{DSC}
Virgin Nylon-6	47.9	19.9%
2.1wt% MWCNTs in nylon-6	47.4	19.7%

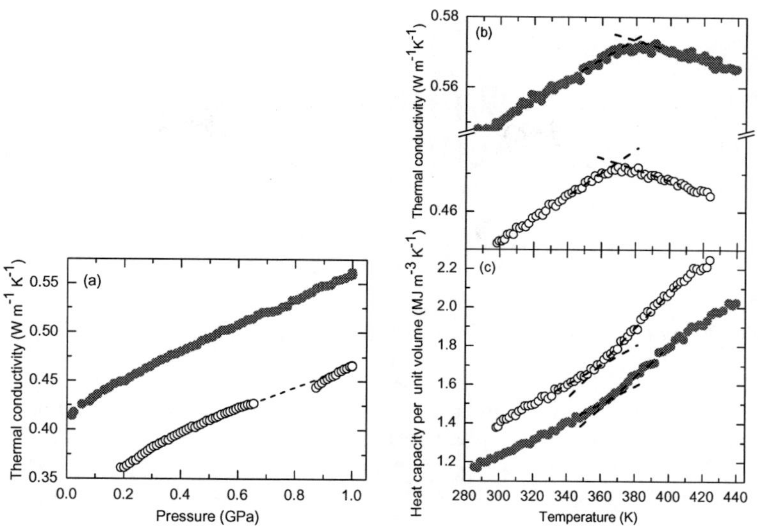

FIGURE 2. (a)Thermal conductivity for: (●) 2.1 wt% nylon-6/MWCNT composite and (○) nylon-6 measured on increasing pressure at 298 K. The dashed line represents a range where the data for nylon-6 were affected by a transition in the surrounding Teflon. (b) Thermal conductivity and (c) heat capacity per unit volume of: (●) 2.1 wt% nylon-6/MWCNT composite and (○) nylon-6 at 0.8 GPa measured on heating. The dash lines in both panels represent linear functions fitted to the data below and above the distinct change of slope. The point of intersection represents T_g.

CONCLUSIONS

In summary, κ and ρc_p for nylon-6 and its 2.1 wt% MWCNT composite have been measured, *in-situ*, under high pressure by the transient hot-wire method. The high κ of the MWCNTs increased κ of the composite by 22%. This increase was almost independent of pressure, which shows that the densification was insufficient to cause thermal percolation through the MWCNTs. Simultaneously, the MWCNTs increased T_g (11 K) but decreased ρc_p (10%) of nylon-6, which reveal MWCNT induced effects on the bulk properties of the nylon-6 matrix. These effects may have a similar origin and seems to be associated with the MWCNT-nylon-6 interfacial interaction.

ACKNOWLEDGMENTS

This work was financially supported by Swedish Research Council.

REFERENCES

1. M. Moniruzzaman and K. I. Winey, *Macromolecules* **39** (16), 5194-5205 (2006).
2. M. Fujii, X. Zhang, H. Xie, H. Ago, K. Takahashi, T. Ikuta, H. Abe and T. Shimizu, *Phys. Rev. Lett.* **95** (6), 065502 (2005).
3. P. Kim, L. Shi, A. Majumdar and P. L. McEuen, *Phys. Rev. Lett.* **87** (21), 215502 (2001).
4. W. Yi, L. Lu, Z. Dian-lin, Z. W. Pan and S. S. Xie, *Physical Review B* **59** (14), R9015 (1999).
5. D. J. Yang, Q. Zhang, G. Chen, S. F. Yoon, J. Ahn, S. G. Wang, Q. Zhou, Q. Wang and J. Q. Li, *Physical Review B* **66** (16), 165440 (2002).
6. B. Tonpheng, J. Yu and O. Andersson, *Macromolecules* **42** (23), 9295-9301 (2009).
7. J.-M. Moon, K. H. An, Y. H. Lee, Y. S. Park, D. J. Bae and G.-S. Park, *The Journal of Physical Chemistry B* **105** (24), 5677-5681 (2001).
8. J. Gao, M. E. Itkis, A. Yu, E. Bekyarova, B. Zhao and R. C. Haddon, *Journal of the American Chemical Society* **127** (11), 3847-3854 (2005).
9. H. Bjorn and G. R. Russell, *Journal of Applied Physics* **68** (7), 3285-3292 (1990).
10. D. W. Van Krevelen, *Properties of polymers: correlations with chemical structure*. (Elsevier publishing company, 1972).
11. V. K.-H. Illers and H. Haberkorn, *Die Makromolekulare Chemie* **142** (1), 31-67 (1971).
12. N. R. Pradhan, H. Duan, J. Liang and G. S. Iannacchione, *Nanotechnology* **20** (24), 7 (2009).

Hemp reinforced composites: surface treatment, manufacturing method and fabric type effects

G.Cicala, G.Cristaldi, G.Recca

Dipartimento di Metodologie Fisiche e Chimiche per l'Ingegneria (DMFCI), Università di Catania,

Abstract. Hemp mats and weaved fabrics were used as received and after surface treatment as reinforcement for composites. Mercerization and amino silane surface treatments improved fibre/matrix adhesion and, as results, the mechanical properties of the composites were also improved. However, if surface treatment was too severe degradation of the mechanical properties of the single fibre was observed and this resulted in a reinforcing efficiency loss. Weaved fabrics obtained from twisted fibres in unidirectional and 0/90° architecture were used. The use of weaved fabrics lead to high improvements of composite mechanical properties despite the absence of fibre's surface treatment. The specimens manufactured by LRTM (Light Resin Transfer Moulding) showed enhanced mechanical properties compared to specimens made by hand lay up. Mechanical models were also used to predict the mechanical properties of the composites.

Keywords: Natural fibres, LRTM, Twisted fibres, Mechanical properties.
PACS: 62. Mechanical and acoustical properties of condensed matter

INTRODUCTION

The use of natural fibres as reinforcements for composite is attracting interest for a wide range of industries. Natural fibre mats are widely used in the automotive sector in interior and exterior components[1,2]. Other uses include, as examples: building components, furniture and packaging [3,4]. Among natural fibres for composites the bast fibres, extracted from the stems of plants such as jute, kenaf, flax, ramie and hemp are widely accepted as the best candidates due to their very good mechanical properties. Hemp was shown to have very promising tensile properties for applications where mechanical properties are a requisite [5-8]. Natural fibres are lighter and cheaper than glass fibres and these factors make usually them attractive despite the lower absolute mechanical properties compared to glass fibres [5]. Most of the studies reported in literature are focused on the use of mat fabrics which are the cheap alternative [9] among technical fabrics. However, several studies showed that the presence of short and randomly oriented fibres, which characterize mats, limits the performance efficiency of the fibres [10]. Andersons et al. reported a review of the models which can be used to predict the mechanical properties of composites reinforced with random distributed fibers [11]. These models are based on the rule of mixture model and take into account the presence of short and randomly distributed fibres introducing some correction factors. Overall, for the case of mat fabrics, a reduction of about half or more can be considered compared to the values predicted from the rule of mixture. An additional problem is the hydrophilic character of the fibre's surface which can lead to poor adhesion between fibres and hydrophobic matrix polymers [12,13].

In the present study hand layup technique and untreated hemp mats as reference manufacturing process and reinforcing fabrics were selected respectively. Incremental modifications to the manufacturing process and to the raw materials were considered separately such as: the use of LRTM (Light Resin Transfer Moulding) technique instead of hand layup, the pre-treatment of the fibre surfaces and the use of twisted fibres as alternative to mats. Test laminates were manufactured after each modification and mechanical properties measured and compared.

CP1255, *Vth International Conference on Times of Polymers (TOP) and Composites,*
edited by A. D'Amore, D. Acierno, and L. Grassia

MATERIALS AND METHODS

A general purpose unsaturated polyester resin was used as thermoset matrix. Methyl ethyl ketone peroxide (MEKT) and cobalt naphthenate were used as catalyst and accelerator respectively. 3-aminopropyltriethoxysilane (A1100) was purchased from Aldrich, Italy, and used without further purification. Several hemp fabrics were used in this study, varying from random mat fabric, purchased by Hempcore Ltd. (United Kingdom) to unidirectional [0°] and bidirectional [0°/90°] woven fabrics purchased by Canipificio Italiano, Italy. The woven fabrics were obtained weaving yarns of natural fibres made of stable filaments twisted together.

Hand layup was used to prepare the laminates for mechanical testing unless differently specified. Each composite was cured at room temperature for 48 h. Light Resin Transfer Moulding (LRTM) was selected as an alternative manufacturing technique to hand lay up.

RESULTS AND DISCUSSIONS

Fig. 1 reports the data on tensile testing for laminates manufactured by hand layup and reinforced with hemp mats before and after surface treatment. The results of the prediction of the cured laminates properties according to three models rule of mixture, Cox-Krenchel [11] and Madsen [21] are also reported.

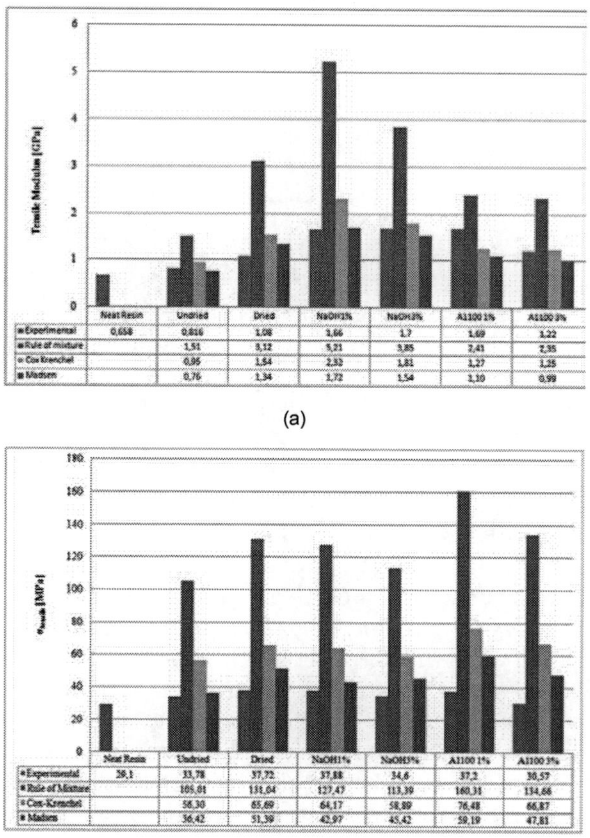

(a)

(b)

FIGURE 1 Tensile testing of surface treated mats: Modulus (a), Strength (b)

Fig. 1a reports the prediction for tensile modulus calculated for 10% fiber volume fraction. A porosity fraction of 5% was assumed when Madsen model was applied. The data show that the rule of mixture overestimates both the tensile modulus and the tensile strength for the composites. This is a consequence of the fibre orientation which, for the mats, is random and not unidirectional with the applied load as it is assumed with the rule of mixture model. The introduction in Cox-Krenchel model of correction factors to account for the random distribution of the fibres and for the fibre length efficiency improves the fitting with the experimental data. The fitting is further improved with Madsen model where the porosity fraction (5%) is considered. This finding supports the conclusion that the sample manufactured by hand layup contains some voids. Microscopy images confirmed the presence of voids both in the resin rich area and inside the fibre bundles.

If the laminates were manufactured by LRTM higher and uniform compaction pressure and reduced porosity were achieved which lead to higher mechanical properties.

As reported above the mechanical properties of the laminates reinforced with mats were reduced by a factor of about 3/8 because of the random distribution of the fibres. To overcome this limitation the use of weaved fabrics made of twisted yarns was considered. Two architectures, namely, unidirectional (UD) and 0/90 were considered (Fig. 2). The results of tensile testing obtained for laminates prepared with these fabrics are summarized in Fig.3.

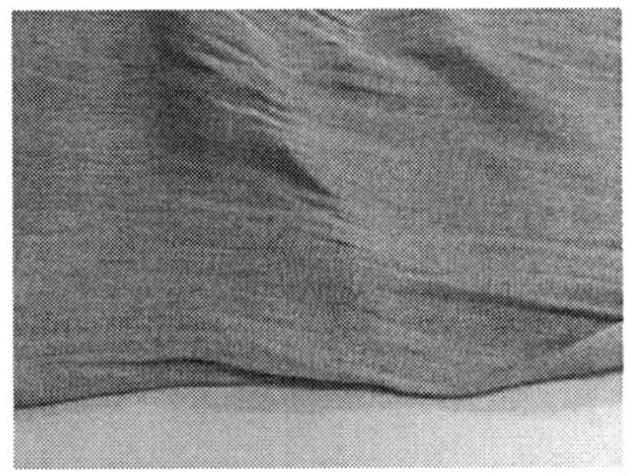

FIGURE 2 Hemp fabric made of twisted yarns

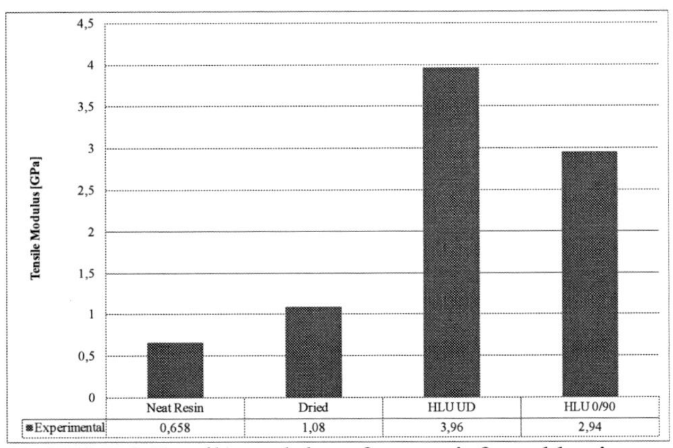

FIGURE 3 Tensile modulus of yarn reinforced laminates

Fig. 3 clearly shows that both modulus and strength are greatly enhanced when twisted yarns are used despite their low mechanical properties in dry form compared to single fibres extracted from hemp mats. This finding is the outcome of the impregnation of the yarns with the resin which, upon curing, stabilizes the yarn reducing the sliding effect of the filaments. Similar results were presented by Goutianos et als [14] when comparing impregnated and dry rowing strength. The good properties measured for the composites reinforced with hemp is the results of the favourable orientation, along the loading direction, of the staple fibres of the yarns. As it can be expected the 0/90 fabrics present lower mechanical performances compared to unidirectional fabrics. This result is due to the presence in the 0/90 fabric of yarns directed transversely compared to loading tensile direction. The modulus reported in Fig.3 are slightly lower than the values found in literature because of the manufacturing method (ie. hand layup) selected and of the low fibre volume fraction achieved.

CONCLUSIONS

The study presented showed that even with the application of simple hand lay up techniques the natural fibers can lead to reinforcement effects. The fiber efficiency is greatly enhanced if surface treatment and proper fiber architecture are used. The manufacturing method can also have profound effects and LRTM, despite being a low cost techniques, can allow to achieve good performances.

REFERENCES

1. Magurno A. *Vegetable fibres in automotive interior components.* Die Angew Makromol Chem 1999;272:99–107.
2. De Bruijn J.C.M. *Natural fibre mat thermoplastic products from a processor's point of view.* Appl Compos Mater;7: 415–20, 2000.
3. Soury E., Behravesh A.H., Rouhani Esfahani E., Zolfaghari A. *Design, optimization and manufacturing of wood-plastic composite pallet.* Material and Design, 30:4183-4191, 2009.
4. Singh B., Manorama Gupta, Anchal Verma, *Mechanical behaviour of particulate hybrid composite laminates as potential building materials*, Construction and Building Materials, vol.9, No.1, pp.39-44, 1995.
5. Saheb DN, Jog JP. *Natural fiber polymer composites: A review* Adv Polym. Technol. 1999;18(4):351–63.
6. Bledzki AK, Gassan *Composites reinforced with cellulose based fibres* J. Prog. Polym. Sci. 1999;24:221–74.
7. Eichhorn SJ, Baillie CA, Zafeiropoulos N, Mwaikambo LY, Ansell MP, Dufresne A, et al. *Current international research into cellulosic fibers and composites* J.Mater.Sci. 2001;36:2107–31.
8. Prasad BM, Sain M. *Mechanical Properties of Thermally Treated Hemp Fibres in Inert Atmosphere as Potential Composite Reinforcement* Mater. Res. Innov. 2003;7(4):231–8.
9. Madsen B. *Properties of plant fibre yarn polymer composites – an experimental study.* Ph.D. Thesis. Technical University of Denmark, Denmark, 2004, Series R, No. 82.
10. Aziz S.H., Ansell M. P., *The effect of alkanization and fibre alignment on the mechanical and thermal properties of kenaf and hemp bast fibre composites: Part 1 – polyester resin matrix*, Composites Science and Technology, 64: 1219-1230, 2004.
11. Andersons J., Sparnins E., Joffe R., *Stiffness and Stregth of Flax Fiber/Polymer Matrix Composites*, Polymer Composites, vol.27, issue 2, 221-229, 2006.
12. Kalia S., Kaith B.S., Kaura I., *Pretreatments of Natural Fibers and their Application as Reinforcing Material in Polymer Composites – A Review*, Polymer Engineering and Science, vol.49, issue 7, 1253-1272, 2009.
13. M.J.John, B.Francis, K.T. Varughese, S.Thomas, *Effect of chemical modification on properties of hybrid fiber biocomposites, Composites: Part A – Applied Science and Manufacturing*, 39: 352-363, 2008.
14. Goutianos S., Peijs T., Nystrom B., *Development of Flax Fibre based Textile Reinforcements for Comnposite Applications*, Appl. Compos. Mater., 13: 199-215, 2006.

Analysis of the creep behavior of thin polymer films by finite elements simulation of micro-bubble inflation

Sylvie Castagnet, Eric Lainé

Institut P'- Départment de Physique et Mécanique des Matériaux – 1 Avenue Clément Ader, BP40109, 86961 Futuroscope cedex, France ; sylvie.castagnet@ensma.fr

Abstract. Finite elements simulations have been performed to investigate the various effects bound to influence the viscoelastic response of nano-confined polystyrene films, among which surface tension, residual stresses, properties gradient. The work is based on the micro-bubble inflation experiment developed by McKenna et al. [1,2]. The preliminary results presented here focus on the analysis of the stress gradients into the film, the influence of the boundary conditions and some aspects of the film preparation.

Keywords: finite elements calculations; thin-films; creep; amorphous polymers
PACS: 68.60.Bs; 81.05.Lg; 66.20.Hg

INTRODUCTION

The viscoelastic properties of amorphous polymers are known to be modified by the nano-scale confinement enhanced in various situations, e.g. in ultra-thin films, nanopores, in nano-reinforced polymers or in interlamellar amorphous layers of semi-crystalline polymers. A better understanding of confinement effects at the nano-scale is necessary to improve the mechanical modelling of such systems but experimental access is very difficult. Several dedicated experiments have been developed to this aim, for instance based on calorimetry [3] or mechanical testing in ultra-thin films –by embedding of particles at the surface [4], dewetting [5], buckling instability [6] or creep inflation of micro-bubbles into an atomic force microscope (AFM) [2]. Most of them focus on the nano-confinement influence on the glass transition. However, results obtained from these techniques are very complex to be analyzed, and very dependent on the experimental protocol and the tested material: for instance, a decrease or increase of the glass transition temperature is measured in free-standing and supported films respectively. This seems to be due respectively to interactions between the film and the substrate, or to a possibly increased molecular mobility near the film surface, but both effects are still rather badly understood. Many open issues to explain the whole set of experiments and more generally nano-scale confinement effects.

The aim of the present work, based on the micro-bubble inflation experiment developed by McKenna et al. [1], is to use Continuum Mechanics and Finite Elements Simulations to contribute to highlight separately the influence of simultaneous effects, among which surface tension, residual stresses and molecular mobility gradients.

NUMERICAL MODELING

In the experiment, spin-coated nanometre thick films are laid on a substrate with 5 micron diameter channels. The film is initially drawn into the holes because of capillary forces, and after a one hour relaxation stage at the testing temperature, a constant pressure is applied: then, the bubble profile is tracked by and the inflation of the bubble is imaged by the AFM. It was shown that the bubble profile keeps a circular shape so that results can be analyzed into a membrane framework [1]: circumferential stress and strain can be calculated from the curvature radius of the bubble, and the biaxial creep compliance can be deduced. By applying a time-temperature superposition principle to curves obtained at different temperatures, the glass transition temperature can be determined: it was shown to decrease from the bulk value for the thickest films (about 100 nm) down to Tg-50°C for about 10 nm thick films. The present work is based on polystyrene (PS) data obtained by Plazek et al. (cited by Ferry

CP1255, *Vth International Conference on Times of Polymers (TOP) and Composites,*
edited by A. D'Amore, D. Acierno, and L. Grassia

[7]). It is crucial to identify the viscoelastic constitutive law from the glassy to the rubbery state to avoid any risky extrapolation when further applying time-temperature superposition. The material is assumed to obey a linear viscoelastic law modelled by Prony series of the shear modulus G and the bulk modulus K, as given in Equation (1).

$$G(t) = G_0\left[1-\sum_{i=1}^{50} g_i\left(1-\exp\left(-\frac{t}{\tau_i}\right)\right)\right] \quad ; \quad K(t) = K_0\left[1-\sum_{i=1}^{50} k_i\left(1-\exp\left(-\frac{t}{\tau_i}\right)\right)\right] \tag{1}$$

The deviatoric part of the constitutive law was identified by minimizing the difference between the shear strain evolution calculated from the compliance law displayed by Plazek at 100°C, and the simulated creep strain calculated in Abaqus® in a single quadratic element undergoing pure shear at the same stress (0.001 MPa). Comparison is plotted in Figure 1(a). It corresponded to a glassy modulus E_0=3000 MPa and a Poisson ratio of 0.4.

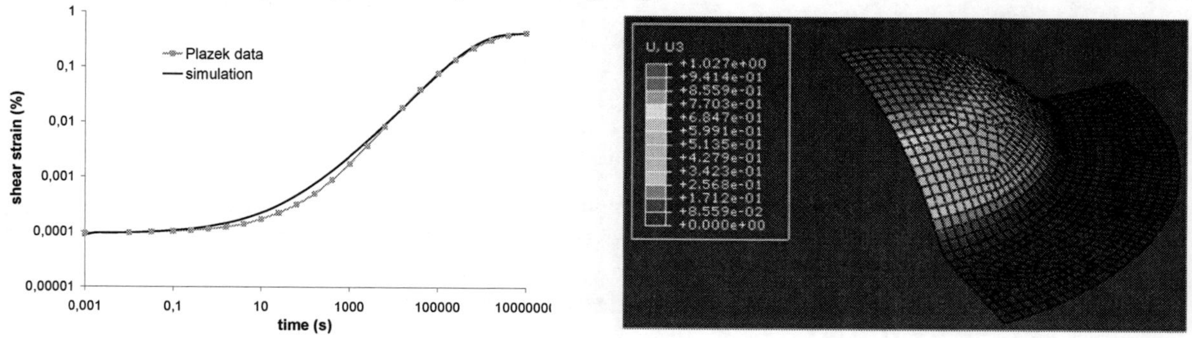

(a) (b)

FIGURE 1. (a) Identification of the deviatoric part of the constitutive from creep data by Plazek, and (b) numerical model used in Abaqus® simulations.

Since the hydrostatic part of the constitutive law cannot be identified from bulk experiments in the same polymer, the Poisson ratio was assumed to be constant. Two situations have been tested: an identical distribution for the (g_i) and (k_i) parameters (elastic Poisson ratio ν=0.4), and a constant bulk modulus K (k_i=0 and ν=0.49). Comparison between the bubble height and circumferential stress-strain curves in both cases support a negligible influence of the Poisson ratio. A comparison between simulations performed with shell and solid elements has been performed displaying similar results, so that axy-symmetric conditions were applied finally. The film (112 nm thick in the present paper) was discretized using 20 nodes iso-parametric quadratic bricks (C3D20) with 27 integration points. The film thickness was divided into 6 layers of elements. Besides the symmetry conditions, nodes at the right part of the lower surface were clamped in the vertical direction. Pressure (4.5 psi) was applied linearly during the first 5 seconds on the lower surface of the film (i.e. the 2.5 μm length corresponding to half diameter of the hole) and maintained constant for 10^5 seconds.

RESULTS

First, the deformation mode and the mechanical gradient along the film thickness have been investigated. Figure 2 displays the local circumferential stress-strain for several integration points across the film thickness at the middle of the bubble. As carefully explained by McKenna et al. [2], films first deform according to a flexural mode: the lower surface undergoes compression loading while the upper one is loaded in tension. After a few hundreds of seconds (about 500 in this case), the film behaves as a membrane. The profile is part of a circle (see in Figure 3(b)) which curvature radius allows calculating the local biaxial stress and strain (as made after 1240 and 2470 s and plotted with square symbols). At the end of the pressure loading step, a significant stress gradient can be observed in the thickness, followed by a rather quick stress homogenization.

As a first attempt, the simulation was compared to McKenna et al. creep experiments in 112nm-thick films at 4.5 psi and 99°C. Figure 3 shows that the film compliance is much lower in simulation than measured in the experiment. Several factors could explain the observed difference, among which a difference between the tested PS (despite close molecular weight), a consequence of thermo-mechanical history not described in the simulation or an influence of unrealistic boundary conditions applied in the simulation.

Consequence of stretching prior to the pressure application has been investigated first. The same film as previously simulated was stretched up to 20%, relaxed over one hour and inflated under the same pressure. The time evolution of the bubble height is plotted also in Figure 3. First, the bubble inflation appears to be slower than in the "virgin" film down, suggesting that the tensile residual stresses could act as a surface tension effect. However, this phenomenon cannot explain the difference observed with the experiment.

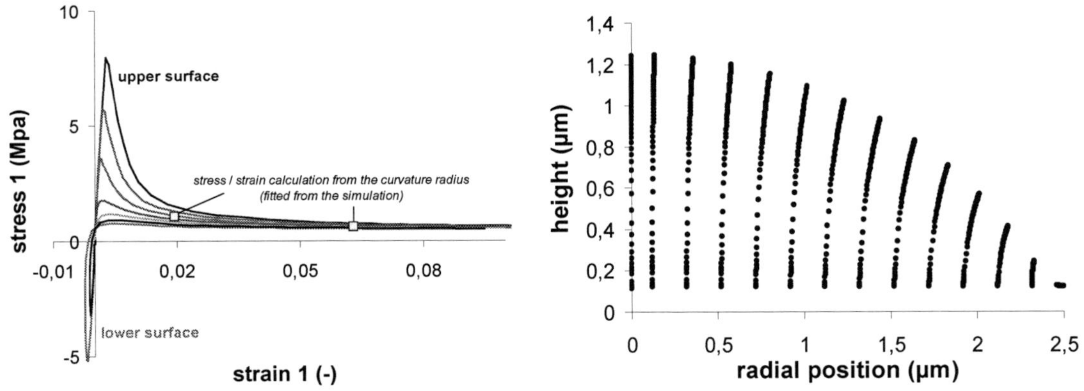

(a) (b)

FIGURE 2. (a) Time evolution of the local circumferential stress-strain behavior into the thickness and (b) evolution of the profile of a 112 nm-thick PS film undergoing creep at 4.5 psi and 100°C.

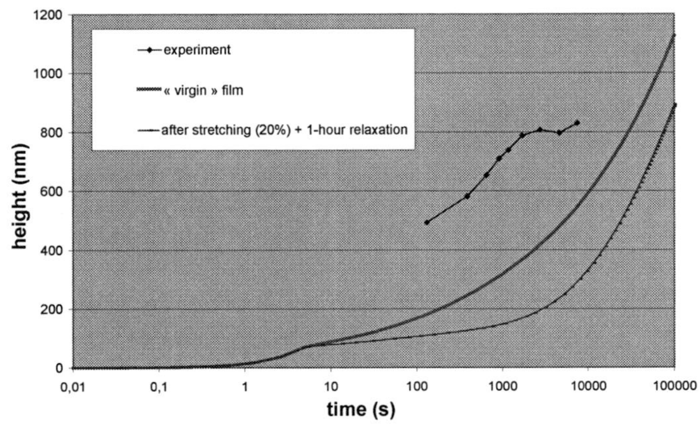

FIGURE 3. Comparison between the simulated inflated of a "virgin" 112 nm film, a pre-elongated and relaxed film and an experiment performed in the same pressure condition [2].

Work is in progress to make the numerical model more realistic and to investigate, qualitatively first, the relative influence of on the behavior of variable thickness films.

ACKNOWLEDGMENTS

This paper presents the first results from a study in progress: authors would like to gratefully acknowledge G. McKenna and P. O'Connell (Texas Tech University, USA) for their collaboration to it, and thank J.C. Grandidier (Institut P' - Poitiers, France) for fruitful discussion.

REFERENCES

1. P. O'Connell, G.B. McKenna, *Rev. Sci. Instr.* **78**, 013901 (2007).
2. P. O'Connell, S.A. Hutcheson, G.B. McKenna, *J. Polym. Sci. Part-B* **46**, 1952–1965 (2008).
3. Y.P. Koh, G.B. McKenna and S.L. Simon, *J. Polym. Sci. Part-B* **44**, 3518–3527 (2006).
2. Z. Fakhraai, J.A. Forrest, *Phys. Rev. Lett.* 95, 025701 (2005).
5. H. Bodiguel, C. Fretigny, *Macromolecules* **40,** 7291-7298 (2007).
6. C.M. Stafford, B.D. Vogt, C. Harrison, D. Julthongpiput, R. Huang, *Macromolecules* **39**, 5095-5099 (2006).
7. J.D. Ferry, *Viscoelastic Properties of Polymers*, New York: John Wiley & Sons, 1980, pp. 605.

Thermal aging of electrical wire jacket

Ines Mkacher[a,b], Yves Brument[a] and Xavier Colin[b]

[a]*EDF R&D Renardières, EDF - Laboratoire des Matériels Electriques, 77250 Moret-sur-Loing, France*
[b]*PIMM, ART et Métiers ParisTech, 151 boulevard de l'hôpital.75013 Paris, France*

Abstract The thermal degradation of high density polyethylene (PE) films, stabilized by a common commercial synergistic blend of antioxidants, were studied at 120, 140, 150 and 160°C in air under atmospheric pressure. Antioxidants depletion has been monitored by UV and FTIR spectrophotometry and induction oxidation time measurements. The peculiar shape of kinetic curves suggests that the physical loss of antioxidants cannot be neglected, but also that classical evaporation-reaction models are inadequate to predict such a kinetic behavior. A model assuming the existence of two distinct antioxidant phases in equilibrium into the PE matrix is tentatively proposed. The insoluble (dispersed) phase, corresponding to the antioxidant excess relatively to the solubility threshold, supplies in antioxidant the soluble phase, until its complete exhaustion, whereas the soluble phase evaporates and is consumed by the chemical reaction.

Keywords: Polyethylene, antioxidants, thermal ageing, physical loss, chemical consumption.

PACS: 82.35.-x

Introduction

In France, about 23% of the synthetic electric cables, especially low and medium voltage cables, are buried. Most of them are used already for 20 years, although their lifetime has been estimated to about 30 years **[1]**. They are constituted of an external polymer barrier (the jacket), made of carbon black filled high density polyethylene (PE) or plasticized poly(vinyl chloride) (PVC), protecting the cable against the mechanical and chemical aggressions (see Figure 1). It is suspected that, if this essential component degrades and losses its functionality, the remaining cable lifetime would be drastically shortened. Main possible causes of degradation are: physical loss and chemical consumption of antioxidants followed by polymer thermal oxidation in the case of PE **[2]**, and physical loss of plasticizers in the case of PVC **[3]**.

The aim of the present study is to evaluate the thermal stability of a high density PE jacket.

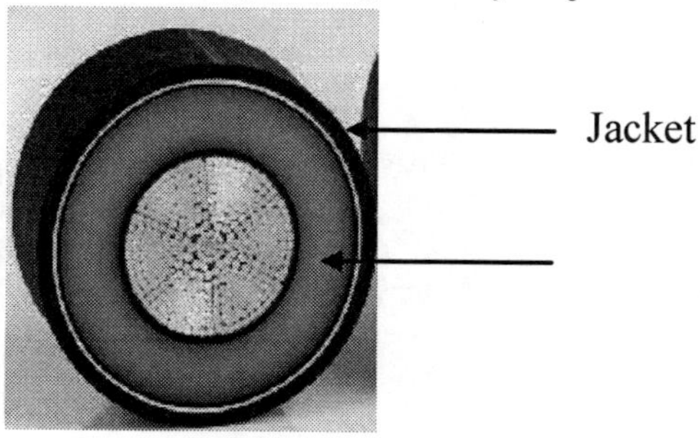

Figure 1 : View of a synthetic electrical cable

Experimental

Material

The high density PE grade (the stenungsund SWPE) was supplied by Borealis. Its main characteristics are: melting temperature of 129.5 ± 2°C, crystallinity ratio of 62 ± 5 % and density of 944 kg.m^{-3}. It is stabilized by a common synergistic blend of antioxidants (B225 of Ciba Geigy): about 0.22% by weight of hindered phenol (Irganox 1010) plus about 0.13 % by weight of organic phosphite (Irgafos 168). The presence of a secondary antioxidant of the phosphite type was confirmed by elemental analysis (detection of about 62 ppm of phosphorus).

Thin PE films of 140 ± 30 μm thickness were compression molded at 180°C under 3MPa pressure.

Methods

PE films were exposed in air-ventilated ovens at 120°C, 140°C, 150°C and 160°C during several thousands of hours. They were periodically removed from ovens in order to be analyzed by different analytical methods. The antioxidant depletion and degradation product build-up were monitored by UV (Perkin Elmer, Lambda 35, UV/visible spectrometer) and Fourier Transform IR spectrophotometries (Brüker, IFS 28 spectrometer, 4 cm^{-1} resolution) in a transmission mode. The residual oxidation time was determined at 200°C by Differential Scanning Calorimetry (TA Instruments Q10 and Q1000) with the following program: heating a sample of 4-8 mg weight from room temperature to 200°C at a 10°C.min^{-1} rate under a nitrogen flow of 50 ml.min^{-1}. After 5 min at 200°C to equilibrate the system, pure oxygen was admitted (50 ml.min^{-1}) and the DSC signal was recorded. The induction time OIT corresponds to the onset of the oxidation exotherm.

CP1255, *Vth International Conference on Times of Polymers (TOP) and Composites,*
edited by A. D'Amore, D. Acierno, and L. Grassia

Results and Discussions

Examples of kinetics curves of antioxidant depletion and degradation products build-up, obtained at 150 and 160°C in air under atmospheric pressure, are presented in Figures 2 and 3. These results call for the following comments:

- Irgafos 168 is a high temperature antioxidant used to prevent oxidation in processing conditions. It decomposes POOH groups by a non radical way **[4]**:

POOH + Dec $\rightarrow$ inactive products (k_{Dec}) where Dec represents a phosphite group.

On the contrary, Irganox 1010 is a low temperature antioxidant used to prevent oxidation at long term in use conditions. Since it contains hydrogen atoms most labile than those of any hydrocarbon substrate **[5]**, it scavenges efficiently radicals **[4]**:

$PO_2^{\bullet}$ + AH $\rightarrow$ POOH + inactive products (k_{AH}) where AH represents a phenol group.

Since both antioxidant types act in a distinct location of the "closed-loop" oxidation mechanism, they constitute, in principle, a synergistic blend of antioxidants **[6]**.

- Kinetic curves of antioxidant depletion display a peculiar shape constituted of three distinct domains: i) an initial period during which the antioxidant concentration decreases pseudo-exponentially with exposure time; ii) a pseudo-steady-state period during which antioxidant concentration remains apparently constant; iii) and a final period during which the antioxidant concentration vanishes rapidly.

Thus, antioxidant depletion can be characterized by four distinct quantities: initial antioxidant concentration $[X]_0$, initial rate of antioxidant concentration depletion v_{X0}; pseudo-plateau antioxidant concentration $[X]_p$ and time at which the antioxidant concentration vanishes t_X, where X refers to Dec or AH antioxidant. These quantities have been reported in Table 1.

It is clear that, in the temperature range under study (typically, for T $\leq$ 160°C), Irgafos 168 is noticeably less efficient than Irganox 1010.

- As a consequence, the kinetic curves of degradation product build-up are also constituted of three distinct domains: i) an initial period during which the degradation product concentration increases rapidly up to a maximal value; ii) a pseudo-steady-state period during which the degradation product concentration remains apparently constant; iii) and a final period during which the degradation product concentration decreases pseudo-exponentially and vanishes.

- In a recent publication **[7]**, we demonstrated that such a kinetic behavior cannot be explained by a classical evaporation-reaction model. This model can be applied in the present case for each type of antioxidant. It is considered three distinct antioxidant concentrations: a) concentration of dissolved, soluble antioxidant $[X]_{sol}$; concentration of excess, insoluble antioxidant $[X]_{ins}$; and saturation threshold above which the excess antioxidant becomes insoluble $[X]_{sat}$. The evaporation loss and chemical consumption of antioxidant during thermal exposure are expected to occur only for the soluble antioxidant phase.

According to this model, if the initial total antioxidant concentration is higher than its saturation threshold into the PE matrix: $[X]_0 > [X]_{sat}$, then the antioxidant exists in a insoluble phase (for instance, as a reservoir) and an antioxidant transfer from the insoluble to the soluble phase will equilibrates exactly the physical loss and chemical consumption in this latter, which leads to the formation a pseudo-plateau on the kinetic curves. As soon as the concentration of the insoluble phase vanishes, the total antioxidant concentration $[X] = [X]_{sol} + [X]_{ins}$ begins to decrease. In other words, according to this model, the pseudo-plateau antioxidant concentration corresponds exactly to the saturation threshold: $[X]_p = [X]_{sat}$.

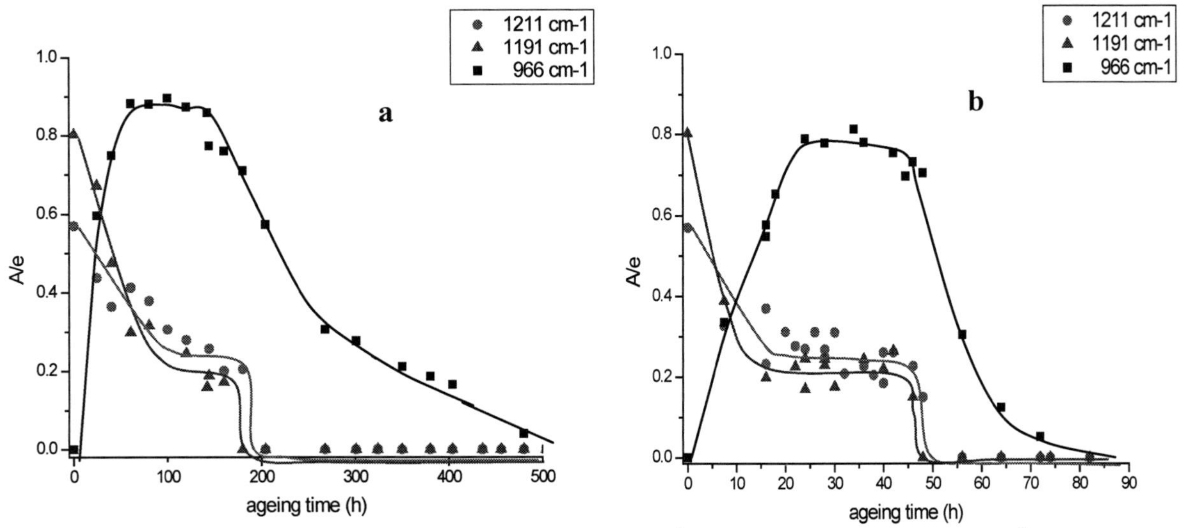

Figure 2: Kinetics curves of phosphite depletion (at 1211 and 1191 cm^{-1}) and phosphate build-up (at 966 cm^{-1}) determined by FTIR spectrophotometry after thermal exposure at 150°C (left) and 160°C (right) in air-ventiled ovens.

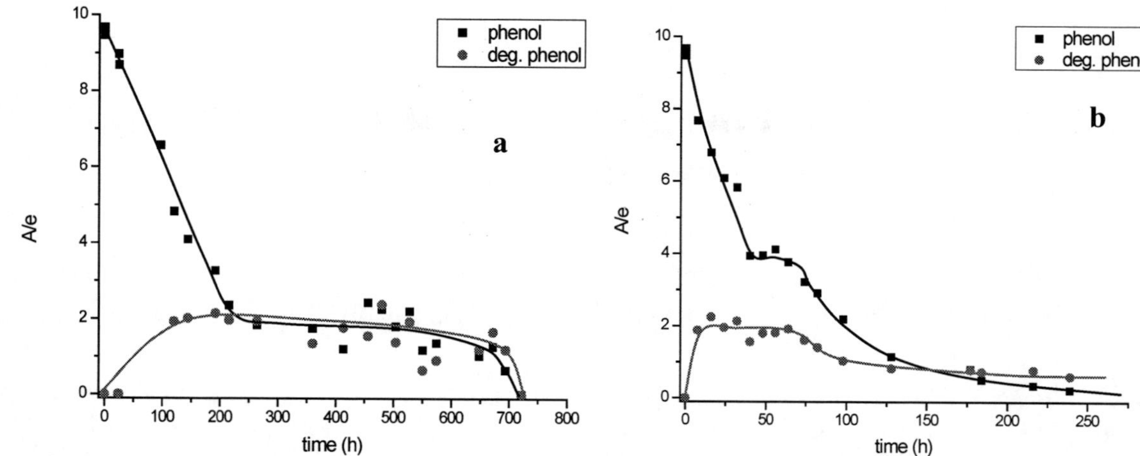

Figure 3: Kinetics curves of phenol depletion (at 275 nm) and quinoneic product build-up (at 283 nm) determined by UV spectrophotometry after thermal exposure at 150°C (left) and 160°C (right) in air-ventiled ovens.

Antioxidant	T (°C)	$[X]_0$ (mol.l^{-1})	v_{X0} (mol.l^{-1}.h^{-1})	$[X]_P$ (mol.l^{-1})	t_X (h)
Irgafos 168	150	1.9 10^{-3}	1.8 10^{-5}	6 10^{-4}	180
Irgafos 168	160	1.9 10^{-3}	1.2 10^{-4}	5 10^{-4}	48
Irganox 1010	150	7.0 10^{-3}	2.6 10^{-5}	1.2 10^{-3}	700
Irganox 1010	160	7.0 10^{-3}	1.7 10^{-4}	2.8 10^{-3}	250

Table 1: Main quantities used to describe the kinetics curves of antioxidant depletion at 150 and 160°C in air under atmospheric pressure.

Examples of kinetics curves of oxidation induction time (OIT), obtained at 150 and 160°C in air under atmospheric pressure, are presented in Figure 4. These curves display the three distinct domains already evidenced for each type of antioxidant in kinetics curves 2 and 3. So, in a first approximation, the OIT value can be considered as the sum of contributions of each antioxidant type:

$OIT = K_1 \times [Dec] + K_2 \times [AH]$

where K_1 and K_2 are constants which will be determined by OIT measurements on PE matrices stabilized by a single type of antioxidant.

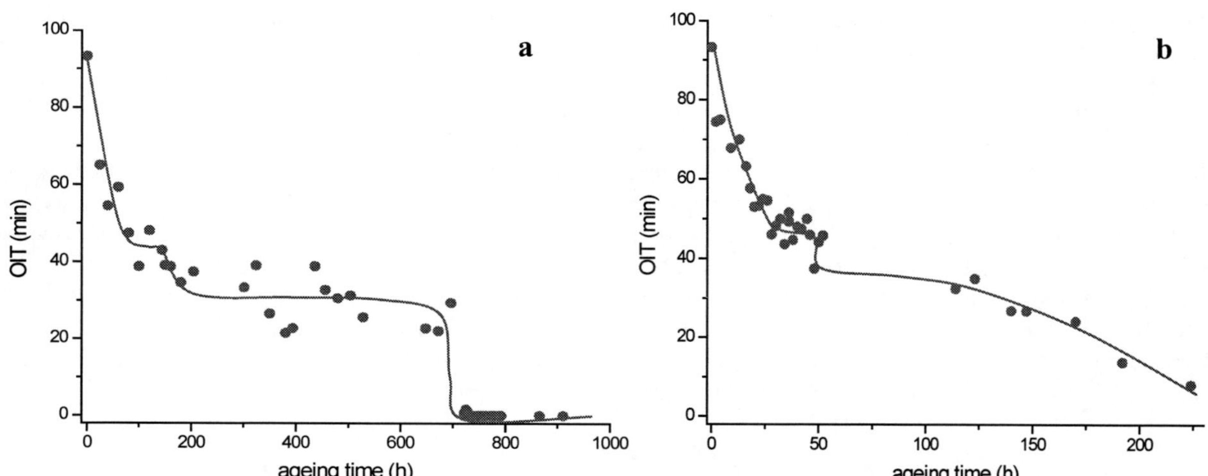

Figure 4: Kinetics curves oxidation induction time determined by DSC at 200°C in pure oxygen after thermal exposure at 150°C (left) and 160°C (right) in air-ventiled ovens.

REFERENCES

1. R. Grasset, *Sycabel*. 2005: EDF
2. Y.G. Hsuan, R.M. Koerner, *Journal of Geotechnical and Geoenvironmental Engineering*, 1998. **124**(6): 532.
3. L. Audouin, A. Andre, J. Verdu, *Journal of Vinyl Technology*, 1994. **16(1)**: 57.
4. H. Zweifel, *Plastics Additives Handbook*, 5th Edition, Hanser Publishers, Munich, p. 10, 2001.
5. F.G. Bordwell, X.M.A.N. Zhang, *Journal of Physical Organic Chemistry*, 1995. **8**: 529.
6. J. Verdu, J. Rychly, L. Audouin, *Polymer Degradation and Stability*, 2003. **79**: 503.
7. E. Richaud, C. Monchy Leroy, X. Colin, L. Audouin, J. Verdu, *Polymer Degradation and Stability*, 2009. **94**(11): 2004.

Tailoring Mater-Bi properties by the use of a biowaste-derived additive

Pierfrancesco Cerruti[a], Gabriella Santagata[a],Giovanna Gomez d'Ayala[a], Mario Malinconico[a], Veronica Ambrogi[b], Cosimo Carfagna[a] and Paola Persico[a]

[a] *Institute of Polymer Chemistry and Technology (ICTP-CNR), via Campi Flegrei 34, 80078 Pozzuoli (Na), Italy*
[b] *Department of Materials and Production Engineering, University of Napoli "Federico II", p.le Tecchio 80, 80125 Napoli, Italy*

Abstract. In this work, a polyphenol-containing extract from winery bio-waste (EP) has been used as additive to tailor Mater-Bi properties. EP was able to efficiently modulate both polymer processing and mechanical, thermal and biodegradation properties. EP decreased the melt viscosity, behaved as a Mater-Bi plasticizer and delayed the Mater-Bi crosslinking process occurring upon thermal aging. Finally, the biodisintegration rate of doped Mater-Bi decreased, thus indicating that EP interfered with the microbial digestion of the polymer films.

Keywords: Mater-Bi, bio-waste, additive, thermal crosslinking, biodisintegration.
PACS: 68.60.Dv, 81.70.Pg, 82.35.Lr, 83.80.Uv, 88.20.R.

INTRODUCTION

The employment and industrial development of biodegradable polymers is often limited because of their poor chemical-physical properties, unsuitable mechanical performances and difficult processability. Therefore, research has focused a growing attention to biodegradable polymer systems, such as composites and blends, in which the biodegradable polymeric matrix is associated with reinforcing fibers, organic and inorganic additives and degradable or non-degradable polymers [1,2].

Among biopolymers, starch is one of the most widely investigated, as it is widely available and easily modified to get a thermoplastic polymer; nevertheless, due to the hydrophilic nature responsible for fast degradation via hydrolysis, thermoplastic starch applications are limited. To overcome this experimental drawback, starch is generally modified by blending with synthetic polymers, such as polyesters or vinyl alcohol copolymers. This approach has been adopted by Novamont under the Mater-Bi trademark [3].

The use of suitable additives could provide both upgrading of processing and manufacturing performance and the enhancement of physico-chemical properties of biodegradable polymers. As a matter of fact, thermoplastic polymers rely on specific additives to obtain products that fulfill the market needs. Several categories of natural and synthetic additives are available, such as processing aids, plasticizers, stabilizers and antibacterial additives [4].

In this contribution, an extract from winery waste (named EP) has been used as a natural additive for Mater-Bi. The use of a natural processing by-product can be attractive to tailor Mater-Bi properties in a cost-effective manner, to further improve its performance in agricultural and packaging applications. The effect of EP on processability, thermal stability, mechanical properties and biodegradability of Mater-Bi has been evaluated and discussed.

EXPERIMENTAL

The polymer employed was Mater-Bi CF03A grade. The grape-based additive (EP) was obtained through hydroalcoholic extraction from grape pomace. Doped polymer pellets were obtained by mixing EP and Mater-Bi in methanol (16% w/v). Polymer films were obtained using a single screw extruder equipped with a film blowing unit. Two film samples were prepared, namely neat Mater-Bi (MB), and Mater-Bi containing 4 wt.% of EP (MB4)

CP1255, *V[th] International Conference on Times of Polymers (TOP) and Composites,*
edited by A. D'Amore, D. Acierno, and L. Grassia

Apparent viscosity of both plain and doped unprocessed Mater-Bi pellets was measured by means of a capillary rheometer. Tests were performed at 150 and 180 °C.

In order to assess the effect of aging on neat and doped Mater-Bi, the polymer films were thermo-oxidized at 70 °C in a forced air oven up to 1000 h.

Infrared spectra were recorded in transmission mode. The hydroxyl concentration was measured as the ratio of areas under the absorbance bands in the range 3650–3000 cm^{-1} and 3000-2720 cm^{-1}, related to -OH and –CH stretching, respectively.

Thermogravimetric measurements were carried out as follows: samples were heated at 10° C/min, from 30 to 600 °C in nitrogen. To eliminate small amounts of water absorbed, a 30-minute isothermal treatment at 90 °C was carried out prior to the heating run.

Dynamic mechanical analysis was performed with oscillation frequency of 1 Hz at a heating rate of 5 °C/min from -70 to 130 °C under nitrogen. Glass transition temperatures (T_g) were taken as the maximum values of tanδ curves. Tensile tests were performed using a dynamometer, according to ASTM D882-02 standard test method, with a 50 mm/min clamp separation rate.

The biodisintegration degree of the samples was evaluated according to ISO 20200:2004 test method. The plastic films were inserted into testing frames, then buried in a solid composting material and subjected to aerobic degradation at 58 °C. The composting process was monitored by visual inspection of the buried films: samples were collected and photographed periodically.

RESULTS AND DISCUSSION

Viscosity of plain and doped unprocessed Mater-Bi pellets was measured through capillary viscometry. Apparent viscosity data are shown in Fig. 1 (a). Both Mater-Bi based melts displayed a decrease of viscosity with increasing shear rate. This trend is commonly referred to as shear thinning behaviour and it is ascribed to the gradual loosening of polymer intermolecular interactions. The presence of EP was responsible for lower viscosity values over the entire range of explored shear rates, the difference being higher at lower rates (from 50 to 500 s^{-1}). Moreover, increasing temperatures led to a reduction in viscosity, as the samples were more prone to flow.

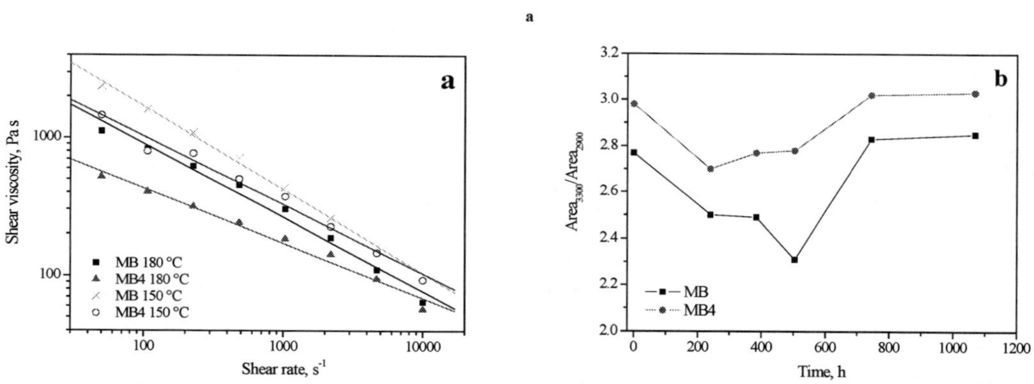

FIGURE 1. (a): Melt viscosity versus shear rate for MB and MB4 at 180 °C and 150 °C. Lines correspond to the theoretical linear fits. (b): Ratio of -OH to -CH stretching peak areas versus aging times, as calculated from FTIR spectra.

Fig. 1 (b), shows the ratio of the peak areas of -OH and -CH stretching vibrations as a function of the aging time. Significant variations of the hydroxyl concentration can be observed. More specifically, a reduction up to t = 504 h was followed by an increase of the band area at longer aging times. This tendency may be explained in terms of two opposed phenomena occurring in the course of the aging process: hydroxyl consumption due to their condensation reaction and -OH groups formation associated to thermal oxidation.

From the analysis of the thermogravimetric curves (not reported), it was observed that degradation of Mater-Bi was composed of two weight-loss steps. The first one, occurring from about 280 to 340 °C, is due to the degradation of the starch fraction, while at higher temperatures (> 350 °C), the thermal decomposition of the polyester portion was observed. In Table 1, the temperatures of 5% weight loss ($T_{5\%}$), the maximum decomposition rate temperature for both degradation steps (T_{max1} and T_{max2}), and the char yield at 600 °C ($Char_{600}$) are shown. From $T_{5\%}$ values, it can be noticed that the aged MB samples were more thermally stable than the untreated ones in the first weight loss stage. The increased stability can be due to starch crosslinking [5]. It is also worth to notice that the temperature of maximum degradation rate of starch (T_{max1}) in the first 504 h increased up to 320 °C, then dropped to 315 °C after 1000 h, confirming the occurrence of crosslinking followed by thermal oxidation due to oven aging. On the other hand, T_{max2} was not affected by the thermal aging.

TABLE 1. Temperature of 5% weight loss ($T_{5\%}$), maximum decomposition rate temperatures (T_{max1} and T_{max2}), and char yield at 600 °C ($Char_{600}$) for MB and MB4.

Aging time (h)	MB				MB4			
	$T_{5\%}$ (°C)	T_{max1} (°C)	T_{max2} (°C)	$Char_{600}$ (%)	$T_{5\%}$ (°C)	T_{max1} (°C)	T_{max2} (°C)	$Char_{600}$ (%)
0	289	313	410	4.5	262	318	407	5.2
240	303	317	409	9.1	296	317	410	6.3
504	301	320	410	7.3	297	319	407	6.6
1000	305	315	408	8.0	299	319	410	5.8

Table 2 lists the glass transition temperatures of polyester (T_{g1}) and starch (T_{g2}) components [6] for MB and MB4 at different aging times. The aging process significantly affected T_g of starch in MB, as a sharp increase in T_{g2} was noticeable after 504 h. This finding could be ascribed to starch crosslinking, which retards molecular relaxations, leading to a more rigid system. At t = 1000 h, T_{g2} decreased to 70.7 °C, as an indication of partial disruption of the starch network caused by thermal oxidation. As for the synthetic thermoplastic portion, a small decrease in T_{g1} was noticed after 1000 h aging, probably due to a reduction of the molecular weight as a consequence of the degradation process. In the case of MB4, the increase in T_{g2} after 504 h was less significant than that observed for MB. This suggests that EP may interfere with starch crosslinking. By comparing T_{g1} of unaged MB and MB4, it can be inferred that EP acts as a plasticizing agent since MB4 showed a slightly lower T_g value. This hypothesis is supported by the results of tensile tests, which show that elongation at break (ε) for unaged MB4 was about 14% higher than that of MB. As far as the tensile strength (σ) is concerned, the presence of EP did not affect the mechanical performance of Mater-Bi.

The effect of the thermal oxidation on MB and MB4 was also evaluated. Elongation at break was influenced by the aging treatment, and a significant drop was observed for both samples at 240 h, as a consequence of the partial crosslinking of starch. Tensile strength was slightly dependent on the thermal-oxidation times for both the investigated samples.

TABLE 2. Glass transition temperatures (T_g), tensile strength (σ) and elongation at break (ε) for MB and MB4 at different aging times.

Aging time (h)	MB				MB4			
	T_{g1} (°C)	T_{g2} (°C)	σ (MPa)	ε (%)	T_{g1} (°C)	T_{g2} (°C)	σ (MPa)	ε (%)
0	-17.8	64.3	13.6±1.7	518±55	-19.5	65.9	13.2±1.3	589±38
240	-	-	13.7±0.3	368±15	-	-	13.5±2.7	387±27
504	-16.7	82.0	15.9±2.0	410±67	-19.2	67.8	14.6±0.9	426±10
1000	-20.3	70.7	15.9±1.7	416±36	-19.5	62.3	14.6±2.3	413±33

In order to evaluate the modulating effect of EP on Mater-Bi degradation, biodisintegration tests were carried out. Fig. 2 shows the evolution of degradation of MB and MB4 films buried under solid composting material.

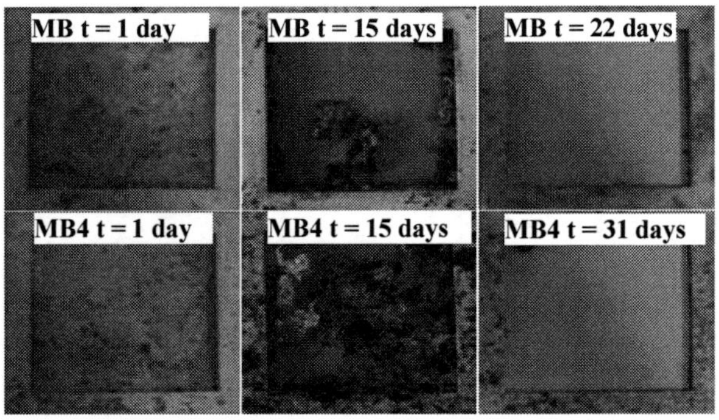

FIGURE 2. Evolution of degradation of MB and MB4 films buried under solid composting material.

It can be observed that both samples achieved complete biodisintegration within 30 days, thus well before 90 days that is the period by the end of which plastic materials are considered disintegrable, according to ISO 20200:2004. Moreover, a 15-day composting period brought about slight degradation of MB4, whereas MB exhibited a considerable consumption of the area exposed to soil. MB was completely biodisintegrated after 22 days, when the compost was sieved and no residual fragments were recovered. The same results were evident for MB4 after 31 days, indicating that biodegradation was efficiently retarded in presence of EP.

Therefore, EP probably has a broad-spectrum antibiotic activity [7] which may interfere with microbial digestion of polymer films, slowing down the rate of Mater-Bi biodisintegration.

CONCLUSIONS

A polyphenol-containing extract (EP) derived from winery waste was used as a Mater-Bi natural additive. EP was found to cause a decrease in melt viscosity, thus suggesting that it could act as a Mater-Bi processing aid. These effects are particularly significant to determine the processing features of Mater-Bi based systems and also relevant to the evaluation of rheology-dependent parameters aimed to scaling up the extrusion process, from a lab level to a high output equipment. As evidenced by thermal and viscoelastic characterization, EP affected the polymer crosslinking which occurred in the starch fraction upon thermal aging. Furthermore, tensile tests indicated that EP exerted a plasticizing effect on Mater-Bi. Also the polymer biodegradability was influenced by the additive, as neat Mater-Bi biodisintegrated earlier than the loaded polymer.

Overall, the use of a natural agricultural processing by-product showed to be attractive due to its ability to tailor Mater-Bi properties in a cost-effective manner.

ACKNOWLEDGMENTS

Dr. Maurizio Tosin (Novamont, Italy) is gratefully acknowledged for supplying Mater-Bi and for performing the biodisintegration tests. Dr. Gianni Gasperoni (CTAEX, Badajoz, Spain) is thanked for providing EP.

REFERENCES

1. X. F. Ma, J. G. Yu, Y. B. Ma, *Carbohydrate Polym.* **60**, 111-116 (2005).
2. N. Wang, J. Yu, X. Ma, *Polym. Int.* **56**, 1440-1447 (2007).
3. C. Bastioli, *Polym. Degrad. Stabil.* **59**, 263-272 (1998).
4. W. Soroka, "Fundamentals of Packaging Technology, 3rd edition", Institute of Packaging professionals, Naperville, IL, 2002.
5 E. Rudnik, G. Matuschek, N. Milanov, A. Kettrup, *J. Therm. Anal. Calorim.* **85**, 267-270 (2006).
6. R. Mani, M. Bhattacharya, *Eur. Polym. J.* **37**, 515-526 (2001).
7. G. Özkan, O. Sagdiç, N. G. Baydar, Z. Kurumahmutoglu, *J. Sci. Food Agric.* **84**, 1807-1811 (2004).

PVC/$CaCO_3$ Nanocomposites: Influence of nanoparticle surface treatment on properties

Irene Bonadies[a], Maurizio Avella[b], Roberto Avolio[a], Cosimo Carfagna[b], Maria Emanuela Errico[b], Gennaro Gentile[b]

[a]*Università di Napoli "Federico II"– Dipartimento di Ingegneria dei Materiali e della Produzione*
P.le Tecchio 80 – 80125 Napoli, Italy
[b]*Istituto di Chimica e Tecnologia dei Polimeri – Consiglio Nazionale delle Ricerche*
Via Campi Flegrei 34 – 80078 Pozzuoli (NA), Italy

Abstract. The influence of nanosized calcium carbonate on the gelation process, thermal stability and mechanical properties of rigid poly(vinyl chloride) has been studied. The gelation time of rigid PVC/calcium carbonate nanocomposites depends on the percentage and superficial treatment of $CaCO_3$. The results of thermogravimetric analysis underline an improvement of the thermal stability as a function of the surface modifier. Finally, the flexural modulus, the flexural strength as well as the critical stress intensity factor increase with the addition of poly(acrylic acid) (PAA) or poly(butadiene-co-acrilonitrile-co-acrylic acid) (PBAA) coated nanoparticles.

Keywords: nanocomposites, PVC, calcium carbonate
PACS: 81.05.Qk, 82.35.Np, 82.35.Gh, 81.70.Bt

INTRODUCTION

Poly(vinyl chloride) (PVC) is a low cost thermoplastic polymer which finds applications in several industrial sectors. Its versatility depends on the possibility to incorporate various additives able to impart different properties to PVC obtaining, as an example, flexible or rigid materials. Nowadays the continuous efforts to realize high performances, easy processability and low cost materials meet with the new opportunities related to nanotechnology leading to the development of a new class of materials: nanocomposites. In this study unplasticized polyvinylchloride (U-PVC) based nanocomposites containing calcium carbonate nanoparticles as filler were prepared and characterized. In order to improve polymer/filler interactions polyacrylic acid (PAA) and polybutadiene-co-acrilonitrile-acrylic acid (PBAA) were tested as surface modifiers. The influence of the nanoparticle surface treatment on final properties was investigated and compared with materials containing commercial uncoated and stearic acid coated $CaCO_3$ nanoparticles.

EXPERIMENTAL

Materials and Techniques

U-PVC with K-value viscosity index 67 was supplied by Solvin. 40 nm sized $CaCO_3$, either neat or modified with 3 wt% of stearic acid, was supplied by Nachmann.

The surface modification of $CaCO_3$ with poly(acrylic acid) (PAA) and with poly(butadiene-co-acrilonitrile-co-acrylic acid) (PBAA) was carried out in chloroform for 2 hrs at 80°C. The amount of acrylic acid onto nanoparticle

CP1255, *V[th] International Conference on Times of Polymers (TOP) and Composites,*
edited by A. D'Amore, D. Acierno, and L. Grassia

surface was evaluated by thermogravimetric analysis (TGA) performed on purified samples and estimated being around 2 wt%.

Materials containing 1 and 3 wt% of different $CaCO_3$ nanoparticles were prepared by melt blending in a twin roller mixer (Plastograph EC of Brabender GmbH & Co. KG) for a period of 10 min at 170°C and at 65 rpm. Morphological analysis of composites was performed by using a scanning electron microscope (SEM), Philips XL20 series, on the fractured surface of samples coated with an Au-Pd alloy. Thermogravimetric analyses were carried out using a Perkin Elmer Pyris Diamond TG/DTA in nitrogen atmosphere from 30 to 800°C at a heating rate of 10°C min^{-1}. Fracture tests were carried out with a Charpy Ceast Resil Impactor equipped with a DAS 4000 Acquisition System, using an impact energy of 3.6 J and an impact speed of 1 m/s. Three-point flexural testing was performed by an Instron 4504 machine at a test speed of 1mm min^{-1}.

RESULTS AND DISCUSSION

In this paper, polyacrylic acid (PAA) and polybutadiene-co-acrilonitrile-acrylic acid (PBAA) have been selected as $CaCO_3$ surface modifiers. This choice depends on the possibility either to improve PVC/nanoparticle compatibility or to transfer the intrinsic elastic properties of the modifier, through the interphase to materials. It is well known that an effective optimization of PVC processing requires the understanding of the PVC gelation process. During this process the initial particulate structure, containing a small percentage of primary crystallites (5÷10%), partially disappears, promoting the formation of a three-dimensional macromolecular network. In this network, new formed crystallites (secondary crystallites) act as crosslinking points. The amount of residual primary and produced secondary crystallites depends on processing conditions and strongly influences the final physical and mechanical properties of the product [1,2,3]. The analysis of the torque curve recorded during melt mixing has shown a decrease of the gelation time as a function of nanoparticle content and surface modifier. This finding can be justified considering that the size of $CaCO_3$ is compatible with the dimension of PVC primary particle; in this way a good contact between particles is obtained and a more frictional heat is developed [4]. In this respect, the gelation time is shortened and the homogenization time is increased thus assuring a more effective filler/matrix mixing. As concerning the influence of surface modifier, it must be to take into account that stearic acid could also act as lubricant influencing further the gelation process. In Table (1) materials with the corresponding code and gelation time values are resumed.

TABLE (1). Samples, codes and gelation times

Sample	Code	Gelation time (min)
PVC	PVC	1,87
PVC/1wt% neat $CaCO_3$	A1	1.28
PVC/3wt% neat $CaCO_3$	A3	0.93
PVC/1wt% stearic acid-$CaCO_3$	B1	1.11
PVC/3wt% stearic acid-$CaCO_3$	B3	0.78
PVC/1wt% PAA-$CaCO_3$	C1	1.61
PVC/3wt% PAA-$CaCO_3$	C3	1.39
PVC/1wt% PBAA-$CaCO_3$	D1	1.69
PVC/3wt% PBAA-$CaCO_3$	D3	1.53

Morphological characterization has proved the effectiveness of PAA and PBAA as surface modifiers allowing to achieve the typical nanostructure; nanoparticles appear finely and homogeneously dispersed into matrix. Moreover, the absence of voids along the surface suggests strong interfacial adhesion between polymer and nanophase. Otherwise, in materials containing neat $CaCO_3$ serious agglomeration phenomena occur demonstrating that particle/particle interactions result dominant. At last, the stearic acid treatment permits to reduce the clustering area but it is not able enough to improve polymer/filler interactions.

TGA analysis has underlined a very significant improvement of the thermal stability as a function of $CaCO_3$ surface treatment. In fact the addition of neat and stearic acid modified nanoparticles increases the temperature of degradation, T_{onset}, up to 16°C. On the contrary, in the case of materials containing acrylic acid modified nanoparticles, the PVC thermal stability is almost the same. These results can be explained considering that calcium ions onto nanoparticle surface are able to scavenge part of produced HCl reducing the extent of its autocatalytic effect in dehydrochlorination process [5]. In this respect, the effect of neat $CaCO_3$ can be considered. In the case of modified nanoparticles, although surface calcium ions are hided as a consequence of surface treatment, the thermal stability of surface modifier must to be taken into account. The degradation of the stearic acid starts at temperature lower than PVC. In this way, nanoparticle surface results uncoated and able to interact with hydrochloric acid. As concerning PAA and PBAA, the relative degradation processes are almost overlapped to that of PVC. For this reason they have not influence on the degradation stability in the corresponding materials.

In Table (2) mechanical parameters of PVC/$CaCO_3$ nanocomposites are resumed.

TABLE (2). Mechanical Properties

Code	Flexural modulus (MPa)	Flexural strength (MPa)	Kc [MPa m$^{1/2}$]
PVC	2309 ± 216	76 ± 4	2.30 ± 0.05
A1	2296 ± 124	74 ± 3	2.49 ± 0.14
A3	2343 ± 148	78 ± 3	2.55 ± 0.18
B1	2268 ± 150	81 ± 3	2.66 ± 0.13
B3	2321 ± 108	73 ± 3	2.78 ± 0.23
C1	2600 ± 79	92 ± 3	2.96 ± 0.15
C3	2692 ± 48	92 ± 2	2.82 ± 0.05
D1	2518 ± 83	92 ± 3	2.88 ± 0.07
D3	2517 ± 53	93 ± 3	3.02 ± 0.10

As reported, both the flexural modulus and flexural strength are not influenced by the addition of neat and stearic acid coated nanoparticles. On the contrary as far as PAA and PBAA coated nanoparticles, these mechanical parameters increase independently from the filler percentage. As above discussed, neat and stearic acid coated $CaCO_3$ result mainly agglomerated. In this respect, the corresponding materials are more similar to microcomposites for which only a larger amount of filler (20 ÷ 40 wt%) assures the improvement of final properties.

In the case of PAA and PBAA coated nanoparticles, the increase of the flexural modulus can be ascribed to the good dispersion and distribution of the nanophase. Conversely with respect to the modulus, the flexural strength is particularly sensitive to the interfacial adhesion. Therefore, the improvement of the flexural strength can be considered as a result of strong interactions established between PVC and PAA or PBAA coated nanoparticles.

Finally, also the toughness of PVC, measured as the critical stress intensity factor (K_c), was improved with the addition of $CaCO_3$ particles coated with PAA and PBAA.

This finding can be justified taking into account both the good interfacial adhesion and the elastic properties of acrylic coatings: the elastic coatings yield prior to the matrix and the yielding is transferred to the matrix in a great number of sites [6,7] thanks to the fine dispersion of the filler. In this respect, during the crack propagation a large amount of the energy is required.

REFERENCES

1. T. Kuriyama, I. Narisawa, R. Shina, M. Kotaki, *Journal of vinyl and additive technology* **4**, 164-168 (1998).
2. R.P. Marques, J.A. Covas, *Processing characteristics of U-PVC compounds*, CIRES, 2003.
3. L. A. Fillot, P. Hajji, C. Gauthier, K. Masenelli-Varlot, *Journal of vinyl and additive technology* **12**, 98-107 (2006).
4. D. W. Cornwell, *Industrial Minerals*, 35-37 (2001)
5. A. Marongiu, T. Faravelli, G. Bozzano, M. Dente, E. Ranzi, *J. Anal. Appl. Pyrolysis* **70**, 519-553 (2003).
6. Fu Qiang, Guiheng Wang, Chunxiao Liu, *Polymer* **36,** 2397-2401 (1995).
7. Zeng X., Wang W., Wang G., Chen J., *Journal of Material Science* **43**, 3505-3509 (2008)

Acoustic Emission Source Location in Unidirectional Carbon-Fibre-Reinforced Plastic Plates Using Virtually Trained Artificial Neural Networks

G. Caprino, V. Lopresto, C. Leone, I. Papa

Department of Materials and Production Engineering, University of Naples "Federico II", Piazzale Tecchio, 80, 80125, Naples, Italy

Abstract. Acoustic emission source location in a unidirectional carbon-fibre-reinforced plastic plate was attempted employing Artificial Neural Network (ANN) technology. The acoustic emission events were produced by a lead break, and the response wave received by piezoelectric sensors, type VS150-M resonant at 150 kHz. The waves were detected by a Vallen AMSY4 eight-channel instrumentation. The time of arrival, determined through the conventional threshold crossing technique, was used to measure the dependence of wave velocity on fibre orientation. A simple empirical formula, relying on classical lamination and suggested by wave propagation theory, was able to accurately model the experimental trend. Based on the formula, virtual training and testing data sets were generated for the case of a plate monitored by three transducers, and adopted to select two potentially effective ANN architectures. For final validation, experimental tests were carried out, positioning the source at predetermined points evenly distributed within the plate area. A very satisfactory correlation was found between the actual source locations and the ANN predictions.

Keywords: acoustic emission; carbon fibre reinforced plastics; Artificial Neural Network (ANN); source location.
PACS: 43.60.Np; 43.60.Np; 07.05.Mh

SCOPES

The engineering problem faced in this paper is schematically depicted in Fig. 1. A square composite plate, 300 mm in side and 2 mm in thickness, is monitored by three AE sensors, P_i, located as shown in the figure. When an acoustic signal is generated at point S, the wave travels within the material, achieving the probes after times depending on both the length d_i and the orientation of the lines connecting the source point to each sensor. Knowing the Time of arrival, TOA, [1, 2] at each sensor, the source location (i.e. the coordinates x, y of point S) is wanted, using ANN which have found widespread application in many scientific domains, providing reliable solutions [3].

If a model able to yield TOA as a function of fibre orientation and distance from the source is available, the TOA at each probe in Fig. 1 can be easily calculated. Therefore, the training set necessary for network learning can be developed without carrying out experimental tests. In principle, this offers an attractive means to virtually train the candidate ANN architectures, finding the most effective solution.

A simple empirical law to provide TOA in an orthotropic composite plate, was here assessed by experimental tests. Using this law, a virtual training set was constructed and used to design two potentially efficient ANNs. The precision of the networks was estimated by comparing the ANN predictions with the results of appropriate AE tests.

Since the difficulty in source location is increased with increasing the anisotropy ratio of the material under concern [4,5], a unidirectional Carbon-Fibre-Reinforced Plastic (CFRP), characterized by a very high anisotropy ratio, was purposely selected in this work.

RESULTS AND DISCUSSION

To verify the consistency of TOA values provided by the AE system, acoustic waves were excited by lead breaks and detected by three sensors, located at known distances from the source along a straight line oriented at θ with respect to fibre direction. The distance of the source from the centre of first sensor was approximately 20 mm, and the distance between two subsequent sensors 100 mm. Frothe differences Δt in TOA between two subsequent probes, two independent values of the velocity ν_e, defined as:

CP1255, *Vth International Conference on Times of Polymers (TOP) and Composites,*
edited by A. D'Amore, D. Acierno, and L. Grassia

$$v_e = \frac{\Delta d}{\Delta t} \tag{1}$$

with Δd being the distance between the transducers, were calculated for each experimental test.

As anticipated, v_e was found to be strongly dependent on orientation, steadily decreasing with increasing θ. Besides, the wave velocity is independent of distance, assessing the consistency of the TOA measurements; therefore, if a suitable expression $v_e = v_e(\theta)$ is available, eq. (1) can be used to theoretically predict TOA for whichever orientation and distance from the source. The equation, obtained by a well known relationship [2]:

$$v_e = \sqrt{\frac{E_x}{\rho(1 - \nu_{xy}\nu_{yx})}} \tag{2}$$

where E, ν are the Young's modulus and Poisson's ratio, and the indexes x, y affecting them designate the propagation direction and its perpendicular, respectively, with appropriate constant values varied until a satisfactory correlation of the curve with the experimental trend was reached, was found to good describe the above mentioned dependence. Indeed, the correlation between the experimental points and eq. (2) is reported in fig. 2 is outstanding. Therefore, a simple formula is available for the accurate prediction of TOA.

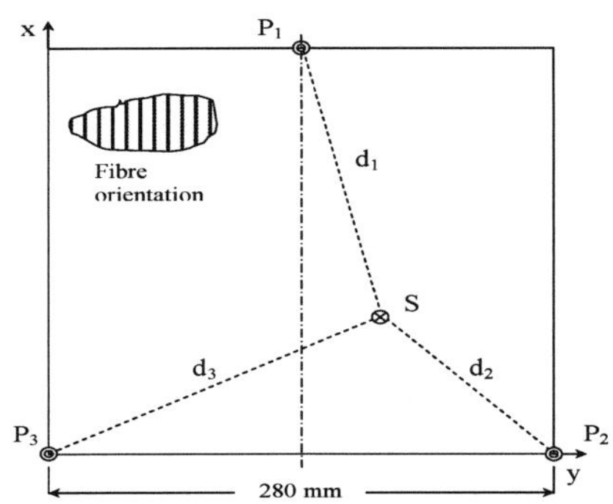

FIGURE 1. Composite plate monitored by three AE sensors.

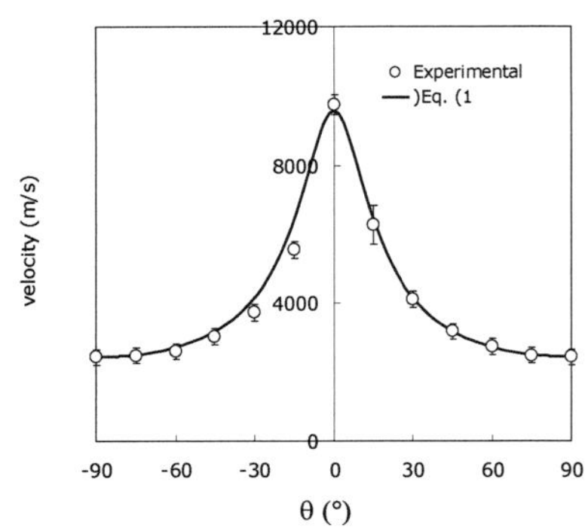

FIGURE 2. Wave velocity, v_e, against fibre orientation, θ.

The ability of an ANN to solve an engineering problem must be exploited in a learning stage: a training data set, consisting of input values and corresponding desired output values, must be available and analyzed by the network.

To build the training set a sufficient number of experimental tests, suitably varying the location of the source S (Fig. 1), and measuring the associated TOA, is necessary to perform. Here, 2500 different location points were selected by a random number generator; for each, the TOA at the three sensors was theoretically evaluated as the ratio d_i/v_e (Fig. 1), using eq. (1) to calculate v_e; finally, 15% of the location points theoretically evaluated were used for cross validation during the training stage. Training was stopped after a maximum of 5x104 iterations, or when the error signaled by the learning curve dropped below 0.005%, whichever of these conditions was met first.

After training, the ANNs examined were subjected to a validation test. A testing set, consisting of 14 unknown location points, was built following the same procedure of the training set, and was presented to the network.

The best ANN network, consisted of a 3/10/8/2 structure, i.e. was made of 10 and 8 nodes in the first and second hidden layer, respectively, besides having three nodes in the input layer and two nodes in the output layer. For this solution, a characteristic distance D_c=2.0 mm, with D_m=1.24 mm and SD=0.76 mm, was computed (Table I), after about 4x104 epochs, requiring a training time of approximately 16', were accomplished.

TABLE I – Accuracy parameters characterizing the ANNs considered in this work.

Data source	ANN architecture	D_m (mm)	SD (mm)	D_c (mm)	D_{max} (mm)
Testing set	3/10/8/2	1.24	0.76	2.00	2.62
	2/10/8/2	0.94	0.37	1.30	1.58
Experimental	3/10/8/2	2.18	1.99	4.17	6.95
	2/10/8/2	1.85	2.04	3.89	7.03

A comparison of the points included in the testing set and their positions as predicted by the 3/10/8/2 ANN is shown in Fig. 3, where the network is labeled "3IN" (three input nodes). The black arrow in the figure indicates the point in correspondence of which the maximum distance (D_{max}=2.62 mm) of the predicted from the real source location was found.

An identical network but with two input nodes instead of three was trained to verify the effect of the redundant input parameters. The white diamonds in Fig. 3 are the predictions of this ANN (labelled as "2IN") fed. The white arrow indicates the location point where the maximum error (D_{max}=1.58 mm) was verified. The main accuracy parameters are shown in Table I. From them, the 2/10/8/2 network is foreseen to perform better than its 3/10/8/2 counterpart. Therefore, it seems that the redundant input parameter does not result in a more effective response of ANN. Further, this finding suggests that an even more efficient solution might be individuated by an accurate optimization procedure.

To assess the two ANN architectures selected, experimental tests were performed, using as source locations the points represented by the black circles in Fig. 4. Acoustic waves were generated by lead breaks, and the measured TOA were given in input to the trained ANNs, which provided the predicted locations (open symbols in the figure).

The correlation between actual and predicted source location is very satisfactory.

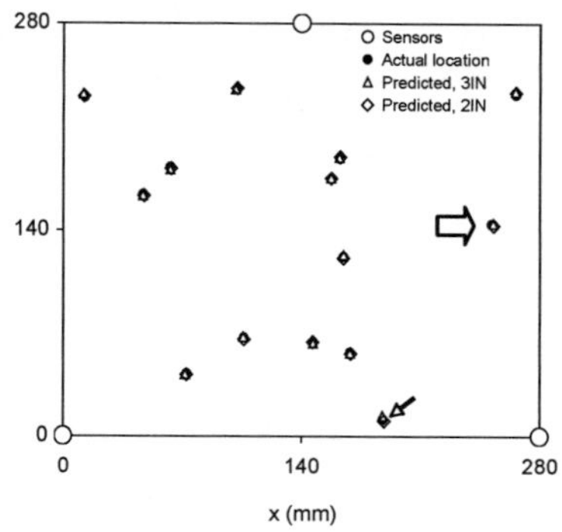

FIGURE 3. Composite plate monitored by three AE sensors.

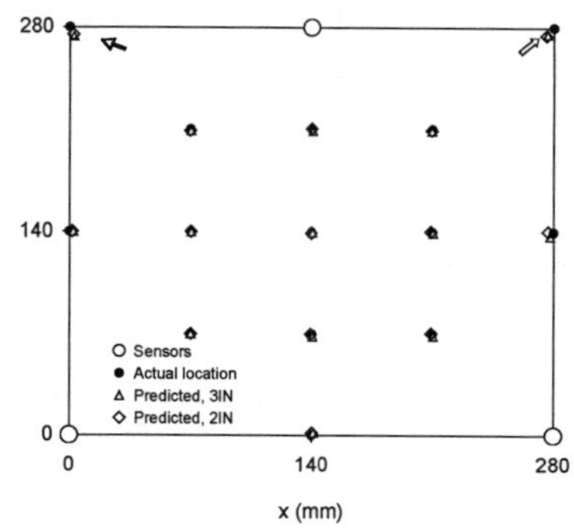

FIGURE 4. Experimental and predicted source locations.

Using theory, the differences in TOA between the sensors P_i in Fig. 1 were calculated, and compared with the experimental ones. The results are shown in Fig. 5, where the label ti-j designates the difference in TOA between probe i and j: the agreement between theory and experiments was excellent. This explains why the virtual ANN training was successful in yielding an effective architecture, directly usable to interpret the experimental results.

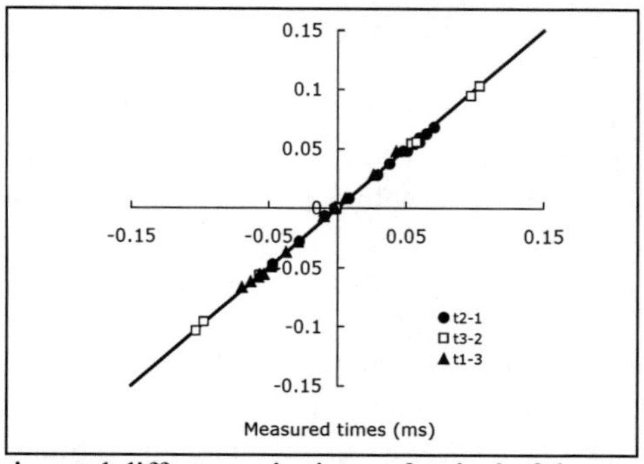

FIGURE 5. Theoretical and experimental differences in times of arrival of the source waves at the three AE sensors.

REFERENCES

1. A Tobias, *Non-Destruct. Test.* 9, 1976, pp. 9-12.
2. SM Ziola and MR Gorman, *J. Acoust. Soc. Am.* 90, 1991, pp. 2551-8.
3. HKDH Bhadeshia, *ISIJ Int.* 39, 1999, pp. 966-79.
4. H Jeong and Y-S Jang,, *Compos. Struct.* 49, 2000, pp. 443-50.
5. S Banerjee, F Ricci, E Monaco and A Mal, *J. Sound Vibrat.* 322, 2009, pp. 167–83

Rheological behaviour, mechanical properties and morphological aspects of thermoplastic polyurethane reinforced with multiwalled carbon nanotubes

Pietro Russo, Domenico Acierno and Paola Spena

Department of Materials and Production Engineering – University of Naples Federico II
P.le Vincenzo Tecchio, 80 – 80125 Naples (I)

Abstract. Melt blended compounds based on a film grade thermoplastic polyurethane resin filled with relatively low contents of multiwalled carbon nanotubes have been investigated. Materials, prepared through the production of a masterbatch containing 3 wt% of nanotubes and subsequent dilution of the same by addition of matrix pellets, were analyzed in the form of tapes and films always taking the neat matrix, processed in the same conditions, as the reference. Improvements of the matrix extrudability and mechanical benefits showed for all investigated composite samples have been satisfactorily interpreted by morphological observations carried out in both transmission and scanning electron microscopy.

Keywords: Thermoplastic polyurethane, carbon nanotubes, rheological behavior, mechanical properties, morphology.
PACS: 82.35.Np

INTRODUCTION

Thermoplastic PolyUrethanes (TPUs) are linear segment block copolymers characterized by hard segments and soft segments: the former are made from diisocyanate while the latter consist of the long flexible polyether or polyester chains which interconnect two or more hard segments. In particular, the hard segments act as multifuncional tie points working both as physical crosslinks and reinforcing fillers, while the soft segments form an elastomer matrix which accounts for the elastic properties of TPU.

Thanks to this particular chemical structure, the materials have several interesting properties, such as wide range of service temperatures, excellent tear and abrasion resistances and very good compression and tensile strength, and therefore current applications already cover many industrial fields from textiles to shoes, from synthetic leathers to mechanical tools and so on.

However the ever increasing interest in the use of plastics in place of traditional materials also for extreme applications, leaves, even in this case, enough space to the research for improvements. In this frame, recent interest has been addressed to the inclusion of carbon nanotubes in order to combine benefits of both matrix and fillers.

Carbon nanotubes, as witnessed by the huge amount of available literature, are largely diffused being optimal reinforcing fillers for composites because they have low density, excellent electronic and mechanical strength, high electrical and thermal conductivity and thermal stability.

Besides these individual properties, a number of potential benefits are expected when carbon nanotubes are employed as reinforcing agents in nanocomposites as long as an optimal dispersion of the filler is ensured.

In this light, main aim of the research was to investigate the influence of carbon nanotube inclusions on extrudability and physical properties of thermoplastic polyurethane based compounds. In particular interest was devoted to highlight influences related to relatively low loadings of multiwalled carbon nanotubes on rheological behaviour of melts and mechanical performance of tapes and films based on the examined formulations and to relate these features with morphological observations.

CP1255, *Vth International Conference on Times of Polymers (TOP) and Composites,*
edited by A. D'Amore, D. Acierno, and L. Grassia

EXPERIMENTAL

Materials

All investigated composites were based on a film grade Thermoplastic PolyUrethane supplied by the Elastogran Gmbh under the trade name Elastollan 1185A and multiwalled carbon nanotubes (MWNTs) from Shenzen Nanotechport Co. Ltd used in powdery shape. In particular, used MWNTs have the outer diameter ranging from 60-100 nm while their length is between 5 and 15 μm.

All materials were adequately dried before blending.

All products were prepared by a two-step melt compounding procedure. In the first one a masterbatch containing 3.0 wt% of nanotubes has been prepared by using a HAAKE twin screw extruder with a screw speed of 90 rpm and a flat temperature profile from the hopper to the die at 200 °C with the care of use of a constant flow of nitrogen in the hopper.

Compounds, pelletized and re-dried, were subsequently diluted with matrix pellets at 0.2, 0.5 and 1.0 wt% during the tape or film preparation by employing a *Brabender DCE 330* (L/D: 20) and under the same processing conditions.

The reference material has been obtained processing the neat TPU resin with the same procedure.

Experimental techniques

Rheological measurements of all materials were carried out in oscillatory way but also in steady shear mode by using a parallel plate geometry and a capillary rheometer to extend evaluations at shear rates covering the processing conditions, too.

Tensile tests were performed by using an Alpha Technologies Tensometer (Mod. 2020) on rectangular specimens sized 4 mm x 150 mm and 1 mm thick (useful length 100 mm). All tensile experiments were carried out at ambient temperature at a strain rate of 300 mm/min. Load cell: 5 kN, gauge length: 15 mm.

For each sample at least ten tests were performed.

Morphological observations of extrudates were performed by scanning electron microscopy on surface fractured in liquid nitrogen along the transverse direction. Further studies were carried out on 110-120 nm-thick microtomed sections derived from samples previously embedded in an epoxy resin by using a Transmission Electron Microscope Philips EM 208 equipped with a high definition data acquisition system Olimpus Soft Imaging Solutions.

RESULTS AND DISCUSSION

Melt compounding procedures performed by discontinuous or more often in continuous by twin screw extruders are normally monitored considering the change in time of some processing parameters. For example, it is generally assumed that a steady state level of mixing is achieved when the decreasing trend of the torque motor of the used equipments reaches a plateau value. In this context, a preliminary consideration in terms of processability may be drawn by the revealed effect of filler content on this value. As highlighted from Fig. 1, it seems that the higher the filler loading the lower the torque. This effect is explainable in terms of a plasticizing action of the included filler and promptly confirmed by the evidence of the capillary rheology.

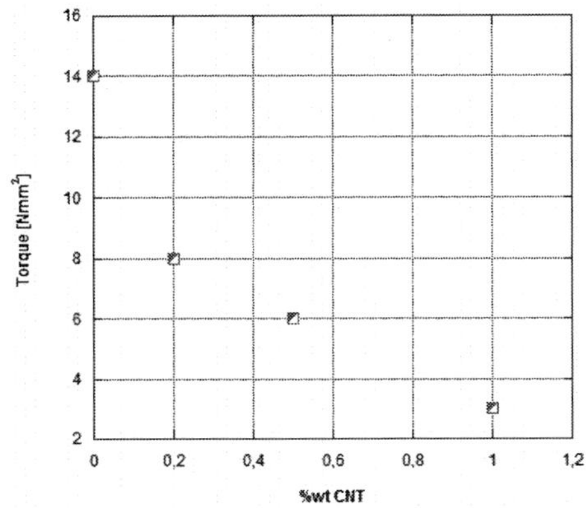

FIGURE 1. Torque values as a function of the nanotube content during the dilution step.

Rheological measurements performed by oscillatory dynamic tests indicate an increase of the storage modulus of composite melts for frequency over than 0.1 (see Fig. 2). At very low frequency this increase seems to be verified only for the 1.0 wt% content of nanofiller.

In terms of complex viscosity, for all composite formulations that parameter appears to be increased compared to the pure matrix at least on frequencies range considered by now. This effect clearly gives no indication about the influence of filler on processability of the matrix but indicates, again especially for the system containing 1 wt% of filler, the occurrence a yield strength phenomena, an increase of viscosity to decrease the frequency of oscillation, typical of nanostructured materials.

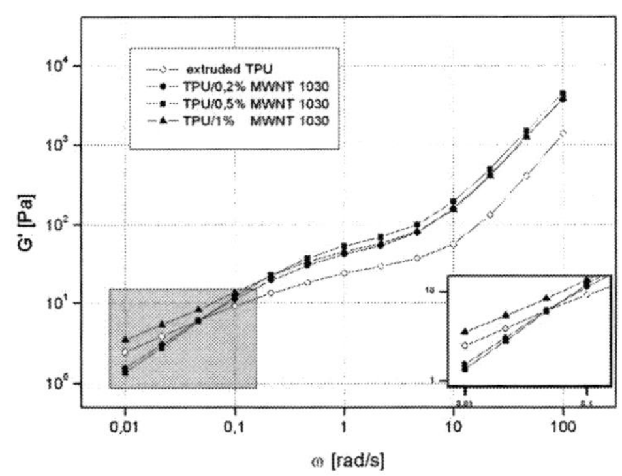

FIGURE 2. Frequency dependence of the storage modulus G' as a function of the nanotube content.

Steady shear tests showed a reduction of the viscosity of the matrix for all composite materials as a sign of the occurrence of significant melt morphology modifications.

Mechanical properties of the extruded composites were investigated for both tape or film products. Typical stress-strain curved referred to the first type of samples are compared in Fig. 3 while the average tensile properties including elastic modulus, failure strain and stress have been summarized in Table 1.

In practice we see a relevant effect of the reinforcement on the stiffness that appears to be almost doubled for the weight loading equal to 0.5%. Anyway, this trend does not significantly sacrifice the product performances at high deformations where an increase of the strain hardening phenomena can be detected by increasing of nanotubes loading with a reduction of the strain at break approximately equal to 20 wt% for contents of nanofiller higher than 0.5 wt%.

Finally, the morphological analysis performed by electron microscopy in scanning and transmission modes has confirmed both the homogeneous distribution included nanofillers and a satisfactory degree of adhesion at interface.

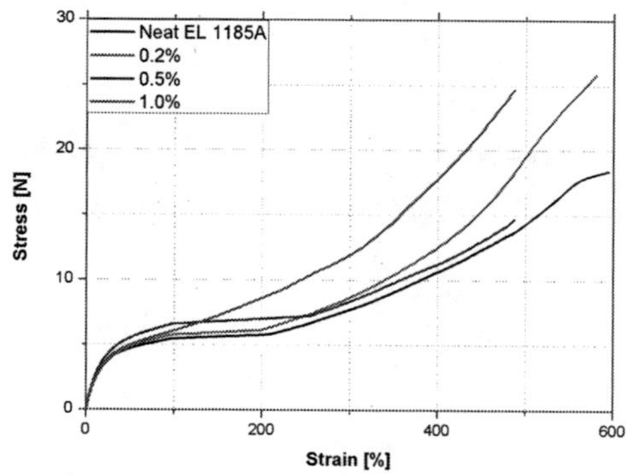

FIGURE 3. Comparison of typical stress-strain curves of all investigated materials.

TABLE 1. Mechanical properties of composite tapes based on TPU/MWNTs and its corresponding pure TPU.

Sample	*Young Modulus (MPa)*	*Strain at break (%)*	*Stress at break (MPa)*
Pure TPU	8.84	597.5	23.25
0.2 wt% MWNT	10.26	585.5	32.53
0.5 wt% MWNT	14.27	486.0	40.86
1.0 wt% MWNT	13.84	487.8	36.42

CONCLUSION

Strips based on TPU/MWNTs compounds have been preliminarly investigated in terms of rheological behaviour, mechanical properties and morphological features as a function of the included carbon nanotubes content and taking the neat TPU based ones as the reference.

The processability of the studied composites with respect to the matrix appeared to be improved by the inclusion of nanofiller.

Static tensile mechanical tests have shown the influence of the reinforcement, both at low and at high deformations, where the higher the content of nanotubes the higher the breaking strength of the material.

Morphological observations highlighted homogeneity of nanotube dispersion and an adequate level of adhesion at the interface with the hosting TPU matrix.

REFERENCES

1. F. Jiang, G. Hu, S. Wu, Y. Wei and L. Zhang, *Polymers and Polymer Composites* **16**, 471-480 (2008).
2. F. Jiang, G. Hu, S. Wu, Y. Wei, L. Zhang and G. Hu, *Polymers and Polymer Composites* **16**, 509-518 (2008).
3. W. Chen and X. Tao, *Macromol. Rapid. Commun.* **26**, 1763-1767 (2005).
4. W. Chen, X. Tao and Y. Liu, *Composite Sci. and Tech.***66**, 3029-3034 (2006).
5. W. Chen and X. Tao, *App. Surf. Sci.* **252**, 3547-3552 (2006).
6. M.T. Hunley, P. Pötschke, T.E. Long, *Macromol. Rapid. Commun.* **30**, 2102-2106 (2009)
7. U. Almeida Pinto, L.L.Y. Visconte and R.C.Reis Nunes, *Eur. Polym. J.* **37**, 1935-1937 (2001)
8. A. Frick and A. Rochman, *Polymer Testing* **23**, 413-417 (2004)
9. P. Pötschke, L. Häuβler, S. Pegel, R. Steinberger and G. Scholz, *KGK* 432-437 (September 2007)

Effect of silica particles concentration on the physical aging of PMMA/silica nanocomposites

Virginie M. Boucher[a], Daniele Cangialosi [b], Angel Alegrìa[b,c], Juan Colmenero[a,b,c]

[a]*Donostia International Physics Center, Paseo Manuel de Lardizabal 4, 20018 San Sebastián, Spain,* [b]*Centro de Física de Materiales Centro Mixto (CSIC-UPV/EHU), Apartado 1072, 20080 San Sebastián, Spain,* [c]*Departamento de Física de Materiales, Universidad del País Vasco (UPV/EHU), Apartado 1072, 20080 San Sebastián, Spain,*

Abstract. We have monitored the physical aging of poly(methyl methacrylate) (PMMA)/silica nanocomposites, namely the slow evolution occurring to the structure of glasses below the glass transition temperature (T_g), following the time evolution of the enthalpy as measured by differential scanning calorimetry (DSC). We have systematically varied the concentration of silica to provide insight about the influence of the ratio area of silica/volume of PMMA (A_{sil}/V_{PMMA}) on the physical aging process. Our results clearly indicate that physical aging speeds up with increasing silica concentration. Furthermore, the amount of recovered enthalpy decreases with increasing silica concentration, indicating that part of the enthalpy is recovered during the delay time before the first data are collected.

Keywords: Accelerated physical aging, polymer nanocomposites, enthalpy recovery.
PACS: 61.20.Lc, 64.70.P-, 66.70.Hk

INTRODUCTION

Physical aging of amorphous glassy polymers has been the subject of intense study.[1] More recently, increasing emphasis has been devoted to the study of physical aging in polymer nanocomposites and other nanostructured glasses.[2] In this case, the presence of an interface may deeply affect the physical aging of the polymer. Within the classical idea that physical aging is intimately related to the segmental mobility of the polymer, two effects can be invoked to explain the change of the physical aging rate in comparison to the bulk polymer: finite size and interfacial effects. In both cases the effect on the segmental mobility of the glass former should only be visible at length scale shorter than several nanometers.[3-4] Notwithstanding these observations, accelerated physical aging was measured in nanostructured systems with typical length scale (the interparticle distance in nanocomposites; the thickness in thin films) in the micrometer range.[6-7]

In this work, we followed the enthalpy recovery of PMMA/silica nanocomposites in a wide range of silica concentrations. The aim is to clarify whether the tendency of physical aging to speed up with increasing the ratio area of silica/volume of PMMA, previously observed in PMMA/silica nanocomposites,[7] is maintained.

MATERIALS AND METHODS

The silica particles, displaying an average particle size of 100 nm, were provided by ESA, Inc., under the name NaOsil. PMMA (M_w=M_n= 120000 g mol^{-1}) was purchased from Polymer Source, Inc.

The silica particles were surface-treated with octadecyltrimethoxy silane (OTMS), in a Tetrahydrofuran (THF) solution. To prepare the PMMA/silica nanocomposite films, the appropriate amount of the previous silica particles solution was added to a solution of dissolved PMMA, then cast onto a clean glass substrate for solvent evaporation.

The weight fraction of silica particles in the nanocomposite samples was measured by thermogravimetric analyses (TGA), on a thermogravimeter TA Q-500, with a heating ramp of 10 °C.min^{-1} up to 750 °C under flowing nitrogen (20 cm^3.min^{-1}). The values of silica weight fraction are reported in table 1.

CP1255, *V*[th] *International Conference on Times of Polymers (TOP) and Composites,*
edited by A. D'Amore, D. Acierno, and L. Grassia

Thermal analyses of the samples were carried out by means of a DSC-Q2000 (TA-Instruments), temperature-calibrated with melting indium. All experiments were performed under nitrogen atmosphere, by following the thermal procedure sketched in the inset of Fig. 2a. The amount of enthalpy relaxed during aging of a glass for a period of time t_a at the temperature of aging, T_a, was evaluated by integration of the difference between aged and unaged samples thermograms, as described elsewhere.[9]

RESULTS AND DISCUSSION

The values of T_g recorded by DSC, at 10 °C.min^{-1}, on cooling, for pure PMMA and the investigated nanocomposites, are reported in table 1. For all the samples, the addition of silica does not affect the position of the glass transition. This implies that PMMA segmental mobility is unaffected by the presence of silica nanoparticles.

TABLE 1. Summary of the samples characteristics : silica weight fraction, theoretical interparticle distance, glass transition temperature (T_g).

Sample	Silica weight fraction, W_f (%)	Interparticle distance, l (nm)	Tg (°C)
PMMA	-	-	121.3
M10	10.6	380	120.0
M17	16.7	250	121.6
M24	23.9	165	120.9
M38	37.5	85	119.5
M52	52	40	120.3

Figure 1 displays the normalized heat capacity, C_p, versus temperature, measured after annealing at T_a= 80 °C (40 °C below the nominal T_g) for various aging times, for PMMA and for the nanocomposite M17, as an example. In both samples, as aging time increases, the magnitude of the endothermic overshoot increases, as well as the temperature of its maximum, T_p, as expected. However, some differences can be observed between pure PMMA and the nanocomposite M17: (i) for the shortest aging times (t_a < 124 h), the nanocomposite sample scans exhibit broader and more important endothermic overshoots, that appear to develop for shorter aging times than neat PMMA; (ii) the nanocomposite sample shows smaller endothermic overshoots than that of the bulk for long aging times ($t_a \geq$ 124 h). Similar observations were made on the other nanocomposite samples thermograms obtained after annealing (not displayed here). Moreover it was found that, for the same aging time, the higher the amount of silica particles in PMMA, the more important the endothermic overshoot at very short aging times, and the smaller the magnitude of the overshoot at long aging times. This latter observation indicates that the total amount of recovered enthalpy decreases with increasing silica particles amount in PMMA. This result suggests that the initial enthalpy state is not the same in the different samples, the nanocomposites having a lower initial enthalpy than pure PMMA.

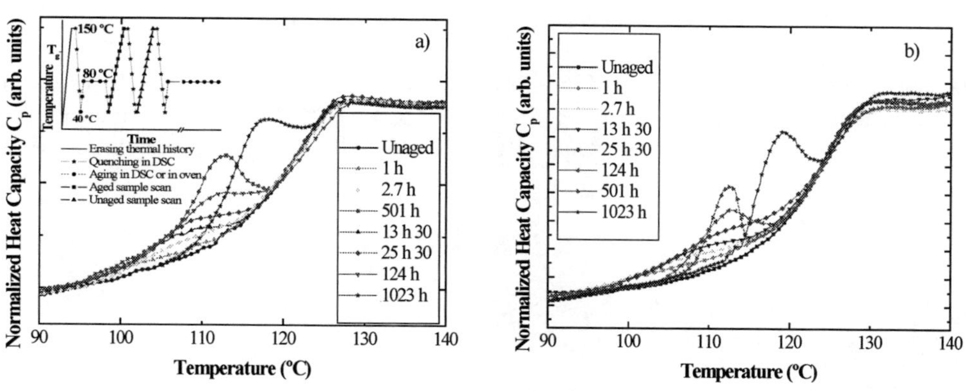

FIGURE 1. Enthalpy recovery profiles upon aging at 80 °C for different aging times for (a) PMMA, and (b) M17. The inset shows the thermal procedure for the enthalpy relaxation measurements.

These results are highlighted in Fig. 2a, where the enthalpy relaxation function, defined as: $\phi(t) = (\Delta H(t) - \Delta H(\infty)) / (\Delta H(0) - \Delta H(\infty))$, is displayed as a function of aging time. Figure 2a confirms that physical aging is accelerated in the nanocomposite samples in comparison to PMMA. Furthermore, due to the already partly relaxed state of the nanocomposites, their relaxation functions present initial values smaller than the unity. This is schematically displayed in Fig. 2b, where typical relaxation functions are presented with particular emphasis to the available time window for aging measurements. For the nanocomposites of the present study, the time delay inherent to the experimental technique does not allow to follow the initial evolution of the enthalpy recovery.

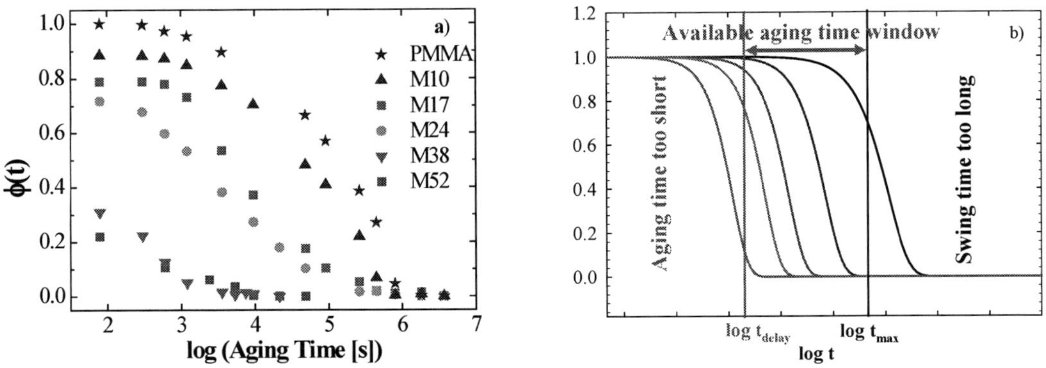

FIGURE 2. a) Structural relaxation function *ϕ(t)* for $T_a = 80$ °C, for all investigated systems. **b)** Typical variation of different relaxation functions ϕ(t) with the the aging time and schematical representation of the experimental available aging window. The time t_{delay} corresponds to the inherent delay time needed to reach the selected aging temperature; the time t_{max} corresponds to the maximum time employed in conventional physical aging experiments.

CONCLUSIONS

The effect of the concentration of silica particles on structural recovery of PMMA was investigated by DSC. The comparison of DSC thermograms of PMMA and its nanocomposites showed that: i) at very short aging times, the endothermic overshoot is larger for the nanocomposites; ii) at very long aging times, the endothermic overshoot is reduced for the nanocomposites and stabilises at shorter aging times than neat PMMA.

The use of a relaxation function, taking into account the different total amounts of recovered enthalpy among the different samples, showed an acceleration of the physical aging of PMMA in presence of silica particles and a correlation between the area/volume ratio of silica in PMMA and the physical aging rate of PMMA. This latter observation suggests that the physical aging process is driven by the diffusion of free volume holes towards polymer interfaces, as previously proposed.[6-8]

ACKNOWLEDGMENTS

The authors acknowledge the University of the Basque Country and Basque Country Government (Ref. No. IT-436- 07, Depto. Educación, Universidades e Investigación) and Spanish Minister of Education (Grant No. MAT 2007-63681) for their support. We gratefully acknowledge ESA, Inc. for kindly providing the silica nanoparticles.

REFERENCES

1. J. M. Hutchinson, *Prog. Pol. Sci.* **20**, 703-760 (1995).
2. R. D. Priestley, *Soft Matt.* **5**, 919-926 (2009).
3. D. Cangialosi *et al.*, *Phys. Rev. E* **76**, 011514 (2007).
4. E. Donth *et al.*, *J. Phys: Cond. Matter* **12**, 281–286 (2000).
5. P. Scheidler *et al.*, *Europhys. Lett.* **59**, 701-707 (2002).
6. M. S. McCaig, D. R. Paul, *Polymer* **41**, 629-637 (2000).
7. D. Cangialosi *et al.*, *Phys. Rev. B* **70**, 224213 (2004).
8. V. M. Boucher *et al.*, *Soft Matt.*, submitted for publication.
9. I. M. Hodge, *J. Non-Cryst. Solids* **169**, 211-266 (1994).

Design of novel polymeric materials by controlled self-assembly

Stefano P. Piotto[a,b], Simona Concilio[b,c], Pio Iannelli[a,b]

[a]*Department of Pharmaceutical Sciences, University of Salerno, via Ponte Don Melillo, I-84084 Fisciano (Salerno), Italy*

[b] *Research Centre for Nanomaterials and Nanotechnology, University of Salerno, via Ponte Don Melillo, I-84084 Fisciano (Salerno), Italy*

[c]*Department of Food and Chemical Engineering, University of Salerno, via Ponte Don Melillo, I-84084 Fisciano (Salerno), Italy*

Abstract. In the last years the role of confinement in selecting conformations of polymers and in accelerating the convergence has been clearly stated. The choice of proper confining systems is crucial in designing new materials. In this contribution, the consequences of tailored mesostructures in controlling a series of physical-mechanical-electronical properties are discussed with a particular attention to magnetic materials and photonic crystals.

Keywords: DDFT, mesostructures, confinement, magnetic materials
PACS: 62.20.-x

INTRODUCTION

The ultimate goal of nanotechnologies is to hierarchically assemble molecules into objects, along several length scales, and to disassemble objects into molecules. This task is superbly mastered by living cells that use confinement to select proper assembly and/or proper folding of natural polymers. The need to control folding and assembly is crucial in cells, and small variations in the structure assisting folding, namely chaperons and chaperonins, can lead to structures without the minimum ability to let cell survive. Famous examples of improper folding of proteins, which are polyamides, are Alzheimer's, Mad Cow (BSE), CJD, ALS, Huntington's, Parkinson's disease among many others.

By means of confinement, cells can also speed up the assembly of order of magnitude. It is reasonable to assume that proteins may experience different energy landscapes when they fold *in vivo* context rather than in solution bulk, and these differences may constitute a significant piece of the folding puzzle. In a recent work [1,2] dynamic mean-field density functional method has been applied to the dynamics of block copolymer melts in three dimensional lattice model. The focus of the work is on the possibility to design new materials with the desired morphologies. In the present work the assisted assembly of polymers is analyzed in terms of mechanical, electronics and photonic properties.

METHODS: DYNAMIC DENSITY FUNCTIONAL THEORY

MesoDyn [3] is an important mesoscopic simulation technique, which is based on dynamic mean field density functional theory (DDFT). This technique will be applied to simulate a series of polymers. The dynamics of the system is described by a set of functional Langevin equations that represent diffusion equations in the component densities, which take account of the noise in the system. Each bead represents a group of atoms corresponding to a structural units of a polymer chain. The chain topology depends on the degree of coarsening of the original system.

CP1255, *V[th] International Conference on Times of Polymers (TOP) and Composites*,
edited by A. D'Amore, D. Acierno, and L. Grassia

Several block copolymers are simulated: polystyrene (PS), polyacryloamides (PAAM), polyacrylonitrile (PAN), poly(methyl methacrylate) (PMMA).

All the mesoscale simulations and analyses are done with the MesoDyn module in Materials Studio 4.1. The dimensions of the cage is set in the range of 20÷100 nm. To ensure a stable numerical algorithm, as an approximation, all bead diffusion coefficients are 10^{-7} cm^2s^{-1}.

The simulation temperature is from 298K to 550K, the time step is 50.0 ns and the noise-scaling parameter is 100. The mean-field interaction energy (a_{ij}) between different chemical components is related to the Flory–Huggins interaction parameters.

METHODS: FINITE ELEMENTS

As previously reported [1,2] a wide variety of new morphologies appear if the self-assembly of block copolymers take place under spatial confinement. An example is given in figure 1 where PAN-b-PS systems is let to spontaneously converge under spherical confinement. The output of the previous simulation is used to calculate the set of properties of the list:

- Elastic constants (stiffness and compliance)
- Thermal expansion coefficients
- Thermal conductivity
- Electrical conductivity
- Dielectric constants
- Gas diffusion constants
- Refractive index
- Magnetic moments
- Swelling in solvents
- Liquid diffusion
- Liquid permeability

FIGURE 1. PAN-b-PS system inside a sphere of diameter equal to 40 nm. The boundary between PAN and PS regions is represented with isosurfaces. In brown is represented PS, in yellow PAN.

RESULTS

Two particular applications of novel materials are taken into account: the design of novel polymeric photonic crystals and the use of polymeric template to control the synthesis of magnetic materials.

Photonic crystals are of large interest both for fundamental and applied research. They are periodic dielectric structures that are designed to control the propagation of electromagnetic (EM) waves by defining allowed and forbidden energy bands in the photon dispersion spectrum. The absence of propagating EM modes inside the structures gives rise to distinct optical phenomena such as inhibition of spontaneous emission. Three-dimensional photonic crystals (PCs) grant a number of optical properties such as optical confinement [4] and the superprism effect, [5] which may be exploited to fabricate high-density optical integrated circuits. In these synthetic crystals, the artificial periodic dielectric structure of the PC creates a photonic band gap.[6] Complete gaps between bands correspond to forbidden wavelengths, a feature of particular utility for passive photonic devices such as filters, waveguides, and resonant cavities. Structures with partial gaps (i.e., in restricted dimensions) may also have application in the emerging field of PC-based optical elements. Photonic band structure calculations predict partial and complete band gaps for a number of periodic dielectric lattices including 3D simple cubic, face-centered-cubic, and hexagonal-close-packed lattices. Unfortunately, the fabrication of such PCs with visible band gaps is problematic and time consuming. In the present contribution, photonic bands are predicted for all novel morphologies.

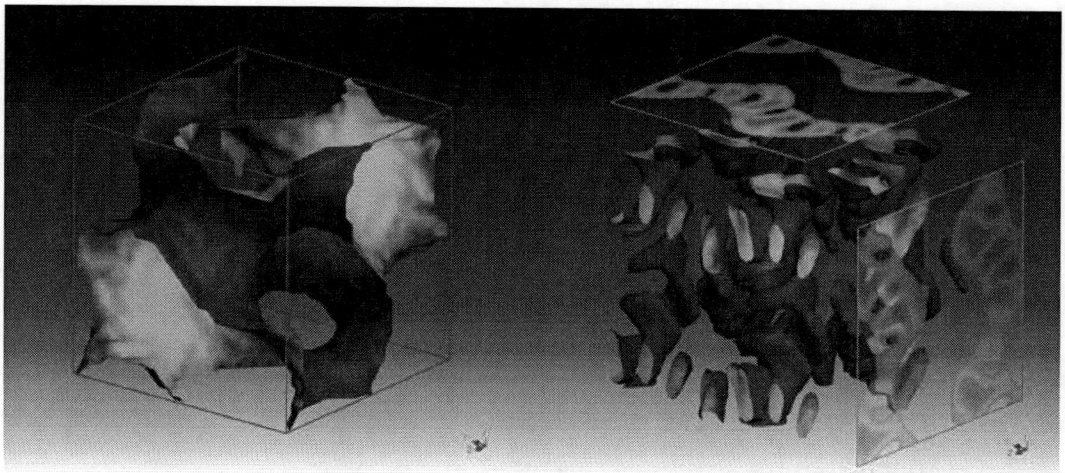

FIGURE 2. Spatial confinement of PAN-PS polymers inside a gyroid scaffold induce the organization of blocks of PAN in chiral nano helices. On the right the inner organization is shown. The two slide planes represent electron density. The formation of nano helices, which can host ferromagnetic nanoparticles, can lead to the formation of novel magnetic materials.

The other important application, the self-assembly nanostructured magnetic materials resulting from the association of complex mesostructures and inorganic magnetic nanoparticles, is nowadays of tremendous interest for applications (see figure 2). Original magnetic may be determined from such an association. The relative ease of preparation of well known and characterized polymers and the high degree of versatility of self-assembled polymeric materials based on diblock copolymers enable the synthesis of a large variety of diverse applications. Self-assembled nanostructured films are eminently suitable for the construction of devices based on molecular recognition. Molecules or supramolecular nanoparticles can be addressed spontaneously either by changing the temperature, or the size of the cage, or by the application of an electric or magnetic field. These characteristics allow the formation of superlattices with the desired symmetry. Hence, the fabrication of photonic, electronic, magnetic, and nonlinear optical devices can be possible. The self-assembly of nanoparticles of metals, magnetic, and semiconductor nanoparticulate films, in any desired order, allows the construction of molecularly novel mesostructures. As previously demonstrated [1,2] the size and the morphology of the self-assembled material can be finely controlled.

REFERENCES

1. S. Piotto, S. Concilio, P. Iannelli, F. Mavelli, IV INTERNATIONAL CONFERENCE TIMES OF POLYMERS (TOP) AND COMPOSITES. AIP Conference Proceedings, **1042**, 44-45 (2008)
2. S. Piotto, S. Concilio, F. Mavelli, P. Iannelli, Macromol. Symposia, 2**86**, 1, 25-33 (2009)
3. J.G.E.M. Fraaije, Dynamic density functional theory for microphase separation kinetics of block copolymer melts, J. Chem. Phys. **99**, 9202-9213 (1993)
4. A. Chutinan and S. Noda, Appl. Phys. Lett. **75**, 3739-3742 (1999)
5. H. Kosaka, T. Kawashima, A. Tomita, M. Notomi, T. Tamamura, T. Sato, S. Kawakami, Phys. Rev. B **58**, 10096-10099 (1998)
6. E. Yablonovitch, J. Phys.: Condens. Matter **5**, 2443 (1993)

Synthesis and Characterization of New Photoluminescent Oxadiazole/Carbazole Containing Polymers

Simona Concilio [1,*], Valeria Bugatti [1], Pio Iannelli [2], Stefano P. Piotto [2]

[1] *Department of Chemical and Food Engineering, University of Salerno, via Ponte Don Melillo, I-84084 Fisciano (Salerno), Italy*

[2] *Department of Pharmaceutical Science, University of Salerno, via Ponte Don Melillo, I-84084 Fisciano (Salerno), Italy*

Abstract. We report on the synthesis and the characterization of a new class of segmented polyethers POC containing the oxadiazole and carbazole units. The polymers exhibit a high thermal stability, high glass transition temperatures and good solubility in common organic solvents, despite the extended aromatic portion in the main polymer chain. According to previous reports on oxadiazole/carbazole containing materials, POC show high photoluminescence activity in the blue region of the visible spectra. The good solubility in chlorinate solvents allows the preparation of films with homogeneous thickness by spin coating. Glass transition temperature in the range 115-125°C ensures good stability of film morphology at room temperature. For these features, polymers POC are potential candidate materials for fabricating blue light emitting devices

Keywords: photoluminescence; polymers; oxadiazole; carbazole.
PACS: 78.66.Qn

INTRODUCTION

Polymer light-emitting diodes (PLEDs) have attracted considerable attention for their convenient preparation from solution by spin-coating or inkjet printing methods. In order to have a balanced carriers recombination in the emitting layer, the polymer should possess good carrier transport properties, as well as energy-level matching with electrodes for effective charge injection.[1-5]

To improve the electron transfer in PLEDs, researchers usually spin a p-type polymer on an n-type polymer with electron transport properties. But this kind of multilayer system may suffer difficulties in spinning a polymer solution on an organic film, which can easily dissolve. To deal with this issue, polymers containing both hole- and electron-transporting moieties have been prepared. Recently, polymers with either electron-withdrawing (such as oxadiazole, diphenylquinoline etc.) or electron-donating groups (such as triphenylamine, carbazole etc.) have been prepared. [6-9]

Oxadialzole/carbazole derivatives having good solubility, high quality film formation and a strong blue emission can play a very important role in organic/polymeric optoelectronic materials, and they can be used as single layer materials for OLEDs.

In this work we report on the synthesis and characterization of a series of oxadiazole/carbazole containing polymers (POC), having formula:

CP1255, *V[th] International Conference on Times of Polymers (TOP) and Composites,*
edited by A. D'Amore, D. Acierno, and L. Grassia

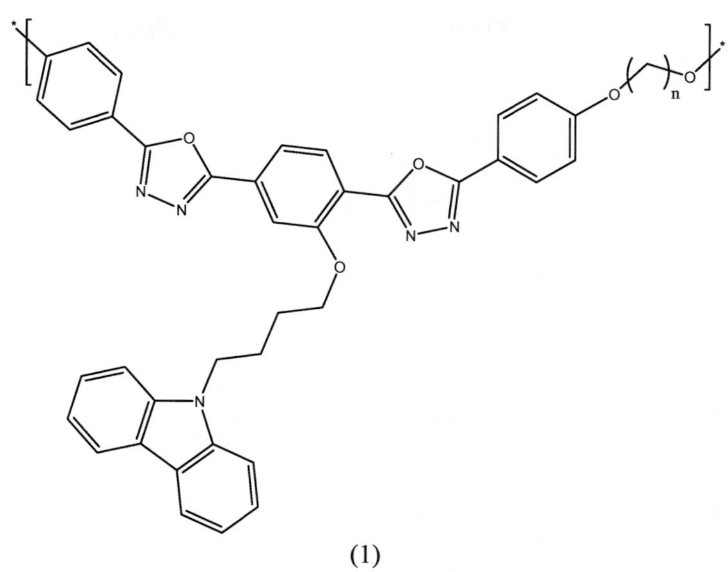

(1)

Where n is 12, or 6 for the two homopolymers, and 12 and 6 or 8 and 4 for the two copolymers. All the polymers show high glass transition temperatures and good solubility in common organic solvents. POC(n,m) show high photoluminescence activity in the blue region of the visible spectrum and they can be processed in high quality thin films by solution casting or spin-coating.

SYNTHESIS AND DISCUSSION

The synthetic route to obtain polymers POC is illustrated in figure 1:

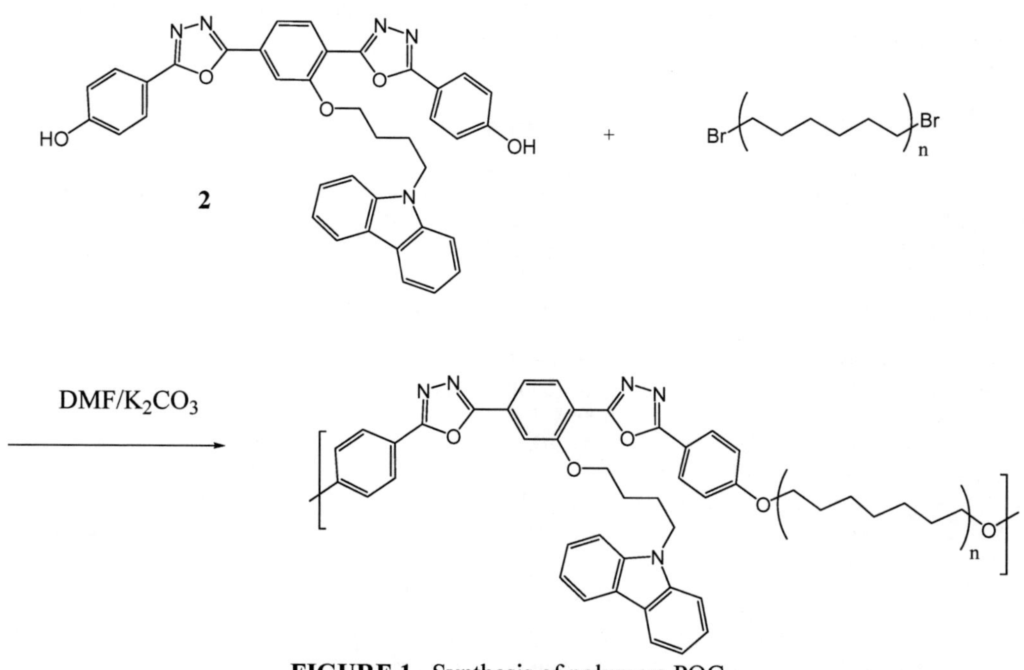

FIGURE 1. Synthesis of polymers POC.

The structure of all polymers POC was confirmed by proton NMR analysis. Thermodynamic data of POC are listed in Table 1. From the DSC analysis only the glass transition temperature is observed, which indicates that POC, as obtained from the synthesis, are amorphous polymers, with glass transition temperatures depending on the length of aliphatic segments in the main chain. Glass transition temperature in the range 115-125°C ensures good stability of film morphology at room temperature. The only exception is POC(6), which shows a poor solubility and higher T_g (170°C), due to its shorter aliphatic chain portion.

TABLE 1. Features of POC(n,m)

Polymer	Tg (°C) *	Td (°C) **
POC(12)	122	314
POC(6)	169	359
POC(12,6)	115	374
POC(8,4)	124	318

* Glass transition temperature, from DSC, N_2 flow at 10 °C/min rate; ** 5% weight loss temperature, from TGA, air flow at 20 °C/min rate.

In analogy with our previous paper on low molecular weight carbazole/oxadiazole compounds [10], and on oxadiazole containing polymers [11-13], the carbazole moiety absorbs at lower wavelength (strong peak at 292 nm) with respect to the oxadiazole unit (two strong and broad peaks at 322 and 345 nm) and emits in the wavelength range (two strong peaks at 354 and 369 nm) where the oxadiazole unit strongly absorbs. The result is that POC show a strong blue emission, identical to that of analogue oxadiazole containing polymers; in particular, the emission occurs at 407 nm (main peak, sharp) and at 460 nm (lower peak, broad band).

The good solubility in chlorinate solvents and DMF allows the preparation of films with homogeneous thickness by spin coating.

CONCLUSIONS

The new synthesized carbazole/oxadiazole containing polymers (POC) are blue photoluminescent materials. Taking advantage of the steric constraints due to the side insertion of carbazole unit to the oxadiazole moiety, and to the insertion of flexible aliphatic segments in the main polymer chain, POC are soluble in chlorinate organic solvents or DMF and they can be easily processed to homogenous films. The synthesized polymers show strong photoemission in the blue region of the visible spectrum. Taking advantage from these features, POC could be good candidate materials for the realization of single layer organic LED

REFERENCES

1. J.H. Burroughes, D.D.C. Bradley, A.R. Brown, R.N. Marks, K. Mackay, R.H. Friend, P.L. Burns, A.B. Holmes, Nature 347 (1990) 539–541.
2. A.B. Holmes, D.D.C. Bradley, A.R. Brown, P.L. Burns, J.H. Burroughes, R.H. Friend, N.C. Greenham, R.W. Gymer, D.A. Halliday, Synth. Met. 57 (1993) 4031–4040.
3. M. Berggren, O. Inganas, G. Gustafsson, J. Rasmusson, M.R. Andersson, T. Hjertberg, O. Wennerstrom, Nature 372 (1994) 444–446.
4. G. Grem, G. Leditzky, B. Ulrich, G. Leising, Adv. Mater. 4 (1992) 36–37.
5. C. Adachi, T. Tsutsui, S. Saito, Appl. Phys. Lett. 56 (1990) 799–801.
6. X. Li, Y. Liu, M. Liu, A. Jen, Chem. Mater. 11 (1999) 1568–1575.
7. Z. Peng, Z. Bao, M. Galvin, Chem. Mater. 10 (1998) 2086–2090.
8. J. Lu, A. Hlil, Y. Sun, A. Hay, T. Maindron, J. Dodelet, M. D'Iorio, Chem. Mater. 11 (1999) 2501–2507.
9. M. Casu, P. Imperia, S. Schrader, B. Schulz, M. Jandke, P. Strohriegl, Synth. Met. 121 (2001) 1397–1398.
10. V. Bugatti, S. Concilio, P. Iannelli, S. P. Piotto, S. Bellone, M. Ferrara, H. C. Neitzert, A. Rubino, D. Della Sala, P. Vacca, Synthetic Metals, 2006,156(1), 13-20.
11. D. Acierno, E. Amendola, S. Bellone, S. Concilio, L. Ferrara, P. Iannelli, H.-C.Neitzert, A. Rubino, F. Villani, Journal of Non-Crystalline Solids, 2004, 338–340, 278–282.
12. D. Acierno, S. Concilio, A. Diodati, P. Iannelli, S.P. Piotto, P. Scarfato, Liquid Cryst. 29 (2002) 1383–1392.
13. D. Acierno, E. Amendola, S. Bellone, S. Concilio, P. Iannelli, H. Neitzert, A. Rubino, F. Villani, Macromolecules 36 (2003) 6410–6415.

The Effect of processing on the PVC/Clay Nanocomposites Structure

A. Kalendova[a], J. Zykova[a], L. Kovarova[a], M. Slouf[b], J. F. Gerard[c]

[a] *Department of Polymer Engineering, Faculty of Technology, Tomas Bata University in Zlín, Nám. TGM 275, 762 72 Zlín, Czech Republic,* kalendova@email.cz
[b] *Institute of Macromolecular Chemistry AS CR, Prague, Czech Republic*
[c] *Laboratoire des Matériaux Macromoléculaires IMP INSA de Lyon, Bât Jules Verne, 20 Avenue A. Einstein, 69621 Villeurbanne Cedex – France*

Abstract. Reported nanocomposites of poly(vinyl chloride) have been prepared using bentonite-based clay, Na-montmorillonite (nature clay) and organophilic clay 30B. Polymer nanocomposites of differing compositions were produced using Buss KO-kneader via melt intercalation method. The effect of different type of plasticizer (both low molecular and high molecular) and compounding conditions on the structure of PVC/clay nanocomposites was investigated. Different compounding conditions were tested to study their influence on nanoparticles dispersion, orientation and exfoliation in PVC/clay nanocomposites. The structure of PVC/MMT nanocomposites was observed using X-ray diffraction, transmission electron microscopy (TEM) and AFM (Atomic Force Microscopy). It was found that the Na-montmorillonite offer low exfoliation level, while 30B modified by plasticizer exhibits fine dispersion of partial to nearly full exfoliated MMT. Moreover the processing conditions play also important role in nanocomposite production.

Keywords: Polyvinylchloride, nanocomposite, morphology, microscopy.
PACS: 81.07.Pr Organic-inorganic hybrid nanostructures

INTRODUCTION

Polymer-layered silicate nanocomposites (PLS) have experienced an important development over the past 20 years. PLS has been the focus of academic and industrial attention because the final composites often exhibit a desired enhancement of mechanical, barrier, thermal or other properties relative to the original polymer matrix, even at very low clay contents.[1-4]

The most commonly used layered silicate in PLS is montmorillonite (MMT), a layered aluminosilicate smectite clay. A sheet structure consists of octahedral alumina sandwiched between two silica tetrahedral layers. Negatively charged, massive silicate sheets are separated by interlayer regions of 5-10 Å, which contain charge-compensating mobile cations..[5] Moreover, montmorillonite has a low cation-exchange capacity (CEC) and, therefore, does not have a large number of ionic interactions holding the clay plates together.[6]

Polyvinylchloride (PVC) is a polymer, which has a lot of characteristics suitable for industrial applications such as good mechanical properties, fire retardancy, good processability and, in addition, it offers the advantage of low cost.[7] Despite the fact that PVC is one of the major thermoplastic polymers, PVC/clay nanocomposites are still in the introductory phase. This work is based on the study of the influence of different processing conditions, and various plasticizers on the morphology of PVC/clay nanocomposites.

EXPETIMENTAL

Materials and PVC/Clay Processing

The suspension type of PVC Neralit 652 (K = 65) from Spolana Neratovice, Czech Republic was used as polymer matrix. The samples were stabilized with 3 % Lankromark LZB 968. As a filler was employed bentonite-based clay, Na-montmorillonite and organophilic montmorillonite 30B from Southern Clay Products, USA. Na-

CP1255, *V[th] International Conference on Times of Polymers (TOP) and Composites,*
edited by A. D'Amore, D. Acierno, and L. Grassia

montmorillonite is a sodium type with *d* spacing 1,2 nm. Montmorillonite 30B is organophilic clay with *d* spacing 1.8 nm. Dioctylphthalate (DOP), dioctyladipate (DOA) and polyester Lankroflex PLA (PLA) were used as PVC plasticizer and montmorillonite modifiers.

The PLS were prepared in a Buss KO-kneader MKS 30 (L/D = 18, D = 30 mm). The screw speed was at 80, 50, 35, 25 rpm. The operating temperature was at range 130–160 °C. Selected blends were compounded two or three times. Table 1 shows overview of tested samples and conditions.

TABLE 1. PVC/Clay Composition and Processing Conditions

Composition		v_{screw} (rpm)	t (times)	Composition	Filler (%)	v_{screw} (rpm)	t (times)
PVC/DOP 50	0	50	t				
PVC/DOA 50	0	50	t				
PVC/PLA 50	0	50	t				
PVC/30B/DOP 80	5	80	t	PVC/Na/DOP 80	5	80	t
PVC/30B/DOP 80	5	80	2t	PVC/Na/DOP 80	5	80	2t
PVC/30B/DOP 50	5	50	t	PVC/Na/DOP 50	5	50	t
PVC/30B/DOP 50	5	50	2t	PVC/Na/DOP 50	5	50	2t
PVC/30B/DOP 50	5	50	3t	PVC/Na/DOP 50	5	50	3t
PVC/30B/DOA 50	5	50	t	PVC/Na/DOA 50	5	50	t
PVC/30B/PLA 50	5	50	t	PVC/Na/PLA 50	5	50	t
PVC/30B/DOP 35	5	35	t	PVC/Na/DOP 35	5	35	t
PVC/30B/DOP 35	5	35	2t	PVC/Na/DOP 35	5	35	2t
PVC/30B/DOP 25	5	25	t	PVC/Na/DOP 25	5	25	t
PVC/30B/DOP 25	5	25	2t	PVC/Na/DOP 25	5	25	2t

v_{screw}.....screw speed, t.....retention time

Structural Study

The structure of obtained materials was observed by XRD and microscopic analysis. X-ray diffraction data were obtained at room temperature by an URD-6 Diffractometer, the X-ray beam was nickel-filtered CuKα_1 (λ = 1,54 Å) radiation.

The dispersion and delamination of the clays in PVC matrix were observed through microscopic technique. For the transmission electron microscopy (TEM, Phillips PM 120) the specimens were cut using an ultramicrotome equipped with a diamond knife. For the atomic force microscopy (AFM, Explorer^{TM}) were used the same samples as for TEM. It was used non-contact AFM method and for profile determination was applied software SPM Lab 5.01.

RESULT AND DISCUSSION

X-Ray Diffraction and Microscopy

X-ray reflectivity is one of the most powerful techniques to study the structural aspects of intercalates and nanostructures. Figer 1 represent the XRD patterns PVC/30B nanocomposites produced with Buss KO-kneader.

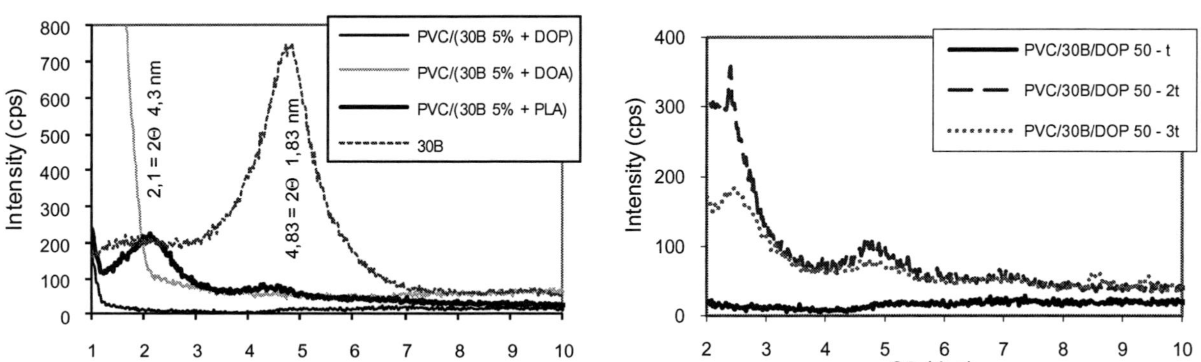

FIGURE 1. XRD Patterns a) PVC/30B 50 – t, b) PVC/30B/DOP 50 – t, 2t, 3t

The XRD measurement of PVC nanocomposites (Figure 1a) confirms good exfoliation by samples with montmorillonite 30B/DOP, 30B/DOA (low molecular plasticizers). It can indicate good level of filler exfoliation to polymer matrix, and this result was confirmed by TEM (Figure 2). For the PVC/30B/PLA, changes in the structure of clay 30B in presence of plasticizer PLA (high molecular plasticizer) were observed, the *d* spacing increased from 1.83 to 4.3 nm. On the other hand, the compositions base on Na-montmorillonite does not show changes in XRD pattern and low level of exfoliation was also confirmed by TEM.

In Figure 1b are presented XRD patterns PVC/30B/DOP compositions prepared by different retention time t. It was noticed, that the composition shows some ordered structures, which could be probably connected with stacks of several layers which were not fully exfoliated. In Figure 3, it can be seen that the silicate layers are dispersed evenly in the polymer matrix. It was observed that the longer time of mixing in the Buss KO-kneader caused regular dispersion and orientation of the nanoparticles in the PVC matrix. On the other hand too long retention time leads to the destruction of exfoliated layers. For various retention time t for PVC/Na composites, it was not noticed dramatic changes in the structure of PVC/Na/DOP. The *d* spacing moved from 1.2 to 1.3 nm.

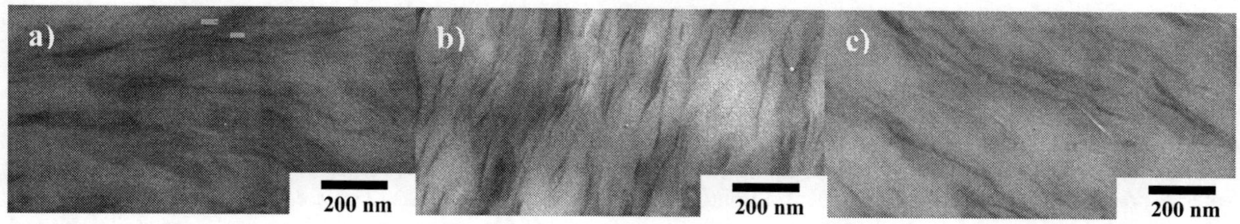

FIGURE 2. TEM Pictures of PVC/30B 50 a) PVC/30B/DOP – t, b) PVC/30B/DOA – t, c) PVC/30B/PLA – t

FIGURE 3. TEM Pictures of PVC/30B/DOP 50 – t, 2t, 3t

CONCLUSION

PVC/clay nanocomposites were obtained via melt compounding of PVC with clay in the presence of low and high molecular plasticizers. The type of plasticizer play important role in the process of intercalation or co-intercalation of montmorillonite and could be helpful during the nanocomposite processing. Better results give low molecular plasticizers.

Next the different processing conditions were tested. It could be concluded, that the lowest screw speed had the most positive effect on the reduction of nanofiller agglomerates and exfoliation process. On the other hand, this process is hold only for successful intercalate agents. If the intercalation is not fruitful then different conditions could not affect the exfoliation process and could not lead to nanocomposite systems.

ACKNOWLEDGMENTS

This project was supported by the internal grant of TBU in Zlín No. IGA/16/FT/10/D funded from the resources of specific university research.

REFERENCES

1. R. A. Vaia, S. Vasudevan, W. Krawiec, L.G. Scanlon and E.P. Giannelis, *Adv. Mater.* **7**, 154 (1995).
2. E. P. Giannelis, *Adv. Mater.* **8**, 29-35 (1996).
3. Naoki Hasegawa, Hirotaka Okamoto, Makoto Kato, Arimistu Usuki, Norio Sato, *Polymer* **44**(10), 2933-2937 (2003).
4. A. Kalendova , L. Kovarova, Z. Malac, J. Malac, J. Vaculik, J. Hrncirik, J. Simonik, Annual Technical Conference - Society of Plastics Engineers, Conference proceedings, 60th (Vol. 2), Dallas, USA, 2002, pp. 2250-2254.
5. A. Essawy Hisham, A. Abd El-Wahab Nourelhoda, A. Abd El-Ghaffar Mahmoud, *Polymer Degradation and Stability* **93**(8), 1472-147 (2008).
6. N. Yarahmadi, I. Jakubowicz, T. Hjertberg, *Polymer Degradation and Stability* **95**, 132-137 (2010).
7. L. I. Nass, Encyclopedia of PVC, Marvel Dekker Inc., USA, 1976.

Reaction in the Melt of Post-consumer Poly(ethylene terephthalate) (PET) with Ester Functionalized Polyolefins

Maria-Beatrice Coltelli[a], Stefania Savi[a], Mauro Aglietto[a], Francesco Ciardelli[a]

[a]Dipartimento di Chimica e Chimica Industriale, via Risorgimento 35, 56126, Pisa, Italy
e-mail: fciard@dcci.unipi.it

Abstract- The blending of PET from post-consumer packaging products with polyolefins can provide toughened materials. PET and polyolefins are immiscible polymers and to the above purpose, different method of reactive compatibilization are investigated to achieve a fine dispersion of polyolefin domains into a PET matrix. The use of a functionalized polyolefin, bearing reactive groups toward PET terminals, is a promising route to obtain grafted copolymers acting as interface stabilizers. In particular the use in the melt blending of ester or hydroxyl functionalized polyolefins in the presence of transesterification catalysts and or anhydride functionalized polyolefins as compatibilizer precursors were both investigated by focusing onto the relevant chemical aspects. The prepared blends were analyzed through suitable fractionation methods, such as selective extractions, and spectroscopic analysis in order to identify the molecular architecture of the PET-polyolefin copolymers macromolecules resulting from the process and study their effectiveness at the interface region. Moreover the phase morphology and the thermo-mechanical properties were investigated and correlated to structure of the macromolecular species in the system..

Keywords: polyester, blending, recycling, reactive processing
PACS: 81.05.Lg

1. INTRODUCTION

The low density plastic fraction is particularly abundant in solid wastes in terms of volume and contains a significant amount of Polyethylene terephthalate (PET) extensively used for the production of fibers, films, and bottles for water and other beverages (especially carbonated drinks). In this regard, it is particularly fortunate because it is easily separable from other city waste and allows a relatively economical collection as the PET mean stream comes from easy identified post-consumer beverages bottles. As an engineering plastic, it offers, among others, excellent high-temperature properties, creep, and solvent resistance, in addition to unique mechanical, and permeability properties, as well as processability

PET recycling can however be sustainable only by developing strategies leading to materials and products with high technical value according to an up-cycling approach.

2. FUNCTIONALIZED POLYOLEFINS AS COMPATIBILIZER PRECURSORS

Functionalized polyolefins have been diffusely studied and used as possible compatibilizer precursors.

The advantages of these compatibilizer precursors consist of the possibility of preparation by modifying commercial pre-made polymers [1-2] (Scheme 2, a), so that the process is particularly economical and versatile. In fact polyolefins bearing various reactive groups can be easily obtained with a satisfactory control of structure modification by radical functionalization in the melt using a peroxide and a functionalizing monomers (M) [3-6].Groups grafted include ester group by Diethylmaleate grafting.

The functionalized polyolefin ester groups are able to react with PET hydroxyl and/or carboxyl terminals to yield a PO-PET graft copolymer (Scheme) in situ during the melt blending.

CP1255, *V[th] International Conference on Times of Polymers (TOP) and Composites,*
edited by A. D'Amore, D. Acierno, and L. Grassia

3. Zn AND Ti CATALIZED TRANSESTERIFICATION IN THE MELT

For the reaction among PET and dialkyl maleate functionalized polymers (PO-g-DAS) ,transesterification catalysts can be used in the reactive blending in order to improve the compatibilizer formation.. The study was mainly focused on the mechanism of the transesterification reactions occurring in the melt during blending (Scheme), but also morphological and mechanical properties of blends were kept into account.

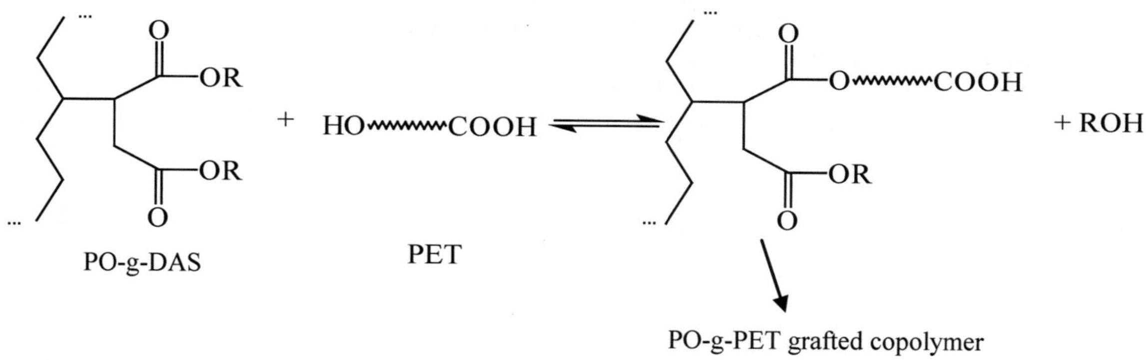

SCHEME. Reaction scheme illustrating the formation of PO-g-PET grafted copolymer during reaction ikn the melt of dialkyl maleate functionalized polyolefins (PO-g-DAS) and PET.

The yield of the transesterification between functionalized poly(ethylene) and PET was determined by the analysis of FT-IR spectra of blends residual fraction after selective extractions of unreacted polyester. A neat increase in grafting yield was observed when a dibutyl maleate functionalized polyolefin (POF) was used. Probably this was not the consequence of difference in macromolecular structure but in rheological properties of the functionalized polyolefins. In fact POF has a lower melt viscosity of POG and this could produce a morphology characterized by an higher interfacial area, as confirmed by SEM results, that favour interphases processes.

In a successive work the chemical reactions occurring during the melt blending between dibutyl maleate functionalized polyolefins (POF) and poly(ethylene terephthalate) (PET) in the presence of $Zn(OOCCH_3)_2*2H_2O$ were studied as a function of blending conditions, such as the preliminary drying of PET, the presence of a nitrogen flow during blending and the time of blending [7]. The selective extraction of POF/PET 30/70 by weight blends to remove unreacted PET and the determination of its molecular weight by viscosimetric measurements allowed to evaluate both the grafting points number and the molecular weight of PET. In this way the impact of degradation of PET onto its grafting by transesterification onto POF was investigated.

Differential scanning calorimetry (DSC), Scanning Electron Microscopy (SEM) and tensile measurements confirmed the trend deducted from chemical analysis, showing a good agreement between evidences about macromolecular structure and final properties of materials.

The results of this study can give some important indications, based on chemical reactivity and structure, for the preparation of blends by an innovative, sustainable and cheap process from post-consumer PET, with an alternative reactive blending approach, encouraging its recycling. The time of processing plays a very crucial role and it should be also optimised to achieve the right balance between PET degradation and grafting onto functionalized polyolefins by transesterification.

4. ANHYDRIDE FUNCTIONALIZED POLYOLEFINS AS COMPATIBILIZER PRECURSORS

The good reactivity of maleic anhydride modified polyolefins towards PET during the reactive compatibilization of very low density poly(ethylene) (VLDPE) /post-consumer PET blends was evidenced [8] in blends having a PET content of 20% by weight. In particular post-consumer PET was more reactive than high molecular weight virgin PET probably because of the presence of a larger number of terminal hydroxyl groups able to react with anhydride groups grafted onto the polyolefin [9].. Overall, the experimental results highlighted the possibility of producing materials with modulated and new properties from recycled PET by overcoming the lack of interfacial adhesion of polyolefin-PET blends typical of an immiscible system. Maleic anhydride functionalized ethylene-octene (PO-g-MAH) was employed as compatibilizer precursors for composition PET/LDPE/PO-g-MAH 70/20/10 whereas the content of maleic anhydride grafted in the ethylene-octene copolymer was varied. The SEM analysis showed dispersed phase diameter value decreasing as the anhydride content of PO-g-MAH increased. The tensile tests showed an increase in stiffness as a consequence of the addition of PO-g-MAH, in agreement with a more effective formation of chemical linkages between the macromolecules during the extrusion as the consequence of PO-g-PET copolymers formation. On the other hand the elongation at break showed a maximum value at 1% of anhydride content. The Izod impact strength showed a plateau-like trend. The rubber was distributed as dispersed domains in the brittle polymer matrix. The mechanism of toughening was investigated by Wu [9], who evidenced the fundamental role of inter-particle distance (Tc). In particular the general condition for toughening requires an inter-particle distance lower than Tc. This parameter depends on blend volume fraction ϕr and critical particle diameter dc through the equation :

$$d_c = \frac{T_c}{\left(\frac{\pi}{6\phi_r}\right)^{1/3} - 1}$$

The critical inter-particle distance is a property of the matrix, independent of rubber volume fraction and particle size, whereas the critical particle diameter depends on composition, being higher by increasing the rubber volume.

CONCLUSIONS

The possibility of improving the graft copolymer formation through a new process based on the transesterification between ester functionalized poly(ethylene) and PET in the presence of proper metal catalysts was performed and discussed. The success of the transesterification reaction to give a PO-g-PET copolymer depended on both the structure of the metal derivatives and compatibilizer precursor[10.11]. From a chemical point of view the less nucleophilic feature of the metal derivative ligands was effective in limiting the PET degradation. Moreover the Zn derivatives showed a slightly higher tendency than Ti derivatives to give PET degradation, but with a higher tendency to give copolymer formation. The modification of the functional groups of the compatibilizer precursor with hydroxyl group appeared as the most promising strategy to improve the copolymer formation by the transesterification approach.The use of anhydride modified poly(ethylene) as compatibilizer precursor allowed to obtain toughened blends, very attractive in applications requiring technical grade materials. Hence this last method provided a further valuable up-cycling perspective to post-consumer based materials.

REFERENCES

[1] G. Moad, *Progr. Polym. Sci.* **24**, 81 (1999)
[2] G. H. Hu and H. Cartier, *J. Appl. Pol. Sci.* **71**, 125 (1999)
[3] M. B. Coltelli, S. Savi, I. Della Maggiore, V. Liuzzo, M. Aglietto and F. Ciardelli, *Macromol. Mater. .Eng*. **289,** 400 (2004
[4] F. Ciardelli, M. Aglietto, E. Passaglia and G. Ruggeri, *Macromol. Symp.* **129,** 79 (1998)
[5] E. Passaglia, P. Siciliano, F. Ciardelli and G. Maschio, *Polym. Int.* **49**, 949 (2000)
[6] E Passaglia, L Corsi, M Aglietto, F Ciardelli, M Michelotti and G Suffredini, *J. Appl. Pol. Sci.* **87**, 14 (2003)
[7] M. B. Coltelli, S. Bianchi and M. Aglietto, *Polymer* **48**, 1276 (2007)
[8] M. Aglietto, M. B. Coltelli, S. Savi, F. Lochiatto, F. Ciardelli and M. Giani, *Jour. Mater. Cycles Waste Manag.* **6**, 13 (2004)
[9] S. Wu, *J. Pol. Sci.: Pol. Phis. Edn.* **21**, 699 (1983)
[10] M. B. Coltelli, C. Harrats, M. Aglietto, G. Groeninckx, *Polym. Eng. Sci.* **48**,1424 (2008)
[11] M. B. Coltelli, M. Aglietto, F. Ciardelli, *Eur. Polym. Jour.* **44**, 1512 (2008)

Effect of the Compounding Procedure on the Structure and Viscoelasticity of Polymer Nanocomposites

G.Capuano[a], G.Filippone[a], G.Romeo[a], D.Acierno[a]

[a]*Department of Materials Engineering and Production, University of Naplesi FedericoII, p.le Tecchio 80 80125 Napoli, Italy*

Abstract. We investigate the relation between the structure and the viscoelastic properties of a model polymer nanocomposite system based on a mixture of inorganic particles and poly(ethyleneoxide). Hydrophilic fumed silica nanoparticles were used as fillers and PEO-based nanocomposites were prepared by melt compounding and freeze-dryng. In both cases, dynamic oscillatory measurements in the melt state highlighted an increase of the frequency-dependent linear viscoelastic moduli with the filler content and a solid-like behaviour above a critical volume fraction. The freeze-dried samples exibhited a significant enhancement of the elasticity at lower filler contents as a result of the effect of particle dispersion on polymer chain dynamics.

Keywords: nanocomposites; filler dispersion; linear viscoelasticity
PACS: 83.60.Bc

INTRODUCTION

Adding nano-sized solid particles to commercial polymers can alter their flow behaviour and significantly improve their mechanical, thermal and transport properties. The degree of property enhancement achieved in the composite depends on the state of the particle dispersion and the ability of the particle surface to alter polymer segment dynamics. Even at very low filler loadings, the high interface between the phases results in micro and meso-structures, due to adsorption of polymer chains to the particle surface and interactions between individual particles. Since interface characteristics, as well as the particle size and shape, play a crucial role in the reinforcement mechanism of filled polymers, the development of new compounding procedures, that provide an efficient dispersion of nanoparticles, can potentially lead to the production of advanced materials, with unique features and competitive manufacturing cost [1].

The linear viscoelastic response of polymer nanocomposite systems (PNS) is extremely sensitive to their internal structure and the use of rheology to characterize the morphology of the dispersed phase in the melt state has been already validated, since small amplitude oscillatory measurements enable to probe the at-rest structure of these complex fluids.

Colloidal particles ($d<1\mu m$) suspended in viscoelastic media are favorable to self-aggregation and, above a critical volume fraction Φ_c, can rearrange and form a space spanning three-dimensional network. During this process the suspension is converted into a self-supporting material which exhibits a predominant elastic character, with the appearance of a secondary plateau of G'at low frequencies. This solid-like behaviour has been attributed to either particle-particle or polymer-particle interactions, both occurring at the nanoscale. The first mechanism originates from the formation of spatial agglomerated structures [2] of small particles (due to electrostatic and van der Waals' forces) and the evolution of these structures during loading. When fillers are well dispersed in the matrix, reducing the filler size while preserving the filler volume fraction, leads to a dramatic increase of interfacial area and a reduction of the average distance between particles or aggregates. Under such circumstances, a large fraction of polymer chains are in contact with at least one filler, some of them forming bridges between neighboring particles [3].

Although rheology, as a macroscopic characterization technique, cannot specifically elucidate what the nature of these mesostructures is, it provides an easily accessible way to inquire about the state of dispersion of the filler and its interactions with the polymer matrix.

CP1255, *Vth International Conference on Times of Polymers (TOP) and Composites,*
edited by A. D'Amore, D. Acierno, and L. Grassia

Fumed silica is an amorphous form of silicon dioxide made of primary spherical particles (d=14nm) fused together in branched aggregates, whose native surface is hydrophilic, the silanol groups (Si-OH) being the characteristic functionality. The hydrophilic character of these particles promotes their dispersion in water-soluble polymers such as poly(vinyl alcohol) and poly(ethylene oxide).

Poly(ethylene oxide), also known as poly(ethylene glycol), is a homopolymer of ethylene oxide monomers and one of the most well studied water-soluble polymers both in terms of its solution and interfacial properties. Interactions between PEO and fumed silica are dictated by hydrogen bonding and the silanols are thought to be the principal adsorption sites for the ether groups.

Fumed silicas have already been employed as a gelling agents in polyether liquids to yield composite polymer electrolytes for use in rechargeable lithium batteries.

EXPERIMENTAL SECTION

Fumed silica (SiO_2 Aerosil® A150 by Degussa) nanoparticles were used as fillers. Poly(ethyleneoxide) based nanocomposites at different filler volume fractions, Φ, were obtained by mixing the nanonparticles and the pure polymer (PEO, M_w=100KDa by Sigma Aldrich) following a melt compounding procedure and a freeze-drying method.

The melt compounded samples were prepared using a counter-rotating intermeshing twin-screw extruder (thermohake minilab) and compression moulded into disks to allow rheological tests. The pure polymer was processed under the same conditions to have a base for comparisons.

To prepare freeze-dried samples, nanoparticles were first dispersed in bidistilled water and the suspension was stirred and sonicated. Then the polymer was added and dissolved through vigorous stirring for 24 hours. After mixing the solution was poured in glass dishes, suddenly quenched in acetone/CO_2 and freeze-dried for 24-48 hours to remove all the water. The porous appearing nanocomposites were disk-shaped in a vacuum oven under slight pressure, to obtain compacted samples for rheological measurements.

Dynamic oscillatory shear tests were carried out by means of a stress-controlled rotational rheometer (TA instruments ARG2) using parallel plates geometry. All the measurements were conducted in a dry nitrogen atmosphere. Preliminary strain-sweep tests were carried out to probe the linear regime of the samples at each volume fraction. Frequency-sweep tests in the linear regime were carried out after samples had been equilibrated until reaching a steady state value of the viscoelastic moduli.

To determine the effective filler content of each sample, thermogravimetrical analyses (TGA) were performed at the end of rheological tests.

RESULTS AND DISCUSSION

The presence of the inorganic phase greatly affects the dynamic properties of the nanocomposites. This is reflected in the frequency dependence of both G' and G'', shown in Fig.1 for both the PEO/FS samples and the neat polymer.

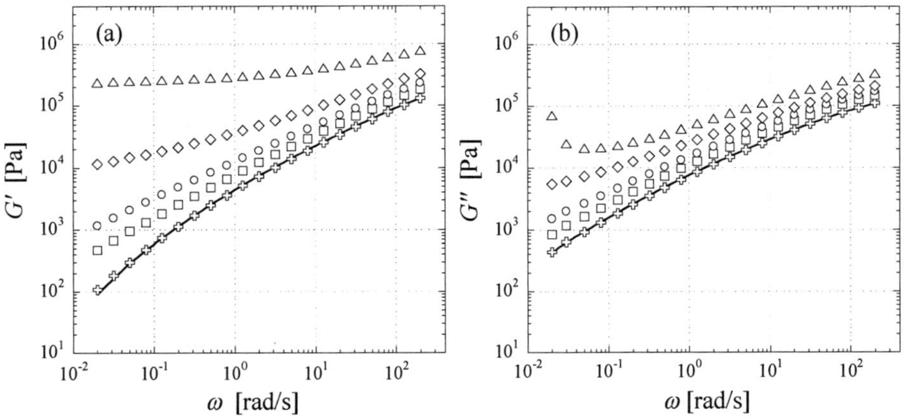

FIGURE 1. Frequency dependence of G' (*a*) and G''(b) for the neat PEO (*solid lines*) and *melt compounded* PEO/FS samples at different compositions: Φ = 1.1% (*crosses*), 3.2%(*squares*), 4.1% (*circles*), 5.1% (*diamonds*), 7.4% (*triangles*).

Irrespective of the frequency, both moduli increase monotonically with increased filler loading. The faster relaxation modes resemble those of the unfilled polymer and the slight increase in the moduli at high frequencies could be a consequence of the increase in the internal strain in the polymer phase with increased filler content.

The weak fequency dependence of G'and G''exhibited by the PEO/FS samples in the low frequencies region can be explained considering a networking mechanism based on polymer-particle interactions: above a critical volume fraction a tridimensional polymer-filler network is formed and G' becomes nearly frequency independent as a consequence of the arrest of polymer dynamics.

The freeze-dried samples exibhit solid-like dynamics at lower filler loadings (Fig.2) and a more effective dispersion of particles, whit smaller aggregates, can be superimposed.

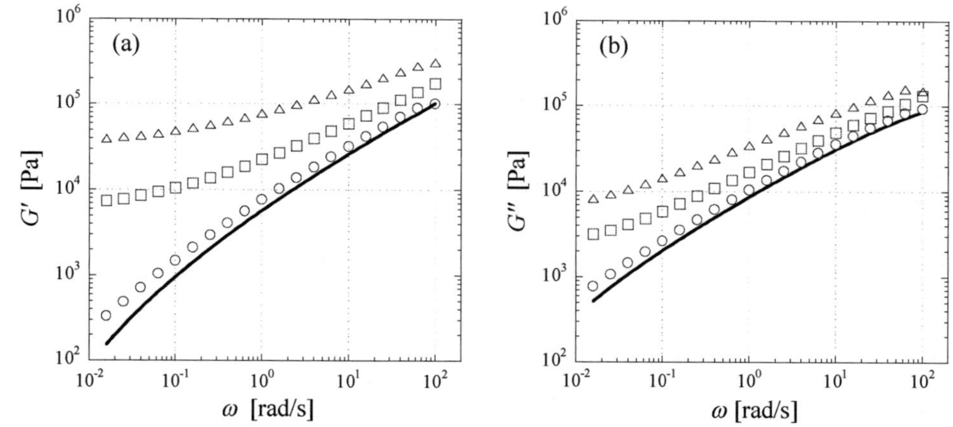

FIGURE 2. Frequency dependence of G' (*a*) and G''(b) for the neat PEO (*solid lines*) and *freeze dried* PEO/FS samples at different compositions: Φ = 1.4% (*circless*), 3.4%(*squares*), 4% (*triangles*).

When filler particles become (at least in one dimension) as small as the polymer coil, polymer molecules can simultaneously adsorb to the surfaces of many particles and connect them, even when they are quite far apart in comparison with their size. Detachment of physisorbed chains is needed to allow relaxation, thus enhancing the elastic response of filled samples (Fig.3).

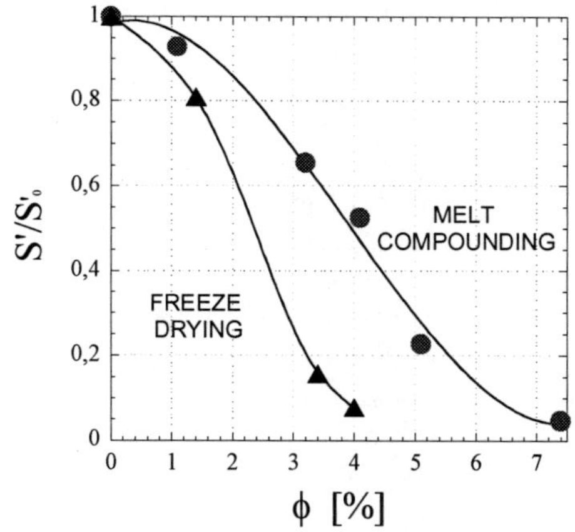

FIGURE 3. Low frequency slope of G' as a function of fumed silica volume fraction for samples prepared using two different compounding procedures: melt compounding (circles) and freeze dying (triangles).

REFERENCES

1. Ren J, Silva AS, Krishnamoorti R, *Macromolecules* 33(10), 3739-3746J, (2000).
2. G. Romeo, G. Filippone, A.F. Nieves, P. Russo, D. Acierno, *Rheol. acta* 47,989-997, (2008)
3. Zhang Q and Archer LA *Langmuir* 18(26), 10435-10442, (2002)

Impact of Nanoparticles on the Microstructure and Properties of Immiscible Polymer Blends: Preliminary Investigations

G. Filippone and D. Acierno

Dipartimento di Ingegneria dei Materiali e della Produzione,
Università di Napoli FedericoII, P.le Tecchio 80, 80125 Napoli, Italy

Abstract. The control of the morphology represents one of the most important aspects in designing polymer blends. A particular arrangement of the phases known as co-continuity can be promoted within a narrow range of compositions and using appropriate expedients during the mixing process. The distinguishing feature of co-continuous morphologies is the mutual interpenetration of the phases, which is often desirable as it may result in a remarkable combination of functional and structural properties of the blend constituents. Besides acting on the composition and processing conditions, adding nanoparticles in polymer systems with an existing phase-separated morphology such as polymer blends represents an innovative way to promote co-continuity in blends with low amounts of either phase. In the present work we focus on this topic, investigating the ability of nanoparticles to affect the morphology and properties of different kinds of immiscible polymer blends. In addition, the implications of the microstructural changes promoted by the filler on the high-temperature mechanical behavior of the blends are discussed.

Keywords: polymer blends; nanoparticles; structure-properties relationships; creep resistance.
PACS: 83.80.Tc, 82.35.Np

INTRODUCTION

Blending polymers represents a simple and economical way to improve the final properties of polymeric materials without investing in new chemistry. Because of a small mixing entropy, most polymer pairs are immiscible and tend to phase separate when mixed together. As a consequence, the macroscopic properties of immiscible blends are to a large extent determined by the micron scale arrangement of the phases, which control can promote performances even better than those of the single blend constituents [1]. Within a narrow range of compositions and using appropriate expedients during the mixing process, a particular space arrangement of the phases can be promoted which is known as co-continuity. The distinguishing feature of co-continuous morphologies is the mutual interpenetration of the phases, which may result in a remarkable combination of functional and structural properties of the blend constituents. Co-continuity can be achieved through the quenching of non-equilibrium morphologies attained during intense mechanical mixing. However, once stopped the flow the material quickly moves toward equilibrium loosing its co-continuous structure. Such a tendency can be minimized by using compatibilizing agents such as copolymers, but such an approach requires to be chemically tailored to each specific polymer pair. This restricts the use of compatibilizers to a relatively small number of systems. In the last years it has been proposed to substitute the traditional compatibilizing agents with small amounts of nanoparticles [2]. Typically, one of the polymer preferentially adsorb onto the particles, which can be used as a medium to assist the formation of co-continuous morphologies [3]. We recently exploited this phenomenon by promoting co-continuity in a blend of high density polyethylene (HDPE) and polyamide 6 (PA6) (HDPE/PA6 75/25) filled with organomodified clay (5 phr). The nanocomposite blend exhibited structural integrity up to temperatures much higher than the melting temperature of the major HDPE phase due to the subtle but continuous framework of PA6 [4]. Starting from these results, in the present work we try to gain further insight into the nature of this ability of nanoparticles in altering the microstructure of immiscible blends. In particular, we prepared and characterized different kinds of nanocomposite blends with the purpose of ascertain the possibility of extending our approach to convenient compounds for technologically relevant applications.

CP1255, *Vth International Conference on Times of Polymers (TOP) and Composites,*
edited by A. D'Amore, D. Acierno, and L. Grassia

EXPERIMENTAL SECTION

Two kinds of blends were prepared by blending a hydrophobic and a hydrophilic polymer: atactic polystyrene (PS N2982 by Polimeri Europa, Italy; molecular weight M_w=125 kDa; polydispersity index M_w/M_n=1.98; density ρ=0.95 g/cm^3) was mixed together with polyamide 6 (PA6 Radilon® S from Radici Group, Italy; ρ=1.13 g/cm^3; intrinsic viscosity [η]=1.5 dL/g at 30°C in 80 vol-% formic acid); high-density polyethylene (HDPE by Sigma Aldrich; ρ=0.95 g/cm^3; melt flow index MFI=12 g/10′) was blended with polyethylene oxide (PEO by Sigma Aldrich; ρ=1.13 g/cm^3; M_w=100 kDa). Two kinds of organomodified montmorillonite (Cloisite® 15A and 25A from Southern Clay Products) and a fumed silica (SiO_2 Aerosil® A150 by Degussa, primary particle size ~14 nm, specific surface ~150 m^2/g; ρ=2.2 g/cm^3) were used as filler. Since previous analyses have shown that the filler locates preferentially inside the more polar constituent of the blends, we kept constant the weight ratio between the nanoparticles and the hydrophilic polymer. The composition of the blends are summarized in Tab. 1.

TABLE 1. Composition of the blends.

Blend	Composition [wt/wt]	Filler type	Filler content
PS/PA6	50/50; 60/40; 70/30; 80/20	Cloisite 15A	Filler/PA6=0.05
HDPE/PEO	50/50; 55/45; 60/40; 65/35; 70/30; 75/25; 80/20; 85/15	Cloisite 25 A	Filler/PEO=0.05
HDPE/PEO	50/50; 55/45; 60/40; 65/35; 70/30; 75/25	Aerosil 150	Filler/PEO=0.05

The blends were prepared by melt-compounding the constituents using a co-rotating intermeshing twin-screw extruder (Minilab Microcompounder by Thermohaake) equipped with a capillary die (diameter 2 mm). A feedback chamber allowed for an accurate control of the residence time, which was set to four minutes for all the samples. The extrusions were performed in nitrogen atmosphere to prevent thermo-oxidative degradation phenomena, and the screw speed was 150 rpm, corresponding to shear rates of order of ~75 s^{-1}. The extrusion temperatures were T=240°C for the PS/PA6 based blends and T=150°C for the HDPE/PEO based blends. Prior to mixing, the hygroscopic polymers and the powders were dried under vacuum over night. Finally, the extruded pellets were compression molded in order to obtain disks (diameter 25 mm, thickness ~1.2 mm) for the subsequent analyses.

Quantitative extraction experiments were performed in order to estimate the degree of phase continuity of the minor constituent of the blends. Three disks for each composition, previously weighed, were immersed into a selective solvent for the minor polymeric constituent of the blend. In particular, formic acid was used to remove the PA6 phase from the PS/PA6 blends, while water was used to dissolve the PEO in the HDPE/PEO blends. Once dried, the samples were weighed again and the weight loss was recorded. The procedure was repeated until a constant mass was attained. Hence, the extent of phase continuity of the soluble polymer i, φ_i, was computed as $\varphi_i=(m_{i0}-m_{if})/m_{i0}$, where m_{i0} and m_{if} represent the nominal masses of the extractable polymer phase in the blend before and after the extraction experiments respectively. All the samples remained self-supporting at the end of the extractions.

Dynamic-mechanical analyses (DMA) were carried out using a Tritec 2000 DMA (Triton technology, UK). The dynamic moduli were measured as a function of temperature in single cantilever bending mode at a frequency ω=1 Hz and total displacement of 0.05 mm, which is small enough to be in the linear regime. The sample bars were heated at 2°C min^{-1} from about -90°C to either ~130°C or ~200°C depending on their composition.

RESULTS AND DISCUSSION

In the case of blends where a polar polymer is blended with a hydrophobic one, the filler generally tend to distribute in the former due to more favorable polymer-particle interaction [5]. This is what happens in our systems, in which all the fillers considered have been found to locate preferentially inside the PA6 and PEO phases. The uneven distribution of the filler brings about drastic changes in the microstructure of the blends, favoring the formation of co-continuous morphologies at lower amounts of the minor phase with respect to the case of unfilled blend. In addition, we found that the shape factor of the filler is an important parameter for co-continuity to occur, and the platelet-like structure of clays is more effective than the spheroidal shape of the fumed silica aggregates. This is shown in Fig. 1, where the extent of continuity of the PA6 and PEO phases in the filled blends is compared with that of the unfilled systems

The high extent of $\varphi_{PA6+Clay}$ in the PS/PA6 blend at 70% of PS has an interesting implication in terms of mechanical resistance at high temperature. The neat PS softens just above its glass transition temperature (T_g~100°C), and the same occurs when 30% of PA6 is added because of the distributed morphology of the blend.

Conversely, the corresponding filled blend keeps structural integrity at temperatures much higher than the softening of the PS due to the continuous framework of filled PA6, which melts down at ~220°C.

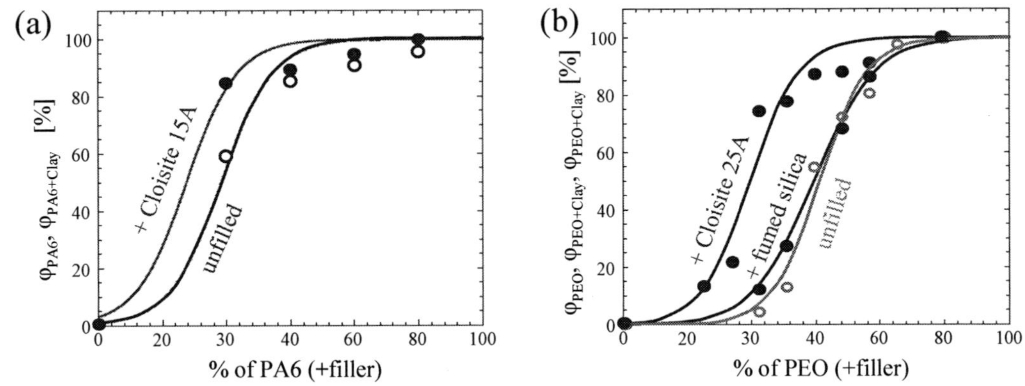

FIGURE 1. Extent of phase continuity, φ, of the PA6 phase in PS/PA6 based blends (a) and of the PEO phase in the HDPE/PEO based blends (b).

This is shown in Fig. 2, where the storage moduli of unfilled and clay-filled PS/PA6 blends are shown as a function of temperature as measured through DMA analyses. We emphasize that such a result was obtained without using any expedient during the preparation of the sample, suggesting that a targeted choice of the materials and compositions may result in optimized nanocomposite blends for specific technological applications.

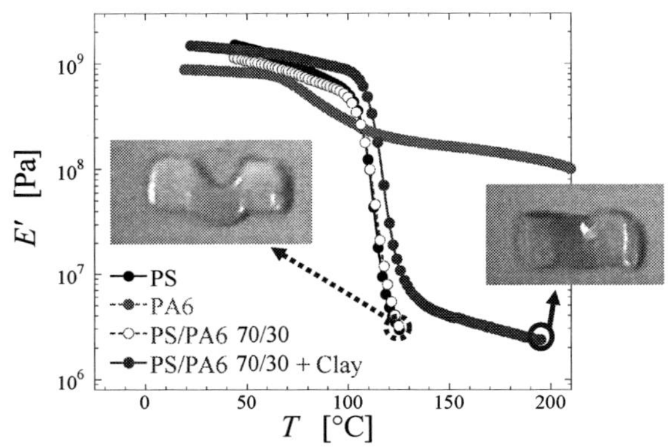

FIGURE 2. Storage modulus, E′, of the neat polymers and the unfilled and clay-filled PS/PA6 blends at 70% of PS.

CONCLUSIONS

Our results show that the synergism between the reinforcing action of the filler and its ability to affect the blend microstructure can be exploited in order to enhance relevant technological properties of the materials, such as their high temperature structural integrity. Given the simplicity and generality of the approach adopted, we believe that it can be extended to a wide variety of systems, representing a viable route for the obtaining of nanocomposite polymeric materials with optimized behaviors.

REFERENCES

1. Paul DR and Newman S in "Polymer blends", Academic Press, New York (1978)
2. Elias L, Fenouillot F, Majesté JC, Cassagnau P. Polymer 2007;48:6029-6040
3. Filippone G, Dintcheva NT, Acierno D and La Mantia FP, Polymer 2008;49(5):1312-1322
4. Filippone G, Dintcheva NT, Acierno D and La Mantia FP, "Impact of Organoclay on the Microstructure and Properties of High Density Polyethylene/Polyamide 6 blends – Effect of Filler Content and Polymer Matrix Composition", submitted to Polymer
5. Fenouillot F, Cassagnau P, Majesté J.-C. Polymer 2009;50:1333-1350

Rubber Characterization with Fatigue and Crack Growth Analysis

Laurent Perier* and Arnaud Favier*

* 01dB-Metravib, 200 chemin des Ormeaux, 69578 Limonest - France

Abstract. Dynamic Mechanical Analysis brings advanced capabilities to generate quantitative dynamic property data of rubber compounds for modelers and end users. Additionally to the recognized DMA testing benefits, a new METRAVIB Analyzer DMA+300 includes new testing solutions for analyzing rubbers mechanical property data under extended testing conditions closer to the real product's life conditions. The DMA+300 is specifically designed and dimensioned for fatigue tests on elastomers and crack growth tests. Combined with new MULTITEST software, DMA+300 makes possible to generate multi harmonics excitation and to control the application of a specific wave form to the specimen of rubber. This paper presents the main benefits available with this machine in the domain of rubber and elastomer applications.

Keywords: Dynamic Mechanical Analysis DMA, Elastomers, Rubber, Fatigue, Crack growth.
PACS: 81, 83

INTRODUCTION

The new METRAVIB machine testing platform for Dynamic Mechanical Analysis brings advanced capabilities to generate quantitative dynamic property data of rubber compounds for modelers and end users.

The description of the machine with its software is described in a first time. In a second time, description of applications and results are presented.

TEST MACHINE

The DMA+300 with a 300N force range, consists of a floor-standing mechanical frame, control and acquisition electronics and a computer station equipped with dedicated software (cf. figure 1).
The dynamic mechanical excitation is delivered by an electrodynamic actuator which includes a high performance guiding and anti-rotation system maintening the excitation within the specimen plane.

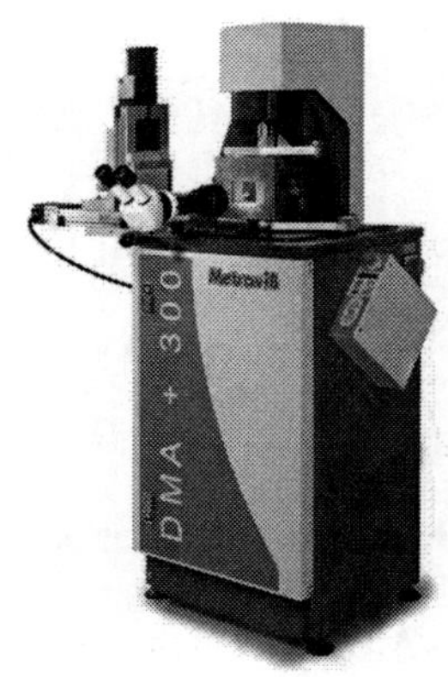

FIGURE 1. DMA+300 Machine

CP1255, *Vth International Conference on Times of Polymers (TOP) and Composites,*
edited by A. D'Amore, D. Acierno, and L. Grassia

The specimen is mounted on a specific mounting template according to a strict protocol, ensuring the jaws are accurately positioned on the same plane and parallel to the specimen plane. Once the specimen is pre-mounted on its jaws and the mounting template, it can be placed on the test machine with no risk of degradation or unwanted stress.

The dynamic and static strain/ stress are measured respectively at each end of the specimen.

The excitation signal is programmed using a new software, Multitest, applied to fatigue and crack growth test.

The control principle is based on the comparison of the measured waveform to the target waveform. To do so, MULTITEST software compares the FFT of the measured displacement signal to the FFT of the displacement set point. The command signal is modified according to the measured deviations

After a few iterations, the command signal is no longer sinusoidal, but the deformation applied to the specimen is compliant with the defined setting. This new approach makes possible to control both the amplitude and the waveform of the distortion actually applied to the specimen.

The thermal chamber (from -150°C up to 600°C) includes a large window that enables the entire surface of the specimen to be observed. The optical quality of this window allows for highly accurate control and optical analysis of the specimen, paticularly in the frame work of crack growth tests.

APPLICATION TO RUBBER CHARACTERIZATION

Characterization versus the applying wave form

The multi harmonic control principle makes possible to handle the deformation really applied to the specimen with very high accuracy, but also to control various waveforms such as sine, pulse, triangle, haversine and user defined specific ones: it makes then possible elastomers testing under conditions as close as possible of the final material application.

Heat build up

If no specific frequency value is required for the test, DMA+300 machine makes possible to optimize the test duration by increasing the test frequency. However, when imposing high strain or stress and high frequency to an elastomer specimen, the applied energy is transformed into heat inside the specimen: this phenomenon is known as heat build up [1].

The heat build up depends on strain ε or stress σ, frequency f, loss factor tan (δ),storage Young modulus E' and specimen volume Vs as shown in (1)

$$P_d = 2\pi f.\tan(\delta)E'\varepsilon^2 V_s, \qquad P_d = 2\pi f.\frac{\sigma^2}{E'\tan(\delta)}V_s \qquad (1)$$

If the frequency is increased in order to optimize the test duration, it is necessary to verify that the heat build up phenomenon is negligible. Actually, the model DMA+300, is especially well adapted to study this phenomenon, in making possible applying high strain excitation. An additional thermal probe may be used to accurately measure the specimen temperature changes. This additional probe can be located at different places depending on testing requirement (standard): close to specimen surface, inside specimen, inside specimen's supports).

Crack Growth Test

Better understanding of crack growth behaviour in elastomer or the study of the material resistance against crack growth can be another important point in product knowledge. DMA+300 is particularly well adapted to this type of test when equipped with the crack growth test module which incorporates cutting and optical systems (cf. figure 1).

The operator initiates cracks in the specimen on left and right sides by using a dedicated tool of the machine. The initial crack length is about few millimetres in order to initiate the cracking.

The crack growth test consists in following the crack growth according to solicitation and environmental test conditions. In the case of the improvement of new rubber compound exhibiting advanced crack propagation features, the cracks can be deviated in order to avoid the propagation from the tire tread to the tire (cf. photo Figure 2)

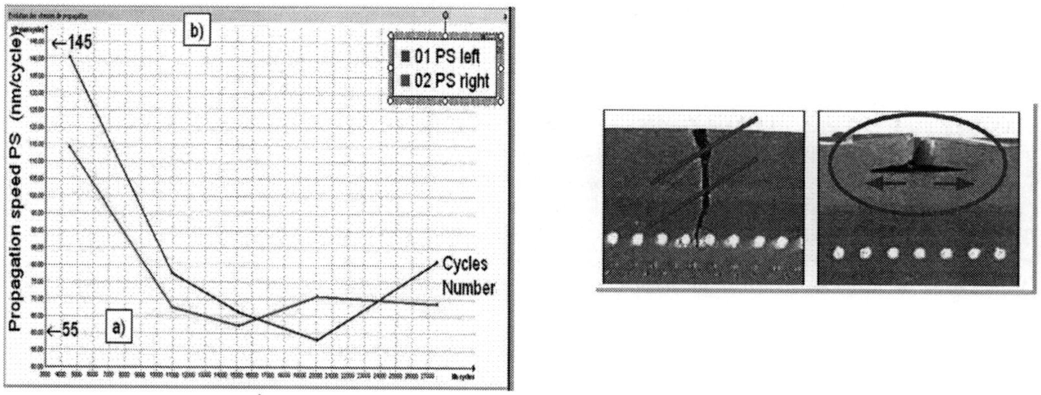

a) b)

FIGURE 2. Elastomer – a) Crack Growth Speed in nm/cycle vs. Cycles Number – b) Photo: Improvement of A New Rubber Compound for a Tire Tread (Michelin's Source)

CONCLUSION

Rubber compounds are complex materials that require accurate testing. The good knowledge of the performances of the rubber compounds, are essential to compounders, researchers and modelling engineers.

The data obtained with METRAVIB DMA+300 can be used as well in research domain to understand, characterize and compare new compounds, as in quality control to ensure that the production process is producing materials featuring the expected properties.
The data obtained with DMA+300 can be used as well in research domain to understand, characterize and compare new compounds, as in quality control domain to ensure that the production process is producing material with expected properties.

Thanks to new MULTITEST software, DMA+300 allows to perform on one single instrument, fatigue testing, crack growth tests, in order to understand the material's mechanical properties evolution under different excitation conditions and different excitation waveforms. The possible application on the instrument of excitation corresponding to vibrations encountered by the manufactured component allows providing more accurate data on the lifetime of the product.

REFERENCES

1. Ferry, John D., Viscoelastic Properties of Polymers. Third edition Wiley – pp 575-576

A DMA Instrument with Versatile Capabilities For Rubber Testing

Laurent Perier* and Arnaud Favier*

* 01dB-Metravib, 200 chemin des Ormeaux, 69578 Limonest - France

laurent.perier@areva.com; arnaud.favier@areva.com;
www.dma-instruments.com

Abstract. Following a development of 40 years, 01dB-Metravib is proposing a set of powerful instruments, including innovative and unique capabilities. The different instruments are covering a unique frequency range from static up to 10,000Hz. Additionnaly to standard DMA tests, thanks to high force capabilities (up to +/-450N), it is possible to understand strain dependence of the material up to very high dynamic strain (+/-300% and higher) and also to propose on the same instrument, complementary tests such as: fatigue, heat build up, crack growth, excitation waveform control, automated glass transition detection and optimization of measurement, … .
This presentation illustrates some of the capability of the DMA+ range of instrument applied to different kind of rubbers and elastomers material.

Keywords: Dynamic Mechanical Analysis, DMA, glass transition, complex modulus, E*, G*, damping properties, tangent delta, Payne effect, Mullins effect, heat build up, fatigue, high force, high frequency, high strain, cross linking degree.
PACS: 81, 83

INTRODUCTION

Contrarily to most other polymer materials, rubbers exhibit very particular behaviour. The properties of such materials are sensitive to temperature, frequency, strain, heat build up and even to the dynamic history of the sample itself! Measuring with high accuracy and high reproducibility dynamic properties of rubbers (E*, G*, tangent delta) requires specific care.
The sample preparation and its handling, the specifications and capabilities of the testing instrument, and the efficiency of the algorithm used to control the excitation are key points for the quality of the final measurement.
Measuring dynamic mechanical properties over a wide range of parameters (frequency, strain, temperature, excitation cycle, …) is becoming more and more important, in order to understand and predict the behaviour of the material in its final configuration of excitation. Taking into account the dynamic behaviour of materials is becoming a key point to improve the design of products, and to ensure a perfect correspondence with the performances requirements.

The very high rigidity of the mechanical frames, the powerful electro dynamic shaker are some of the key points of the DMA+ instruments, that enables its high level of performances (force up to +/-450N, frequency from static up to 1000Hz, temperature from -150°C up to 450°C, wide range of accessible strain (up to +/-300%) which is particularly appreciated for rubber testing …

DMA+ series, propose together with the instrument a complete range of specimen holder that allows adapting different testing mode to perform, tension, compression, shear and bending tests on the instrument. Considering elastomer material, the adequate testing modes are tension/compression and shear. Shear is particularly interesting because allows reaching very high strain deformation. For who wants to understand non linearity of the material, this capability is essential. Depending on the geometry of the specimen (Goodrich like cylinder, sheet of few mm thickness, …), these specimen holders allows to adapt easily the specimen to the instrument, and to characterize its properties.

Temperature dependency

Figure 1 presents some typical measurement on elastomers materials. This test is consisting in a temperature ramp (with adjustable rate), and following constantly the evolution of the properties while the temperature is increasing.

CP1255, *Vth International Conference on Times of Polymers (TOP) and Composites,*
edited by A. D'Amore, D. Acierno, and L. Grassia

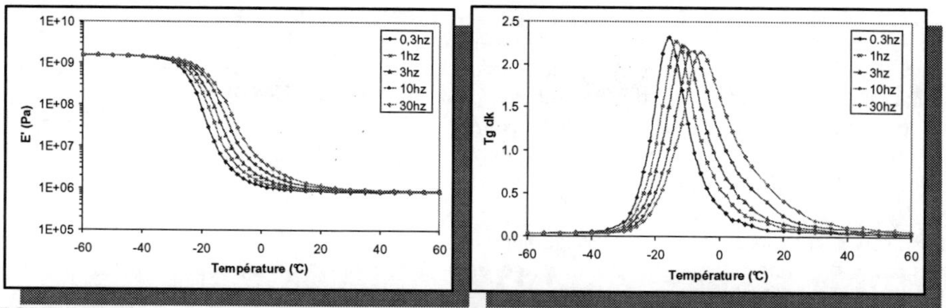

FIGURE 1: Elastomer (E' and Tg delta versus temperature) temperature and frequency dependency

The DYNATEST software used on DMA+ series includes several valuable capabilities, such as adapting automatically the temperature rate. In temperature area were the evolutions of properties are slow, the temperature rate can be fast, while in area were the material is presenting high temperature dependence, the software can automatically reduce the temperature rate, to have the best accuracy of the measurement. This functionality allows minimizing the duration of the test, without any compromise on the quality of the measurement.

Strain dependency

Rubbers and elastomer materials are material which presents some non linear behaviour with strain, which means that depending on the strain level applied, the modulus and tangent delta of the compound are changing. It is then very important to be able to measure and quantify the dependency of the material with the strain. In order to make possible high dynamic strain tests (+/-50%, +/-100% or even higher), it is necessary to perform measurement in the shear testing mode. Shear testing mode is the only mode that allows making some high level of strain.

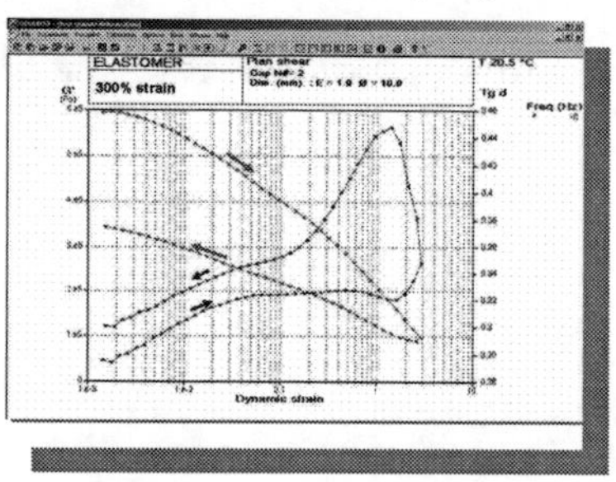

A single test performed on DMA+ allows characterizing accurately the dependency of the material to the strain known as Payne and Mullins effect.

Figure 2 displays an example of measurement with a strain sweep up and down, up to 300%. The non linearity, that is to say the dependency of dynamic properties to the strain is important (factor 6 on the modulus from low to high strain).The Mullins effect (value back) is presenting an important difference, with a factor 2 on the modulus between the starting and the ending point, which were performed at the same strain level.

FIGURE 2: Elastomer strain sweep up and down (G' and tan delta versus dynamic strain)

Frequency dependency

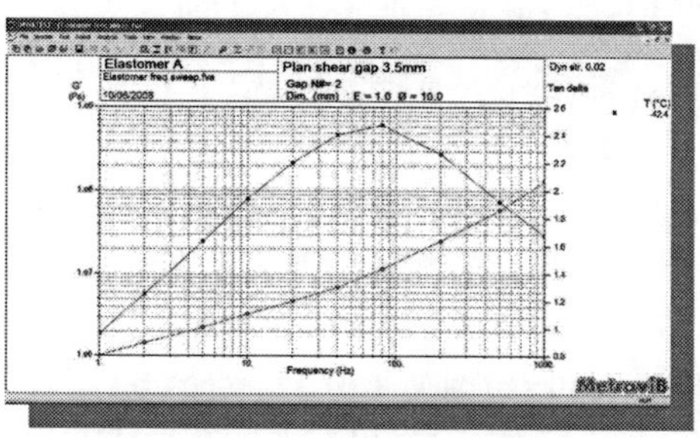

The frequency is one of the key parameter in rubber and elastomer testing. The location of the glass transition is strongly dependent on the frequency of excitation. It is also possible to run wide frequency scan at stabilized temperature, to understand better the influence of frequency.

Figure 4 is presenting such a test, with a frequency sweep from 1Hz up to 1000Hz, at a temperature of -42°C; that is to say in the transition area. Results displays that when going from low to high frequency, it is going through the maximum of the tangent delta values. Only because of the frequency excitation change, the storage modulus is presenting an evolution of more than two decades!

FIGURE 3: Elastomer frequency sweep, from 1 up to 1000Hz (G' and tan delta versus frequency)

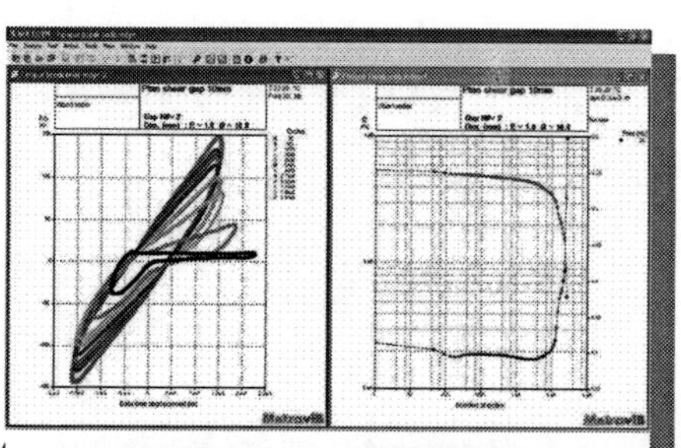

Fatigue testing

DMA+ series of instrument are also proposing some complementary capabilities of analysis, that overpass standard DMA testing. Thanks to the additional software named MULTITEST, it is possible to perform some fatigue tests, and so to understand about the evolution of the dynamic properties along with the number of cycle applied to the sample.

FIGURE 4: Fatigue test up to rupture. Lissajous curves (left), modulus and tangent delta vs number of cycles (right)

When looking at lifetime prediction and fatigue testing, the waveform apply to the sample is an important parameter. MULTITEST software is including some predefined usual wave form, such as: sinus, Haversine, pulse, Triangle. MULTITEST is also proposing to import some customized waveform. It means that the operator define onto ASCII files the waveform that correspond to the real final excitation of the product in the ASCII file. Then it is possible to import it in MULTITEST software and to apply this excitation waveform on the sample. Operator keeps the capability to define the frequency of excitation, in order to possibly accelerate the fatigue phenomenon. The more similar are the excitation waveform on the instrument compared to the real life excitation of the product, the more accurate will be the lifetime prediction.

Applying the adequate waveform allows to obtain information much closer to the real life of the product. For example, failure of the material will not occur after the same amount of cycle, in case of sinus excitation or in case of pulse excitation. The measurement of the heat build up can also be very related with the waveform applied.

Heat build up

DMA+ series is proposing as option capabilities to plug some additional input measurement channel, which the operator can use for its own purpose, depending on its project requirement. Example presented on figure 5 below is illustrating the use of this functionality, by measuring the temperature directly inside the sample. When this sample is submitted to high strain and high frequency excitation, parts of the energy injected is absorbed by the sample. This energy is transformed into heat, which increases the temperature of the sample, which modifies its properties.

In this example, the frequency of excitation is 30Hz, and the applied dynamic force is 75N. During the first 2 minutes of the test, the temperature in the heart of the sample is increasing by 15°C, which leads to a decrease of the modulus of approximately 20%. After 10 minutes, the increase of temperature is reaching more than 35°C, and the modulus decrease around 35%. After a certain period of time, the temperature in tending to stabilized (equilibrium between the injected energy due to heat build up and the dissipated energy in air and specimen holder).

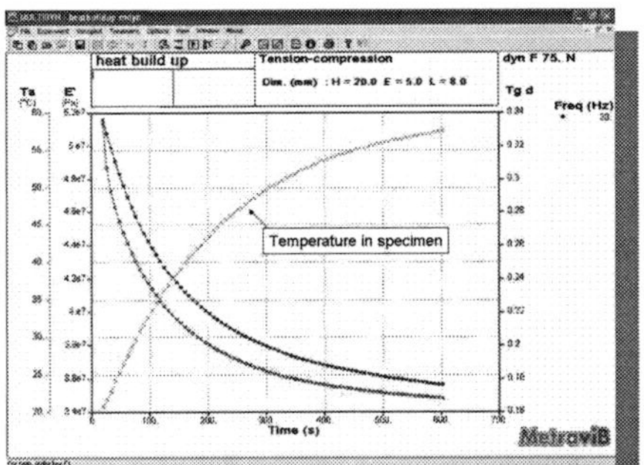

FIGURE 5: Heat build up measurement inside a sample

Conclusions

High quality and high performance tests on rubbers and elastomers, which lead to good knowledge of the performance of the material, are some of the key to success for compounder, researcher and modelling staff.

The results obtained with DMA+ series of instruments are used in research and compounding area to understand, characterize, compare and develop new compound, as well as in quality control domain to ensure that the production process is in line with the expected properties.

DMA+ series is today gathering a unique range of possibility of tests in one single instrument. Thanks to its wide range of specifications (force up to +/-450 Newton, frequency from static up to 1000Hz, temperature from -150°C to 450°C ...), DMA+ series enables to characterize accurately dynamic and static properties of material and their dependency to temperature, strain (Payne and Mullins effect) and frequency.

With the full range of specimen holders available, it is easy to adapt the different geometry of sample (sheet, cylinder, fibres ...), as well as the different state of material (liquid, pasty, gel, powder and solid), which makes the instrument applicable for any kind of materials (plastics, thermoplastics, thermosets, composites....)

Thanks to MULTITEST software, the same instrument allows to perform fatigue testing, in order to understand the properties evolution under different excitation conditions and different excitation waveform. Being able to reproduce on the instrument the real excitation, allows providing more accurate data on the lifetime of the product. When controlled accurately, increasing the frequency of excitation enables to get fatigue information faster, and so to save significantly testing time.

DMA+ series also propose additional options such as hygrometry control, customized additional channel, which open the instrument to further kind of simultaneous analysis. It makes then the operator free to measure together with the standard DMA or fatigue tests some additional measurement channel of its requirement.

Production and Characterization of Active Transparent PET Films for Oxygen Sensitive Foods Packaging

Maria Rosaria Galdi[a] and Loredana Incarnato[a]

[a]*Chemical and Food Engineering Department. University of Salerno*
via Ponte Don Melillo, CAP 84084, Fisciano, Salerno, Italy. lincarnato@unisa.it

Abstract. The aim of this work is to investigate possible solutions to realize active, transparent PET film suitable for packaging oxygen sensitive foods. At this purpose, monolayer active PET films at different oxygen scavenger concentrations and multilayer active ones were produced by cast extrusion laboratory scale equipments. To assess their activity and to verify the efficacy of such solutions, O_2 absorption analyses were carried out in continuous by an innovative oxygen meter.

Keywords: Oxygen scavenger films, active films properties, flexible packaging.

INTRODUCTION

Innovation in food packaging are related to the development of functional and eco-compatible packages able to satisfy the products needs and shelf-life requirements. Active technologies represent a very promising solution to improve polymeric materials properties. In fact, they allow to realize packages that interact with the foods and their head space to optimize the product conservation during the time. Nowadays, the tendency is to add active phases and, in particular, oxygen scavengers (OS) into polymeric matrices, in order to produce active packages by means of conventional processes. Most of the patents and applications for active packages are focused on polyester based packaging (1-3). The main PET active applications are relative to rigid packaging, in particular to bottles. The OS films applications are still rare due to the difficulty of realizing functional and efficient active films, by guaranteeing their activity over time. In particular, thin films (up to 100 microns) can promote a fast scavenger exhaustion, compromising the package performances.

The purpose of this work is to investigate possible solutions to obtain active PET films suitable for oxygen sensitive food packaging.

In our previous work (4), PET active films were realized using low concentration of an oxygen scavenger (up to 5% in weight). Films with 5% of AMS showed significant scavenging activity, but their exhaustion time was limited to 26h. In this work, the effect of higher concentrations of OS (up to 20% in weight) in monolayer films were evaluated. Active films were produced by cast extrusion laboratory equipment and were analyzed to determine their scavenging capacity and activity time. Moreover multilayer active structures were also realized by coextrusion process. In particular, a multilayer films, comprising active PET as central layer and pure PET as external layers were produced to form a tri-layer structure. Oxygen absorption analyses, performed in continuous on produced films, have demonstrated the efficacy of these solutions in prolonging the scavenger activity of active PET films.

EXPERIMENTAL

The materials selected as matrix is Polyethilentereftalate (PET) Cleartuf P60 (M&G Polimeri Italia). The oxygen scavenger selected is Amosorb (AMS) DFC 4020, a copolyester-based polymer. Monolayer PET films at different AMS concentration (0%wt, 5%wt,10%wt and 20%wt) were produced by a laboratory cast film extrusion equipment. The extruder was equipped with a screw diameter of 12 mm and an L/D ratio of 24. The extrusion flat die had lip width of 200 mm and thickness of 0.25 mm. The extruder temperature profiles were: 280°C along the extruder and 270°C at the die. The chill roll temperature was about 10°C and its speed was 2.5 m/min. Multilayer active PET films were produced by cast coextrusion, using two single screw extruders with a coextrusion feedblock technique process. The extruders were equipped with a screw diameter of 20 mm and an L/D ratio of 25. The extrusion flat die had lip width of 200 mm and thickness of 0.25 mm. The extruders temperature profiles were: 280°C along the

CP1255, *V[th] International Conference on Times of Polymers (TOP) and Composites,*
edited by A. D'Amore, D. Acierno, and L. Grassia

extruder and 275°C at the die. The chill roll temperature was about 10°C and its speed was 7 m/min. The final dimension of films were: about 150 mm wide, and about 30μ thick. The overall process was carried out under nitrogen flow. The oxygen absorption measurements were carried out by means of Fibox-3 Trace (Minisensor Oxygen Fibox 3-Trace V3, PreSens GmbH) at ambient temperature. The measurements were conducted on samples with a defined geometry, introduced into glass cells (volume 9 ml) containing the spot sensor PSt3, used for the analyses, and then the cell was hermetically capped. The oxygen concentration inside the closed cells was measured and recorded by the Fibox 3-Trace in continuous, in order to detect O_2 variation over time. The oxygen permeability was measured by GDP-C, Brugger at 23°C temperature, according to ASTM D 1434. The percentage of crystallinity (x_c) was evaluated by Differential Scanning Calorimeter (mod. DSC 822, Mettler Toledo). The thermal scan consisted of a first heating at a rate of 10°C/min, from 25°C to 300°C, an isotherm of 5 minutes at 300°C. a cooling from 300°C to 25°C at a rate of 10°C/min and a second heating up to 300°C at a rate of 10°C/min.

RESULTS AND DISCUSSION

In order to evaluate the activity of monolayer films, oxygen absorption analyses were performed in continuous to determine their scavenging capacity (μ) and the exhaustion time (t_e), as reported in Table 1. Permeability (P) and diffusivity (D) coefficients obtained by permeability measurements, performed on exhausted films, are also reported in Table 1.

Looking at the scavenging properties of active films, it is possible to observe that films with 5% AMS show a significant scavenging activity, but their exhaustion time is low. Increasing the AMS content it is possible to improve both scavenging properties. Moreover, the best performance were measured for films at 10% of AMS. These results can be attributed to a better dispersion and distribution of active phase in 10% of AMS samples, as shown in Fig. 1.

TABLE 1. Scavenging Capacity, Exhaustion time, Permeability and Diffusivity Coefficients and Cristallinity of produced films

Films	μ (cc O_2/g)	t_e (h)	P ($cm^3cm/m^2day\ bar$)	D ($cm^3cm/m^2\ day\ bar$)	x_c (%)
PET	/	/	0,23	$4{,}7*10^{-9}$	12
PET + 5% AMS	0,86	26	0,22	$4{,}1*10^{-9}$	13
PET + 10% AMS	4,68	170	0,19	$2{,}4*10^{-9}$	16
PET + 20% AMS	2,60	90	0,17	$2{,}3*10^{-9}$	20

Increasing the scavenger content up to 20%, an increase of AMS domains can be observed. Moreover, a worsening of active phase distribution within the PET matrix occurs (Fig. 1 (b)).

The permeation results reported in Table 1, show that for concentration higher than 5% of AMS, P and D coefficients show a significant reduction, probably related to higher crystallinity degree. In fact, the thermal analyses results underline that the scavenger presence promotes the film crystallization. This phenomenon could be due to nucleant effect of the scavenger and to molecular weight reduction enhanced increasing the AMS content, as verified by rheological analyses and intrinsic viscosity measures (5).

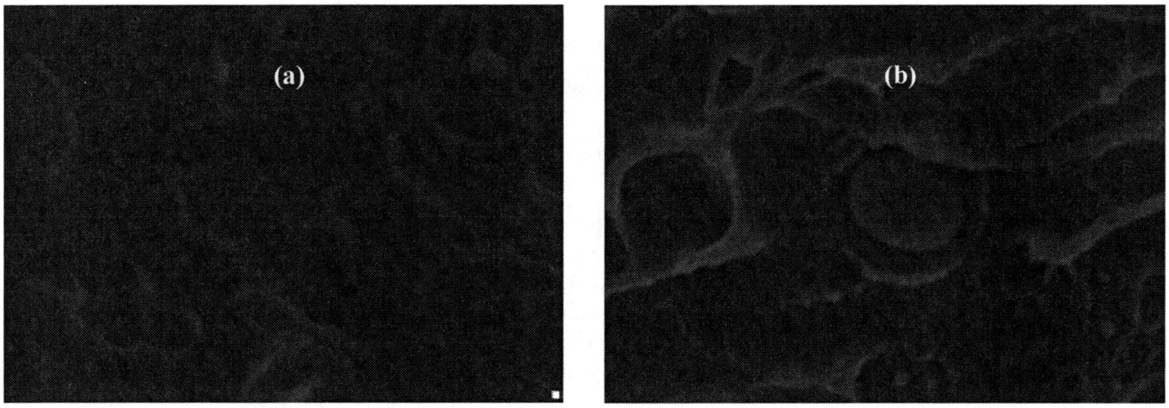

FIGURE 1. SEM images at 25kx of magnification: (a) active film at 10% AMS, the active domains size about 1 micron; (b) active film at 20% AMS, the active domain are larger than 4 microns.

In order to further increase the activity time of the oxygen scavenger, a tri-layer ABA structure was designed to protect the active layer (90% in weight of PET and 10% AMS) by two external layer of pure PET. Multilayer active films were realized by coextrusion process. O_2 absorption curves of monolayer at 10% of AMS and multilayer films are reported in Fig 2. It is possible to observe that multilayer film shows longer exhaustion time compared to monolayer active film. In fact, for these samples, pure PET external layers hinder the OS reactivity thus protracting the scavenging effect of AMS during the time. The exhaustion time evaluated by oxygen absorption measures was also compared to theoretical data obtained by applying the Solovyov-Goldman Model for active films (6-7).

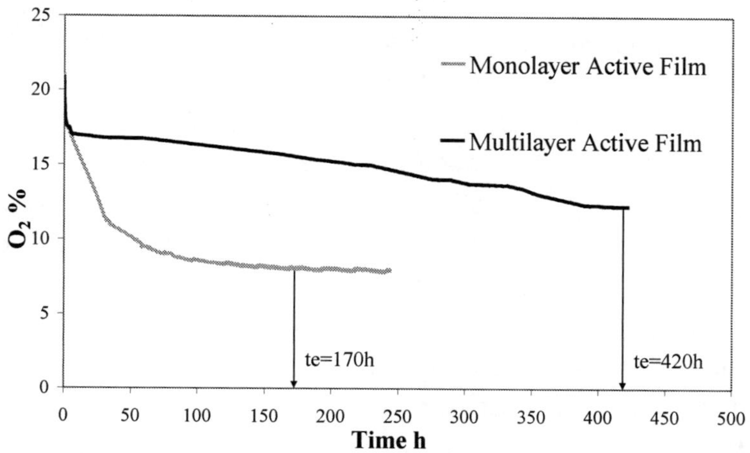

FIGURE 2. O_2 absorption curves of active monolayer films compared to active multilayer films, evaluated by Fibox-3 Trace in continuous.

Packaging tests with fresh oxygen sensitive food, not here reported, have confirmed the above results and have verified the efficacy of multilayer films to preserve these products by oxidation, increasing their shelf life (5).

CONCLUSIONS

The results obtained on monolayer active PET films at different AMS concentrations highlighted that the film at 10% of AMS content was the most performing, showing a high scavenging capacity and a prolonged exhaustion lag time. Increasing the OS content up to 20% a significant worsening in the state of dispersion and distribution of AMS has been observed. As a consequence the scavenger performances are significantly reduced. Moreover, O_2 absorption measurements pointed out that multilayer active films, as designed, further prolong the activity time of PET films with AMS. These results and shelf life tests performed on oxygen sensitive fresh foods assessed that the multilayer active structures realized can be suitable for food packaging applications.

ACKNOWLEDGMENTS

This research was partially supported by the Scientific Research Council of Campania (L.R.n.5 of 28/03/2002).

REFERENCES

1. P.J. Cahill, S.Y. Chen. US Patent No. 6,083,585, (2000).
2. J..M. Tibbitt, P..J. Cahill, G.E. Rotter, D.P. Sinclair, G.T. Brooks, T.R. Behrends. US Patent No.7,214,415 (2007).
3. D. Tung, E. A. Sisson, R. A Leckonby. US Patent No. 6,780,916 (2004).
4. M.R. Galdi, V. Nicolais, L .Di Maio, L. Incarnato. *Pack. Tech. and Sci.* **21**, 257-268 (2008).
5. M. R. Galdi, L. Incarnato. *Pack. Tech. and Sci.* Sent Dec 2009.
6. S. E. Solovyov, A. Y. Goldman. *International Journal of Polymeric Materials.* **54**, 71-91 (2005).
7. S. E. Solovyov, A. Y. Goldman. *International Journal of Polymeric Materials.* **54**, 93-115 (2005).

From rheology of nanocomposites to rheology of polymer melts: step back or forward?

Valery Kulichikhin, Alexander Semakov, and Alexander Malkin
A.V.Topchiev Institute of Petrochemical Synthesis, Russian Academy of Sciences
29 Leninsky Pr. Moscow 119991 Russia
e-mail address: klch@ips.ac.ru

1. Introduction:

For all heterophase systems including polymer ones the relation between rheology and stream morphology becomes very important. The stable tendency to combine rheological tests with "in situ" or at least "post factum" morphology or structure exists for this kind of objects. Rheooptics (including light scattering, birefringence, microscopy), as well as such scattering methods as X-ray or neutron-scattering are using for estimation of interrelationship between evolution and transformation of structure in flow and rheological response [1-3]. This approach helped us to obtain unusual morphological pictures for flowing nanocomposites based on polymer matrices, consisting in self-arrangement of filler particles [4]. These results forced us to find the driving force of such ordering, coming back to rheology of viscoelastic polymer melts behavior at high shear rates. Our vision of interrelationships between rheological properties of nanocomposite precursors and polymer melts are considered in this paper.

The main driving force predetermining a full set of morphological patterns is elasticity of complex fluids. The matter is that many real biological and technological liquids, besides irreversible deformation at flow, can store elastic energy and slowly relax after cessation of flow. The dual viscoelastic behavior is observed for numerous colloid systems, including proteins, liquid crystals, polymer melts and solutions, as well as for such biological media as sinovial liquid, saliva and even blood. Elasticity inherent to such liquids leads to rather unusual behavior and two effects of such kinds –die-swell in capillary flow [5] and the Weissenberg effect [6] consisting in raising up along a rotating unit. Here we shall demonstrate two other phenomena originated from elasticity of a liquid.

One of them is the already mentioned arrangement of solid particles initially randomly dispersed in an elastic liquid to form the regular ring-like morphology, as shown in Fig. 1. The main feature of this kind of reorganization is strong flow (high shear rates and stresses). It may be two main reasons of this effect: either intrinsic disturbances of polymer stream in presence of particles or something has happened with polymer matrix leading to arrangement of particles in definite positions. The second effect consists in very similar morphological pattern obtained for unfilled polymer melt at the same conditions. Manifestation of this effect is also shown in Fig.1, where one can see formation of regular ring texture of a molten polymer after intense shear deformation. Comparison of pictures inherent for nanocomposite and neat polymer melt says in favor of the second mechanism of arrangement, i.e., polymer melt is suffered the regular instability, and causes arrangement of impurities (particles).

It is well-known that in critical regimes of shear polymer melts the relaxation "liquid-like to solid-like" transition takes place. Effects of regular instability were mentioned for Newtonian liquids in strong shear fields where they are attributed as secondary flows and related to inertial input to flow [7]. Meanwhile, the effects under discussion here are of different nature. They are related to the loss of fluidity of viscoelastic medium due to impossibility of quick relaxation of

CP1255, *Vth International Conference on Times of Polymers (TOP) and Composites,*
edited by A. D'Amore, D. Acierno, and L. Grassia

deformed structures. So, self-assembling proceeds at condition of shear-induced relaxation transition from a liquid to rubber-like state.

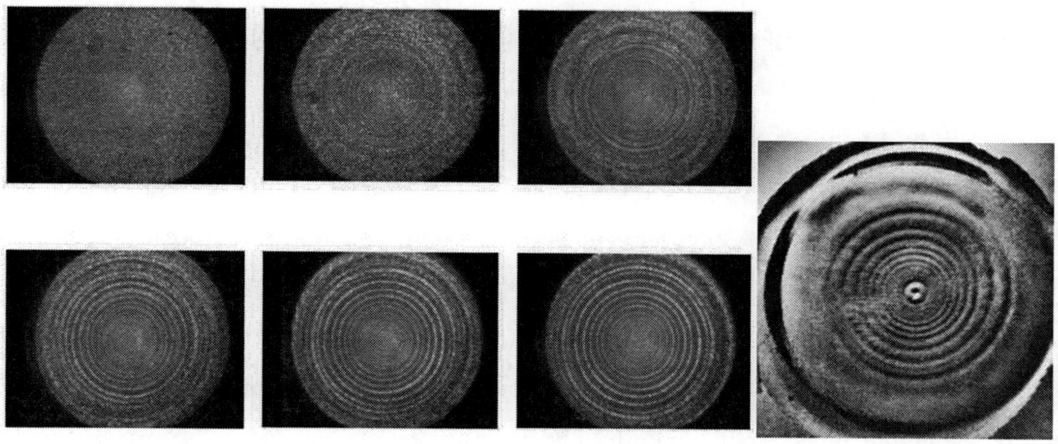

(A) (B)

Fig. 1. (A) Stages of morphology evolution of PIB solution filled with 7% of clay in strong flow in time (from left to right and from top to bottom); (B) The texture of individual polymer melt in similar conditions.

The threshold of this transition is covered by the critical value of the Weissenberg Number, Wi, which is the analogue of the Reynolds Number but with the substitution of inertial forces by elastic forces. So, at the critical ratio of elasticity to viscosity (the Weissenberg Number corresponding to this threshold is of the order of several units) prevailing input of elasticity leads either to the real turbulent flow (as the analogue of the Reynolds inertial turbulence) or to the transition from chaos to regular structures as we observed and demonstrated in Fig 1. The rate of this process depends on the level of the Weissenberg Number as well as on the elasticity of a matrix and dispersity degree of the internal phase.

Theory

The effects observed at high shear rates cannot be explained by standard constitutive equations, which relate to low shear rates (linear viscoelastic region). In the cases described above, the strong non-linear effects dominate and it requires to develop a new model of viscoelastic behavior. A model adequately corresponding to the effect of self-arranging at high deformation rates can be developed based on the renunciation of treating a viscoelastic material as a continuous media and transition to a model of interacting elastic structure elements (corpuscular or granular model) [8]. The content of the model is the theoretical analysis of mechanically interacting elements ("particles") in shear fields of different configuration. According to this model, shearing results in two elastic effects – deformation of initially symmetrical elements leading to their orientation in space and interaction of particles. Both are the source of the elastic energy storage, w_1 and w_2, respectively. The input of the first and the second items is expressed as

$$w_1 = -\beta E\left(1 + |S_n||S_{n+1}|\right)\log\frac{1 + S_n S_{n+1}}{1 + |S_n||S_{n+1}|} \text{ and } w_2 = -\alpha\tau\log\left(1 + h_e S_n\right)$$

where S_i are vectors coinciding with main axes of ellipsoids, E is elastic modulus, τ is shear stress of the external field with direction determined by vector h_i, α and β are constants of the order of 1.

The Hamilton function of this system can be written as the sum of two above listed components. So, the problem under discussion can be treated as the solution of nonlinear Schrödinger equation, which describes the process of spreading of solitary waves or their superposition (multi-soliton solutions) [10].

Computer simulation has demonstrated that this model really predicts the main experimentally observed effect – the "chaos-to-order" transition. The initial state of the structural elements on the lattice is assumed random. The increase of the number of iterations leads to an attractor, i.e., some utmost configuration. The attractor under discussion is a pseudo-spherical surface with self-intersections. The lines of self-intersection are zones of bifurcation, where phase streams separate. So, bifurcations create the necessary prerequisites for elastic instability of flow. The calculations also showed the effect of the separation of the cooperative behavior domains of particles. It actually means that space localization of the movement modes for a system of particles takes place.

Rings visually observed in Fig. 1 are limiting stage of spiral evolution. To prove this statement the micrograph obtained at flow of polyisobutylene solution is presented in Fig. 2. One

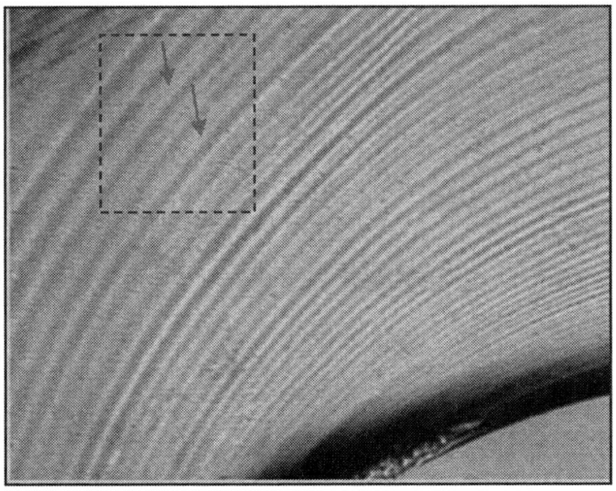

Fig. 2. Spiral texture of PIB in transparent operating unit. Twinning effect is shown in selected fragment by arrows.

can see the spiral structure, and in the selected fragment a zone of helix twinning. This indicates on bifurcation mechanism of spiral structure formation in axial shear field.

References:

[1] G.Schmidt, A.I.Nakatani, P.D.Burler, A.Karim, C.C.Han, *Macromolecules*, **33**, 7219 (2000).

[2] C.de Ruiter, E.Mendes, H.Boerstoel, S.Picken, *Polymer*, **47**, 8517 (2006)

[3] V.Kulichikhin, E.Plotnikova, A. Subbotin, N. Plate', *Rheol. Acta*, **40**, 49 (2001).

[4] V.G.Kulichikhin, A.V.Semakov, V.V.Karbushev, N.A.Plate, S.J.Picken, *Polymer Sci.*, **A51**, 2044 (2009).

[5] A.C. Merrington, *Nature*, **152**, 663 (1943).

[6] K. Weissenberg, *Nature*, **159**, 310-311 (1947).

[7] E.S.G. Shaqfeh, *Ann. Rev. Fluid Mech.*, **28**, 129 (1996).

[8] A.V. Semakov, V.G. Kulichikhin, *Polymer Sci.*, **51**, 2054 (2009)

Thermo-mechanical and electrical properties of epoxy composites containing ceramic nanofillers

A.M. Scamardella [a,b], E. Amendola [b,c], C. Petrarca [d] and D. Acierno [a]

[a] *Department of Materials and Production Engineering, University of Naples "Federico II, P.le Tecchio 80, 80125 Naples, Italy*
[b] *Institute of Composite and Biomedical Materials, CNR-Italy's National Council of Research, P.le E. Fermi1, 80055 Portici (Na), Italy*
[c] *Technological District of Polymer and Composite Materials Engineering and Structures, IMAST S.C.A.r.L., P.e E. Fermi 1, 80055 Portici, Naples, Italy*
[d] *Department of Electrical Engineering, University of Naples "Federico II, Via Claudio 21, 80125 Naples, Italy*

Abstract. Epoxy nanocomposites with ceramic nanoparticles, i.e. Al_2O_3, TiO_2 and SiO_2, were prepared by combined dispersion techniques, namely mechanical mixing and sonication. The influence of fillers addition on thermal, mechanical and dielectric properties have been investigated by means of Differential Scanning Calorimetry (DSC), Dynamic-mechanical analysis (DMA) and dielectric spectroscopy to measure the dielectric permittivity, tan δ and volume resistivity. Morphological investigations by Scanning Electron Microscopy (SEM) has been also performed.

Keywords: Epoxy nanocomposites, Al_2O_3, dielectrics, dc resistivity, thermal conductivity.
PACS: 65.80.-g

1 INTRODUCTION

Epoxy nanocomposites are widely used in industrial applications as structural composites, adhesives, surface coatings and as insulating materials in electrical motors. The insulation of rotating machines is subjected to multiple and concomitant stresses during operation, e.g. electrical, mechanical and thermal[1]. The generated heat, if not dissipated, might potentially affect the structural integrity of the final product, thus it is necessary to improve thermal conductivity of dielectric insulators. Epoxy resins are a class of thermosetting polymers widely used due to their ease of processability, good mechanical properties and thermal resistance. Unfortunately, the relatively disordered polymeric structure is also responsible of very poor thermal conductivity[2].
The use of epoxy nanocomposites as dielectrics requires optimal dielectric properties, therefore in this direction the addition of insulating ceramic nanoparticles to epoxy resin can provide a promising material with good dielectric and thermal properties[3].

2 EXPERIMENTAL

2.1 Materials and methods

A diglicydyl ether of bisphenol-A epoxy resin (EC01) with an epoxide equivalent of 187g/equiv. cured with methyl tetrahydrophthalic anhydride hardener (IG850) supplied by Elantas Camattini S.p.a. was used for the investigations. The addition of 0.25 phr of methyl imidazole (supplied by Elantas Camattini S.p.a. with the trade name of IG847) catalysed the crosslinking reaction. Alumina nanoparticles with an average particle sizes smaller than 50 nm (Al_2O_3 supplied by Sigma Aldrich), silica (SiO_2, Aerosol® 200) and titanium dioxide (TiO_2, Aeroxide P25) produced by Degussa with average particle of 12 and 21nm respectively, are used for composites preparation. The amount of ceramic particles added is 1 and 3 weight percent.

CP1255, *V[th] International Conference on Times of Polymers (TOP) and Composites*,
edited by A. D'Amore, D. Acierno, and L. Grassia

A scanning electron microscope (SEM, type Leica S440) was used to analyze the inorganic filler dispersion. A thin layer of gold was applied to the specimens surface, prior to SEM investigation, for providing electrical conductivity.

The thermal conductivity of neat epoxy resin and nanocomposites has been estimated by modulated DSC *(MDSC)*, at 10°C/min heating rate, ±0,5°C amplitude temperature modulation and a period of 80sec (5) according to ASTM E 1952 .

Dynamic-mechanical analysis was carried out by a *TA Instruments DMA 2980* at a heating rate of 5°C/min in a range of temperature from 0 to 200°C. Single cantilever geometry has been adopted, with a deformation amplitude of 15μm at the frequency of 1 Hz.

The electrical properties under evaluation are: relative dielectric permittivity (ε_r), dissipation factor (tan δ) and dc volume resistivity (ρ_v). Permittivity and tan δ have been measured in the frequency rango of 10^2-10^6 Hz by using an impedence analyzer HP 4192A. The volume resistivity has been measured at 20°C, by holding the specimens in a suitable shielded cell and using a stabilised dc source and a picoammeter HP 4140B. The data presented in this paper are an average value of 5 samples.

2.2 Nanocomposites preparation

Combined dispersion techniques have been used for mixture preparation, namely mechanical mixing and sonication[4]. The suitable amount of nanoparticles has been added to IG850 hardener, due to its low viscosity. The system has been mechanically mixed with a high speed homogeneizer "*Turrax T25*" at 6500 rpm for 10 min. The mixture has been subsequently sonicated in a ultrasonic processor *"Hielscher UP200S"* with a 20% of maximum amplitude (200W) for about 30 min. Temperature has been controlled during sonication at 35°C by means of a cooling bath. After addition of the suitable amount of EC01, the system has been further sonicated for 10min. Finally, the samples has been cured according to the thermal cycle suggested by technical data sheet (4h at 80°C, 3h at 100°C and 4h at 140°C).

2.3 Experimental results and discussion

DMA. Addition of ceramic particles determines, as expected, an increase in elastic modulus from the value of ~2700MPa observed for the unfilled resin, reaching the maximum value of ~3250MPa for the composite containing 3% of TiO_2 (figure 1).

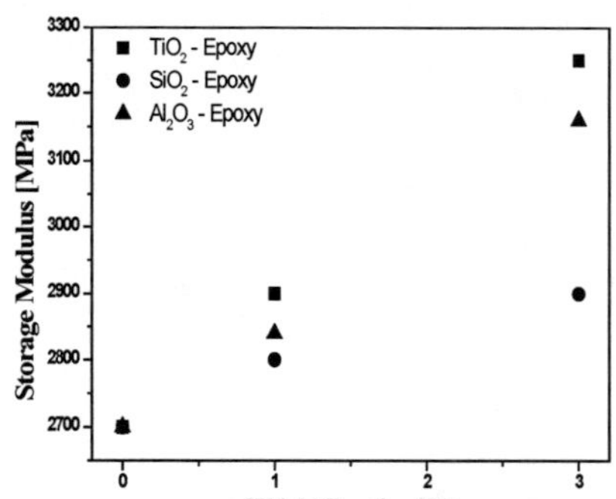

FIGURE 1. Storage Modulus of epoxy nanocomposites.

The glass transition temperature of composites, reported in table 1 increases of about 7°C for sample containing 3%wt of SiO_2 filler. The addition of rigid nanoparticles with a very large surface area and the formation of the interface boundary region between matrix and solid particles reduces the polymer chains mobility resulting in an increased glass transition temperature.

TABLE 1. T_g values assessed by DMA analyses for the nanocomposites and the neat epoxy resin.			
Filler [%wt]	Al_2O_3	2	2
0	139	139	139
1	138	145	141
3	145	147	144

DSC. The thermal conductivity of neat epoxy and its various hybrids has been evaluated by modulated DSC method. At lower filler content, the thermal conductivity remains approximately constant for all specimens containing different fillers. Increasing Al_2O_3 content up to 30%wt, the λ value is roughly doubled compared to the neat resin. This improvement, from 0,19 W/mK, for neat resin, to 0,29 W/mK for 30% Al_2O_3 sample, is due to the formation of a percolative pathway for heat conduction.

Electrical tests. dc volume resistivity has been measured by applying a dc voltage of 1400 V on thin samples characterize by a thickness ranging between 0.6 and 0.8 mm, corresponding to an applied electrical field included in the interval from 15.5 to 23.0 kV/cm. The values of dielectric current have been recorded after 120 s from the application of the dc voltage. In Figure 2 dc resistivity ρ_v vs. the filler content % wt is reported for the tree different typologies of samples. It can be noted a decrease of ρ_v with respect to the pure epoxy resin, wich is moe evident in the TiO_2 samples (in the 3% wt samples the resistivity is one order of magnitude lower with respect to pure epoxy resin).

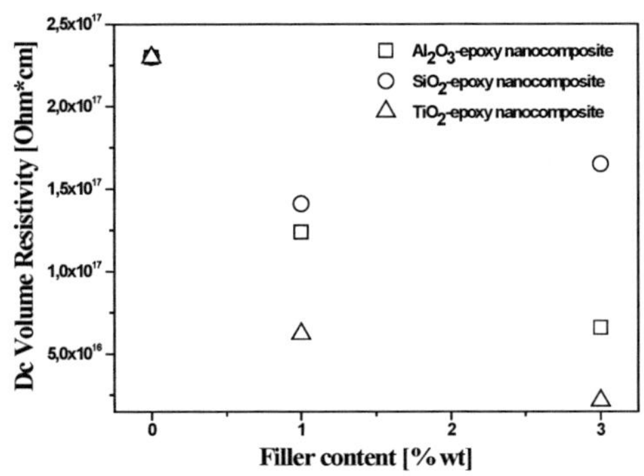

FIGURE 2. Dc volume resistivity of epoxy nanocomposites.

3 CONCLUSION

The paper presents results on the investigation of epoxy-nanocomposites containing ceramic nanoparticles. At low filler content (1% and 3%wt), the threshold of countinuous patway is not reached, and thermal conductivity λ values are only slightly reduced with respect to neat matrix. At high filler content (30%wt) the formation of a percolative heat sink pathway increases thermal conductivity.

The elastic moduli and the Tg of the composites increased compared with neat resin. A slight reduction in the epoxy dc volume resistivity is observed by incorporating a small amount of nano-filler into epoxy matrix.

ACKNOWLEDGMENTS

The financial support of IMAST in the frame of research project PIROS is greatly acknowlwdged. The authors would like to thank Mr. Mario De Angioletti and Mrs. Mariarosaria Marcedula for their technical support.

REFERENCES

1. A. Omrani, L. C. Simon, A. A. Rostami, *Mat. Chem.and Physics,* **114**, 145-150 (2009).
2. G. Lee, M. Park, J. Kim, J. Ik Lee, H. Yoon, *Composites A* **37**, 727-734, (2006).
3. S.Singha, M. Joy Thomas, *IEEE Transactions on Dielectrics and Electrical Insulation* **15**, 12-23, (2008).
4. C. Lekakou, A.K. Murugesh, P.A. Smith, *Polymer Engineering and Science*, 216-222, (2008).
5. S.Singha, M. Joy Thomas, IEEE Transactions on Dielectrics and Electrical Insulation , 1-5 , (2009).

Two timescales in polymer solidification: processing vs polymer crystallization

Stefano Piccarolo, Anesh M. Poulose, Domenico Carbone[§]

Dipartimento di Ingegneria Chimica dei Processi e dei Materiali, Università di Palermo, Viale delle Scienze, 90128 Palermo, Italy & INSTM Udr Palermo, E-mail: piccarolo@unipa.it,
§ Istituto di Chimica e Tecnologia dei Polimeri, CNR, Catania

Abstract. In this work we compare the influence of a plasticizer on the density drop, and therefore on the disappearance of any long range order crystalline phases, observed after fast controlled quenching of PET and PBT. The plasticizer enhances significantly the maximum cooling rate at which PET still shows long range order crystalline phases while barely modifies the situation for PBT.
The results are discussed on the basis of an interpretation of polymer crystallization strongly determined by the topology of the polymer, a view unconventional with respect to most of the well established interpretations of polymer crystallization all based on data obtained in a time scale much longer with respect to solidification in processing

Keywords: polymer crystallization, topology, processing, nucleation
PACS: *61.20, 61.25.hk, 81.05*

INTRODUCTION

Relevance of polymer crystallization in materials science does not need to be further pointed out if one thinks that out of the overall synthetic polymers production the most important materials and materials classes are those of crystalline polymers.This situation has certainly given rise to a continuous interest on the understanding and gathering of relevant information whenever this was made possible by the experimental approaches. By far the most important is certainly Differential Scanning Calorimetry able to collect, in different modes, data on crystallization and melting with characteristic times down to 10-100s, certainly the "fastest" method if one compares with the majority of those available. A high accuracy is obtained and one can collect all sort of information related to temperature dependence of overall crystallization kinetics or identify peculiar mechanisms recently summarized [1]

Although this work is not aiming to examine all these efforts in detail it is clear that the time range explored is by far different with respect to processing conditions where characteristic times are of the order of 1-10ms at least 4 orders of magnitude smaller. A situation encountered only on studying another dynamic process: that of glass transition.

DISCUSSION

In the polymer processing environment two situations determine a dichotomy between two classes of crystallisable materials depending on whether they are amenable to processing by injection moulding or not, i.e. fast or slow crystallizing polymers. Where, surprisingly enough, the role played by the possibility to add a nucleating agent to modify this situation becomes less significant under processing conditions.At first glance it is even more surprising that crystallization kinetics of fast crystallizing polymers is not easily experimentally available. The latitude of cooling rates or temperatures under which non isothermal or isothermal crystallization kinetics, respectively, is available is relatively modest. The upper bound of cooling rates is determined by instrumental dynamics while the bounds for studying crystallization under isothermal conditions depend on instrumental sensitivity the closer the temperature to the equilibrium melting one or by instrumental dynamics on decreasing temperature, i.e. in this case

CP1255, *Vth International Conference on Times of Polymers (TOP) and Composites,*
edited by A. D'Amore, D. Acierno, and L. Grassia

the material might already partially crystallize upon dropping to the crystallization conditions if the cooling rate had not been large enough.
On the opposite for slowly crystallizing polymers it is even possible to inhibit crystallization by either increasing cooling rate or decreasing isothermal crystallization temperature to such an extent that the typical pattern obtained by WAXD is replaced by a broad diffuse halo, apparently related to the onset of the fully amorphous phase. In the case of such polymers it is possible to study under a broad range of conditions the crystallization kinetics eventually being able to discriminate the underlying mechanisms. It is for this reason that those materials are often used as a model for interpreting and quantifying those mechanisms.

Table 1 The critical cooling rate, CR*, at which long range crystalline phases disappear for several polymer compositions

Polymer	**Additive** (% w/w)	**Mw (MWD),** Da	**[η]**, dl/g	**CR***, K/s
PET	- ?	1.9 E4		1.8
PET	- ?+ PEGDME (5)	1.9 E4		~50
PBT	- ?	1.7 E4		~250
PBT	- ?+ PEGDME (5)	1.7 E4		~350
PEN	- ?		0.51	0.12
iPP1	pure	4.8 E5 (6)		80
iPP3	Catalyst (~0.1)	4.9 E5 (9.7)		100
iPPM7N	Catalyst +Talc (~0.1+0.1)	3.8 E5 (5.3)		~1000

Notes: PET: Polyethyleneterephtalate, PBT: Polybuthyleneterephtalate, PEN: Polyethylenenaphtalate, PEGDME: poly (ethylene glycol) di-methyl ether, iPP: isotacticPolypropylene. ?: commercial sample

The solidification behaviour of several polymers was also investigated under conditions emulating polymer processing by a heuristic approach whereby a thin sample is solidified in a mould by a controlled continuous cooling approach very similar to that adopted in metallurgy to investigate the morphology developed in steels [2]. In contrast to the usual approaches adopted to study crystallization, nothing is known on the evolution of the morphology development during cooling but its thermal history since suitable design of experimental conditions [3] determine the onset of a homogeneous morphology throughout a sample that is therefore amenable for use of macroscopic probes for characterization of its overall structure. One of these, sample density, once plotted versus a suitable cooling rate describes the non isothermal crystallization behaviour, it is the solidification curve and it is shown in figure 1 for two polyesters: polybutylene terephthalate, PBT, and polyethylene terephthalate, PET. Common feature to all curves is a smooth decrease of density with cooling rate but for a density sudden drop at a cooling rate above which crystalline phases giving rise to distinct WAXD peaks disappear. It is this critical cooling rate, CR*, we shall examine and compare in the following of the discussion and some actual figures for a few typical polymers with/without nucleating agents is reported in Table 1.
The difference between the so called slow and fast crystallizing polymers is apparently related with CR*, fast are those with CR* on the order or above 100K/s, slow the ones with CR* below 1-2K/s. Between the two classes there are two orders of magnitude in the cooling rate determining the disappearance of phases carrying the fingerprint of long range order, which does not necessarily mean that above such cooling rates the materials are amorphous. Although this is not within the subject of the present discussion.
In Fig 1 it is also shown the influence of a plasticizer on the solidification curve of PET and PBT. Its influence on PET crystallization is very significant since CR* increases by one and a half order of magnitude becoming close to that of PBT. On the other side the same plasticizer does not affect to any significant extent the CR* of PBT. The different behaviour of the plasticizer in this context, i.e. on determining a faster crystallization of PET such as to make it almost comparable to PBT while barely affecting the latter is surprising.
Crystallization in polymers is a complex phenomenon where the mechanisms involved might not simply be those reported for simple substances, i.e. nucleation and growth. Considering however this simplifying approach there is only a difference in the surface energy contribution between primary and secondary nucleation so that nucleation is in any case the rate determining step.
At least three approaches interpret the nucleation in polymers. The first due to Turnbull and Fisher [4] and then directly borrowed by Hoffman and Lauritzen in what has been the most established viewpoint of polymer crystallization [5] accounts of a thermodynamic balance. The second due to Strobl [6] is also based on a

thermodynamic interpretation albeit it postulates, on the basis of experimental evidence, the presence of a mesomorphic precursor which should overcome steps like the conformational regular arrangement of chain sequences necessary for stem deposition on the growing surface.

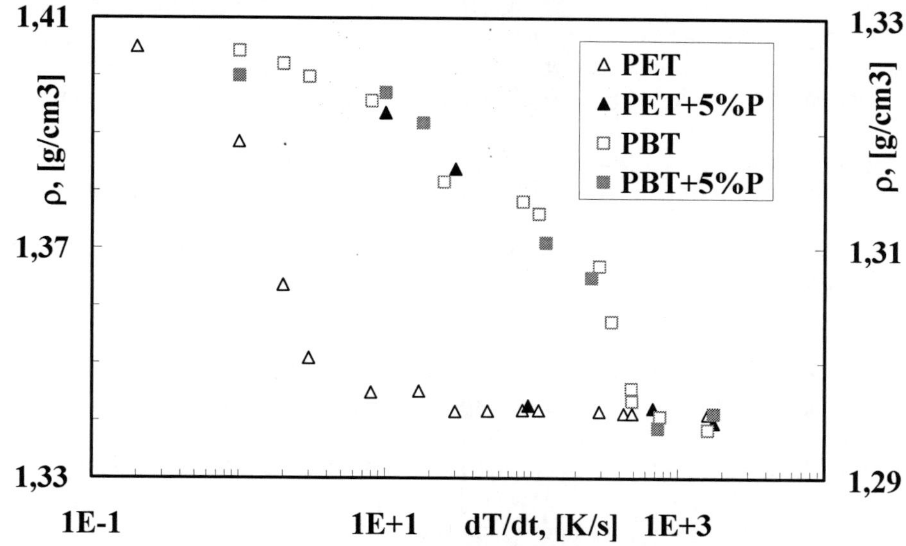

FIGURE 1. Density vs cooling rate, the solidification curve [3], for some of the polymers listed in Table 1 and the influence of a plasticizer PEGDME in the figure shown as P, see Table 1.

Finally the third due to Wunderlich, recently reviewed in [7], assumes a molecular nucleation to occur during stem deposition although the evidence for its occurrence is inferred from experimental indirect observations. Finally another approach is due to Rault [8] bearing many similarities to the one described by Strobl.
In all the cases cited the experimental tests of these approaches are based on solidification conditions not much departed from equilibrium, i.e. conditions which give the possibility to follow the solidification in real time by some macroscopic method of characterization, therefore the timescale of the experiment is above the one quoted above of ca 10s.
The reason for invoking a timescale stands on an unstated hypothesis which has never been given sufficient evidence: the conformational rearrangement for stem deposition occurs freely without constraints whatsoever, quite hard to digest if one thinks that the stem length is of the order of 5nm while the path distance between two entanglements is between 20 to 40nm depending on the polymer. This issue has been recently taken in due consideration by Hikosaka [9] where the disentanglement mechanism is postulated to occur as a prerequisite for nucleation to take place. Also in this case the experimental verification is but at timescales available for a realtime investigation.
Certainly disentanglement occurs in crystallization and it may even be pushed to such an extent that separation of different molecular weight fractions in a process called Crystaf[10] is obtained, the acronym standing for crystal fractionation. Again, however, the timescale of the crystallization fractionation is much longer than the figure of 10s given above.
Finally a beautiful experiment in this context should be acknowledged: it starts from the observations of Lemstra and coworkers who obtained high modulus PE fibers from very diluted solutions of UHMWPE, a process patented by DSM as Dyneema [11]. Very high draw ratios can be obtained upon drawing the solid precipitate although if molten the solid cannot be drawn anymore to the same extent. These results led Rastogi [12] to observe the onset of a metastable melt once the disentangled solid polymer, whatever the route adopted for its development, is slowly heated. This melt will eventually crystallize much faster than the normal one obtained by fast heating above the melting point. This is a striking evidence of the importance entanglements have on crystallization.
These observations help on interpreting the results shown in Fig 1 related to the influence of a plasticizer on the processing conditions crystallization kinetics of two very similar polyesters, PET and PBT. Certainly the influence of entanglements cannot be only ascribed to dilution effects since entanglement concentration is scaled with polymer volume fraction to -1 [13].
From a purely heuristic point of view the dramatic influence of mobility on nucleation can be grasped by the dependence of the self diffusion coefficient of different polymers in several solvents, an invaluable information reported in the review of Irzhak [14]. The diffusion coefficient decreases abruptly in the vicinity of a volume

fraction equal to 1, i.e. for very small concentrations of the solvent, scaled with an exponent, on volume fraction, often above 10. This information should be compared with the notion of tube diameter, which should in principle be independent of the notion of number of segments between entanglements, ν_e, and it is qualitatively clear that, as well as the resulting mobility, it is influenced by long range interactions and therefore by the presence of other chemical species with respect to segments of the same polymer.

Although this observation would help to explain the strong influence of the plasticizer on the CR* for PET as shown in Fig 1, does not improve the view in the case of PBT, since in this case the results are in contradiction with these observations.

In order to interpret such different behaviour one should abandon the general framework of understanding of the phenomena involved and focus on the specific data of the two polymers pertinent to local mobility as eventually influenced by a plasticizer, plausibly a good local solvent.

In the literature there are several sources of information regarding entanglement density and tube diameter for several polymers but, at least to our knowledge, not that many. One major source is reported in the series of papers of Fetters et al where several quantities are reported including Me and its relationship with other molecular parameters [15-16] another in a review by Aharoni on Mc, the critical molecular weight for the onset of entanglement dynamics, for a broad range of materials[17]. Although PET is certainly among the materials listed, no mention is made of PBT, therefore for this polymer one may infer a trend on the basis of polymers with similar chemical structure or on the basis of a comparison with the parent polymer PET attributing to PBT a larger mobility due to the four methylene groups. The collection of several data on other materials which is supported by the packing model [18], albeit criticized in a recent work [19], mentions that ν_e decreases with chain stiffness because a stiffer chain pervades a larger volume that can accommodate the same number of chains at a shorter chain length. This observation should be compared with the Kuhn length for several polymers [16] which also, for the same reason, decreases for stiffer chains. The different behaviour of PET and PBT, if one excludes specific interactions with the plasticizer adopted in this work, should therefore come from being respectively a stiffer and a softer chain polymer respectively. Although this statement reconciles with the experimental evidence, the results account of a threshold like behaviour difficult to interpret on the basis of such understanding.

These remarks are but some hints to reconsider the role of topology in polymer crystallization whose influence can remarkably influence the kinetics particularly under relevant experimental conditions.

It is clear that the early works on this subject, collected in the famous Faraday Discussions of 1979 [20], led to several misconceptions on polymer crystallization, one certainly being the idea that the time for disentanglement is much smaller than the time for selecting a stem in order to: a) obtain a conformation compatible with the crystalline state, b) aggregate on the growing embrio, what has been referred as the process of "reeling in" a chain to the crystalline phase.

Only new methods for studying polymer crystallization can give perspectives on the understanding of polymer solidification with crystallization under the drastic conditions met in processing.

REFERENCES

1. Wunderlich, B. J Polym Sci Part B 2008 **46**, 2647
2. Piccarolo S., E. Vassileva, Z. Kiflie, in Lecture Notes in Physics, 2002, **609**, p326
3. Brucato V., S. Piccarolo, V. La Carrubba, Chem Eng Sci, 2002, **57**, 4129
4. Turnbull, D.; Fisher, J. C. J Chem Phys 1949, **17**, 71
5. Lauritzen, J. I., Jr.; Hoffman, J. D. J Res. Natl Bur Stand A 1960, **64**, 73
6. Strobl, G. in Lecture Notes in Physics 2007, **714**, 481
7. Hu, W. in Lect Notes Phys 2007, **714**, 47
8. Rault, J. J Macromol Sci Phys, 1978, **B15**, 567
9. Hikosaka, M., Watanabe, K., Okada, K. & Yamazaki, S. Adv Polym Sci 2005, **191**, 137
10. Monrabal, B. Macromolecular Symposia, 1996, 110, pp. 81
11. Smith, P. & Lemstra, P. J Mater Sci 1980, **15**, 505
12. Rastogi, S., Lippits, D., Terry, A. & Lemstra, P in Lect Notes Phys 2007, **714**, 285
13. Cheng, G., Graessley, W. & Melnichenko, Y. Phys. Rev. Lett. 2009, **102**, 157801
14. Irzhak, V. Russ Chem Rev 2000, **69**, 261
15. Fetters, L., Lohse, D., Richter, D., Witten, T. & Zirkel, A. Macromolecules 1994, **27**, 4639
16. Fetters, L., Lohse, D., Milner, S. & Graessley, W. Macromolecules 1999, **32**, 6847
17. Aharoni, S.M. Macromolecules, 1986, **19**, 426
18. Heymans, N. Novel Macromolecules 2000, **33**, 4226
19. Wang, S. Macromolecules 2007, **40**, 8684
20. Faraday Discussions "Organization of Macromolecules in the Condensed Phase" 1979, **68**

Predicting The Photoageing And Photostabilization Of Polymer Nanocomposites

S Therias[a,b], S Lonkar[e], F Leroux[c,d], JL Gardette[a,b], NT Dintcheva[f], FP La Mantia[f]

[a]Clermont Université, Université Blaise Pascal, Laboratoire de Photochimie Moléculaire et Macromoléculaire, F-63177 Aubière Cedex
[b]CNRS, UMR 6505, F-63177 Aubière Cedex
sandrine.therias@univ-bpclermont.fr
[c]Clermont Université, Université Blaise Pascal, Laboratoire des Matériaux Inorganiques,F-63177 Aubière Cedex,
[d]CNRS, UMR 6002, F-63177 Aubière Cedex
[e]Division of Polymer Science and Engineering, National Chemical Laboratory, Pune -411 008 India.
[f]Dipartimento di Ingegneria Chimica dei Processi e dei Materiali, Università di palermo, Viale delle Scienze, 90128 Palermo, (Italy).

Abstract. The influence of organo-clays on the durability of the nanocomposite materials finds expression in the efficiency of stabilizers. With the two kinds of stabilizers tested that are conventionally a processing one with a phenolic function and a long term stabilizer which is a redox antioxidant (hindered amine light stabilizer), the main conclusion is that the presence of organo-montmorillonite decreases the efficiency of the stabilizer. However, with specific stabilization additives the prodegradant effect of nanoclays can be cancelled and it offers a new insight in the stabilization strategy of nanocomposites.
The photochemical study of PP/LDH nanocomposites revealed that the divalent cations of LDH nanofiller (Mg or Zn) have an influence on the durability of the polymer and also on the photostabilization with an opposite behaviour between phenolic and redox antioxidants.

Keywolds: Photodegradation, polymer, nanocomposite, clay, LDH.stabilization
PACS: 81.05.Qk

INTRODUCTION

Smectite-type materials (montmorillonite) are widely used as nanoclays in polymer nanocomposites. However, few attention has been paid to another class of filler, the layered double hydroxide (LDH), which are described according to the ideal formula, $[M^{II}_{1-x}M_x^{III}(OH)_2]^{x+}_{intra}[A^{m-}_{x/m}\cdot nH_2O]_{inter}$, ($M^{II}$ and M^{III} are the metallic cations, A the anions, *intra* and *inter* denote the intralayer and interlayer domains, respectively). The structure consists of brucite-like layers constituted of edge-sharing octahedra. The presence of trivalent cations induces positive charges in the layers, which are counterbalanced by interlamellar anions (**FIGURE 1**).

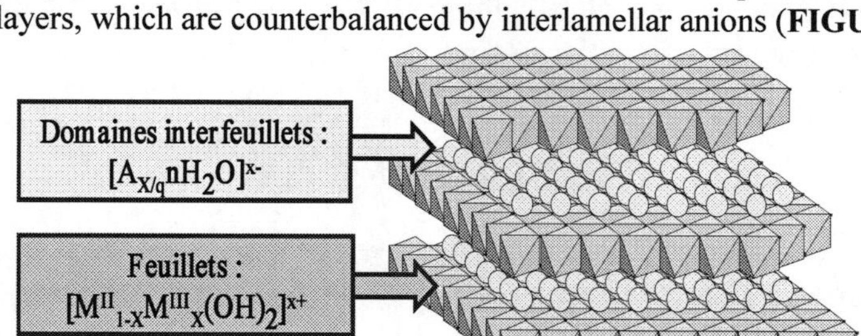

FIGURE 1: Chemical structure of LDH.

CP1255, *V[th] International Conference on Times of Polymers (TOP) and Composites,*
edited by A. D'Amore, D. Acierno, and L. Grassia

Depending on the nature of the added filler, properties such as fire resistance, thermal and electrical conductive properties can been enhanced. Among the tremendous literature devoted to nanocomposite materials, most of the articles concern the synthesis, the processing and the characterization of nanocomposites, and the evaluation of their properties. Only few papers deal with the long term behaviour of these materials.

Previous studies concerning the photooxidation of polypropylene (PP) containing organo-montmorillonite (oMMt)[1-2] or boehmite [3] have shown that the nanofiller did not modify the oxidation photoproducts of PP but that the additives usually used for the photostabilization of PP had only a very poor efficiency in presence of the nanofiller [1-3]. Various stabilizers were tested to improve the durability of PE/clay nanocomposites [4]. The influence of LDHs has been recently studied in the case of polystyrene nanocomposites [5], and it has been shown that the durability of the polymer was not modified. The photochemical behaviour of new polypropylene nanocomposites with hybrid LDHs, (the organic anion being dodecylsulfate (DS)) was studied. Three different LDHs were synthesized, with either Mg or Zn as divalent cation or both cations in [ZnMg-Al-DS]. These LDHs were characterized by XRD and the corresponding nanocomposites by XRD and DSC [6]. Different formulations with phenolic and/or redox antioxidants were also studied and compared under photooxidative conditions. Photooxidation was produced by exposure of the sample in accelerated photoageing conditions carried out irradiation at $\lambda > 300$ nm and 60 °C. The UV-light induced modifications of the polymer structure were followed by infrared and UV-visible spectroscopies. The oxidation products were identified and the rates of oxidation of the various samples were compared.

RESULTS AND DISCUSSION

Polymer/Clay Nanocomposite

The obtained results indicated that the mechanism of the polypropylene photooxidation was not modified by the interactions with the nanoclay and the compatibilizing agent. The same photoproducts were formed, in the same relative concentrations.

Photooxidation of the organo-modified montmorillonite revealed that the alkylammonium cations can be degraded and that the presence of iron species in the clay might play a role in the degradation mechanism. The presence of the clay and of the compatibilizer modified dramatically the kinetics of oxidation, leading to a shortening of the induction period. This behaviour can be attributed to an initiation of the photooxidation by active species generated by photolysis or photooxidation of the organoclay and/or by interactions between the anti-oxidant, montmorillonite and maleic anhydride.

With specific stabilization additives (such as metal deactivator) the prodegradant effect of nanoclays can be cancelled.

Polymer/LDH Nanocomposite

The modifications of the IR spectra of PP/LDH nanocomposites resulting from photoxidation are similar to those of PP, with the same PP photoproducts. However, a new absorption band develops at 1600 cm^{-1} in presence of LDH, whereas it is not observed in the case of organo-montmorillonite

In order to compare the rates of oxidation of nanocomposites with the three different LDH and with phenolic or redox stabilizers, and also with montmorillonite, the increase of absorbance at 1715 cm^{-1} has been plotted versus irradiation time. The rate of oxidation of PP is lower with LDH than in the presence of organo-clay and the results reveal differences between the three different LDHs depending on the intra-lamellar cations (Mg or Al). Moreover, the best stabilizer (phenolic or HALS) depends on the divalent cation of the LDH nanofiller.

REFERENCES

1. S. Morlat, B. Mailhot, D. Gonzalez and J.L. Gardette, *Chem. Mater.*, **16(3)** 377 (2004).
2. S. Morlat-Therias, B. Mailhot, D. Gonzalez, J.L. Gardette, *Chem. Mater*, **17(5)** 1072 (2005).
3. S.Bocchini, S. Morlat-Therias, J- L. Gardette, G. Camino, *Polym. Degrad. Stab.*, **92** 1847 (2007)
4. S. Morlat-Therias, E. Fanton, JL Gardette, N. Tzankova Dintcheva, FP La Mantia and V. Malatesta, *Polym. Degrad. Stab.*, **93** 1776 (2008).
5. F. Leroux, L. Meddar, B. Mailhot, S. Morlat-Therias, J.-L. Gardette, *Polymer*, **46(11)** 3571 (2005).
6. S.P. Lonkar, S. Morlat-Therias , N. Caperaa , F. Leroux , J.L. Gardette , R.P. Singh ,*Polymer*, **50** 1505 (2009).

Twinkling Fractal Theory of the Glass Transition and Yield Stress

Richard P. Wool

Department of Chemical Engineering and Centre for Composite Materials, University of Delaware, Newark DE 19716-3144 USA

The recent Twinkling Fractal Theory (TFT) of T_g provides new insight and a method of analyzing fracture and yield in amorphous materials. In particular we show that the following three aspects of yield in amorphous polymers are simply interrelated: (a) the rate dependence of T_g; (b) the temperature dependence of yield and (c) the rate dependence of yield.

1.0 Introduction

The Twinkling Fractal Theory (TFT) of the glass transition temperature T_g provides a new method of analyzing rate effects and time-temperature superposition in amorphous materials [1,2]. The rate dependence of the glass transition temperature was examined [3] in the light of new experimental and theoretical evidence for the nature of the dynamic heterogeneity near T_g.[2] As T_g is approached from above, dynamic solid fractal clusters begin to form and eventually percolate rigidity at T_g. The percolation cluster is a solid fractal and to the observer, appears to "twinkle" as solid and liquid clusters interchange in dynamic equilibrium with a vibrational density of states $g(\omega)\sim\omega^{df-1}$ as explicitly shown by Stanzione et al [2].

The solid-to-liquid twinkling frequencies ω_{TF} are controlled by the Boltzmann population of intermolecular oscillators in excited energy levels of their anharmonic potential energy functions U(x) such that $\omega_{TF} = \omega\exp -B(T^{*2}-T^2)/kT$ in which $T^*\approx 2T_g$. An oscillator changes from a solid to a liquid when a thermal fluctuation causes it to expand beyond its inflexion point in the anharmonic potential. This leads to a continuous solid fraction P_s near T_g given by $P_S \approx 1-[(1- p_c)\ T/T_g]$ where $p_c\approx 1/2$ is the rigidity percolation threshold. Since $g(\omega)$ is continuous from very low to very high frequencies, the complex twinkling dynamics existing near T_g produces a continuous relaxation spectrum with many different length scales and times associated with the fractal clusters.

The twinkling frequencies control the kinetics of T_g such that for a given observation time t when the rate $\gamma>1/t$, only those parts of the twinkling spectrum with $\omega>\gamma$ can contribute to relaxation or percolation up to time t. The most important results in this paper are as follows: The TFT describes the rate dependence of T_g, both for DSC thermal heating/cooling rates and DMA frequencies as the classic T_g-lnγ law as [3]

$$T_g(\gamma) = T_{go} + (k/2B)\ \ln \gamma/\gamma_o$$

in which the constant B = 0.3 cal/mol $^oK^2$. The constant B appears quite universal for the 17 thermosets polymers investigated in this study and 18 linear polymers investigated by others. Many other amorphous metal and ceramic glass materials exhibited the same rate law but required a new B value approximately half that for polymers. The same B=0.3 value was also used to successfully describe the TTS shift factors using the twinkling fractal frequencies $\omega_{TF} = \omega\exp -B(T^{*2}-T^2)/kT$, as $\ln a_T(TFT) = \exp B(T_R^2-T^2)/kT$, which gave comparable results with the classical WLF equation, $\log a_T = [-C_1\ (T-T_R)]/[C_2 + (T-T_R)]$. The advantage of the TFT over the WLF is that C_1 and C_2, are not universal constants and must be determined for every material, whereas the TFT uses one known constant B which appears to be the same for all polymers. The TFT has also been found to describe the rate and temperature depndence of the yield stress in polymers

CP1255, *Vth International Conference on Times of Polymers (TOP) and Composites,*
edited by A. D'Amore, D. Acierno, and L. Grassia

In this paper, we show that the TFT uniquely predicts and interrelates (a) the rate dependence of T_g as, measured by dynamic mechanical analysis (DMA), (b) the temperature dependence of the yield stress of model thermosets, as reported by A. Lesser et al [4] and (c) the rate dependence of yield in glassy polymers, as reported by M. Boyce et al [5]. Thus, the TFT offers a new method to predict the outcome of elaborate mechanical experiments (e.g. high rate impact at varying temperatures) by simple DMA tests and is a useful material design tool.

2. Rate and Temperature dependence of Yield Stress σ_y.

For an amorphous solid below T_g, subjected to triaxial stresses σ_1, σ_2, σ_3, the Distortional Strain Energy Function U_d is determined by (Von Mises) as:

$$U_d = (1+\nu)[(\sigma_1 - \sigma_2)^2 + (\sigma_2 - \sigma_3)^2 + (\sigma_2 - \sigma_3)^2]/6E \qquad (1)$$

in which ν is the Poisson ratio and E is the isotropic Elastic Young's modulus. In the simple uniaxial case, we obtain the stored strain energy with $\nu = 0.5$ as $U = \sigma^2/2E$. This stored energy must be utilized to overcome a percolation number of interatomic anharmonic oscillator bonds of energy $U(x_c)$. For a Morse oscillator $U(x_c) = 0.08\ D_o$, which is the energy necessary to reach the Lindemann bond expansion at x_c. The number of such oscillators per unit volume is $1/V_m$, where V_m is the oscillator molar volume. At the critical stored energy, we have yielding when the amorphous solid is raised to the twinkling fractal state by the release of the mechanical energy and flow begins in accord with:

$$\sigma_y^2/2E \geq 0.08\ D_o\ [p_s - p_c]/V_m \qquad (2)$$

in which the solid fraction p_s is given by Eq (5) and is both temperature and rate dependent. Thus, the yield stress is obtained as [2]:

$$\sigma_y = [0.16\ E\ [p_s - p_c]\ D_o/V_m]^{1/2} \qquad (3)$$

Note that the term D_o/V_m corresponds to the traditional Cohesive Energy Density (CED). For example, if E = 1 GPa, p = 1 (high rate of deformation), $p_c = 1/2$, $D_o = 3$ kcal/mol, $V_m = M_o/\rho$, $\rho = 1$ g/cc, the monomer molecular weight $M_o = 100$ g/mol, then Eq 13 predicts that $\sigma_Y = 100$ MPa. This value is typical for high performance amorphous polymers. The modulus E ~ CED such that a plot of σ_y/CED should normalize all yield data for many polymers, as noted by Lesser et al [4]. Since the modulus E(T) behaves as $E(T) = E_o\ (1 - T/T_g)$ in the vicinity of T_g [6] and the solid fraction behaves similarly, then we obtain the temperature dependence of the yield stress as:

$$\sigma(T)/\sigma_\infty = [1 - T/T_g] \qquad (4)$$

where σ_∞ is the value of yield stress at T = 0°K. Similarly, combining Eq 1 and Eq 4, we obtain the rate dependence of yield as:

$$\sigma(\gamma)/\sigma_\infty = 1 - T/\{T_{go}[1+(k/\beta T_{go}) \ln \gamma/\gamma_0\} \qquad (5)$$

where σ_∞ is the yield stress at infinite rate and is equivalent to σ_∞ in Eq 4 These results are in excellent agreement with data reported by Lesser and Calzia [4] (Fig 1)and Mulliken and Boyce [5] (Fig 2) for the temperature dependence and rate dependence of the yield stress, respectively.

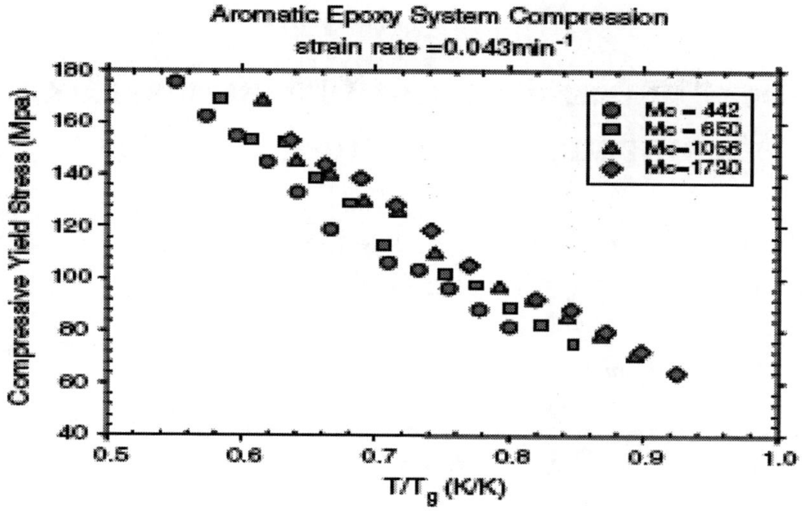

Fig 1. Yield Stress vs T/Tg (A. Lesser data [4])

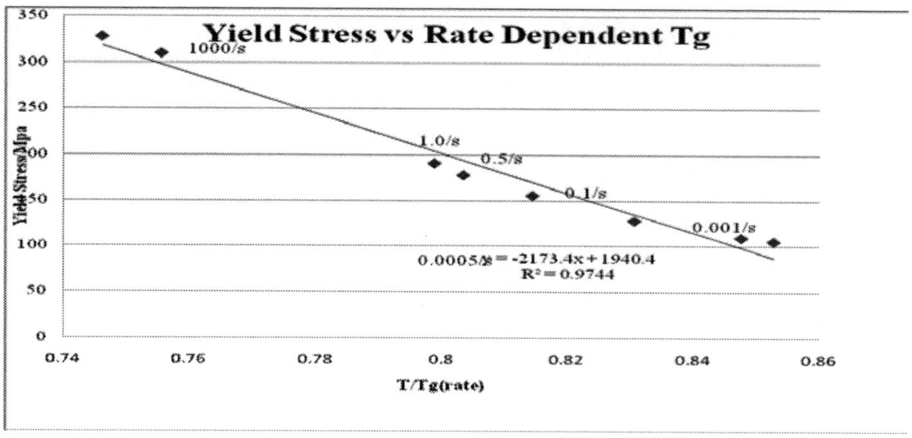

Fig 2: Yield vs rate dependent Tg (M Boyce data [5])

In summary, the rate and temperature dependence of yield and fracture of glassy polymers are conveniently interrelated through the rate dependence of T_g, as described by the TFT. This allows the results of highly complex experiments such as ballistic impact to be understood in terms of a more simple DMA experiment and provides meaning to the WLF equation.

Acknowledgement: The author is grateful to the Army Research Lab at Aberdeen MD for financial support of this research under the Army Materials Center of Excellence Program, W911NF-06-2-011 .

References

[1] Wool R.P., J. Polym. Sci., Part B: Polym Phys. 46, 2766 (2008)
[2] Stanzione, J.F., Strawhecker K.E., Wool, R.P., J. Non-Cryst Solids(in press) (2010)
[3] Wool R.P., Campanella A., J Polym. Sci., Part B: Polym Phys. 47, 2578 (2009)
[4] Lesser A.J; Calzia K.J; J. Polym. Sci. Part B: Polym. Phys., 42, 2050 (2004)
[5] Mulliken A.D: Boyce M.C: Proceedings of 13th International Conference DYFP, Kerkrade, Nederland, page 61 (2006)
[6] Wool, R.P; Sun X.S; *Bio-Based Polymers and Composites*, Elsevier (2005)

Mass Transport in Nanocomposite Materials for Membrane Separations

Michele Galizia,[a] Ilaria Puccini,[b] Massimo Messori,[b]
Maria Grazia De Angelis[a] and Giulio C. Sarti[a]

[a] *Dipartimento di Ingegneria Chimica, Mineraria e delle Tecnologie Ambientali*
Via Terracini 28, Università di Bologna, I-40131 Bologna
[b]*Dipartimento di Ingegneria dei Materiali e dell'Ambiente*
Via Vignolese 905 Università di Modena e Reggio Emilia, I-41125 Modena

Abstract. The vapor transport properties of nanocomposite materials obtained with different techniques and based on a high free volume glassy polymer suitable for membrane separations, poly[1-(trimethylsilyl)-1-propyne] (PTMSP), have been determined and modeled. The simple mixing in solution of hydrophobic fumed silica nanoparticles with PTMSP leads to *mixed matrix* membranes, which show higher free volume and higher values of diffusivity and permeability than the pure polymeric material. If a sol-gel route is followed, with PTMSP and Tetraethoxysylane (TEOS) as precursor of the silica phase, one obtains *hybrid matrices* characterized by lower vapor diffusion and sorption values with respect to the pure polymer. Although the trends observed are very regular functions of the silica content in the composite, none of the behavior observed obeys traditional models for composites permeability, such as the Maxwell's one. Both types of behaviors were modeled considering the variation of polymer fractional free volume induced by the inorganic phase: in the mixed matrices the poor interactions between silica and polymer chains favor the formation of nanovoids at the interface, increasing the free volume and the vapor diffusivity, while in the more interconnected hybrid matrices the inorganic domains act as constraints, reducing the volume occupied by the polymeric phase, which is naturally endowed with a very high excess free volume.

Keywords: Membrane separation, mixed matrix membranes, hybrid materials, vapor, transport.
PACS: 81.07.Pr ; 82.35.Np ; 66.30.-h; 82.35.Lr ; 82.45.Mp ; 82.60.Lf

INTRODUCTION AND BACKGROUND

Nanocomposite materials obtained combining inorganic fillers with high free volume glassy polymers have been investigated in the last decade with the aim to improve the permselectivity of polymeric membranes for gas and vapor separation.[1-4]

The glassy matrix chosen for the present study is poly[1-(trimethylsilyl)-1-propyne] (PTMSP), for which relatively large information about the gas and vapor transport and the microscopic morphology is available in the literature. In particular, this polymer has higher permeability for vapors than for gases, a property that is common in rubbers but less usual in glassy polymers, and this behavior is due to very high fractional free volume. [5-6]

The object of this study is an experimental analysis of n-C_4 and n-C_5 sorption and diffusion at room temperature into:

i) mixed matrices formed by PTMSP and hydrophobic nanometric silica added in solution (mass fractions ranging from 10 to 50wt %);

ii) hybrid matrices formed by PTMSP and SiO_2 obtained from tetraethoxysilane (TEOS) via sol-gel reaction, with silica content ranging from 10 to 30%.

The first aim of the study is to compare the transport behaviour of the two types of matrices at similar silica content and to understand the role of the interactions between polymer and silica domains in driving the transport behavior. In the second place, we looked for a comprehensive model that can represent the transport in mixed and/or hybrid matrices based on a small set of physical and empirical parameters.

CP1255, *V[th] International Conference on Times of Polymers (TOP) and Composites,*
edited by A. D'Amore, D. Acierno, and L. Grassia

The studies dealing with the transport properties of mixed matrices of PTMSP and silica,[3] as well as of hybrid matrices obtained by combining PTMSP with a precursor of the silica phase such as an alkoxysilane,[8] did not provide a definite and unique structure-property correlation to link the permeability and diffusivity to a single physical parameter of the composite matrix.

In this work we apply a model which describes the gas diffusion as a function of one single parameter, i.e. the fractional free volume of the polymeric matrix. The estimation of the fractional free volume in the glassy polymer phase of mixed (or hybrid) matrix state is a non trivial problem, which can be afforded, however through reasonable assumptions. To this aim we considered that silica maintains in the composites the same density and adsorption capacity for vapors as in the state of pure solid, while the polymer density and sorption capacity do vary with the silica content. The vapor sorption is uniquely related to the polymer density through the Non Equilibrium Lattice Fluid (NELF) model for gas solubility in glassy polymers.[10]

Such approach requires the experimental determination of sorption data of one test penetrant in the filler, when it is not negligible as in the present case and in one, or more, composite matrices with filler volume fraction Φ_F. By assuming that the filler sorption capacity remains unvaried upon mixing with the polymer, one obtains the sorption amount in the polymeric phase of the composite, C_P as a function of the pure filler and mixed matrix sorption capacity, C_F^0 and C_M, respectively:

$$C_P = \frac{C_M - \Phi_F \cdot C_F^0}{(1-\Phi_F)} \tag{1}$$

This quantity is then modeled by the NELF theory,[10] based on which different sorption capacities of the same polymer are due to different initial density, ρ_P^0 values, so that the value of ρ_P^0 is retrieved for each matrix. This procedure requires one adjustable parameter, i.e. the binary energetic parameter for the gas in the polymer, k_{ij}, that is generally available for the systems under study. For swelling penetrants a second parameter has to be fitted, which represent the amount of swelling induced by the penetrant at fixed pressure, k_{sw}:

$$\rho_P(p) = \rho_P^0 (1 - k_{sw} p) \tag{2}$$

Knowledge of ρ_P^0 allows to estimate the FFV of the polymer in the mixed matrix state, FFV_M^0 as:

$$FFV_M^0 = \frac{\rho_P^{vdW} - 1.3 \cdot \rho_P^0}{\rho_P^{vdW}} \tag{3}$$

where ρ_P^{vdW} is the van der Waals density of the polymer, available from the literature. Finally, the FFV value can be used to estimate, with 2 adjustable parameters A and B, the infinite dilution diffusivity of gases in the mixed matrix materials, by means of the free volume theory:

$$D_{i,M}(0) = \frac{1}{\tau} \exp\left(A - \frac{B}{FFV_M^0} \right) \tag{4}$$

with the tortuosity τ given by $(1+\Phi_F/2)$ as in Mawell's model.

The approach has been previously tested on mixed matrices based on various glassy polymers, [9] and it is proven here that it can describe also the transport properties in the hybrid matrices obtained from PTMSP and TEOS. Moreover, the data relative to hybrid and mixed matrices lie on a same mastercurve when the diffusivity in the matrix is plotted against the fractional free volume (FFV).

RESULTS AND DISCUSSION

The experimental data collected show that vapor solubility and diffusivity in the mixed matrices increases monotonously with increasing silica content, while the transport rates in hybrid matrices are lower than in the pure polymer and much lower than those observed in mixed matrices at equal content of silica.

None of these trends can of course be described by the Maxwell's model,[7] which assumes that the polymer phase is not altered by interactions with the filler: the permeability in mixed matrix is higher than predicted by this model, while the permeability in hybrid matrices is lower than the estimated value. Clearly, therefore, the interactions between polymer and silica play an important role in both cases.

In this work, we used experimental sorption capacity as input to evaluate, from the NELF model, the polymer density and, from it, the fractional free volume (eq 3). From this analysis it has been seen that, for mixed matrices, the FFV is higher than that of the pure polymer while hybrid matrices show lower FFV values with respect to the unloaded PTMSP. This result actually explains the different transport trend in the composite matrices, and is consistent with the correlation based on the free-volume theory (eq 4) which is shown in Figure 1, that is able to depict in a single curve the data of hybrid and mixed matrices. We believe that in mixed matrices additional free volume is created at the interface between polymer and filler due to poor interactions between the phases; for hybrid matrices the inorganic phase is deeply interconnected to polymeric chains and acts as an additional constraint on the open structure of PTMSP, lowering its fractional free volume.

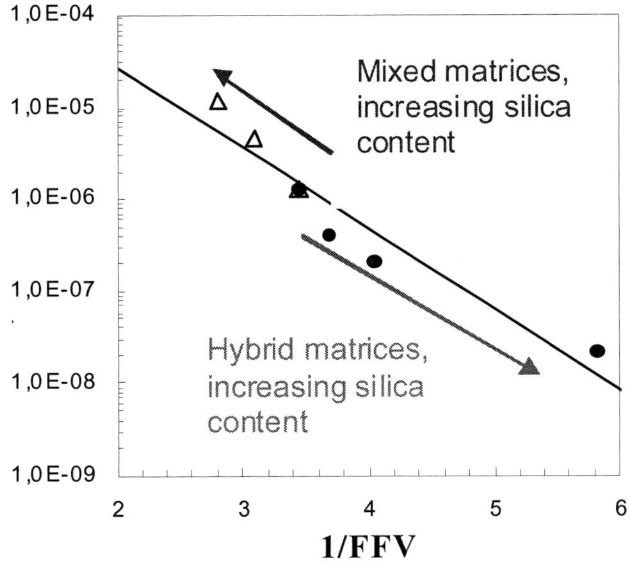

FIGURE 1. Infinite dilution diffusivity of n-C_4 in PTMSP-based composites at 25°C, in cm^2/s, versus the reciprocal of fractional free volume (FFV).

REFERENCES

1. T. C. Merkel et al., *Science* **396**, 519-522 (2002).
2. T. C. Merkel et al., *Macromolecules* **36**, 8406-8414 (2003).
3. T. C. Merkel et al., *Macromolecules* **36**, 6844-6855 (2003).
4. T. C. Merkel et al., *Chem. Mat.* **15**, 109-123 (2003).
5. K. Nagai et al., *Prog. Polym. Sci.* **26**, 721-798 (2001).
6. I. Pinnau and L.G. Toy, *J. Membrane Sci.* **116**, 199-209 (1996).
7. C. Maxwell, *Treatise on Electricity and Magnetism*, London: Oxford University Press, 1873.
8. D. Gomes et al., *J. Membrane Sci.* **246**,13-25 (2005).
9. M.G. De Angelis and G.C. Sarti, *Ind. Eng. Chem. Res.* **37**, 5214-5226 (2008).
10. F. Doghieri and G.C. Sarti, *Macromolecules***29,** 7885-7896 (1996).

Vibrational Properties Of A Reactive Mixture Investigated During A Chemical Vitrification Process

S. Caponi[a,b], S. Corezzi[c], D. Fioretto[c], A. Fontana[a], G. Monaco[d], F. Rossi[a]

[a] *Dipartimento di Fisica, Università di Trento, 38050 Povo, Trento, Italy*
[b] *IOM-CNR, c/o Università di Perugia, I-06123, Perugia, Italy*
[c] *Dipartimento di Fisica, Università di Perugia, I-06123, Perugia, Italy*
[d] *European Synchrotron Radiation Facility, BP 220, F-38043 Grenoble, France*

Abstract. Raman and Brillouin light scattering measurements were performed to follow the modification of the vibrational properties of a reactive epoxy-amine mixture during isothermal polymerization. We analyze the variations of the vibrational density of states while the system changes from liquid to glass upon increasing the number of covalent bonds among the constituent molecules. The spectral shape of the boson peak follows a precise and universal scaling law. Surprisingly, bond-induced modifications of the structure do not affect this master curve. Moreover the Brillouin scattering results here reported for a curing epoxy system provide evidence for the validity of a Cauchy-like relation.

Keywords: Boson Peak, Glass transition.
PACS: 63.50.Lm:, *64.70.pj.*

INTRODUCTION

The glass transition is a general phenomenology, common to a large number of materials ranging from simple liquids to systems with a growing level of complexity. Decreasing the temperature or increasing the pressure are the most common routes to the glassy state. In these cases the glass transition is a reversible process and a physical glass is obtained. On the other hand, a glass can also be the result of the formation of chemical bonds between the constituents monomers in a reactive mixture. In this case one obtains a chemical glass by an irreversible process [1,2,3].

In this paper the chemical vitrification process is investigated. The evolution of the vibrational properties of a reactive mixture has been studied by Brillouin light scattering and Raman scattering experiments.

It is well known that compared to their crystalline counterparts, disordered materials present a non-Debye behavior of the vibrational density of states $g(\omega)$. The existence of an excess over the prediction of the Debye theory shows up as a broad band, the so-called boson peak (BP), in the reduced density of states $g(\omega)/\omega^2$ [4]. The characterization of the BP evolution in a chemical glass, where this evolution is connected with bond-induced modifications of the structure, may help clarify the nature of the vibrational modes giving rise to this peak.

The system that we have chosen for this study is an epoxy-amine mixture consisting of diglycidyl ether of bisphenol-A (DGEBA) and diethylenetriamine (DETA) in the molar ratio of 4:3, undergoing a polymerization reaction at a constant T = 295 K. At this temperature the monomers bind slowly to each other and the total reaction time is about 10 hours. The reaction proceeds by stepwise addition of the amino hydrogen to the epoxy group, without elimination of by-product molecules. Starting from a liquid sample the final product is a glassy network polymer.

RESULTS AND DISCUSSION

Figure 1 shows the experimental depolarized Raman intensity. The spectra are collected sequentially during the reaction. In the high-frequency region (inset fig.1a) intensity variations only occur for those peaks that are related to

CP1255, *Vth International Conference on Times of Polymers (TOP) and Composites,*
edited by A. D'Amore, D. Acierno, and L. Grassia

molecular vibrations affected by the reaction. The spectra shown in the main panel of Fig. 1 can be divided in three parts: (i) an optical band centered at 80 cm^{-1}; (ii) the BP band centered at 15 cm^{-1}; and (iii) the quasielastic scattering (QES) region for frequencies lower than 5 cm^{-1}, that appears as a broadened elastic line. Figure 1a shows that, as the reaction proceeds, the BP shifts toward higher frequencies and decreases its intensity, while the QES significantly decreases until it becomes negligible. We can effectively model the Raman intensity in the frequency region of interest as the superposition of two contributions, one (I_{QES}) ascribed to relaxation processes, and one (I_{BP}) related to vibrations, and then ultimately to the vibrational density of states g(ω). Since in this work we are only interested in the BP variations, we subtract from the total signal the QES contribution. Following the procedure suggested in ref. [5], the evolution of the BP can be related to the modification of the elastic medium and a master curve for the BP, whose existence has been theoretically predicted, can be obtained. The master curve, shown in Fig 1b, is obtained by rescaling the frequency of the spectra with the BP position, $\nu=\omega/\omega_{BP}$, and imposing the conservation of the total number of vibrational states, $g(\omega)d\omega=g(\nu)d\nu$. This master curve demonstrates that the spectral shape is unchanged during the polymerization process, the number of modes remains the same and hence the decrease of the BP intensity is apparent.

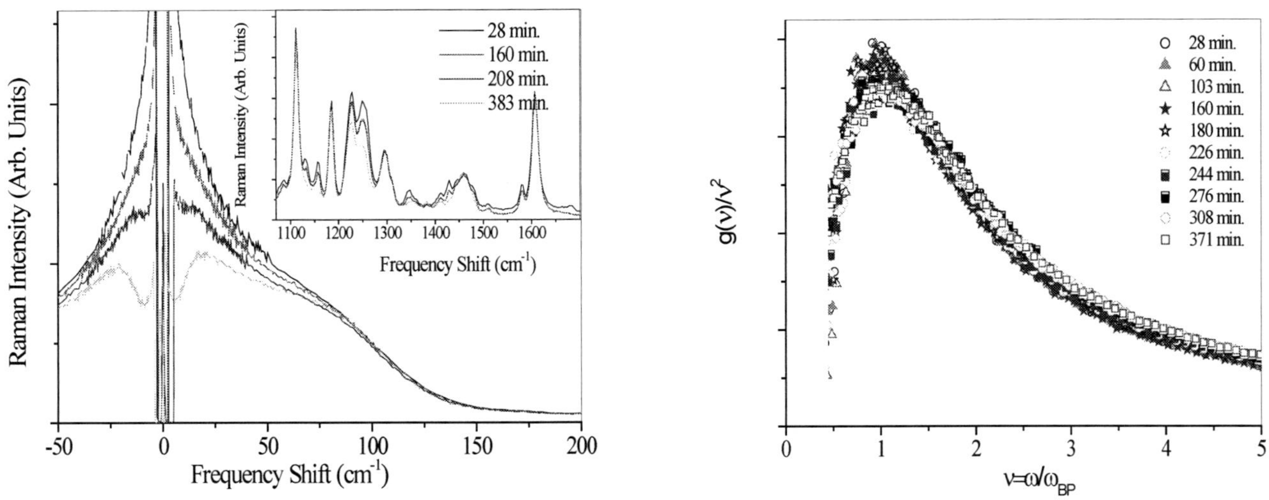

a) b)

FIGURE 1. a) Depolarized Raman spectra at some selected reaction times as indicated in the legend, during the isothermal reaction DGEBA-DETA 4:3 at T = 295 K. The inset shows the same spectra in the frequency range of the molecular peaks. The peak at 1610 cm^{-1}, corresponding to the aromatic quadrant stretch, is invariant during the reaction, while the intensity decrease of the band at 1260 cm^{-1}, related to the breathing of the epoxy ring, corresponds to the consumption of epoxy groups. b) Master curve of the BP.

Changes of the elastic constants of the medium during the polymerization process are clearly visible in the GHz frequency range probed by Brillouin light scattering measurements. The frequency position of the longitudinal, ν_L, and transverse, ν_T, Brillouin peaks are reported in the insets of fig.2. The considerable increase in both ν_L and ν_T during the polymerization reaction is due to the increase in the stiffness of the system, which evolves from the liquid to the glassy state on approaching the end of the reaction. Moreover, these measurements permit a quantitative comparison between the variation of the longitudinal and transverse sound waves throughout the reaction, as shown in the main panel of fig.2. The real part of the longitudinal modulus M' at the frequency of the longitudinal mode $\omega_L=2\pi\nu_L$ and the real part of the shear modulus G' at the frequency of the shear mode $\omega_T=2\pi\nu_T$ are calculated through the formulas

$$M'(\omega_L)=\rho\frac{\omega_L^2}{q^2}$$ and

$$G'(\omega_T) = \rho \frac{\omega_T^2}{q^2}$$

where ρ is the mass density and q is the exchanged wavevector. Figure 2 shows that a linear relationship M'(ω)=A+BG'(ω) as predicted by the Chauchy-like relation between M' and G' is fulfilled within the experimental uncertainty during the whole reaction (see ref. [6]). The obtained value of the parameter B is very close to 3. It is worth noticing that most of these data points are obtained in the viscoelastic region above the glass transition, where Brillouin light scattering is affected by the structural relaxation. This evidence confirms those accumulated on polymerizing epoxy systems, and strongly supports the idea that the Cauchy-like equation suitably describes the relation between apparent longitudinal and shear moduli.

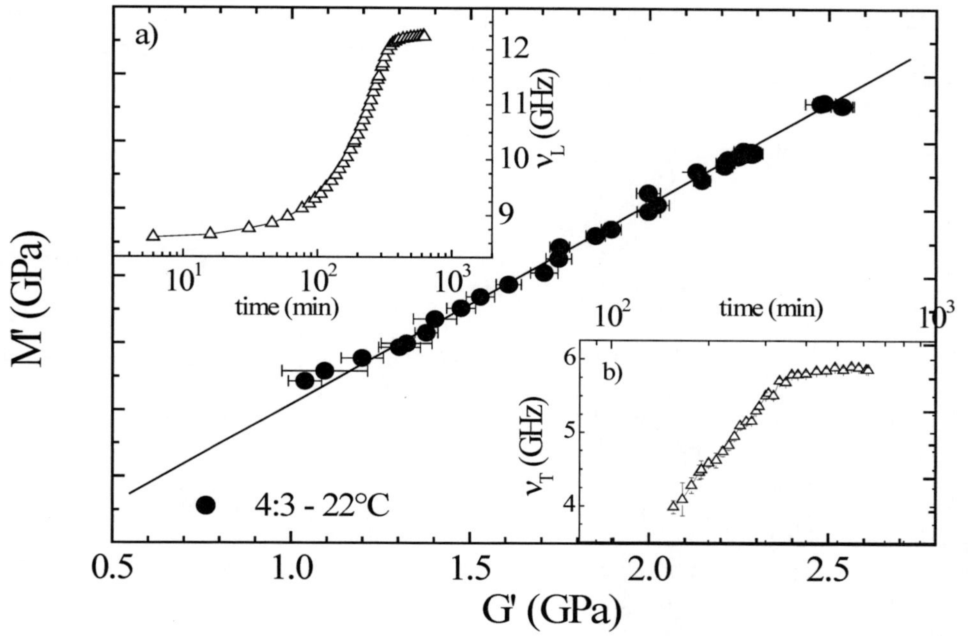

FIGURE 2. Main panel: Real part of the longitudinal modulus M' vs the real part of the shear modulus G'. Insets: a) Frequency shift of the longitudinal acoustic mode as a function of the reaction time. b) Frequency shift of the transverse acoustic mode as a function of the reaction time.

REFERENCES

1. S. Corezzi, D. Fioretto, P.A. Rolla, Nature 420, 653-656 (2002).
2. S. Corezzi, D. Fioretto, J.M. Kenny, Phys. Rev. Lett., 94, 065702 (2005).
3. S. Corezzi, D. Fioretto, D. Puglia, J.M. Kenny, Macromolecules 36 (14), 5271-5278 (2003).
4. S.Caponi, A.Fontana, F.Rossi, G.Baldi, E.Fabiani, Phys. Rev. B 76, 092201 (2007) and references therein.
5. S. Caponi S. Corezzi, D. Fioretto, G. Monaco, A. Fontana, F.Rossi, Phys. Rev. Lett. 102, 027402 (2009).
6. D. Fioretto S. Corezzi, S. Caponi, F. Scarponi, G. Monaco, A. Fontana, L. Palmieri, J. Chem. Phys. 128, 214502 (2008).

Effect of Solvent Viscoelasticity on the Stress Induced Demixing

Mario Minale[a]

[a]*Seconda Università di Napoli - Dipartimento di Ingegneria Aerospaziale e Meccanica.*
Via Roma 29 – 81031 Aversa (CE)

Abstract. Stress induced demixing of a polymer blend is studied in the mainframe of the Two-fluid theory. The free energy of mixing is modified so to account for the variation of the mixing entropy due to chain stretch. Both the blend constituents are non Newtonian and the effect of the solvent viscoelasticity on polymer demixing is here investigated and it is concluded that it tends to stabilize the solution.

Keywords: Emulsion, Multi-phase flow, Demixing, Polymer Blends.
PACS: 82, 83

INTRODUCTION

Demixing has been repeatedly observed [1-3] and different theoretical approaches have been followed to describe this behavior. From one hand it was assumed that during shear an actual phase separation took place and this was studied with thermodynamic theories applied at the steady state. The Gibbs free energy was modified by adding a term describing the stored elastic energy coming from the long polymer chain stretch due to flow. From the other hand, the observed onset of turbidity during flow was explained with a kinetic theory describing the flow induced enhancement of concentration fluctuations [4]. This approach has been also formalized within the framework of the two-fluid theory ([5]-[6]).

Further optical measurements, together with rheological data have been interpreted as the evidence for actual phase separation, rather than temporary fluctuations of concentration [7]. The thermodynamic theories were able to describe phase separation, but not the early stage of the process where only concentration fluctuations appear. Conversely the kinetic theories were able to describe the enhancement of concentration fluctuations, but not the actual phase separation. Recently, a new formulation of the two-fluid theory was proposed [8] reconciling the two phenomena, indeed also the actual phase separation of a polymer solution was predicted in nice agreement with experimental results.

We here study the effect of the solvent elasticity to the stress induced demixing modifying the free energy of mixing appearing in the two-fluid theory and choosing an appropriate constitutive equation for both polymer and solvent.

THE TWO FLUID THEORY

Minale [8] reformulated the two-fluid theory so to be able to write the appropriate boundary conditions at the interface that is formed during the demixing process. The spirit of two-fluid theory is that in addition to the classical individual mass balances also the individual momentum balances can be written:

$$-\zeta\left(\mathbf{v}_p-\mathbf{v}_s\right)+\phi\nabla\Pi_p-\nabla p_p+\nabla\cdot\boldsymbol{\sigma}_p=0 \tag{1}$$

$$\zeta\left(\mathbf{v}_p-\mathbf{v}_s\right)+\left(1-\phi\right)\nabla\Pi_s-\nabla p_s+\nabla\cdot\boldsymbol{\sigma}_s=0 \tag{2}$$

CP1255, *Vth International Conference on Times of Polymers (TOP) and Composites,*
edited by A. D'Amore, D. Acierno, and L. Grassia

where ζ is the friction coefficient per unit volume, $\mathbf{v}_p$ and $\mathbf{v}_s$ are the polymer and solvent velocity, ϕ is the polymer volume fraction, Π_p and Π_s are the osmotic pressures of the polymer and the solvent; p_p and p_s are the polymer and solvent pressure respectively and $\boldsymbol{\sigma}_p$ and $\boldsymbol{\sigma}_s$ their stresses. The solvent and polymer pressure are given by:

$$\begin{aligned} p_p &= p\frac{\nabla\cdot\left[\phi\left(\mathbf{v}_p-\mathbf{v}_s\right)\right]}{\nabla\cdot\left(\mathbf{v}_p-\mathbf{v}_s\right)} = p\phi + p\frac{\left(\mathbf{v}_p-\mathbf{v}_s\right)\cdot\nabla\phi}{\nabla\cdot\left(\mathbf{v}_p-\mathbf{v}_s\right)} \\ p_p &= p\frac{\nabla\cdot\left[\left(1-\phi\right)\left(\mathbf{v}_p-\mathbf{v}_s\right)\right]}{\nabla\cdot\left(\mathbf{v}_p-\mathbf{v}_s\right)} = p\left(1-\phi\right) - p\frac{\left(\mathbf{v}_p-\mathbf{v}_s\right)\cdot\nabla\phi}{\nabla\cdot\left(\mathbf{v}_p-\mathbf{v}_s\right)} \end{aligned} \tag{3}$$

Eqs. (3) represent the frame invariant constitutive equations for the pressure p_s and p_p, the terms correcting the partial pressure (p_i) vanish both in the limit of no flow and when the solution is homogeneous. Eqs (1,2) with eq. (3) sum up to the classical momentum balance for the whole solution.

To find the demixed stratified solution it was proven that the following equations must be solved for the concentration of the two layers; thus constitutive choices are required:

$$\sigma_{yy}^{(p)}{}' + \int d\phi'\left[\phi'\frac{d\Pi_p}{d\phi'}\right] = \sigma_{yy}^{(p)}{}'' + \int d\phi''\left[\phi''\frac{d\Pi_p}{d\phi''}\right] \tag{4}$$

$$\sigma_{yy}^{(tot)}{}' = \sigma_{yy}^{(tot)}{}'' \tag{5}$$

CONSTITUTIVE CHOOICES

The polymer and solvent stress are deviatoric thus the required stresses can be obtained from the first and the second normal stress difference, N1 and N2. We here restrict our attention to the linear viscolelastic regime. We thus consider a polymer solution where the polymer can be entangled and the solvent not. The solvent will thus obey to a Rouse dynamics, while the polymer either to the reptation, if enough concentrated, or to the Rouse dynamics. In the linear regime the contribution of the polymer and the solvent linearly superimpose, thus, knowing the spectrum G(t) of the polymer and the solvent, the whole solution spectrum is obtained adding the two. From G(t) the first and the second normal stress difference can be obtained:

$$N_1 = 2J(\phi)\sigma_{xy}^2 = 2J(\phi)T_0^2 \qquad N_2 = -\frac{1}{7}N_1 \tag{6}$$

where the compliance $J(\phi)$ is:

$$\mathrm{J} = \int_0^\infty tG(t)dt \Big/ \left(\int_0^\infty G(t)dt\right)^2 \tag{7}$$

The elastic modulus of the solvent and of the polymer can be expressed as

$$\begin{aligned} G_S &= \left(1-\phi\right)\ \rho\frac{RT}{M_S}\sum_{p=1}^{n->\infty} e^{\frac{-2p^2t}{\tau_S}} \\ G_P &= \phi\ \left[\rho\frac{RT}{M_P}\sum_{p=1}^{n->\infty} e^{\frac{-2p^2t}{\tau_L}} + \rho\ \phi\ \frac{RT}{M_E}e^{\frac{-t}{\tau_d}}\right] \end{aligned} \tag{8}$$

where ρ is the density, M_S and M_P are the solvent and the polymer molecular weight respectively, τ_P and τ_S are the solvent and the polymer Rouse relaxation time, τ_d is the disengagement relaxation time and M_E is the molecular weight above which the pure polymer is entangled. In this paper we limit the summations in eqs. (17, 18) to n = 2. The relaxation times are $\tau_S = \left(M_S/M_E\right)^2\tau_E$; $\tau_P = \left(M_P/M_E\right)^2\tau_E$; $\tau_d = 3\ \left(M_P/M_E\right)^3\phi\ \tau_E$.

The free energy of mixing per unit volume is given by Flory Huggins theory that is modified so to account for the evidence that mixing stretched molecules is different from mixing molecules at equilibrium (random coil), thus the entropic contribution to the free energy of mixing may vary during flow. We have only accounted for the variation of the entropy of mixing, discarding the entropic contribution due to the elastic stored energy caused by stretching, since this contribution was already accounted for. With a combinatorial analysis we get

$$f_{mix} = \frac{RT}{V}\left[\frac{\phi\ln(\phi)}{N_P}+\frac{(1-\phi)\ln(1-\phi)}{N_S}+\chi\phi(1-\phi)+\phi\frac{N_P-1}{N_P}g_P+(1-\phi)\frac{N_S-1}{N_S}g_S\right] \tag{9}$$

where g_P and g_S are known functions of the polymer and solvent degree of stretching, respectively.

RESULTS

In Figure 1, the minimum stress (T_{min}) above which the demixed stratified solution can exist is shown for a solution with a viscous solvent and for a solution with a viscoelastic solvent.

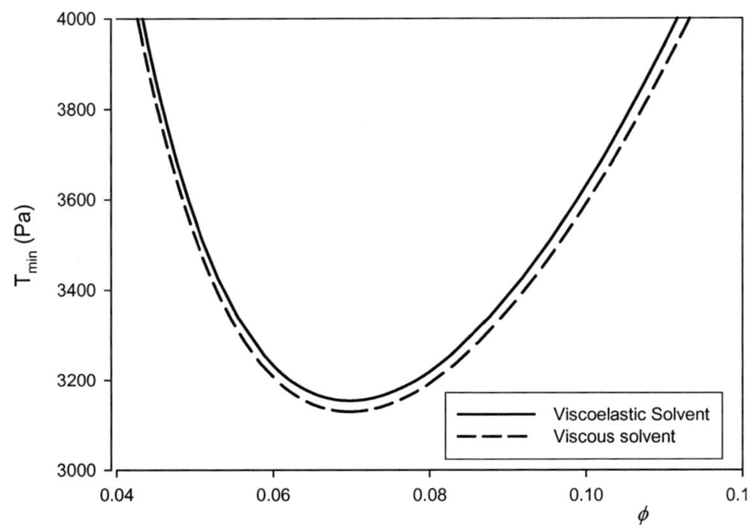

FIGURE 1. Prediction of Tmin for a solution with a viscous solvent (broken line) and a solution with a viscoelastic solvent (solid line)

It is then clear the solvent viscoelasticity tends to stabilize the homogeneous solution

REFERENCES

1. G. Ver Strate and W. Philippoff *J. Polym. Sci., Polym Lett. Ed.*, **12**, 267 (1974).
2. C. Rangel-Nafaile, A. B. Metzner and K. F. Wissbrun *Macromolecules* **17**, 1187, (1984)
3. R. Horst and B. A. Wolf *Macromolecules* **25**, 5291, (1992)
4. H. Helfand and G. H. Fredrickson *Phys. Rev. Lett.* **62**, 2468, (1989).
5. M. Doi in *Dynamics and Patterns in Complex Fluids*, edited by A. Onuki, K. Kawasaky, Berlin: Springer, 1990; pp. 100-112.
6. H. Ji, and E. Helfand *Macromolecules* **28**, 3869, (1995).
7. H. Yanase, P. Moldenaers, J. Mewis, V. Abetz, J. van Egmond and G.G. Fuller *Rheol. Acta* **30**, 89, (1991).
8. M. Minale *Macromolecules* **41**, 4471-4478, (2008).

Dispersion and characterization of Thermoplastic Polyurethane/Multiwalled Carbon Nanotubes in co-rotative twin screw extruder

Adolfo Benedito[a], Ignacio Buezas[a], Enrique Giménez[a,b], Begoña Galindo[a].

[a]*Instituto Tecnológico del Plástico – AIMPLAS. C/Gustave Eiffel, 4. València Parc Tenològic 46980 Paterna (Valencia) SPAIN.*
[b]*Universidad Politécnica de Valencia– Instituto de Tecnología de Materiales, Camino de Vera s/n 46022 (Valencia) SPAIN*

Abstract: The dispersion of multi-walled carbon nanotubes in thermoplastic polyurethanes has been done in co-rotative twin screw extruder through a melt blending process. A specific experimental design was prepared taking into account different compounding parameters such as feeding, temperature profile, screw speed, screw design, and carbon nanotube loading. The obtained samples were characterized by thermogravimetric analysis (TGA), light transmission microscopy, dynamic rheometry, and dynamic mechanical analysis. The objective of this work has been to study the dispersion quality of the carbon nanotubes and the effect of different compounding parameters to optimize them for industrial scale-up to final applications.

Keywords: Multiwalled carbon nanotubes, twin screw extruder, polyurethane, dispersion, screw.
PACS: 81.07.-b

INTRODUCTION

Using carbon nanotubes has become an interesting way to improve some properties such as electrical and thermal conductivity, mechanical strength [1,2], fireproof protection or shape memory properties for sensor applications [3, 4]. The suitable balance of properties is achieved with low nanotube loadings, replacing some traditional non-nanoscale additives or fillers. Several methodologies to obtain nanocomposites are well known, like *in situ* polymerization [5], solution mixing [6], electrospinning [7] or melt blending. Last methodology is interesting from industrial point of view, because it allows producing products with lower costs. Most of the works have been done using conical laboratory microcompounders [8,9], with some difficulties to scale-up the process because the compounding control variables in industrial compounding extruders are quite different. Other authors have used laboratory co-rotative twin screw extruders with high mixing power [10] closer to industrial practices.

Thermoplastic polyurethane (TPU) is a copolymer that contains soft and hard segments. The soft ones are obtained from polyesther or polyether macrodiol and the hard ones from diisocyanate and diol or diamine. It is a very versatile material with great mechanical and memory shape properties used in a lot of industrial applications such as tubing, footwear, sealants, adhesives, sports, or automotive parts. An important issue in the preparation of TPU/MWCNT nanocomposites is the capacity to filler agglomeration. The CNTs usually form high structured particles that reduce the properties enhancement. It is essential to obtain a good interfacial interaction between the nanotubes and the polymeric matrix. This work has studied the quality of the dispersion by means of the density of carbon nanotube agglomeration. Processing parameters and the selection of suitable compounding parameters are critical to achieve a good dispersion and therefore the most optimized properties.

CP1255, *V[th] International Conference on Times of Polymers (TOP) and Composites,*
edited by A. D'Amore, D. Acierno, and L. Grassia

EXPERIMENTAL

Commercial MWCNT (Nanocyl 7000) has been used for the experimental work. This is a MWCNT produced via catalytic carbon vapor deposition (CCVD) process with an average diameter of 9.5 nanometers, length of 1.5 microns and carbon purity around 90%. The selected commercial thermoplastic polyurethane is a PEARLTHANE D12F75 grade supplied by Merquinsa. To obtain the dispersed nanocomposites a co-rotative twin screw extruder (COPERION W&P ZSK25) has been used. The thermoplastic material was previously dried in an oven at 80ºC during 4 hours. The pellets were cut, dried and then compressed in a hot press to obtain probes for the different characterization techniques.

The characterization techniques include light transmission and reflection microscopy (LEICA model Leitz DMRX) with particle size analysis software from the same manufacturer, and field emission scanning electron microscopy (JEOL JSM-70001F). Also it was used dynamic mechanical analysis (DMA 2980 from TA Instruments), dynamic rheometry (AR G2 TA Instruments), and themogravimetric analysis (TGA Q 5000 from TA Instruments) for morphological, structural, thermal and mechanical characterization.

Based on the experimental design several CNTs loadings were prepared (1%, 2%, 3%, 4%, 10% and 15% in weight), varying some compounding parameters such as feeding zone, screw rates (200-600 rpm), and preparing different shear screw designs. The low loading levels were prepared from dilution of previously prepared 15wt%. masterbatch. The preparation of nanocomposites from commercial masterbatches seems an excellent solution to avoid safety problems related to the handling of carbon nanotubes at industrial level.

ANALYSIS OF RESULTS

Light transmission images showed the presence of microscopical agglomerated structures around 20-40 microns. The agglomerates seem to be well distributed. No empty or nanofiller over-concentrated regions have been detected. Comparing with the results observed in other thermoplastic polymers, like polyolefins, the quality of the dispersion in TPU materials in terms of average size particle and agglomerated density is quite good. With regard to compounding parameters, the screw speed and the screw design seem to have major influence on the break of agglomerate structures. The light transmission microscopy images show similar agglomeration but bigger sizes in the case of low screw speeds (see figure 1). The histograms of particle size showed agglomerated around 80 microns of length and demonstrate that high screw speeds caused high shear forces and then the breakage of these big agglomerates. The calculated agglomeration density was 0.49% for low screw speeds and 0.33% for high screw speeds. This parameter is defined as the relation between measured agglomerated area and total area.

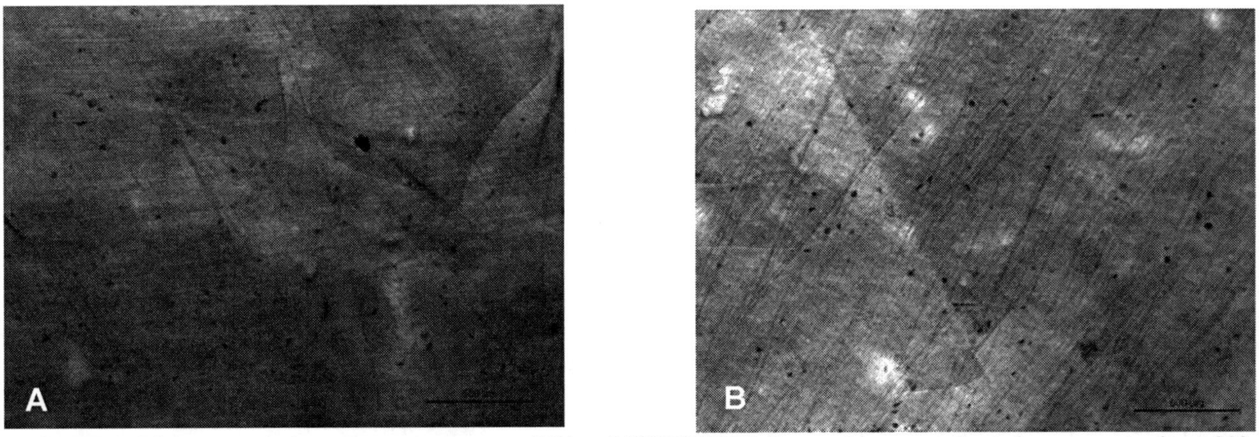

FIGURE 1. Light transmission microscope images of 1%wt. MWCNT/TPU dispersed in co-rotative twin screw extruder at 200 rpm (A) and 600 rpm (B).

FEG-SEM analysis also showed a suitable dispersion and distribution of CNTs (see figure 2), despite some agglomeration regions in high concentrated compounds.

FIGURE 2. FEG-SEM of nanotubes dispersed regions in 15wt.% of TPU/CNT.

Dynamic rheology is an interesting characterization technique to detect morphological changes. Storage modulus (G') at low frequencies is related to the structure stiffness and the presence of intermolecular soft electrostatic bonds. Well dispersed carbon nanotubes show a stabilized net structure that gives stiffness to the melt. It is the rheological percolation point, around 2% and very similar to conductivity percolation threshold [11]. Figure 3 shows the modulus G' at low frequencies at different CNTs loadings. The reference material shows fully relaxed polymer chains at low frequencies. With dispersed CNTs the terminal behaviour disappears and this means that tube-tube interactions dominate over polymer-particle interactions, as a nanofiller network is established.

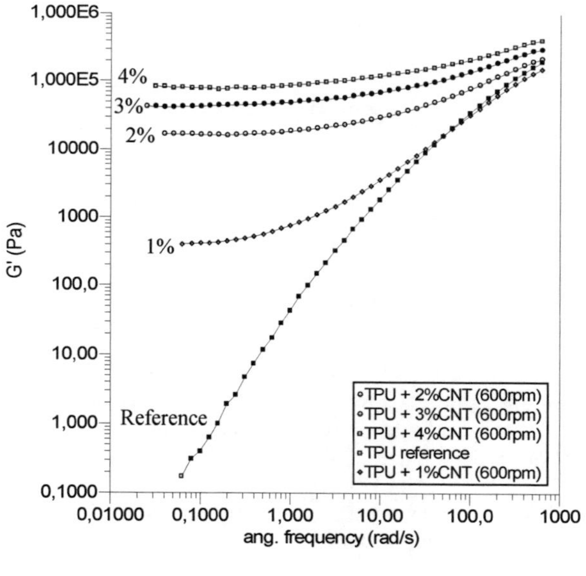

FIGURE 3. Dynamic rheograms for MWCNT/TPU compounds at fixed screw speed and the rest of processing parameters.

The effect of the carbon nanotubes loading on the mechanical properties can be studied through the DMA (dynamic mechanical analysis). The storage modulus (E') remains constant at low temperatures. It seems that the CNTs do not modify mechanical properties at low temperatures. Above 0°C, environmental temperature, the CNTs loading increases the storage modulus in a critical way (see figure 4).

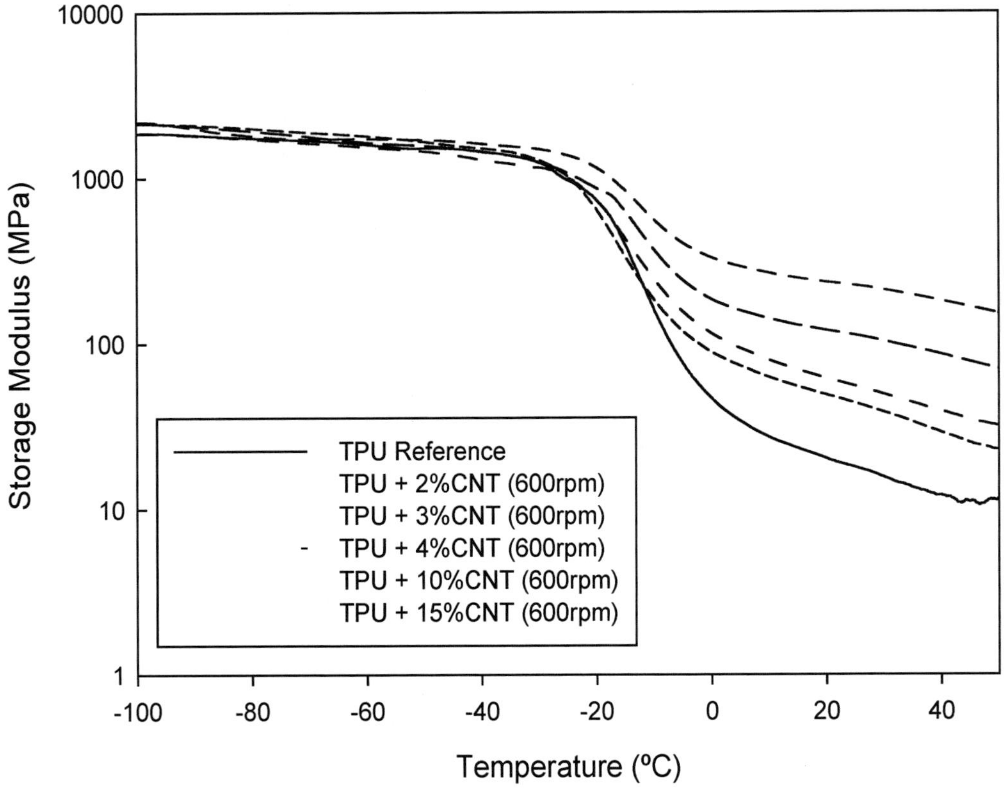

FIGURE 4. DMA for nanocomposites prepared at 600 rpm and different CNT loadings.

ACKNOWLEDGMENTS

This work has been co-funded by the *Ministerio de Industria, Comercio y Turismo* through the *Plan Nacional de Investigación Científica, Desarrollo e Innovación Tecnológica (IAP-600100-2008-26),* by the program *Fomento de la Innovación* from IMPIVA and the *European Regional Development Fund* (FEDER).

REFERENCES

1. J.N.Coleman, U.Khan, W.J.Blau, Y.K.Gun'ko. Carbon. Vol.44, Issue 9, August 2006, 1624-1652.
2. M.A.López Manchado, L.Valentini, J.Biagiotti, J.M.Kenny. Carbon. Vol.43, Issue 7, June 2005, 1499-1505.
3. A.Ramaratnam, N.Jalili. Journal of Intelligent Material Systems and Structures, Vol. 17, No. 3, 2006, 199-208.
4. T.Xu, Z.Wang, J.Miao. Nanoelectronics Conference, 2008, INEC 2008, 2nd IEEE International, 555-558.
5. H.Xia, M.Song. Soft Matter (2005) 386.
6. H.Koerner, W.Liu, M.Alexander, P.Mirau, H.Dowty, R.A.Vaia. Polymer 46 (2005), 4405.
7. Leslie Y. Yeo, James R.Friend. Journal of Experimental Nanoscience, Vol.1, No.2, June 2006, 177-209.
8. F.Jiang,G.Hu,S.Wu,Y.Wei,L.Zhang. Polymers & Polymer Composites, Vol.16, No.8, 2008, 471-480.
9. W.Chen,X.Tao,Y.Liu. Composites Science and Technology, 66, 2006, 3029-3034.
10. P.Pötschke, L.Häubler,S.Pegel, et all. KGK. September 2007, 432-437.
11. M.Fernández, M.Landa, J.Peña, M.E.Muñoz, S.Santamaría. NanoSpain2010 (March).

Microcellular Foams Based on High Performance Thermoplastic Nanocomposites

Luigi Sorrentino[a,b], Marcella Gargiulo[a], Giuseppe Pezzullo[c], Salvatore Iannace[a,b]

[a]*Institute for Composite and Biomedical Materials – National Research Council*
Piazzale V. Tecchio, 80 – 80125 Napoli – Italy
[b]*IMAST - Technological District in Polymer and Composite Engineering*
P.le Fermi 1 80055 Portici (NA) (Italy)
[c]*Department of Materials and Production Engineering – University of Napoli Federico II*
Piazzale V. Tecchio, 80 – 80125 Napoli – Italy

Abstract. Foams from engineering thermoplastics nanocomposites based on Polyethersulphone and Polyethylene-2,6-naphthalate were prepared by using two different nanofillers (Silica nanoparticles and Graphite nanosheets). The effects of the nanofiller type and content on the foaming process was investigated and related to the density and cellular morphology of foams. The nanocomposite foams based on PES matrix exhibited improved nucleated cells both with SiO_2 and Graphite nanosheets, but the density increased at all temperatures. On the contrary, nanocomposite foams based on PEN matrix showed different behaviors with the filler type. In this case, in fact, silica nanoparticles allowed lower densities when compared to the unfilled polymer foams, without influencing cells density. The Graphite nanosheets extended towards higher temperatures the foaming window of PEN nanocomposites, allowing densities as low as 0.15 at 260°C.

Keywords: Foams, PES, PEN, nanocomposites
PACS: 82.70.Rr, 81.05.Qk, 81.07.Wx

INTRODUCTION

Foams are low densities materials widely used in industrial commodity applications, such as in packaging and thermal or acoustical insulation, for cushion, such as in car seats, or in case high mechanical energy absorption properties are required with low allowable stress, such as in helmets. Recently, transportation industries, such as automotive, nautics and aeronautics, is definitely pointing to structural foams, mainly for weight reduction and for their high specific mechanical properties. Among the candidate materials, thermoplastic polymers are preferred as foam matrices due to their recyclability, higher productivity and reduced environmental impact of their manufacturing processes.

The service temperatures of the conventional foams are near room temperature and this require the use of high performance microcellular foams based on thermoplastics, such as polyethersulphone, polyphenylensulphone, polyetherimide, as recently prepared and characterized [1-3]. The expansion process of neat polymers was performed by using CO_2 in the supercritical state as blowing agent. Very few works were done on semicrystalline PEN [3]. In order to improve the mechanical response of these engineering polymers and to further extend the temperature working range, nanocomposites were considered, due to their several advantages over neat polymers and conventional composites [4]. The aim of this work was to study microcellular foams prepared from nanocomposites based on PES and PEN engineering thermoplastics.

EXPERIMENTAL

The materials used for the experimental activities were the PES (Ultrason E3010; Basf) and PEN Homopolymer (TN8065S; Teijin). The polymer resins were dried at 120°C for about 24 h in an vacuum oven before being tested or

CP1255, *V[th] International Conference on Times of Polymers (TOP) and Composites,*
edited by A. D'Amore, D. Acierno, and L. Grassia

processed. The silica nanoparticles (Aerosil R380, mean average primary particle size = 7nm, specific surface area = 380 m^2/g) were furnished by Degussa (Germany). Proprietary expanded graphite particles (platelets width smaller than 65 μm, platelets thickness smaller than 1 μm) were supplied by GrafTech International (TG-741 and its availability exclusively from GrafTech International Holdings, Inc.). Four concentrations were used for each nanofiller (0.1, 0.5, 1 and 2.5%). Nanocomposites were prepared by melt blending in an internal mixer. An high mixing speed was employed to promote a good dispersion of nanofillers in the polimeric matrix.

The foaming agent used was carbon dioxide (with a purity of 99.9%) supplied by from S.O.N. (Italy) and used as received. Solid state foaming technique was employed to prepare foam samples. They were solubilized with CO_2 at 80bar and 50°C in a pressure vessel and weighted to evaluate the gas uptake after 24 hours. All different nanocomposites were foamed in an oil bath at temperatures ranging from 140 and 260°C. After being polished, the density of samples was measured according to the ASTM D792 standard. The cellular morphology was evaluated by analyzing SEM images and measuring the number of cells for cubic centimeter.

RESULTS AND DISCUSSION

The blending process allowed to obtain PES and PEN nanocomposites. The dispersion was verified by rheological tests and XRD analysis. The PES exhibited a strong enhancement of the viscosity with both nanofillers, while PEN showed an opposite behavior (results not shown). This was attributed to some extent of macromolecular degradation occurred during the melt blending process.

Nanosilica did not reduced the gas uptake in PES, while in PEN a strong reduction was induced at high filler concentrations. Graphite nanoplatelets reduced the gas uptake in both matrices, more and more as the concentration rised (Table 1). This behavior was probably due to reduction of gas diffusivity in graphite nanocomposites, in which the planar shape of the filler increased the tortuosity and gas diffusion was slowed.

Filler Content	Gas uptake [%]			
	PEN-SiO_2	PEN-Graphite	PES-SiO_2	PES-Graphite
Neat polymer	5.2	5.2	6.5	6.5
0.10%	5.2	4.8		
0.50%	5.2	4.1	6.7	6.2
1.00%	4.3	4.4	6.4	5.8
2.50%	4.4	3.5	6.7	

TABLE 1. Gas uptake after 24 hours for all nanocomposites prepared

It's worth to note that the foaming temperatures were near the PES T_g (220°C) and higher than the PEN T_g. As a consequence, the foaming process of PES was strongly influenced by the very high viscosity, which led to low expansion ratios and high foam density when compared to PEN foams. The nucleation of cells, on the contrary, was positively influenced, and microcellular foams on PES nanocomposites were obtained in quite all conditioins.

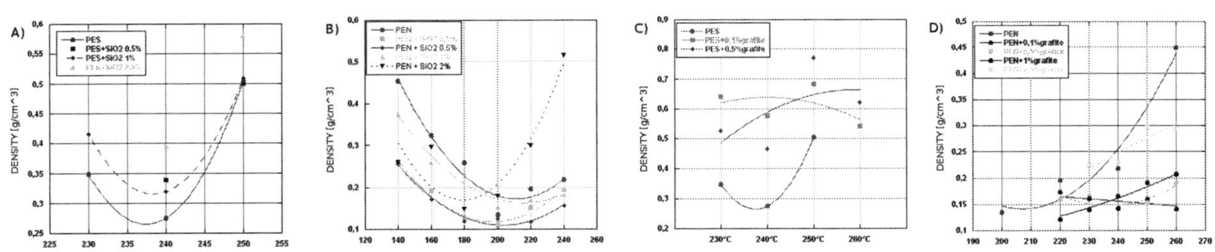

FIGURE 1. Foam densities for a) PES-SiO_2, b) PEN-SiO_2, c) PES-Graphite, d) PEN-Graphite nanocomposite foams

The effects of nanofillers on foam properties are showed in Figure 1 (foam density) Figure 2 (cell density). The foaming behavior was different for PES and PEN nanocomposites. In particular, PES-SiO_2 and PES-Graphite samples showed an increase of the foam density with respect to the neat polymer foams, related to the strong increase of the viscosity. PEN-SiO_2 systems exhibited lower densities in the entire temperature range, in particular for low filler concentrations. PEN-Graphite samples showed improved foam stability at high temperatures (above 230°C) allowing to obtain low density foams whereas neat PEN was not able to be regularly foamed, exhibiting few and collapsed cells. This was attributed to the enhanced crystallization kinetics allowed by graphite sheets in PEN (data not shown), which contributed to stabilize the cellular structures at high temperatures.

The cells density was increased of one order of magnitude in PES-SiO_2 nanocomposites, meaning that more cells were nucleated, while PEN-SiO_2 nanocomposites did not show a regular behavior, and cell densities were about the same in quite all conditions. This behavior could be related to the cell collapse after nucleation due to the very low polymer viscosity, but further investigations should be performed to clarify the phenomenon.

Graphite was more effective than SiO_2 for nucleating cells in PES samples. An increase of two order of magnitude of cell density was exhibited by PES-Graphite nanocomposites. Again, the PEN-Graphite samples showed an opposite behavior and a lower cell density was detected with respect to the neat polymer. This was attributed to the low viscosity of PEN-Graphite nanocomposites, which prevented the cell stabilization and the cell collapse took place.

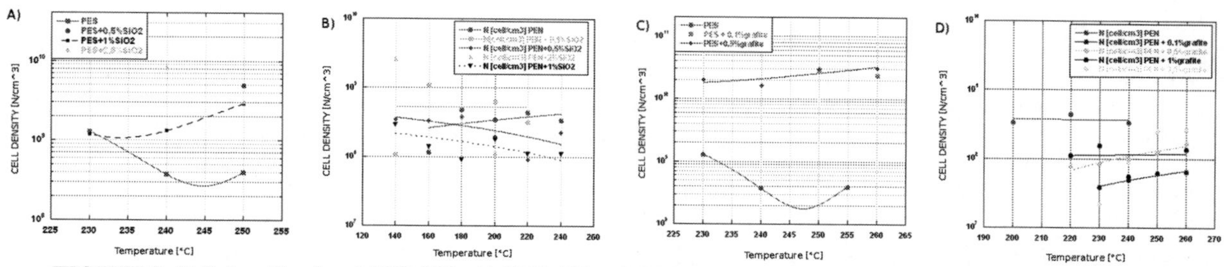

FIGURE 2. Cell densities for a) PES-SiO_2, b) PEN-SiO_2, c) PES-Graphite, d) PEN-Graphite nanocomposite foams

CONCLUSIONS

Nanocomposites of PEN and PES high performance thermoplastics were successfully prepared by the melt mixing technology. Silica nanoparticles and graphite nanosheets were used as fillers. The presence of the nanofillers heavily affected the sorption behavior of PEN nanocomposites, reducing the diffusivity of the blowing agent (CO_2) and the gas uptake. The rheological behavior of PES based nanocomposites was strongly enhanced, exhibiting an increase of zero shear viscosity and a shear thinning behavior for all particle contents. PEN based nanocomposites, on the contrary, showed minor influences, and for PEN-Graphite systems a lowering of the viscosity curve, probably due to some molecular degradation during the melt mixing process.

The crystallization behavior of PEN based nanocomposites was evaluated by means of nonisothermal DSC scans. Nanoparticles increased the crystallization kinetics, in particular for graphite nanosheets .

Foams based on PES nanocomposites showed a microcellular morphology but an increased foam density with respect to the neat polymer. PEN nanocomposites showed larger cells and lower densities. In the PEN-Graphite foams the temperature range for foaming was strongly extended. The different behavior was related to the polymer viscosities at the foaming temperatures: very high in the PES nanocomposites and low in PEN nanocomposites.

Foams based on PES nanocomposites exhibited a very pronounced increase of the cell density (from one to two order of magnitude for SiO_2 and Graphite, respectively), indicating that nanofillers had a strong nucleating effect. On the contrary, nanofilled PEN showed quite constant, when not lower, number of cells. This was probably due to the reduced melt strength of PEN nanocomposites at the foaming temperatures which provoked the collapsing of nucleated cells during the cell growth.

ACKNOWLEDGMENTS

This study has been carried out with financial support from MIUR (Italy) within the projects FIRB “MANTA”, Prot. RBIP065YCL. The authors would like to thank Vincenzo Scognamiglio for the assistance during the lab activities.

REFERENCES

1. Hongliu Sun, Gil S. Sur, James E. Mark, *European Polymer Journal* 38 (12), 2373-2381 (2002)
2. B. Krause, R. Mettinkhof, N.F.A. van der Vegt, M. Wessling, *Macromolecules* 34, 874-884 (2001)
3. L. Sorrentino, M. Gargiulo, S. Iannace, Adv Polym Tech, submitted
4. Suprakas Sinha Ray, Masami Okamoto *Prog Polym Sci* 28, 1539–1641 (2003)

Nanotechnology and Food Contact Materials

Constantine (Costas) D. Papaspyrides

Laboratory of Polymer Technology, School of Chemical Engineering
National Technical University of Athens, Athens 157 80 Greece

Abstract. The use of nanotechnology in food contact materials is highlighted in relation to novel applications and potential implications for consumer safety and regulatory controls. Nanotechnology applications are expected to bring a range of benefits to the food sector, including improved packaging, antimicrobial properties, traceability and security of food products. The toxicological nature of hazard, likelihood of exposure and risk to consumers from nanotechnology-derived food/food packaging are largely unknown and this work highlights the benefits of nanotechnology in food contact materials but also the gaps in knowledge regarding consumers safety that require further research.

Keywords: Food Packaging, Barrier Properties, Nanocomposites
PACS: 82.35.Np

Traditionally, polymeric materials have been filled with synthetic or natural inorganic compounds in order to improve their properties, or simply to reduce cost. However, although conventionally filled or reinforced polymeric materials are widely used in various fields, it is often reported that the addition of these fillers imparts drawbacks to the resulting materials, such as weight increase, brittleness and opacity. Nanocomposites, on the other hand, are a new class of composites, for which at least one dimension of the dispersed particles is in the nanometre range. Nanocomposites represent a new alternative to conventional technologies for improving polymer properties. They exhibit increased barrier properties, increased mechanical strength, and improved heat resistance compared to their neat polymers and conventional composites [1,2].

Depending on how many dimensions are in the nanometer range, one can distinguish isodimensional nanoparticles when the three dimensions are on the order of nanometers, nanotubes or whiskers when two dimensions are on the nanometer scale and the third is larger and layered crystals or clays, present in the form of sheets of one to a few nanometers thick and hundreds to thousands nanometers in extent. Among all the potential nanocomposite precursors, those based on clay and layered silicates have been most widely investigated [1,3]. Regarding their structure, two types can be obtained; intercalated and exfoliated or delaminated nanocomposites (Figure 1) [4]. Intercalated structures are formed when a single (or sometimes more) extended polymer chain is intercalated between the silicate layers. The result is a well ordered multilayer structure of alternating polymeric and inorganic layers, with a repeat distance between them. On the other hand, exfoliated or delaminated structures are obtained when the clay layers are well separated from one another and individually dispersed in the continuous polymer matrix.

FIGURE 1. Schematic illustration of two different types of thermodynamically achievable polymer/layered silicate nanocomposites: Intercalated (left) and Exfoliated (right).

CP1255, *Vth International Conference on Times of Polymers (TOP) and Composites,*
edited by A. D'Amore, D. Acierno, and L. Grassia

Nanomaterials have been demonstrated to possess novel characteristics that can be applied in developing new packaging with better properties than conventional packaging. When used in food packaging, nanocomposites are better able to withstand the stress of thermal food processing, transportation and storage. In addition to mechanical properties, nanotechnology leads to the production of packaging with improved barrier properties, which when applied to the food industry will extend the shelf life of a food, thereby expanding its marketing potential. For example, polymers ranging from epoxies and good sealants (like siloxanes) to semi-permeable (e.g. polyureas) and highly hydrophilic (e.g. PVA) are all improved up to an order of magnitude by low clay loadings [1]. The dramatic improvement of barrier properties can be explained by the concept of tortuous paths [5,6]. That is, when impermeable nanoparticles are incorporated into a polymer, the permeating molecules are forced to diffuse around them in a random walk by a tortuous pathway, as shown in Figure 2 [7].

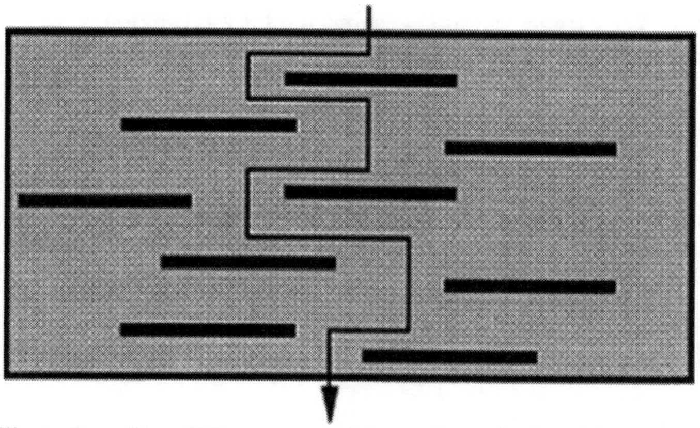

FIGURE 2. Schematic illustration of two different types of thermodynamically achievable polymer/layered silicate nanocomposites: Intercalated (left) and Exfoliated (right).

In addition to improving mechanical and barrier properties of plastic food packaging, several nanoparticles can provide active and/or intelligent properties to food contact materials, such as antimicrobial properties, oxygen scavenging ability, enzyme immobilization, or indication of the degree of exposure to some degradation related factor. Thus, nanocomposites can not only passively protect the food against environmental factors like humidity or oxygen, but also incorporate properties to the packaging material so it may actually enhance stability of foods. Polymer nanocomposites incorporating metal or metal oxide nanoparticles have been developed for antimicrobial ''active'' packaging, abrasion resistance, UV absorption and/or strength. The nanomaterials used as UV absorbers (e.g. titanium dioxide) can prevent UV-degradation in plastics such as PS, PE or PVC [8]. In addition based on the antimicrobial action of nanosilver, a number of ''active'' food contact materials have been developed that are claimed to preserve the food materials within longer by inhibiting the growth of microorganisms [8,9].

However, as with most new and evolving technologies, potential benefits are emphasized while little is known on safety of the application of nanotechnologies in the food contact materials sector. Due to very large aspect ratios, a relatively low level of nanoparticles is sufficient to change the properties of packaging materials without significant changes in density, transparency and processing characteristics [10]. Yet, the drastically greater surface area of nanoparticles also means potentially greater reactivity with biological or chemical materials around them—such as increased reactivity with the immune system [11]. The main risk of consumer exposure to nanoparticles from food packaging is likely to be through potential migration of nanoparticles into food and drinks. However, such migration data are not currently available, despite the fact that a number of food contact materials containing nanomaterials are already available and in commercial use in some countries. To date, there is only one published study from Avella et al.[12] that has determined the migration of minerals (Fe, Mg, Si) from biodegradable starch/nanoclay nanocomposite films. The results of this study showed an insignificant trend in the levels of Fe and Mg in packaged vegetables, but a consistent increase in the amount of Si (the main component of nano-clay). This study, however, provided only a small piece of information for a biodegradable material and not for the plastic nanocomposites more likely to be used in food and drink containers, such as PET, PE and PP. Thermoplastic matrices, are generally expected to be better barriers than biodegradable materials in the migration of nanoparticles. Mathematical predictions from Simon et al. indicate that any detectable migration of nanoparticles from packaging to food will only take place in the case of very small nanoparticles with a radius in the order of 1 nm, from polymer matrices that have a relatively low dynamic viscosity, and that do not interact with the nanoparticles. These conditions could be met for example in the case of composites of nanosilver with polyolefins.[13].

In general, nanocomposite materials, introduce significant improvements in polymers and thus in the plastic food packaging by rendering plastics light, fire-resistant and stronger in terms of mechanical and thermal performance. In addition, barrier properties –a topic of great interest in the food packaging sector- are greatly improved as plastics have become less permeable to substances due to the incorporation on nanomaterials even in low quantities. Nanotechnology applications also bring a range of novel benefits to the food contact materials like antimicrobial properties, traceability and security of food products. On the other hand, the toxicological nature of hazard, likelihood of exposure and risk to consumers from nanoparticles in food contact materials are largely unknown. This work synopsizes the current knowledge regarding nanotechnology in polymers and food contact plastic materials and also highlights major gaps that remain unexplored, regarding consumers safety and exposure to nanotechnology.

REFERENCES

1. S.Pavlidou and C.D.Papaspyrides. *Progress in Polymer Science* **33** (2008)
2. P. Kiliaris and C.D. Papaspyrides. *Advanced Research in Polymer Science* **10** (2009).
3. G. Gorrasi, M. Tortora ,V. Vittoria, G. Galli G, E. Chiellini. *J Polym Sci Polym Phys* **40** (2002)
4. S. Ray and M. Bousima. *Prog Mater Sci.* **50** (2005)
5. J. Nielsen. *J. of Macromolecular Science A* **1**(5) (1967)
6. D. Adame and G. Beall *Applied Clay Science* **42** (2009)
7. K. Yano, A. Usuki , A. Okada, T. Kurauchi, O. Kamigaito . *J Polym Sci Polym Chem* **31** (1993)
8. Q. Chaudhry et al. *Food Add and Cont* A **25**:3 (2008)
9. C. Damm, H. Munstedt, A. Rosch. *Materials Chem Phys* **108** (2008)
10. S. Lei, S. Hoa, M. Ton-That. *Composites Science and Technology* **66** (2006)
11. H. Bouwmester et al. *Reguatoryl Toxicology and Pharmacology* **53** (2009)
12. Avela et al. *Food Chemistry* **93** (2005)
13. Simon et al. *Journal of Food and Nutrition Research* **47** (2008)

Predicting the ageing and the long-term durability of organic polymer solar cells

Jean-Luc Gardette[a,b], Agnès Rivaton[a,b], Sandrine Thérias[a,b], Sylvain Chambon[a,b], Matthieu Manceau[a,b], Julien Gaume[a,b]

[a]*Clermont Université, Université Blaise Pascal, Laboratoire de Photochimie Moléculaire et Macromoléculaire, F-63177 Aubière Cedex*
[b]*CNRS, UMR 6505, F-63177 Aubière Cedex*
luc.gardette@univ-bpclermont.fr

Abstract. Organic solar cells based on conductive polymers exhibit a unique combination of properties which include low cost, flexibility and large surface processability. Organic photovoltaic could then prevail for some applications alongside silicon, such as nomad or indoor. To achieve this objective, the sustainability of the initial properties in conditions of use of the cell is required, since it could be a lock to the emergence of these devices in the market. The polymers used in solar cells are indeed known to exhibit low resistance to environmental constraints, in particular to the combined action of sunlight, oxygen and water.
We present recent results on both the accelerated artificial and the natural outdoors ageing of MDMO-PPV (Poly[2-methoxy-5-(3',7'-dimethyloctyloxy)-1,4-Phenylenevinylene) and P3HT/PCBM blends poly(3-hexylthiophene) (P3HT) (methano-fullerene[6,6]-phenyl C61-butyric acid methyl ester) ([60] PCBM). The influence of various parameters such as the temperature and the presence of oxygen were studied. The modifications of the chemical structure of both the components of the blend were monitored by spectroscopic analysis (infrared, UV-visible), the morphology of the blends was analysed by AFM and XRD and the photovoltaic performances all along the exposure were recorded.
Two important results have been pointed out: on one hand, the Achilles heel of the chemical structure of MDMO-PPV and P3HT under the impact of light has been evidenced. On the other hand, it has been shown that P3HT:PCBM blends are much more stable than MDMO:PCBM blends whatever the conditions of ageing are. Results show that a convenient encapsulation can ensure a promising lifetime of P3HT/PCBM blends in real conditions of use. This work also focuses on this last point and proposes to study and try to understand the behavior of the materials used in the active layer when submitted to photoaging and thermal aging in the absence of oxygen. To fulfil very good encapsulation, glass substrates can be used. However, as a result, organic devices lose some of their attractive properties, such as flexibility. Organic PV cells finally need a complex layer stack as a permeation barrier to be protected from water and oxygen of the environment. Multilayer barrier coatings are required: on the one hand, there are inorganic materials applied as thin films. Those films show a preferential permeation at defects. For the most part, these inorganic materials are then used in combination with a polymeric substrate. Polymers can be considered as being homogeneous with regard to the usually utilized substrate or film thicknesses. Their barrier properties can be reinforced by nanoclays. The impermeable clay layers force a tortuous pathway for a permeant transversing the nanocomposite. It is reported that gas permeability through polymer films can be reduced by 50–500 times even with small loadings of nanoclays.

INTRODUCTION

The discovery of conducting polymers during the 1970s resulted in intense research, greatly motivated by the possible replacement of conventional inorganic or metallic components in a wide range of applications. The

CP1255, *V[th] International Conference on Times of Polymers (TOP) and Composites,*
edited by A. D'Amore, D. Acierno, and L. Grassia

ultimate goal is to combine the metallic behaviour of π-conjugated systems with the lightness, processability and low-cost production of polymers. Among all of the known conducting polymers, polyphenylenevinylene derivative, MDMO-PPV and poly(3-alkylthiophenes) (P3ATs) have been the focus of particular attention due to their good electrical and mechanical properties [1,2]. These polymers have a high conductivity, and the presence of the alkyl side chain favours their solubility in common organic solvents. As a result, they can be processed from solution and then used in a large variety of technologies involving thin films: light emitting diodes, field effect transistors, electrochromic devices, solar cells, etc. Since most of these applications are optical in nature, it is surprising that the photochemical behaviour of these polymers remains so scarcely documented, especially in the solid state [3,4,5]. Similar to numerous other conjugated macromolecules, they have low photochemical stability, which leads to reduced operating lifetimes. Understanding the degradation processes that result from the interaction of conducting polymers with solar light seems relevant to improve the stability of devices that integrate such polymers.

RESULTS AND DISCUSSION

Our lab has been involved for a few years in this research area. In a first time, the behaviour of a polyphenylenevinylene derivative, MDMO-PPV has been studied in detail [6,7,8,9]. The degradation mechanism was determined, which permitted us to evaluate the stability of MDMO-PPV-based devices. In the case of P3HT, based on observations from irradiated solutions of this polymer [10,11], the literature reports a dual mechanism to account for the solid state photodegradation of P3ATs [3]. On one hand, oxidation of the thiophene ring has been proposed to be induced by singlet oxygen O_2 ($^1\Delta_g$). $^1O_2$ can be obtained by energy transfer from the triplet excited state of P3HT and would undergo a 1,4 Diels–Alder addition to the thienyl units, ultimately forming an unstable endo-peroxide. This species would then decompose into carbonyl, olefinic and sulfine derivatives, while breaking the macromolecular backbone. This disrupts the π-conjugation and thus lowers the UV–visible absorbance. This mechanism was proposed on the basis of experiments involving chemically generated $^1O_2$ [11]. The reaction of P3HT with chemically-produced $^1O_2$ resulted in polymer photobleaching, and it was shown that the presence of a quencher, such as anthracene, reduced the rate of P3HT degradation. The proposed reaction products were also consistent with the mechanisms reported for functionalized thiophene [12,13]. A second pathway involving the oxidation of the side chain by ground state oxygen has also been proposed [3]. The oxidation would proceed via a free-radical chain reaction route, thus forming carbonyl and hydroxyl adducts, and may include cross-linking. This reaction scheme was proposed using infrared (IR) band assignments and size exclusion chromatography of irradiated samples. An identical radical mechanism has also been suggested for thermooxidative degradation of poly(3-octylthiophene) (P3OT) [4], supporting the idea that $^1O_2$ is not involved in the side-chain oxidation. In a previous paper, we have clearly shown that $^1O_2$ is not the main intermediate responsible for the photodegradation of P3HT [14]. Therefore, the photodegradation mechanism of solid-state P3HT has to be reconsidered. In the present study, we monitored the modifications of the chemical structure of P3HT that resulted from photo- and thermal ageing. P3HT thin films were submitted to light irradiation ($\lambda > 300$ nm) or to thermal degradation (60 and 100 °C) in ambient air. Particular attention was paid to the identification of sulfuroxidized species, which was performed by transposition of the techniques used to characterize carbonyl compounds. The obtained data allowed us to propose a degradation mechanism that accounted for the formation of the oxidative species due to the impact of light or heat on the polymer matrix.

Our results show that a convenient encapsulation can ensure a promising lifetime of P3HT/PCBM blends in real conditions of use. The extrapolation of the data obtained in conditions of artificial accelerated ageing to natural ageing suggested that the P3HT:PCBM blend would be chemically stable for at least three years under use conditions if well protected from oxygen. We clearly observed that the irradiation of P3HT in the absence of oxygen provoked modifications of its chemical structure that slowly lead to absorbance decrease. However, we showed that P3HT is much more stable than MDMO-PPV. This improved durability was explained by the different chemical structures of the two materials and, more precisely, by their respective side chains. We also observed that the degradation rate is strongly attenuated, and even nearly suppressed, when P3HT is blended with PCBM. This behaviour has been attributed both to the radical scavenging properties of the fullerene and to its ability to quench the P3HT singlet state. Over the time range investigated (10000 h), this system appeared to be chemically stable. On the basis of natural ageing data, and regarding materials chemical structure, one can extrapolate the lifetime of the P3HT:PCBM system to at least three years of outdoor exposure.

REFERENCES

[1] Roncali J. Conjugated poly(thiophenes): synthesis, functionalization, and applications. Chem Rev 1992;92(4):711–38.

[2] McCullough RD. The chemistry of conducting polythiophenes. Adv Mater 1998;10(2):93–116.

[3] Abdou MSA, Holdcroft S. Solid-state photochemistry of π-conjugated poly(3- alkylthiophenes). Can J Chem 1995;73:1893–901.

[4] Ljungqvist N, Hjertberg T. Oxidative degradation of poly(3-octylthiophene). Macromolecules 1995;28(18):5993–9.

[5] Caronna T, Forte M, Catellani M, Meille SV. Photodegradation and photostabilization studies of poly(3-butylthiophene) in the solid state. Chem Mater 1997;9(4):991–5.

[6] Chambon S, Rivaton A, Gardette J-L, Firon M, Lutsen L. Aging of a donor conjugated polymer: photochemical studies of the degradation of poly[2-methoxy-5-(30,70-dimethyloctyloxy)-1,4-phenylenevinylene]. J Polym Sci Part A Polym Chem 2007;45(2):317–31.

[7] Chambon S, Manceau M, Firon M, Cros S, Rivaton A, Gardette J-L. Photooxidation in an 18O2 atmosphere: a powerful tool to elucidate the mechanism of UV–visible light oxidation of polymers – application to the photodegradation of MDMO-PPV. Polymer 2008;49(15):3288–94.

[8] Chambon S, Rivaton A, Gardette J-L, Firon M. Durability of MDMO-PPV and MDMO-PPV: PCBM blends under illumination in the absence of oxygen. Sol Energy Mater Sol Cells 2008;92(7):785–92.

[9] Chambon S, Rivaton A, Gardette J-L, Firon M. Photo- and thermal degradation of MDMO-PPV:PCBM blends. Sol Energy Mater Sol Cells 2007;91(5):394–8.

[10] Holdcroft S. A photochemical study of poly(3-hexylthiophene). Macromolecules 1991;24(17):4834–8.

[11] Abdou MSA, Holdcroft S. Mechanisms of photodegradation of poly(3- alkylthiophenes) in solution. Macromolecules 1993;26(11):2954–62.

[12] Skold CN, Schlessinger RH. The reaction of singlet oxygen with a simple thiophene. Tetrahedron Lett 1970;11(10):791–4.

[13] Wasserman HH, Strehlow W. Reactions of singlet oxygen with thiophenes. Tetrahedron Lett 1970;11(10):795–6.

[14] Manceau M, Rivaton A, Gardette J-L. Involvement of singlet oxygen in the solid-state photochemistry of P3HT. Macromol Rapid Commun 2008;29(22):1823–7

On Sliding Wear of Nanoparticle Modified Polymer Composites

Klaus Friedrich[a,b] and Li Chang[c]

[a] *Institute for Composite Materials (IVW), University of Kaiserslautern, 67663 Kaiserslautern, Germany*
[b] *CEREM, College of Engineering, King Saud University, Riyadh, Saudi Arabia*
[c] *School of Aerospace, Mechanical and Mechatronic Engineering,*
The University of Sydney, Sydney, NSW 2006, Australia

Abstract. The objective of this study was to achieve an in-depth understanding of the role of the nanoparticles in modifying the sliding wear behavior of SFRPs. In particular, the effects of nanoparticles on the contact mechanics and the wear behavior of the transfer film were investigated. It was found that the additional nanoparticles do not directly contribute to the formation of a high performance transfer film. However, the presence of nanoparticles in the contact region can effectively reduce the adhesion between the transfer film and the polymeric specimen, resulting in a lower coefficient of friction. In particular, the rolling behavior of nanoparticles can significantly enhance the tribological performance of SFRPs, especially under extreme sliding conditions.

Keywords: nanoparticles, polymeric composites, wear and friction, contact mechanics.

INTRODUCTION

Short fiber-reinforced polymers (SFRPs) have been widely used as dry sliding materials, with the attractive advantages of self-lubrication and superior cleanliness [1]. The current trends in development of SFRPs are to tailor their properties for more extreme loading and environmental conditions and to explore new fields of application for these materials. In particular, with the booming of nanophased materials, nano-sized inorganic particles have come under consideration. For instance, Bahadur et al [2] reported that the addition of 2vol.% nano-CuO could generally enhance the wear resistance of short fiber reinforced polyphenylene sulfide. Chang et al [3] systematically studied the effect of nano-TiO_2 on short fiber reinforced epoxy under different loading conditions. A rolling effect of nanoparticles was proposed to explain the lower friction and wear properties of the nanocomposites. Nevertheless, the improvement mechanisms by using nanoparticles have not been deeply understood, although the pioneer researchers have addressed a number of significant factors affecting the wear behavior of these materials. It was, therefore, the objective of the present work to bring further light into the open questions mentioned above.

EXPERIMENTAL

The materials inverstigated were based on an epoxy (EP) or a polyamide 66 (PA 66) matrix, and short carbon fibers (SCF) and two solid lubricants i.e. graphite and PTFE, served as conventional tribo-fillers. Nano-sized TiO_2 particles were used as the additional filler, at a volume content of 5%. Their average diameter amounted to 300 nm.

The wear tests were conducted on a pin-on-disc apparatus on which the specimen pin rotated against a polished steel disc. All tests in this study were conducted for 20 hours under dry condition at room temperature. During the test, the friction coefficient was recorded. The reduction in specimen's height was measured by a displacement transducer, which could be used to characterize the development of the wear process. An additional monitoring of the temperature rise during testing was carried out by an iron-constantan thermocouple positioned at the edge of the steel disc.

SPECIFIC WEAR RATES

Figure 1 summarizes the wear test results of the polymeric composites in comparison with those of the neat polymers. The tribo-fillers enhanced the wear resistance of the polymers by about one order of magnitude at 1 MPa·m/s. It can also be noticed that the wear rates of the SFRP composites are in the range of 10^{-6} mm^3/Nm. With an increase in *pv*, the wear rate of the composites filled

CP1255, *V[th] International Conference on Times of Polymers (TOP) and Composites,*
edited by A. D'Amore, D. Acierno, and L. Grassia

with traditional fillers progressively increased, suggesting changes in the dominant wear mechanisms. For the composite with additional nanoparticles, however, the wear rate of the nanocomposite was relatively stable (within a range between 1 to 3×10^{-6} mm^3/Nm), even under high *pv* conditions.

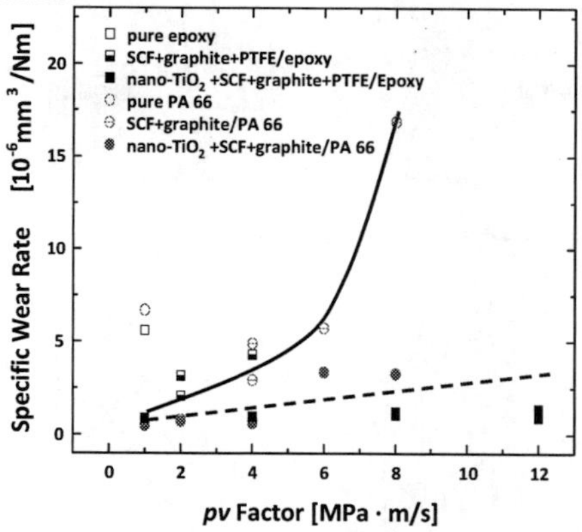

Figure 1: Specific wear rate of various composites. Dashed line: with nanoparticles; full line: without nanoparticles.

NANOPARTICLE RELATED WEAR MECHANISMS

Figure 2 shows an atomic force microscope view of a fiber exposed on the worn surface of an epoxy based SFRP composite filled with nanoparticles. It can be observed that the fiber clearly stands out from the polymeric matrix and is fully exposed to the counterpart. Hence, during the wear process, the short fibers had to carry most of the load.

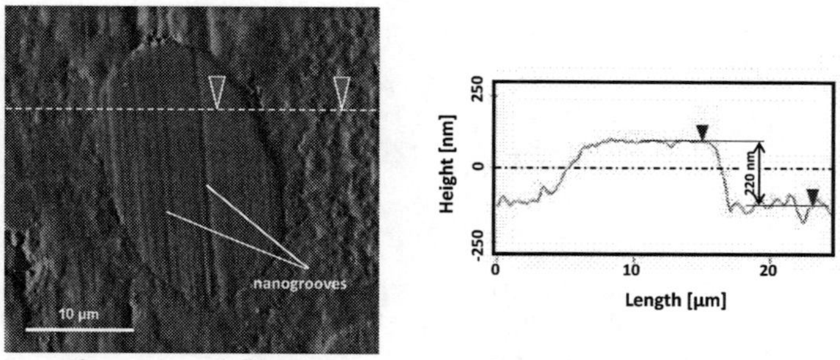

Figure 2: AFM picture of an exposed fiber in the worn surface of nano-TiO_2 +graphite +SCF +PTFE/epoxy (left), and a cross-sectional measurement (right). Loading conditions: normal pressure = 1MPa; sliding velocity = 1m/s.

Figure 3 compares the worn surfaces of two epoxy based composites, without and with nanoparticles. For the composite without nanoparticles, the worn surface is relatively smooth under the low loading condition (Figure 3a), which indicates a gradual removal process of the fibers, normally occuring in the sequence of fiber thinning, fiber fracture, and final removal of broken fiber pieces [1]. However, with an increase in the applied contact pressure, breakage of the epoxy matrix occurs, especially at the interfacial region around the fibers (Figure 3b). As a result, the fibers are removed more easily, because the local support of the matrix is missing. With the addition of nanoparticles, on the other hand, the situation is quite different. As shown in Figures 3c and d, the worn surfaces appear much smoother, even at more severe wear conditions of 4 up to 12 MPa. The fibers fully contributed to the wear resistance of the composites, so that the load carrying capacity of the material was significantly improved. Similar observations were made for the PA- composites.

On the basis of the microscopic observations, Figure 4a gives a schematic illustration of the failure mechanisms of SFRP composites without nanoparticles. Although a continuous transfer film was formed during the running-in stage, the fibers are exposed to most of the normal load. Moreover, due to the frictional heating induced by the fibers, the temperature of the matrix around the fibers is relatively high. As a result, failure of the material occurs firstly in the interfacial region, e.g. brittle fracture occurred in case of the epoxy based composites (cf. Figure 3b), whereas thermal softening/melting was evident for the PA 66 based composites.

For the nanocomposites (Figure 4b), a three-body contact condition was induced by the additional nanoparticles between the contacting surfaces, which is evidenced by the nano-grooves/indents on the fiber surfaces. In this way, the distance between the steel and the composite material was also enhanced i.e. the particle acted as ‘spacers’. This, in turn, can cause a reduction in the adhesion between the contacting surfaces. Moreover, as the nanoparticles were free to move (partially by a rolling motion), they tend to be dispersed uniformly over the transfer film during the wear process, which would result in a more uniform contact stress between the partners, and thus minimize the stress concentration on the individual fibers.

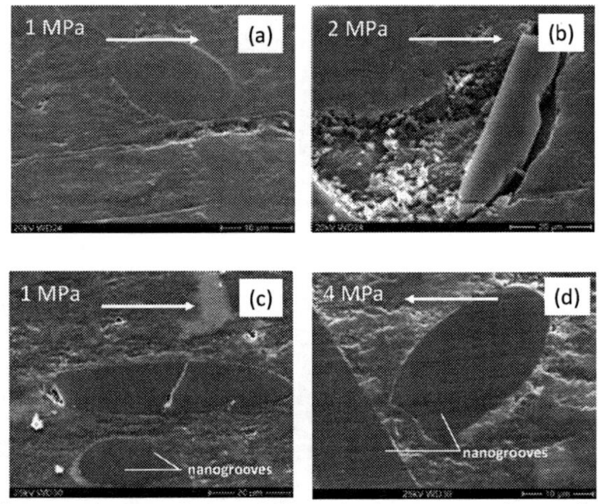

Figure 3: Comparisons of the damage characteristics of fibers in the worn surfaces of epoxy-based composites without and with additional nanoparticles: (a), (b) are SEM images of graphite +SCF +PTFE/epoxy tested under 1MPa and 2MPa, respectively; (c), (d) are SEM images of nano-TiO_2 +graphite +SCF +PTFE/epoxy tested under 1MPa and 4MPa, respectively. The sliding velocity was remained constant at 1m/s. The white arrow indicates the sliding direction of the counterpart.

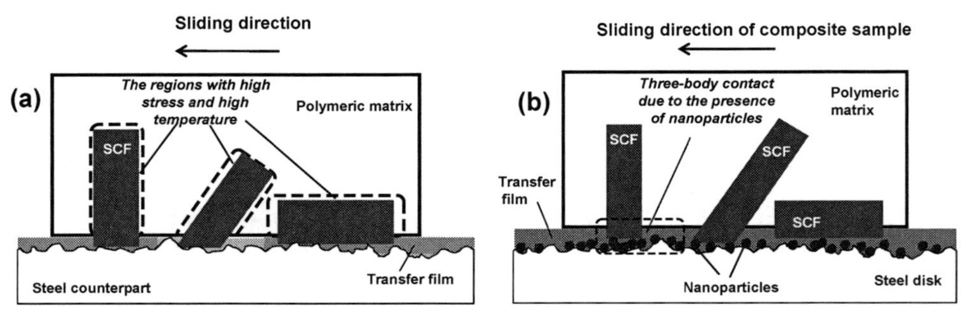

Figure 4: Schematic illustration of the failure mechanism during sliding wear of SFRP composites without (a) and with (b) nanoparticles (for better illustration purposes: fiber diameter too small relative to roughness of steel counterpart and particle size).

CONCLUSIONS

The tribological performance of SFRPs can be significantly improved by using nanoparticles due to their friction reducing abilities, especially under extreme loading conditions. In detail, the following effects were observed:

1. The adhesion between the contact surfaces was reduced by the presence of nanoparticles.
2. The stress concentration on the individual fibers was minimized with the dispersed nanopaticles in the contact region, which consequently protected the polymer matrix in the interfacial regions from the thermal-mechanical failure.
3. The rolling ability of nanoparticles could significantly restrict the increase of the friction force, especially under extreme loading conditions.

ACKNOWLEDGMENTS

This project was partly supported by the Australian Research Council (DP0877750). In addition, Dr.-Ing. L. Chang greatly appreciates the support of the Alexander von Humboldt - Foundation for his research fellowship at IVW in 2009.

REFERENCES

1. K. Friedrich (Ed.), Friction and Wear of Polymer Composites, Elsevier Science Publishers, B.V. (1986).
2. M.H.Cho, S. Bahadur, Wear 258 (2005) 835-845.
3. L.Chang, Z. Zhang, C. Breidt, K. Friedrich, Wear 258 (2005) 141-148.

Water Fluxes in Polymeric Membranes for Desalination via Membrane Distillation

G. A. Mannella, V. Brucato and V. La Carrubba

D.I.C.P.M., University of Palermo, Viale delle Scienze, Ed. 6, 90128, Palermo

Abstract. Membrane distillation is an emerging technique for seawater desalination. Hydrophobic polymeric membranes are used to separate the solute-free water vapour from the hot solution. Vapour fluxes of commercial polymeric membranes were measured in various conditions, i.e. natural and forced convection and vacuum. Vapour fluxes were also predicted with models and compared with experimentals. Higher fluxes were recorded in vacuum conditions.

Keywords: Polymeric membranes, hydrophobic membranes
PACS: 47.56.+r

INTRODUCTION

Membrane distillation (MD) is an emerging technology for seawater desalination at moderate temperatures and pressures [1]. A consistent energy recovery can be achieved, and solar energy can be used to generate the heat needed for the process [2].

MD consists in contacting a hot solution with a hydrophobic membrane, which retains the liquid and allows the diffusion of water vapour through the pores. The driving force for mass transfer is the difference in vapour pressure between the feed side and the permeate side [3].

Polymeric membranes used in MD are mainly made of PP, PTFE or PVDF [4]. Pore dimension of membranes for MD applications ranges from 10^{-1} μm to 1 μm.

EXPERIMENTAL

To measure the water vapour permeability of various commercial membranes, a batch MD apparatus was designed-on purpose. It consisted in a cylindrical vessel ($V \simeq 1\ l$) with an electrical resistance ($P = 1200\ W$) inserted. One end is open, to allow the membrane positioning and sealing via o-rings.

A capillary tube was inserted on the lateral surface: the water level variation in the tube was recorded and related to the vapour flux across the membrane. The temperature was controlled with a PID controller, and experiments were carried out in the range 20-80 °C of temperature of water. The thermocouple was located in the liquid phase near the membrane surface. Distilled water was used to avoid the salt concentration vapour pressure depletion.

When the membrane is exposed to the atmosphere, two configurations can be reproduced: air gap membrane distillation (AGMD), where the vapour is removed by natural convection, and sweeping gas membrane distillation (SGMD), inducing forced convection with a fan. When the membrane is closed into a chamber, using a vacuum pump can be reproduced the vacuum membrane distillation (VMD).

Three membranes of different pore sizes and materials were investigated. Table 1 reports materials and nominal pore sizes.

TABLE 1. Membrane properties

Name	Material	Support	Nominal pore size [μm]
Gore	PTFE	Spunbonded PP	0.45, 0.2
GVS	PVDF	Non-woven polyester	0.2

CP1255, *V*[th] *International Conference on Times of Polymers (TOP) and Composites,*
edited by A. D'Amore, D. Acierno, and L. Grassia

MODELLING

Vapour flux for the MD process is [5]:

$$N = k_P \left(P_f^0 - P_p^0\right) \tag{1}$$

where k_P is the global mass transfer coefficient, P_f^0 and P_p^0 are respectively the vapour pressure at the feed and permeate side.

Mass transfer coefficients were estimated for both natural and forced convection conditions and for two possible mass transfer mechanisms through the membrane: Knudsen diffusion and coupled Knudsen-molecular diffusion (the most reliable for forced convection [6]). As regards to feed side, the mass transport resistance was assumed negligible.

The expressions of mass transfer coefficient for Knudsen and molecular diffusion are [5]:

$$k_P^k = \frac{2\varepsilon r}{3\tau}\sqrt{\frac{8RT}{\pi M}}\frac{M}{RT}\frac{1}{\delta} \tag{2}$$

$$k_P^d = \frac{\varepsilon}{\tau}\frac{P}{P_a}D_{ij}\frac{M}{RT}\frac{1}{\delta} \tag{3}$$

where ε is the porosity, τ the tortuosity, δ the membrane thickness, r the pore radius, D_{ij} the water vapour diffusivity in air, P the pressure, P_a the air pressure, M the molecular weight.

In the case of natural and forced convection, empirical correlation for Nusselt number estimation were used [7]:

$$\mathrm{Nu} = 0.523(\mathrm{Pr}\cdot\mathrm{Gr})^{0.25} \quad \text{(natural convection)} \tag{4}$$

$$\mathrm{Nu} = 0.595\cdot\mathrm{Re}^{0.5} \quad \text{(forced convection)} \tag{5}$$

Assuming valid the Chilton-Colburn analogy [8]:

$$\frac{\mathrm{Nu}}{\mathrm{RePr}^{\frac{1}{3}}} = \frac{\mathrm{Sh}}{\mathrm{ReSc}^{\frac{1}{3}}} \tag{6}$$

values of k_P^{nc} and k_P^{fc} can be evaluated.

RESULTS AND DISCUSSION

The comparison between vapour flux predictions for natural and forced convection cases and data (figure 1), shows that the main resistance to mass transfer is related to permeate side conditions. In other words, membranes should allow fluxes of 1-2 order of magnitude higher than experimental one, as predicted by Knudsen or Knudsen-Molecular diffusion coefficients. Therefore, process effectiveness can be only improved by enhancing mass transport in the permeate side.

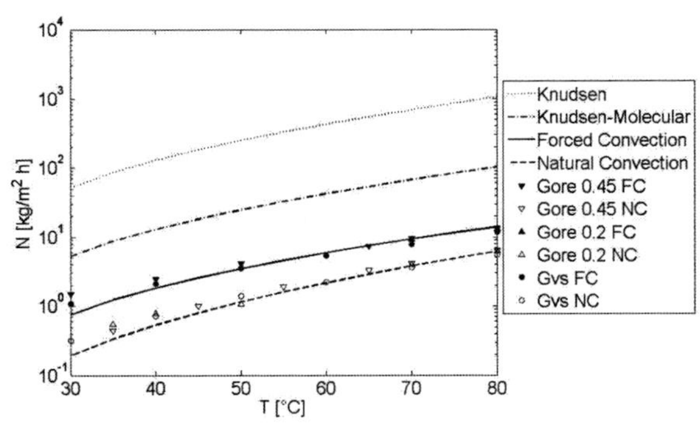

FIGURE 1. Experimental vapour fluxes (dots) compared with predictions (lines)

Vacuum conditions in permeate side induced a vapour flux proportional to (partial) pressure difference between feed and permeate side (figure 2). All fluxes tend to zero at a permeate pressure near to water vapour pressure at experimental temperatures. These results strictly agree to equation 1 predictions.

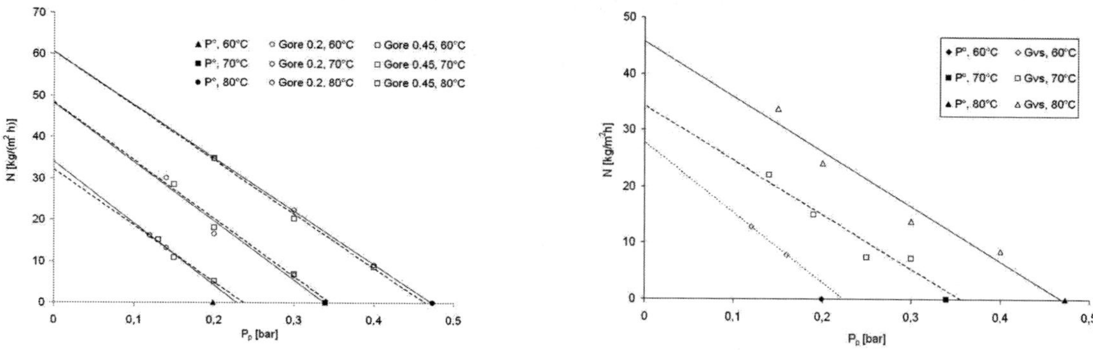

FIGURE 2. Experimental VMD fluxes vs permeate chamber pressure. Filled points at $N = 0$ are water vapour pressures at experimental temperatures

Gore membranes, although are different in nominal pore size, showed similar fluxes under the same conditions. This result can be related to various factors, as porosity, or inferring that the membranes have a distribution of pore size instead of a single value. Gvs membrane showed fluxes lower than Gore membranes, as can be easily noticed by the intercept at $P_p = 0$. This difference can be related to the different nature of the membrane, which shows a different morphology, that presumably affects the vapour diffusion.

CONCLUSIONS

The NC conditions showed a higher mass transfer resistance, which make the MD process independent on the membrane used. For FC conditions, membrane properties as materials and morphology affect the vapour flux, whereas the pore size does not influence the mass transfer through the membrane. The highest vapour flux (36 kg/m^2h) was reached in vacuum conditions.

Prediction on mass transfer coefficients showed the main resistance is ascribable to permeate-side conditions, thus addressing future works on MD desalination unit to be focused on permeate conditions, to enhance mass transfer.

ACKNOWLEDGMENTS

This work has been carried out with the financial support of Lympha project by Regione Siciliana, Assessorato alla Presidenza, Ufficio Speciale per la Cooperazione Decentrata allo Sviluppo ed alla Solidarietà Internazionale. Authors acknowledge the precious help of Salvatore Montesanto, Giuseppe Scaglione and Francesco Guttilla for their contribution in carrying out experiments and assembling the batch MD system.

REFERENCES

1. N. N. Li, A. G. Fane, W. S. W. Ho, and T. Matsuura, *Advanced membrane technology and applications*, John Wiley & Sons, Inc., Hoboken, New Jersey, 2008.
2. S. Al-Obaidani, E. Curcio, F. Macedonio, G. D. Profio, H. Al-Hinai, and E. Drioli, *Journal of Membrane Science* **323**, 85–98 (2008).
3. R. W. Schofield, A. G. Fane, and C. J. D. Fell, *Journal of Membrane Science* **33**, 299–313 (1987).
4. E. Curcio, and E. Drioli, *Separation and Purification Reviews* **34**, 35–86 (2005).
5. K. W. Lawson, and D. R. Lloyd, *Journal of Membrane Science* **124**, 1–25 (1997).
6. M. Khayet, P. Godino, and J. I. Mengual, *Journal of Membrane Science* **170**, 243–255 (2000).
7. D. M. Gates, *Biophysical ecology*, Springer-Verlag New York, 1980.
8. C. J. Geankoplis, *Transport Processes and Separation Process Principles*, Prentice Hall, Upper Saddle River, NJ, 2003.

Using polarized Neutrons for elastic and dynamic studies on protein systems

W. Häussler[a,b*] and J. Kindervater[a]

[a]Forschungsneutronenquelle Heinz Maier-Leibnitz (FRM II), Technische Universität München, D-85747 Garching, Germany
[b]Physik-Department E21, Technische Universität München, D-85747 Garching, Germany

Abstract. We report about measurements at the new Neutron Resonance Spin Echo (NRSE) instrument RESEDA (FRM II, Munich). We have performed polarization analysis, providing information about the fraction of neutrons scattered without and with spin flip at the sample. From comparison with the resolution function measured on a standard sample, the dynamics of the incoherent fraction is detected by NRSE measurements at low temperatures.

Keywords: neutron spin echo; resonance spin echo; proteins.
PACS: 87.14.E-;

1. INTRODUCTION

Neutron Spin Echo[1,2] (NSE) is a high-resolution inelastic neutron scattering technique, that measures the difference between the initial neutron energy and the energy of the neutron after having been scattered from the sample. The NSE technique relies on Larmor precession of the neutron spin in magnetic fields. In NSE, long-distance field coils traversed by the neutron beam are used as spin precession devices. Neutron Resonance Spin Echo (NRSE)[3,4], is a variant of NSE. In contrast to other methods like triple axis spectroscopy, the energy transfer between sample and neutron is obtained in the Fourier time regime by means of NSE. The reachable Fourier time does not dependent to first order on the wavelength spread of the incoming neutron beam.

As NSE relies on polarized neutrons, changes of the polarization due to sample scattering are detected. This fact can be used to combine NSE with an elastic polarization analysis measurement. For this, the polarization of the neutrons scattered from the sample system is compared with the polarization of a standard sample, that does not lead to polarization loss due to the scattering process. In this paper, we report on recent polarization analysis and NSE measurements performed at RESEDA[5] at the FRM II, Munich being commissioned recently. In section 2, we remind some useful characteristics of NSE and of polarization analysis. We present the results of first measurements on a protein system in section 3, and finish with the conclusions in 4.

2. NEUTRON SPIN ECHO

The Neutron Spin Echo Signal

Spin echo spectrometers consist of two spectrometer arms, one before and one after the sample, where precession of the neutron spin ($s = \hbar/2$) in magnetic fields takes place. The spin phase accumulated in each spectrometer arm is proportional to the time-of-flight needed by a neutron for passing through the arm. If the neutron exchanges energy in a sample scattering process in-between the two spectrometer arms, its time-of-flight by traversing the second spectrometer arm is different compared to the first arm. The spin echo technique allows to detect such sample scattering with energy transfer ("inelastic scattering") by comparing the spin phase accumulated in the two spectrometer arms. The main goal in spin echo experiments is to measure this spin phase difference and the energy difference very exactly. The measure for the spin echo resolution is given by the spin echo time

CP1255, *V[th] International Conference on Times of Polymers (TOP) and Composites,*
edited by A. D'Amore, D. Acierno, and L. Grassia

$\tau = \gamma/2\pi \cdot (m^2/h^2) \cdot \lambda^3 \cdot J_B$, (1)

where γ is the neutron gyromagnetic ratio, m the neutron mass, λ the neutron wavelength and J_B the magnetic field integral. τ is defined as the proportionality constant between the energy transfer $\omega = \Delta E/\hbar$ and the resulting spin phase difference $\phi = \tau \cdot \omega$.

The measured polarization p=(up-down)/(up+down) averaged over the neutron beam, contains the desired information about the normalized scattering function $S(Q,\omega)$ through:

$$p(\tau) = \frac{S(q,\tau)}{S(q,0)} = \int S(q,\omega) \cdot \cos(\omega\tau)\, d\omega, \qquad (2)$$

where Q is the scattering vector.

Polarization Analysis

Using a polarized incident neutron beam and counting separately the neutrons scattered with and without spin-flip with regards to the incident beam polarization, coherent and the incoherent nuclear scattering processes from non-polarized materials can be detected. Coherent neutron scattering events do not involve a flip of the neutron spin, whereas only 1/3 of the incoherent neutron scattering events of a non-polarized sample are without spin-flip, the other 2/3 being with spin flip[6-8]. The sample signals, $I_{NSFm}(Q)$ and $I_{SFm}(Q)$, were obtained by correcting the measured intensities for the polarization efficiency measured by a standard coherent elastic sample. From the latter, the instrument flipping ratio R[9] was calculated, and the sample signals were derived as follows:

$$\begin{cases} I_{NSFc}(Q) = I_{NSFm}(Q) + \frac{1}{1-R}\left(I_{NSFm}(Q) - I_{SFm}(Q)\right) & \text{(3a)} \\ I_{SFc}(Q) = I_{SFm}(Q) + \frac{1}{1-R}\left(I_{SFm}(Q) - I_{NSFm}(Q)\right) & \text{(3b)} \end{cases}$$

and the coherent and incoherent terms

$$\begin{cases} I_{coh}(Q) = I_{NSFc}(Q) - \frac{1}{2} I_{SFc}(Q) & \text{(4a)} \\ I_{inc}(Q) = \frac{3}{2} I_{SFc}(Q) & \text{(4b)} \end{cases}$$

were obtained.

The Spectrometer RESEDA

Fig. 1. The instrument RESEDA with the primary (foreground) and the two secondary spectrometer arms, separated by the sample region (for detailed description see text). The black detector shieldings contain the ^{3}He counters.

The Resonance Spin Echo spectrometer RESEDA (Fig. 1) located at the end position of the neutron guide NL5 at the Forschungsneutronenquelle Heinz Maier-Leibnitz (FRM II, Munich) consists of two secondary spectrometer arms, in addition to the primary arm. Two NRSE coils are located at the beginning and end of this flight region, and in between, the NSE coil is found. The secondary arms after the sample region consist of the same components. At the end of this spectrometer arms, the helium detectors are fixed within a shielding reducing the radiation background present in the neutron guide hall. The wavelength band is defined by a velocity selector mounted upstream of the instrument within a gap in the neutron guide. At the end of the neutron guide, a polarizing cavity is fixed (length 2 m). The spin manipulation devices used as described in the previous section consist of the NSE coils.

3. EXPERIMENTAL RESULTS

The sample system under study consists of a powder of the protein Ferritin[10]. The polarization analysis results are shown in Fig. 2. The fraction of coherent and incoherent neutrons varies only very weakly in the Q-range under consideration. The fraction of incoherently scattered is larger than the coherent fraction, as the proteins consist to a big extent of incoherently scattering protons.

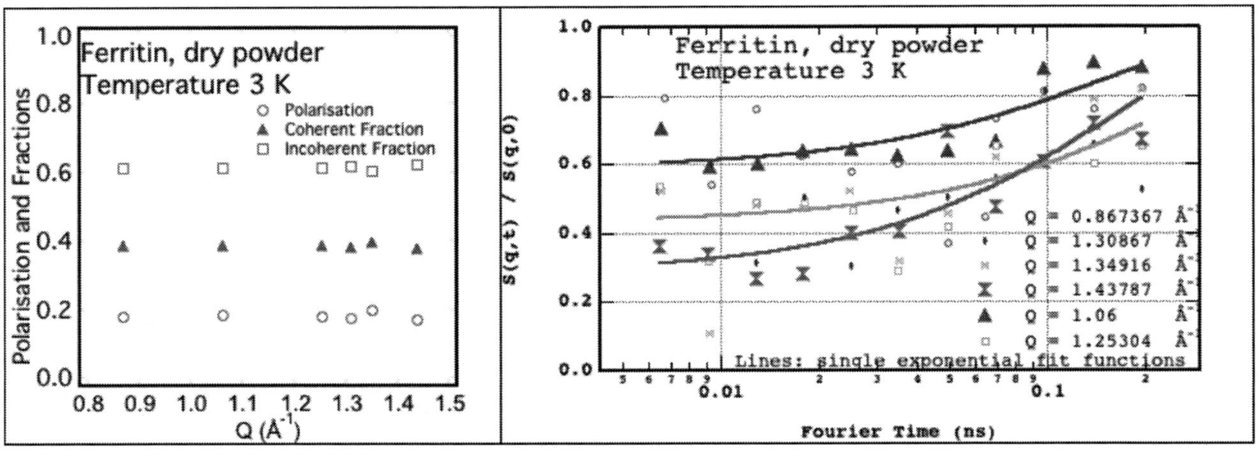

Fig. 2. The polarization analysis (left) and NSE (right) results (for detailed description see text).

NSE data have been acquired by measuring the polarization while varying the magnetic field integral and, thereby, the resolution of RESEDA. The raw polarization data have been corrected for the instrumental resolution measured on a elastic coherent standard sample.

The NSE data show clear evidence of dynamics, even at the low temperature of 3 K. Together with the elastic data, these dynamics can be ascribed to mobile hydrogen in the protein shell. As the self correlations between the hydrogen decay, the impact of the neutrons scattered from hydrogen on the value of the measured $S(Q,t)$ increases. Thus, the impact of the incoherent fraction on the polarization decreases. As the coherent part, leading to positive contribution in the measured polarization, can be assumed to be constant, the total measured signal $S(Q,t)$ increases with increasing Fourier time, the latter being a measure of the resolution. The amplitude and the relaxation rate of the dynamics increase with increasing scattering vector Q being a measure for the spatial resolution.

4. CONCLUSIONS

We have performed elastic neutron spin polarization analysis and achieved information about the fraction of neutrons scattered without and with spin flip from a Ferritin sample. At the same time, the dynamics of the incoherent fraction steaming from mobile hydrogen is detected by NRSE measurements at low temperatures. Such combined polarization analysis and NSE measurements can perfectly be measured by the new instrument RESEDA.

REFERENCES

1. F. Mezei, Z. Phys. **255** (1972) 146.
2. F. Mezei (ed.), *Neutron Spin Echo* (Springer, Heidelberg, 1980).
3. R. Gähler, R. Golub, Z. Phys. B **65** (1987) 269. R. Golub, R. Gähler, Phys. Lett. A **123** (1987) 43.
4. D. Dubbers, P. El-Muzeini, M. Kessler, J. Last, Nucl. Instr. and Meth. A **275** (1989) 294.
5. W. Häussler, D. Streibl, P. Böni, Meas. Sci. Technol. **19** (2008) 034015.
6. R.M. Moon, T. Riste, W.C. Köhler, Phys. Rev. **181** (1969) 920-931.
7. O. Schärpf, Physica B **182** (1982) 376-388.
8. B.J. Gabrys, Physica B **267-268** (1999) 122-130.
9. O. Schärpf, Experience with spin analysis on a time-of-flight multidetector scattering instrument. In Neutron scattering in the 'nineties', Proceedings of the IAEA Conference, Jülich (1985) 85-97.
10. W. Häußler, Eur Biophys J **37** (2008) 563–571.

Recycling of Chrome Tanned Leather Dust in Acrylonitrile Butadiene Rubber

Salwa H. El-Sabbagh[a] and Ola A. Mohamed[b]*

[a]Polymer and Pigments Department, National Research center, Cairo, Egypt.

[b] Chemistry of Tanning materials and Leather Technology, National Research center, Cairo, Egypt.

Abstract. Concerns on environmental waste problem caused by chrome tanned leather wastes in huge amount have caused an increasing interest in developing this wastes in many composite formation. This leather dust was used as filler in acrylonitrile butadiene rubber (NBR) before treatment and after treatment with ammonia solution and sod. formate. Different formulations of NBR/ leather dust (untreated-treated with ammonia solution – treated with sod. formate) composites are prepared.

The formed composite exhibit a considerable improvement in some of their properties such as rheometric characteristics especially with composites loaded with treated leather dust. Tensile strength, modulus at 100 % elongation, hardness and youngs modulus were improved then by further loading start to be steady or decrease.Cross linking density in toluene were increased by incorporation of leather dust treated or untreated resulting in decreases in equilibrium swelling. Distinct increase in the ageing coefficient of both treated and untreated leather with drop in NBR vulcanizates without leather dust. Addition of leather dust treated or untreated exhibit better thermal stability

Keywords: rubber, leather dust, composite, chrome tanned, thermal stability.
PACS: 60

INTRODUCTION

Leather is a natural polymer consisting of collagen fibers cross linked in a three dimensional structure, leather tanning results in improved appearance, physical properties and chemical and biological residue of leather but the process leads to emission of solid and liquid wastes into the environment.(1)

The idea of using leather dust as reinforcement in composite materials is not a new or recent one.

On the other hand, the disposal of municipal solid waste has become an environmental issue of growing concern (2). Leather dust waste management is a global environmental problem. One method, to get ride leather waste is to use it as a filler in rubber. When waste leather fibers are added to elastomer, the former could function as short fiber reinforcing the matrix provided the inherent fibrous nature of the former is retained during processing.

The addition of the leather wastes caused that density and hardness of the composites increased, whereas their tensile properties were considerably deteriorated with the incorporation of leather. In order to improve mechanical properties, the granulated leather was modified with vinyl acetate to form an adhesive layer on the granules surface, which considerably improved the tensile strength of composite by 30%. Leather dust with silica was used as a filler of PVC (7).

This additive decreased the density of the composite by 30-40 %, but increased its melting point and maximal elongation.

The aim of the present work is to evaluate the effect of leather dust waste on the mechanical ,physical and thermal properties of acrylonitrile butadiene rubber (NBR), before and after surface modification of this waste by ammonia and sodium formate which are used to enhance the surface interaction between the leather dust waste and rubber.

CP1255, *Vth International Conference on Times of Polymers (TOP) and Composites,*
edited by A. D'Amore, D. Acierno, and L. Grassia

RESULTS AND DISCUSSION

Filler play an important role in enhancing favorable physical properties of rubber. They can also modify processing characteristics and reduce the cost of compounds.

The influence of untreated and treated leather on mechanical properties of NBR mixes.

Figure 1-2 shows the mechanical properties of composites as a function of leather waste loading before and after treatments of the leather. The addition of untreated leather waste to NBR decrease the tensile strength and elongation at break (Fig. 1-2) until reaching the loading 6 phr. Further loading lead to the value of tensile strength approximately constant. While the addition of leather waste treated with ammonia to NBR improves the tensile strength in the first and then decrease and became approximately the same level for the other ratio. Also by adding leather waste treated with sodium formate to NBR, the tensile properties increase with increasing leather content and then decreases. The decrease in tensile strength and elongation at break can be attributed to the interference of leather wastes as a filler in NBR causes reduction in the mobility or deformation of the rubber matrix. This interference created through the physical interaction, there by immobility of the rubber matrix by imposing the mechanical restraints so the composite tends to form a weak structure. On the hand, the gradual increase of tensile strength could be attributed to the improvement in the interfacial bonding between leather waste treated ammonia or sodium formate and rubber matrix. These properties were confirmed by the modulus at 100 % elongation, hardness and young's modulus .

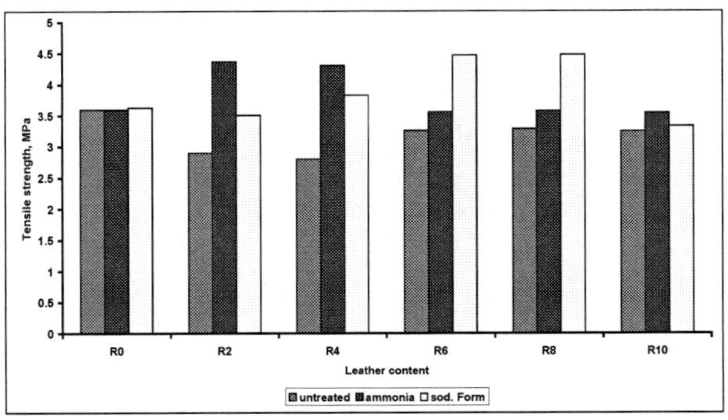

Figure (1): The tensile strength of NBR loaded with untreated and treated leather

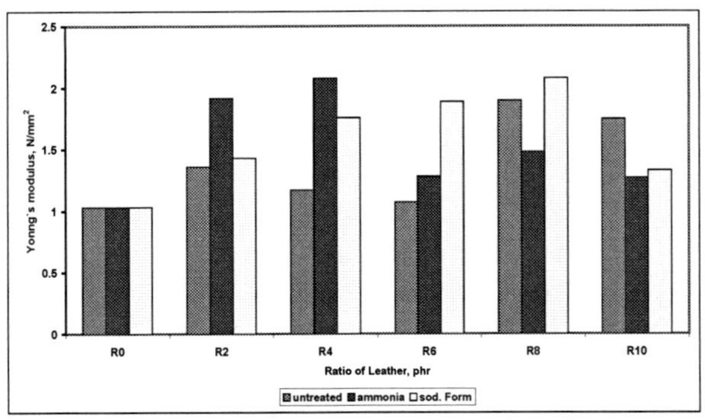

Figure (2): The modulus at 100% elongation of NBR treated leather and untreated leather

The Influence of Leather Waste on Equilibrium Swelling and Crosslinking Density of the NBR Vulcanizates

This is confirmed by the value of crosslink density calculated from the swelling in toluene by using Flory-Rehner equation given earlier. On the other hand, the value of equilibrium swelling of composite containing untreated leather is found to be relatively higher than the corresponding compositions containing treated leather.

Table 1: Swelling characteristics for NBR vulcanizates in Toluene with untreated and treated leather.

Rating of leather phr	Equilibrium swelling Q_m, %	Soluble of reaction %	Q_f/Q_g	Molecular weight between crosslinking M_c (g/mole)	$\upsilon \times 10^4$, mole/cc (crosslinking density)
Untreated leather					
0	189	1.24	----	2063	2.42

2	183	0.92	0.97	1946	2.57
4	176	0.78	0.934	1817	2.75
6	165	0.54	0.87	1624	3.08
8	156	0.48	0.83	1474	3.39
10	147	0.66	0.79	1332	3.75
Treated leather with ammonia					
2	133	3.07	0.70	1125	4.44
4	127	2.41	0.67	1042	4.799
6	120.3	2.18	0.64	949	5.27
8	120	1.66	0.63	892	5.60
10	119.9	1.41	0.634	948	5.27
Treated leather with sodium formate					
2	154	2.67	0.81	1439	3.47
4	128	2.2	0.679	1061	4.71
6	126.5	1.91	0.669	1041	4.83
8	118	1.77	0.624	922	5.42
10	116	2.04	0.613	959	5.21

CONCLUSION

The above research can be summarized as follows: chrome tanned leather was disintegrated into dust and incorporated into acrylonitrile butadiene rubber (NBR) before and after neutralization with ammonia solution and sodium formate with different concentration. The resultant NBR /leather composites show the following characteristics:

1- Improvement in rheological properties and reduce the cost of compounds.
2- Addition of leather dust treated with ammonia and sodium formate improve the physic mechanical properties, tensile strength in the first then decrease and become approximately stable, modulus at 100 % elongation, hardness and youngs modulus have the same trend.
3- Increase the cross link density of volcanizates for both treated or untreated leading to reduce in equilibrium swelling.
4- The composites system show enhancement in the ageing coefficients which indicate that resistance of both untreated or treated leather to thermal ageing.

REFERENCES

1. Ravichandran, K., Natchmuthn, N., *Polimeros* ,2005, **15**, 12-29.
2. Wark, D.T., May 1991, Blends and Alloys to 200 – Structural Performance and Feed stock Implications, ECM International Conference, Advances in High Performance Polymer Blend and Alloys.
3. Panov, A. K., Minskar, K. S., IL'ina, T. F., Panov, A.A., , *Int. Polym. Sci. Technol.,* 2001 , **28**, 84-85.

Luminescent Polymer Materials on the Basis of Aromatic Isocyanates and Open-chain Analogues of Crown Eithers

A.M Gumerov, I.M. Davletbaeva, A.I. Ahmetshina

Kazan State Technological University, 420015, K.Marks Str., 68, gumerov_a@mail.ru

Abstract. One of the promising trends in optics is the search of polymers possessing luminescent and electroluminescent properties[1-3]. Research in this field is both of fundamental and practical value as it gives the possibility to produce active environments of reconstructed lasers, optical sensor controls, limiters of optical radiation, various screens and information indicators on the flexible polymeric basis. Luminescent materials along with luminescent activity should possess sufficient stability. The important point is that phosphor groups should be a component of the polymeric matrix.

INTRODACTION

In our works it was determined [4] that polyisocyanate groups of acetal nature formed as the result of the aromatic isocyanates opening which was initiated by alcoholate endgroups of open-chain analogues of crown ethers.

In the current work the influence of reaction conditions of synthesis on the phosphor properties of polyisocyanates was investigated. The interaction of polyisocyanate groups of acetal nature with specially introduced compounds having their own system of delocalized electron was studied. It was expected that interchain hydrogen links would appear in the system and it would be possible that the quasiconjugated system of intermolecular links would appear parallel with the complex formation.

RESULTS AND DISCUSSION

Data on the polymers containing polyisocyanate links of acetal nature in their composition are not numerous. It is known, however, that in unstabilized condition they are far from being steady. Therefore the problem is not only their stabilisation, but also obtaining more or less lengthy chain consisting of these links. In our works it was shown, that by regulating such reaction conditions as temperature, the nature of solvent and the molar proportion of initial reagents it is possible to achieve the increase of the constant of reaction speed leading to the primary formation of polyisocyanate links of acetal nature. In connection with it the influence of these conditions on the nature of luminescence spectra of polyisocyanate films was studied.

Photoluminescence spectra for the polymers obtained at various molar proportions of initial reagents – 2,4-tolyilendiisocyanat (TDI) and polyoxyethylene glycolate (PEG).

. It appeared that maxima of spectral dependences of luminescence correspond at least to two areas (FIGURE 1). When surplus of tolyilendiisocyanat is not large the stripe is observed at 425 nanometers and 550 nanometers. When the molar surplus of tolyilendiisocyanat is increased further the displacement of not only long-wave edge of absorption of the stripe of 425 nanometers in the area of smaller energies, but also of the maximum up to 450 nanometers is observed. The stripe of 550 nanometers does not shift to the large lengths of waves, but as a result of overlapping it by the stripe of 450 nanometers it is revealed in the form of a shoulder. Such changes of the spectrum can be the consequence of the increase of degree of π-conjugation of the structural units which are a component of polymer. The consideration of the chemical structure of polyisocyanate links of acetal nature permits to explain the strengthening of π -conjugation by increasing their number in a chain.

CP1255, *Vth International Conference on Times of Polymers (TOP) and Composites,*
edited by A. D'Amore, D. Acierno, and L. Grassia

The further research was directed at modifying polymeric matrix by the compounds having their own system of delocalized electron. For this case phenanthroline (FT) was used. While synthesizing modificated polymer the small surplus of tolyilendiisocyanat (2,5 и 3,5) was taken. It appeared that phenanthroline influences considerably the change of the spectrum form. The stripe of 550 nanometers intensifies, and the shoulder in the area of 500 nanometers starts appearing in the form of an individual stripe (FIGURE 2).

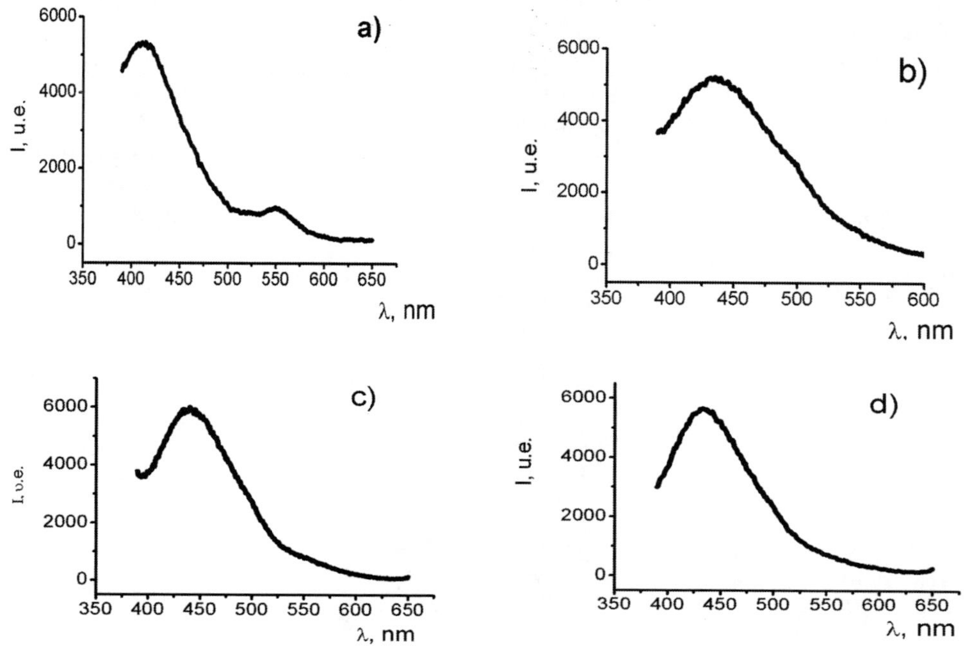

FIGURE 1. Luminescence Spectrum of Polymer Films, obtained at the interacting in the system : [TDI] : [PEG]= a) 2; b) 4; c) 5; d) 6

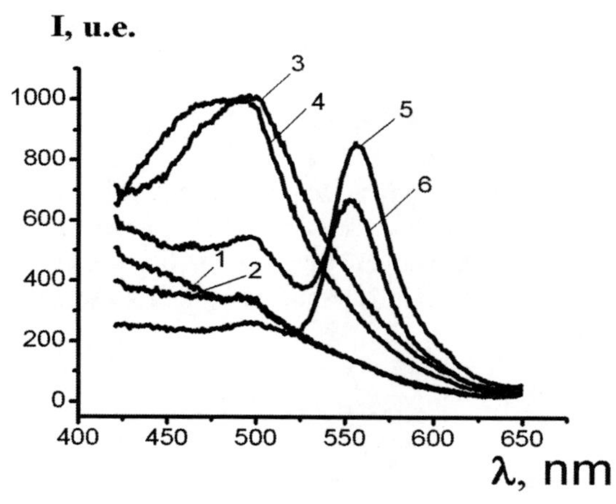

FIGURE 2. Luminescence Spectrum of Polymer Films, obtained at the interacting in the system : [PEG] : [TDI] : [FT]=
1-1:2,5:0,01; **2**-1:2,5:0,05; **3**-1:2,5:0,1; **4**-1:2,5:0,3; **5**-1:2,5:0,5; **6**-1:2,5:1

As opposed to the area of 420 to 450 nanometers the stripes of 550 and 500 nanometers for base polymer do not vary at increasing of surplus of tolyilendiisocyanat. The last fact permits to connect the revealing of invariable stripes of luminescent glowing with the response of end polyisocyanate groups. As phenanthroline does not reveal its own luminescent activity, strengthening of intensity of these stripes for polyisocyanates modified by it can be the consequence of the formation of the complex of quasiconjugated system of intermolecular links(FIGURE 3).

By using kinetic methods of the analysis it was established that introduction of lanthanoid ions into reaction system in the process of polymerisation leads to strong linkage of the centre of anion polymerisation and even to the stop of the process of polymerization.

Introducing lanthanoid ions upon termination of the process was appeared to be the means of stabilisation of the active centres and the formation of additional knots of a spatial polymeric network.

The lanthanoid luminescence in polymeric complexes is usually connected with the electron excitation by transferring them the energy from the triplet level of ligands and the subsequent transition of electrons onto the lower permissible conditions. In this case ligand is the channel transferring excitation energy to lanthanoid. The availability of intrinsic lines of lanthanoid radiation depends on the efficiency of the transmission of energy from ligand to lanthanoid. The latter is defined by the distance between the centre of ligand and lanthanoid, and also by the proportion of energies of the triplet condition and the luminescent condition of lanthanoid.

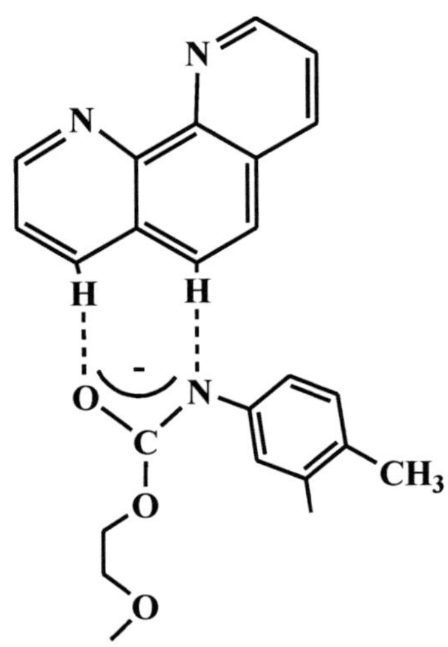

FIGURE 3. Formation of intermolecular links in polymers, obtained at the interacting PEG, TDI and FT

At modifying polyisocyanates by lanthanoid ions the luminescent glowing became apparent when ions Eu (III) were used. On the spectrum the stripe appeared resulted from energy transfer onto Eu (III) ion in the area of 618 nanometers. The spectrum arises at relatively low content of Eu (III) (1-2 %).

CONCLUSIONS

Thus, the formation of intermolecular links of an end polyisocyanate group and a molecule with the system of the conjugated links can lead to the occurrence of supramolecular ensembles. It is known that the change of properties of such intermolecular formations in the basic and activated conditions in comparison with properties of individual molecules as a result of co-operative character of the process can lead to strengthening and even to occurrence of new photophysical properties of polymers.

REFERENCES

1. E. I. Krivchenko, R. V. Matvienko, V. N. Pavlova, et al.,*Plast. Massy*, No. 4, 26 (1988).
2. C. J. Twomey, J. Polym. Sci., Part B: *Polym. Phys*. 29, 859 (1991).
3. C. J. Twomey, J. Polym. Sci., Part B: *Polym. Phys*. 31, 647 (1993).
4. I.M. Davletbaeva, V. F. Shkodich, E. O. Ekimova, A. M. Gumerov, *Polymer Science,* Pleiades Publishing, Ser. A, 2007, Vol. 49, No. 8, (896).

Metalcoordinated Polyurethanes And Their Properties

I.M. Davletbaeva, R.S. Davletbaev

Kazan State Technological University, 420015, K.Marks Str., 68, davletbaeva09@mail.ru

Abstract. The synthesis of metal coordination polymers is a way of affecting the processes of crosslinking of macrochains; interchain and intraionic interactions; and, thereby, preparing polymer materials with special properties [1-6]. From the standpoint of designing materials with electric and magnetic properties, it is promising to form, in a polymer matrix, chains of transition metal ions bound by exchange interaction. Conventional methods for the creation of interactions of this type in a polymer are primarily based on the presence in a macromolecule of certain units, e.g., those including the phthalocyanine and azomethine moieties, or on conformational potentialities of macrochains that contain coordination-active groups in their branches.

INTRODACTION

Metal atoms in metal phthalocyanine LC complexes are bound to one another by chloride bridges and play the role of a spacer between phthalocyanine units, thus promoting overlap of electronic orbitals of parallel molecules. As a result, the electric conductivity of metal-coordination polymers obtained on the basis of these complexes is increased by a few orders of magnitude relative to undoped systems. The coordination bonding of comblike LC polymers can also give rise to stacked structures. Interaction between metal ions was revealed in such polymers, which is realized owing to the association of metal ions in an indirect manner, through ester oxygen bridges. However, this approach is seriously limited and cannot be used for the creation of stacked metal-coordinated fragments in a disordered polymer matrix. With the aim of creating such structures in heterochain disordered polymers (including network polymers), we prepared modifier metal complex compounds with a preformed chain of exchange-interaction-bound transition metal ions that were subsequently fixed in a polymer matrix.

RESULTS AND DISCUSSION

In the early studies devoted to imparting special properties to polyurethanes by their coordination bonding, it was shown that the interaction of aromatic isocyanates with transitive ion metal chlorides in the presence of trace amounts ofwater proceeds as a sequence of chemical transformations including simple addition and decomposition reactions, redox processes, and subsequent complexation.

The ultimate products are polynuclear complexes of azaaromatic compounds, in which metal ions occurring in two variable oxidation states are connected by chloride bridges. As it is seen from the stuctural formulas, some transitive-metal ions are stabilized at the initial degree of oxidation due to the formation of heterovalent pairs connected by halogenid bridges. Free isocyanate groups present in these compounds were reacted with oligoether diols, and polyurethane was crosslinked by coordination bonding of urethane and formed azo groups by metal ions.

It was shown that the chloride bridges are replaced by heteroatoms that are present in polymer chains while metal ions are coordinatively bound to macromolecules to crosslink them and, occurring in two interacting variable oxidation states, form local centers of exchange interactions. As a result of electron transfer from one local coordinated unit to another, which is mediated by electron-donating groups, such an ester group, the conductivity of polyurethane increases by several orders of magnitude.

Further studies established that similar structural units could be formed even in the polyurethane matrix itself as a result of its modification with metal complexes synthesized for this purpose. A characteristic feature of these coordination compounds is the presence of chloride-bridged metal ions in their structure.

CP1255, *Vth International Conference on Times of Polymers (TOP) and Composites,*
edited by A. D'Amore, D. Acierno, and L. Grassia

There was possible an interaction of these coordination connections with aromatic isocyanate, leeding to formation of azaaromatic structures. It was found that, in the progressively sophisticated sequence of metal complex systems made from Fe(III) chloride and monoethanolamine, Cu(II), Mo(V) chloride and N,N`-dietilgidroksilamin the transitive metal ions are connected in a long chain of atoms interacting with one another via hloride bridges.

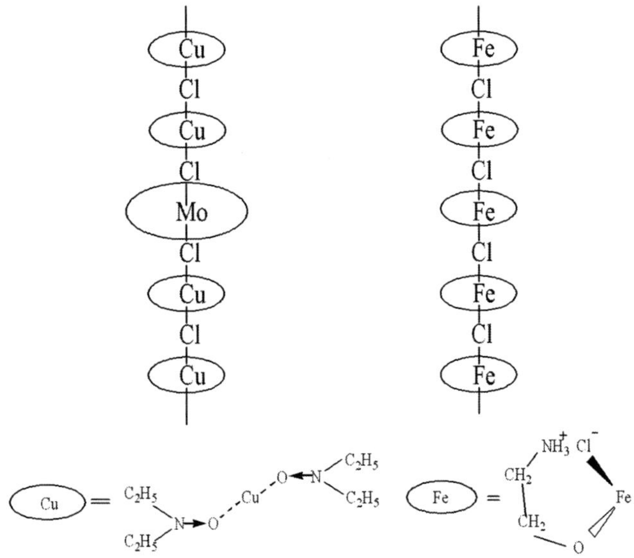

The Mössbauer study confirms the columnar structure of the metal complexes, obtained from Fe(III) chloride and monoethanolamine, the possibility of their fixation in a flexible-chain polymer matrix containing electron-donating groups, and the existence of magnetic ordering at a temperature below 70 K. Sizes of very thin magnetic fields correspond to high-spin state of iron(III) ions (S=5/2). It has been established, that metal-complex modification of polyurethanes results in the change of their physicomechanical properties and spasmodic reduction of volumetric and superficial electric resistance (FIGURE 1).

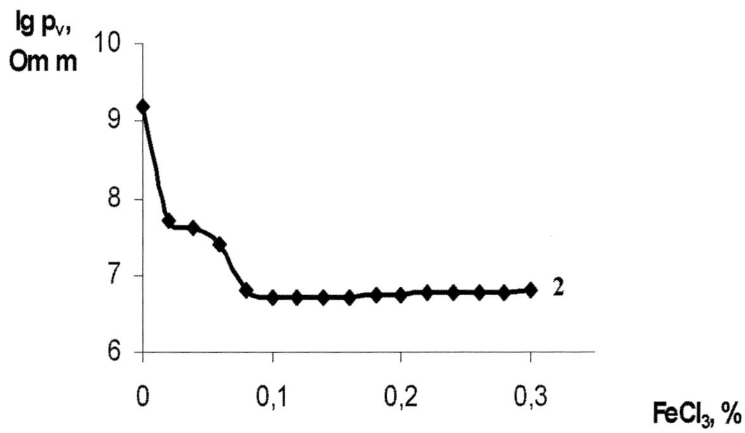

FIGURE 1. Dependence of lg ρ_v metalcoordinated polyurethanes on iron(III) chloride contents

The reduction of the specific bulk electrical resistance is the most significant effect accompanying the metal-complex binding of polyurethanes. In this case the electrical resistance falls spasmodically depending on the nature of flexible chains of polyurethanes, the range of transitive-metal ions concentrations. If to increase further the content of metal-complex modifying agent the electrical resistance increases to some extent. The main role in the mechanism of charge transmission in metal-coordinated polyurethanes was assigned to electronodonar groups which are included in the structure of flexible chains of polyurethane matrix and the presence of transitive-metal ions having two degrees of oxidation in it.

On the base of data of the thermomechanical analysis, IR-spectroscopy and electronmicroscopic research was established that a growth of physical-mechanical and temperature characteristies is consequence of the strongly pronounced microphase division, which is going on in a result of the unification of hard segments (as well as inhomogeneous dimensions) into account of the coordination tying of transitive-metal ions of cruel aromatic segments by urethane connects.

CONCLUSIONS

The structure of a metal-complex system based on Fe(III) chloride and monoethanolamine, Cu(II), Mo(V) chloride and N,N`-dietilgidroksilamin and the products of interaction of this system with urethane prepolymer was studied. It was found that the transitive metal ions are connected in a long chain of atoms interacting with one another via hloride bridges. The columnar structure of metal complexes, and the feasibility of formation of metal complexes ordered in the same manner in a flexible-chain polymer matrix containing electron-donating groups were revealed.

REFERENCES

1. Thuichida E., Nichide H., *Andvanses in Polymer Science*, 1977, 24, pp.2-87.
2. Shirai H., Kobayashi K., Takemae V., Hirabaru O., Hoji N., *Macromol. Chem.*, 1979, v.180, pp.2073-2084
3. Tobita, S. Solid polymers doped with rare earth metal salts, *J. Phys. Chem,* 1984,pp. 697.
4. J.L. Serrano, L. Oriol, *Adv. Mater.,* 1995,v 7, № 4, pp. 348-436.
5. Klink, S. Intensities 4f – 4f transitions in glass materials, *J. Appl. Phys,* 1998, pp.1181.
6. Davletbaeva I.M., Kirpichnikov P.A., Rakhmatullina A.P., *"Macromolecular Symposia Nano-Structures in Polymer Systems",* 1996, v.4, pp. 87-90.
7. Klink, S, Red electroluminescence of a europium complex dispersed in poly, *J. Phys. Chem. A,* 2000, pp.457.

Investigation the effect of time detention on the removal of chromuim from aqueous solution by using polyaniline/polystyrene nanocomposite

Hossein Eisazade[a] , Behzad Davodi [b]and Mohammad Soleimani lashkenari[c]

[a] *Prof in chemical engineering department of Babol University of Technology, P.O.Box 484, Babol, Iran*
[b,c] *Ms students of Babol University of Technology, P.O.Box 484, Babol, Iran*

Abstract: In this study Polyaniline/polystyrene nanocomposites were prepared in the aqueous solution by using ammonium peroxy disulfate as an oxidant in the presence of poly (vinyl pyrrolidone) as surfactant. Morphology and chemical structure was characterized by using scaning electron micrograph(SEM), fourier transform infrared (FTIR) spectrometer and removal percentage obtained by using mass atomic absorption.
Keywords: nanocomposite, surfactant, morphology, chemical structure

INTRODUCTION

Hence it becomes necessary to remove these toxic metal ions from municipal and industrial effluents to protect plants, animals and humanbeings from their adverse effect before discharging into natural water bodies or onto land. Although alarge number of separation methods are available, the process of adsorption is the most versatile and plays a significant role in removal of heavy metal contamination from water bodies [1-3]. Several polymers, clay minerals, dead biomass and metal oxides have been used as adsorbents [4-7]. Considerable attention was given in recent years for the removal of heavy metal toxic ions such as chromium, nickel, cadmium, zinc, etc. by polymeric beads [8-9].

In recent years, considerable attention has been devoted to fine new adsorbents. Chromium is widely used in chromium-plating, leather tanning, metal finishing, and chromate preparation. Chromium exists in two stable oxidation states Cr (III) and Cr (VI). The Cr (VI) state is of particular concern because of its toxicity. The recommended limit of chromium in potable water is 0.05 mg/L. The chromium ion Cr (III) can significantly affect the ecological environment when present in large amounts. Therefore, chromium pollution has recently received much attention in industries [10].

EXPERIMENTAL

Instrumentation

A magnetic mixer model MK20, digital scale model FR200, scanning electron microscope (SEM) model XL30, fourier transform infrared (FTIR) spectrometer model shimadzu 4100 and inductively coupled plasma (ICP) model vista-pro were employed.

Reagents and standard solutions

Materials used in this work were styrene and aniline monomer from Merck. All reagents were used as received without further purification, unless stated otherwise. Distilled deionized water was used throughout this work. Aniline monomer was purified by simple distillation.

Preparation of polyaniline/polystyrene nanocomposite

PAn/PS nanocomposite was prepared using ammonium peroxydisulfate as an oxidant. The results indicate that the PAn/PS nanocomposite is able to remove chromium from aqueous media. The chromium removal percentage was increased in alkaline solution. Also, the structure of obtained product was determined by FTIR spectroscopy. The microemulsion polymerization **of polystyren** was occurred at 75°C for 4 h. For the preparation of PS 6 mL of styrene monomer was added to a stirred aqueous solution (100 mL) containing 0.5 g poly(vinyl pyrrolidone) (PVP) and 0.06 g of ammonium peroxydesulfate (APS). The obtained PS latex was highly stable and precipitation was not observed even after centrifugation at 8000 rpm for 20 min. Aniline monomer was added to solution containing PS

CP1255, *V[th] International Conference on Times of Polymers (TOP) and Composites,*
edited by A. D'Amore, D. Acierno, and L. Grassia

latex, APS and PVP. Polymerization was carried out for 4 h at room temperature. Product was centrifuged several times and then dried in a vacuum oven at 60◦C for 6 h.

RESULTS AND DISCUSIONS

The morphology of nanocomposite was studied using scanning electron microscopy. The effect of contact time in the removal of chromium is shown in Table 1.

As can be seen in table the removal of chromium related to contact times. Also removal percentage increased by increasing the detention time. In Figs.1 SEM micrographs of polyaniline/polystyrene nanocomposite was shown.

Table1. Effect of contact time in the removal of chromium by PAn/PS nanocomposite

Type of adsorbent	Contact time (min)	Initial conc. (ppm)	Final conc. (ppm)	Removal percentage (w/w %)
	5	100	19.31	80.69
PAn/PS	10	100	18.73	81.27
nanocomposite	15	100	18.59	81.41
	20	100	18.38	81.62
	30	100	18.28	81.72

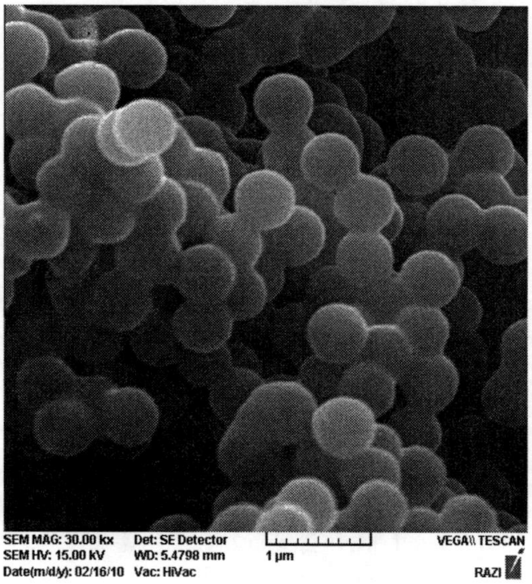

FIGURE1. Scanning electron micrograph of polyaniline/polystyrene

The structure of obtained product was determined by FTIR spectrum. The FTIR spectroscopy has provided valuable information regarding the formation of polyaniline/polystyrene nanocomposite. FTIR analysis has been done to identify the characteristic peaks of product.

FTIR spectra in the 4000-400 cm^{-1} region, for PAn/PS nanocomposite is shown in Fig 2. As can be seen characteristic absorption bands at 1598 cm^{-1} (C=C stretching vibration of the quinoid ring), 1467 cm^{-1} (stretching vibration of C=C of the benzenoid ring), 1304 cm^{-1} (C-N stretching vibration), 1180 cm^{-1} (C-H in-plane deformation) , 800 cm^{-1}(C-H out-of-plane deformation) selected to polyaniline. Also characteristic absorption bands at 3046 cm^{-1}(aromatic C–H stretching) ,2920 cm^{-1} (methylene C–H stretching) and 1600 cm^{-1} (C–H out-of-plane deformation bending of aromatic ring) selected to polystyrene .

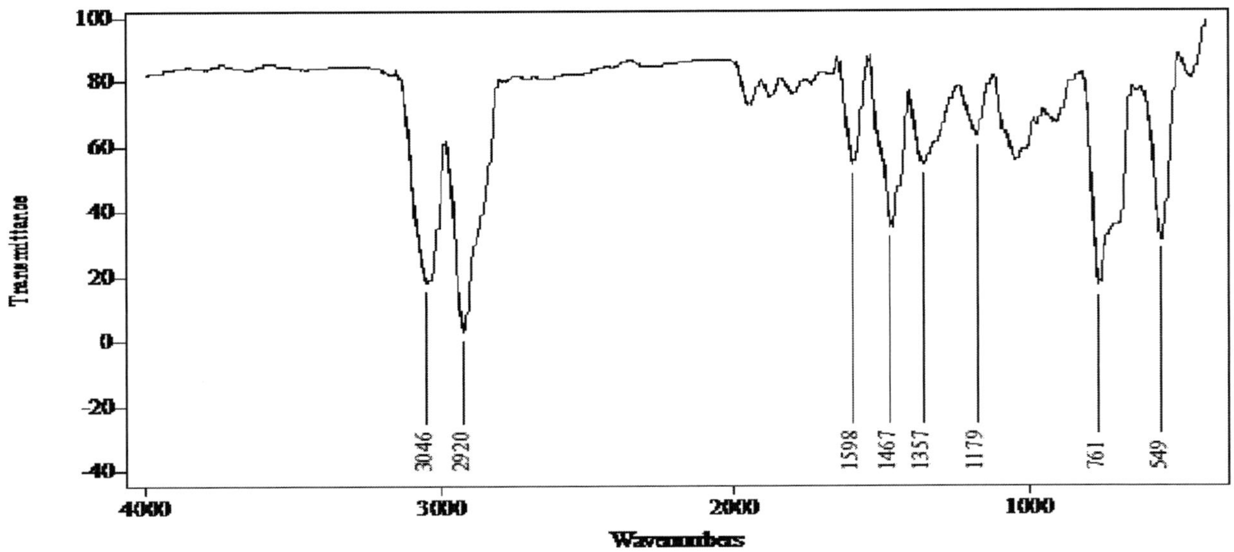

FIGURE 2. FTIR spectra of polyaniline/polystyrene in aqueous media

CONCLUSIONS

In this research, the effect of PAn/PS nanocomposite and detention time on the removal of chromium from aqueous media also as shown in table was studied. As can be seen the PAn/PS nanocomposite can be used for the removal of chromium from aqueous media. The equilibrium was reached within 20 minutes.

REFERENCES

1. J.B. Stankovic, S.K. Milonjic, M.M. Kopenci and T.S. Ceranic.,Colloids Surf. **46** 283-296 (1909)
2. S.S. Krishnan, A. Cancilla, E. Jervis and J. Radioanal., Nucl.Chem. Art. **110** 379-384 (1987)
3. D.K. Bhattacharya, N.C. Datta and J. Nucl., Sci. Technol. **128** 1014-1112 (1991)
4. S.P. Mishra, V.K. Singh and D. Tiwari., Radiochim. Acta. **73** 49-53 (1996)
5. B. Volesky, Biotechnology **2** 137-142(1988)
6. W. Mocka, H. Wihlidal, G. Stehlik, J. Washuttl and E. Bancher.,Chemosphere. **10** 787-796 (1979)
7. M.A. Spitsyn and V.N. Andreev., Elektrokhimiya. **25** 1171-1179 (1989)
8. A. Denizli, C. Arpa, S. Bektas and O. Genc., Adsorp. Sci. Technol. **20** 91-106 (2002)
9. N. Akiko, Y. Masao, I. Tomoyuki, O. Taiji, T. Kunihiko and Kagaku Kogaku Ronbunshu . **28** 501-509 (2002)
10. M. H. Liu, Zhang, J. S. Zhang . Fang, Y. C. Fang and Y. Chin. Deng . Southwest Inst. Nationalities.**34** 283 (1998)

Natural Fiber Reinforced PLA Composites

Young Seok Song[a], Jung Tae Lee[b], and Jae Ryoun Youn[b]

[a]*Department of Fiber System Engineering, Dankook University, Yongin, Korea*
[b]*Department of Materials Science and Engineering, Seoul National University, Seoul, Korea*

Abstract. We investigated the physical properties of natural fiber reinforced PLA composites, which were prepared by using PLA films and fibers. The impact, tensile, and dynamic mechanical properties of the composites were observed to examine the reinforcing effect of natural fibers. In addition, numerical analysis was carried out to more clarify the experimental results. The natural fibers/PLA composites shows a good potential for substituting traditional composites such as glass fiber or carbon fiber reinforced composites.

Keywords: PLA, natural fiber, numerical analysis

INTRODUCTION

Fully biodegradable green composites which are composed of biodegradable matrix and fibrous biomass have received enormous attention due to their complete degradation, light weight, and aesthetics. Many researchers have examined flax, cordenka, hemp, jute, ramie, kenaf, bamboo, caraua, and miscanthus as natural fibrous biomass for the reinforcement of PLA matrix [1-6]. The main objective of this study is to improve the mechanical and thermal properties of PLA resin by using denim woven fabrics. Impact and tensile tests were performed to identify the mechanical properties and elastic stiffness of the composites, and they were also predicted by using a numerical analysis code.

EXPERIMENT

Transparent PLA sheets (Toraysaehan Inc.) were used as the polymer matrix and twill weaved denim fabrics were used as fibrous biomass. PLA sheets were cut into square sheets, and each square sheet was dried at 80°C in a convection oven for 4 hr. Denim fabrics were also cut into a size of 14.5 x 14.5 mm^2 and dried at 80°C for 4 hr. The denim fabrics were placed between the PLA sheets to prepare denim reinforced composite specimens as described in Fig. 1.

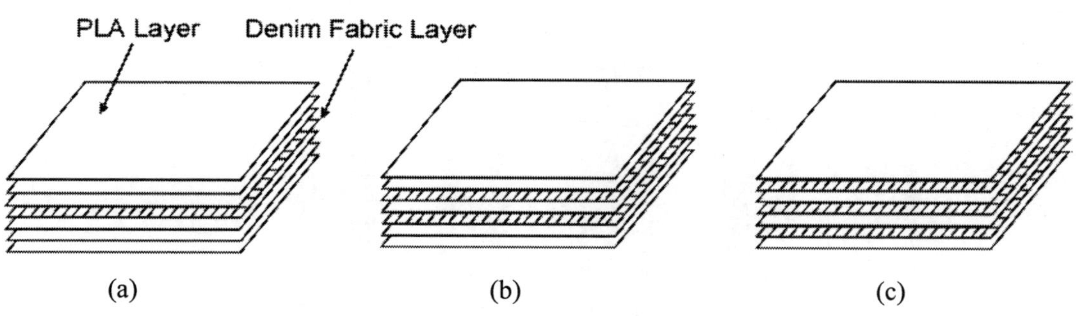

FIGURE 1. Description of the method used for preparing test samples.

CP1255, *Vth International Conference on Times of Polymers (TOP) and Composites,*
edited by A. D'Amore, D. Acierno, and L. Grassia

The impact, tensile, and dynamic mechanical properties of the composites were observed with increasing denim layers to examine the reinforcing effect of denim fabrics. Numerical analysis was carried out to model the elastic modulus of the composite by using a commercial software. Three-dimensional geometry of the denim fabric reinforced PLA composite was generated through a CAD program.

RESUTS AND DISCUSSION

Figure 2 shows the impact strength of the denim fabric reinforced PLA composites. As more denim layers were laminated, higher impact strength was achieved. Superior impact property of denim/PLA composite can be explained by fiber/matrix interaction and crack propagation. Measured tensile strength of the denim fabric PLA composite is given in Fig. 3. Tensile strength of the denim reinforced composite was improved but the improvement is not satisfactory. Interfacial adhesion between fibers and matrix is mainly responsible for tensile strength.

Young's modulus of the PLA/denim fabric composite is shown in Fig. 4 with respect to the number of denim layers. Tensile modulus was also increased as more denim fabric layers were reinforced. Elastic modulus of the composite was predicted by assuming the iso-strain condition in the numerical simulation and compared with the experimental results.

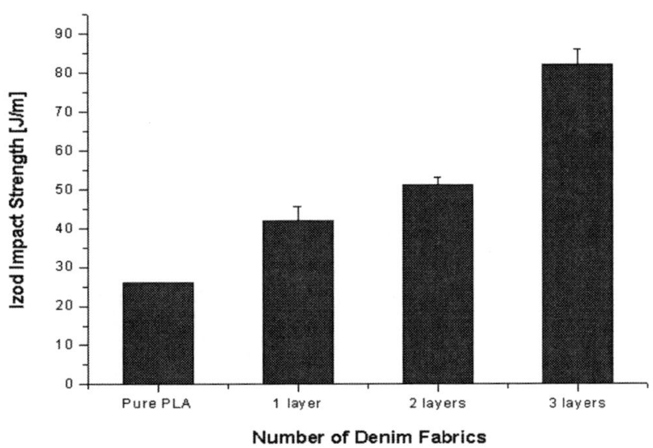

FIGURE 2. Plot of Izod impact strength with respect to the number of denim fabrics.

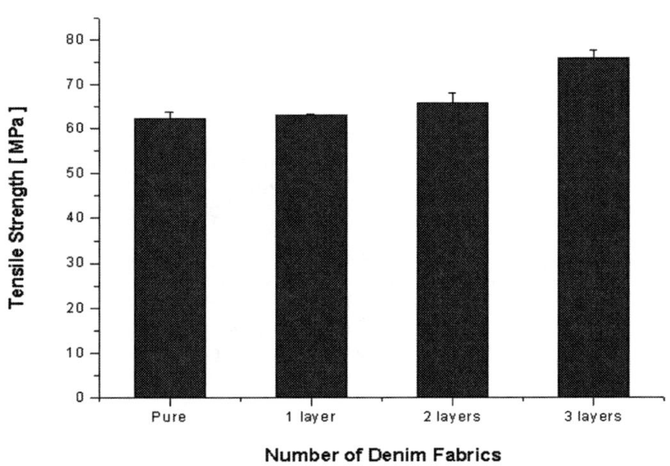

FIGURE 3. Plot of tensile strength with respect to the number of denim fabrics.

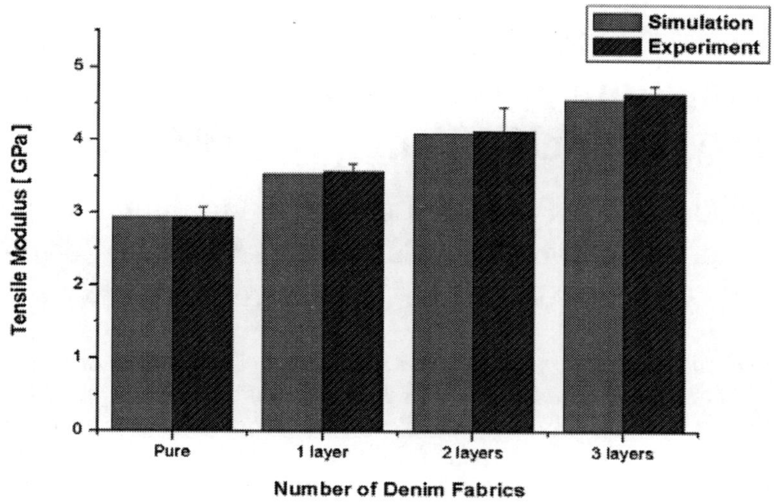

FIGURE 4. Plot of tensile modulus with respect to the number of denim fabrics.

CONCLUSION

It was demonstrated that the denim fabric is a good reinforcing material for PLA composites. The poor interaction between the fibers and the matrix strongly influences both impact and tensile strength and should be the reason for comparatively small increase of tensile strength. Geometry modeling and stress analysis on the tensile moduli of the composites were carried out to compare with experimental results. Tensile moduli calculated through the stress analysis of the RVE by assuming perfect bonding between the matrix and fibers were well fitted to the experimental results since the poor interaction between the PLA matrix and the denim fabric would affect the initial elastic moduli of the composites slightly.

Acknowledgments

This study was supported by the Korea Science and Engineering Foundation (KOSEF) grant funded by the Korean government (MEST) (R11-2005-065) through the Intelligent Textile System Research Center (ITRC). The authors are grateful for the support.

REFERENCES

1. E. Bodros, I. Pillin, N. Montrelay, and C. Baley, *Compos. Sci. Technol.*, **67**, 462 (2007).
2. K. Oksman, M. Skrifcars, and J. F. Selin, *Compos. Sci. Technol.*, **63**, 1317 (2003).
3. B. Bax and J. Mussig, *Compos. Sci. Technol.*, **68**, 1601 (2008).
4. R. Hu and J-K Lim, *J. Compos. Mater.*, **41**, 13 (2007).
5. D. Plackett, T.L. Andersen, W.B. Pedersen, and L. Nielsen, *Compos. Sci. Technol.*, **63**, 1287 (2003).
6. M. S. Huda, L. T. Drzal, A. K. Mohanty, and M. Misra, *Compos. Sci. Technol.*, **68**, 424 (2008).

FT-IR Investigation of Hoveyda-Grubbs' 2nd Generation Catalyst in Self-Healing Epoxy Mixtures

Liberata Guadagno[a,d*], Pasquale Longo[b,d], Marialuigia Raimondo[a], Carlo Naddeo[a,d], Annaluisa Mariconda[a], Vittoria Vittoria[a,d], Generoso Iannuzzo[c], Salvatore Russo[c]

[a] *Dipartimento di Ingegneria Chimica e Alimentare Università di Salerno, Via Ponte Don Melillo1, 84084 Fisciano (Salerno) Italy (UE)*
[b] *Dipartimento di Chimica Università di Salerno, Via Ponte Don Melillo 84084 Fisciano (Salerno) Italy (UE)*
[c] *Alenia Aeronautica SpA Viale dell'Aeronautica 80038 Pomigliano D'Arco (Napoli) Italy (UE)*
[d] *NANO_MATES, Research Centre for Nanomaterials and Nanotechnology Università di Salerno Via S. Allende, 84081 Baronissi (Salerno), Italy (UE).*

**corresponding author: lguadagno@unisa.it, tel. +39 0890964142, fax +39 089 964057*

Abstract. The development of smart composites capable of self-repair on aeronautical structures is still at the planning stage owing to complex issues to overcome. A very important issue to solve concerns the components' stability of the proposed composites which are compromised at the cure temperatures necessary for good performance of the composite. In this work we analyzed the possibility to apply Hoveyda Grubbs' second generation catalyst (HG2) to develop self-healing systems. Our experimental results have shown critical issues in the use of epoxy precursors in conjunction with Hoveyda-Grubbs II metathesis catalyst. However, an appropriate curing cycle of the self-healing mixture permits to overcome the critical issues making possible high temperatures for the curing process without deactivating self-repair activity.

Keywords: Self-Healing Materials, Epoxy Matrix, Aircraft Materials.

INTRODUCTION

Polymeric composites in aeronautic applications are used to replace the traditional metal alloys. These materials, however, are subject to weakening through impact damage which can lead to substantial matrix micro-cracking, delamination and fibre-matrix de-bonding, resulting in the reduction of structural capability. If damage is not detected and repaired, the result will be premature failure. The idea to put into action self-healing composites in aeronautical applications is therefore very attractive. Many strategies were formulated in the development of self-healing materials [1]. The right strategy to manufacture a self-repair material seems to be the storage of healing agents inside composites that restore the strength of the materials after damage. A very interesting autonomic self-healing system for aeronautic applications, in terms of design, was recently proposed by White et al. [2]. This system consists of incorporating a microencapsulated healing agent and a catalytic chemical trigger within an epoxy matrix. An approaching crack ruptures embedded microcapsules releasing a polymerizing agent into the crack plane through capillary action. Polymerization of the healing agent is triggered by contact with the embedded catalyst, bonding the crack faces. Thermosetting auto-repair polymers, which have been proposed so far, include Grubbs' first-generation catalyst (G1); and currently, the possibility of applying other ruthenium catalysts for ring-opening metathesis polymerization-based self-healing applications is being evaluated. In a recent paper, the dissolution properties, initial polymerization kinetics, chemical and thermal stabilities were analyzed for three catalysts: Grubbs' first (G1) and second generation (G2) catalysts and Hoveyda-Grubbs' second generation catalyst (HG2). Regarding Hoveyda catalysts, until now the experimentation was focused on the second generation catalyst because it has demonstrated impressive chemical stability and recyclability. In this work we analyzed the possibility to apply Hoveyda Grubbs' second generation catalyst (HG2) to develop self-healing systems. Our experimental results have shown critical issues in the use of epoxy precursors in conjunction with Hoveyda-Grubbs II metathesis catalysts. In particular, it has been found that a reaction takes place between the oxirane rings of the epoxy precursors and the metathesis catalyst, which thereby was not subsequently capable of promoting the polymerization of the reactive

CP1255, *V[th] International Conference on Times of Polymers (TOP) and Composites,*
edited by A. D'Amore, D. Acierno, and L. Grassia

monomer, thus deteriorating the self-healing ability. Such phenomena could strongly limit the use in the practice, in self-healing composite materials, of Hoveyda-Grubbs II catalysts, preventing a proper exploiting of the excellent intrinsic properties of the latter. In fact, such catalysts have a very high decomposition temperature, of the order of 232 ° C, and therefore it would be suitable to promote the reactive monomer polymerization up to such high temperatures, as well as at low temperatures even below -50 ° C [3]. Our experimental results have shown that by choosing the appropriate curing cycle as well as the specific chemical formulation, unprecedented control of the catalytic activity is obtained. In particular, the catalyst remains intact in the formed epoxy matrix during the curing process, and is thus capable of subsequently performing its catalytic activity of the polymerization of the reactive monomer consisting in the cyclic olefin (ENB), when the latter comes out of a microcapsule affected by a crack.

EXPERIMENTAL SECTION

Materials. T*he epoxy matrix composite* (EHA) was prepared by mixing an epoxy with an epoxy flexibilizer which was used in small percentage to improve the toughness of the material and consequently crack growth stability. These resins, both containing an epoxy, were obtained by Sigma-Aldrich. *The curing agent* investigated for this study is an anionic initiator Phenol, 2,4,6-tris[(dimethylamino) methyl] *The catalyst* used in the epoxy matrix is Hoveyda-Grubbs' 2nd generation catalyst. A self-healing mixture was obtained homogeneously dispersing finely pulverized Hoveyda-Grubbs' catalyst HG2 (5 wt %) into the EHA sample. This last mixture was cured by a single or two-stage curing cycle: a first isothermal stage at lower temperature for 2 hours and a second isothermal stage with higher temperatures up to 180 °C; these samples are named EHA-HG2(Y) where Y is the temperature of the second stage.

RESULTS AND DISCUSSION

Evaluation of Catalyst Thermal Stability by FT/IR. Thermal stability of HG2 catalyst was evaluated under conditions of actual use for applications in self-healing structural systems. The catalyst is dispersed in the epoxy matrix in solid state and both the epoxy precursor and the composite were cured (with the curing agent) by choosing processing conditions that contemplate permanence at high temperatures. Infrared spectroscopy was also used to evaluate the catalytic activity, directly inside the epoxy resin, after the cure processing.

Infrared Analysis. Infrared spectroscopy provides a useful way to identify metathesis products and therefore catalyst activity. The manufactured EHA-HG2 (170) sample was cut by a serrated blade; the powder which was produced from the sample was collected in two separate mortars and a drop of healing agent (5-ethylidene-2-norbornene - ENB) was added before dispersing the powder samples into the KBr disks for FTIR investigation.
Figure 1 shows FTIR spectra of EHA-HG2(170) powder treated with ENB. From the absence of the peak at 966cm^{-1}, characteristic of ring-opened poly(ENB), we deduce that the HG2 catalyst shows an inability to polymerize the ENB monomer, providing evidence that the embedded catalyst is inactive for a curing temperature of the epoxy formulation of 170 °C.

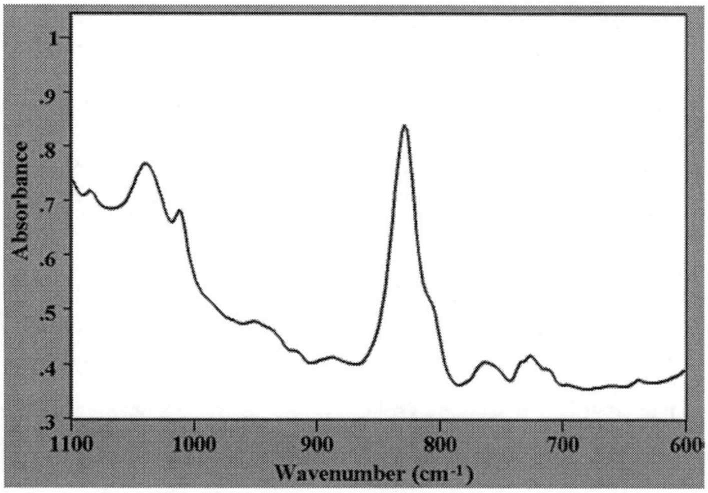

FIGURE 1. FTIR spectrum of EHA-HG2 (170) powders treated with ENB.

In light of the TGA results reported in literature, which show for HG2 a high temperature of beginning degradation (~232 °C), the unexpected results obtained from FT/IR investigation have spurred us to undertake an investigation aimed at understanding the reactivity of the catalyst inside the epoxy resins. Figure 2 shows FTIR spectra of the formulation EHA with the HG2 catalyst after two different curing processes: a) after a curing process directly carried out at high temperatures (2 h a 150 °C), b) after a two stage curing process composed of a pre-treatment at 70 °C for 7 hours and a following treatment at 150 °C for 2 hours. The spectra reported in figure 2 show that a direct treatment at high temperatures deactivates the catalyst, whereas, a pre-treatment at lower temperatures preserves the activity of the catalyst, as it can be deduced from the peak at 966 cm^{-1} characteristic of ring-opened poly(ENB).

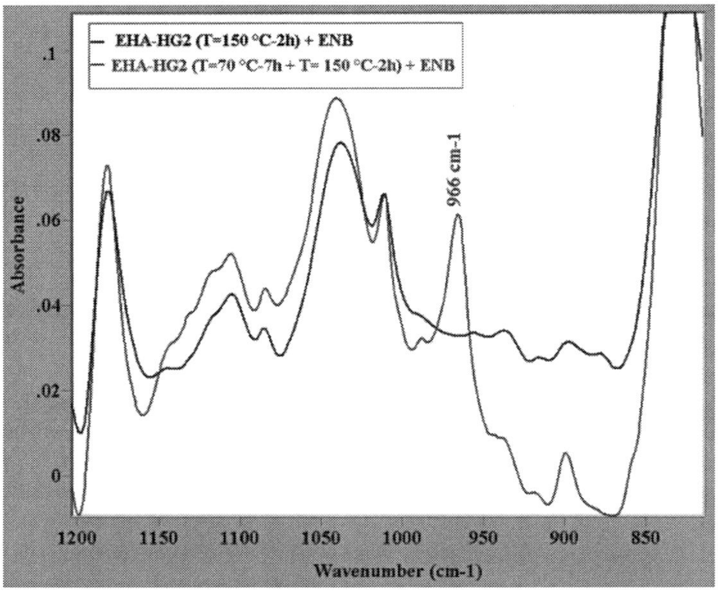

FIGURE 2. FTIR spectra of EHA-HG2 sample after two different curing process: a) a single stage directly at high temperature (2 h at 150 °C) b) a pre-treatment at 70 °C for 7 hours and the following treatment at 150 °C for 2 hours.

Our experimentation has shown that, with a pre-treatment at lower temperatures, the catalytic activity of the HG2 catalyst can be preserved up to temperatures between 150 °C and 160 °C; other treatments at temperatures higher than 160 °C deactivated the catalyst.

CONCLUSIONS

We have investigated the possibility to apply Hoveyda Grubbs' second generation catalyst (HG2) to develop self-healing systems. Our data have shown that by choosing the appropriate curing cycle, the catalytic activity of HG2 in the ROMP reactions can be kept under control. In particular, the catalyst remains intact in the formed epoxy matrix during the curing process, and is thus capable of subsequently performing its catalytic activity of the polymerization of the healing agent (ENB) in applications where the healing agent comes out of the damaged microcapsules.

REFERENCES

1. Self Healing Materials: An Alternative Approach to 20 Centuries of Materials Science (*Springer* Series in Materials Science) by *Sybrand van der Zwaag* (2007).
2. White SR, Sottos NR, Geubelle PH, Moore JS, Kessler MR, Sriram SR, et al.., Nature, 2001;409:794.
3. L. Guadagno, P. Longo, M. Raimondo, C. Naddeo, A. Sorrentino, V. Vittoria, G. Iannuzzo, S. Russo, E. Calvi, PCT/IB2009/051005.

Self-Healing Structures in Aerospace Applications

Liberata Guadagno[a], Marialuigia Raimondo[a],
Generoso Iannuzzo[b], Salvatore Russo[b]

[a] *Dipartimento di Ingegneria Chimica e Alimentare Università di Salerno, Via Ponte Don Melillo1, 84084 Fisciano (Salerno) Italy (UE)*
[c] *Alenia Aeronautica SpA Viale dell'Aeronautica 80038 Pomigliano D'Arco (Napoli) Italy (UE)*

**corresponding author:* lguadagno@unisa.it, *tel. +39 0890964142, fax +39 089 964057*

Abstract. The present paper relates to autonomically-healing composites which function actively and quickly at temperatures reaching as low as -50°C while maintaining the performance of the current structural composites. This is achieved thanks to the high temperatures possible for the formulation cure without deactivating self-repair activity.

Keywords: Self-Healing Materials, Epoxy Matrix, Aircraft Materials..

INTRODUCTION

Advances in the growing use of polymer composites in aerospace applications explore the development of smart materials capable of self-repair. Many realistic strategies have been formulated in the development of self-healing materials [1-5]. One of the most promising self-healing systems was proposed by White et al [1]. This system consists of incorporating a microencapsulated healing agent and a catalytic chemical trigger within an epoxy matrix. An approaching crack ruptures embedded microcapsules releasing a polymerizer agent into the crack plane through capillary action. Polymerization of the healing agent is triggered by contact with the embedded catalyst, bonding the crack faces. Aeronautic and aerospace applications require that the function of self-healing must be active also under extreme environmental conditions and with materials having high mechanical performances. It is critical that the self-healing activity functions at low working temperatures which can reach values as low as – 50 °C. Also, another problem concerns the catalyst stability which is compromised at the cure temperatures necessary for good performance of the composite.

We have developed self-healing composites exhibiting high levels of healing efficiency under severe operational conditions [6]. These systems are active in the self-healing process even at -50 °C and the self-repair function is fast and efficient within a wide temperature range. At room temperature, for example, these composites repair themselves in less than 30 seconds. As in the White design, our composite includes monomer-filled microcapsules and a solid catalyst embedded in a thermosetting polymeric matrix. The wide and elevated efficiency is due to the chemism of new components. The polymerizer selected for the micro-encapsulation is 5-ethylene-2 norbornene (ENB) which gives a faster kinetic reaction than those proposed up to now. The microcapsules containing this polymerizer agent can be synthesized to become highly heat-proof. The amazing self-healing efficiency in our systems have been obtained thanks to the catalysts developed by Grubbs and Hoveyda.

EXPERIMENTAL SECTION

Materials. T*he epoxy matrix composite* was prepared by mixing an epoxy (trade name EPON 828 – Acronym DGEBA) with an epoxy flexibilizer (Dimer Acid Diglycidyl Ester - trade name HELOXY 71 – Acronym DADGE) which was used in small percentage to improve the toughness of the material and consequently crack growth stability. These resins, both containing an epoxy, were obtained by Sigma-Aldrich.
The curing agent investigated for this study is an anionic initiator Phenol, 2,4,6-tris[(dimethylamino) methyl] (Trade name Ancamine K54). *The catalysts* used in the epoxy matrices are Grubbs' 1st and 2nd generation and

CP1255, *Vth International Conference on Times of Polymers (TOP) and Composites,*
edited by A. D'Amore, D. Acierno, and L. Grassia

Hoveyda-Grubbs' 1st and 2nd generation catalysts. All the catalysts were purchased from Aldrich. These catalysts, in the form of thin powders, were reduced before dispersion into the epoxy matrix.

Polymerizations. The monomer and the catalyst in two different vials were kept for 1h at the temperature of the runs (table 3). Then the polymerizations were activeted by addition of the monomer (ENB or DCPD) to the vial containing the catalyst. The runs were stopped by introducing a few drops of ethyl vinyl ether. For conversion percentages (Yields %) in the ROMP reaction less than 100%, the polymers were coagulated in methanol, recovered by filtration and dried in vacuum at room temperature.

Microcapsules containing ENB, with the outer shell composed of poly(urea-formaldehyde) and the inner shell of ethylene maleic anhydride copolymer (EMA), were prepared by *in situ* polymerization in an oil-in-water emulsion in accordance with a previously described procedure [8-9] .

Epoxy specimens manufacture. Sample EHA was obtained by mixing Epon 828 with Heloxy 71 flexibilizer at a concentration of 63%: 37% (by wt) epoxide to flexibilizer. Ancamine K54 was added at a concentration of 10:100 (by wt) hardener to mixture (Epon 828 and Heloxy 71).

Self-healing epoxy specimen manufacture. A self-healing mixture was obtained homogeneously dispersing finely pulverized Hoveyda-Grubbs' catalyst HG1 (5 wt%) and microcapsules containing ENB (10 wt%) into the epoxy specimen (EHA). This last mixture was cured by a two-stage curing cycle: a first isothermal stage at 125 °C (2 hours) and a second isothermal stage at 180 °C (2 h).

RESULTS AND DISCUSSION

Control of the metathesis polymerization.

Our self-healing composites are based on the ring-opening metathesis polymerization (ROMP). In this work, the metathesis activity of the Grubbs catalysts with dicyclopentadiene (DCPD) is compared to those of the Hoveyda-Grubbs 1 and Hoveyda-Grubbs 2 catalysts with ENB as substrate. For all the tests, a molar ratio between catalyst and reactive monomer of 1:1000 was used. The obtained data demonstrate that the catalyst's heterocyclic ligand significantly enhances the catalytic activity, and the best results were obtained with ENB as polymerizer agent and Hoveyda-Grubbs 1 and Hoveyda-Grubbs 2 as catalysts for which the metathesis reaction is also active at - 50 °C. In addition, in the temperature range where the DCPD is active, the polymerization reaction rate of ENB is progressively faster with respect to DCPD.

TABLE(1). Conversion percentage (Yields %) in the ROMP reaction of ENB and DCPD with different catalysts.

Catalyst	Monomer	Monomer/ catalyst molar ratio	Temperature (°C)	Time (m)	Yields (%)
GRUBBS 1	DCPD	1000	25	15	34
GRUBBS 1	DCPD	1000	10	30	0
GRUBBS 1	ENB	1000	25	0.5	100
GRUBBS 1	ENB	1000	0	13	100
GRUBBS 1	ENB	1000	- 30	1440	74
GRUBBS 1	ENB	1000	- 40	1440	0
HOVEYDA-GRUBBS 1	ENB	1000	25	0,5	100
HOVEYDA-GRUBBS 1	ENB	1000	-50	420	100
HOVEYDA-GRUBBS 2	ENB	1000	25	8	100
HOVEYDA-GRUBBS 2	ENB	1000	-50	420	100

Self-healing systems properties and morphology.

The self-healing mixture was obtained using a curing cycle up to 180 °C, as described in the experimental section. For these self-healing systems, it is important to resolve the problem of catalyst stability which can be seriously compromised because of the chemical environment and the cure temperatures necessary for good performance of the composite.

A cross-sectional view of our self-healing specimens (A, B, C,) is reported in the SEM images of figure 2 which shows: healed crack faces (A) with a magnification on the metathesis product inside the crack (B); an ampler view of the self-healing specimen sections where the metathesis product, inside of microcapsules after damage, can also be observed (C, D).

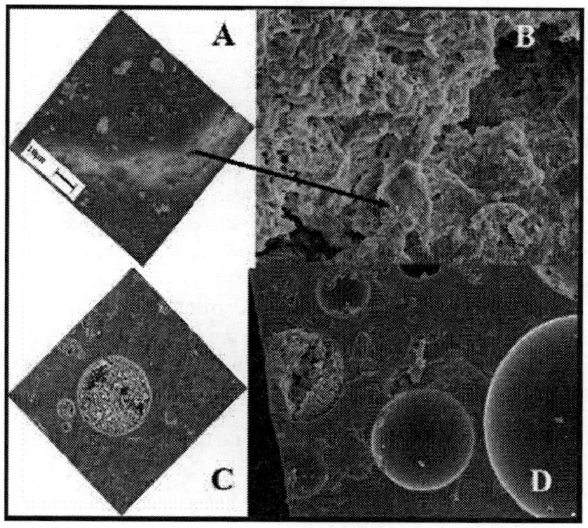

FIGURE 2. (A) healed crack faces closed by means of the metathesis product; (B) magnification on the metathesis product inside the crack; (C, D) self-healing specimen sections with metathesis product inside of the microcapsules after damage.

SEM images of a cross-sectional view of our self-healing specimen have allowed us to verify the formation of the metathesis product inside the crack.

The interesting results obtained from the analysis of the Ring-Opening Metathesis Polymerization inside of the epoxy resin using HG1 catalyst are of general validity. These can be applied also for another proposed bioinspired coating/substrate design[7-8] which delivers a healing agent to cracks in a coating via a three-dimensional microvascular network in the substrate.

REFERENCES

1. S. R. White, N. R. Sottos, P. H. Geubelle, J. S. Moore, M. R. Kessler, S. R. Sriram, E. N. Brown, S. Viswanathan, *Nature* **409**, 794 (2001).
2. P. Gould, *Materials Today* **6**, 44 (2003).
3. M. Trau, D. A. Saville, I. A. Aksay, *Langmuir* **13**, 6375 (1997).
4. X. Chen, M. A. Dam, K. Ono, A. Mal, H. Shen S. R. Nutt, K. Sheran, F. Wudl, *Science* **295**, 1698 (2002).
5. C. Dry, *Smart Matls & Structs.* **3**, 118 (1994).
6. L. Guadagno, P. Longo, M. Raimondo, A. Mariconda, C. Naddeo, A. Sorrentino, V. Vittoria, G. Iannuzzo, S. Russo, E. Calvi; Italian Patent **n. TO2008A000194** (03/13/2008).
7. K.S. Toohey, N.R. Sottos, J.A. Lewis, J.S. Moore, S.R. White, R. Scott, *Nature Materials* **6**, 581 (2007).
8. K. S. Toohey, N.R. Sottos, S. R. White, *Exp. Mech.* 2008, online at http://www.springerlink.com/content/p844pj0181×8874/fulltext.pdf.
9. E.N. Brown, M.R. Kessler, N.R. Sottos, S.R. White, *J. Microencapsul.* **20,** 719 **(2003).**

Hydrogen at Polymer Surfaces

J. Zemek[a], J. Houdkova[a], B. Lesiak[b]

[a] *Institute of Physics Academy of Sciences of the Czech Republic, Cukrovarnicka 10, 162 53 Prague 6, Czech Republic*

[b] *Institute of Physical Chemistry, Polish Academy of Sciences, ul. Kasprzaka 44/52, 01-224 Warszawa, Poland*

Abstract. We present a relatively simple method for analyzing H content on surfaces of selected commercially available polymers, *i.e.* polyethylene, polypropylene, polystyrene, and poly(methylphenylsilylene). The applied method, *i.e.* the Elastic Peak Electron Spectroscopy (EPES), is based on quasi-elastic scattering of primary beam electrons with atoms of solid sample constituents. Electron energy, scattering angle and difference between electron mass and atomic mass of sample constituents induces an electron recoil energy loss. Recoil effect results of splitting the elastic peak into components, by their energy shifting and broadening. For above-mentioned polymers, the EPES method indicates the surface H content close to the nominal composition. Sampling depth of the method does not exceed inelastic mean free path of electrons used in the scattering experiment. Therefore, the technique is useful to monitor H surface content during or after various surface modifications as UV, electron and ion beam modification of polymer surfaces or surface plasma treatments. Limitations of the technique are mentioned.

Keywords: Electron quasi-elastic scattering, Recoil effect, Elastic Peak Electron Spectroscopy, EPES, Hydrogen content
PACS: 68.35.bm, 68.35.Dv, 68.49.Jk, 81.65.Cf, 81.70.Jb

INTRODUCTION

Evaluation of hydrogen content at the surface from the EPES spectra is based on an electron recoil from atom, in which electron is undergoing a quasi-elastic scattering from target atoms of different atomic mass. The recorded elastic peak splits into components of different recoil energy shift. Within the classical model for quasi-elastic electron scattering on an atom, the energy transferred from an electron to a target atom, *i.e.* the recoil energy shift, E_r, is expressed as [1]:

$$E_r = \frac{4m}{M} E \sin^2\left(\frac{\theta}{2}\right), \tag{1}$$

where m is the mass of electron, M is the mass of the scattering atom, θ is the scattering angle (the angle between the direction of the primary beam and the direction towards the analyzer) and E is the incident electron energy.

Quantification of the H content is based on the measured ratio of H-induced elastic peak area and the area(s) induced by other elements found near a sample surface combined with calculations of corresponding electron backscattering probabilities. For binary C-H system, homogeneous in composition over the analyzed volume, the corresponding ratio of electron backscattering probabilities, and η_H, calculated for a given geometry of the measurement and electron energy, considering the differential backscattering cross sections $d\sigma/d\Omega$ [2] for C and H atoms, as a function of hydrogen percent content, x, can be estimated from the equation [3,4]:

$$\frac{\eta_H}{\eta_C} = \frac{(d\sigma/d\Omega)_H\, x}{(d\sigma/d\Omega)_C (100 - x)}, \quad 0 \le x < 100. \tag{2}$$

CP1255, *Vth International Conference on Times of Polymers (TOP) and Composites,*
edited by A. D'Amore, D. Acierno, and L. Grassia

Eq. (2) derived within the single scattering approach [1] is based on assumption that an electron undergoes only one large angle elastic scattering event along the trajectories and enters the analyzer. Other elastic collisions are not significant. The derivation of a formula for multi elemental sample is straightforward.

EXPERIMENTAL

EPES method requires electron source, sample manipulator, and electron energy analyzer placed in a vacuum chamber and can be conducted in a commercial electron spectrometer. A mono-energetic electron beam strikes upon a sample surface, electrons undergo quasi-elastic events and some of them enter the electron energy analyzer.

Several commercially available polymers as polyethylene (PE), polypropylene (PP), polystyrene (PS) were cleaned mechanically. Poly(methylphenylsilylene) (PMPSi) thin films were deposited from a toluene solution by casting on Au covered silicon substrates under protecting He atmosphere in a small preparation chamber attached to a photoelectron spectrometer ADES-400 (VG Scientific, UK) [5]. Cleanness and surface composition were checked by XPS (Mg Kα radiation).

The EPES spectra were recorded using a defocused electron beam of energy 1500 eV and 2000 eV, the beam current of 1×10^{-6} A, and a spot diameter at the sample surface of 3 mm. The quasi-elastically backscattered electrons were collected at pass energy of 20 eV within a small conical acceptance angle of ±4° of electron energy analyzer. Typical width of the elastic peak was about 0.5 eV (FWHM). Particular values of electron energy were chosen owing to optimal energy shift of about 3 eV. Results of the shift calculation for H, C and Si are shown in Fig. 1 (a). The mean penetration depths of quasi-elastically backscattered electrons in PE at 1500 eV and 2000 eV are 2.1 and 2.7 nm, respectively [4].

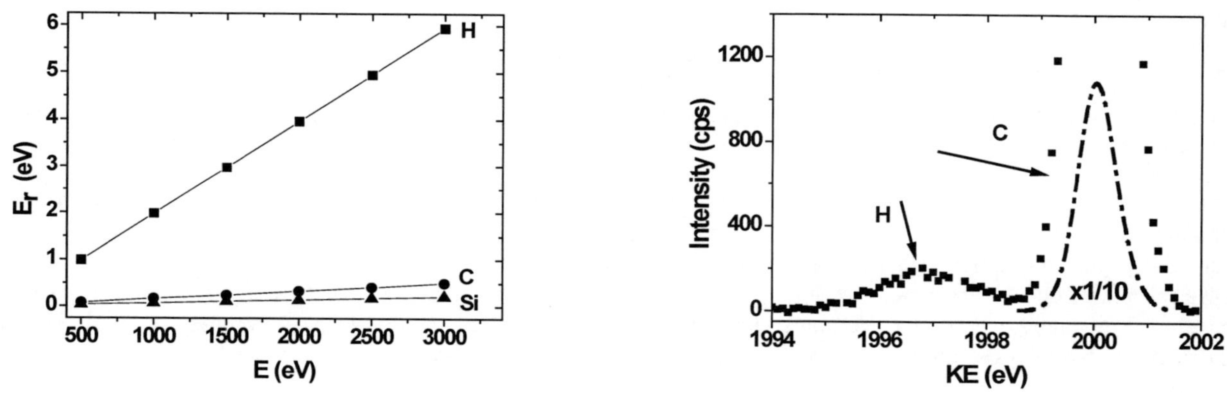

(a) **(b)**

FIGURE 1. (a) Recoil energy vs. electron energy calculated for H, C, and Si atoms. (b) Typical EPES spectrum recorded from PE surface at electron energy 2000 eV and scattering angle of 145°. More pronounced elastic peak intensity originates from quasi-elastic scattering of impinging electrons on carbon atoms, a weak elastic peak comes from electron-hydrogen interaction.

RESULTS and DISCUSSION

Typical EPES two peak spectrum recorded from PE surface at 2000 eV electron energy is shown in Fig. 1 (b). Due to very different cross sections for quasi-elastic scattering of electrons on hydrogen and carbon atoms [2], hydrogen-induced intensity is rather weak. The energy difference between the peaks is slightly smaller than that calculated by Eq. (1), *i.e.* 3.2 eV instead of 3.6 eV, likely due to surface charging [6].

Results of surface composition of analyzed polymers are summarized in Table 1. Carbon and silicon percentage were evaluated from the C 1s and Si 2p XPS spectra using a common quantification procedure [7], while hydrogen concentration was evaluated from the EPES spectra by Eq. (2). Hydrogen percentage is nearly approaching the nominal value except for the PMPSi surface where the measured value is much smaller than the nominal one. This finding is caused mainly by electron-induced degradation associated with hydrogen depletion from a surface region of the material under analysis.

TABLE (1). Nominal and measured composition by XPS and EPES techniques in atomic %

Sample	Nominal Composition			Measured Composition		
	C	Si	H	C	Si	H
PE	33.3	-	66.7	28	-	72
PP	33.3	-	66.7	27	-	73
PS	50.0	-	50.0	46	-	54
PMPSi	43.75	6.25	50.0	51	8.5	40

The hydrogen depletion from an analyzed volume of PE is illustrated in Fig. 2. Hydrogen percentage steeply falls from nearly nominal value to about 15% for electron dose of 1000 Cm^{-2}.

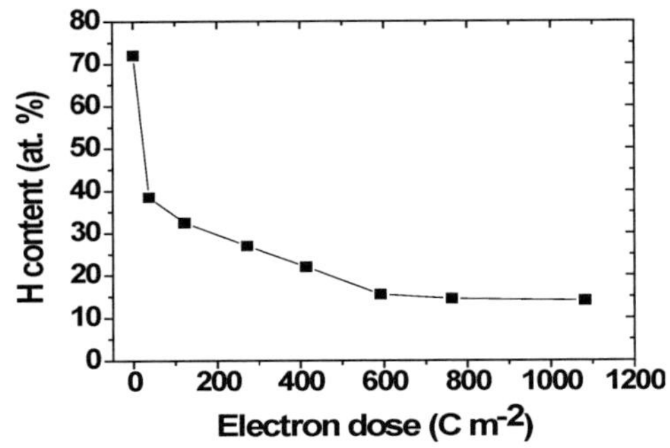

FIGURE 2. Analyzed hydrogen content in a surface region of PE, evaluated by Eq. (2), and its decrease due to electron induced degradation.

The technique presented here can be useful for controlling of various surface modifications of materials containing hydrogen. However, there are several limitations one should keep in mind, namely (i) relatively low sensitivity caused by a rather low value of cross-section for electron-hydrogen quasi-elastic interaction, (ii) possible spectral interference with signal intensity of another origin, *e.g.* from π-π^* transitions and (iii) surface charging in case of low electrical conductive materials. Therefore, it is recommended to use the method with caution, particularly the analysis should be done at several electron energies and scattering angles.

ACKNOWLEDGMENTS

The authors acknowledge the support of exchange between Polish Academy of Sciences and Academy of Sciences of the Czech Republic, Institutional Research Plan No AV0Z10100521 and Czech Science Foundation Project 202/09/0428 and Polish Science Foundation Project NN507378235.

REFERENCES

1. D. Laser and M. P. Seah, *Phys. Rev. B* **47**, 9836-9840 (1993).
2. F. Salvat and A. Jablonski, *NIST Electron Elastic Scattering Cross-Section Database,* (SRD 64), v. 3.1, NIST, Gaithersburg MD, 2003; http://www.nist.gov/srd/nist64.htm.
3. M. Filippi and L. Calliari, *Surf. Interface Anal.* **40**, 1469-1474 (2008).
4. B. Lesiak, J. Zemek and J. Houdkova, *Polymer* **49**, 4127-4132 (2008).
5. I. Kuritha, F. Schauer, P. S. Saha, J. Zemek, P. Jiricek, S. Nespurek, *Czech. J. Phys.* **56**, 41-50 (2006).
6. M. Vos and M. R. Went, *J. Electron Spectrosc. Relat. Phenom.* **155**, 35-39 (2007).
7. J. Zemek, *Acta Physica Slovaca*, **56**, 543-553 (2006).

On the Efect of the Oxidative Reagents on the Conductivity of Polyaniline/MMT Nanocomposites

A. Garcia-Bernabé[a], M. Gil-Agustí[a], G. Ortega[a], P. Llovera[b], A. Almarza[c], S. Vázquez[c], D. Amantia[c] and L. Aubouy[c].

[a] *Energy Technological Institute (ITE), Av. Juan de la Cierva 24, Parque Tecnológico de Valencia, 46980 Paterna-Valencia (Spain)*

[b] *Instituto de Tecnología Eléctrica, Universidad Politécnica de Valencia, Camino de Vera s/n 46022-Valencia (Spain)*

[c] *LEITAT Technological Center, Passeig 22 de Juliol, 218, 08221 Terrassa-Barcelona (Spain)*

Abstract. The synthesis of polyaniline has been reported using three different oxidative reagents: ammonium persulfate, potassium iodate and potassium iodate+sodium hypochlorite. This polyaniline has been used to prepare several nanocomposites with different percentage of Montmorillonite. The DC conductivity of the nanocomposites was determined by impedance spectroscopy. The oxidative reagent that gives higher conductivity is ammonium persulfate. The temperature dependence of the conductivity was studied.

Keywords: polyaniline; MMT; nanocomposite, conductivity.
PACS: 72.80 Le, 81.05 Fb, 82.35 Cd

INTRODUCTION

The polyaniline (PANi) is a conducting polymer with important potential applications in electronic devices, chemical sensors and antistatic coating. The conductivity of PANi depends upon the oxidation state: the fully reduced, leucoemeraldine base (LEB) and fully oxidized states, pernigraniline (PNB), are not conductive with conductivity values around 10^{-9} S/cm. However the half oxidized/half reduced polyaniline emeraldine base (PANiEB) is good conductor showing conductivity values ranging between between 10^{-7} and 10^{1} S/cm. This great divergence depends on the experimental conditions in the synthesis and the doping level of the PANi. In this way, the HCl-doped polyaniline at pH equal to 7 has a conductivity of 2.7×10^{-11} S/cm and at pH equal to 0 the conductivity falls down to 0.8 S/cm.[1] On the other hand, the conductivity of the H_3PO_4-doped polyaniline varies between 10^{-7} and 10^{-1} S/cm at different doping levels.[2]

Composites of PANi and inorganic materials, especially different kind of clay, has been studied and proposed to several applications as secondary battery, electrodes, pH sensors, electro-rheological materials and anti-corrosion materials. Particularly, the montmorillonite (MMT) is good filler in the case of polianiline-clay composites, the addition of MMT leads to a modification of the mechanical and conducting properties. Joo *et al* reported the nanocomposite of polyaniline and sodium-MMT doped with docecylbenzenesulfonic acid. In this nanocomposite , it was observed the intercalation of PANi-DBSA between the MMT layers at the nanoscale. The clay layers interrupt the delocation of charge carrier, they weaken the interchain interaction and they induce a more disordered state, resulting in lower conductivity.[3] Yu *et al* analyzed the factors affecting the conductivity of PANi/MMT composite.[4] In this paper, the authors studied the influences of reaction time, concentration of hydrochloric acid, oxidizing content and MMT content. They found the highest conductivity for PANi/MMT composite at values of 11.5 S/cm in determined conditions. However, they did not consider the influence of oxidative reagents on conductivity of PANi/MMT composite.

In this study it was investigated the effect of three different oxidative reagents on the conductivity of the polymer. It was determined which of these reagents gives the highest conductivity to this polymer. Conductivity was measured by impedance spectroscopy.

CP1255, *V[th] International Conference on Times of Polymers (TOP) and Composites,*
edited by A. D'Amore, D. Acierno, and L. Grassia

EXPERIMENTAL PART

Mechanical intercalation of PANi to montmorillonite: To 500 mL of 1M hydrochloric acid was added aniline (10 mL; 100 mmol). Upon the addition of the aniline it appears a cloudy white precipitate that instantly disappeared. The clear solution was stirred for 15 minutes at room temperature. An aqueous solution of ammonium persulfate (APS; 4.9 g; 21 mmol in 500 mL) was added and left to stir for 10 minutes until the colour changed from light grey to slight purple and dark blue. The solution was left overnight without stirring at room temperature. The resulting green precipitate was filtered through by Buchner with paper filter and rinsed with distilled water (3 x 100 mL) and acetone (3 x 100 mL). The polyaniline (PANi) powder was dried at 60°C in oven for 12 hours. The same synthesis was repeated using potassium iodate (KIO_3) and potassium iodate (KIO_3) + sodium hypochlorite (NaClO) as oxidative agent instead of ammonium persulfate (APS).

The montmorillonite (MMT) was mechanically introduced to polyaniline with the use of mortar and pestle. Typically, the nanocomposites were made as followed; it was used 400 mg of PANi for each samples and added to it 44.4, 50 and 57.1 mg of MMT to obtain mixture of 10, 20 and 30% in weight of nanoclay.

The impedance spectroscopy was carried out using a Dielectric Novocontrol Concept 80 Analyzer of the Department of Applied Thermodynamics at the Polytechnic University of Valencia in the range 10^{-1} to 3×10^{6} Hz isothermal conditions at 10° steps from -100 to 30°C.

RESULTS AND DISCUSSION

For the synthesis, aniline was used and is previously protonated in 500 mL of 1M hydrochloric acid. The different oxidative agents (a) Ammonium persulfate, (b) Potassium iodate and (c) potassium iodate + sodium hypochlorite were dissolved in 500 mL of distilled water and mixed with the anilinium cation to start the polymerization reaction. The mixtures were stirred 10 minutes where it can be observed a change of colour (colorless to grey and then dark blue) and heating of the solution indicating the beginning of the polymerization reaction. The coloured solutions were left without stirring overnight at room temperature to reach completion of polymerization reaction. The obtained green precipitates were washed with 1M HCl and acetone in order to protonate the polymeric chains and remove impurities. In the case of the reaction with the potassium iodate + sodium hypochlorite the change of colour was instantaneous with the addition of the oxidant to the aniline solution, this can be explained by a fast polymerization reaction giving short polymeric chain and poor conductivity as it was observed in figure 2. The resulting PANi polymers were mixed with the correct amount of montmorillonite with a mortar and pestle to give 10, 20 and 30% in weight of PANi/MMT nanocomposites where the polymer chains of PANi are intercalated in between the exfoliated nanoclay.

The real part of AC conductivity of the PANi-MMT nanocomposites using ammonium perdisulfate as oxidative reagent is shown in the figure 1. The AC conductivity is constant in frequency domain, which can be identified with DC conductivity, $\sigma^*(\omega)=\sigma_{DC}$. This non-dependent with frequency character indicates an electronic conduction.

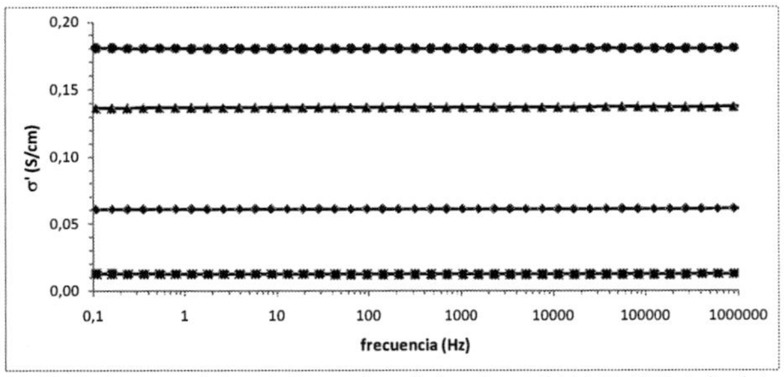

FIGURE 1. Frequency dependence of real part of the AC conductivity for 100% PANi (circle), 90% PANi/10 %MMT (triangle), 80 % PANi/20 % MMT (diamond), and 70% PANi/30 % MMT (square) using ammonium perdisulfate as oxidative reagent.

The conductivity increases with the contents of the PANi in the nanocomposite. The minimum conductivity is obtained at 70 % of PANi/30 % of MMT. In the figure 2, the % PANi dependence of the conductivity is shown. Ammonium perdisulfate is the oxidative reagent that provides higher values of DC conductivity.

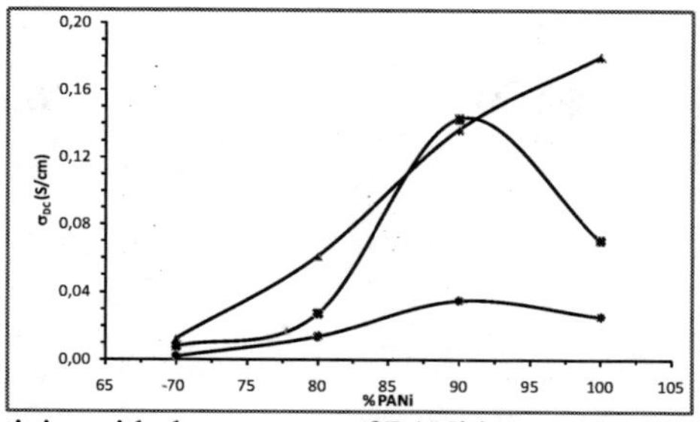

FIGURE 2. Variation of de DC conductivity with the contents of PANi in nanocomposite using ammonium perdisulfate as oxidative reagent (triangle), potassium iodate as oxidative reagent (square) and sodium hypochlorite and potassium iodate as oxidative reagent (circle)

In order to deeply analyze the conduction mechanism, the study of the temperature dependence of DC conductivity has been carried out. The analyzed nanocomposites were PANi/MMT using ammonium perdisulfate as oxidative reagent. The experiment was isothermal conditions at 10° steps from -100 to 30°C. The conductivity increases with temperature. A mechanism very often proposed to explain the DC conductivity in disordered materials is 3D-Variable Range Hopping (3D-VRH) model[5]. This model proposes that the electronic states of such material are overlapping between the neighboring chains. The model has this form.

$$\sigma_{DC} = \frac{\sigma_0}{\sqrt{T}} \exp\left(-\frac{T_0}{T}\right)^{1/4} \tag{1}$$

where σ_0 and T_0 are two parameters. The plot of $\ln\left(\sigma_{DC}\sqrt{T}\right)$ as a function of $T^{-1/4}$ is shown in figure 3.

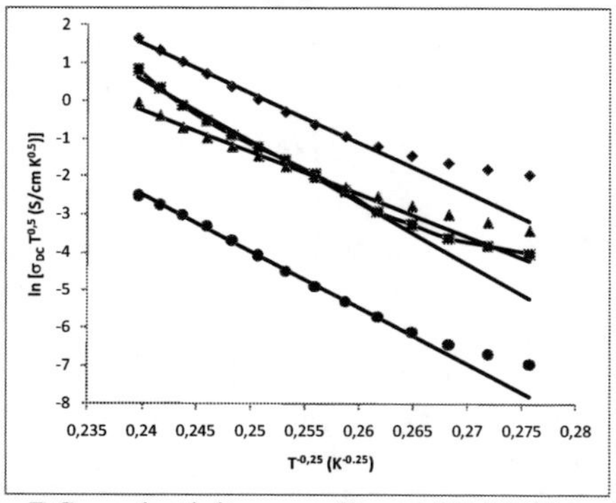

FIGURE 3. Temperature dependence of the DC conductivity according to 3D-VRH model of the PANi-MMT nanocomposites using ammonium perdisulfate as oxidative reagent. 100 %PANi (diamond), 90% PANi/10% MMT (square), 80% PANi/20% MMT (triangle), 70% PANi/30% MMT (circle). Continuous line 3D-VRH model.

DC conductivity data can support the 3D-VRH model in the range of temperature from -60 to 30°C. At lower temperature, another model must be used to explain the conductivity of these nanocomposites.

ACKNOWLEDGMENTS

This work was supported by the Ministry of Science and Innovation (Spain) in the context of the National Plan for Scientific research, Development and Innovation 2008-2011 (project ref. IAP-600200-2008-49).

REFERENCES

1. R. Singh, V. Arora, R. P. Tandon and S. Chandra. *J. Mater. Sci.* **33**, 2067-2072 (1998)
2. J. F. Rouleau, J. Goyette, T. H. Bose, R. Singh and R. P. Tandon. *Phys. Rev. B* **52**, 4801-4805 (1995)
3. B. H. Kim, J. H. Jung, S. H. Hong, J. Joo, A. J. Epstein, K. Mizoguchi, J. W. Kim and H. J. Choi. *Macromolecules* **35**, 1419-1423 (2002)
4. C. Liu, Y. Zhou, S. Li, L. Li and M. Yu. *Polym. Adv. Technol.* **19**, 1693-1697 (2008)
5. N. F. Mott, E. A. Davis, *Electronics Processes in Noncrystalline Materials.* Clarendon, London, 1979.

Blends of PEO in Poly(ethylene oxide)/Poly(2-vinyl pyridine)-b-poly(ethylene oxide)

E. Araneda[a], L. Gargallo[a], A. Leiva[a], N. Hadjichristidis[b], I. Mondragon[c], D. Radic'[a]

[a]*Laboratorio de Química Fisica de Macromoléculas, Departamento de Química Fisica, Facultad de Química, Pontificia Universidad Católica de Chile. Casilla 306, Correo 22, Santiago, Chile.*
[b]*Laboratory of Industrial ChemistryDepartment of Chemistry National and Kapodistrian University of Athens, Panepistimioupolis, Zografou, 15771 Athens, Greece.*
[c]*'Materials + Technologies' Group, Dpto. Ingeniería Química y M. Ambiente, Escuela Politécnica, Universidad País Vasco/Euskal Herriko Unibertsitatea Pza Europa 1. 20018 Donostia - San Sebastián, Spain.*

Abstract.
The crystallization behavior of homopolymer Poly(ethylene oxide) PEO and block copolymer Poly(2-vinyl pyridine)-b-poly(ethylene oxide) (P2VP-b-PEO) blends was investigated by differential scanning calorimetry (DSC), polarized optical microscopy (POM), thermogravimetric analysis (TGA) and atomic force microscopy (AFM). Different compositions of two homopolymer/block copolymer blends were studied. The formation of the spherulites was observed for the systems in study by POM. A diminishing of the growing rate of spherulites G (μm/s) for these blends can be observed as the copolymer content increases

Keywords: Crystallization, Poly(ethylene oxide), Poly(2-vinyl pyridine)-b-Poly(ethylene oxide).
PACS: 83.80.Tc

INTRODUCTION

Polymer blends have been extensively studied due to their application for the development of new materials with new chemical and physical properties (1,2). Crystallization behavior and morphology developments of block copolymers have also been investigated widely for several years in the last decades (3,4).

Studies on synthesis, characterization, isothermal crystallization and spherulitic morphology of different types of copolymers can be found in the literature (6,7). In multiblock copolymers, Petrova et al. (5) have reported the synthesis and thermal analysis of poly(ethylene oxyde)/poly(caprolactone) (PEO/PCL) multiblock copolymers by differential scanning calorimetry. They found two melting peaks, reflecting the presence of two crystalline domains.

The majority of studies on polymer blends have been focused on systems containing mainly homopolymers. Nevertheless in the last years the study of blends containing copolymers have been a matter of current interest due to the versatility of chemical compositions of the samples that can be able to enhance interpolymer miscibility and then miscibility or at least compatibility conducting to a new material with new properties.

The aim of this work is the analysis of the thermal behavior of blends containing poly(ethylene oxide)(PEO) and a diblock copolymers of poly(2-vinylpyridine)-b-poly(ethylene oxide) (P2VP-b-PEO), and the analysis of the influence of each one on the crystallization rate. The variation on crystallinity and changes in the morphology have been investigated using differential scanning calorimetry, polarized optical microscopy (POM), thermogravimetric analysis (TGA) and atomic force microscopy (AFM).

RESULTS AND DISCUSSION

TABLE 1. Polymers characterizations

Polymers	M_n
PEO(200M)	200000
PEO(40M)	41500
P2VP-b-PEO	80000 ; 50% wt of P2VP

CP1255, *V[th] International Conference on Times of Polymers (TOP) and Composites,*
edited by A. D'Amore, D. Acierno, and L. Grassia

The spherulitic growth through crystallization of PEO on the block copolymers was studied by polarized optical microscopy. The formation of the spherulites was observed for the systems in study by POM, Figure 1

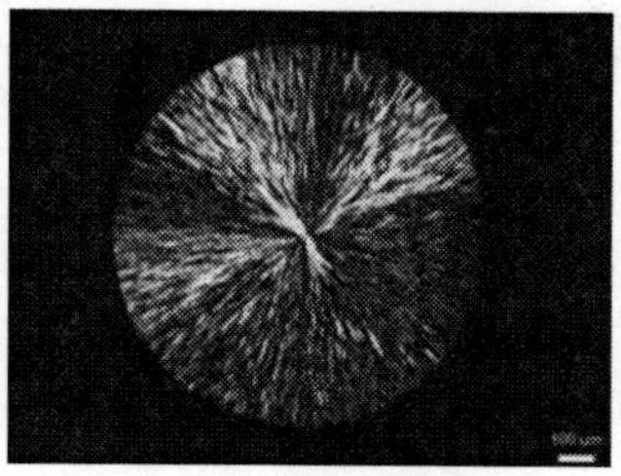

FIGURE 1. Micrograph of 80% POE(200M)/P2VP-b-POE blend, isothermal crystallization at 45°C

Tables 2 and 3 compile the results of the growth rate G (μm/s) for PEO(200M)/P2VP-b-PEO and PEO(40M)/P2VP-b-PEO blends. The value of growth rate G (μm/s) of PEO are in the range expected according to results found in the literature. A diminishing of the growing rate of spherulites G (μm/s) for these blends can be observed as the copolymer content increases. This behavior is enhanced as the temperature of the isothermal crystallization or the molecular weight of the homopolymer in the blends increases.

TABLE 2. Crystallization growth rate as a function of homopolymer content in PEO(200M)/P2VP-b-PEO blends at three isothermal crystallization temperatures.

System	40 °C	43 °C	45 °C
PEO(200M)	55.9	29.7	20.2
80 % PEO	5.5	1.92	0.37
50 % PEO	0.40	0.20	0.18
20 % PEO	0.06	0.03	0.03
P2VP-b-PEO	0.02	0.008	0.006

TABLE 3. Crystallization growth rate as a function of homopolymer content in PEO(40M)/P2VP-b-PEO blends at three isothermal crystallization temperatures.

System	40 °C	43 °C	45 °C
PEO(40M)	65.6	46.2	28.1
80 % PEO	13.2	14.9	5.7
50 % PEO	0.99	0.89	0.50
20 % PEO	0.10	0.12	0.05
P2VP-b-PEO	0.01	0.008	0.006

ACKNOWLEDGMENTS

DIPUC, Fondecyt project 1080007, Conicyt project 24090082, Conicyt for a doctoral fellowship

REFERENCES

1. Li, C.; Zhang, Y.; Zhang, Y.; Zhang, C. *Eur. Polym. J.*, **39**, 305 (2003).
2. Urzua, M.; Leiva, A.; Alegria, L.; Gargallo, L.; Radic, D. *Int. J. Polym. Mater.* **56**, 687 (2007).
3. Albuerne, J.; Marquez, L.; Müller, A.; Raquez, J.; Degée, Ph.; Dubois, Ph.; Castelletto, V.; Hamley, I. *Macromolecules*, **36**, 1633 (2003).
4. Hamley, I.; Castelletto, V.; Castillo, R.; Müller, A.; Martin, C.; Pollet, E.; Dubois, Ph. *Macromolecules*, **38**, 463 (2005).
5. Petrova, Ts.; Manolova, N.; Rashkov, I.; Li, S.; Vert , M. *Eur. Polym. J.* , **40**, 1461 (2004).
6. Fragouli, P; Iatrou, H.; Hadjichristidis, N. *Polymer,* **43**, 7141 (2002).
7. Arnal M.; López-Carrasquero F.; Laredo E.; Müller A. *Eur. Polym. J.* **40**, 1461 (2004).

CALORIMETRIC CHARACTERIZACION OF PLLA- MWCNTs AND PLLA-MWCNTsf

C. Albano[1,*], G. Gonzalez[2,*], A. Karam[3], M. Covis[1]
[1]Facultad de Ingeniería, UCV, Caracas-Venezuela.
[2]Centro de Ingeniería de Materiales y Nanotecnologia, and [3]Centro de Química, IVIC, Caracas-Venezuela.
E-mail: carmen.albano@ucv; gemagonz@ivic.ve

Abstract
In the present work the thermal characterization of composites of polylactide (PLLA) with pristine or functionalized multiwalled carbon nanotubes (MWCNTs) prepared by the method solution mixing. The effect of % MWCNTs added on the thermal transitions was evaluated.

Introduction

Carbon nanotubes (CNTs) have a very high aspect ratio and very high Young's modulus; therefore, it is expected to be an excellent reinforcement for polymeric composites [1-2]. To have effective reinforcing effects, CNTs are normally surface-modified before adding to a polymer matrix, although the unmodified CNTs could enhance crystallization. On the other hand, PLLA is biodegradable and biocompatible aliphatic polyester and is widely used in biomedical applications. Since crystalline morphologies of PLLA can affect its physical properties and thus its applications, a thorough understanding of the crystallization and melting behaviors of PLLA is crucial. In this work, PLLA-MWCNTs and PLLA-MWCNTsf nanocomposites were prepared and the effects of the CNTs on melting and crystallization behaviors of PLLA were investigated by DSC.

Experimental

The MWCNTs were thoroughly washed in a mixture of nitric acid and sulphuric acid, and then functionalized with HNO_3 (60%) for 2 h. The composites were prepared in a chloroform solution at room temperature. Characterization was carried out by FTIR, Raman spectroscopy, MET, TGA and DSC.

Results and Discussion

The structural characterization of the pristine and funtionalized multiwalled carbon nanotubes (MWCNTs and MWCNTsf) by FTIR and Raman spectroscopy resulted in a $I_{D/G}$ of 1.15 for MWCNTsf and 1.06 for the pristine MWCNTs. The concentration of acid sites was 3.04 x 10^{-3} mol/g for the MWCNTCsf. The MET characterization showed a decrease in the density of aggregates due to the reduction of surface energy, as result of the acid sites creating a high electrostatic repulsion.

To evaluate the degree of interaction between PLLA and MWCNTs the thermal behaviour was analyzed for the different composites. Figure 1 shows the variation on crystallization enthalpies with % of MWCNTs. It can be observed that the enthalpy of the cold crystallization peak presents higher values for the composites PLLA-MWCNTsf than for the composites PLAA-MWCNTs. This difference is more drastic for the the composite materials with 5% MWCNTs, where the cold crystallization peak disappears when pristine MWCNTs were used but reaches a

CP1255, *V[th] International Conference on Times of Polymers (TOP) and Composites,*
edited by A. D'Amore, D. Acierno, and L. Grassia

maximum when the same proportion of functionalized nanotubes was employed. On the other hand, in general the values of cold crystallization enthalpies decreases with the increase of pristine MWCNTs content and increases with the increase of functionalized MWCNTsf. This is an evidence of the strong interaction matrix-filler due to the functional groups created by the functionalization treatment of the MWCNTs.

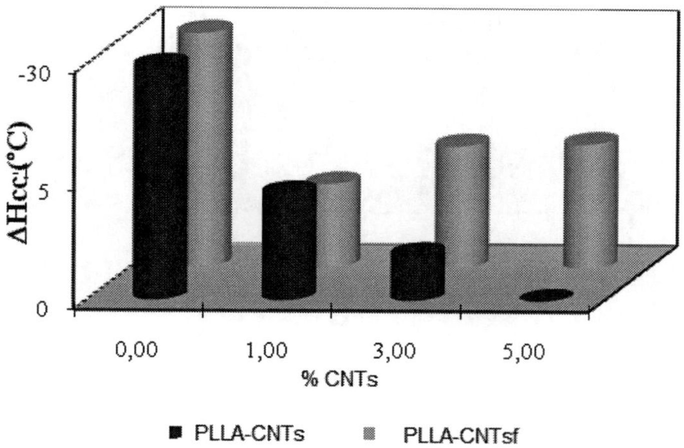

Figure 1: Enthalpy associated to the cold crystallization of the composite materials

Figure 2 shows the variation of the temperatures of the melting peak for the different composite materials with the addition of different proportion of nanotubes. An increase in the melting temperature is observed with the increase in the proportion of nanotubes. This effect is more pronounced in the composites with the MWCNTsf, implying that the addition of MWCNTsf contributes to the formation of more perfect crystals.

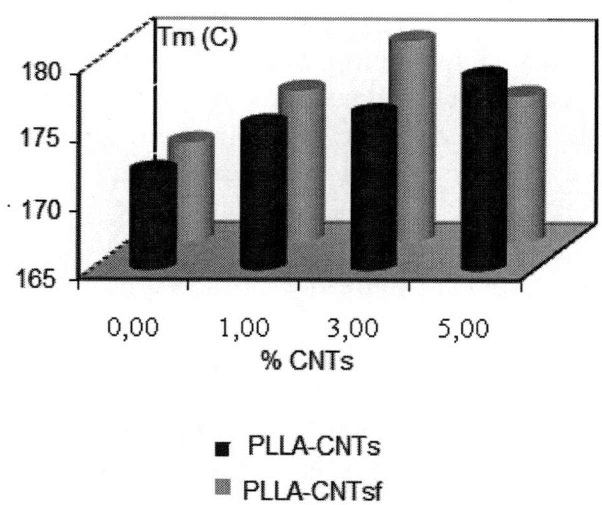

Figure 2: Melting Temperatures of composite materials vs. % nanotubes.

Finally Figure 3, compares the crystallinity of PLLA in the different composites. The crystallinity of PLLA is higher when functionalized nanotubes were incorporated. On the other hand, it is observed that the addition of pristine nanotubes in concentrations higher than 1wt% inhibits this process. Therefore, this implies that different important mechanisms of interaction matrix-filler are present, depending on the surface characteristics of the nanotubes. These effects affect the thermal behaviour of the composites studied.

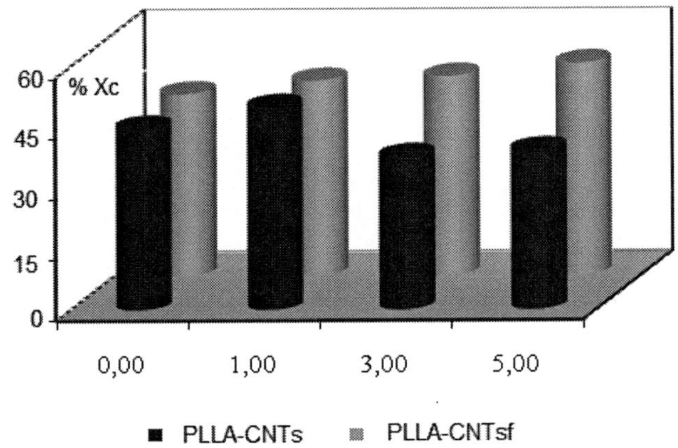

Figure 3: Crystallinity of the different composite materials.

This can be explain considering that the pristine carbon naotubes have a high surface energy that induces agglomeration and impedes their dispersion in common solvents. The process of functionalization is very efficient to modify the surface energy creating polar groups at the surface of nanotubes that contribute to the repulsion force between them facilitating dispersion.

The composites PLLA- MWCNTs and PLLA- MWCNTsf show a slight increase in the polymer crystallization fraction for concentrations of 1wt% of nanotubes. However, when higher nanotubes quantities are incorporated to the matrix a decrease in the crystallization process is observed. This behaviour can be explained considering that when the filler concentration is increased the formation of agglomerates is favoured limiting not only the density of nucleation sites given by the nanotubes but also the auto-nucleation of the polymer chains. On the contrary, when increase proportions of functionalized nanotubes are used, due to their surface energy, the repulsion energy keeps them apart and a good dispersion in the polymer matrix is obtained, resulting in a better crystallization process, since additional effective nucleation sites for the polymer crystallization are accessible.

Conclusions

Carbon nanotubes inhibit the cold crystallization of PLLA, this effect is dependent of the surface chemistry of the nanotubes. On the other hand the melting temperature increases with the addition of nanotubes, suggesting the formation of larger and more perfect crystals.

Also, the melting enthalpy and crystallinity of PLLA decreases with the addition of 3 and 5wt% of pristine MWCNTs, this is attributed ti the formation of agglomerates. On the contrary, for the composites PLLA-MWCNTsf these parameters increase proportional to the load content due to the higher dispersion of the MWCNTsf exerting a more effective nucleation effect.

References

[1] Coleman J., Balu W., Gunko Y. (2006). Carbon, 44, 1624-1652.
[2] Liu P. (2005). Eur. Polym. J., 41, 2693-2703.

Effects of electron beam irradiation on polyamide 12 with fiberglass reinforcement

Joon-Pyo Jeun, Bum-Sik Shin, Hyun-Bin Kim, Young-Chang Nho and Phil-Hyun Kang*

Radiation Research Division for Industry & Environment, Korea Atomic Energy Research Institute, 1266 Shinjeong-dong Jeongeup-si Jellabuk-do 580-185, Korea

Abstract. In the present study, the effects of electron beam irradiation of polyamide 12 (PA12) with fiberglass reinforcement on the thermal and wear properties were investigated. Electron beam irradiation of PA 12 was carried out over a range of irradiation doses (100-600 kGy) in air. The gel formation in the presence of a curing agent was dependent on the radiation doses. The thermal properties of irradiated PA 12 were studied in the temperature region 50-250 °C to observe the changes in the melting point with radiation dose. The dimensional stability was significantly increased by electron beam irradiation and the related crosslinking of the PA 12.

Keywords: Polyamide 12, Electron beam irradiation, Wear property, Thermal property
PACS: 68.55.am

INTRODUCTION

Polyamides are characterized by their possession of a high tensile strength, elasticity, tenacity and resistance to abrasion. These mechanical properties are maintained even under high temperatures and therefore, the polyamides can be used in automotive under the hood applications – the driving force being greater fuel economy and weight reductions coupled with the critical requirements of higher continuous use temperature and resistance to thermal degradation.

Radiation process is a well-known technique for the modification of polymers. The irradiation of polymeric materials with ionizing radiation such as gamma rays, X-rays, electron and ion beams leads to the formation of very reactive intermediates. These intermediates are able to enter into chemical reactions leading to chemically reactive products that initiate the crosslinking reaction. In Particular, the electron beam technology speeds up production, improves productivity, lowers costs, makes news and often better products.

During the last few decades, the industrial employment of the ionizing radiation has been growing for crosslinking thermoplastics, with excellent results in several sectors. The aim of this study is to study the effect of electron beam irradiation with different doses on the mechanical and thermal properties PA 12 with 15% fiberglass reinforcement and compare these results with those of non-irradiated samples. The change in thermal behavior of the prepared unirradiated and E-beam irradiated sample was investigated using DSC and TGA. In order to evaluate the sliding wear behavior, the wear performance of the samples was evaluated experimentally in the abrading test with the Taber Abraser.

EXPERIMENTALS

PA12 pellets with 15% of fiberglass reinforcement with a density of 1.16 g/cm^3 was kindly supplied by Mando cor., Republic of Korea. Trially cyanurate (TAC), triallyl isocyanurate (TAIC) and trimethylopropane trimethacylate (TMPTMA) as a crosslinking agent were obtained from Aldrich Chemical Company and used without further purification.

The PA12 pellets were vacuum dried at a temperature of 60 °C for 12 h and then mixed with 5w% of crosslinking agent by means of a lab-scale Brabender. Sheet of 1 mm thickness were then obtained by compression

CP1255, *Vth International Conference on Times of Polymers (TOP) and Composites,*
edited by A. D'Amore, D. Acierno, and L. Grassia

molding at 250 °C for about 5 min. The sheets kept in polyethylene zipper packets were subjected to e-beam irradiation in air at room temperature and in ambient humidity. The integral irradiation dose levels varied from 100 kGy to 600 kGy. 10 kGy dose was given in each pass and the samples in packets were cooled in air before the next pass.

The gel content of the samples was evaluated using a solvent extraction method according to ASTM D2765. 2g of irradiated samples of PA12 at various doses were accurately weighed and put into the Soxhlet extractor. After extracting with *m*-cresol for 36 h and with methanol for 24 h, the residue was dried in the vacuum oven untill the constant weight was obtained, the calculation was based on with the gel content. The DSC traces were recorded on a SDT Q600 (TA instrument, USA) at the rate of 10 °C/min. Thermogravimetric analysis was performed on a DSC Q100 (TA instrument, USA) with heating rate of 20 °C/min and nitrogen was used as the purge gas.

The wear tests according to ISO 5470 and IEC60068-2-70 were carried out using a linear abraser Model 5750 (TABER industries) in air at 25 °C with a relative humidity of 50 %. The samples cut had dimensions of 30 × 30 × 2 mm made by diamond cutter as wear partners of the slider (6 mm in diameter, H-18). The slider was loaded with about 9.8 N and the sliding rate was fixed at 75 cycle/min until a maximum cycle of 10,000 were reached. After testing, the loss weight of samples were measured and compared.

RESULTS AND DISCUSSION

The effects of various additives such as TAC, TIAC, TMPTMA on the formation of gel of PA12 during e-beam irradiation are shown in Fig. 1(a). In the absence of additives, there was almost no formation of gel up to 100 kGy indicating that the crosslinking and scission occurred simultaneously during irradiation. While the addition of 5w% additives led to a significant increase of the gel content by up to about 90%, which demonstrated that the crosslinking predominated over the scission reaction. On comparing PA12, it can be clearly seen that the overall tendency in density of the network is in the order of PA12-TAC >PA12-TAIC >PA12-TMPTMA >PA12. Thus, TAC is the most effective additive for PA12 under irradiation.

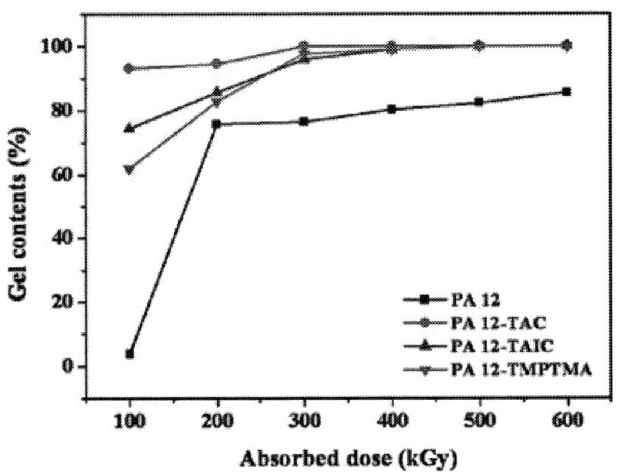

FIGURE 1. The effects of additives on gel formation as a function of radiation dose.

The thermal behavior of the un-irradiated PA12 and its blends with additives is displayed in Table 1. The presence of additives decreased the melting point, decomposition temperature, and residue content. However, the melting temperatures showed a decrease by e-beam irradiation for all samples. The decrease in melting temperature with an exposure dose has a different reason for the heating run. The behavior reflects the scission of molecules and chain folds at the crystal. Also, the decrease in melting temperature could be explained by the Gibbs-Thomson relationship, which correlated the decrease in melting temperature to the decrease in the crystallity in the blend. In this case, radiation causes a decrease in the length of the macromolecular chains, crosslinking, and other modifications in the crystal domains and, as a result, a decrease in melting temperature takes place.

The degree of crosslinking can be influenced by the radiation dose and additives species. The effects of e-beam irradiation and additives on the wear resistance of each samples were shown in Fig. 2. All e-beam irradiated samples showed significantly lower loss weight in comparison to the neat PA12. In particular, the wear resistance of the irradiated PA12 is about 3.75 greater than that of the un-irradiated PA12. PA12-TMPTMA showed the lowest changes.

TABLE 1. Thermal parameters of un-irradiated and irradiated PA12.

Sample composition	Dose (kGy)	T_m (°C)	T_d (°C)	Residue (%)
PA12	0	178.9	453.6-	15.1
	100	177.6	-	-
	200	176-9	-	-
	300	175.8	456.4	16.8
	400	174.5	-	-
	500	173.7	-	-
	600	173.0	-	-
PA12-TAC	300	166.9	453.48	15.6
PA12-TAIC	300	167.8	451.71	15.6
PA12-TMPTMA	300	173.2	449.65	15.6

T_m : Melting temperature determined by DSC.
T_d : Deformation temperature determinded by TGA.

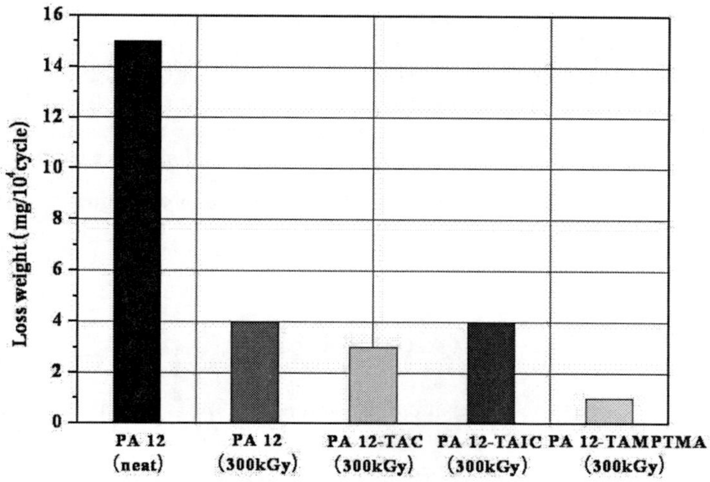

FIGURE 2. The abrasive wear data of PA samples as a function of additive species.

CONCLUSIONS

Radiation crosslinking of thermoplastic polymers offers an opportunity to increase the continuous operating temperature. However, the crosslinking and scission occur simultaneously during irradiation of PA12. The presence of additives was beneficial to radiation crosslinking of PA12. The addition of additives into PA12 led to a decrease of thermal properties, such as melting point and decomposition temperature. The melting temperatures showed a decrease by e-beam irradiation due to the scission of molecules and chain folds. Radiation crosslinked PA12 has a higher wear resistance compared to un-irradiated PA12.

ACKNOWLEDGMENTS

This research was supported by the Nuclear R&D program through the Korea Science and Engineering foundation funded by the Ministry of Education, Science and Technology, Republic of Korea.

REFERENCES

1. R. Sengupta, V. K. Tikku, A. K. Somani, T. K. Chaki and A. K. Bhowmick, *Rad. Phys.Chem.* **72**, 751-757 (2002).
2. M. M. Hassan, N. A. Badway, A. M. Gamal, M. Y. Elnaggar and E. A. Hegazy, *Nucl. Instrum. Meth. B ,In Press* (2010).
3. R. Feulner, Z. Brocka, A. Seefried, M. O. Kobes, G. Hulder and T. A. Osswald, *Wear,* **268**, 905-910 (2010).
4. R. Sengupta, S. Sabharwal, A. K. Bhowmick, T. K. Chaki, *Polym. Degrad. Stabil.* **91**, 1311-1318 (2006).
5. W. Feng, F. M. Hu, L. H. Yuan, Y. Zhou and Y. Y. Zhou, *Rad. Phys.Chem.* **63**, 493-496 (2002).
6. C. Pinto and L. G. Andrae e Silva, *Rad. Phys.Chem.* **76**, 1708-1710 (2007).

Preparation of Ag nanoparticles on PAN/TiO_2 nanofibers and their photocatalytic and antibacterial properties

Joon-Pyo Jeun, Dong-Kwon Seo, Hyun-Bin Kim and Phil-Hyun Kang*

Radiation Research Division for Industry & Environment, Korea Atomic Energy Research Institute, 1266 Shinjeong-dong Jeongeup-si Jellabuk-do 580-185, Korea

Abstract. Polyacrylonitrile/titania (PAN/TiO_2) nanofibrous mats were prepared by an electrospinning method and silver nanoparticles were deposited by electron beam (e-beam) irradiation of silver ions in aqueous solutions. The morphology of the electrospun fibers was characterized by SEM analysis. The deposition of silver particles was not of a uniform coverage on the PAN/TiO_2 surface, but of a heterogeneous growth of the silver particles on the PAN/TiO_2 surface. The photocatalytic efficiency of the composites against methylene blue dye in an aqueous solution improved and the actibacterial activity against E. *coli* and S. *aureus* was found to increase with the silver nanoparticles being induced.

Keywords: Silver nanoparticles, Electron beam irradiation, Photocatalytic activity, Antibacterial property
PACS: 62.23.Pq

INTRODUCTION

Nanomaterials often show unique and considerably changed physical, chemical and biological properties compared to their macroscaled counterparts. Therefore, the application of nanoparticles is an emerging area of nanoscience and nanotechnology. Nanoparticles such as silver, gold and copper may provide the solution to technological and environmental challenges in the areas of catalysis, electronics, optics, biological labeling, medicine and water treatment. Most of all, silver nanoparticle is of particular interest due to its distinctive properties, such as chemical stability, good conductivity and antibacterial activity.

Nanocomposite materials composed of silver nanoparticles contained in polymers have attracted much attention due to their noticeable catalytic, optical and electric properties which have potential application in the fields of catalysis, photonics and electronics. Polymers are considered as good host materials for silver nanoparticles as well as other stabilizing agents such as citrates, organic solvents, long chain alcohols, and surfactants. Various silver-polymer nanocomposites such as silver-polyacrylonitrile, silver-polyacylamide, silver-polyvinyl alcohol, silver-poly(N-vinyl-2-pyrolidone), and silver-polyimide have demonstrated their usefulness in these roles. Silver/TiO_2 surfaces have many advantages such as photocatalysis, antibacterial activity, and biological compatibility. Immobilization of silver nanoparticles on TiO_2 to form hybrid composites with various structures and surface functionalities have been reported to show improved photocatalytic activities.

In this contribution, the aim of this study was to compare the effect of the silver deposition on their properties, especially on the photocataltic activity and antibacterial property. PAN/TiO_2 nanofibrous mats were prepared via electrospinning method, and the mats were irradiated by an e-beam to enhance the photocatalytic activity. The deposition of silver nanoparticle was performed by using the e-beam irradiation method, and the resulting composite structure was characterized by scanning electron microscopy and energy-dispersive spectroscopy. Photoefficiency against methylene blue dye were carried out in an aqueous solution as the adsorption cycle varies and antibacterial properties again E. *coli* and S. *aureus* were investigated.

EXPERIMENTALS

PAN (average molecular weight = 1.5 X 10^5 g/mol, density 1.18g /cm^3, Aldrich, USA) and $Ti(OBu)_4$ (97 %) were used as the precursor. N,N-dimethyl formamide (DMF, Showa, Japan) was used as a solvent without any

CP1255, *Vth International Conference on Times of Polymers (TOP) and Composites,*
edited by A. D'Amore, D. Acierno, and L. Grassia

further purification. Acetic acid (CH_3COOH, 97 %, Showa, Japan) was used as a complex agent to control the hydrolysis rate. TritonX-100 (Aldrich, USA), Isopropyl alcohol (99.5% Aldrich, USA) and $AgNO_3$ (99.8%, Showa, Japan) were used as silver deposit materials. Staphylococcus aureus KCTC 1916 (S. aureus, Gram positive) and Escherichia coli KCTC 1682 (E. coli, Gram negative) and agar (Difco, USA) and distilled water were used.

Polymer solution was prepared with weight ratio [PAN: $Ti(Obu)_4$: DMF = 2:1:18]. $Ti(OBu)_4$ was allowed to react with acetic acid. Acetic acid added with the ratio of acetic acid and $Ti(OBu)_4$ was 2:1. The Polymer solution was drawn horizontally from the needle tip (ID 0.241 mm), driven by the electrostatic force generated form from the high voltage between the needle tip and the collector. The polymer concentration was controlled at 9 wt% (w/w). The electrospun PAN/TiO_2 nanofiber mat was collected on a collector drum which was placed at a distance of 10 cm from the needle tip. The electrospun nanofibrous membrane was dried under a vacuum. The PAN/TiO_2 nanofiber mat was placed in a stainless still chamber and sealed after the introduction of N_2 gas for electron beam irradiation in N_2 atmosphere. Electron beam irradiation was carried out at a dose rate 14,000 kGy/h. The electron beam was generated with a 1.14 MeV acceleration voltage, 4 mA of current, and 10,000 kGy absorbed dose.

In order to deposit Ag nanoparticles, the samples of PAN/TiO_2 nanofiber mats were dipped in TritonX-100 (9 ml), isopropyl alcohol (45 ml), silver nitrate (1.53 g) aqueous solution (300 ml). The solution with PAN/TiO_2 mats were irradiated by the electron beam up to 50 kGy. After electron beam irradiation, the mats were washed thoroughly by de-ionized water and dried at 60°C for 24h. Study of the morphology was conducted by operating scanning electron microscope (SEM, Sirion, FEI, Netherlands).

Photocatalytic activities of Ag nanoparticles deposited on PAN/TiO_2 nanofiber mat and bare PAN/TiO_2 nanofiber mat were determined by the decomposition rate of methylene blue in a dark aqueous chamber. A 50 W lamp as the UV light source (365 nm) was exposed in a cylindrical glass vessel (diameter: 8 cm, height: 10 cm). Changes of the maximum absorption at 355-360 nm in UV-vis absorption spectra versus irradiation time were recorded, which reflected the change of methylene blue concentration during the photocatalysis.

The antibacterial activity of the prepared silver nanoparticles was tested on Staphylococcus aureus KCTC 1916 (S. aureus, Gram positive) and Escherichia coli KCTC 1682 (E. coli, Gram negative) by the disk susceptibility test. The bacteria were cultured at 37°C for 24 h. A zone of inhibition test was done on the Ag nanoparticles deposited on nanofiber mat.

RESULTS AND DISCUSSION

The SEM images of silver nanoparticles deposited PAN/TiO_2 fibers were shown in Fig. 1. It can be seen from the figure that silver containing fibers appeared to have roughened and show a non-uniformed surface at the silver particles containing places but otherwise the surfaces are smooth. The amount of silver nanoparticles was increased by increasing the e-beam irradiation dose.

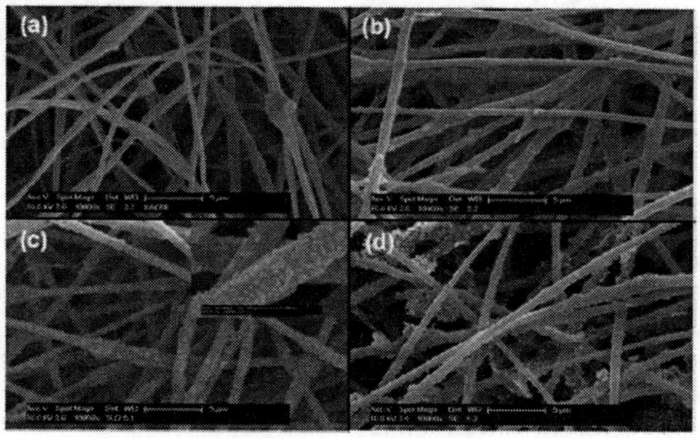

FIGURE 1. SEM micrographs of (a) PAN/TiO_2 mat and PAN/TiO_2 mat containing Ag nanoparticles with different e-beam dose; (b) 10 kGy, (c) 20 kGy, and (d) 50 kGy.

The concentration of aqueous methylene blue dye solution was reduced by increasing the UV irradiation time, as shown in Fig. 2. After e-beam irradiation, the photoactivity of PAN/TiO_2 mat increased to some degree. However, a significant increase in the photodegradation of methylene blue dye were observed in silver particles depositd on PAN/TiO_2 mat. The photocatalysis mechanism involves adsorption of methylene blue dye molecules in water on the surface of PAN/TiO_2 fiber and silver nanoparticles, which are then excited by the UV energy to generate the electron and hole pairs within the particle volume. However, the work function of silver particles deposited on PAN/TiO_2 fibers is much higher than that of PAN/TiO_2 fibers. When silver particles deposited on PAN/TiO_2 are excited by UV irradiation, a portion of photo-induced electrons will transfer to the silver particles and will be scavenged by the electron acceptor. Therefore, more photo-induced holes will have an opportunity to escape the geminated recombination of electron-hole pairs and take part in the following oxidation reactions that will results in the enhanced catalytic efficiencies. Deposition of silver on TiO_2 surface has been reported to enhance the photo-induced electron-hole lifetimes as a result of the effective trapping of the photo-induced electrons to the surface deposited silver nanoparticles.

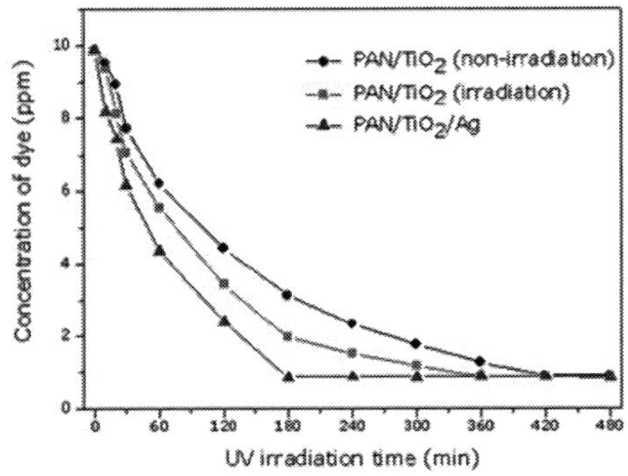

FIGURE 2. Photodegradation of methylene blue dye solution as a function of UV irradiation time.

The silver nanoparticles deposited on PAN/TiO_2 fibrous mat were tested for antibacterial properties on E. *coli* and S. *aureus*. A parallel test of the electrospun PAN/TiO_2 without the silver deposition has also been carried out for comparison. The zone of inhibition observed for the silver coated mat confirms the antibacterial action of the silver nanoparticles, whereas there was no zone of inhibition for the PAN/TiO_2 mats, as shown in Fig. 3a and b, respectively. These observations clearly demonstrate the antibacterial properties of silver nanoparticles.

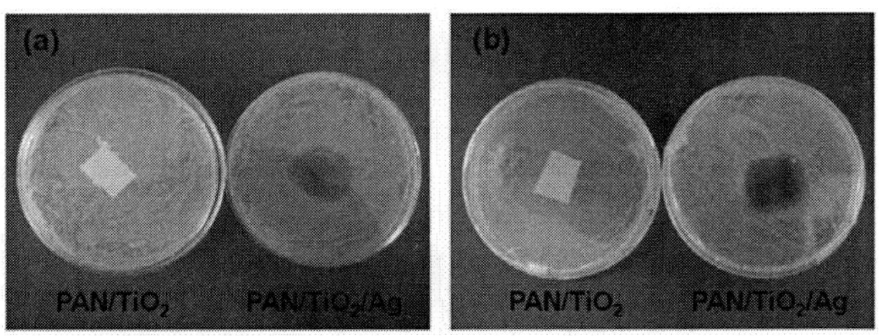

FIGURE 3. Comparison of inhibition zone test on E. *coli*(a) and S. *aureus*(b) between PAN/TiO_2 mat and Ag on PAN/TiO_2 mat.

CONCLUSIONS

PAN/TiO_2 nanofibrous mats containing silver nanoparticles were prepared by electrospinning method and e-beam irradiation process. The prepared fibers were nano-scale sized and contained silver particles on the surface of the fibers. The method of radiation-induced silver deposition is an effective way to produce nanoparticles on the PAN/TiO_2 mat. The photocatalytic efficiency is enhanced by treating e-beam irradiation. Furthermore, the degradation of methylene blue dye was more effective in the case of the suspending silver nanoparticles. The silver particles deposited on PAN/TiO_2 fibers showed antibacterial activity against both E. *coli* and S. *aureus*. These antibacterial silver nanoparticles, with can be prepared in a highly stable state by e-beam irradiation, may be suitable for the formulation of new types of bactericidal materials.

REFERENCES

1. V.K. Sharma, R.A. Yngard, and Y. Lin, *Adv. Colloid Interfac.* **145**, 83-96 (2009).
2. W.C. Lin, C.N. Chen, T.T. Tseng, M.H. Wei, J.H. Hsieh, and W.J. Tseng, *J. Eur. Ceram. Soc. In press* (2010).
3. J.S. Im, M.I. Kim, and Y.S. Lee, *Mater. Lett.* **62**, 3652-3655 (2008).
4. X.G. Hou and A.D. Liu, *Rad. Phys.Chem.* **77**, 345-351 (2008).

Crystallization of PCL/Clay Composites

Ivica Janigová, Daniela Jochec Mošková and Ivan Chodák

Polymer Institute, Slovak Academy of Sciences, Dúbravská cesta 9, 842 36 Bratislava, Slovak Republic

Abstract. Biodegradable polymer/clay nanocomposites and/or composites based on poly(ε-caprolactone) (PCL) were prepared by conventional melt mixing. Three kinds of clays, organomodified Cloisite 10A, Cloisite 15A and unmodified Cloisite with Na cations were used for preparation in different concentrations (from 1 to 10wt%). The isothermal crystallization of the samples were studied by DSC at temperatures from 41 to 47°C. The obtained results of overall crystallization rate v_c and half-time of crystallization $t_{0.5}$ indicated that the filler presence in the PCL matrix leads to superposition of two opposing effects – nucleation and retardation. The different tendency of the changes of v_c and $t_{0.5}$ in the dependence of filler content was showed between samples with organomodified and unmodified fillers. It can be explained by different surface areas of modified and unmodified clays and proportional to an extent of filler exfoliation.

Keywords: Poly(ε-caprolactone), Clay, Isothermal crystallization, DSC, Composites, Filler.
PACS: 82.35.Np

INTRODUCTION

It is known that a study of crystallization of nanocomposites and/or composites helps us to understand the changes of mechanical and physical properties. It has been recently reported that well dispersed organoclay platelets act as nucleating agent in PCL matrix[1,2] resulting in an increasing of the overall crystallization and reducing of half-time of crystallization. This effect is remarkable for the low concentration of the clay.

The overall crystallization rate can be obtained from the slope of the dependence[3]

$$x(t) = f(t) \tag{1}$$

in which the amount of the crystalline phase x(t) measured by DSC can be expressed as follows

$$x(t) = A_t/A \tag{2}$$

where A_t is the area under the DSC curve corresponding to an amount of the heat released during crystallization on time t and A is the area under the whole DSC curve related to overall heat released during crystallization.

In the present paper PCL nanocomposites and/or composites filled with organomodified and unmodified clays prepared by conventional melt mixing have been investigated regarding isothermal crystallization.

EXPERIMENTAL

Materials

PCL (CAPA 6800, Solvay Interbox Ltd. , UK) was used as the polymeric matrix. Three kinds of montmorillonite (Southern Clay Products, Inc., USA) were chosen for nanocomposite and or composite preparation. Two of them are organomodified materials containing different quarternary ammonium cations Cloisite 15A (dimethyl dehydrogenated-tallow ammonium, C15A) and Cloisite C10A (dimethyl benzyl hydrogenated-tallow ammonium, C10A). Natrified clay Cloisite Na (CNa) was used for comparison as a filler without hydrophobic surface modification.

CP1255, *V[th] International Conference on Times of Polymers (TOP) and Composites,*
edited by A. D'Amore, D. Acierno, and L. Grassia

The composites were prepared by conventional melt mixing at 75°C for 10 minutes at 30 rpm in a 50 ml mixing chamber of Plastograph Brabender PLE 330. The filler content in the samples varied between 1 to 10 wt %. After mixing, the samples were compression molded at 70°C for 2 minutes to 1 mm thick slabs.

Method

Isothermal crystallization was measured at four different temperatures for each sample in temperature range from 41 to 47°C by Mettler-Toledo DSC 821^e calorimeter. The samples were heated to 100°C (heating rate 10°/min) and kept at this temperature for 5 min. They were subsequently cooled at a rate 80°/min to the crystallization temperature.

RESULTS AND DISCUSSION

Prepared nanocomposites and/or composites characterized by X-ray diffraction, electron microscopy were described in the previous study[4]. The study showed intercalation of clay layers by polymer matrix and confirmed a formation of nanostructure and also existence of the exfoliated layers of organomodified clays.

The values of crystallization parameters overall crystallization rate v_c, half-time of crystallization $t_{0.5}$ and overall heat released during crystallization ΔH_c in the dependence of crystallization temperatures and filler content in PCL matrix for each type of filler are summarized in Tables 1-3.

TABLE 1. Crystallization parameters of overall crystallization rate v_c, half-time of crystallization $t_{0.5}$ and heat released during crystallization for virgin PCL and nanocomposite PCL/C10A for filler content 1, 3 and 10 wt%.

PCL/C10 (wt%)	T_c (°C)	$V_c.10^2$ (s^{-1})	$t_{0.5}$ (s)	ΔH_c (J/g_{PCL})
0	41	0.587	317	57
	43	0.348	570	57
	45	0.167	1239	58
	47	0.067	2864	50
1	41	0.695	224	62
	43	0.546	302	63
	45	0.221	672	56
	47	0.098	1661	62
3	41	0.582	247	59
	43	0.350	418	61
	45	0.191	753	69
	47	0.099	1650	65
10	41	0.435	358	64
	43	0.266	483	63
	45	0.187	878	73
	47	0.050	2282	73

TABLE 2. Crystallization parameters of overall crystallization rate v_c, half-time of crystallization $t_{0.5}$ and heat released during crystallization for nanocomposite PCL/C15A for filler content 1, 3 and 10 wt.%.

PCL/C15A (wt%)	T_c (°C)	$V_c.10^2$ (s^{-1})	$t_{0.5}$ (s)	ΔH_c (J/g_{PCL})
1	41	0.704	223	62
	43	0.421	363	62
	45	0.245	663	63
	47	0.124	1478	60
3	41	0.569	249	62
	43	0.358	392	59
	45	0.164	819	58
	47	0.112	1582	63
10	41	0.425	334	66
	43	0.246	573	66
	45	0.132	1095	69
	47	0.066	2607	67

The comparison of v_c and $t_{0.5}$ values for the samples filled with organomodified and unmodified clays with increasing filler content leads to different results. The samples filled with C15A and C10A showed the decrease of v_c and the increase of $t_{0.5}$ for all isothermal temperatures (Tables 1,2). The opposite effect was observed at the samples with CNa as a filler (Table 3). The values of v_c increased and $t_{0.5}$ decreased with increasing filler content. This can be explained by differences of surface area of the filler and the degree of filler intercalaction or exfoliation in the matrix. At lower filler content organomodified clays act as nucleation agents proportional to the exfoliation. At higher filler content a diffusion of polymer chains is reduced and crystallization is retarded. This effect is pronounced at the fillers with high surface area. On the other hand unmodified filler CNa with low degree of intercalation, smaller and hydrophilic surface area acts as a nucleation agent at the samples with higher filler content where values of v_c significantly increase and $t_{0.5}$ decrease. From this point of view we can assume the superposition of two contrary effects during crystallization: nucleation and retardation. The values of heat released during crystallization are higher for all prepared samples at the different isothermal temperatures and filler content in comparison with the values of unmodified PCL. This correlates with the nucleation activity of the filler. High value of ΔH_c for sample PCL/10wt%CNa (Table 3) is caused by double peak of crystallization, probably due to secondary crystallization.

TABLE 3. Crystallization parameters of overall crystallization rate v_c, half-time of crystallization $t_{0.5}$ and heat released during crystallization for nanocomposite PCL/CNa for filler content 1, 3 and 10 wt%.

PCL/CNa (wt%)	T_c (°C)	$V_c.10^2$ (s^{-1})	$t_{0.5}$ (s)	ΔH_c (J/g_{PCL})
1	41	0.772	246	58
	43	0.461	422	59
	45	0.271	818	63
	47	0.139	1793	71
3	41	0.649	236	54
	43	0.472	383	58
	45	0.219	706	58
	47	0.132	1655	64
10	41	0.988	169	59
	43	0.747	244	64
	45	0.509	419	75
	47	0.265	882	290*

*The double peak of crystallization

CONCLUSSIONS

The study of isothermal crystallization PCL/clay nanocomposites and/or composites prepared by conventional melt mixing with three kinds of commercial Cloisite (C10A, C15A-organomodified, CNa-unmodified) leads to the assumption of the superposition of two effects during the process of crystallization. Nucleation or retardation of crystallization prevails in the dependence of filler content in the samples. The intensity of these phenomena correlated with the extent of surface area and degree of clay exfoliation in the matrix. The presence of all clay kinds accelerates the crystallization of PCL. The values of overall crystallization rate and heat of released during crystallization are higher and the values of half-time of crystallization for all filled samples are lower than those for the origin PCL matrix.

ACKNOWLEDGMENTS

The authors are grateful for financial support provided by Slovak Research and Development Agency (Grant No APVV 51-050505 and Slovak Grant Agency for Science (Grant No VEGA 2/0185/10).

REFERENCES

1. E. Di Maio, S. Iannace, L. Sorrentino and L. Nicolais, *Polymer* **45**, 8893-8900 (2004).
2. D. Homminga, B. Goderis, i. Dolbnya and G. Groeninckx, *Polymer* **47**, 1620-1629 (2006).
3. P. C. Vilanova, S. M. Ribas and G. M. Guzman, *Polymer* **26**, 423-428 (1985).
4. I. Janigová, F. Lednicky, D. Mošková, K. Csomorová and I. Chodák, "Morphology and properties of poly(ε-caprolactone) /montmorillonite nanocomposites" in *Nanostructured and Functional Polymer-Based Materials and Nanocomposites*, Proceedings of 4th International Conference, Roma, Italy, 2008, pp. 227-235.

Effect of Diisocyanate Structure on Thermal Properties and Microstructure of Polyurethanes Based on Polyols Derived from Renewable Resources

Mª Angeles Corcuera, Lorena Rueda, Borja Fernández d'Arlas, Ainara Saralegui, Cristina Marieta, Aitor Arbelaiz, Iñaki Mondragon, Arantxa Eceiza

"Materials + Technologies" Group, Department of Chemical and Environmental Engineering, Escuela Politécnica, University of the Basque Country, Pza Europa 1, 20018, Donostia-San Sebastián, Spain

Abstract. Polyols derived from renewables resources are good candidates to obtaining segmented polyurethane elastomers. Diisocyanates with different chemical structure, aliphatic and aromatic, have been used to synthesize by a two step polymerization procedure polyurethane elastomers with different hard segment content. Microphase separation and thermal stability have been studied using attenuated total reflection Fourier transform infrared spectroscopy (ATR-FTIR), differential scanning calorimetry (DSC) and thermogravimetric analysis (TGA). The analysis of the H-bonded and non H-bonded urethane carbonyl stretching vibration in the amide I region, the glass transition temperature of the soft and hard segments and the melting temperature and enthalpies of hard segment reveal that aliphatic diisocyanate based polyurethanes present higher phase separation degree and harder segment crystallinity and also superior thermal stability than aromatic diisocyanate-based polyurethanes.

Keywords: polyurethane, castor oil, microphase structure, thermal properties.
PACS: 81.05.Lg, 81.05.Zx, 81.70.Pg, 82.80.Gk, 88.20.ft

INTRODUCTION

Polyols derived from vegetable oils are new materials from renewable sources (1, 2). Combined with diisocyanates, these polyols produce polyurethanes (PU) that can compete in many aspects with PU derived from petrochemical polyols, reducing the environmental impact. These PU can be used in many applications such as abrasion-resistant components, biomaterials, durable coating, and foams (3, 4). Segmented polyurethane elastomers are block copolymers formed by soft segments (SS) and hard segments (HS). The soft segment, the polyol, provides extensibility and the hard segment, derived from diisocyanate and chain extender, plays the role of physical crosslinker and acts as high modulus filler. The incompatibility between both segments causes phase separation determined by factors such as chemical structure of PU precursors, segmental lengths, nature of interfacial region between the segments, hydrogen bonding (intramolecular interactions) and crystallization extent (5). The nature of hydrogen bonding in the HS causes a strong mutual attraction leading to hard and soft segments domain formation (6, 7). Interesting properties of segmented polyurethane elastomers are strongly dependent on their phase separated morphologies (8).

The choice of used monomers to synthesize elastomeric polyurethanes is dependent on the final applications of the material. It is well-known that the properties of polyurethanes are remarkably affected by the content, type, and molecular weight of the soft segments (5, 9). The nature of diisocyanate also concerns the polyurethane properties (5, 10). The study of the effect of the diisocyanate structure on microstructure and thermal properties of polyurethane based on polyols from renewable resources has been the main aim of this work. Segmented PU elastomers based on castor oil as soft segment have been synthesized using 1,4-butanediol (BD) as chain extender and two diisocyanate, one aliphatic, 1,6-hexamethylene diisocyanate (HDI), and other aromatic, 4,4′-diphenilmethane diisocyanate (MDI), as components of the hard segment by a two-step synthetic procedure. Microphase structure and properties have been analyzed using attenuated total reflexion Fourier transform infrared spectroscopy (ATR-FTIR) and differential scanning calorimetric (DSC) as complementary techniques, used to

CP1255, *Vth International Conference on Times of Polymers (TOP) and Composites,*
edited by A. D'Amore, D. Acierno, and L. Grassia

characterize thermal transitions of the polyurethane and to assign them to functional groups interactions. The thermal stability has also been analyzed by means of thermogravimetric analysis (TGA)

EXPERIMENTAL

Materials and Synthesis

Diisocyanates used HDI (Desmodur H) and MDI (Desmodur 44) were kindly provided by Bayer. The polyol, was kindly supplied by BASF (Lupranol Balance 50), being derived from castor oil. The polyol number-average-molecular weight is 3366 g/mol, as determined using ASTM-D 4274-88 standard for measuring the hydroxyl number. Chain extender BD was provided by Aldrich.

PU were prepared using two-steps bulk polymerization procedure. Before reaction, the polyol and BD were dried under vacuum for 12 h at 90 °C and 40 °C, respectively. HDI and MDI were used as received. Dried polyol and an excess of diisocyanate were placed in a three-necked round-bottom flask fitted with a dry nitrogen inlet, condenser, and mechanical stirrer, and heated in a thermoregulated silicon bath for 5 h at 90 °C for the HDI-based systems, PU-HDI, and 5 h at 80 °C for the MDI-based systems, PU-MDI, according to kinetics measurements, with stirring to end-cap the macrodiol with isocyanate. Chain extender was added to the prepolymer at 90 °C for the HDI-based systems and at 60 °C for the MDI-based systems, while rapidly stirring with a stainless steel paddle for 10 min to homogenize the mixture and for evacuating the air trapped within it. The resulting viscous liquid was quickly poured between two domain coated metal plaques separated by 1 mm and left to cure under pressure for 10 h at 100 °C. PU are named as shown in Table 1.

TABLE 1. Molar Ratio, Hard Segment Content and Designation of the Synthesized Polyurethanes.

Polyurethane Code	Molar Ratio Polyol:Diisocyanate:Chain Extender	Hard Segment (wt % HS)
PUHDI-1	1:3:2	21
PUHDI-2	1:5:4	32
PUHDI-3	1:7:6	40
PUHDI-4	1:11:10	50
HDI-BD	0:1:1	100
PUMDI-1	1:3:2	27
PUMDI-2	1:5:4	39
PUMDI-3	1:7:6	48
PUMDI-4	1:11:10	59
MDI-BD	0:1:1	100

Characterization Techniques

Infrared Spectroscopy

Attenuated total reflexion Fourier transform infrared spectroscopy was used to characterize the functional groups of the synthesized polyurethanes and also to investigate hydrogen bond formation in the hard segments. Measurements were performed with a Nicolet Nexus FT-IR spectrometer equipped with a MKII Golden Gate accessory, Specac, using diamond crystal as ATR element at a nominal incidence angle of 45° with a ZnSe lens. Single beam spectra of the samples were obtained after averaging 20 scans in the range from 4000 to 600 cm^{-1} with a resolution of 2 cm^{-1}.

Differential Scanning Calorimetry

Differential scanning calorimetry thermograms of the low-temperature behaviour of the synthesized polyurethanes was accomplished cooling the sample from room temperature to -90 °C with a intracooler and then heating up to 0 °C at 20 °C/min under dry helium atmosphere using a Perkin-Elmer DSC-7 instrument. The thermal behaviour above room temperature of PU-MDI and PU-HDI polyurethane systems was investigated by scanning the specimens from -60 °C up to 200 °C under dry nitrogen atmosphere at a scan rate of 20 °C/min. Sample weights employed were 5-10 mg.

The mid-point of the heat capacity change was chosen to represent the glass transition temperature. Melting points refer to the maximum of the endothermic peak temperatures.

Thermogravimetric Analysis

Thermogravimetric analysis at a scanning rate of 10 °C/min from room temperature up to 700 °C under nitrogen atmosphere was performed to investigate the polyurethane thermal stability using a TGA/SDTA 851 Mettler Toledo.

RESULTS and DISCUSSION

ATR-FTIR spectroscopy was used to investigate the effect of HDI and MDI diisocyanate chemical structure and hard segment content on the microstructure of polyurethanes, based on a polyol derived from castor oil as soft segment, by means of hydrogen bonding formation in the amide I region. Figure 1a-b show typical infrared spectra of HDI and MDI-based polyurethanes and also pure polyol in the carbonyl group stretching vibration region. Polyol shows a characteristic triglyceride ester carbonyl band at 1743 cm^{-1}. It is well known that the infrared absorbance of H-bonded urethane carbonyl groups in ordered (crystalline) hard domains, Hb_{ord}, appears at lower wavenumbers than that of H-bonded urethane carbonyl groups in disordered (amorphous) domains, Hb_{disord}. On the other hand, the urethane carbonyl band appears at different wavenumbers depending if the free electron pair of the amine group is conjugated with the π bond of carbonyl group in HDI-based polyurethanes or with the aromatic ring in MDI-based polyurethanes. In both polyurethane systems, while the band associated with Hb_{disord} at 1728 and 1714 cm^{-1} for HDI and MDI-based polyurethanes, respectively, decreases with increasing hard segment content, the Hb_{ord} at 1700 and 1685 cm^{-1} for MDI and HDI-based polyurethanes, respectively, increases with hard segment content increasing and, in this way, it provokes the formation of large HS domains that contribute to the phase separated microstructure. Nevertheless the relative intensity of Hb_{disord} respect to Hb_{ord} is higher for MDI-based polyurethane, which means that less phase separated microstructure is obtained. In this way, the band associated with polyol carbonyl group in MDI-based polyurethanes is not observed, as it can be masked with the Hb_{disord}. The higher intensity observed for Hb_{ord} of HDI-based polyurethanes could be related with a higher hard domain crystallinity degree.

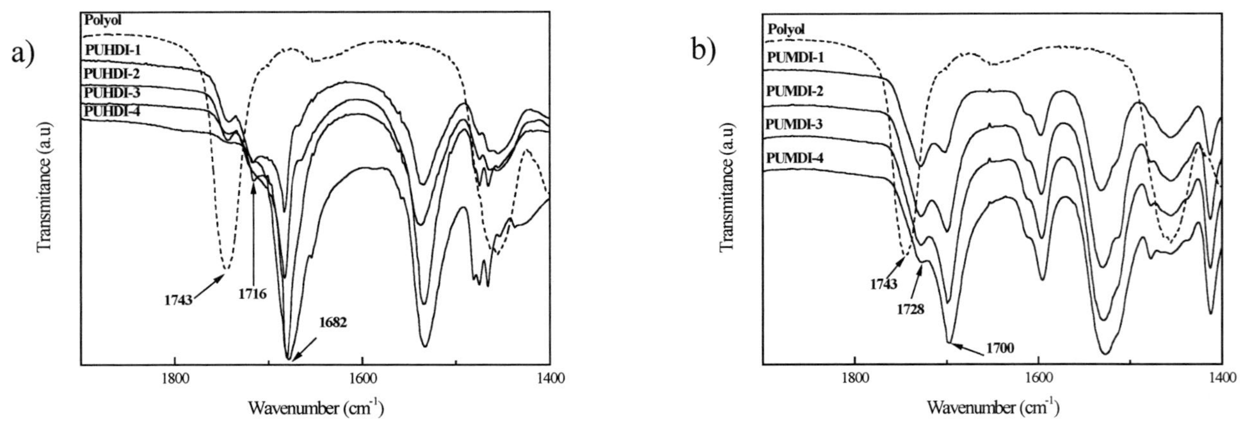

FIGURE 1. ATR-FTIR Spectra of Pure Polyol and PU Systems in the Stretching Vibration Carbonyl Group Region: a) PU-HDI Systems b) PU-MDI Systems.

Thermal properties of PU-HDI and PU-MDI systems have been analyzed by means of DSC. The measured values are given in Table 2. Segmented PU generally display several thermal transitions which can be related with microdomain purity or microdomain separation degree i.e., the presence of the hard segments in the soft domain and viceversa. The soft segment SS and hard segment HS glass transition temperatures, T_{gSS} and T_{gHS}, respectively, showed that HDI-based PU are more phase separated than MDI-based PU, suggesting that a relatively lower amount of HS is mixed within the soft domain and viceversa. On the other hand, although MDI-based PU systems showed higher melting temperature, they presented lower values of melting enthalpy (ΔH_m), indicating that this system is less likely to crystallize, which could be related with the steric hindrance that aromatic ring introduces on the hard segment structure lowering hydrogen bond formation, as shown above by FT-IR.

TGA has been employed to study polyurethane thermal degradation which occurs in three stages; a first stage related with the depolymerization of urethane bond, a second stage related with soft segment polyol degradation and a third stage related with the degradation of advanced fragments produced after the second stage (11). The first stage of polyurethane degradation depends on diisocyanate structure. Polyurethanes based on alkyl diisocyanates show higher thermal stability than PUs based on aryl diisocyanates (12). As observed, while both polyurethane systems showed similar second and third degradation stages, differences in the first degradation stage were observed. Indeed, MDI-based PU showed a first degradation stage between 260-345 °C and HDI-based PU between 270-370 °C. The results indicate that HDI-based PU present higher thermal stability than MDI-based PU.

TABLE 2. Thermal Properties of the Pure Polyol, Pure Hard Segment and Polyurethanes Systems.

Polyurethane Code	T_{gSS} (°C)	T_{gHS} (°C)	T_m (°C)	ΔH_m (J/g)
Polyol	-62.8	-	-	-
PUHDI-1	-50.1	58.2	149.9	3.1
PUHDI-2	-50.9	53.7	158.8	13.2
PUHDI-3	-50.9	46.7	159.8	27.9
PUHDI-4	-51.2	50.2	164.1	40.8
HDI-BD	-	52.1	189.0	122.3
PUMDI-1	-42.1	69.8	176.7	0.9
PUMDI-2	-42.7	62.2	178.0	7.6
PUMDI-3	-41.8	58.8	182.5	15.3
PUMDI-4	-42.1	62.2	186.0	26.7
MDI-BD	-	100.0	223.9	39.6

CONCLUSIONS

Polyurethanes derived from castor oil with HDI/BD and MDI/BD as hard segments have been synthesized, analyzed and compared. FTIR and DSC results show that independently of diisocyanate used, the phase separation degree and crystallynity increase with increasing hard segment content, being higher for the HDI-based polyurethanes. The degradation analysis reveals that HDI-based polyurethanes present higher thermal stability.

ACKNOWLEDGMENTS

The authors would like to thank Gobierno Vasco 'Saiotek 2008 (S-PE08UN11) and 2009 (S-PE09UN07)', and 'Grupos de Investigación Consolidados (IT-365-07)'.

REFERENCES

1. C. K. Williams and M. A. Hillmyer, *Polym. Rev.* **48**, 1-10 (2008).
2. A. Gandini, *Macromolecules* **41**, 9491-9504 (2008).
3. H. Uyama, M. Kuwabara and S. Kobayashi, *Biomacromolecules* **4**, 211-215 (2003).
4. P. Krol and B. Pilch-Pitera, *Eur. Polym. J.* **37**, 251-266 (2001).
5. I. Yilgor, E. Yilgor, I. G. Guler, T. C. Ward and G. L. Wilkes, *Polymer* **47**, 4105-4114 (2006).
6. M. M. Coleman, K. H. Lee, D. J. Skrovanek and P. C. Painter, *Macromolecules* **19**, 2149-2157 (1986).
7. L. Ning, W. De-Ning and Y. Shenk-Kang, *Macromolecules* **30**, 4405-4409 (1997).
8. L. Rueda-Larraz, B. Fernandez d'Arlas, A. Tercjak, A. Ribes, I. Mondragon and A. Eceiza, *Eur. Polym. J.* **45**, 2096-2109 (2009).
9. L. T. J. Korley, B. D. Pate, E. L. Thomas and P. T. Hammond, *Polymer* **47**, 3076-3082 (2006).
10. K. Kojio, S. Nakashima and M. Furukawa, *Polymer* **3**, 153-158 (2002).
11. I. Javni, Z. S. Petrovic, A. Guo and R. Fuller, *J. Appl. Polym. Sci.* **77**, 1723- 1734 (1999).
12. D.K. Chattopadhyay and D.C. Webster, *Prog. Polym. Sci.* **34**, 1068-1133 (2009).

EXPERIMENTAL DESIGN ON LAMINATED VENEER LUMBER FIBER COMPOSITE: SURFACE ENHANCEMENT

U. Meekum [1,2] and Y. Mingmongkol[1,2]

[1] *School of Polymer Engineering, Institute of Engineering, Suranaree University of Technology, Nakorn Ratchasima, 30000, THAILAND*

[2] *Centre of Excellent for Petroleum, Petrochemical and Advanced Material, Chulalongkorn University, Bangkok, THAILAND*

Thick laminate veneer lumber(LVL) fibre reinforced composites were constructed from the alternated perpendicularly arrayed of peeled rubber woods. Glass woven was laid in between the layers. Native golden teak veneers were used as faces. In house formulae epoxy was employed as wood adhesive. The hand lay-up laminate was cured at 150°C for 45 mins. The cut specimen was post cured at 80°C for at least 5 hours. The 2^k factorial design of experimental(DOE) was used to verify the parameters. Three parameters by mean of silane content in epoxy formulation(A), smoke treatment of rubber wood surface(B) and anti-termite application(C) on the wood surface were analysed. Both low and high levels were further subcategorised into 2 sub-levels. Flexural properties were the main respond obtained. ANOVA analysis of the Pareto chart was engaged. The main effect plot was also testified. The results showed that the interaction between silane quantity and termite treatment is negative effect at high level(AC+). Vice versa, the interaction between silane and smoke treatment was positive significant effect at high level(AB+). According to this research work, the optimal setting to improve the surface adhesion and hence flexural properties enhancement were high level of silane quantity, 15% by weight, high level of smoked wood layers, 8 out of 14 layers , and low anti termite applied wood. The further testes also revealed that the LVL composite had superior properties that the solid woods but slightly inferior in flexibility. The screw withdrawn strength of LVL showed the higher figure than solid wood. It is also better resistance to moisture and termite attack than the rubber wood.

Keywords: Design of Experiment, Laminate Veneer Lumber, Surface Treatment and Fiber Reinforced Composite .
PACS: 81

INTRODUCTION

LVL is one kind of engineered wood. It is commonly made by bonding thick or thin veneers with adhesives resulted a large billet. It is mainly used for both heavy and low load construction materials. The properties of the LVL are depended on the constituents used such as type of wood, adhesive, processing parameter and reinforcements. The outstanding mechanical property of the LVL is bending strength. It can be stronger than the wood lumber by more than two folds. There were several works reported on the properties dependency of the LVL components[1-4]. From our previous work that studied on wood, epoxy adhesive and reinforcement types by using Design of Experiment(DOE) approach. It was found that para rubber wood, room temperature epoxy system and glass woven fiber are superior constituent for making the LVL by mean of economic and engineering aspects. In this publication, the surface treatment in order to improve the bonding strength of LVL and then enhancement overall bending properties of the material. Silane, smoking and termite resistance substance were examined. The 2^K factorial design method was used to conclude the effectiveness of experiment results.

EXPERIMENTAL PROCEDURE AND RESULTS DISCUSSION

Specimen Preparation and Testing

Pre-dried peeled rubber wood with average thickness of 2.5, thin teak veneer at 0.5 mm, plain woven E-glass fabric with aerial density of 821 g/m^2 and in house room temperature cure epoxy system were the main composition of LVL. Surface smoking process was achieved by constructed smoke house, using rice husk as the resource, for two days. The 3-aminopropyl silane and commercial termite resistance solution were employed. The rubber wood sheets

CP1255, *V[th] International Conference on Times of Polymers (TOP) and Composites,*
edited by A. D'Amore, D. Acierno, and L. Grassia

and fiber glass woven were alternately stacked and the epoxy adhesive was applied between the layers. Total of 14 layers of wood sheet were required. The faces of sample were covered with teak/glass woven for the purpose of finishing and termite protection. Curing process was achieved 150°C for 45 minutes under compression press at 30 bars. The bending specimen was cut, sand polished and post-cured at 80°C for at least 5 hours.

Flexural testing, ASTM D790, was adapted. The strength, modulus and toughness were recoded. The specific values were also calculated by dividing the flexural figures with the sample density. The 2^K design matrix used is shown in table 1. Plus(+) and minus(-) are referred to high and low level, respectively. Each level was further divided into two sub-level. Varying the silane addition by % weight in epoxy hardener used, number of layers of smoked rubber sheet and also layers of the termite treated wood were assigned as parameters A, B and C, respectively.

TABLE 1. The Design matrix for 2^K- factorial and the flexural test results

Run No.	Factors			Flexural Properties			Specific Flexural Properties		
	A	B	C	Strength(MPa)	Modulus (MPa)	Toughness(kJ/m²)	Strength	Modulus	Toughness
T1	+10	+6	+6	31.81±6.37	636.99±134.30	92.65±14.28	36.48±6.94	730.91±149.57	106.46±16.86
T2	+15	+8	-2	45.55±5.26	873.87±68.68	125.10±13.65	53.09±7.10	1018.21±101.25	145.57±15.41
T3	+10	-2	+8	41.69±8.03	890.00±106.91	128.29±35.18	47.05±8.67	1004.43±106.97	144.53±37.58
T4	+15	-4	-4	54.01±7.75	1286.36±73.75	119.83±28.92	57.50±6.81	1371.74±50.26	127.67±28.89
T5	-2	+6	+6	44.53±5.73	1160.83±84.95	70.52±7.84	50.28±6.22	1311.34±94.08	79.56±7.42
T6	-5	+8	-2	52.46±8.45	1003.37±134.68	91.83±40.83	50.54±10.08	1047.58±104.56	95.42±41.26
T7	-2	-2	+8	48.66±9.92	966.83±75.44	143.15±31.75	57.20±3.45	1041.95±90.14	154.01±31.81
T8	-5	-4	-4	45.50±3.64	934.01±128.13	111.01±19.31	50.22±4.83	1031.57±155.64	122.56±22.56

Results and Discussion

Table 1 is summarized flexural testing result and their calculated specific values. At 95% degree of confidential(α= 0,05) and using the statistical ANOVA testing on the calculated effect values of the above parameters, they were found that the designed model had been significantly effect in accordance with modulus and toughness responds and their specific figures, respectively. Other responds shown no significant effected by those designed parameters. Table 2(***a***) and 2(***b***) are selective summarized the ANOVA results for flexural modulus and specific toughness. Considering only the ANOVA, it is shown that interacted AB significantly effect on modulus. Parameter A, B and interaction AB, AC, and BC show that significantly influence by mean of specific toughness of the glass woven reinforced LVL.

TABLE 2. Summary of the ANOVA result for (a) Flexural Modulus and (b) Specific toughness of the glass woven reinforced LVL

Source	Sum of Squares	df	Mean Square	F-value	*p*-value
AB	107848.85	1	107848.85	7.40	0.0417
AC	84772.87	1	84772.87	5.82	0.0607
Residual	72830.49	5	14566.10		
Cor Total	265452.21	7			

(***a***)

Source	Sum of Squares	df	Mean Square	F-value	*p*-value
A	660.22	1	660.22	88.48	0.0111
B	1853.09	1	1853.09	248.35	0.0040
AB	828.51	1	828.51	111.03	0.0089
AC	178.87	1	178.87	23.97	0.0393
BC	1333.09	1	1333.09	178.66	0.0056
Residual	14.92	2	7.46		
Cor Total	4868.69	7			

(***b***)

Further investigation using the Pareto plots the t-value of the effect shown in figure 1(***a***) and 1(***b***) confirm that the interacted AB- which is significantly negative effect on flexural modulus and parameters A+, B-, BC-, AB+ and AC- are significantly effect on toughness, especially, specific values. Based on this research outcome, the silane content(A+) is positive effect on the toughness, low modulus, of LVL, smoking process had negative effect(B-) on toughness and modulus. The combined parameters had an effect by mathematical mean. These analysis reveal that the optimal composition of glass reinforced LVL would be using high silane ingredient in epoxy hardener with high layers of smoke rubber wood sheet but minimal or without termite treated wood.

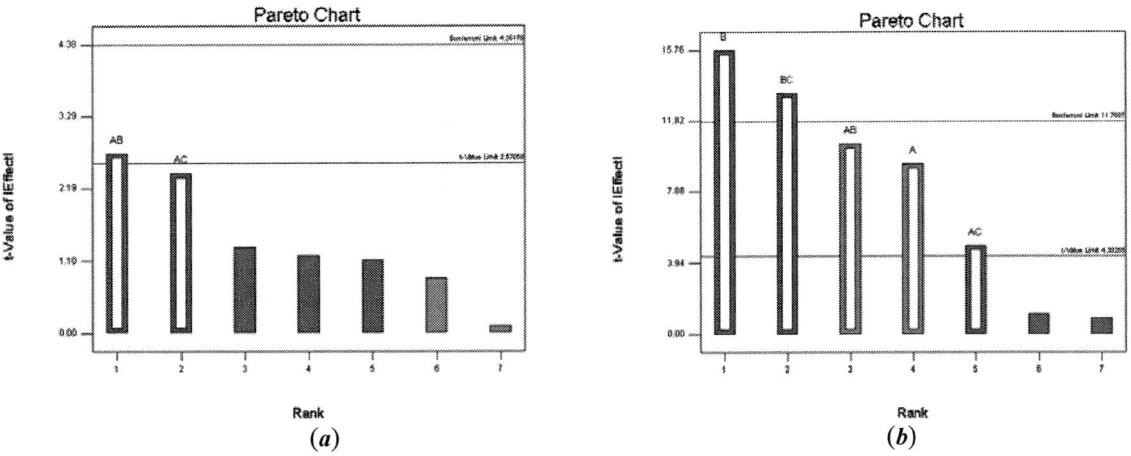

FIGURE 1. The Pareto chart of (***a***) flexural modulus and (***b***) specific toughness of glass woven reinforced LVL

Complementary durability testing were conducted. It was observed that screw nail withdrawing strength, water absorption and termite attack resistance of glass reinforced LVL had superior strength properties than hard wood but slightly inferior in toughness and lightness.

ACKNOWLEDGMENTS

This work was conducted as master thesis. The all financial support and appreciation for both tuition fee and material expenses during study was compulsory given by Centre of Excellent for Petroleum, Petrochemical and Advanced Material that is part of Office of Higher Education Commision, Ministry of Education, Royal Thai Government.

REFERENCES

1. C.A. Eckelman, Forest *Production J.* **43**, 19-24(1993).
2. T. Hayashi and S. Oshiumi, *MokuZai Gakkaishi J. Japan Wood Res. Soc.*, **39**, 985-992(1993).
3. B. Ozaeska, *Wood Sci. Tech.*, **33**, 341-351(1999).
4. E.D. Wong, A.K. Razali and S. Kawai, *Wood Research*, **83**, 8-16(1996).

Thermal properties of PMMA and PS filled with organo-silylated Montmorillonite

A. De Maria[a], C. Borriello[a], M. Schwarz[b]

[a]*ENEA, UTTP Nano, CR. Portici, P.le Fermi n.1, 80055 Portici (Naples) Italy*

[b]*ENEA, MATTEC, CR. Casaccia, 00060 S. Maria di Galeria (Roma), Italy*

Abstract. Clay minerals dispersed in a polymer matrix at the nanoscale show an improvement of many polymer properties at relatively low loadings. However a modification of hydrophilic clay surface is necessary to increase the compatibility with the hydrophobic polymer matrix. In this study the modification by organosilane of a commercial montmorillonite, containing alkylammonium cations, has been investigated. These organo-modified montmorillonites (OMMTs) have been also used to prepare Polystyrene (PS) and Polymethylmethacrylate (PMMA) nanocomposites by in-situ polymerization and their structural and thermal properties have been examined.

Keywords: nanocomposites, silylation, montmorillonite, thermal stability, PS, PMMA.
PACS: 81.07.Pr; 82.35.Np

INTRODUCTION

The potential application of polymeric nanocomposites are strictly connected with the clay exfoliation, so many researchers have been working on modification of inorganic clay using organic additives in order to improve the compatibility between hydrophobic polymer and hydrophilic clay. Traditional methods for the preparation of hydrophobic clays involve the treatment with alkylammonium cations by ion-exchange reaction [1] but broken bonds on the edges are common phenomena for layered silicates and lead to the formation of hydroxyl groups, which can be used for chemical modification [2]. A possible one is the reaction with a reactive organosilane of the type RSiX3 where X is either Cl or a methoxy group and R is alkyl, aryl or organo-functional group [3]. Silane treatment can improve processing, performance and durability of mineral-modified products.

In this research it has been studied the effect of silylation of a commercial alkylammonium montmorillonite on the properties of PS and PMMA nanocomposites.

RESULTS AND DISCUSSION

Synthesis and Characterization of organo-silylated Montmorillonite

The commercial montmorillonite (MMT-72T), modified by dimethyl dihydrogenated tallow ammonium ions, has been silylated by chloro-dimethyl-octadecylsilane. The silane has been added to a suspension of MMT-72T in boiling acetone, obtaining MMT-OS (figure 1).

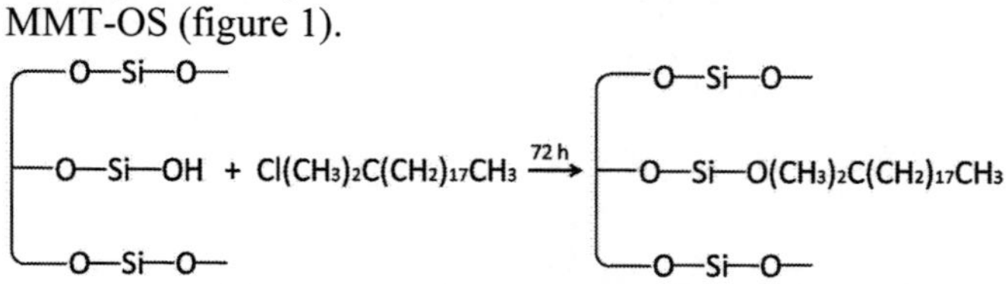

FIGURE 1. Scheme of silylation of MMT-72T by chloro-dimethyl-octadecylsilane

CP1255, *Vth International Conference on Times of Polymers (TOP) and Composites,*
edited by A. D'Amore, D. Acierno, and L. Grassia

Comparison between IR spectra of precursor (MMT-72T) and modified MMT (MMT-OS) has been useful to confirm the success of silylation. Indeed in the spectrum b of figure 2 it is the complete absence of a large band at 3500 – 3400 cm^{-1} attributed to the stretching of hydrogen bonded silanol group modes and trapped water molecule presenting a well known band at 3420 cm^{-1}.

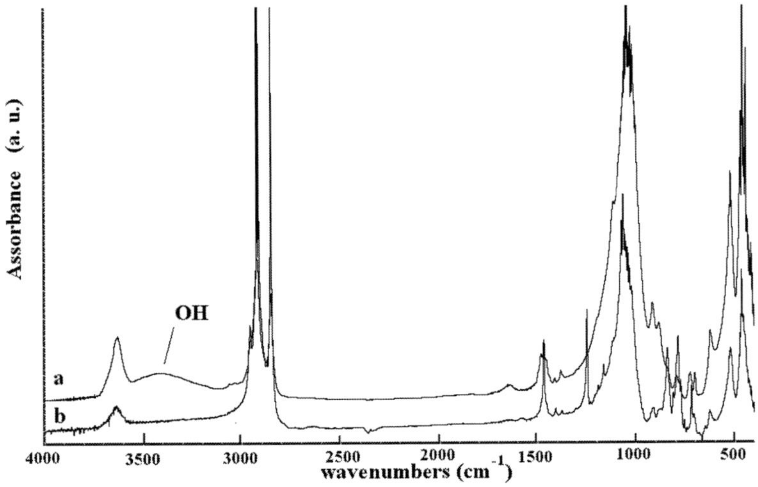

FIGURE 2. FTIR spectra of MMT 72T before (spectrum a) and after (spectrum b) reaction of silylation.

The thermogravimetric analysis of OMMTs have shown that the silane treatment increases up to 17% the onset of thermal decomposition.

In the XRD diffractograms it has been observed the shift of peak d(001) position from 3.60° for MMT-72T to 3.42° for MMT-OS, corrisponding to a small increase of basal space from 2.45 to 2.58 nm for the silylated MMT.

Synthesis and characterization of nanocomposites

Synthesis of nanocomposites by in-situ polymerization has involved the dispersion of the organoclay in styrene or methyl methacrylate followed by free-radical polymerization initiated by the addition of AIBN [4]. The monomers containing the desired amount of organoclay (1, 3 or 5 % wt.) have mechanically mixed for 24 h at 40 °C. The AIBN has been then added to the mixture to initiate the polymerization which took place at 100°C for 24 hours. The composites obtained have been kept overnight in an oven at 120°C to complete polymerization and remove the remaining monomer.

The diffractograms of the nanocomposites prepared have revealed a decrease of 2θ values, indicating expansion of the clay structure due to the polymer insertion into the silicate galleries.

The thermal properties of nanocomposites and homopolymers has been studied by TG and DSC analysis.

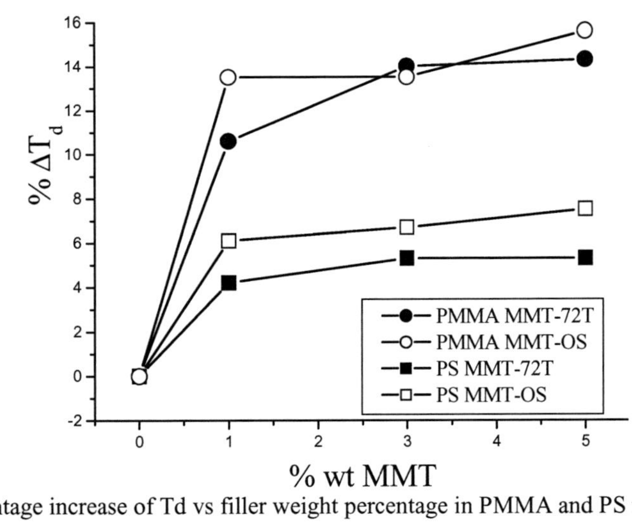

FIGURE 3. Percentage increase of Td vs filler weight percentage in PMMA and PS nanocomposites.

It has been observed (figure 3) that the T_d measured to 5% weight loss presents higher values for PS and PMMA nanocomposites, compared with homopolymer ones. The best results have been obtained in presence of MMT-OS, in particular the nanocomposites containing 5% wt show an increase of 15.6 % for PMMA and 7.5% for PS.

DSC analysis have shown that the glass transition temperature (T_g) of nanocomposites is higher than homopolymer ones (figure 4). Also in this case the best results have been realized in presence of 5% wt of MMT-OS. Indeed, it has been observed an increment up to 13% for PMMA and 7.8% for PS.

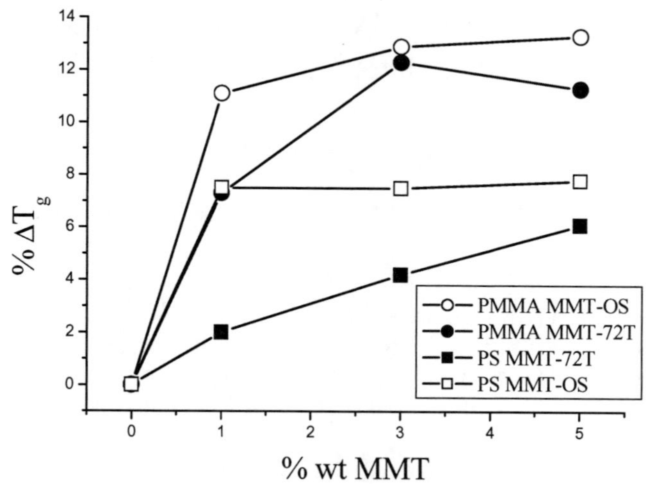

FIGURE 4. Percentage increase of Tg vs filler weight percentage in PMMA and PS nanocomposites.

The nanocomoposites glass transition temperatures tend to a plateau with the increase of the clay content. This result can be attributed to a lower molecular weights of polymers due to the high viscosity of MMT-monomer mixture that inhibits the diffusion of initiator molecules and of the chain propagation during polymerization.

CONCLUSIONS

It has been modified a commercial montmorillonite, MMT-72T, containing quaternary ammonium ions, by silylation reaction. Both MMT-72t and MMT-OS have been used to prepare PS and PMMA nanocomposites.

It has been observed a thermal properties improvement, compared with homopolymer ones, in particular in presence of MMT-OS. This probably is due to its higher thermal stability and to better compatibility between polymer matrix and silylated clay.

ACKNOWLEDGMENTS

This research was conducted in the "MAC- Nuove Metodologie di processo e materiali Avanzati per la Cantieristica da diporto" project supported by MIUR of Italy.

REFERENCES

1. (a) M. Alexandre, P. Dubois *Mat. Sci. and Eng.* 2000, *28*, 1; (b) S. Sinha Ray, M. Okamoto *Prog. Polym. Sci.* 2003, 28, 1539; (c) S. Sinha Ray, M. Bousmina *Progress in Materials Science*, 2005, 50, 962.
2. J. A. Rausell-Colom, J.M. Serratosa *Chemistry of clays and clay mineral,* ACD Newman, Ed.Wiley-Interscience, New York, 1987; 371.
3. Jinguo Zhang, Rakesh K. Gupta, and Charles A. Wilkie *Polymer* 2006, 47, 4537.
4. C. Borriello, A. De Maria, M. Schwarz *Energia, Ambiente ed innovazione*, 2007, 4, 62.

Poly(3-alkoxyphenylthiophenes) For VOCs Detection

E. Massera [a], M.L. Miglietta [a]*, T. Polichetti [a], G. Di Francia [a], F. Borbone [b], A. Carella [b], S. Pappalardo [b], A. Roviello [b]

[a] *ENEA Portici Research Center, Piazzale E. Fermi 1, 80055 Portici (NA), Italy*

[b] *Department of Chemistry, University of Naples "Federico II", Via Cinthia, 80126 Naples, Italy*

Abstract. In this work, poly(3-(4-octyloxyphenyl)thiophene has been synthetized by an oxidative coupling polymerization reaction and employed in a chemiresistor structure to probe its potential as VOCs sensor. The devices has been exposed to toluene, acetone and ethanol vapours. The chemiresistor has exhibited fast response time (< 1 min), good stability over more than two weeks under ambient conditions and good sensitivity to VOCs, demonstrating to be a viable option for detection of VOCs.

Keywords: polythiophenes, volatile organic compounds, chemical sensors.
PACS: 82.35.Cd Conducting polymers

INTRODUCTION

Conjugated polymers have been extensively investigated since they were demonstrated to become electrically conductive through a chemical doping reaction which involves partial reduction or oxidation (n-type or p-type doping respectively) of its spatially extended π-bonding system [1]. The interest on conductive polymers arises from the suitability of these materials for applications in organic electronics, solar cells, batteries, supercapacitors, biological and chemical sensors [1-5].

Among the conjugated polymers, polythiophenes and, in particular, poly(3-alkylthiophenes), are attracting an outstanding interest of researchers in sensing applications because of their good environmental stability, easy processability and their particular sensing ability which is not restricted to a single sensing modality as regards to polymer/nanofiller composite materials or metal oxide sensors.

The mechanism of interaction between analyte and polymer can be either physical or chemical: nucleophilic gases (NH_3, ethanol, methanol, etc.) interact chemically with the polymer chain dedoping the polymers and decreasing the conductivity, while electrophilic (NOx, SO_2, etc.) gases have an opposite effect. Other gases, including VOCs, interact via Van der Waals forces with the polymer chain and upon absorption by polymer film, a change in resistance, correlated with the concentration of analyte, can be observed [1].

In this work, the potential of the poly(3-(4-octyloxyphenyl)thiophene (PO) as VOCs sensor has been probed; exposing a PO-based chemiresistor to toluene, acetone and ethanol vapours. The device shows good responses in terms of sensitivity and response time.

RESULTS AND DISCUSSION

The poly(3-(4-octyloxyphenyl)thiophene (Fig. 1) has been synthetized by an oxidative coupling polymerization reaction that leads to a highly regioregular polymer (up to 80%) that has then lightly doped with $FeCl_3$ [6]. The chemiresistors have been obtained by drop casting a 10 mg/ml chloroform solution of PO onto a 5x5 mm^2 Al_2O_3 substrate with gold interdigitated electrodes.

CP1255, *Vth International Conference on Times of Polymers (TOP) and Composites,*
edited by A. D'Amore, D. Acierno, and L. Grassia

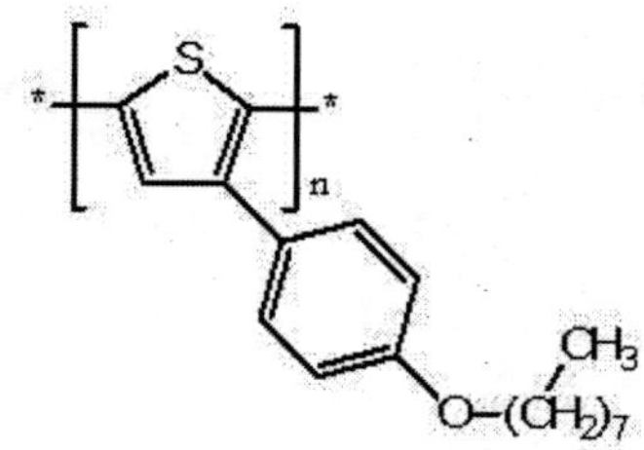

FIGURE 1. Poly-3 (4-octyloxyphenyl)thiophene chemical structure.

The device has been mouted in a Gas Sensor Characterization System where VOCs vapors, such as toluene, acetone and ethanol, are introduced by means of a bubbler system in controlled by mass flowmeters amounts. The gas mixture and concentration output were checked by a Thermo Antaris IGS FTIR gas analyzer.

The PO chemiresistor, exposed to different concentrations of those analytes (Fig. 2), shows good responses in terms of sensitivity and response time (see table 1). It is worthwhile to note that the response time of these devices is comparable or even lower than the filling time of the test chamber. These performances persist even after more than two weeks under ambient conditions; after this period the polymer's electrical conductivity is drammatically reduced by the prolonged exposure to the environmental water humidity.

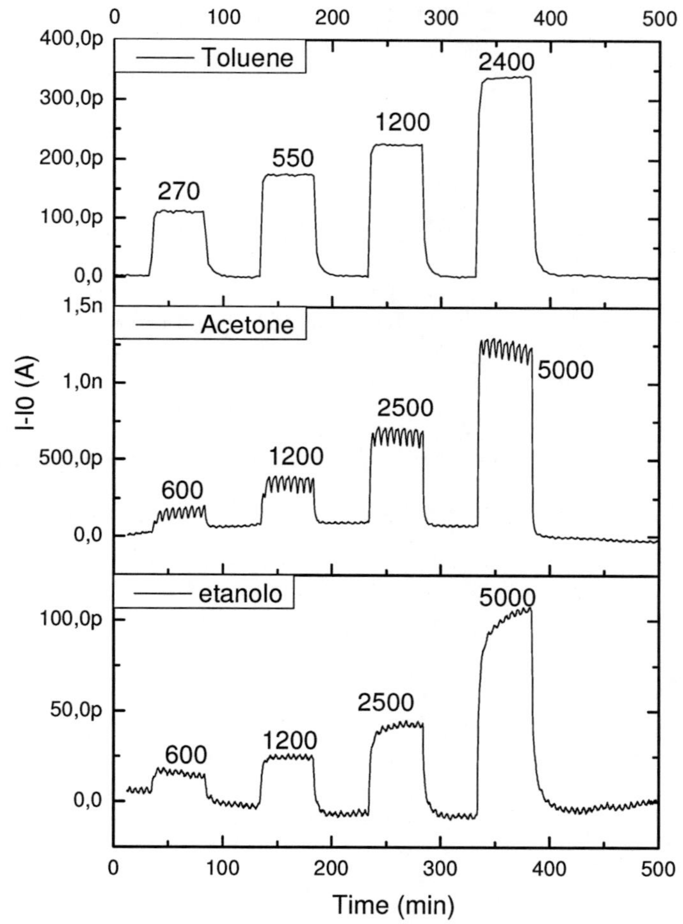

FIGURE 2. PO sensor response upon exposure to increasing concentration of toluene, acetone and ethanol.

Table 1			
Analyte	**Sensitivity (S/ppm)**	**Lower Detection limit (ppm)**	**Response time (10%-90% sec)***
Toluene	$7.5x10^{-13}$	10	<120
Acetone	$2.3x10^{-13}$	40	<120
Ethanol	$2.3x10^{-14}$	250	<120

*VOC chamber filling time is about 120 sec.

CONCLUSIONS

Concluding, sensors based on poly(3-(4-alkoxyphenyl)thiophenes) exhibited fast response time (< 1 min), good stability over more than two weeks under ambient conditions and good sensitivity to VOCs, demonstrating to be a viable option for detection of VOCs. Besides, the material stability at room operation condition has been checked over a long observation period, showing that the main parameter that affects the PO sensor conductivity is the moisture.

This sensing material will be further tested towards different chemical environments and the effect of the chemical structure of the alkyl substituent in the polymer backbone on the sensor selectivity will be also evaluated.

ACKNOWLEDGMENTS

The authors gratefully acknowledge the Ministry of Universities and Research (MUR) for financial support as part of Public-Private Laboratory TRIPODE (Technologies and Research for the application of polymers in electronics devices).

REFERENCES

1 Lange, N. V. Roznyatovskaya, V. M. Mirsky, Analytica Chimica Acta 614 (2008) 1-26
2 Lizhi Zhan, Zhiping Song, Ning Shan1, Jingyu Zhang, Jing Tang, Hui Zhan, Yunhong Zhou, Zaoying Li, Caimao Zhan , Journal of Power Sources 193 (2009) 859–863
3. Hongliang Guan, Peng Zhou, Shang Zeng, Xianglei Zhou, YunWang, Zhike He, Talanta 79 (2009) 153–158U
4. V. Dua, S. P. Surwade, S. Ammu, X. Zhang, S. Jain, S. K. Manohar, Macromolecules (2009), 42, 5414-5415
5. B. Li, S. Santhanam, L. Schultz, M. Jeffries, M. C. Iovu, G. Sauvè, J. Cooper, R. Zhang, J. C. Revelli, A. G. Kusne, J. L. Snyder, T. Kowalewski, L. E. Weiss, R. D. McCullough, G. K. Fedder, D. N. Lambeth, Sensors and Actuators B 123 (2007) 651-660
6. M. Barra, M. Biasucci, A. Cassinese, P. D'Angelo, A.C. Barone, A. Carella, A. Roviello, J.Appl. Phys. 102 (2007) 093712

Infrared Thermography: Some Applications of Microencapsulated PCMs as Investigative Tool for Heat Transport Evaluation

Paola Persico[a], Adolfo Izzo Renzi[b] and Cosimo Carfagna[a,b]

[a] *Institute of Chemistry and Technology of Polymers, CNR, Via Campi Flegrei 34, 80078 Pozzuoli (Na), Italy*
paola.persico@ictp.cnr.it
[b] *Department of Materials and Production Engineering, University of Naples "Federico II", P.le Tecchio 80, 80125 Naples, Italy*

Abstract
Microcapsulated Phase Change Materials (PCMs) were additivated onto the surface of natural leathers and mixed in the formulation of composites tailored for secondary indoor applications. The thermal response to external temperature dumping has been evaluated using an experimental equipment composed of a Peltier-effect device linked to an electrical wave generator and an infrared thermocamera to record images.

Keywords: Phase Change Materials (PCMs), Infrared Thermal Analysis, Natural Leather, Building Materials
PACS: 44 (heat transfer)

INTRODUCTION

Latent heat storage of a phase change material (PCM) is obtained during a change in phase. When a substance melts, vaporizes and sublimates, heat is absorbed. Heat is released when a material solidifies and condenses. During a phase transition, the temperature remains practically constant. Paraffin wax and salt hydrates are examples of such materials. The observation of mammals, whose mechanism of heat balance is based on transpiration, led to the production of functional textiles such as those in which phase change materials (mainly paraffins) are microencapsulated onto a polymeric shell. The spun fibers contain thousands of microspheres, each of which represents a reservoir of energy to be buffered on request. The effect of thermoregulation is most noticeable in technical clothing worn by workers who frequently move from cold to warm environments. The presence of PCMs contributes to reduced thermal damping due to oscillating temperature making the clothing more comfortable

PCMs can be also incorporated into building materials used in passive indoor applications. The purpose is to store thermal energy and then subsequently release it to aid in the reduction of the heating load.

Several testing methods have been developed during the last years for measuring the temperature-regulating ability of PCMs. Infrared – thermal investigation techniques have been used successfully in several applications, i.e. inspection of sub-surface defects and features, identification of thermo-physical properties, detection of coating thickness and hidden structures . Infrared (IR) thermographic systems provide images that represent surface temperatures, or thermograms, by measuring the magnitude of infrared radiation emitted by the surface of an object [1-4]. To perform the temperature distribution analysis, a thermal vision camera can be used to detect IR radiation and convert this information into an image, where each pixel corresponds to a temperature value. Modern IR imagers resolve surface temperature differences of 0.1°C or less. With this high sensitivity, they can provide a non-destructive evaluation of thermal phenomena, which are only revealed in the form of slight temperature gradients.

Infrared thermography can be used as both a qualitative and a quantitative tool. In the active approach, a thermal excitation source, modulated (lock-in) or pulsed, is employed with the intention of inducing thermal contrasts. The surface under investigation is heated by thermal waves using one or more modulated heating sources, and the resulting thermal transient at the surface is monitored using the thermal camera [5]. Basically the thermographic system is coherently coupled with a thermal wave source operating in such a way that a sinusoidal temperature modulation is generated by a non-linear electrical signal [6].

Evaluation of Thermo-Regulated Performance

Thermoregulation Effects in Natural Leather

Microencapsulated PCMs were coated onto natural leathers using a polymer binder to add the thermoregulating properties demanded by modern consumers [7]. The testing apparatus was designed to assess the average Heating Time Ratio (HTR) which can be calculated as [8]:

$$HTR = HT/HC \times 100\%$$

CP1255, *V[th] International Conference on Times of Polymers (TOP) and Composites,*
edited by A. D'Amore, D. Acierno, and L. Grassia

where HT is the average heating time of each cycle, HC is the average time of each cycle, as defined in Figure 1, that reports the typical thermogram obtained by comparing the reference sample (coated leather without microPCMs) and the activated one with 40wt% microcapsules Ty65.

The left side of the peak is related to the heating step and the right to the cooling run. At around 18°C and 12.5°C two shoulders on the signal of the additivated sample can be observed, caused respectively by the melting and crystallization process of microPCMs. These transitions are accountable for the delayed thermal response of additivated leathers, since part of the heat generated is used to melt the paraffin inside the microcapsules. During the cooling cycle the crystallization of PCMs is regained.

The HTR factor can be associated with the ability of thermally regulated systems to prolong the time required to reach an established temperature due to the melting process.

The HC of microPCMs additivated leather is about 255s, which is indeed not significantly longer than that of treated leather without PCMs (209s). Hence, the HTR in the presence of microcapsules is 42% but it is 44% for the sample not containing microPCMs. Probably, the fibrous leather is able to trap warm air within its low density structure, thus restricting the microcapsules melting effect.

The action time of microPCMs is limited by phase change, because once the external temperature goes over the phase change range, the PCMs coated leather is no longer effective as an active thermal adjuster.

The thermal qualitative effect due to the presence of microPCMs can be evaluated also by recording colour images using the infrared thermocamera, as displayed in Figure 2.

The leather sample containing microPCMs (reported on SX side of each frame) placed on the surfaces of the Peltier cell is able to delay and soften the thermal response in comparison with the sample without PCMs (reported on DX side of each frame), since microcapsules absorb heat during the melting process of paraffin waxes [9].

Temperature magnitude in a thermograph can be visualized by a colour scale in which each colour is related to a temperature range. From the hottest to the coldest temperature, the colours are so scaled: white, red, yellow, green and blue; the exact value depending on the experimental set fixed. In this experiment, the image was equivalent to the temperature range from 7°C to 19°C.

When the cooling/heating cycle starts, the samples are at the same temperature (step 1), both showing red colour (T = 17°C). During the cooling cycle, the sample coated with PCMs (SX side) turns green, while reference leather (DX side) is cyan (step 2). Such colour differences indicate that the temperature of the sample containing PCMs is higher than that of leather not containing microcapsules. In this process, PCMs act as a thermal buffer material by releasing stored heat delaying the leather cooling. When the cooling cycle stops, the samples are at the same temperature as confirmed by the blue colour for both (step 3, T = 11°C). During the heating cycle, the sample with PCMs is green, while uncoated leather is yellow (step 4). This colour difference indicates that the temperature of the sample with PCMs is lower than that of uncoated leather, since part of the available heat is used for wax melting. When the cooling cycle stops, samples reach the same temperature and their colour is red (step 5).

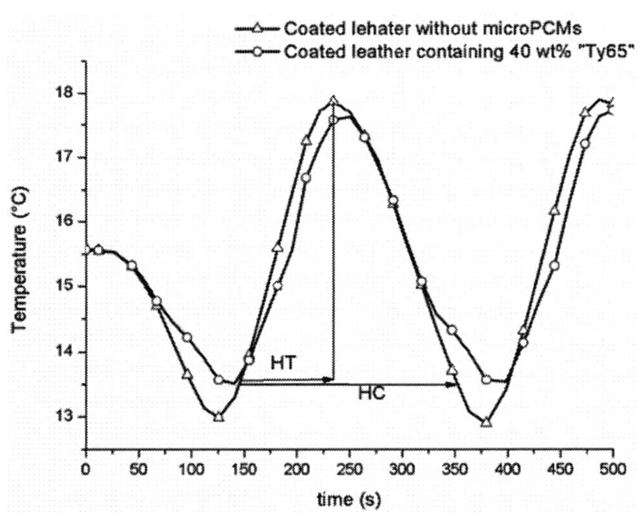

FIGURE 1. Thermogram of dry coated leather without PCMs and dry coated leather with 40wt% Ty65

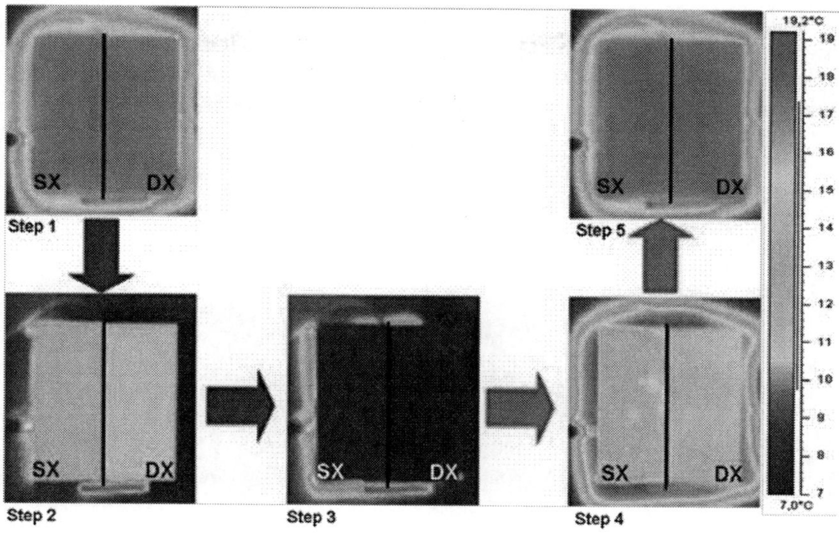

FIGURE 2. Infrared camera images of dry coated leather without PCMs (DX side of each frame) and coated leather with 40wt% Ty65 (SX side of each frame)

Thermoregulation Effects in Building Materials

Passive solar systems integrated into residential structures significantly reduce heating energy consumption. Taking advantage of latent heat storage has further increased energy savings. This is accomplished by the incorporation of phase change materials into building materials used in passive applications. Trombe walls, ceilings and floors can all be enhanced with phase change materials. Increasing the thermal storage of floor tile by the addition of encapsulated paraffin wax is the proposed topic of research.

In figure 3 the cooling/heating cycle for the composite containing 20%wt microcapsules (left side) is compared to neat resin (right side). When the cooling/heating cycle starts, the samples are at the same temperature (step1), showing both yellow colour (T = 20°C). During the heating cycle, the sample containing PCMs (left side) turns to orange (step2), while neat resin (right side) is red. Such colour difference indicates that the temperature of the composite containing PCMs is lower than that of neat resin. In this process, PCMs act as a thermal buffer material by absorbing free heat and delaying the resin heating. When the heating cycle stops (step3), the samples are at the same temperature as confirmed by the red colour for both (T = 33.7°C). This active temperature compensation mechanism enables the obtained composites to contribute to more comfortable room conditions miming the thermal capacity of conventional building materials, without additional space.

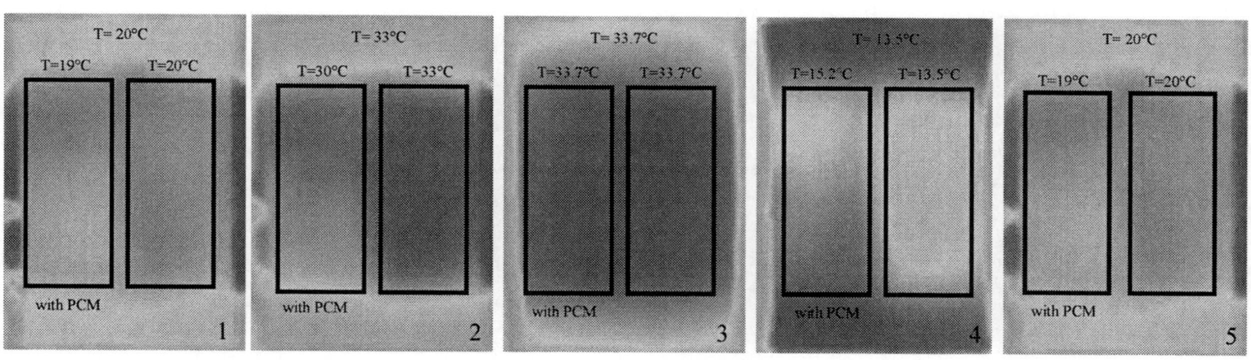

FIGURE 3. Infrared camera images of floor tiles containing microencapsulated PCMs

CONCLUSIONS

In this study thermally regulated natural leathers have been produced by dry coating of phase change microcapsules (PCMs). The thermal response to external temperature dumping has been evaluated using an experimental equipment composed of a Peltier-effect device linked to an electrical wave generator and an infrared thermocamera to record images. The average Heating Time Ratio associated with the thermal ability of the system to prolong the time required to reach a fixed temperature has been calculated.

The proposed materials and the related dry coating technique can represent an interesting tool to functionalize leather and improve its thermal performance without sacrificing comfort and softness of the garment.
Application of microencapsulated PCMs is demonstrated to be effective as well in those fields for which an active response to thermal dumping is required, such as architectural and civil engineering saving energy solutions.

AKNOWLEDGMENTS

The Authors gratefully thanks CARISMA S.p.A., Solofra (AV), Italy for leather supply and Dr Cimino Francesca for the preparation of composites.

REFERENCES

1. K. Koo, Y. Park, J. Choe, E. Kim, *J. Appl. Pol. Sci.*, **108**, 2337–2344 (2008).
2. C. Maierhofer, R. Arndt, M. Rőlling, C. Rieck, A. Walther, H, Scheel, B. Hillemeier, *Cement & Concrete Composites*, **28**, 393-401 (2006).
3. D. Bates, G. Smith, D. Lu, J. Hewitt, *Composites Part B*, **31**, 175-185 (2000).
4. H. Wiggenhauser, *Infrared Phys Tech*, **43**, 233-238 (2002).
5. N. P. Avdelidis, D. P. Almond, Z. P. Marioli-Riga, A. Dobbinson, B. C. Hawtin, "Pulsed termography: philosophy, qualitative & quantitative analysis on aircraft materials & applications" in *Advances in Signal Processing for Non destructive Evaluation of Materials*; Proc. Vth International Workshop, ISBN 2-9809199-0-X.
6. C. Meola, G. M. Carlomagno, L. Giorleo, *J. Materials Processing Technology*, **155/156**, 1132–1137 (2004).
7. A. Izzo Renzi, C. Carfagna, P. Persico, *Appl. Therm. Eng.* (2010), doi:10.1016/j.applthermaleng.2010.02.026.
8. S. X. Wang, Y. Li, J.Y. Hu, H. Tokura, Q. W. Song, *Polymer Testing*, **25**, 580–587 (2006).
9. C. Carfagna, P. Persico, *Macromolecular Symposia*, **245/246**, 355–362 (2006).

Effect of Carbon Nanotubes (CNTs) on Mechanical and Thermal Properties of Polypropylene Matrix

J. Pascual[a], E. Fages[a], R. Balart[b], O. Fenollar[b] and T. Boronat[b]

[a]*Textile Research Institute (AITEX), Plaza Emilio Sala 1, 03801, Alcoy, Alicante (Spain)*
[b]*Institute of Materials Technoloy (ITM), Polytechnic University of Valencia (UPV), Plaza Ferrandiz y Carbonell s/n, 03801, Alcoy, Alicante (Spain)*

Abstract. In the last years, a remarkable increase in the use of nanoparticles as fillers of polymer materials has been detected. It is mainly due to the excellent properties that these materials can provide when added to polymer matrices as a consequence of the high specific surface, so that, small additions can lead to a remarkable increase in overall properties. Small additions of carbon nanotubes (CNTs) in a polypropylene matrix promote an increase in mechanical and thermal performance. Presence of CNTs leads to formation of interlock points with polypropylene chains and this has a positive effect on mechanical properties; thus an increase in elastic modulus and tensile strength are obtained for relatively small amounts of CNTs in the 1-3 wt. %. Regarding thermal properties, it is important to note that a remarkable increase of thermal stability is detected at medium temperatures (oxidation onset measured by differential scanning calorymetry) and high temperatures (degradation measured by thermogravymetric analysis). So, we can conclude that the use of small additions of CNTs is a useful technique to improve polypropylene performance from both technical and economical point of view since good balanced properties are obtained. Furthermore, the cost of these nanoparticles has decreased considerably in the last years and

Keywords: Carbon Nanotubes (CNTs), Polypropylene, Mechanical Properties, Thermal Properties, Composites.
PACS: 62.23.Pq, 81.07.De, 82.35.Np.

INTRODUCTION

Polypropylene is highly used at industrial level due to excellent balance between easy processing and overall properties (mechanical, thermal, chemical...). In the last years, with the growing development of nanotechnologies, many research works have been focused on the addition of different particles at a nano-scale range in order to change a wide variety of properties: mechanical resistance, chemical barrier, electrical conductivity...[1-5]

Small additions of nanoparticles can lead to a remarkable increase in some properties due the high specific surface of these particles. In this work we have used carbon nanotubes (CNTs) to improve mechanical and thermal performance of PP-CNTs compounds. Polypropylene is characterized by interesting mechanical properties and this is responsible for its wide use as a commodity plastic and, even, in technical applications. On other hand, it is important to remark that polypropylene is highly sensitive to thermo-oxidative degradation (in fact, processing additives must be added in order to avoid thermal degradation during processing). For this reasons, it is useful to increase the overall properties of polypropylene by addition of nanoparticles in order to increase mechanical performance and thermal behavior.[6-9]

The aim of this work is to improve mechanical and thermal performance of polypropylene by addition of carbon nanotubes (CNTs). Tensile strength, elastic modulus and elongation at break have been determined from tensile tests and thermal degradation has been followed by thermogravymetric analysis (TGA). In particular, degradation temperature at 1% and 5% weight loss has been determined as a representative parameter of the degradation improvement.

CP1255, *V[th] International Conference on Times of Polymers (TOP) and Composites,*
edited by A. D'Amore, D. Acierno, and L. Grassia

EXPERIMENTAL

Commercial grades of polypropylene and carbon nanotubes were use to prepare PP-CNTs compounds with a twin screw co-rotating extruder Werner & Pfleiderer. The amount of CNTs varied in the 0.2-10 wt. %. After compounding, standard samples were injection molded with a Babyplast for sample characterization. Mechanical properties were obtained by tensile tests using a universal test machine Elib 50 (SAE Ibertest, Madrid, España) following the guidelines of the UNE EN ISO 527-1. Thermal stability of PP-CNTs compounds was determined by thermogravymetric analysis with a temperature program from 30°C to 600°C at 20°C min^{-1} with air atmosphere. Degradation temperatures were determined at different degradation stages: 1%, 5%, 50%, 95% and 99% weight loss).

RESULTS AND DISCUSSION

Tensile strength of unfilled polypropylene is around to 27.9 MPa. As we can see in Figure 1, a remarkable increase is tensile strength of PP-CNTs compounds is observed as CNTs content increases. For intermediate CNTs content (1-5 wt. %), the tensile strength is raised up to values in the 34-35 MPa. Even for lower CNTs additions, a significant increase in tensile strength is detected; thus, for PP-CNTs compounds with 0.2 wt. % CNTs, the tensile strength reaches values close 32 MPa. These results are interesting since the effect of particle addition to polymer matrices could lead to two different behaviors: if particles do not interact with polymer chains, a decrease in tensile strength could be expected due to the lack of particle-polymer interactions so that, particles act as fillers and promote stress concentration. On other hand, when added particles interact with polymer chains, some interlock points are formed and this has a positive effect on overall reinforcing properties. It is the case of PP-CNTs compounds since addition of CNTs leads to an increase in tensile strength which is a mechanical property highly sensitive to presence of particle fillers.

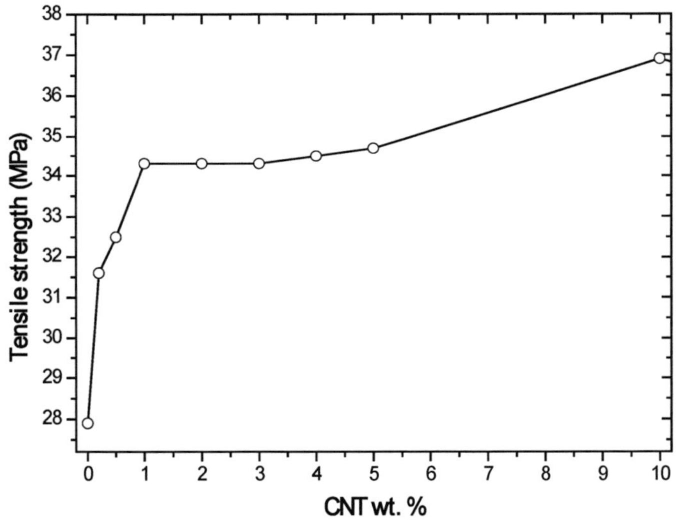

FIGURE 1. Plot of the evolution of tensile strength of polypropylene-CNTs compounds in terms of the weight percentage of CNTs.

Regarding thermal properties, presence of CNTs also leads to a remarkable increase in thermal stability at high temperatures. Figure 2 shows the evolution of the degradation temperatures (temperatures corresponding to 1% and 5% weight loss) for the PP-CNTs system, as representative of the degradation processes at initial stages. We can clearly observe that these temperatures are raised to higher values thus indicating a noticeable increase in thermal stability. Similar behavior is observed at the final stages (95% and 99% weight loss) of the degradation process but the improvement on thermal stability is more pronounced at initial stages.

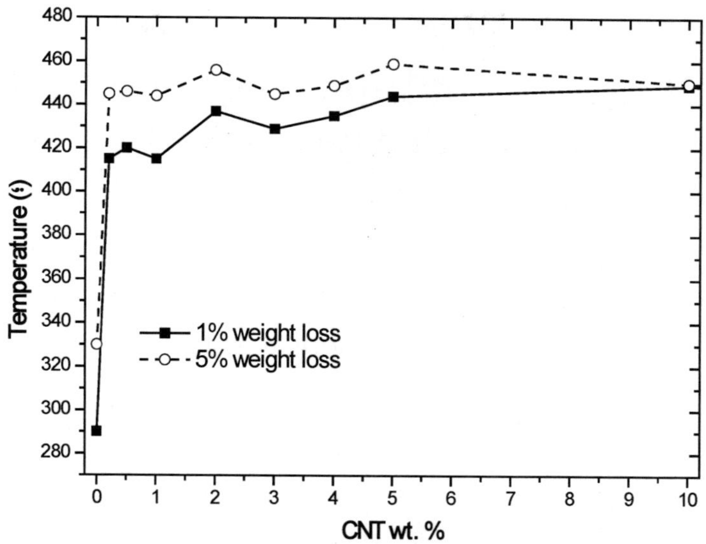

FIGURE 2. Plot of the evolution of the degradation temperatures (1% weight loss) of polypropylene-CNTs compounds in terms of the weight percentage of CNTs.

CONCLUSIONS

The addition of small amounts of CNTs leads to an increase in both mechanical and thermal stability (at intermediate and high temperatures). Presence of CNTs promote formation of interlock points with interaction with polypropylene chains and this has a positive effect on mechanical properties, specially on tensile strength which is more sensitive to presence of particles. The initial tensile strength of unfilled polypropylene (27.9 MPa) is increased up to values in the 34-35 MPa range for intermediate CNTs content (1-4 wt. %). With regard to thermal stabilization effect of CNTs, the addition of CNFs (even at low concentrations, 0.5-2 wt. %) promotes an increase in the thermal degradation at high temperatures (in the degradation range) and the typical thermogravymetric (TGA) curves are shifted to high temperatures. In some cases the typical degradation temperatures are shifted 70-80°C so that the effect of CNF is clearly observable. The overall effect of CNTs addition is positive and represents and interesting technique to increase performance of polypropylene for uses in technical applications.

ACKNOWLEDGMENTS

Authors thank "Ministerio de Ciencia y Tecnología", Ref: DPI2007-66849-C02-02 and Generalitat Valenciana FPA/2010/027 & ACOMP/2010 for financial support.

REFERENCES

1. S.P. Bao and S.C. Tjong, *Mater. Sci. Eng. A-Struct. Mater. Prop. Microstruct. Process.*, **485**, 508-516 (2008).
2. M. Ganss, B.K. Satapathy, M. Thunga, R. Weidisch, P. Potschke and D. Jehnichen, *Acta Mater.*, **56**, 2247-2261 (2008).
3. D.T.G. Katerelos, R. Joffe, D. Labou and L. Wallstrom, *Mech. Compos. Mater.*, **45**, 423-434 (2009).
4. A.A. Kovalchuk, V.G. Shevchenko, A.N. Shchegolikhin, P.M. Nedorezova, A.N. Klyamkina and A.M. Aladyshev, *J. Mater. Sci.*, **43**, 7132-7140 (2008).
5. N.G. Sahoo, N.T. Thet, Q.H. Tan, L. Li, S.H. Chan, J.H. Zha and S.Z. Yu, *J. Nanosci. Nanotechnol.*, **9**, 5910-5919 (2009).
6. S. Choi, Y. Jeong, G.W. Lee and D.H. Cho, *Fiber. Polym.*, **10**, 513-518 (2009).
7. M. Hemmati, G.H. Rahimi, A.B. Kaganj, S. Sepehri and A.M. Rashidi, *J. Macromol. Sci. Part B-Phys.*, **47**, 1176-1187 (2008).
8. J.H. Ko, J.C. Kim and J.H. Chang, *Polym.-Korea*, **33**, 333-341 (2009).
9. K. Prashantha, J. Soulestin, M.F. Lacrampe, P. Krawczak, G. Dupin and M. Claes, *Compos. Sci. Technol.*, **69**, 1756-1763 (2009).

Mechanical Properties of Injection Moulded Binary Blends of Polyethylene with Small Additions of a Liquid Crystalline Polymer

Tatjana Ivanova[a], Janis Zicans [a], Ilze Elksnite [a], Martins Kalnins[a] and Robert Maksimov[b]

[a]*Institute of Polymer Materials, Riga Technical University, 14/24 Azenes street, Riga, LV-1048, Latvia*
[b]*Institute of Polymer Mechanics, University of Latvia, 23 Aizkraukles street, Riga, LV-1006, Latvia*

Abstract. Results of the investigation of the mechanical properties of polyethylene blend with small (no more than 3 wt.%) additions of thermotropic liquid crystalline polymer (LCP) are reported. Data on the influence of LCP additions on the stress-strain behaviour in tension, elastic modulus, and long-term (up to a 1000-h) creep are obtained. Elastic properties of the materials are analyzed taking into account the effect of self-reinforcement due to the formation of extended LCP inclusions.

Keywords: LCP, PE, stress-strain characteristics, creep characteristics, modelling
PACS: 45.20.D-; 62.20.-x; 81.05.Lg; 83.80.Ab; 83.80.Tc; 83.80.Xz

INTRODUCTION

There are several purposes, which are considered when developing theromtropic liquid crystal polymer (LCP) modified thermoplatics. One of these reasons is decrease of viscosity of some polymers thus facilitating their processing, while some other is increase of mechanical properties of the materials due to formation of the extended LCP fibrils in the used thermoplastic matrix [1, 2]. Self reinforcing effect is especially pronounced at extrusion and subsequent drawing of the compositions containing considerable amounts of LCPs (up to 30 wt. %) [2].

The current work, however, is devoted to the investigation of the effect of small additions of LCP on short-term and long-term mechanical properties of polyethylene (PE).

EXPERIMENTAL

In the current work polyethylene (Egyeuroptene MD 3804U, Egyptian European Company) with density of 0.938 g/cm^3 and melt index of 4.0 g/10 min has been used. Vectra B950 copolymer (Ticona GmBH) has been used as a liquid crystal polymer. LCP/PE blends have been obtained by initial mixing in a Banbury type mixer followed by processing in a twin screw extruder above LCP melting. In total, LCP content has been varied from 0 to 3 wt. %. Specimens for mechanical tests have been obtained by injection molding.

Short-term tensile tests have been performed on a Zwick/Roell BDO FB020TN universal testing machine at 20°C with speed of the moving clamp 50 mm/min. 1000 h tensile creep experiments have been performed at constant 3 MPa static loads, corresponding to the stresses equal to 15-20 % from the yield maxima of the investigated blends.

RESULTS AND DISCUSSION

Stress-strain diagrams of the PE/LCP blends are shown in Fig. 1a. As one can see, by increasing LCP content in the blends, stress-strain curves in the "σ–ε" coordinate system are shifted to higher stress values. Thus at LCP

CP1255, *Vth International Conference on Times of Polymers (TOP) and Composites,*
edited by A. D'Amore, D. Acierno, and L. Grassia

content 3 wt. %, yield stress σ_Y of the composite is almost 20% higher than for neat polyethylene. Besides considerable decrease (approximately two times) of yield strain ε_Y is observed.

Experimental values of elastic modulus E as a function of LCP weight content W_f are demonstrated in Fig. 1b. It is clearly seen, that by increasing LCP content up to 3 wt. %, E has risen 1.4 times in comparison to neat PE.

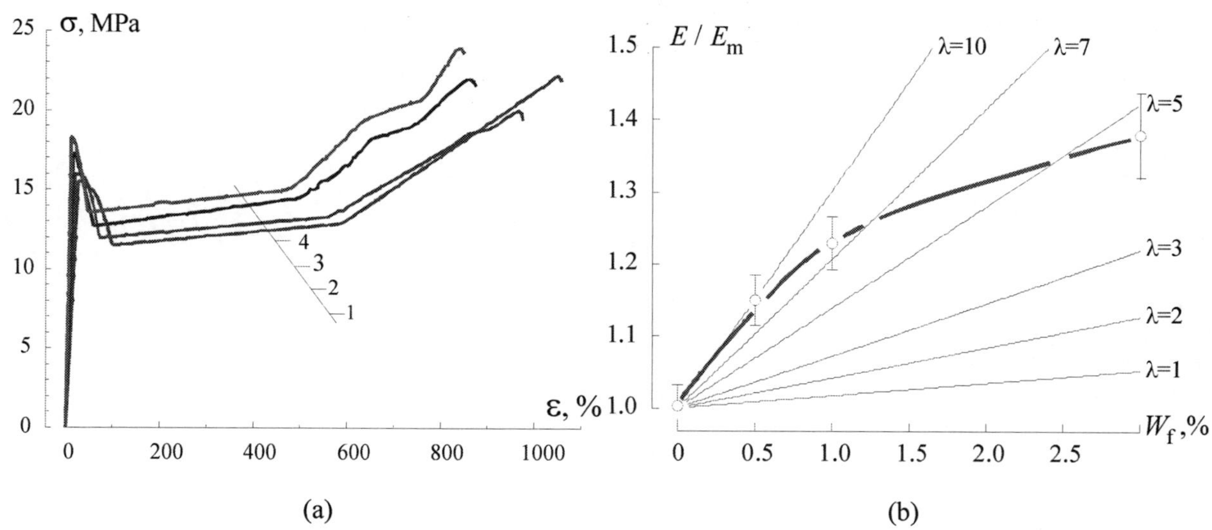

FIGURE 1. Short-term stress-strain characteristics of LCP/PE blends: (a) Stress-strain diagrams of LCP/PE blends with different LCP contents: 0 (1), 0.5 (2), 1 (3) and 3 (4) wt. %; (b) Relative modulus of elasticity E/E_m (E_m- elastic modulus of neat polyethylene) of LCP/PE blends as a function of LCP weight content: points – experimental data; lines – approximations calculated at different draw ratios λ of the LCP inclusions.

Theoretical analysis of the E dependence on volume fraction V_f of LCP is performed, considering provisions about morphology of the LCP modified blends and classical models of the composites with anisometric inclusions. In the calculations we have assumed that LCP inclusions have shape of ellipsoids with long axis oriented in the direction of the melt flow in the injection moulded specimens. Elastic modulus of the composites is calculated according to the famous equation of Halpin-Tsai with consideration of different aspect ratios of LCP inclusions [3]. It is well known that axial deformation (orientation) of the fibrous inclusions increase their aspect ratio and subsequently raises modulus of elasticity of the composite. Therefore in the calculations experimental results on the relationship between elastic modulus and drawing ratio λ, obtained by Choy [2], has been used. Relationship between length l and diameter d of fibrous particles with λ is depicted by following equation:

$$l / d = \lambda^{3/2}, \tag{1}$$

where λ – draw ratio of LCP inclusions equal to the relationship between cross-sectional areas of the particle before and after drawing.

Calculated values of relative modules of elasticity E/E_m of the investigated composites are shown in Fig.1b with lines. From the figure, it is clearly seen that increase in draw ratio at one and the same LCP content causes significant rise in the modulus of elasticity of the investigated composites. Besides it experimental relationships E/E_m of various investigated LCP/PE compositions lay within predictable values characteristic for draw ratios from 5 to 10.

In opposite to short-term stress-strain behaviour, long-term creep effects on the exploitation characteristics of LCP containing thermoplastic composites are studied only scarcely, in spite of the fact that such materials are intensively investigated for more than 20 years.

Published results on the long-term creep investigations (10 000 hours) of PP modified with various amounts of liquid crystalline additives (polyethylene terephtalate/p-hydroxybenzoic acid) [4], show that certain LCP additions allow considerably improve creep resistance of the polymer.

Analogous effect has been observed also for the investigated LCP/PE blends as it is demonstrated in Fig.2a. After duration of tensile creep (at 3 MPa) for 1000 hours, strain of the composite containing 3 wt. % of LCP, is 1.3

times smaller than that of neat PE. In Fig. 2b, possibilities of depicting the experimental creep curves of LCP/PE composites by using Findley power law [5] are shown:

$$\varepsilon(t) = \varepsilon_0 + \varepsilon_c t^n \tag{2}$$

where ε_0 – instantaneous strain; ε_c, n – coefficients.

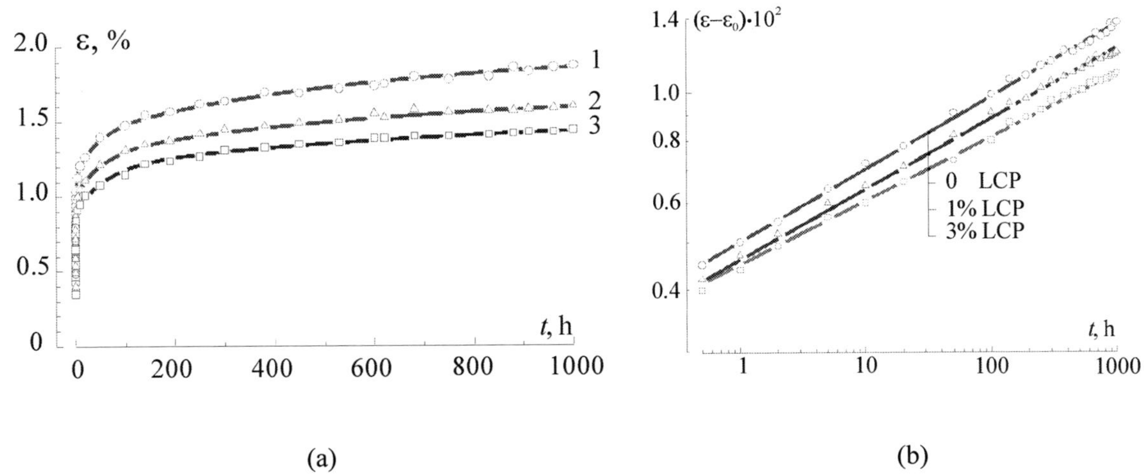

FIGURE 2. Creep characteristics of LCP/PE blends: (a) Creep curves *ε(t)* of LCP/PE composites at various LCP concentrations: 0 (1), 1 (2) and 3 (3) wt. %; (b) Creep strains (total creep strain minus instantaneous strain) of LCP/PE blends in logarithmic coordinates: points – experimental data; lines – approximations according to equation 2.

CONCLUSIONS

In the current research the effects of small additions (up to 3 wt. %) of liquid crystal polymer (LCP) on short-term and long-term (creep up to 1000 hours) tensile properties of polyethylene (PE) are revealed. Experimental results of the investigation show that even small additions of LCP allow considerably rise modulus of elasticity, yield stress, ultimate stress and creep resistance of polyethylene. Theoretical analysis, carried out in the framework of the current investigation, allow assuming that LCP inclusions in the continuous PE matrix have been elongated 5 to 10 times due to their orientation during manufacturing of the test specimens. Besides it theoretical analysis shows that creep of the investigated LCP/PE blends can be sufficiently well described by power law.

REFERENCES

1. W. Brostow, T. Sterzynski, and S. Triouleyre, *Polymer* **37,** 1561-1574 (1996).
2. C. L. Choy, K. W. E. Lau, and Y. W. Wong, *Polymer Eng. Sci*, **36,** 1256-1265 (1996).
3. S. W. Tsai and N. J. Pagano, "Invariant properties of composite materials" in *Composite Materials Workshop*, edited by Tsai, Halpin, and Pagano., Technomic Publishing Co., Stamford, 1968.
4. J. Kubat and R. D. Maksimov, "Creep and stress relaxation" in *Mechanical and thermophysical properties of polymer liquid crystals,* edited by W. Brostow, Chapman & Hall, London 1998, pp. 407-430.
5. W. N. Findley, J. S. Lai, and K. Onaran, "Creep and relaxation of nonlinear viscoelastic materials", Dover Publications, Inc., New York 1989.

Improvement of the T-peel Strength of Polypropylene Adhesion Joints by Surface Photografting Pre-Treatment with Methyl Methacrylate

R. Balart, L. Sánchez-Nácher, J. Balart, V. Fombuena, J.M. España

Institute of Materials Technology (ITM), Polytechnic University of Valencia (UPV), Plaza Ferrandiz y Carbonell s/n, 03801, Alcoy, Alicante (Spain)

Abstract. Although polypropylene is one of the most used polymers at industrial level due to good balanced properties, it presents some restrictions in applications that require good adhesion properties as well as coating and painting. These restrictions are related to its non polar nature which leads to low wetting properties. So that, in most cases, it is necessary a previous surface pre-treatment in order to improve adhesion properties. These surface treatments could be physical or chemical. Among the wide variety of physical processes, plasma technologies are useful from both technical and environmental points of view. If we take into account economic considerations, chemical processes are interesting due to low cost equipment and procedures. In particular, we have used photografting of methyl methacrylate (MMA) monomer on polypropylene substrates with UV radiation and initiators. This process is useful to promote chemical modification of polypropylene surface by grafting MMA monomers into polypropylene polymer chains. Due to polarity of some groups in MMA monomers, it is possible to increase surface wettability thus promoting a remarkable increase in adhesion properties of polypropylene. In this work, changes in wettability of polypropylene surfaces in terms of the exposure time to UV radiation in presence of MMA monomers and initiators has been investigated. Furthermore, chemical changes have been characterized by FTIR analysis and mechanical performance of adhesion joints has been evaluated by T-peel tests.

Keywords: photografting, polypropylene, methyl methacrylate, adhesion.
PACS: 82.35.Gh, 68.35.Np, 68.35.bm, 61.80.Ba

INTRODUCTION

Polypropylene is a "commodity plastic" widely used in low-medium technologic content applications. Nevertheless, due to an excellent balance between properties and procesabillity, it finds growing applications in technical-engineering applications. In some cases, it must appear in adhesion joints and for this reason it is necessary a previous surface pre-treatment in order to increase its low intrinsic wetting properties. These surface pre-treatments are focused on surface activation by insertion of polar species together with promoting changes in surface roughness. This has a positive effect on overall surface free energy thus improving adhesion properties.

Surface modification can be carried out by several physical (plasma-based) and/or chemical processes. Physical processes such as plasma (corona, atmospheric…) are efficient from technical point of view by, usually, they are expensive. On other hand, chemical processes are characterized by low cost and easy implementing, so that, they are more interesting for industrial applications. Among the wide variety of chemical processes, photopolymerization by UV radiation offers interesting properties to polymer surface modification.

By using photopolymerization or photografting[1-4] processes by UV radiation, it is possible to achieve surface activation and changes in surface roughness by grafting a selected monomer into polymer chains. In this work we have used a methyl methacrylate[2, 5] (MMA) monomer (characterized by high polarity) as a base monomer for surface modification of a polypropylene substrate (which is a non polar polymer) by using UV[6, 7] radiation in presence of an initiator (benzophenone, BP)[8] for free radical formation. This research work has focused on the effect of the exposure time to UV radiation in presence of a solution containing optimized monomer:initiator ratio. Changes in wetting properties of polypropylene have been characterized by contact angle measurements and

CP1255, *V[th] International Conference on Times of Polymers (TOP) and Composites,*
edited by A. D'Amore, D. Acierno, and L. Grassia

subsequent surface free energy calculation. Also, a FTIR study has been carried out in order to elucidate chemical changes in polypropylene surface as a consequence of MMA surface grafting. Finally, the effect of the exposure time to UV radiation during the photografting process, on mechanical performance of polypropylene-polypropylene adhesion joints has been evaluated by T-peel tests.

EXPERIMENTAL

A commercial polypropylene grade ISPLEN® PB 180 G2M supplied by REPSOL YPF has been used as base substrate for surface modification. The monomer used for surface activation was a stabilized methyl methacrylate (MMA) monomer with a purity of 99% supplied by Acros Organics and the photoinitiator was benzophenone (BP) provided by Scharlau Chemie S.A.

Photografting process was carried out with a UV equipment UVASPOT 1000RF2 (Honle UV Technology). Yhis equipment is provided with a high pressure mercury lamp. A goniometer model EasyDrop Standard FM140 (Krüss) was used for contact angle measurements.

Mechanical characterization of PP-PP adhesion joints was carried out in a universal test machine Elib 30 (SAE Ibertest) following the guidelines of the UNE EN 1895.

RESULTS AND DISCUSSION

Table i) shows contact angle values for different test liquids. We can see a remarkable decrease in contact angle values for all the test liquids as the exposure time to UV radiation increases. The contact angle values reach minimum values for exposure times in the 180-210 s. Nevertheless, it is important to remark that a noticeable decrease in contact angle values is detected for low exposure times. So that, the initial water contact angle of polypropylene (87.3°) is reduced up to values close to 75.5° for an exposure time to UV radiation of about 30 s. Similar or more pronounced tendency can be observed for the other test liquids. This fact shows that the photografting process is useful to increase surface wettability of polypropylene. After the photografting process, samples were properly washed in order to remove non-reacted monomer and solution. This way, it is possible to observe the grafting process by FTIR analysis which shows the typical absorption peaks and bands of methyl methacrylate (MMA), thus indicating good bonding between polypropylene and MMA monomer.

TABLE i). Contact angle values of polypropylene surface for different test liquids in terms of the exposure time to UV radiation in presence of a solution containing MMA monomer and initiator.

time (s)	θ, water (°)	θ, formamide (°)	θ, diiodomethane (°)	θ, glycerol (°)
0 (untreated)	87.3	71.4	54.0	74.7
30	75.5	47.8	37.1	64.4
60	66.4	46.1	36.7	57.1
90	47.6	45.8	35.7	47.6
120	46.4	40.2	31.8	48.2
150	44.0	39.6	29.9	47.2
180	43.8	32.4	29.5	40.0
210	39.3	31.9	21.8	38.3
240	37.0	30.8	15.9	32.6

The increase in wetting properties on polypropylene has a positive effect on overall adhesion properties. So that, we have prepared PP-PP adhesion joints with a polyurethane adhesive and adhesion joints have been characterized by T-peel tests. Figure 1 shows a plot evolution of the maximum T-peel force as a function of the exposure time to UV radiation in presence of a solution containing MMA monomer and initiator. We can see that a remarkable increase is obtained for relatively short exposure times (about 30 s); nevertheless, as the exposure time increases, T-peel force is increase up to higher values of about 110 – 120 N for exposure times of about 180 – 210 s, thus indicating a remarkable increase in adhesion properties.

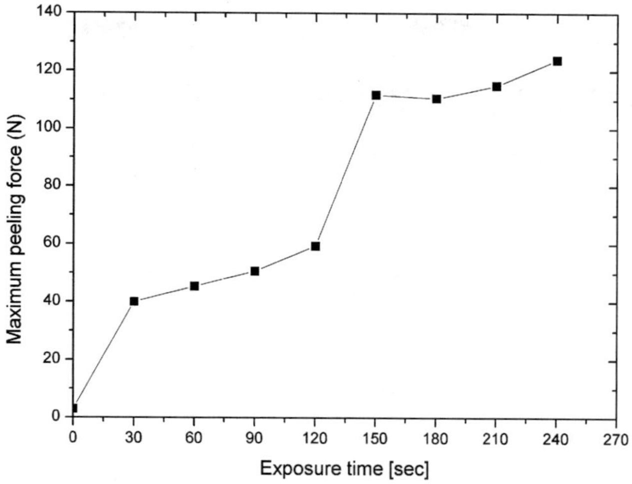

FIGURE 1. Evolution of the maximum T-peel force vs. exposure time to UV radiation in presence of a solution containing MMA monomer and initiator.

CONCLUSIONS

Photografting process with methyl methacrylate (MMA) is a useful technique to increase the low intrinsic adhesion properties of polypropylene. FTIR shows good grafting of methyl methacrylate monomer on polypropylene chains. Wettability is remarkably improved as observed by contact angle measurements for different test liquids. As the exposure time to UV radiation increases, the contact angle decreases. So that, water contact angle is reduced from 87.3° for untreated polypropylene up to values of 37° for 240 s exposure time.

Regarding mechanical properties, the effect of the photografting process is highly positive since T-peel force is increase from values close to 0 (for untreated material) up to values of about 110 – 120 N for exposure times in the 180 – 210 s range. The overall effects of the photografting process are interesting from an industrial point of view since it allows increasing adhesion properties of polypropylene with a low cost and technical efficient process.

ACKNOWLEDGMENTS

Authors thank "Ministerio de Ciencia y Tecnología", Ref: DPI2007-66849-C02-02 and Generalitat Valenciana FPA/2010/027 & ACOMP/2010 for financial support.

REFERENCES

[1] C. Decker and K. Zahouily. Surface modification of polyolefins by photografting of acrylic monomers. Bratislava, Slovakia: Wiley-V C H Verlag Gmbh, 1997. p. 99.
[2] T.K. Maji and A.N. Banerjee, J Appl Polym Sci, 73, 2187 (1999)
[3] L. Tan, J.P. Deng and W.T. Yang, Polym Adv Technol, 15, 523 (2004)
[4] J.W. Zhu, J.P. Deng, S.M. Cheng and W.T. Yang, Macromol Chem Phys, 207, 75 (2006)
[5] Y.X. Wang, W.B. Zhong, N. Jiang and W.T. Yang, Macromol Rapid Commun, 26, 87 (2005)
[6] H.P. Deng and W.T. Yang, Eur Polym J, 41, 2685 (2005)
[7] C.M. Xing, J.P. Deng and W.T. Yang, Macromol Chem Phys, 206, 1106 (2005)
[8] L.B. Kong, J.P. Deng and W.T. Yang, Macromol Chem Phys, 207, 2311 (2006)

Electrospinning of drug-loaded polymer systems: preparation, characterization and drug release

Giuseppina Russo, Vittoria Vittoria, Gaetano Lamberti, Giuseppe Titomanlio.

Dipartimento di Ingegneria Chimica e Alimentare, Università degli Studi di Salerno, 84084 Fisciano (SA) Italy

Abstract. In this paper we formulated and characterized biomedical devices for bone tissue regeneration. These systems were realized homogenously dispersing lamellar Hydrotalcite loaded with Diclofenac Sodium in a polymeric matrix of PCL (Poly-caprolactone). These biomedical devices were obtained through the electrospinning technique that has shown many advantages with respect to others techniques, in particular very interesting micro-fiber loaded with HDik were obtained. Drug delivery results were analyzed considering a mathematical modeling to evaluate the diffusivity coefficients.

Keywords: Poly-caprolactone, drug delivery, Diclofenac Sodium, HDik

INTRODUCTION

In the past years, a variety of materials have been used for replacement and repair of damaged or traumatized bone tissues. These materials include metals, ceramics, polymers (natural and synthetic) and their combinations. Metals and ceramics have two major disadvantages for tissue engineering applications: they are lacking of degradability in a biological environment, and show very limited processability. In contrast, polymers have great design flexibility because the composition and structure can be tailored to the specific needs. They are therefore attractive candidates. Biodegradability can be imparted into polymers through molecular design. Some polymers contain chemical bonds that undergo hydrolysis upon exposure to the body's aqueous environment, and some others can degrade by cellular or enzymatic pathways. For these reasons, polymeric materials have received considerable attention and are widely studied for bone tissue engineering applications **[1]**.

Among the available biopolymers, Poly(e-caprolactone) (PCL) is a very attractive polymer owing its biocompatibility and biodegradability properties.

EXPERIMENTAL PART

Materials. The polymeric matrix chosen to realize biomedical devices is "Poly-Caprolactone" (PCL) which belongs to the polyester polymer group and is biodegradable, thermoplastic, aliphatic and have an hydrophilic nature.

The fillers chosen to design active devices were:

a) <u>Diclofenac Sodium (Dik)</u>, a drug that belongs to the category of the NSAID drugs (Non -Steroidal Anti Inflammatory Drugs). The name derived from its chemical name: 2-(2,6-Dichloranilino) Phenylacetic Acid.

b) <u>HDik</u> (Hydrotalcite loaded with Diclofenac Sodium): an interclay based on Hydrotalcite and Diclofenac Sodium that allows to control the drug delivery phenomena. To realize HDik a chemical synthesis was necessary; in this way Diclofenac Sodium molecules were entrapped in this inorganic compound (Hydrotalcite) and this final structure delays the release of the drug.

HDik chemical formula: [Mg 0.65Al 0.35(OH)2]DIK 0.35.0.86H_2O

Methods. The technique used to realize biomedical devices was the: "Electrospinning". This technique allowed us to

CP1255, *Vth International Conference on Times of Polymers (TOP) and Composites,*
edited by A. D'Amore, D. Acierno, and L. Grassia

create polymeric fibers with diameters in the range of nano-micro scale [2]. The micro-nano scale diameter of electrospun fibers introduced properties like increased surface to volume ratio, and modification of the release rate.

RESULTS AND DISCUSSIONS

Electrospun membranes were characterized using SEM studies and drug delivery phenomena at different temperatures (T=37°C, T= T amb).

Morphological analysis. Figure 1 shows the micrographs of the membrane with 10% of Hydrotalcite loaded with Diclofenac Sodium (sample PCL+10% HDik) before (left image) and after drug delivery experiments (right image) .

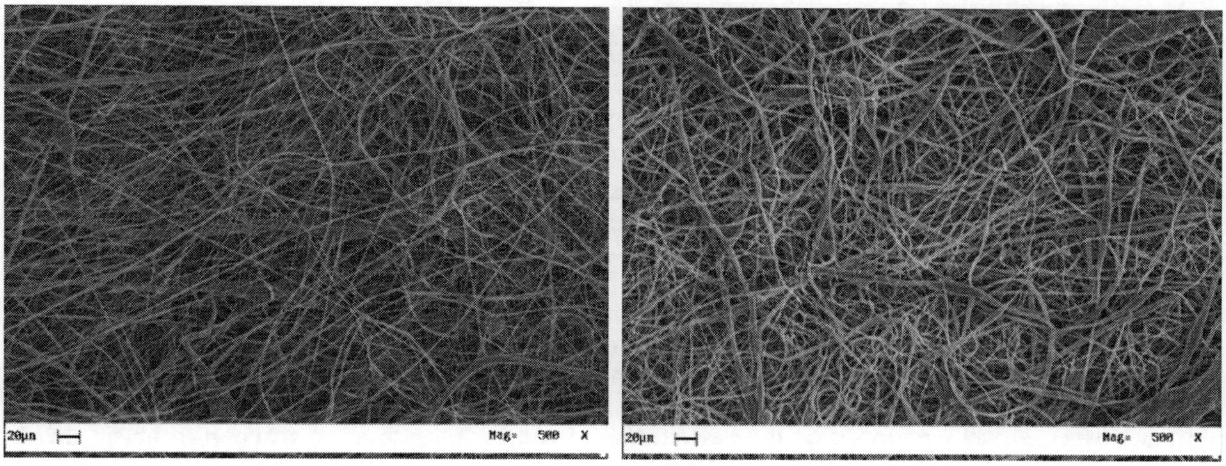

FIGURE 1. SEM: membrane morphology of the sample PCL+10% HDik before (left image) , and after drug delivery (right image) at room temperature.

The images show that the membrane morphology does not change after drug delivery.

Drug delivery phenomena. Drug delivery phenomena were studied using different operative temperatures (T1= Tamb; T2= 37°C) for samples with 5% and 10% of Hydrotalcite loaded with Diclofenac Sodium These samples are named *PCL+10% HDik* and *PCL+5 % HDik* respectively. Here, for comparison we also show two samples where the drug (Diclofenac Sodium – DS) is directly dispersed in the polymeric matrix. The percentages of DS used for these last samples are 2.5% (*PCL+2.5% DS)* and 5.0% (*PCL+5.0% DS)* respectively.
The analyzed samples are listed in table 1 which also shows the operative parameters.

TABLE 1. Drug delivery operative conditions

Samples:	Dissolution medium	Dissolution volume [ml]	Agitation [rpm]	[°C]	Experimental last [hours]
-PCL+2.5% DS	physiological	25	7	T amb	624
-PCL+5% DS	solution			37	
-PCL+5% HDik					
-PCL+10% HDik					

In figure 2 we show the Drug Delivery Curve for the above mentioned samples.

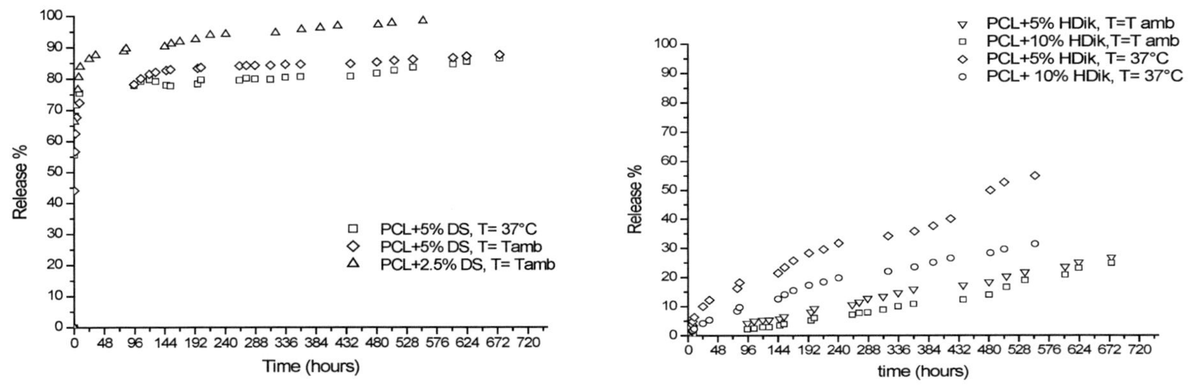

FIGURE 2. Drug Delivery Curve samples loaded with: a) Diclofenac Sodium; b)HDik

The curves of the samples where the DS is directly dispersed in the polymeric matrix show an initial burst followed by a plateau. A very different behaviour is obtained for samples loaded with HDik. In this case, the curve shape is indicative of an ionic interchange between the dissolution medium molecules and the interclay (HDik). As for the effect of the temperature on drug delivery phenomena we observe that it influences the kinetic of the process but not the shape of the Drug Delivery Curves.

Drug delivery experimental results were considered to evaluate the drug delivery diffusivity coefficients (D). In this evaluation the system (membrane in the dissolution medium) is considered as a one dimensional diffusion in medium bounded by two parallel planes in a non steady –state [3]. Solving the mass variation equation and using experimental data, it was possible to calculate the Diffusivity coefficient using the concentration variation law for long time, which is well approximated by the following equation. The use of eq.1 allows to obtain the drug diffusivity.

$$1 - R(t) = \frac{8}{\pi^2}\exp\left[-\pi^2 \frac{Dt}{b^2}\right] \tag{1}$$

Table 2 shows the diffusivity coefficient for the samples loaded with HDik.

TABLE 2. Drug Delivery Study: Diffusivity Coefficients Results

Samples:	**: D/b^2**	**D $[m^2/s]$**	**[°C]**
PCL+ 5% HDIK	$^{-5}$	$2.5 \cdot 10^{-14}$	T amb
PCL+ 10% HDIK	$4.0 \cdot 10^{-5}$	$2.4 \cdot 10^{-14}$	
PCL+ 5% HDIK	$\cdot 10^{-4}$	$4.0 \cdot 10^{-14}$	37
PCL+ 10% HDIK	$6.0 \cdot 10^{-5}$	$\cdot 10^{-14}$	37

These results indicate that the diffusivity coefficient is not influenced by HDik percentage in the samples at room temperature; its value, as expected, increases with temperature.

REFERENCES

1. Burg, K. J. L., S. Porter, and J. F. Kellam. Biomaterial developments for bone tissue engineering. Biomaterials 21:2347–2359, 2000
2. Reneker, D. H. (2007), Sawicka, K.M., Gouma, P. (2006) & Liang, D. et al (2007)
3. Cranck J., "The mathematics of diffusion", Oxford Science Editions, (1975).

Study of the low-pressure plasma effect on polypropylene nonwovens

R. López[a], T. Boronat[b], M. Pascual[a], O. Calvo[a] and R. Balart[b]

[a]*Textile Research Institute (AITEX)*
Plaza Emilio Sala s/n, 03801, Alcoy (Alicante) Spain
[b]*Materials Technology Institute (ITM) Polytechnic University of Valencia*
Plaza Ferrandiz y Carbonell s/n, 03801, Alcoy (Alicante) Spain

Abstract. In this work we have used low-pressure plasma with a gas based on methane and oxygen mixture to improve wettability and durability of a PP nonwoven fabrics. The obtained results show good durability with the use of methane-oxygen plasma mixture gas. The effects of the plasma are similar to a plasmapolymerization process but in this case we obtain hydrophilic properties with high durability. The surface does not suffer important changes and the roughness of the material remains constant.

Keywords: plasma, surface, polymer, nonwovens, wettability.
PACS: 82.35.Gh, 85.40.Sz, 52.40.Hf

INTRODUCTION

Nonwovens textile fabrics are a versatile engineering material, that are used in a great variety of applications ranging from baby diapers to industrial high performance textiles including liquid and gas filtration. On the hand, the nonwoven industry is one of the fastest growing industries in the world. It has been exhibiting an average growth of about 10% over the past twenty years and should continue this rate of growth in the next ten years [1-3].

Nonwoven filtration media has existed for hundreds of years in one form or another, including wool felt and cellulose However appearance of the synthetic polymer like polyester or polypropylene promoted the increasing of its use as coolant and water filtration. These applications need in many cases good adhesive properties and high wettability. Regarding to this polypropylene (PP) has a strong hydrophobicity [4-8].

In this study, hydrophobic nonwoven polypropylene was treated with methane-oxygen gas plasma activation and subsequent plasmapolymerization to modify its surface hydrophilicity [9-13]. The gas methane-oxygen ratio used was 80:20. The effect of the variation of the plasma treatment conditions has been studied in order to optimize the plasma treatment. The film wettability has been analyzed by the study of the variation of the contact angle and free surface energy calculation. The nonwoven surface topography was analyzed by scanning electron microscopy (SEM).

The plasma treatment containing an organic monomer (methane) deposites activated molecules on the surface material. On the other hand the activation/functionalization of the surface was obtained by several ways in function of the processing condition. By this way, the activation is obtained by inserting active species, surface abrasion or cross-linking processes. The final result of plasma treatment is an increase of surface energy of nonwovens and consequently an improvement of wettability properties; in addition an improvement of aging behavior is experimented.

EXPERIMENTAL

The materials used were the following:

a. Nonwoven fabrics PEGATEX®S made from polypropylene supplied by PEGAS NONWOVENS, SA (Luxembourg) with a weight of 16.5 $g.cm^{-2}$.

CP1255, *V[th] International Conference on Times of Polymers (TOP) and Composites,*
edited by A. D'Amore, D. Acierno, and L. Grassia

b. Nonwoven fabrics VICATEX® made from polypropylene supplied by VICATEX NV, S.L. (Onteniente, Spain) with a weight of 30 $g.cm^{-2}$.
c. Nonwoven fabrics MANATEX® made from polypropylene supplied by MANTATEX, S.A. (Bañeres de Mariola, Spain) with a weight of 70 $g.cm^{-2}$.

PP substrates were exposed to radio frequency (RF) low pressure CH_4-O_2 plasma. It was used a glow discharge RF generator (operating at 13.56 MHz with a maximum power of 150 W) type CD 400 MC option PC (Europlasma, Oudenaarde, Belgium). The plasma chamber consists of four aluminium shelves for sample holder and a total volume of 64 l. The gas used for the plasma generation was a mixture of CH_4-O_2 with a 80:20 volume ratio. It was used a gas flow rate of 100 cm^3min^{-1} and the working pressure varied in the 31-32 Pa range. The treatment power varied in the 50-150 W range.

Static contact angle measurements of the plasma treated samples were carried out at room temperature on a KSV CAM 200 using four different test liquids: water, glycerol, diiodomethane and formamide. Al least six different measurements on the plasma treated surfaces were obtained and the average values for contact angles were calculated. Surface energies were calculated using the Owens–Wend method derived from the general theory of adhesion work among solid and liquid phases on which polar and nonpolar (dispersive) contributions are considered to explain interactions among the two phases.[6] In this method, it is possible to determine the solid surface energy (γ) as the sum of polar (γ_p) and dispersive (γ_d) contribution using at least two different test liquids. The relationship between the equilibrium contact angle of the liquid phase deposited onto a solid phase is derived from the general Fowkes expression, which considers the polar and dispersive contributions for both solid and liquid designed as γ_l and γs with a superscript ''d'' or ''p'' for the dispersive and polar contributions, respectively.[7,8]

$$\gamma_l \cdot (1+\cos(\theta))/2(\gamma_l^d)^{1/2} = (\gamma_s^p)^{1/2} \cdot \left[(\gamma_l^p)^{1/2} / (\gamma_l^d)^{1/2}\right] + (\gamma_s^d)^{1/2} \quad (1)$$

TABLE (1). Constant values of test liquids used for the contact angle measurements.

Liquid	$\gamma_l^d(mJ \cdot m^{-2})$	$\gamma_l^p(mJ \cdot m^{-2})$	$\gamma_l(mJ \cdot m^{-2})$
water	22.0	50.2	72.2
glycerol	34.0	30.0	64.0
diiodometano	48.5	2.3	50.8
formamida	32.3	26.0	58.3

SEM analysis was performed on a scanning electron microscope FEI mod. Phenom (FEI, Oregon, USA). Prior to each measurement, all samples were covered with a Au-Pd alloy layer with a Sputter Coater EMITECH mod. SC7620 (Quorum Technologies Ltd. East Sussex, UK).

RESULTS AND DISCUSSION

The exposure time is an important factor to determine the effect of the plasma process on the nonwoven substrate. By this way, with the aim of evaluating the influence of the plasma exposure time on PP nonwovens wettability, contact angles measurements and subsequent surface energy calculation were used. Figure 1 shows the variation of surface energy as a function of the CH_4-O_2 plasma exposure time in the range 0-30 min. As can be observed the surface energy increases with the exposure time in all nonwovens substrates analyzed. However significant differences can be observed for different nonwoven weight.

Regarding to PEGATEX® nonwoven (16.5 $g.m^{-2}$ weight) the initial value of surface energy is around 36 $mJ.m^{-2}$, this value increase abruptly up to 62 $mJ.m^{-2}$ with only 10 min of exposure time and for higher exposure times this value remains constant. The VICATEX® nonwoven (30 $g.m^{-2}$ weight) the results obtained are poor. The surface energy is not increased in the same way as above. For 15 min of exposure time the surface energy increase from 27 $mJ.m^{-2}$ to 40 $mJ.m^{-2}$ and it reach 56 $mJ.m^{-2}$ for 30 min. On the other hand the MANATEX® nowoven (70 $g.m^{-2}$ weight) increase the surface energy up to 57 $mJ.m^{-2}$ with 15 min of time exposure and it remains constant for higher time exposure values.

In general terms the results obtained offers an improvement of wettability with the PP nonwoven with the lower weight, while nonwoven substrates with higher weight show an lower increase of wettability specially the PP nonwoven VICATEX®.

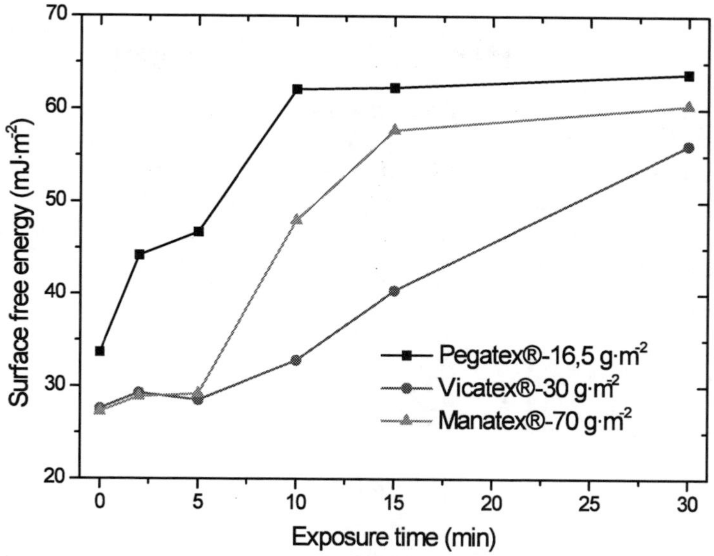

FIGURE 1. Variation of the surface free energy of PP nonwovens as a function of exposure time to CH_4-O_2 plasma.

SEM analysis of nonwovens substrates were examinated to complete the study of the plasma treatment effect over the nonwovens surface. Figure 2 shows SEM images of nonwovens used in this study both without treatment and plasma treated for 30 min of exposure time. No significant changes have been observed on the PP fibers that compose the nonwoven. This fact demonstrates that the effect of the deposition of plasma product typical of plasmapolymerization process is higher than the material removing associated to other conventional plasma gases.

FIGURE 2. SEM micrographs of PP nonwovens: a) PEGATEX® no treated; b) PEGATEX® treated with CH_4-O_2 plasma t=30 min; c) VICATEX® no treated; d) VICATEX® treated with CH_4-O_2 plasma t=30 min; e) MANATEX® no treated; f) MANATEX® treated with CH_4-O_2 plasma t=30 min.

CONCLUSION

In this work we have used low pressure plasma with a gas based on methane and oxygen mixture to improve wettability and durability of a PP nonwoven fabrics. In general terms the results obtained offers an improvement of wettability with the PP nonwoven with the lower weight, while nonwoven substrates with higher weight show an lower increase of wettability specially the PP nonwoven VICATEX®. On the other hand, the effects of the plasma are similar to a plasmapolimerization process but in this case we obtain hydrophilic properties. The surface does not suffer important changes and the roughness of the material remains constant

ACKNOWLEDGMENTS

Authors thank "Ministerio de Ciencia y Tecnología", Ref: DPI2007-66849-C02-02 and Generalitat Valenciana FPA/2010/027 & ACOMP/2010 for financial support.

REFERENCES

1. D. L. Cho, C. N. Choi, H. J. Kim, A. K. Kim and J. H. Go, *J. Appl. Polym. Sci.* **82,** 839-846 (2001)
2. Q. F. Wei, Q. Li, X. Q. Wang, F. L. Huang and W. D. Gao, *Polym. Test* **25,** 717-722 (2006)
3. M. R. Yang, K. S. Chem, J. C. Rsai, C. C. Tseng and S. F. Lin, *Mater. Sci. Eng. C-Biomimetic Supramol. Syst.* **20,** 167-173 (2002)
4. H. S. Choi, V. V. Rybkin, V. A. Titov, T. G. Shikova and T. A. Ageeva, *Surf. Coat. Technol.* **200,** 4479-4488 (2006)
5. H. Demir, D. Balkose and S. Ulku, *Polym. Degrad. Stabil.* **91,** 1079-1085 (2006)
6. C. C. Wang and C. C. Chen, *J. Appl. Polym. Sci.* **98,** 391-400 (2005)
7. H. Y. Yu, Y. Xie, M. X. Hu, J. L. Wang, S. Y. Wang and Z. K. Xu, *J. Membr. Sci.* **254,** 219-227 (2005)
8. S. Nowak, H. P. Haerri, L. Schlapbach and J. Vogt, *Surf. Interface Anal.* **16,** 418-423 (1990)
9. K. S. Chen, J. C. Rsai, C. W. Chou, M. R. Yang and J. M. Yang, *Mater. Sci. Eng. C-Biomimetic Supramol. Syst.* **20,** 203-208 (2002)
10. Y. J. Hwang, M. G. McCord, J. S. An, B. C. Kang and S. W. Park, *Text. Res. J.* **75,** 771-778 (2005)
11. M. M. Radetic, D. M. Jocic, P. M. Jovancic, Z. L. Petrovic and H. F. Thomas, *Environ. Sci. Technol.* **37,** 1008-1012 (2003)
12. Y. Shin, K. Son, D. Il Yoo, S. Hudson, M. McCord, S. Matthews and Y. J. Whang, *J. Appl. Polym. Sci.* **100,** 4306-4310 (2006)
13. J. Verschuren, P. Van Herzele, K. De Clerck and P. Kiekens, *Text. Res. J.* **75,** 437-441 (2005)

Improvement of adhesion properties of low density polyethylene (LDPE) substrate using atmospheric plasma.

L. Sánchez-Nácher, D. García-Sanoguera, O. Fenollar, J. Balart, V. Fombuena

Materials Technology Institute (ITM) Polytechnic University of Valencia
Plaza Ferrandiz y Carbonell s/n, 03801, Alcoy (Alicante) Spain

Abstract. In this work we have used atmospheric plasma technology on polyethylene surface with different treatment conditions. These modify surface pre-treatments on polyethylene, thus having a positive effect on overall surface activity of polymer surface and, consequently, adhesion properties can be remarkably improved. We have evaluated the influence of the nozzle/substrate distance and atmospheric plasma speed on wettability changes and adhesion properties. Wettability changes have been studied by contact angle measurements and subsequent surface energy calculation. Mechanical characterization of adhesion joints has been carried out in two different ways: peel and shear tensile test. The overall results show a remarkable increase in mechanical properties of adhesion joints for low nozzle/substrate distances and low speed. So plasma atmospheric technology is highly useful to increase adhesion properties of polypropylene.

Keywords: Atmospheric plasma, adhesion, polyethylene, wettability.
PACS: 82.35.Gh, 85.40.Sz, 52.40.Hf

INTRODUCTION.

Among the wide variety of thermoplastic materials, low density polyethylene LDPE is one of the most used at industrial level and continuously it finds growing uses. It is classified as a "commodity" plastic with excellent balanced properties (physical, mechanical, electrical and chemical) and easy processing. So that it is not only used in low technology sectors but also in medium-high technology applications such as automotive or medical devices. Low density polyethylene, as other polyolefins, is characterized by high chemical inertness, as a consequence of its non polar nature and this fact results in very low wetting properties. For this reason, LDPE finds enormous difficulties for different industrial processes such as painting and adhesion, which are highly related to surface wetting properties [1-2].

The main problem regarding low density polyethylene adhesion is due to its low surface roughness; this fact, together with its non polar nature, leads to poor adhesion properties; so that, adhesion joints with this polymer show relatively low mechanical performance. Hence, in most cases it is necessary to carry out a surface pre-treatment in order to modify surface wetting properties. There are several technologies to promote surface changes and improve adhesion behaviour [3]. Physical processes such as plasma technologies (corona plasma, cold plasma, atmospheric plasma…) are interesting solutions from a technical point of view. They increase surface activation by the action of the plasma gas, which is able to promote chain scission and subsequent free radical formation. After chain scission, two main processes can occur: one mechanism is surface activation by insertion of polar groups (mainly oxygen-based species) into free radicals. The second mechanism is characterized by changes in surface roughness as a consequence of chain scission that leads to formation of low molecular weight oxidized material (LMWOM), which can be removed from surface thus promoting changes in surface topography. These processes are highly advantageous from a technical and environmental point of view; despite this, they are expensive and this restricts their use at industrial level [4-5].

CP1255, *V[th] International Conference on Times of Polymers (TOP) and Composites,*
edited by A. D'Amore, D. Acierno, and L. Grassia

EXPERIMENTAL

The substrate used in study was low density polyethylene (LDPE) ALCUDIA PE-109 supplied by Repsol YPF (Repsol Chemical, S.A., Madrid, Spain). The LDPE sheets used with the plasma treatment were of 80 x 30 x 2 mm in size.

The plasma treatment was carried out by a Plasmatreat atmospheric plasma equipment mod. FG3001 (Plasmatreat GmbH, Steinhagen, Germany) with a maximum power of 1kW. It was used a rotative nozzle.

Wettability changes on treated polyethylene surface were characterized by contact angle measurements using a EasyDrop Standard goniometer. Four different test liquids were used for contact angle measurements and surface free energy calculation: diiodomethane, formamide, and glycerol. Surface free energy values were calculated using the Owens-Wendt method, which takes into account the dispersive (non polar) and polar contribution to the total surface free energy value. The Owens-Wendt method, considers the following expression:

$$\gamma_l \cdot (1+\cos(\theta))/2(\gamma_l^d)^{1/2} = (\gamma_s^p)^{1/2} \cdot \left[(\gamma_l^p)^{1/2} / (\gamma_l^d)^{1/2} \right] + (\gamma_s^d)^{1/2} \qquad (1)$$

Where θ is the contact angle, γ_l is the total surface free energy of the liquid and γ_s represents the surface free energy of the solid. The terms with subscripts "d" and "p" refer to the "dispersive" and "polar" component of the surface free energy. We can see in an easy way that the previous equation can be expressed as a linear expression ($y = a + bx$). So that, we can represent $(\gamma_l^p)^{1/2}/(\gamma_{ld})^{1/2}$ versus a $\gamma_l\ (1+\cos\theta)/(\gamma_l^d)^{1/2}$. Then, the slope of the line will be $(\gamma_s^p)^{1/2}$ and the y-axis intercept will be $(\gamma_s^d)^{1/2}$. The total solid surface free energy (γ_s) will be calculated as the sum of the polar (γ_s^p) and dispersive (γ_s^d) component.

RESULTS AND DISCUSSION

In general it is observed that the surface modification of LDPE sheets through the application of atmospheric plasma allows optimal functionalization of the polymer surface at low speed and low spacing nozzle/substrate.

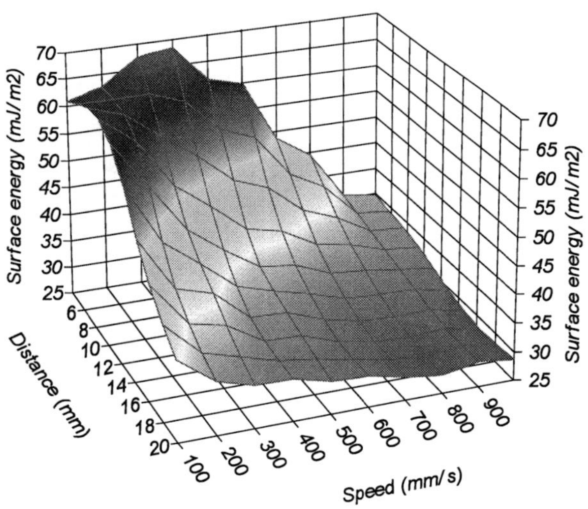

FIGURE 1. Variation of total surface energy (γs) of low density polyethylene surfaces as a function of plasma treatment conditions: speed and distance nozzle/LDPE

The optimum wetting properties by atmospheric plasma treatment on LDPE surface are obtained for distance nozzle/LDPE 6 and 10 mm, and speed 100 mm.s^{-1}. Regarding mechanical properties of LDPE-LDPE adhesion joints, it is important to remark that similar tendency is observed for both shear and T-peel tests. The increase surface activation by atmospheric plasma treatment permits important wettability increase on treated LDPE surface, and then an improvement of adhesion properties. The establishment of strong interactions between the polyurethane adhesive and LDPE substrates plasma treated thus obtaining good adhesion properties. In this case, SEM analysis has revealed a combined adhesive-cohesive fracture type thus indicating good adhesion properties.

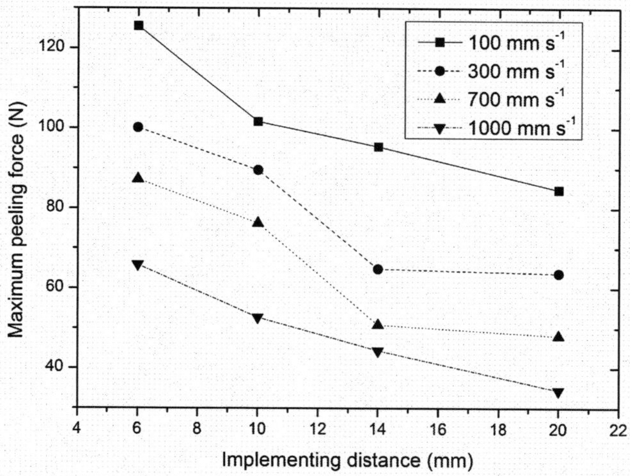

FIGURE 2. Variation of maximum peeling force of low density polyethylene adhesion as a function of plasma treatment conditions: speed and distance nozzle/LDPE

CONCLUSIONS

In this work we have used atmospheric plasma on low density polyethylene (LDPE) substrate. The results obtained demonstrate that the atmospheric plasma treatment is an adequate technique to improve the adhesion properties. The processing conditions play an important role in the final properties of the substrate, which are strongly linked with the functionalization of polymer surface.

ACKNOWLEDGMENTS

Authors thank “Ministerio de Ciencia y Tecnología”, Ref: DPI2007-66849-C02-02 and Generalitat Valenciana FPA/2010/027 & ACOMP/2010 for financial support.

REFERENCES

1. D. L. Cho, C. N. Choi, H. J. Kim, A. K. Kim and J. H. Go, *J. Appl. Polym. Sci.* **82,** 839-846 (2001).
2. Y. Takemura, N. Yamaguchi, T. Hara, *J. Appl. Phys.* **47**, 5644- 5647 (2008).
3. J. Vartiainen, M. Ratto, U. Tapper, S. Paulussen, E. Hurme, *Polym. Bull.* **54**, 343-352 (2005).
4. L. Zhu, W. H. Teng, H. L. Xu, Y. Liu, Q. R. Jiang, C. X. Wang, Y. P. Qiu, *Surf. Coat. Technol.* **202**, 1966-1974 (2008).
5. Y. Ren, C. X. Wang, Y. P. Qiu, *Appl. Surf. Sci.* **253**, 9283-9289 (2007).

Mechanical and morphological characterization of novel vinyl plastisols with epoxidized linseed oil as natural-based plasticizer

O. Fenollar, R. Balart, L. Sánchez-Nácher, D. García-Sanoguera, T. Boronat

Institute of Materials Technology (ITM), Polytechnic University of Valencia (UPV), Plaza Ferrandiz y Carbonell s/n, 03801, Alcoy, Alicante (Spain)

Abstract. Poly(vinyl chloride) (PVC) is one of the most commonly used plastics in the current market due to its low cost and versatility in processing, combined with its satisfactory physical and chemical properties. However, there is an important problem associated to the use of plasticized PVC. This problem is regarding to the toxicity of the most common plasticized used like DOP, DEHP, DINP, due to its possible migration. This problem limits the use of the plasticized PVC in the industry. In this work we have used epoxidized linseed oil (ELO) as a non toxic plasticizer for PVC. This type of natural oil is characterized by acting as both plasticizer and stabilizer of PVC. With this purpose, ELO have been added to PVC. The processing conditions (temperature and time of curing) are vital to determine the final properties of the material. A study of the processing conditions shows the adequate temperature and time to achieve the optimum properties.

Keywords: PVC, plasticizers, ELO, linseed, ftalates, migration.
PACS: 87.85.J, 62.20.mm, 68.37.Hk, 81.05.Zx.

INTRODUCTION

The use of vinyl plastisols at industrial level has acquired great relevance as a consequence of their excellent flexibility properties, adaptability to wide range of industrial processes, low costs, etc. Despite this, the use of traditional plasticizers (mainly based on phthalate structures) are being questioned due appearance of toxicity problems associated to plasticizer migration [1-3]. So, most of the efforts in this research field are focused on the use of alternative low toxicity and low migration plasticizer to reduce in a considerable way toxic risk derived from the use of these materials, mainly attributable to plasticizer migration. On the other hand, nowadays vegetable oils are a resource of several polymers. Concretely vegetable-oil-base like esters and epoxies can be used as plasticizer and additives to commercial polymers [4]. Epoxidized vegetable oils like soybean oil (ESBO) and sunflower oil (ESO), have been used in processing of PVC with good results [5-9]. It is necessary to highlight that the epoxidized vegetable oils are listed as authorized substance in the plastic regulation approved by the toxicological evaluation and the legal status in the EU [10]. The aim of this work is to analyze the morphological and mechanical properties of plasticized PVC with ELO in order to become the ELO in an alternative to traditional plasticizers

EXPERIMENTAL

Materials and procedures

The PVC resin Lacovyl PB 1172 H was supplied by Atofina (Atofina UK, Ltd., Midlands United Kingdom) with a k value of 67 obtained according the standard ISO 1628-2. The epoxidized linseed oil was supplied by Traquisa

CP1255, *Vth International Conference on Times of Polymers (TOP) and Composites,*
edited by A. D'Amore, D. Acierno, and L. Grassia

(Traquisa, S.A., Barcelona, Spain). The poly(vinyl chloride) and the plasticizer have been prepared by mixing different percentages of PVC resin and natural plasticizer in a rotative mixer KAPL mod. 5KPMS (KAPL, St. Joseph, MI). After the mixing process, the pastes were subjected to a vacuum process for 30 min in a MCP Group vacuum chamber model 00ILC (HEK-GmbH, Lübeck, Germany) with a maximum vacuum of -1 bar for removing all of the air bubbles. The curing process of the plastisols was performed in a ventilated oven Carbolite mod. 2416CG (Keison products, Barcelona, Spain) with a maximum temperature of 300 °C. Curing time and temperatures. The tensile tests were carried out using a universal tensile test machine ELIB 30 (S.A.E. Ibertest, Madrid, Spain) following ISO 527.. Morphology analysis of tensile fractured surfaces of plastisols was carried out by means of a scanning electronmicroscope JEOL JSM-6300 (Jeol USA, Peabody) usingsecondary electrons with an acceleration voltage of 15 kV. Samples were covered with Au layer (5–7 nm) by sputteringin vacuum conditions prior to each measurement.

RESULTS AND DISCUSSION

Regarding to tensile test, we can see in Figure 1 that for times less than 8 min, and low temperatures (160 and 180°C), the material breaks easily. This indicates that the curing process has not been completed and then, plastisol remains as a solid paste, so mechanical properties are poor . Higher curing times and temperatures lead to an increase in tensile strength. To completely cure a plastisol, it is necessary to reach temperatures over 180–190 °C to promote fusion of microcrystallites. When PVC plastisol is isothermally cured at 220°C, the process is fully completed with relatively short times.

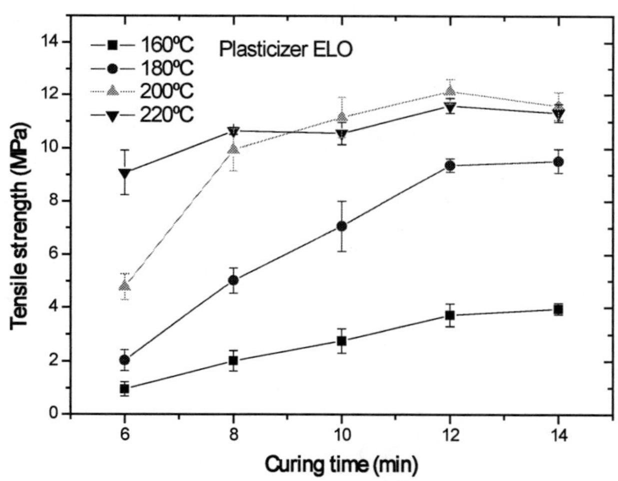

FIGURE 1. Influence of curing conditions (time and temperature) on tensile strength of plastisol prepared with epoxidized linseed oil (ELO).

Figure 2a shows SEM microphotographs of fractured samples isothermally cured at 180 °C and curing time of 10 min). It shows the presence of PVC swollen particles and aggregates as a consequence of plasticizer absorption. This phenomenon indicates that the melting process has started; because of this, mechanical strength is higher, although gelation process has not been completed yet. In figure 2b, we can see a SEM microphotograph of fractured surfaces of vinyl plastisols with a epoxidized linseed oil isothermally cured at 220 °C for t= 10 min. Image indicates that the crystalline phase has fused. This surface morphology shows evidences of elastic and plastic deformation such as waves and threads formed during the deformation-fracture process. Also, with these curing conditions, mechanical properties are increase in a remarkable way as a consequence of particle cohesion.

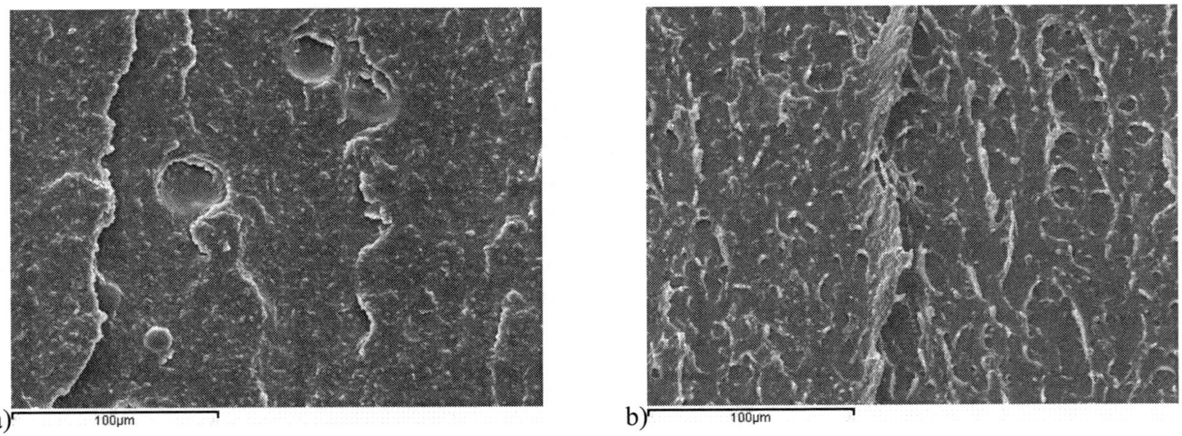

a) b)

FIGURE 2 SEM microphotographs of fractured surfaces of vinyl plastisols with epoxidized linseed oil, a) isothermally cured at 180° C; t = 10 min; a) isothermally cured at 220° C; t = 10 min,

CONCLUSIONS

The use of natural-based plasticizers as ELO rises as an interesting alternative for obtaining plastisols with low toxicity. The curing process may be carried out in a suitable way at isothermal temperatures of 200 or 220 ºC with curing times of about 10 and 14 min, respectively, according to the study of mechanical properties. For these curing conditions, plastisol microstructure is formed by a completely homogenous matrix due to melt of microcrystallites and the plasticizer absorption. So, we can say that the use of epoxidized linseed (ELO) is adequate additive for replacing traditional plasticizers with high toxicity phthalates.

ACKNOWLEDGMENTS

Authors thank "Ministerio de Ciencia y Tecnología", Ref: DPI2007-66849-C02-02 and Generalitat Valenciana FPA/2010/027 & ACOMP/2010 for financial support.

REFERENCES

1. J.W. Summers J Vinyl Addit Technol 1997, 3, 130.
2. D. Braun J Vinyl Addit Technol 2001, 7, 168.
3. A.C. Shah; D.J. Poledna J Vinyl Addit Technol 2002, 8, 214.
4. S.M. Ashraf; S. Ahmad; U. Riaz; M. Alam; H.O. Sharma J. Appl. Polym. Sci. 2007, 103, 1367.
5. M.T. Benaniba; N. Belhaneche-Bensemra; G. Gelbard Polym. Degrad. Stabil. 2001, 74, 501.
6. M.O. Boussoum; D. Atek; N. Belhaneche-Bensemra Polym. Degrad. Stabil. 2006, 91, 579.
7. S.W. Kim; J.G. Kim; J.I. Choi; I.R. Jeon; K.H. Seo J Appl Polym Sci 2005, 96, 1347.
8. M.A. Semsarzadeh; M. Mehrabzadeh; S.S. Arabshahi Eur. Polym. J. 2002, 38, 351.
9. M.A. Semsarzadeh; M. Mehrabzadeh; S.S. Arabshahi Iran. Polym. J. 2005, 14, 769.
10. A. Fankhauser-Noti; S. Biedermann-Brem; K. Grob Eur. Food Res. Technol. 2006, 223, 447.

Mechanical properties of green composites based on thermoplastic starch

F. Fornes, L. Sánchez-Nácher, O. Fenollar, T. Boronat, D. Garcia-Sanoguera

Materials Technology Institute (ITM) Polytechnic University of Valencia
Plaza Ferrandiz y Carbonell s/n, 03801, Alcoy (Alicante) Spain

Abstract. The present work is focused on study of "green composites" elaborated from thermoplastic starch (TPS) as polymer matrix and a fiber from natural origin (rush) as reinforced fiber. The effect of the fiber content has been studied by means of the mechanical properties. The composite resulting presents a lack of interaction between matrix and fiber that represents a performance decrease. However the biodegradability behavior of the resulting composite raise this composite as useful an industrial level.

Keywords: mechanical properties, biocomposites, fiber, TPS
PACS: 81.70.Bt, 81.05.Qk,

INTRODUCTION

In the last years, it has appeared a new type of composites named "green composites". These composites are composed by a biodegradable polymer filled with natural fibers. The composite resulting is biodegradable material with low environmental impact [1]. Nowadays many works are focused on the development of bio-based composites is because of natural fibres biodegradability, light weight, low cost, high specific strength compared to glass and carbon, recycling and renewing natural sources of plants such as coir, sisal, jute, Agave Americana, kenaf, flax, ramie, abaca, kapok and hemp [2, 3]. Especially, natural fibers have a good potential in the future as a substitute for wood- based material in many applications [4, 5].

Rush is a plant, which belongs to the Juncaceae family characterized by a lower lignin content value and high humidity absorption. Regarding to the biodegradable polymer matrix used in the "greencomposites" polyalcaonatederivates as polylactic acid (PLA), polycaprolactone, polyhydroxybutirate (PHB), polyhydroxy butirate-valerate (PHBV) and thermoplastic starch (TPS) are commonly used [7].

Starch is a complex homo-polymer composed of a-D-glucose units linked together in two different forms: the linear form amylose and the highly branched amylopectin, the polymer from starch (TPS) is obtained by solvent precipitation [6].

The objective of this work is to compare the mechanical properties of biopolymers reinforced with natural fibers for obtaining "green composites" which can replace any polymer thermoplastic in optimal conditions. With this aim, mechanical behavior of biocomposite composed by TPS with Rush fibers has been analyzed.

EXPERIMENTAL

The thermoplastic starch used in this study was TPS NF866 supplied by Novamont (Novamon, SPA, Italy). The tensile tests were carried out using a universal tensile test machine ELIB 30 (S.A.E. Ibertest, Madrid, Spain) following ISO 527. A 20 mm min-1 crosshead speed was used to obtain the tensile graph with a load cell of 5 kN. All specimens were tested at room temperature and a minimum of five samples were analyzed and average values of tensile strength and elongation at break were calculated. SEM analysis was performed on a scanning electron microscope FEI mod. Phenom (FEI, Oregon, USA). Prior to each measurement, all samples were covered with a Au-Pd alloy layer with a Sputter Coater EMITECH mod. SC7620 (Quorum Technologies Ltd. East Sussex, UK).

CP1255, *Vth International Conference on Times of Polymers (TOP) and Composites,*
edited by A. D'Amore, D. Acierno, and L. Grassia

RESULTS AND DISCUSSION

The addition of natural fibers in the TPS matrix promotes important changes on the mechanical properties with the fiber content. With respect to elastic modulus as can be expected the addition of rush fiber promotes an increase of stiffness of composite. Thus, for 5 wt% of fiber the increase is around 110 %, for higher amount of fiber (20 wt%) the improvement is close to 320%.

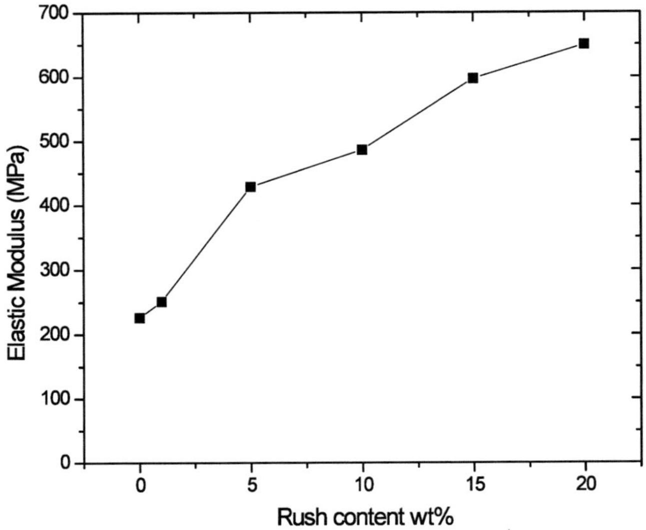

FIGURE 1. Variation of elastic modulus of TPS-rush composite as a function of the fiber content (wt%).

Regarding to tensile strength the rush fiber addition not offers significant changes of this property. The values obtained are in the 13-12 MPa range.

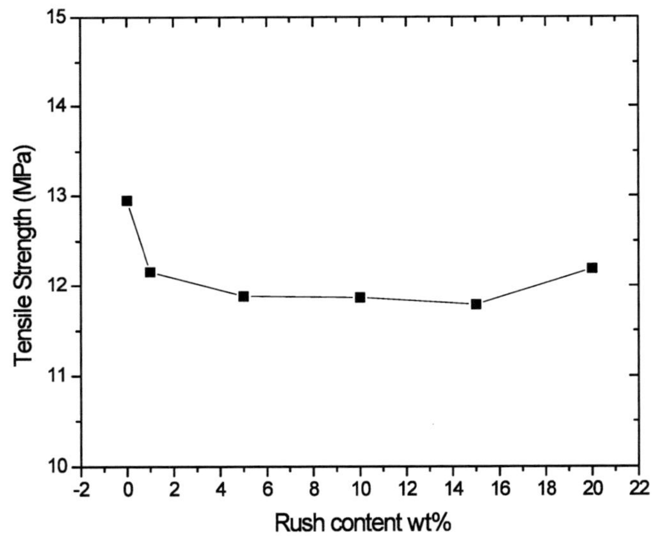

FIGURE 2. Variation of tensile strength modulus of TPS-rush composite as a function of the fiber content (wt%).

Regarding elongation at break, this is a property that is harmed by the addition of fiber rush. The decrease on elongation capability is evident, so it is decrease from 188 to 52 with only 1 wt% of rush fiber.

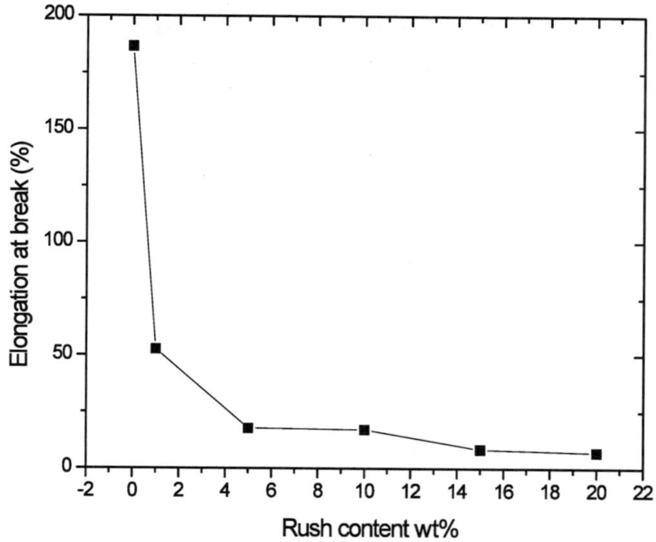

FIGURE 3. Variation of elongation at break of TPS-rush composite as a function of the fiber content (wt%).

Regarding to morphological study of fractured surface of composite after tensile test, the Figure 4a shows the SEM micrograph of composite without fiber, where can be observed the typical morphology of a thermoplastic with high strain. The Figure 4b shows a SEM micrograph of composite with rush fiber. A lack of adhesion between natural fiber and polymer matrix is observed. This fact contributes in a decisive way to the fall in strain at break.

a) b)

FIGURE 4. SEM micrographs of TPS-rush composite: a) TPS without fiber; b) TPS-5 wt% fiber.

CONCLUSION

The results obtained in this study demonstrate that tensile strength of the "green composite" improves with the amount of natural fiber. However the composite presents a lack of adhesion between matrix and fiber that promotes a decrease on the deformation capability. It is expected that an improvement on the interfacial adhesion will result in an important increase on mechanical performance specially regarding ductile properties.

ACKNOWLEDGMENTS

Authors thank "Ministerio de Ciencia y Tecnología", Ref: DPI2007-66849-C02-02 and Generalitat Valenciana FPA/2010/027 & ACOMP/2010 for financial support.

REFERENCES

1. A. Carvalho, M. D. Zambon, A. A. S. Curvelo and A. Gandini, *Polym. Degrad. Stabil.* **79,** 133-138 (2003)
2. S. Kuciel and A. Liber-Knec, *J. Biobased Mater. Bioenergy* **3,** 269-274 (2009)
3. G. Romhany, T. Czigany and J. Karger-Kocsis, *Compos. Sci. Technol.* **66,** 3179-3187 (2006)
4. G. Romhany, J. Karger-Kocsis and T. Czigany, *Macromol. Mater. Eng.* **288,** 699-707 (2003)
5. F. G. Torres, O. H. Arroyo and C. Gomez, *J. Thermoplast. Compos. Mater.* **20,** 207-223 (2007)
6. R. F. T. Stepto, *Macromol. Symp.* **279,** 163-168 (2009)
7. A. Taguet, M. A. Huneault and B. D. Favis, *Polymer* **50,** 5733-5743 (2009)

Modeling and stress-strain characteristics of mechanical properties of carbon nanotube reinforced polyvinylacetate nanocomposites

Remo Merijs Meri [a], Juris Bitenieks [a], Martins Kalnins [a] and Robert Maksimov [b]

[a] *Institute of Polymer Materials, Riga Technical University, 14/24 Azenes street, Riga, LV-1048, Latvia*
[b] *Institute of Polymer Mechanics, University of Latvia, 23 Aizkraukles street, Riga, LV-1006, Latvia*

Abstract. Multiwall carbon nanotube (CNT) modified polyvinylacetate nanocomposites (PVAc) are investigated. PVAc/CNT nanocomposites are obtained by using solution casting technique. Stress-strain characteristics of PVAc/CNT nanocomposites clearly reveal remarkable reinforcing effect of CNT additions already at small weight contents (up to 2 wt. %). Mathematical model for prediction of elastic properties of CNT polymer nanocomposites is proposed.

Keywords: CNT, PVAc, stress-strain properties, modeling
PACS: 45.20.D-; 62.20.-x; 81.05.Lg; 81.05.Qk; 83.80.Ab

INTRODUCTION

Carbon nanotubes, since their discovery in 1991, have attracted considerable interest of scientists due to their unique electrical, thermo-physical, mechanical and other properties. Multiple researches have been devoted to structural, processing, mechanical and other issues [1-3] of carbon nanotube modified polymer nancomposites. The effect of small additions of multiwall carbon nanotubes (CNT) on the structural, calorimetric, thermogravimetric and electric properties of polyvinylacetate (PVAc) matrix nanocomposites is clearly demonstrated also in our previous research [4].

In order to ensure reliable performance of carbon nanotube modified polymer composites, perspective mathematical model for prediction of its exploitation properties should be developed. The current research is devoted to the investigation of mechanical properties of CNT modified PVAc nanocomposites and fitting of a mathematical model for prediction of its elastic characteristics.

EXPERIMENTAL

Finndisp HW 1 polyvinlylacetate (PVAc) is used as a thermoplastic matrix, while *Bayer Baytubes* (CNTs) – as a nanostructured filler for manufacturing of the nanocomposites by solution casting method. Concentration of the CNT in PVAc is varied from 0.01 to 2 wt. %. Sodium dodecylsulphate is used as stabilizing agent in the PVAc/CNT dispersions.

Tensile stress-strain characteristics of PVAc/CNT nanocomposites are determined by using Zwick/Roell BDO020 universal testing machine (speed of the upper moving clamp – 1 mm/min. for Young's modulus determination and 50 mm/min. for measuring of stress-strain characteristics at yield and at break).

RESULTS AND DISCUSSION

Stress-strain characteristics of PVAc/CNT nanocomposites are shown in Fig. 1. As one can see modulus of elasticity E and yield strength σ_r both exponentially increase to certain maximum value along with increasing weight content of CNT from 0.01 wt. % to 2 wt. %. Ultimate elongation ε of the investigated nanocomposites at the same CNT weight contents decrease from more than 500% to 200 %. It is however important to mention that all the

CP1255, *Vth International Conference on Times of Polymers (TOP) and Composites,*
edited by A. D'Amore, D. Acierno, and L. Grassia

investigated nanocomposite compositions in this CNT concentration region retain their deformation behavior, characterized by yield maxima at ca 5 % and subsequent strain hardening until rupture. In order to explain this behavior detailed analysis of elastic constants of PVAc/CNT nanocomposites is presented below.

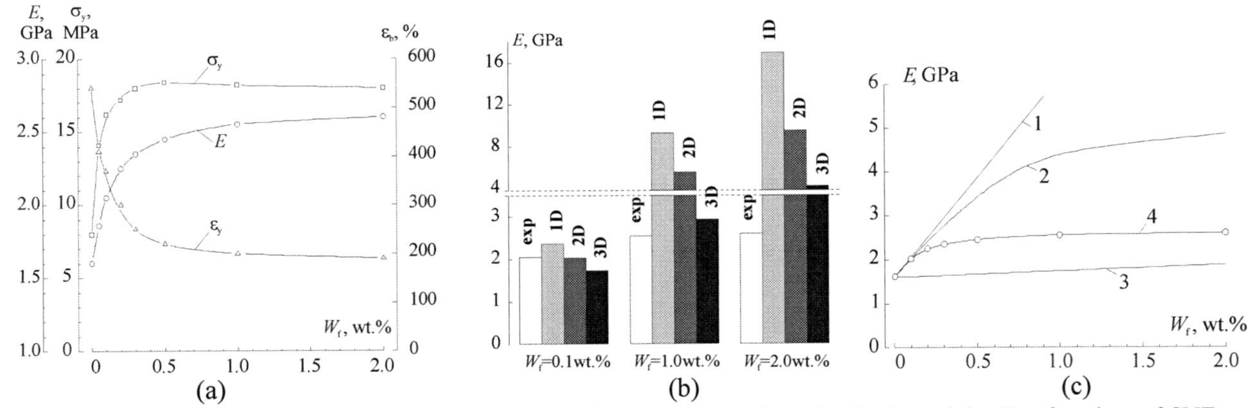

FIGURE 1. (a) Experimental values of yield stress σ, ultimate strain ε, and tensile elastic modulus E as functions of CNTs weight content W_f; (b) Experimental data (exp) and theoretical predictions of E of the nanocomposite, assuming straight CNTs with different orientations: uniaxially aligned (1D) and randomly oriented in 2D and 3D space; (c) E as a function of CNTs weight content W_f in the composite. Dots – experiment, lines – calculations for the cases of straight CNTs (1), zigzag shaped CNTs (2), totally agglomerated CNTs (3) and partly agglomerated and curved CNTs (4).

According to a known generalized Eshelby solution [5] for a composite, containing small concentration of uniaxially oriented (1D) ellipsoidal inclusions, elasticity tensor can be expressed as follows:

$$C^* = C^m + V_f\left(C^f - C^m\right)\cdot\cdot A^f \quad \text{and} \quad A^f = A_{dilute} = \left[I + S\cdot\cdot\left(C - C^m\right)\cdot\cdot\left(C^m\right)^{-1}\right]^{-1}, \qquad (1)$$

where V_f - the volume fraction of inclusions; C_f and C_m - the rigidity tensors of inclusion and matrix, respectively; S - the Eshelby tensor [5]; I - the fourth-rank unit tensor with the components $I_{ijkl} = 1/2(\delta_{ik}\delta_{jl} + \delta_{il}\delta_{jk})$, where δ_{ik} - the Kronecker delta; A_f - the strain concentration tensor.

In the case of finite concentration of the nanofiller and consideration of its elastic interaction in the composite let's use the Mori–Tanaka model [6], which elasticity tensor can be expressed as follows:

$$A^f = A_{MT} = A_{dilute}\left[\left(1 - V_f\right)I + V_f A_{dilute}\right]^{-1}. \qquad (2)$$

Analytical equations for calculations of elasticity constants of composites (tensile modulus E, shear modulus G and Poisson's ratio ν) with uniaxially and randomly oriented (both spatially (3D) and planary (2D)) ellipsoidal inclusions are presented in [7, 8].

In Fig. 2b experimental values of E of various compositions of PVAc/CNT nanocomposite are compared with theoretically calculated values for 1D, 2D and 3D orientations of the nanofiller. In the calculations it has been assumed that E_f=900 GPa, ν_f=0,25, l_f/d_f=200, E_m=1,6 GPa, ν_m=0,35. As one can see there are great differences between experimentally detected and theoretically predicted values of E of the investigated PVAc/CNT nanocomposites. Besides it, theoretical and experimental $E(W_f)$ relationships are also different as it is shown in Fig. 1c. By comparing the experimental data with theoretical ones, calculated for 2D orientation of straight CNT in a uniform thermoplastic matrix (points and curve 1 respectively), sufficient coincidence is observed only at low nanofiller content. The most plausible reasons for such behavior are curvature and agglomeration of CNTs.

In the calculations the effect of the curvature is considered by assuming that in the case of random orientation CNTs can be represented as zig-zags instead of line segments, which leads to decrease of the aspect ratio of CNTs, along with increasing nanofiller content. As one can see from Fig. 1c, by comparing theoretically calculated E values in the case of 2D orientation (curve 2) with experimental data points, there are still considerable differences, which are most probably caused due to agglomeration of CNTs.

On the supposition that part of CNTs are agglomerated, the degree of agglomeration can be expressed by coefficient k ($k=V_{fa}/V_f$, where V_{fa} and V_f represent agglomerated and total amounts of CNTs in the thermoplastic matrix, respectively). Effective elasticity constants of the inclusions with agglomerated CNTs depend on both, aspect ratio l_f/d_f of CNTs and volumetric part μ_{fa} of the nanocomposite, containing agglomerated CNT. In addition let's assume 3D orientation of the agglomerated CNTs in the spherical inclusions. Theoretical values of tensile modulus E of such inclusions, calculated for l_f/d_f range from 0,1 to 0,5 and for μ_{fa} range from 10 to 100, are shown in Fig. 2a.

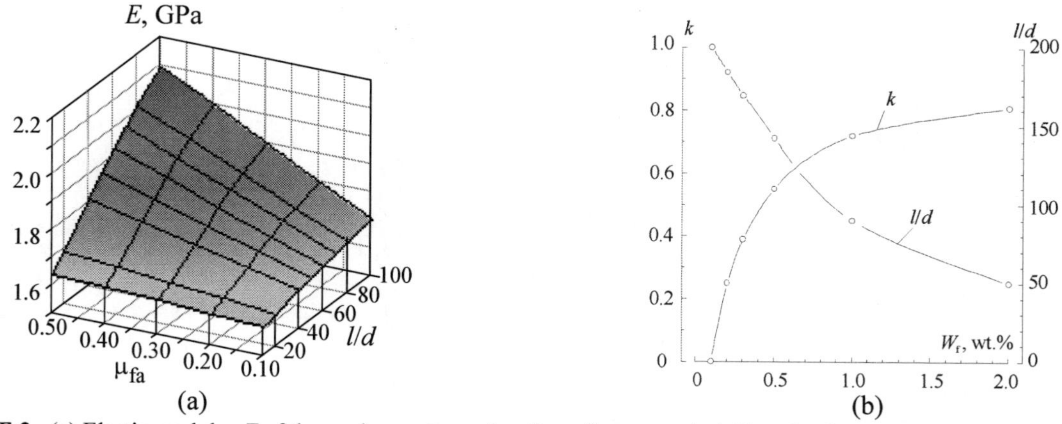

FIGURE 2. (a) Elastic modulus E of the agglomerates as function of aspect ratio l_f/d_f and volume content μ_{fa} of randomly oriented CNTs; (b) Coefficient k and aspect ratio l_f/d_f as functions of CNT weight content W_f.

It is important to mention that volumetric content of spherical inclusions V_a with agglomerated CNTs depends on both, V_{fa} and μ_{fa}:

$$V_a = V_{fa} / \mu_{fa} = kV_f / \mu_{fa} . \tag{3}$$

At small concentrations and total agglomeration, reinforcing effect of CNTs is negligible, as it is shown in Fig. 1c, where the straight monotonously growing relationship 3 represents the case when k=1, μ_{fa}=0,2 and l_f/d_f=50. By comparing experimental data points with theoretically calculated curve 3, one can assume that in real nanocomposite agglomeration degree of CNTs, characterized by k, is changed with the growth of W_f. It is clearly demonstrated in Fig 2b, where calculated k values of the approximation 4, shown in Fig.1c (determined by considering curvature of 2D oriented CNTs with following structural characteristics of the agglomerates - μ_{fa}=0,2 and l_f/d_f=50), are depicted as a function of CNT weight content.

CONCLUSIONS

Results of the investigation on the stress-strain characteristics of multiwall carbon nanotube (CNT) modified polyvinylacetate nanocomposites (PVAc) testify that along with increasing weight content of CNT modulus of elasticity E and yield strength σ exponentially increase to the certain maximum value, while ultimate strain ε correspondingly decrease to the certain minimum value. In addition it has been mathematically proven that elastic constants of CNT containing polymer nanocomposites depend on both curvature and agglomeration of CNTs.

REFERENCES

1. R. Andrews and M.C. Weisenberger, *Current Opinion in Solid State & Materials Science*, **8**, 31-37 (2004).
2. J. N. Coleman, U. Khan, W. J. Blau and Y. K. Gun'ko. *Carbon*, **44**, 1624-1652 (2006).
3. Y. Ganesan and J. Lou. *JOM*, **61**, 32-37 (2009).
3. I. Elksnite, J. Bitenieks, J. Zicans and A.K. Bledzki, *Solid State Phenom.,* 151, 171-175C (2009).
5. J. D. Eshelby, *Proc. Roy. Soc. Series A*., **241**, 376–396 (1957).
6. T. Mori and K. Tanaka, *Acta Metallurgica*, **21**, 571–574 (1973).
7. E. Plume, R. D. Maksimov and A. Lagzdins, *Mech. .Compos. Mater.*, **44**, 341–348 (2008).
8. A. Lagzdins, R. D. Maksimov and E. Plume. *Mech. Compos. Mater.*, **45**, 345-358 (2009).

TiO_2/polymer nanocomposite based inks

F. Loffredo[a], I. A. Grimaldi[a], A. De Girolamo Del Mauro[a], F. Villani[a], R. D'Amato[b] and C. Minarini[a]

[a] *Portici ENEA Research Center, Piazzale E. Fermi I, 80055 Portici (NA), Italy*
[b] *Frascati ENEA Research Center, via E. Fermi 45, 00044 Frascati (Roma) (ITALY)*

Abstract. We report the development and characterization of dielectric inks based on dispersions of TiO_2 in poly(ethylenimine)/ethanol solutions having physicochemical properties suitable to ink-jet printing process. In order to study the effect of polymer dispersant on the printability and stability of inks, we carried out dynamic light scattering analysis of different inks made with and without polymer. Moreover, we compare the curve of distribution of TiO_2 particles size at different aging times. For TiO2polymerwe optimize the inkjet parameters (amplitude and duration of jetting impulse, jetting frequency, substrate velocity) to obtain thin lines based on TiO_2/ poly(ethylenimine) nanocomposite on silicon substrate. Finally, the morphology of films was also investigated.

Keywords: TiO_2 nanoparticles, polymer composite, Ink-jet printing
PACS: 82.35 Np Nanoparticles in polymers

INTRODUCTION

Titanium dioxide nanoparticles represent a promising research subject for various a wide range of applications such as in photocatalysts [1], photovoltaic cells [2] batteries [3], photochromic and electrochromic devices [4], gas sensors [5], thin film transistor (TFT) [6]. For a lot of this applications ink-printing technique could be an interesting low cost and versatile method to make controlled deposition of TiO_2 nanoparticles and their nanocomposites.

Generally, the time stability of nanoparticles in suspension is considered to be one of the most difficult problems to solve in ink-jet deposition of dispersions. Moreover, the difficulty of dispersing nanoparticles of TiO_2 has already been show in several works [7]. The methods commonly used are the simultaneous synthesis and dispersion of TiO_2 nanoparticles [8] and the use of polymeric dispersants and ultrasonic treatments [7]. To aim to investigate the influence of a polymer dispersant on the printability and the stability of inks, two different inks were made optimizing the dispersion of TiO_2 with and without polymer dispersant .

The first suspension is based on water and ammonia (NH_3) and the second on poly(ethylenimine) (PEI) in ethanol. For the last ink, we performed printability tests on silicon substrate at different temperatures.

EXPERIMENTAL PART

TiO_2 nanopowders were synthesised by CO_2 laser pyrolysis by using $Ti(OPr)_4$ as liquid precursor and NH_3 as sensitiser. In this way N-doped nanoparticles with diameter of 10-15nm were obtained. After synthesis, a thermal treatment at 480°C for 4 h was carried out in order to eliminate carbon contamination.

Linear poly(ethylenimine) (M_W=120,000) was supplied from Polyscience. Ammonia (15% v/v) and ethanol (EtOH) were purchased by Aldrich and were used as received.

$TiO_2/H_2O/NH_3$ inks were prepared by dispersing TiO_2 nanoparticles (10mg) in 20mL of solution of NH_3 (4 drops) in deionized water (20mL). For TiO_2/PEI/EtOH inks, TiO_2 nanoparticles (10mg) were dispersed, in a polymeric solution obtained dissolving PEI (10 mg) in ethanol (20mL). Both suspensions were dispersed by ultrasonic bath for 5 min.

The $TiO_2/H_2O/NH_3$ and TiO_2/PEI/EtOH inks were characterized by Dynamic Laser Scattering (DLS) using a HPPS 3.1 (Malvern Instruments) at different times of aging, in order to determine the particle size distribution of filler in suspension and dispersion stability in the solutions.

CP1255, *V[th] International Conference on Times of Polymers (TOP) and Composites,*
edited by A. D'Amore, D. Acierno, and L. Grassia

ENEA has a customized ink-jet equipment with single nozzle print-heads. For this work, a Microdrop piezoelectric dispenser head with 50μm diameter nozzle, was used. Drops can be deposited with frequencies up to 2 kHz. A stroboscopic camera system provided visual control to adjust piezo-voltages and pulse durations for reliable droplet ejection and to avoid satellite drops. Single drops and lines, obtained by drops overlapping, were performed on silicon substrates at different temperatures. The printing parameters chosen were: printing frequency=10Hz and microhead speed=0.5mm/sec.

To investigate the morphology of ink-jet-printed films, the samples were characterized by optical micrography and by scanning electronic microscopy (SEM). The SEM images were obtained after metallization of the samples with a gold thin layer.

RESULTS AND DISCUSSION

To produce TiO_2 inks we performed preliminary tests of solubility in solvents of different polarity: water, ethanol and chlorobenzene. In all cases we obtain stable dispersions with large agglomerations only for 2-3 minutes. A higher degree of dispersion of nanoparticles in water was observed by adding drops of ammonia to the suspension (TiO2/H20/Nh3 ink). The polymer PEI was instead used to improve the dispersion of TiO2 in ethanol (TiO2/pei/etthanol ink.

In Figure 1, the size distribution curves of TiO_2 nanoparticles, obtained by Dynamic Laser Scattering (DLS) analysis performed on $TiO_2/H_2O/NH_3$ and on TiO_2/PEI/EtOH inks, are shown.

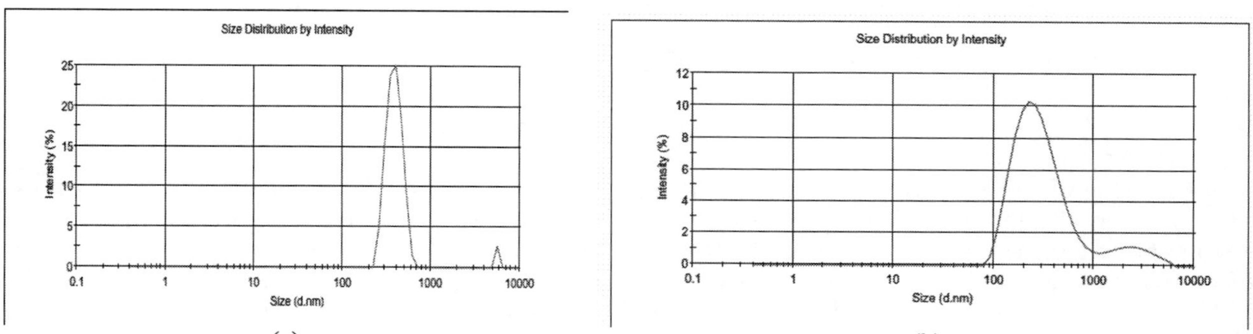

(a) (b)

FIGURE 1. Dynamic light scattering (DLS) analysis of inks obtained with TiO_2 nanoparticles freshly: (a) $TiO_2/H_2O/NH_3$ ink (b) TiO_2/PEI/EtOH ink.

Both types of samples are well dispersed with large amount of TiO_2 particles with sub-micrometer dimensions. DLS analysis shows the presence of a main peak centered around 300nm

In order to test the time stability of the suspensions, we studied the behaviour of inks for different aging times. The longer time stability is observed for TiO_2/PEI/EtOH ink as shown in figure 2. For $TiO_2/H_2O/NH_3$ ink agglomeration is already observed for aging times lower than 1h (2a) and after one day the sample is almost completely sedimented. Instead, TiO_2/PEI/EtOH ink presents only a slightly aggregation and sedimentation which involves a narrowing of the particle size distribution curve and a reduction in their average sizes (2b). For this reason it is chosen as ink for printing tests.

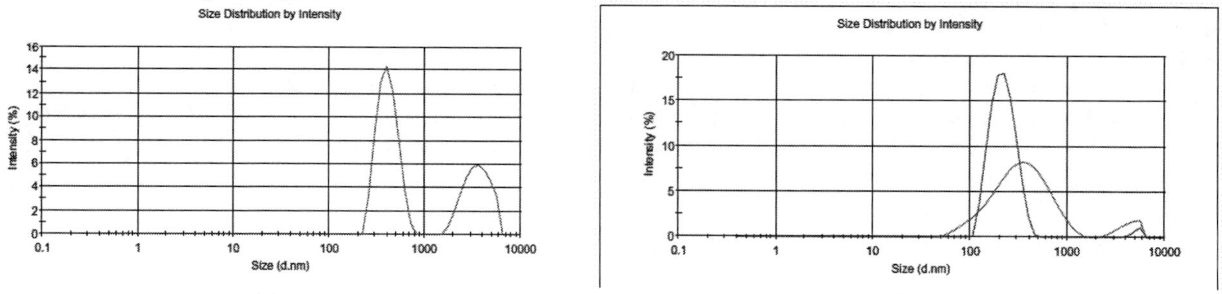

(a) (b)

FIGURE 2. Dynamic light scattering (DLS) analysis of: (a) $TiO_2/H_2O/NH_3$ ink after 20min of aging and (b) TiO_2/PEI/EtOH ink after 1 day of aging (sharp curve).

In figure 3 it is shown the optical micrographs and SEM images of TiO_2/PEI lines printed on a silicon substrate at room temperature and at T=50°C. The printed lines at room temperature (figure 3a) show jagged edges and greater uniformity than those printed at T = 50°C (figures 3b). SEM analysis shows that at room temperature there is less aggregation of nanoparticles (figure 3c).

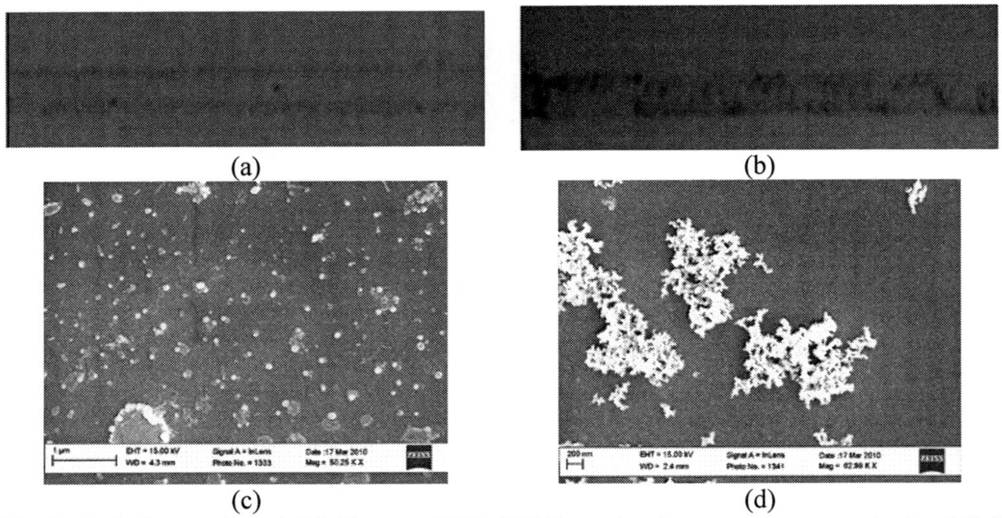

FIGURE 3. Optical micrographs and SEM images of TiO_2/PEI lines printed at room temperature (a,c) and T=50°C (b,d).

CONCLUSIONS

In this study was investigated the possibility to manufacture inks based on TiO_2 nanoparticles printable by ink-jet. It was noted that the introduction of PEI allows to obtain suspensions with average particle size of the order of 300nm having a higher stability in time of the suspension without polymer. Good interaction of PEI with TiO_2 nanoparticles probably allows the polymer intercalation between the nano particles stabilizing by spherical effect.

The improved stability allows to deposit the TiO_2/PEI/EtOH ink by ink-jet printing. Preliminary studies were performed by heating silicon substrates at different temperatures in order to evaluate the morphology variation.

ACKNOWLEDGMENTS

The support by the TECVIM project financed by the Ministero dell'Università e della Ricerca (MUR) is gratefully acknowledged.

REFERENCES

1. A. L. Linsebigler, G. Lu, and J. T. Yates, *Chem. Rev.*, **95**, 735–58 (1995).
2. A. Hagfeldt and M. Gratzel, *Acc. Chem. Res.*, **33**, 269–77 (2000).
3. S. Y. Huang, L. Kavan, I. Exnar, and M. J. Gratzel, *J. Electrochem. Soc.*, **142** , L142–4 (1995).
4. T. Gerfin, M. Gratzel, and L. Walder, *Prog. Inorg. Chem.*, **44**, 345–93 (1997).
5. H. M. Lin, C. H. Keng, and C. Y. Tung, *Nanostruct. Mater.*, **6**, 1001–4 (1995).
6, J. Park, J. W. Lee, D. W. Kim, B. J. Park, H. J.Choi and J. S. Choi, *Thin Solid Films* **518**, 588–590 (2009)
7. S. Peres-Durand, J. Rouviere, and C. Cuizard, *Colloids and Surfaces A: Physicochem. Eng. Aspects*, **98**, 251–70 (1995).
8. K. Sato, J.G. Li, H. Kamiya,T. Ishigaki, *J. Am. Ceram. Soc.*, **91** 2481–2487 (2008)

Investigation of PE blend films through CIE L*C*h color scale

Mohsen Mohammadi*, Ali A. Yousefi, Morteza Ehsani

Faculty of Processing, Iran Polymer and Petrochemical Institute, P. O. Box: 14965-115, Tehran/Iran
* M.Mohammadi@ippi.ac.ir

Abstract
Different ratios of two types of polyethylene (HDPE and LDPE) were blended and then blown films of them were prepared. Color stimuli of the films on CIELCH color scale were investigated. Results on L^* values for films indicated that presence more than 50% LDPE in the blend films causes higher L^*- indices comparing to other corresponding blend films. Comparing CIE coordinates verified that amongst values of L^*, C^*, and h°, in this paper hue angle values has significant magnitude, which showed large and clear differences with changing composition in the blend films. Moreover, thermal analysis confirmed miscibility for blend films comprising more than 75% HDPE and immiscibility of other blend films.

Keywords: PE, lightness, CIEL*C*h color scale, hue angle

INTRODUCTION

For the reason of good ability of polyethylene (PE) to be blown in melt state, producing film using PE has been attracted much attention [1]. Moreover, polyolefin blends and particularly polyethylene blends have been taken into consideration as interesting materials with various properties [2]. Mechanical, thermal, and physical properties of polyethylene blends have been widely investigated [3]. In polymer industry color and appearance is influenced by many factors and depending to type and grade of used materials, a variety of semblance is optically established [4, 5].
In literature, there is not any report on effect of blending on color stimuli of polyethylene blend films and characterization of them through CIE coordinates. In this paper we studied the effect of composition on CIELCH coordinates of polyethylene blend blown films.

EXPERIMENTAL

Materials and Methods

In this work commercial grades of polyethylene (HDPE and LDPE) were used. Some of their physical properties are listed in Table 1.

Table 1- The physical properties of polymers.

Polymer	HDPE	LDPE
Melt Index (g/10 min), 190 °C	0.4	0.8
Density	0.948	0.918
Melting point (°C)	139	112

At first the two polymers were dry mixed and then fed into extruder of a film blowing unit. To attain 50-μm thick blown films, a blow up ratio of 2.5 and a frost line height of 5 cm were used. The composition of the blends is summarized in Table 2. To measure tristimulus values of films in transmission mode under standard illuminant D65, a spectrophotometer 7000A with a diffuse 8° geometry and a 10° standard observer was used and CIEL*C*h° color scale was applied. Thermal behavior of films was studied using a NETZSCH DSC 200 F3 differential scanning calorimeter at a heating rate of 10°C/min.

Measurement of CIE coordinates

The color of an object can be described through the CIELAB color space, by means of L^* (lightness), a^* (the greenness-redness axis) and b^* (the yellowness-blueness axis). Moreover, defining color in cylindrical coordinates is beneficial in

CP1255, *Vth International Conference on Times of Polymers (TOP) and Composites,*
edited by A. D'Amore, D. Acierno, and L. Grassia

most cases and for the intention it is possible to calculate the C^* (chroma parameter) and h° (hue angle) which have more association with visual perception of color. The values are defined based on a^* and b^* values as [6]

$$C^* = \left[a^{*2} + b^{*2}\right]^{\frac{1}{2}}, \qquad (1)$$

$$h^\circ = Arc\tan\frac{b^*}{a^*} \quad , \quad 0^\circ \le h^\circ \le 360^\circ \qquad (2)$$

Table 2- Composition of the blend blown films.

Sample	HDPE (wt %)	LDPE (wt %)
H1	100	0
5L2H	95	5
10L2H	90	10
15L2H	85	15
25L2H	75	25
50L2	50	50
25HL2	25	75
15HL2	15	85
10HL2	10	90
5HL2	5	95
L2	0	100

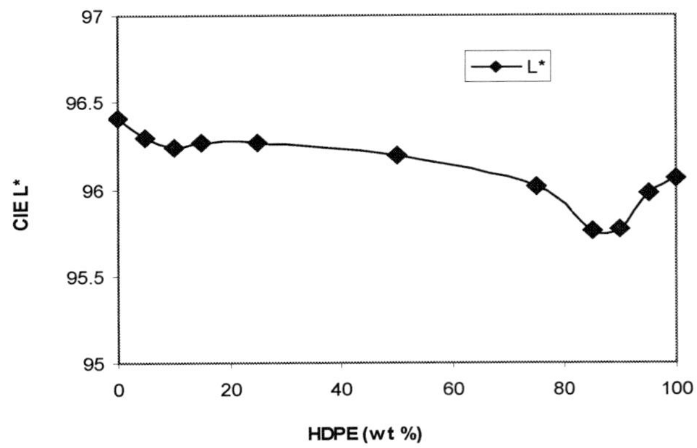

Figure 1- Lightness of the blend blown films with different HDPE weight percent.

RESULTS and DISCUSSION

Lightness

The lightness (L^*) of the films of the original LDPE and HDPE along with the blends are shown in Figure 1. Amongst all samples, LDPE has the highest lightness, while by increasing HDPE content in the film blends, L^* values shows declining trend until the 85% HDPE, but after that up to 100% HDPE the L^* values grow up. Actually, in transmission mode the L^* value of the blend films decreases up to 15% LDPE and then increases but never exceeds L^* of the LDPE. Presence of up to 15% LDPE in the HDPE resin because of more compatibility of the blend in the mentioned composition range bring about abating lightness, whereas because of immiscibility and heterogeneity of the mixture in other compositions, L^* values escalate. It can be also understood that the presence of HDPE into the LDPE matrix has trivial effect on L^* value. Likewise, DSC thermograms of films containing more than 75% HDPE showed only a single melting peak which is a sign of miscibility of blends but for other films of blends two melting peaks were observed reflecting two crystalline phases and therefore immiscibility (Figure 2). It has been reported a large number of polyethylene blends are not compatible or only compatible to rather limit extent [7].

Chroma

Figure 3 illustrates the chart of chroma vs. HDPE concentration. For the samples L2 to 10L2H, C^* value increased from 0.30 to 0.48. Minimum chroma (C^*) was registered for pure LDPE (0 % HDPE) film sample rising with increasing HDPE content, until a HDPE wt% of 90%. On the other hand, a predominant HDPE content (90%) resulted in the C^* parameter augmentation reaching a maximum level at the sample 10L2H. The level of C^* at the sample 50L2 was close to that of LDPE sample.

Hue angle

Hue angle for the samples are located in a part of Cartesian coordinate in which a^* and b^* are both positive i.e. redness and yellowness region (0-90°). Regarding to the magnitude of hue angles they are very close to the b^* axis which corresponds to yellow color and far from a^* axis. It means saturation or hue of samples is predominated by b^* and not a^*. Via incorporating LDPE into HDPE hue angles shifted to less yellowish, in particular for HDPE-rich blend films in comparison with LDPE-rich blend films. The hue angle for blend films with 15 to 85% HDPE ranged from 87 to 75.7. The value for the angle obtained for LDPE film was higher than HDPE film. There was not perceived any great differences in the hue angle values within each samples containing more than 75% HDPE, or within each samples comprising lower than 15% HDPE.

As illustrated in Figure 3 with decreasing HDPE content hue angle (h°) trended towards $+b^*$ (yellowish). Comparing the hue angle values of blends, they generally reached higher levels at lower HDPE contents. Besides, other observations (based on thermal analysis) also support that in concentration region more than 75% HDPE, there is better homogeneity and/or compatibility (Figure 2). So in graphs of L^*, C^*, h° it is obvious that data for this region of concentration is different from other samples. Especially, due to homogeneity of films of HDPE-rich blend films, their hue angles do not have intensive change by increasing LDPE content (or decreasing HDPE content), but for blend films containing lower than 75% HDPE, severe rising of h° happens.

Having performed curve fitting for the plot of hue angle values vs. various HDPE concentrations, which is presented in Figure 3, an equation is established. Using the graph and the fitted_equation, prediction of h° value for any HDPE ratio in the blend films becomes feasible. As a result an implement for predicting hue of two-component film samples is offered. As we can see there is not the same order as described for hue angle comparing with their L^*, C^* counterparts (Figures 1 and 3) and also the changes in h° is more clear and larger.

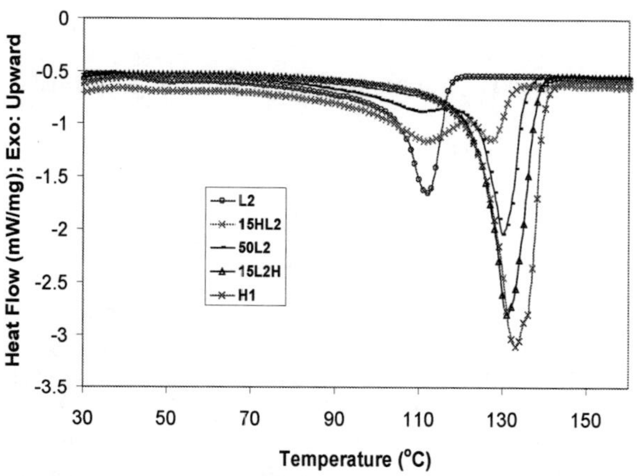

Figure 2 – DSC heating thermograms of PE blend films with different HDPE content.

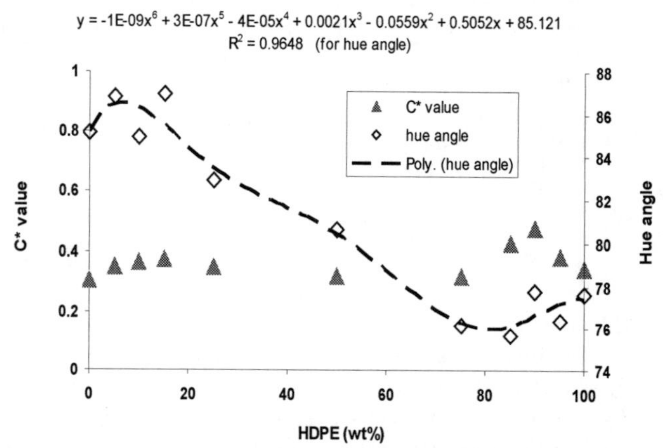

Figure 3- Chroma and hue angle of blend blown films with different HDPE weight percent.

These scrutiny on chroma and h° corroborates the assertion that chroma would not be used as a remarkable parameter when color is used as a yardstick of compatibility of films rather than hue angle.

CONCLUSION

In this paper we prepared blown films of LDPE/HDPE blend and the influence of the parameter blending ratio on CIE coordinates of the resulting films was studied. It was pointed out objective terms obtained from $CIEL^*C^*h$ color space was useful parameters for characterizing blend films and mathematical functions corresponding to CIE coordinates can explain the color as a physical scale. Both color stimuli and thermal behavior supported miscibility of HDPE-rich blend films.

REFERENCES

[1] X. M. Zhang, S. Elkoun, A. Ajji, M.A. Huneault, Polymer **45**, 217–229 (2004).

[2] D. Srivastava, R. Garg, P. Kumar, D. C. Saxena, G. N. Mathur, *Macromol. Mater. Eng.* **283**, 81–87 (2000).

[3] Z. J. Fan, M. C. Williams, P. Choi, Polymer, **43**, 1497-1502 (2002).

[4] J. H. Park, T.W. Lee, Y. C. Kim, O. O. Park, J. K. Kim, *Chem. Phys. Letters* **403,** 293–297 (2005).

[5] T. D. Fornes, P.J. Yoon, D.R. Paul, *Polymer* **44**, 7545–7556 (2003).

[6] S. Westland, C. Ripamonti, *Computational Colour Science using Matlab*, John Wiley & Sons Ltd, 2004, pp. 49-60.

[7] A. Siegmann and Y. Nir, Polym. Eng. Sci. **27**, 1182-1186 (1987).

Hole injection layer containing Polyaniline-Poly (4-styrenesulfonate) for efficient organic light-emitting diodes

Anna De Girolamo Del Mauro[a], Giuseppe Nenna[a], Valentina Bizzarro[b], Immacolata Angelica Grimaldi[a], C. Minarini[a]

[a] *Italian National Agency for New Technologies, Energy and Sustainable Economic Development (ENEA), Portici Research Centre, p.le E. Fermi 1, 80055 Portici (NA), Italy*
[b] *IMAST Portici, Piazzale E. Fermi, 80055 Portici (NA), Italy*

Abstract. Nanocomposite films based polyaniline and polystyrene sulfonate (PANI:PSS) were prepared and characterized by chemical oxidative polymerization. Transparent and uniform films were obtained by spin-coating onto patterned ITO-coated glass substrates and were used as hole injection layer (HIL) in OLED devices. The devices made with HIL-based PANI:PSS were compared with those with only ITO and showed better electrical characteristics, lower turn-on voltage and a three time improved device efficiency.
Keywords: Polyaniline/ polystyrene sulfonate, Hole injection layer, Conducting polymer
PACS: 85.60.Jb Light-emitting devices

INTRODUCTION

Organic light-emitting diodes (OLEDs) have been attracting great attention as a potential flat panel display due to outstanding advantages, such as wide viewing angle, fast response, easily fabricating and processing flexible display. However, OLEDs possess inherently adverse properties: for instance, mismatch of energy level between conventional electrodes and electroluminescent (EL) layer. To improve EL efficiency and reducing driving voltage, the modified layers are usually used, for example, introducing a hole transporting layer and/or the hole injection layer (HIL).

Different layers, such as copper-phthalocyanine [1], polyaniline (PANI) [2], poly(3,4-ethylene dioxy-thiophene) (PEDOT) [3], platinum [4], carbon [5], silicon nitride [6], 4,4V,4W-tris(3 methylphenylphenylamino) triphenylamine [7], Ni-ITO co-sputtered layer [8] or metal oxides [9], were inserted between the ITO anode and the hole transport layer to improve the stability and lower the driving voltage of OLEDs.

Among the above-mentioned materials, PANI is a highly attractive candidate for the electronic devices due to its facile synthesis, environmental/ chemical stability and transparency in the conducting form [10]. The blending of PANI with poly (styrene sulfonate) (PSS) have already been carried out [11] and showed good compatibility as well as doping, resulting in a material that has moderate conductivity and the capability to process by melt or solution.

In this study, PANI:PSS nanocomposites were synthesized via an oxidative dispersion polymerization in acidic aqueous media using PSS as a stabilizer and dopant. The effect of PANI:PSS layer as hole injection layer on the performance of OLED devices was evaluated. Morphology and the electrical conductivity of PANI:PSS films and doped with dimethyl sulfoxide (DMSO) were also investigated.

DEVICES FABRICATION

The water dispersed PANI:PSS was synthesized by chemical oxidation polymerization of aniline in the presence of excess hydrochloric acid according to the method reported in the literature [16]. These dispersions (without and with DMSO) were spin-coated to give a 70 nm-thick film on the patterned ITO glass, then baked on a hot plate at 100°C for 10 min. These layers were used as a polymeric HIL materials for the OLED device. On PANI:PSS HIL films, N, N'-Di-((1-naphthyl)-N,N'-diphenyl)-1,1-biphenyl)-4,4'-diamine (NPD) and aluminium tris(8-hydroxyquinoline) (Alq_3) were sequentially evaporated under vacuum (10-7 mBar) to obtain a 40 nm-thick hole transport layer and 60

CP1255, *Vth International Conference on Times of Polymers (TOP) and Composites,*
edited by A. D'Amore, D. Acierno, and L. Grassia

nm-thick emitting layer. Then a thin lithium fluoride layer (1 nm) and an aluminium (Al) cathode (100 nm) were thermally evaporated to complete the device. Thus, the full device structure is: ITO/PANI:PSS/NPD/Alq_3/LiF:Al. Furthermore, a reference sample of an only ITO layer was used in a conventional OLED with a structure of ITO/NPD/Alq_3/LiF:Al.

RESULTS AND DISCUSSIONS

The properties of PANI:PSS dispersion and of corresponding films are given in table 1 and 2, respectively.

TABLE 1: Physical and chemical properties of PANI:PSS dispersion

	Dispersion
Colour	Dark green
Molar ratio (PANI:PSS)	1:12
Dinamic Viscosity (cP)	2.32
Average size of nanoparticles (DLS) (nm)	200

TABLE 2: Physical and electrical properties of PANI:PSS film

	Film (70nm)
Colour	Green
Conductivity (S/cm)	1.4E-02
	1.31 (doped with DMSO)

The PANI:PSS dispersion is dark green as a conducting emeraldine salt form and, through dynamic light scattering measurements (DLS), it is observed that they are well dispersed in aqueous solution without any phase separation even after 4 months. The addition of DMSO lowered the sheet resistance by two orders of magnitutide.

The morphology and surface roughness of films onto glass substrates were investigated by tapping mode atomic force microscopy (fig.1a and fig.1b).

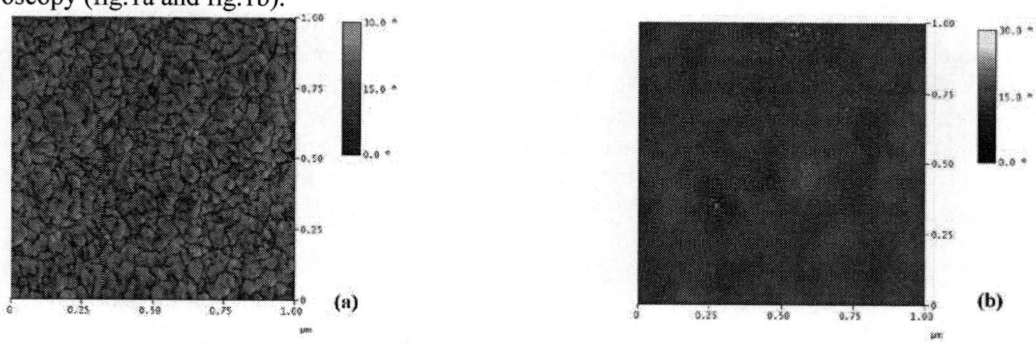

FIGURE 1: AFM images of phase of PANI/PSS (a) and DMSO-doped PANI:PSS (b) nanocomposite films, respectively.

The surface morphology of the films changed adding DMSO with a roughness value that increases from 10nm (fig.1a) to 13.1nm (fig.1b).

Fig. 2 shows the transmission spectra of untreated ITO (150nm thick) and ITO coated in PANI:PSS (50nm) on glass substrate. DMSO-doped PANI/PSS spectra were similar to PANI:PSS and hence only one is shown. All show high transmission over the visible range of the spectrum. The ITO/PANI:PSS film shows a transmission of 84.1 % at the wavelength of 600nm while a transmission of 86% is observed with ITO-coated glass substrate. Thus, the addition of PANI:PSS as HIL in a conventional OLED device does not affect the optical transparency.

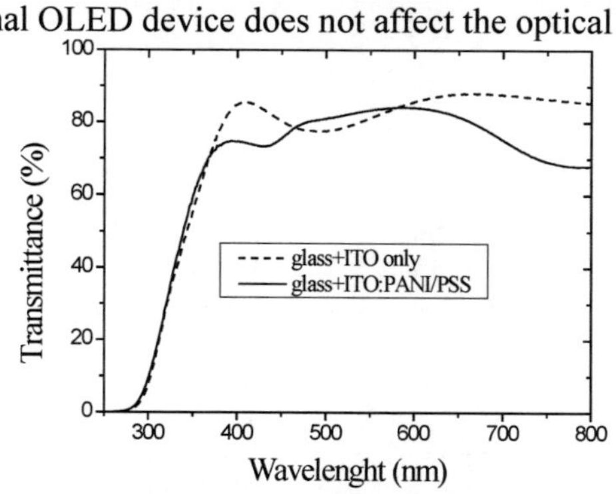

FIGURE 2: Optical transmission spectra of 50nm thick layer of PANI/PSS on ITO-coated glass substrate compared to an only ITO on glass substrate

The devices with HIL-based PANI:PSS were compared with those with only ITO and they showed better electrical characteristics (Fig. 3), lower turn-on voltage (from 5V to 2.5V) (inset fig.4) and an improved devices efficiency (fig.4). Regarding the device efficiency, it grow around two times with a DMSO-doped PANI:PSS film and around three times by using the undoped PANI:PSS layer respect to the device without hole injection layer (fig.4). Films with and without DMSO showed similar trends.

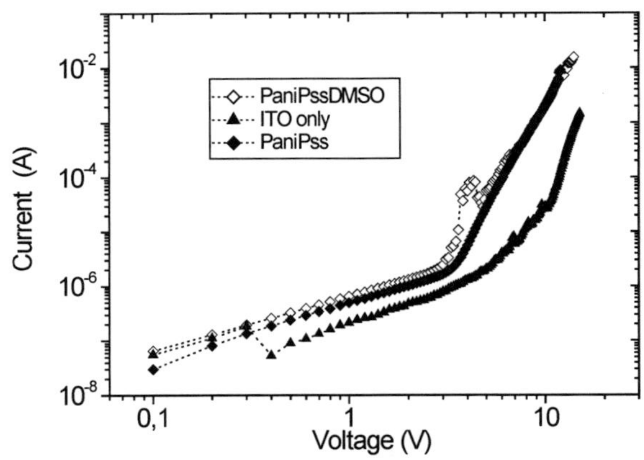

FIGURE 3: I-V characteristics of ITO/PANI:PSS, ITO/DMSO-doped PANI:PSS and only ITO

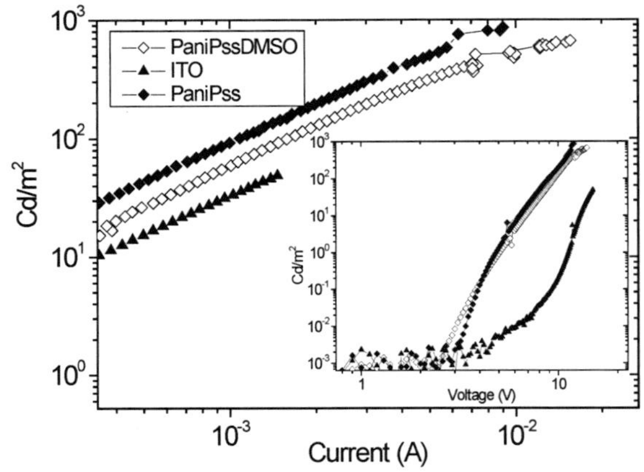

FIGURE 4: Efficiency of OLED devices with PANI:PSS HIL layer and with only ITO.
Inset: Comparison of electroluminescent layer between devices with and without PANI:PSS HIL

Moreover, when a PANI:PSS film with a higher conductivity is employed implies that, to have the same efficiency total current value through the device, an increase in electron current is necessary. For this reason, graphing the device current in function of the luminance (Fig. 4) it is possible to explain the lower luminance obtained through the highest conductive PANI:PSS (DMSO doped) [12]. This probably means that in this case and for this device the right hole injection layer is not necessary the more conductive film. This implies that in the case of PANI:PSS film we already obtain a good increase of recombined electron–hole pairs per time unit and so an increase in device efficiency. Further investigations are necessary in function of DMSO doping and changing the electron injection layer.

CONCLUSIONS

Water dispersed PANI/PSS was synthesized by chemical oxidation polymerization in the presence of excess HCl and the corresponding films were used as HIL layer to improve the performance of OLED devices. The obtained films are homogeneous, have a low surface roughness (about 10-13nm), are transparent (transmittance of about 84%) and have a conductivity of the order of $1.4x10^{-2}$S/cm. The conductivity improves of two orders of magnitude reaching a value of 1,31 S/cm by adding 5% DMSO. The corresponding devices have instead shown better electrical characteristics, lower turn-on voltage (from 5V to 2.5V) and improved efficiency three time respect to the device with only ITO.

ACKNOWLEDGMENTS

Support by the TRIPODE project financed by the Ministero dell'Universita` e della Ricerca (MIUR) is gratefully acknowledged.

REFERENCES

1. F. Nuesch, M. Carrara, M. Schaer, D.B. Romero, L. Zuppiroli, *Chem. Phys. Lett.* 347, 2001, pp. 311-317.
2. Y. Yang, A.J. Heeger, *Appl. Phys. Lett.* 64, 1994, p 1245.
3. A. Elschner, F. Bruder, H.-W. Heuer, F. Jonas, A. Karbach, S. Kirchmeyer, S. Thurm, *Synth. Met.* 111, 2000, p. 139.
4. Y. Shen, D.B. Jacobs, G.G. Malliaras, G. Koley, M.G. Spencer, A.Ioannidis, *Adv. Mater.* 13, 2001, p 1234.
5. A. Gyoutoku, S. Hara, T. Komatsu, M. Shirinashihama, H. Iwanaga, K. Sakanoue, *Synth. Met.* 91, 1997, p. 73.
6. H. Jiang, Y. Zhou, B.S. Ooi, Y. Chen, T. Wee, Y.L. Lam, J. Huang, S. Liu, *Thin Solid Films* 363, 2000, p.25.
7. Y. Shirota, Y. Kuwabara, H. lnada, T. Wakimoto, H. Nakada, Y. Yonemoto, S. Kawami, *Appl. Phys. Lett.* 66, 1994, p.807.
8. C.-M. Hsu, W.-T. Wu, *Appl. Phys. Lett.* 85, 2004, p. 840.
9. I.M. Chan, F.C. Hong, *Thin Solid Films* 450, 2004, p.304.
10. J. Jang, J. Ha, B. Lim, *Chem. Commun.* 15, 2006, p. 1622.
11. W. Liu, J. Kumar, S. Tripathy, K.J. Senecal and L. Samuelson, *J. Am. Chem. Soc.* 121, 1999, p. 71.
12. P. Vacca, M. Petrosino, R. Miscioscia, G. Nenna, C. Minarini, D. Della Sala, A. Rubino, *Thin Solid Films* 516, 2008, pp. 4232-4237.

Improvement of the Inkjet Printed VOCs Sensor Performances through the Sensing Layer Geometry

A. De Girolamo Del Mauro[a], I. A. Grimaldi[a,b], F. Loffredo[a], E. Massera[a], T. Polichetti[a], F. Villani[a], and G. Di Francia[a]

[a]Portici ENEA Research Center, Piazzale E. Fermi I, 80055 Portici (NA), Italy
[b]Department of Physics– University of Naples "Federico II", V. Cintia 1, 80126 Naples, Italy

Abstract. Polymer nanocomposites (PNCs) are attractive as sensing materials in volatile organic compounds (VOCs) chemical sensors because they combine the simple processability of polymers together with the conductivity properties of nanoparticles dispersed in the insulating polymer matrix. Recently, these materials are easily processed by inkjet printing (IJP) technique whose patterning capability is one of its main advantages. In the present work, we fabricated PNCs chemical sensors inkjet printing a polystyrene (PS)/carbon black (CB) composite based ink. The sensor responses have been measured upon exposure to acetone organic vapors and analyzed as function of sensitive material geometry.

Keywords: inkjet printing; polymer nanocomposite; volatile organic compounds; chemical sensor.
PACS: 82.35.Np Nanoparticles in polymers

INTRODUCTION

Polymer nanocomposites (PNCs), consisting of inorganic nanoparticles embedded in a polymer matrix, are well known to be sensitive to different analytes and are extensively used in sensing applications. Chemiresistors based on polymer nanocomposite work on the basis of a change of their electric properties when they come into contact with a specific substance. This mechanism is associated to the polymer matrix volume change due to the solvent absorption (swelling) [1-3], and to the corresponding reduction of the carbon particles volumetric fraction [4].

PNCs can be easily processed by techniques such as spin-coating or drop-casting and, recently, also by inkjet printing (IJP) technology, rapidly emerged as an innovative technique for the deposition of a wide variety of materials [5-10]. The advantages of IJP lie in its potential for printing on both nonflexible and flexible substrates, the efficient use of material, the reduced waste products, low cost of the process, and its patterning capability. As concerning this last item, the IJP technology assumes a key role in the feasibility of modifying the sensitive material geometry in controlled and extremely easy manner in order to optimize the sensor response.

In a previous work [11], we started the study on the correlation between the PNCs pattern and the device response investigating two different geometries. Starting from this point, in the present work we analyze other geometries all realized printing one or more lines differently orientated with respect to the transducer fingers. The sensor responses have been measured upon exposure to acetone organic vapors and quantitatively analyzed. The device performances have been compared in terms of sensitivity, response time and lower detection limit.

EXPERIMENTAL

The employed substrates (5x5 mm^2) are made in alumina on which interdigitated Au electrodes are previously e-beam evaporated and subjected to a lift-off photolithographic standard process.

The sensing material is a polystyrene (PS)/carbon black (CB) composite. The atactic polystyrene, supplied from Sigma-Aldrich, has been chosen for its sensing property to different organic chemical compounds (acetone, ethanol, toluene, benzene, ...); Carbon Black (Black Pearls 2000 - Cabot Co.) (particle average diameter ~ 12nm) has been employed as conductive filler. The PS polymer matrix (80 mg) has been dissolved in n-Methyl-2-pyrrolidinone

CP1255, *V[th] International Conference on Times of Polymers (TOP) and Composites,*
edited by A. D'Amore, D. Acierno, and L. Grassia

(NMP) and the CB nanoparticles (20 mg) have been dispersed in the polymeric solution (0.5 wt%) by means of ultrasonic bath for 90 minutes. The PS/CB suspension has been finally filtered using a 0.2-μm filter with the aim of removing bigger agglomerates. Such a processed dispersion has been used as ink and printed onto 50 °C heated electrodes/substrate system. The inkjet equipment (Aurel S.p.A.) uses piezoelectric Drop on Demand print-heads (Microdrop) with a 70 μm opening nozzle. The patterned sensing material geometries consist of one or more lines, created by overlapped droplet sequences (50% overlap, 0.5 Hz drop emission frequency, 0.05 mm/s print-head speed), differently orientated with respect to the transducer fingers.

The sensor response has been measured upon exposure to acetone organic vapors at different concentrations (600, 1250, 2500, and 5000 ppm) in a test chamber that allows to monitor and control pressure, flow rate, actual gas composition, humidity and environment temperature. The chamber is placed in a thermostatic box at controlled temperature (20 °C) and humidity (30%).

RESULTS AND DISCUSSION

The devices have been fabricated IJ printing the sensing material in five different geometries. Each geometry consists of one or more lines parallel (denominated 'vertical') or transverse (denominated 'horizontal') to electrodes interdigitated fingers; the fifth geometry is a cross. Moreover, we fabricated two different configurations with three- and five-overlapped layers for each pattern with the aim of investigating the thickness effect on the sensor performances.

The 3-layers and 5-layers devices electrical responses to acetone vapors are shown in figure 1 and figure 2, respectively. The parameters that characterized the 3-layers sensor performances are summarized in table 1, while those for the 5-layers are reported in table 2.

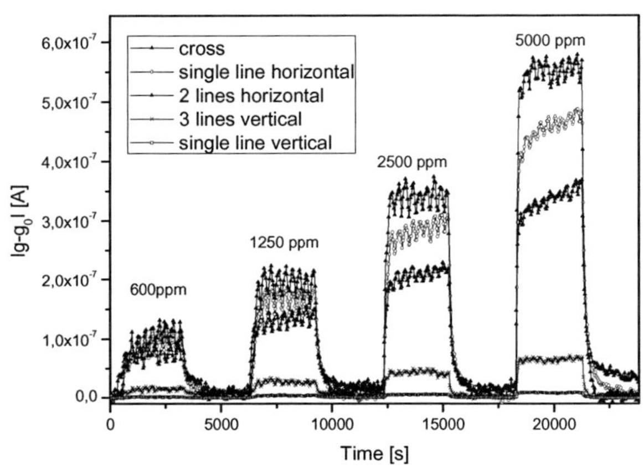

FIGURE 1. 3-layers sensors electrical responses to acetone vapors.

TABLE (1). Electrical responses parameters of 3-layer sensors employing acetone vapors as gas analyte.

Morphology	Resistance [KOhm]	Sensitivity [S/ppm]	Lower detection Limit [ppm]	Time response [s]
Cross	13.3	9.8e-11	150	100
Single line horizontal	57.6	8.3e-11	30	100
2 lines horizontal	38.5	5.8e-11	86	100
3 lines vertical	95.7	1.11e-11	180	100
Single line vertical	1000	1.53 e-12	400	100

The results indicate that the dynamic responses of sensors with different geometries are strongly influenced by the base resistance, i.e. the device resistance measured in absence of the analyte.

In addition, the 3-layers sensors are all faster than 5-layers devices, a side of the geometry. This behavior can be attributed to the different surface/volume ratio of the two configurations. Particularly, the 5-layers devices electrical responses don't reach the saturation condition at each analyte concentration.

Comparing the realized geometries, all patterns with 'horizontal' lines show the best performances in terms of sensitivity and lower detection limit.

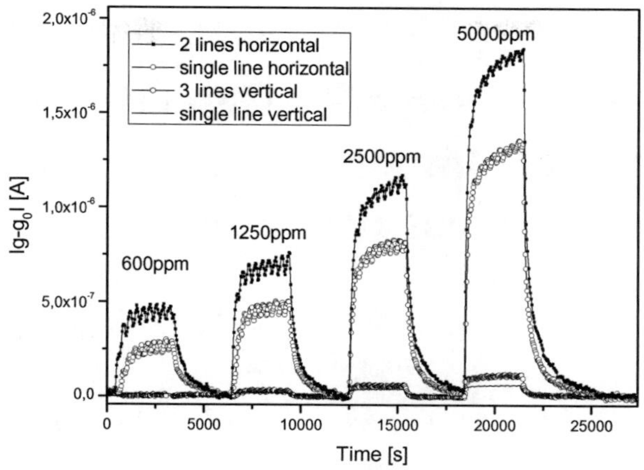

FIGURE 2. 5-layers sensors electrical responses to acetone vapors.

TABLE (2). Electrical responses parameters of 5-layer sensors employing acetone vapors as gas analyte.

Morphology	Resistance [KOhm]	Sensitivity [S/ppm]	Lower detection Limit [ppm]	Time response [s]
2 lines horizontal	8,6	3,05e-10	50	3000
Single line horizontal	8,9	2,29e-10	85	3000
3 lines vertical	50	2,19e-11	200	3000
Single line vertical	175	1,13e-11	100	3000

CONCLUSIONS

In the present work, we fabricated polymer nanocomposites chemical sensors and investigated how their performances are related to sensing material geometry. To this aim, we employed the inkjet printing technology to deposit the nanocomposite material taking advantage of its patterning capability and realized different geometries in extremely easy and controlled manner.

The results point out that the sensing material with less layers improve the device performances thanks to a suitable surface/volume ratio. Moreover, the PNCs chemical sensors fabricated IJ printing lines transverse to interdigitated electrodes finger show the best sensitivity and have got the lower detection limit.

ACKNOWLEDGMENTS

This work was supported by the TRIPODE project (DM 20160) financed by the Ministero dell'Università e della Ricerca (MUR).

REFERENCES

1. S. Chatzandroulis, D. Goustouridis, I. Raptis, *J. of Phys.: Conference Series* **10**, 297-300 (2005).
2. D. Goustouridis, K. Manoli, S. Chatzandroulis, M. Sanopoulou, I. Raptis, *Sensors and Actuators B* **111**, 549-554 (2005).
3. A. Convertino, G.Leo, M. Tamborra, C. Sciancalepore, M. Striccoli, M.L. Curri, and A. Agostiano, *Sensors and Actuators B* **126**, 138-143 (2007).
4. A. Carrillo, I. R. Martin-Dominguez, and A. Marquez-Lucero, *Sensors and Actuators B* **113**, 477-486 (2006).
5. K.E. Paul, W.S. Wong, S.E. Ready, and R.A. Street, *Appl. Phys. Lett.* **83**, 2070 (2003).
6. P. Calvert, *Chem. Mater.* **13**, 3299 (2001).
7. B. Ballarin, A. Fraleoni-Morgera, D. Frascaro, S.Marazzita, C. Piana, and L. Setti, *Synth. Met.* **146**, 201 (2004).
8. Y. Yoshioka, and G.E. Jabbour, *Synth. Met.* **156**, 779 (2006).
9. F. Loffredo, G. Burrasca, L. Quercia, and D. Della Sala, *Macromol. Symp.* **247**, 357 (2007).
10. F. Loffredo, A. De Girolamo Del Mauro, G. Burrasca, V. La Ferrara, L. Quercia, E. Massera, G. Di Francia, and D. Della Sala, *Sensors and Actuators B* **143**, 421 (2009).
11. A. De Girolamo Del Mauro, F. Loffredo, G. Burrasca, E. Massera, and G. Di Francia, Proceed. Aisem 2009.

MWNT-Doped Epoxy Matrix for Detecting Impact Damages in Fiber Reinforced Composites by Electrical Resistance Measurements

Marco Monti, Maurizio Natali, Josè M. Kenny and Luigi Torre

University of Perugia, Civil and Environmental Engineering Department, Strada di Pentima 4, 05100 Terni - Italy

Abstract. In this paper, we report the development of a glass fiber reinforced composite based on a MWNT-doped epoxy matrix. The intrinsic electrical conductivity of carbon nanotubes has allowed the production of a nanocomposite with enhanced electrical properties, respect the neat matrix. Due to the high aspect ratio of carbon nanotubes, very small amounts of these particles were enough for modifying the electrical properties of the obtained glass fiber composites. A low-velocity impact was created on the surface of the laminate, in order to produce a damage, which not involved the catastrophic failure of the sample. In order to evaluate the damage level, a mechanical characterization was performed on pristine samples, on one hand, and impacted The measurement of the electrical resistance, before and after the impact, was exploited as a way for detecting the presence of the damage produced. The results have shown that it is possible to associate the increase in electrical resistance of the composite with the formation of a damage due to an impact.

Keywords: carbon nanotubes, glass fibers, electrical properties, damage tolerance, impact.
PACS: 61.46.Fg, 68.37.Lp, 81.07.De, 82.35.Np

INTRODUCTION

Advanced composites are increasingly used in aerospace, naval and automotive vehicles due to their high specific strength and stiffness. However, the mechanical performances of composite materials may be severely reduced in presence of damages produced by impacts on the surface. Moreover, these damages could not be visible to the naked eye and constitute an important limitation of the employment of the structure [1], since a wide variety of damage modes, such as matrix cracking, delaminations, debonding and fiber failure in composite structures are easily induced by an impact.

The use of techniques that exploit electrical properties of the composite has been widely recognized as a non-invasive way to monitor damage in carbon fibers reinforced plastic, due to the good electrical conductivity of carbon fibers themselves [2-4]. In fact, in this case, fracture of fiber results in a strong increase of electrical resistance. Nonetheless, this approach can not take into account the fractures involving the matrix and the fiber-matrix interface, so it can give just a partial view of what is happening inside the composite during a mechanical solicitation. Moreover, it can not be employed with non-conducting fibers, e.g. glass or aramid. For these reasons, researchers have demonstrated that it is possible to exploit the same technique with non-conducting fibers, using electrically conductive polymer-based nanocomposites as a matrix [5-6].

Due to their high conductivity and to their high length-to-diameter ratio, carbon nanotubes are considered one of the most efficient filler for modifying the electrical conductivity of a polymeric matrix. In fact, a very low concentration is necessary to overcome the electrical percolation threshold of the system in which they are included [7-8].

In this study, the enhanced conductive properties of a epoxy-CNT nanocomposite was exploited to produce a glass fiber reinforced composite, in which the drop in mechanical performance is strictly related to an increase of the electrical resistance.

CP1255, *Vth International Conference on Times of Polymers (TOP) and Composites,*
edited by A. D'Amore, D. Acierno, and L. Grassia

MATERIALS AND PROCESSING

A low-viscosity epoxy resin, supplied by Hexion with the commercial name of RIM 235 was employed as a matrix. Multi-walled carbon nanotubes (MWNTs) were employed as a nanofiller. They were supplied by CNT Co., Ltd, (Korea), under the commercial name of C_{Tube}-100. As claimed by the technical data sheet, and verify by TEM and TGA analysis (fig. 1) they have an average diameter in the range of 10-40 nm, a length between 1 and 25 microns, and a purity of about 93 %. An high-speed mechanical stirring technique was employed to disperse the nanotubes in the matrix.

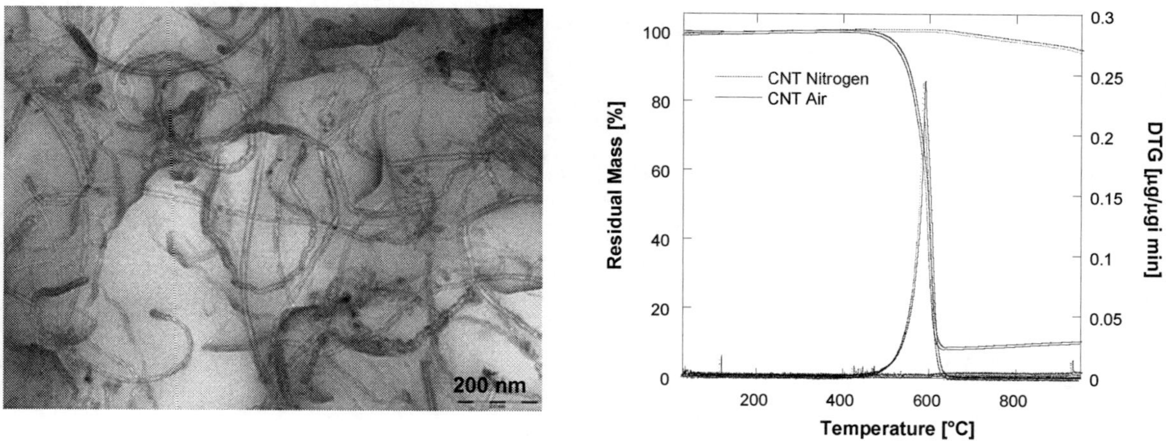

FIGURE 1. (a) TEM micrograph and (b) TGA pattern of the C_{Tube}-100 MWNTs.

RESULTS AND DISCUSSION

Nanocomposite Characterization

The dispersion level of the carbon nanotubes in the epoxy resin was evaluated performing a TEM analysis on the fully-cured samples. As a consequence, it was demonstrated that an homogeneous dispersion of the nanoparticles was reached. As an example, fig. 2 represents a TEM micrograph of the 0.25%-MWNT nanocomposite.

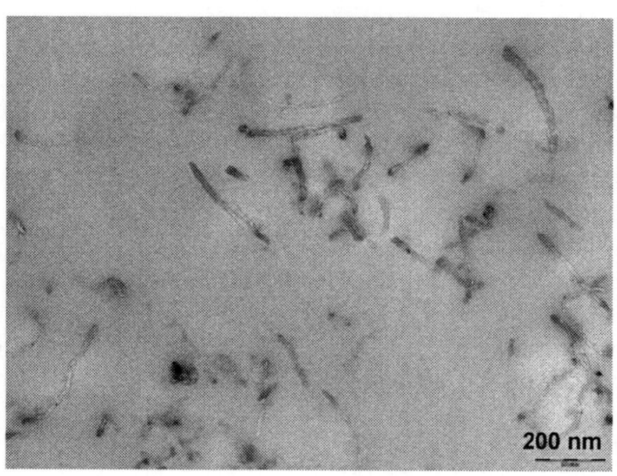

FIGURE 2. TEM micrograph of the 0.25%-MWNT nanocomposite.

In the first part of this study, the electrical behaviour of the epoxy-based nanocomposite, as a function of the CNT content, was analyzed. To this aim, mixtures with different percentages were produced, from 0.1 to 0.5 wt% of MWNTs. The 0.5 wt% content has represented the maximum limit, because over this concentration the viscosity of the mixture was extremely high, and therefore the processing was seriously compromised. As a result, fig. 3 shows the electrical resistivity of the nanocomposites, as a function of the MWNT content. As it is possible to observe, a strong decrease is obtained, even at a MWNT content of 0.1%.

As a consequence of this characterization, the concentration of 0.25% was selected for the production of a nanocomposite-based fiber reinforced composite.

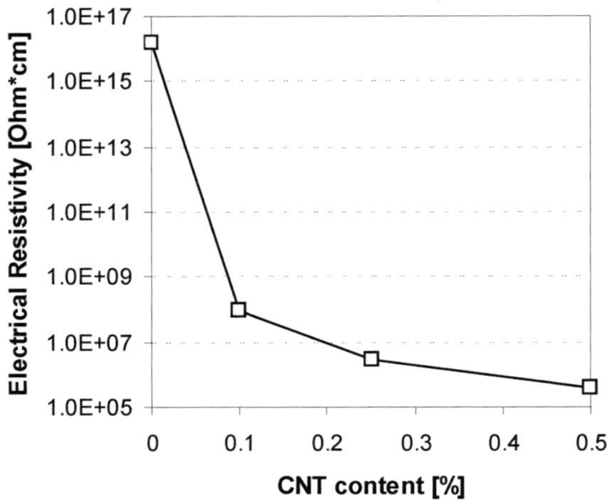

FIGURE 3. Electrical resistivity of the nanocomposites, as a function of MWNT content.

Fiber Reinforced Composite: Impact Damage monitoring

Once the characterization of the nanocomposite was carried out, and the CNT concentration was selected, an amount of this nanocomposite, enough for the fiber composite, was produced. Vacuum infusion was utilized as a technique for the manufacturing of the composite.

Two different sets of specimens were created, from the produced laminate: the first was directly used for the tensile tests, whereas the other was firstly impacted, then characterized by tensile tests, in order to evaluate the drop of the mechanical performances cause by the impact, in terms of elastic modulus, residual tensile strength and failure strain. In order to make the impact tests, a self-made pendulum impactor was realized, with a spherical-shaped impactor of diameter 40 mm.

The mechanical characterization, based on tensile tests, performed on pristine specimens on one hand, and to impacted specimens, on the other, has revealed that the impact, while damaging the composite, strongly affects its mechanical performance. Figure 4 shows how mechanical properties, such as Young modulus, tensile strength and failure strain decrease, as a consequence of the impact.

On the other hand, the measurement of the electrical resistance of the samples significantly rise up as a consequence of the impact: it has been demonstrated that an increase of 7.7% is directly related to the damage produced by the impact, which was produce in our test.

This means that the presence of a damage, which strongly affect the mechanical performance of the composite, but still maintain more than the 50% of the initial properties, is easily detectable by measuring its electrical resistance. Therefore, the measurement of a CNT-doped glass fiber reinforced composite can be successfully exploited as a technique for the real-time monitoring of damage formation.

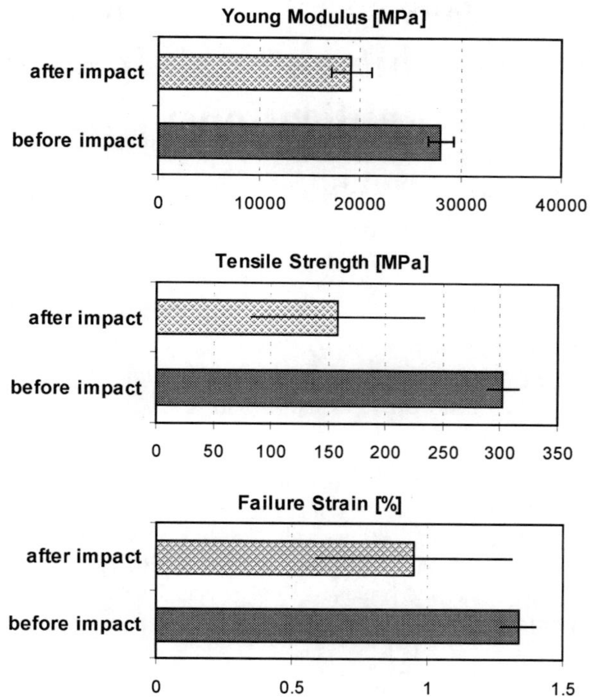

FIGURE 4. Tensile properties of the CNT-doped composite, before and after the impact

ACKNOWLEDGMENTS

The authors wish to thank Dr. Roberto Petrucci for his kindly contribution on manufacturing of the fiber reinforced composite by means of vacuum infusion technique.

REFERENCES

1. U. Polimeno and M. Meo, *Composite Structures,* 2009, Vol. 91, pp. 398–402.
2. A. Todoroki, *Key Eng Mat,* 2004, Vols. 270-273, pp. 1812–1820.
3. A. Todoroki and J. Yoshida, *JSME Int J*, 2004, Series A, Vol. 47, No. 3, pp. 357–364.
4. K. Ogi and Y. Takao, *Compos Sci Technol*, 2005, Vol. 65, pp. 231–239.
5. E.T. Thostenson and T.W. Chou, *Adv Mater*, 2006, Vol. 18, pp. 2837–2841.
6. L.Böger, M.H.G. Wichmann, L.O.Meyer, K. Schulte, *Compos Sci Technol*, 2008, Vol. 68, pp. 1886–1894.
7. J.K.W. Sandler, J.E. Kirk, I.A. Kinloch, M.S.P. Shaffer, A.H. Windle, *Polymer*, 2003, Vol. 44, pp. 5893–5899.
8. J.K.W. Sandler, M.S.P. Shaffer, T. Prasse, W. Bauhofer, K. Schulte, A.H. Windle, *Polymer*, 1999, Vol. 40, pp. 5967–5971.

Silicone containing biomaterials in cardiovascular applications

A. Silvestri[a], S. Sartori[a], P. Serafini[a], P. Ferrando[a], C. Mattu[a], S. Milione[b], F. Boccafoschi[c], G. Ciardelli[a]

[a]*Department of Mechanics, Politecnico di Torino, Corso Duca degli Abruzzi 24, 10129 Torino*
[b]*Department of Chemistry, Università degli Studi di Salerno, Via Ponte Don Melillo, 84084 Fisciano, Italy*
[c]*Department of Clinical and Experimental Medicine, School of Medicine, University of Eastern Piedmont, Via Solaroli 17, 28100 Novara, Italy*

Abstract. A series of biostable polyurethane (PU) formulations, including a composite containing a biocompatible clay as filler, were prepared as new biomaterials for cardiovascular applications. Polydimethylsiloxane (PDMS) and polytetramethylenoxide (PTMO) were selected as macrodiols because of their high hydrolysis resistance[1]. 1,6-Diisocyanatohexane (HDI) and 1,4-cyclohexane dimethanol (CDM) were used as diisocyanate and chain extender, respectively. Chemical and mechanical characterizations of the obtained polymers highlight that they are promising materials for applications in the cardiovascular field.

Keywords: Biostable Polyurethanes, Mechanical Properties, Mitral Device
PACS: 87.85.J, 87.85.Qr, 87.85.Rs

INTRODUCTION

Polyurethanes (PUs) are a class of thermoplastic elastomers possessing excellent mechanical properties and good procesability. They are defined as segmented copolymers and offer high versatility in terms of formulation design. It is possible varying nature and composition of both soft (macrodiols) and hard segment (diisocyanate, chain extender) components in order to tune final polymer's mechanical properties. The incorporation of poly(dimethyl siloxane) (PDMS) into the PU structure has the advantage of imparting some of the attractive properties of PDMS to PUs: high flexibility, good emocompatibility, low toxicity, excellent thermal, oxidative and hydrolytic stability. Accordingly, the use of PDMS based PUs has been proposed in many cardiovascular devices, such as heart valves, catheters and vascular grafts. If only siloxane was incorporated, the resulting PU is characterized by poor mechanical properties[1]. Significant improvements are achieved by incorporating also a polyether in the soft segment.

In this work a series of based PDMS-poly(tetramethylene oxide) (PTMO) PU formulations are proposed, including a composite containing a biocompatible clay as filler, as new biomaterials for cardiovascular applications.

MATERIALS AND METHODS

Materials and Polyurethanes Synthesis

PUs were synthesized by using a two steps procedure[2]. Two kind of macrodioliol were used: polydimethylsiloxane or PDMS (Tegomer H-Si-2311, Evonik, M_w=2500g/mol) and polytetramethyleneoxide or PTMO (Terethane, Aldrich, M_w=2000 g/mol). That used in the larger amount was dissolved in tetrahydrofuran (THF, Aldrich) and the solution was azeotropically dried by refluxing under nitrogen over molecular sivies. Just before the first step reaction, the second macrodiol, previously dried in oven, was added to the solution. In sequence 1,6-hexamethylene diisocyanate (HDI, Aldrich) was reacted (2:1 molar ratio) with the soft segment and the catalyst

CP1255, *V[th] International Conference on Times of Polymers (TOP) and Composites,*
edited by A. D'Amore, D. Acierno, and L. Grassia

(dibutyltin dilaureate, Fluka) to form the prepolymer. At the end of prepolymer formation (150 min, 70°C) cycloexanedimethanol (CDM, Aldrich), previously azeotropically dried, was added at 1:1 molar ratio with respect to the prepolymer at ambient temperature. Chain extension reaction was stopped after 16 hours by the addition of methanol. The polymer was obtained by precipitation in distilled water.

Composite was prepared by using the solution-intercalation film casting technique. The clay (Cloisite 30B, Southern Clay Products, 3% wt), previously dried in oven at 100°C for 24h, was mixed in THF for 8 h. Then PU containing 60% of PTMO (PU_PTMO60), previously dried in oven at 40°C for 16 h, was added to the clay solution, that was mixed for 8 h and then casted in a glass petri dish. The solution was allowed to dry for 1 day. Once dry, the polymer/clay film was placed in a vacuum oven at 40 ◦C for 24 h.

Polymers Characterization

Molecular weights (M_n) and molecular weight distributions (M_w/M_n) of PUs were measured by GPC (Gel Permeation Chromatography) using a Water 1525 binary system equipped with a Water 2414 RI detector. The measurements were performed by using four styragel columns (range 103-106 Å) at ambient temperature, THF as eluent (1.0 mL/min), narrow polystyrene standards were used for calibration.

PU films were prepared by solvent casting and characterized by ATR-IR (Attenuated Total Reflection Infrared Spettroscopy) and DSC (Differential Scanning Calorimetry). These characterizations were performed by using a Perkin Elmer Spectrum GX FTIR System equipped with an ATR attachment and a TA INSTRUMENTS DSC Q20, respectively. ^{1}H-NMR (^{1}H-Nuclear Magnetic Resonance) spectra were recorded on a Bruke Bruker AVANCE 400 operating at 400 MHz for ^{1}H. The ^{1}H chemical shifts are referenced to $SiMe_4$ using the residual protio impurities of the deuterated solvents as external reference.

Tensile tests were performed on PU films to study their mechanical properties. Stress-strain tests and DMA (Dynamic Mechanical Analysis) were performed by using a MTS QTest/10 Elite Controller and a TA ISTRUMENTS DMA Q800, respectively.

RESULTS AND DISCUSSION

Polyurethanes: Synthesis and Characterization

Polyurethanes were successfully synthesized as shown by ATR-IR and ^{1}H-NMR spectra. ATR-IR spectra of the PUs containing PDMS and PTMO in the soft segment show characteristic PU peaks at: 1690 cm^{-1} (urethane C=O), 3335 cm^{-1} (urethane N-H, hydrogen bonded), 1537 cm^{-1} (urethane N-H, not hydrogen bonded). The presence of PDMS is shown by the characteristic double peak of Si-O-Si stretching at 1100 cm^{-1}. C-O-C stretching band is not visible because it is probably covered by the large and strong Si-O-Si stretching band: both vibrations occur at the same wave number value. ^{1}H-NMR was used to confirm the PTMO presence in the polymer chain. In figure 1 the ^{1}H-NMR spectrum of PU containing 40% of PTMO and 60% of PDMS (PU_PTMO40) is shown. The signal at 3,4 ppm corresponds to the -O-CH_2- group and is not present in the ^{1}H-NMR spectrum of PU containing only PDMS.

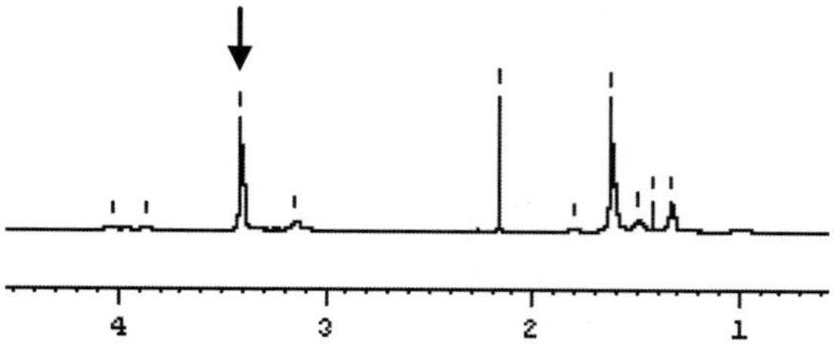

FIGURE 1. Section of ^{1}H-NMR spectrum ($CDCl_3$, 400MHz) of PU_PTMO40: indicated peak is relative to the -O-CH_2- group of PTMO segment.

The experimental data concerning the molar composition, the molecular weights, the polydispersity index (M_w/M_n) and the reaction yields of the obtained PDMS-PTMO based polyurethanes are collected in table 1.

TABLE (1). Experimental results concerning PDMS-PTMO based polyurethanes synthesis reactions.

Sample	**%PTMO in the soft segment**	**M_n (g/mol)**	**M_w/M_n**	**Yield %**
PU_PTMO0	0	22.000	2,03	76
PU_PTMO20	20	20.000	1,41	81
PU_PTMO40	40	26.000	2,03	80
PU_PTMO60	60		1,55	85
80	80	26.000	1,60	90

DSC analysis (not reported in this paper) show the almost complete amorphous nature of PDMS-PTMO based polyurethanes. The study of morphology of this kind of polymers will be deepened by X-Ray and TEM (Transmission Electron Microscopy) analysis, to investigate the presence of a crystalline phase.

In table 2 experimental data about thermal and mechanical characterizations are shown. In particular has been reported: glass transition temperature (T_g) (DMA data), Young Modulus, Maximum Elongation and Maximum Stress (tensile test data). From stress-strain analysis turns out that an higher percentage of PTMO in the soft segment improves polyurethane mechanical properties, especially Young Modulus and Maximum Stress. Mechanical properties of PU_PTMO60 and PU_PTMO80 make these two polymers suitable for cardiac surgery devices. In particular, they can be utilized for annuloplastic rings, that have to be capable of mimicking the physiological 3D motion of the native mitral annulus and accommodating its anatomical saddle shape. DMA analysis show that T_g becomes higher with the increasing of percentage of PTMO. This may indicates a reduction of phase separation, probably due to the PDMS segment (T_g = -110°C).

TABLE (2). Experimental results concerning PDMS-PTMO based polyurethanes DMA and Stress-Strain tests.

Sample	**Tg DMA (°C)**	**Young Modulus (MPa)**	**Average Max % Elongation**	**Average Max Stress (MPa)**
PU_PTMO20	-80	1,5	20	0,2
PU_PTMO40	-67	5,2	30	0,2
PU_PTMO60	-65	5,5	55	1,2
PU_PTMO80	-63	6,8	70	1,1

Mechanical-morphological analysis on the clay-composite and biocompatibility tests on all PU formulations are in progress.

CONCLUSIONS

PDMS-PTMO based PURs have been successfully synthesized, as revealed from spectroscopic analysis. Depending on their chemical and thermo-mechanical properties PU_PTMO60 and PU_PTMO80 seems to be promising materials for biostable mitralic valve applications[3].

ACKNOWLEDGMENTS

Co-financing by the Regione Piemonte through the Bando Regionale - 2006 per la Ricerca e lo Sviluppo Precompetitivo project nr.10-BIADS, and the POR: Piattaforme Innovative (F.E:S.R. 2007-2013) "Active-Advanced Cardiovascular Therapies"References

1. P.A. Gunatillake, D.J.Martin, G.F. Meijs, S.J. McCarthy, R. Adhikari, Aust. J. Chem., 56, 545-557 (2003)
2. G. Ciardelli, A. Rechichi, S. Sartori, M. D'Acunto, A. Caporale, E. Peggion, G. Vozzi, P. Giusti, Polym. Adv. Technol. 17, 786–789 (2006)
3. Arita M. et al., Asian Cardiovasc Thorac Ann, 9, 14-18 (2001)

Development of polypropylene/wood flour ecocomposites. Evaluation of silane as coupling agent.

Bouza R, Barral L, Abad MJ, Montero B

Grupo de Polímeros, Dpto. de Física, E.U.P. Ferrol, Universidad de A Coruña, Avda. 19 de Febrero s/n, 15405 Ferrol, Spain.

Abstract. The effects of *Pinus Sylvestris* wood flour as filler in polypropylene matrix was evaluated. The mechanical properties and the morphology of different wood flour/polypropylene composites (WPC) were studied. The composites materials were prepared with several amounts of wood flour from 10 to 30 % wt. Mechanical properties show that the wood flour incorporation increases the rigidity of the composites. Morphological analysis indicates that agglomerates are formed, with amounts exceeding 30% of wood flour. For the silane – treated composites, the dispersion of the filler into the polypropylene (PP) matrix improved. Shore D hardness of the composites is decreased with the addition of the coupling agent.

Keywords: Polypropylene, wood flour composites, mechanical properties, morphology.
PACS: 81, 89

INTRODUCTION

In the present study, a vinyltrimethoxy silane was used as surface modifier of the filler, to improve the interaction between the polypropylene matrix and wood flour. This product belongs to a new family of coupling agents, synthesized specifically for processing wood – thermoplastic composites. The effectiveness of the coupling agent mainly depends on its chemical nature, and it is therefore necessary to study how the functionalized silane affects the composite properties.

The objectives of this research were to determine the effects of the *Pinus Sylvestris* on the mechanical properties and the microstructure of WPCs. For this, the effects of the amount of filler and the coupling agents on the mechanical properties of the composites were analyzed.

MATERIALS AND PROCESSING

A virgin isotactic polypropylene homopolymer (PP), provided by Repsol – YPF was used as thermoplastic matrix. Wood flour (WF) from *Pinus Sylvestris* specimen, provided by Maderas del Noroeste S.A., Spain, was used as filler, previously characterized [1].

In order to improve adhesion between matrix and filler, the surface of latter can be modified by chemical treatments or by use of coupling agents. The coupling agent used was Geniosil XL 10, a vinyltrimethoxy silane (provided by Wacker). Prior to the extrusion process, the wood flour was dried for 24 h at 100ºC. The extrudate was cooled and pelletized at the die exit, then dried and injection – molded into standard specimens, with a Battenfeld Plus 350 apparatus at an injection pressure of 1400 bar, a barrel temperature of 180ºC, and an injection rate of 56 cm3/s. The nomenclature used for the different composites was a) polypropylene with untreated wood flour: 90/10, 70/30, and 50/50 containing 10, 30 and 50% of wood flour (by weight), respectively; and b) polypropylene with silane – treated wood flour, with the percentage weight of silane indicated e.g. 90/10 – 1 corresponds to the WPC with 10% wood flour treated with 1% silane.

CP1255, V^{th} *International Conference on Times of Polymers (TOP) and Composites,*
edited by A. D'Amore, D. Acierno, and L. Grassia

RESULTS AND DISCUSSION

Tensile tests of the composites were performed at a crosshead speed of 2 mm/min and at room temperature. While the tensile test of the PP was performed at 20 mm/min to check the break point. The typical curves of load/deformation are shown in the figure 1 for the uncompatibilzed composites.

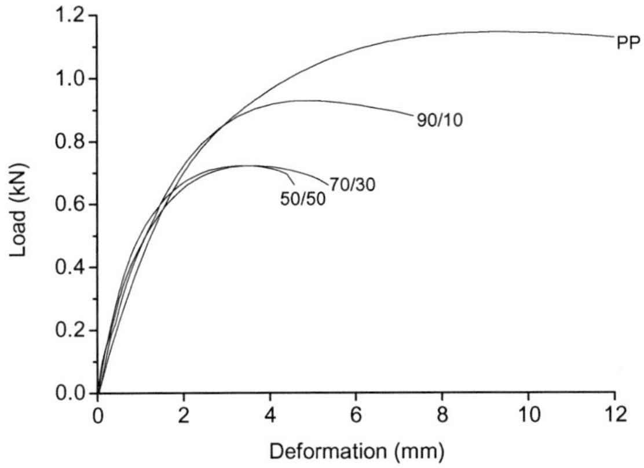

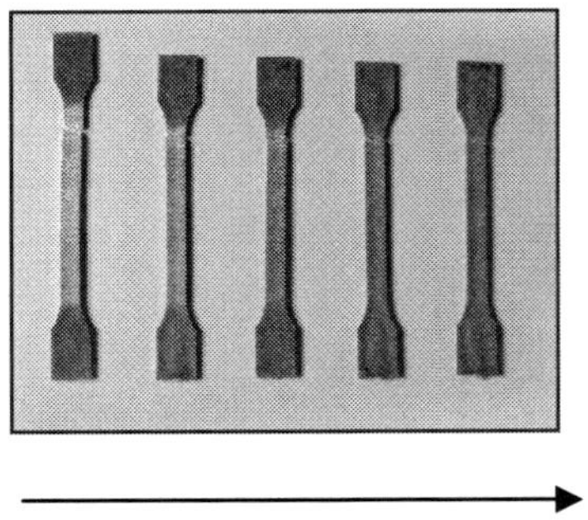

FIGURE 1. Curves load/deformation of the PP, 90/10, 70/30 and 50/50 composites without silane (a) and images of the specimens after the tensile test (b).

In general, the tensile parameters show that the composites become more rigid and have a lower tensile strength as the wood flour content in the materials increased (figure 2). The wood particles are stiffer than the polypropylene matrix; therefore, the result is as expected, since the wood flour restricts the mobility of the matrix. The values of Young modulus exhibit a maximum for the 70/30 sample. This is common behavior for polymer – filler composites, especially composites containing natural fillers [2] and is explained by the tendency for WF to form agglomerates.

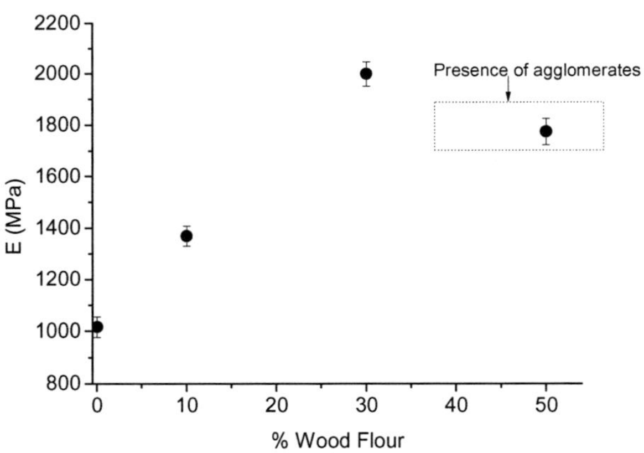

FIGURE 2. Variation of the Young modulus as a function of the wood flour content.

In general, when the filler was treated with the coupling agent, the composite modulus decreased slightly. The coupling agent improved the dispersion of the wood flour into the matrix, and the surface treatment of the WF particles inhibited inclusion of the polypropylene in the wood cells, thereby diminishing the rigidity of the material.

The values at the break point are also related to the interfacial adhesion [3], although the ε_B values can be altered by small imperfections in the specimens (small bubbles, micro crazes ...) which can lead to wrong conclusions.

The variation of the Shore D hardness of WPC as a function of wood flour wt % is shown in figure 3. It can see that the composites have higher hardness compared to the polypropylene matrix due to increased rigidity of the compounds by adding reinforcement. The data also show a slightly higher values for mixtures made with the highest percentage of wood flour and demonstrating a small decrease with the addition of silane.

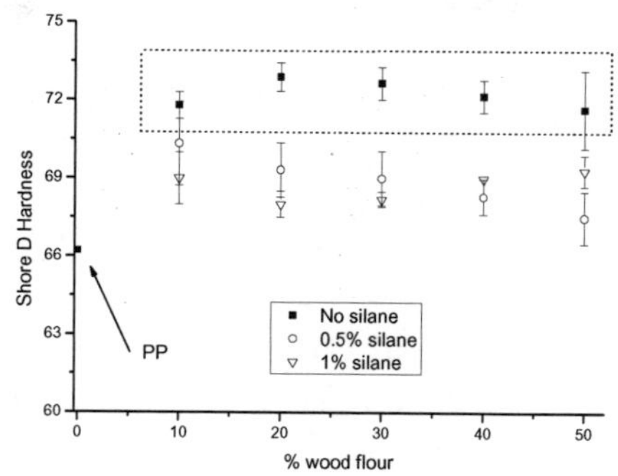

FIGURE 3. Variation of Shore D hardness of composites as a function of WF wt %.

The morphology of the cross sections of samples was observed on metal (Au) – coated fractured samples. The micrographs in Figure 4 show a general view of the fracture surface of the composites. The micrograph of the 90/10 composite shows a rough surface and the wood cells are not visible, whereas they are clearly observed in the 70/30 and 50/50 composites. Furthermore, the micrograph of the 70/30 composite shows that the capillaries of the wood are unaltered and completely filled with PP. In the 50/50 samples, the amount of matrix was not sufficient to fill the wood cells and the empty wood capillaries and WF particle agglomerations can be observed. Moreover, the wood cells were deformed during processing, as observed in the final micrograph in figure 4.

FIGURE 4. Fracture surface of the (a) 90/10, (b) 70/30 and (c) 50/50 composites.

With wood flour treated with 1% of silane, in general, the size of the wood domains decreased and the distribution of filler improved as the quantity of silane increased; this was more evident in the morphology of the composites containing lower amounts of wood flour.

CONCLUSIONS

As regards the tensile properties, the tests demonstrated that the wood flour acted to reinforce the matrix, thereby increasing the elastic modules of the composites up to 30% filler. The tensile strength of the PP decreased when non-treated wood flour was added due to the poor stress transfer at the interphase. In the formulations containing silane, the tensile parameters demonstrated that the dispersion of the wood flour into the PP matrix was improved and less rigid materials were thus obtained. The Shore D hardness of the composites shows similar values and independent of the silane content.

Morphological analysis of the composites revealed that inclusion of silane in the formulations improved the dispersion of the filler, and that surface treatment of the wood particles hindered the inclusion of PP in the cells.

In summary, the use of wood flour modified with the vinyltrimethoxy silane, as filler of polypropylene matrix improves the mechanical properties of the resulting composites in the formulations up to 30% of filler.

REFERENCES

1. R. Bouza, A. Lasagabaster, M.J. Abad and L. Barral *J. Appl. Polym. Sci.*, **109**, 1197 (2008).
2. K, Oksman, C. Clemons *J. Appl. Polym. Sci.* **67**, 1503 (1998).
3. L.E. Nielsen, R.F. Landel. *Mechanical properties of polymers and composites*. New York: Dekker, 1994.

Photo-oxidation Behaviour of EVA Antimicrobial Films

L. Botta, R. Scaffaro and F. P. La Mantia

Dipartimento di Ingegneria Chimica dei Processi e dei Materiali, Università di Palermo, Viale delle Scienze, Ed. 6, 90128 Palermo, Italy

Abstract. In this work the photo-oxidation of neat EVA and antimicrobial EVA/Nisin films was studied. Two EVA samples - containing two different vinyl acetate levels – were added with different amounts of nisin. The influence of the matrix type and of the nisin content on the photo-oxidation behaviour was evaluated. The photo-oxidation has been followed by monitoring the change of the mechanical and spectroscopic properties upon artificial exposure to UV-B light. The results revealed that the films incorporating nisin show a better photo resistance with respect to the neat polymer. This improvement becomes weaker with decreasing the amount of nisin incorporated. Moreover the EVA 28 based films showed a much slower photo-oxidation rate in comparison with the EVA 14 based ones.

Keywords: Photo-oxidation; Antimicrobial film; EVA; Nisin.
PACS: 81.05.Lg; 82.50.Hp

INTRODUCTION

Recently, both industry and academia have shown a growing interest in materials with antimicrobial properties suitable for food packaging.[1] Polymeric materials, due to their flexibility in processing, modification and possible treatments, are excellent candidates to obtain materials with new specific functional characteristics like antimicrobial properties.

For packaging applications, polymeric films must posses several appropriate characteristics and properties such as: good mechanical resistance, improved barrier properties to gases, chemical resistance, etc. Recently, the attention was also focused on the photo-oxidation resistance of packaging films.[2, 3]

In our previous work,[4] antimicrobial films of ethylene-co-vinyl acetate copolymer (EVA) incorporating nisin as antimicrobial agent was prepared by film blowing. Two grades of EVA with two different vinyl acetate content and two commercial formulations of nisin with different activities were used. The results indicated that in all the cases nisin is effective as antimicrobial agent. The nisin with the lowest activity must be added at relatively high levels (5% wt) while less than 1% wt is enough if higher activity nisin is used. In this latter case, neither the mechanical performance nor the transparency were significantly changed.

In this work the photo-oxidation behaviour of neat EVA and antimicrobial EVA/Nisin films was studied. The influence of the matrix and of the amount of incorporated nisin on photo-oxidation behaviour was evaluated. The photo-oxidation has been followed by monitoring the change of the mechanical and spectroscopic properties upon artificial exposure to UV-B light.

EXPERIMENTAL

Materials and Film Preparation

The polymers used in the frame of the present work were two grades of EVA copolymers kindly supplied by Polimeri Europa (Italy): EVA 14 (Greenflex FC 45), containing 14% of vinyl acetate and EVA 28 (Greenflex ML 60) containing 28% of vinyl acetate.

CP1255, *Vth International Conference on Times of Polymers (TOP) and Composites,*
edited by A. D'Amore, D. Acierno, and L. Grassia

A commercial formulation of nisin (Polichimica, Italy) was used as antimicrobial agent: Nisin 8000 (Nisina 8000 IU, nominal activity 8000 IU/mg).

Films of pure EVA or containing 0.2% to 1% (by weight) of Nisin 8000 were prepared by film blowing. EVA and nisin were accurately weighed, premixed at the solid state and then fed to a single screw extruder (Brabender, L/D=25, D=19) preheated at a thermal profile of 95-115-120-120 °C at a speed of 50 rpm, equipped with a film blowing unit.

Accelerated Weathering

The films, thickness of about 100 μm, were photo-oxidized in a QU-V chamber mounting eight UV-B lamps. The weathering conditions were 8 h of light at T = 55 °C and 4 h of condensation at T= 45 °C. The films were exposed up to about 300 hours.

Characterizations

Tensile tests (ASTM D882) were carried out by using a dynamometer Instron 3365 on rectangular shaped specimens (10 × 90 mm) cut off from the film in machine direction.

The Fourier transform infra red (FTIR) spectrometry was performed using an FTIR Spectrum One Instrument (PerkineElmer). The spectra were all normalized using the methylene rocking vibration band at 720 cm^{-1} as a reference.

RESULTS AND DISCUSSION

The progress of the photo-oxidation has been followed by monitoring the changes of the tensile properties of the unfilled and filled EVA films upon exposure to UV-B light.

The dimensionless elongation at break (EB) values of the neat EVA 14 and of the EVA 14 with 1% of Nisin 8000 are plotted at different photo-oxidation times in Figure 1.

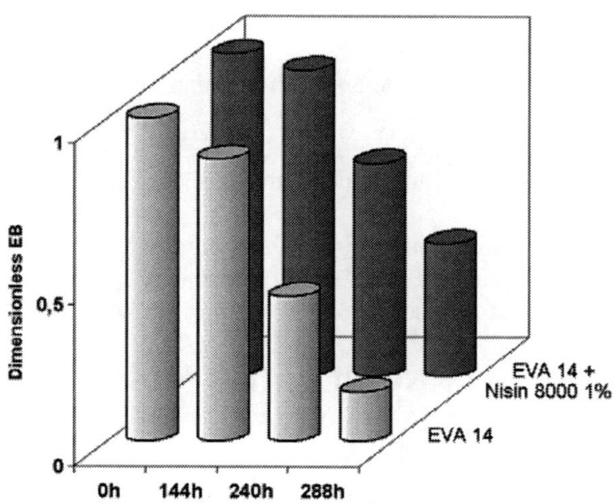

FIGURE 1. Dimensionless elongation at break (EB) values of the neat EVA 14 and of the EVA 14with 1% of Nisin 8000

The elongation at break is an excellent parameter to define the photo-oxidation resistance of polymeric materials as it is very sensitive to molecular and morphological changes undergone by the polymer during the irradiation. Indeed, these changes strongly affect the values of the elongation at break.

The dimensionless EB values highlight that the rate of photo-oxidative degradation of films incorporating the antimicrobial agent is sensibly slower than that of neat EVA 14 films. Moreover, this improvement becomes weaker with decreasing the amount of nisin incorporated.

Similar results are shown by EVA 28 based films, although the process of photo-oxidation is globally much slower for this polymer. Indeed, the higher content of vinyl acetate slows down the photo-oxidation, as shown in our previous work.[5]

The results of FTIR analysis carried out on the photo-oxidized films show a similar trend as that observed in the mechanical tests, although the differences between the neat film and the antimicrobial films are weaker.

In Figure 2 the FTIR spectra in the region 980 - 820 cm^{-1} of neat EVA 14 and of the EVA 14 with 1% of Nisin 8000 after different exposure time are reported.

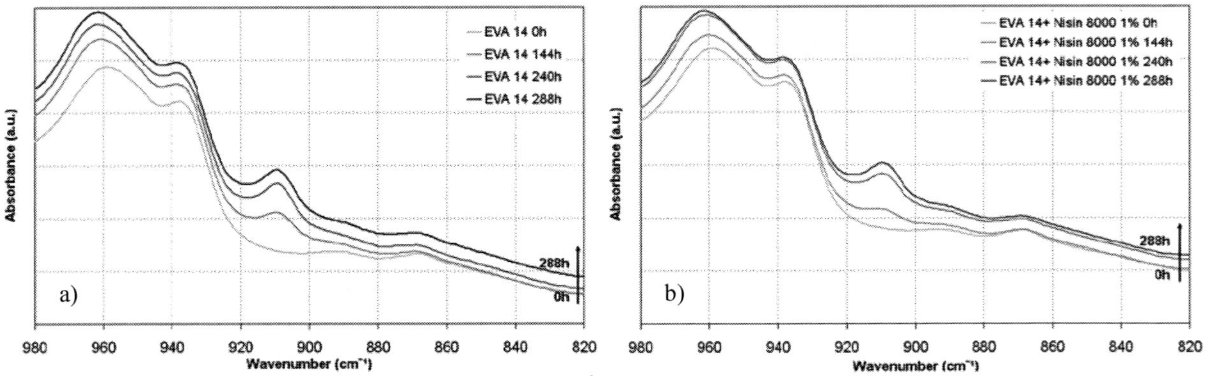

FIGURE 2. FTIR spectra in the region 980 - 820 cm^{-1} of neat EVA 14 and of EVA 14 with 1% of Nisin 8000

UV-light irradiation of EVA produces notable modifications of the FTIR spectra. A broad band appears in the hydroxyl region together with a shoulder of the VA band in the carbonyl region and two other absorption bands around at 910 cm^{-1} and around 960 cm^{-1} (Fig. 2a). These FTIR absorption bands were attributed to the formation of unsaturations corresponding to vinyl bonds ($-CH=CH_2$) at 910 cm^{-1} and to vinylene bonds ($-CH=CH-$) at 960 cm^{-1}. The increase of absorbance in the hydroxyl region was attributed to alcohols associated with acetate groups.[6]

The evolution of FTIR spectra of photo-oxidized antimicrobial films is similar to that of the neat polymer although the absorbance bands corresponding to the photo-oxidation products are slightly less intense as visible in Figure 2b for the vinyl and vinylene bonds.

CONCLUSIONS

In this work the photo-oxidation of neat EVA and antimicrobial EVA/Nisin films was studied. Two EVA samples - containing two different vinyl acetate levels - were added with different amounts of nisin. The influence of the matrix type and of the nisin content on the photo-oxidation behaviour was evaluated. The photo-oxidation has been followed by monitoring the change of the mechanical and spectroscopic properties upon artificial exposure to UV-B light.

The results revealed that the films incorporating nisin show a better photo resistance with respect to the neat polymer. This improvement becomes weaker with decreasing the amount of nisin incorporated. Moreover the EVA 28 based films showed a much slower photo-oxidation rate in comparison with the EVA 14 based ones.

ACKNOWLEDGMENTS

This work has been financially supported by INSTM.

REFERENCES

1. R. D. Joerger, *Packag. Technol. Sci.* **20**, 231-273 (2007).
2. P. A. Tarantili and V. Kiose, *J. Appl. Polym. Sci.* **109**, 674–682 (2008).
3. Li-Xin Lu and Fang Xu, *Packag. Technol. Sci.*, **22**, 107–113 (2009).
4. R. Scaffaro, L. Botta, S. Marineo, A. M. Puglia, *LWT* (2010). Submitted.
5. L. Botta, N.Tz. Dintcheva, F.P. La Mantia, *Polym. Deg. Stab.* **94**, 712-718 (2009).
6. S. Morlat-Therias, E. Fanton, J. Gardette, S. Peeterbroeck, M. Alexandre, P. Dubois, *Polym. Deg. Stab.* **92**, 1873-1882 (2007).

Synthesis and Characterization of Linear Hyperbranched PBA Block Copolymers, as Potential Polymeric Plasticizers.

Ascione L.[a], Ambrogi V.[a], Battiato S.,[c] Carfagna C.[a, b], Persico P.[b], Puglisi C.[c], Samperi F.[c]

[a]*Department of Materials and Production Engineering, University of Naples "Federico II", P.le Tecchio 80, 80125 Naples (Italy).*
e-mail: laura.asc@libero.it
[b]*Institute of Polymer Chemistry and Technology (ICTP)-CNR, Via Campi Flegrei 34, 80078 Pozzuoli (Na) (Italy).*
[c]*Institute of Polymer Chemistry and Technology (ICTP) Sec.Catania-CNR, Via P.Gaifami 18, 95126 Catania (Italy).*

Abstract. Linear hyperbranched block copolymers based on PBA polyesters (HPBA) has been synthesized by a one-pot synthesis procedure and well characterized by MALDI-TOF mass spectrometry, ^{1}H-NMR and SEC tools, to obtain reliable information on its chemical composition and molar mass distribution.

Keywords: Hyperbranched Polyester, Polymeric Plasticizer, Mass Spectrometry.
PACS: 81.05.Lg, 81.16.Be, 81.70.Jb, 82.56.Ub, 82.80.Ms

1. INTRODUCTION

Hyperbranched polymers are a special type of dendritic polymers with a treelike structure, high branching density and high density of terminal functional groups on their surface. The commercial success of hyperbranched polymers is a result of the highly branched and dense but irregular structure that gives them unique physical and chemical properties and leads to excellent solubility, compared to linear polymers, low solution viscosity, modified melt rheology, and high level of terminal end group functionality.[1] They are usually prepared in a one-pot synthesis, which limits the control on molar mass and branching accuracy and leads to heterogeneous products with a distribution in molar mass, chemical composition and branching. Hyperbranched polymers were considered as products suitable for a wide range of application in polymer fields where a perfect structure is sacrificed for an easy and affordable synthetic route. Important commercial applications are: *i)* as a reactive component in coating and resin formulations; *ii)* as polymer additives in linear polymers for improving rheology and flow and for surface modification. Recently, they found also application as polymeric plasticizers, in particular for PVC in substitution of banned phthalate esters (DOP or DEHP).[2] Branched polymers have a higher density of chain-ends, more free volume and mobility compared to linear polymers of the same molar mass. The use of branched polymeric plasticizer increases the mobility of the system compared to a system plasticized with used linear polymers, at the same time the branched polymer's high molar mass and bulkiness significantly reduce its volatility and diffusivity compared to low molar mass plasticizers.[3] Han et al. have synthesized similar hyperbranched PBA samples by reaction of adipic acid terminated PBA with glycerol or pentaerythritol as branching agent.[4]

In the present work we have prepared linear hyperbranched block copolymers in which linear chains of poly(butileneadipate) (PBA), are linked to the focal point of 1,1,1-tris(hydroxymethyl) propane.[1] The progress of reaction was followed by ^{1}H-NMR and MALDI-TOF MS analysis, in order to obtain reliable information about the chemical composition of the initial telechelic PBA pre-polymer and the final hyperbranched products as a function of the reaction parameters (time, temperature and monomers molar ratio). Molar mass distribution ($D=M_w/M_n$) of all polymeric materials were determined by SEC apparatus calibrated with standard poly(styrene)s.

2. EXPERIMENTAL

The linear hyperbranched block copolymers based on PBA polyester used in this work were synthesized through a polycondensation reaction between 1,4-butanediol (BD) and dimethyl ester of adipic acid (DMA), using titanium(IV)isopropoxide (TIP) as catalyst and 1,1,1-tris(hydroxymethyl)propane (TMP) as branching agent, as highlighted in Scheme 1.

CP1255, *V[th] International Conference on Times of Polymers (TOP) and Composites,*
edited by A. D'Amore, D. Acierno, and L. Grassia

In a preliminary study of this work different linear PBA polymers were synthesized, varying reaction parameters such as temperature, time and monomers molar ratio to understand the effect on the structure (i.e. type of end groups) and molar mass distribution of the hyperbranched polymers. From all, we have chosen to describe the synthesis and characterization of linear PBA obtained using an excess of 0,25% of DMA with respect to the other monomer.

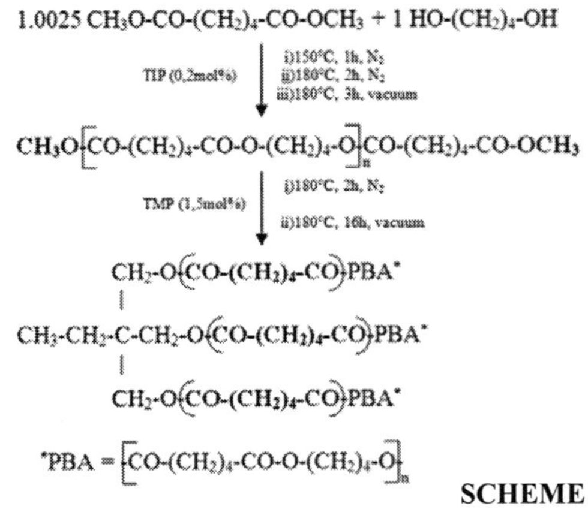

SCHEME 1

In a two-neck round-bottom reaction vessel BD (1;10^{-2} mol), DMA (1,1125·10^{-2} mol) and TIP (0,2 mol%) were mixed at 150°C, under N_2 flow. The reaction was carried out firstly at this temperature for 1 hour and then at 180°C for 2 hours at atmospheric pressure. The methanol formed was continuously distilled from the reaction mixture. The reaction was continued for 3 hours at 180°C under reduced pressure. Successively, the reaction mixture was cooled down to room temperature, the polyesters were dissolved in THF and precipitated into a large excess of methanol. The white polymer was filtrated and dried under vacuum at 40°C for one day.

Linear hyperbranched block copolymers were synthesized in one-pot synthesis according to a two-step procedure. The first step was the same described for linear PBA. In the second step, TMP (1,5 mol%) was added to the reaction mixture. The reaction was carried out for 2 hours at 180°C under N_2 flow at 1 atm, and then at 180°C under reduced pressure for 16 hours. The reaction mixture was cooled down to room temperature. The polyesters were dissolved in chloroform and precipitated into a large excess of methanol and finally filtrated and dried under vacuum.

3. RESULTS AND DISCUSSION

The linear hyperbranched block copolymers based on PBA polyesters formed through the branching reaction between the terminal methyl ester groups (CH_3O–) of linear PBA and the hydroxyl end groups (–OH) of the branching agent TMP as sketched in Scheme 1. Our preliminary studies, in fact, revealed that in the temperature range 180-200°C the (–OH) end groups of TMP react preferentially with (CH_3O–) terminals of polyester rather than with the inner ester functions in the backbone. The intermediate PBA pre-polymer and final white powder HPBA copolymer were characterized by MALDI-TOF mass spectrometry, ^{1}H-NMR and SEC analyses. Figure 1 reports the MALDI-TOF mass spectrum of the PBA pre-polymer. It shows three clusters of homologous peaks corresponding to the sodium and potassium ions of PBA linear chains with *n* repetitive units ranging from 6 up to 33. The most intense peaks correspond to the PBA chains terminated with (CH_3O–) groups at both ends (series C_n). The other two families of peaks correspond one at the linear chains terminated with (–OH) groups at one end and with (CH_3O–) species at the other end (series A_n), and the other one at those having an anhydride group along the chains and terminated with (CH_3O–) moieties at both ends (series D_n). The ^{1}H-NMR spectrum (see Figure 2) confirmed that the linear PBA is mainly terminated with (CH_3O–) groups, since in addition to the peaks corresponding to the protons of the PBA repetitive units it showed a singlet at 3.67 ppm, which is associated to the methyl groups of the methoxy units. Moreover, we calculated the degree of polymerization X_n=24 and the average number molar mass (M_n = 4800 g/mol) for the PBA sample according to equations (1) and (2), respectively.

$$X_n = \left(\frac{A_{4,09}/4}{A_{3,67}/6} \right) + 1 \qquad (1)$$

$$M_n = X_n + 200,2 \qquad (2)$$

Where $A_{4,09}$ is the integrated intensity of methylene protons linked to ester group (–CO-O-CH_2–). The ^{1}H-NMR analysis confirmed also the formation of the hyperbranched HPBA. The corresponding spectrum in Figure 2, in fact, showed new signals at 4,01 ppm, 1,25ppm and 0.84 ppm corresponding to TMP units linked with PBA blocks. Accurate information about the composition of the synthesized HPBA were obtained from their MALDI-TOF mass spectrum, which showed a series of peaks from 1000 up to 10000 Da. Figure 3 reports an enlarged section of this spectrum. It shows mass peaks due the unreacted linear PBA oligomers (series A_n, B_n, C_n and D_n), and those belonging to the formed hyperbranched HPBA macromolecules (species IPA, IPB and IPPB). SEC analysis shows that molar mass distributions of HPBA was higher with respect to the PBA pre-polymers (see Figure 4), confirming that the reaction between PBA and TMP occurred. Average molar masses of both linear PBA and HPBA were calculated from the corresponding SEC curves, calibrated with PS standards, obtaining: M_n=4700 and a M_w=7400 for the PBA; Mn=20500 and a M_w=40900 for the HPBA.

The discussed results indicated that the conversion of the PBA telechelic pre-polymers was incomplete during the branching reaction. In order to obtain HPBA samples with different degree of branching and D, other syntheses have been performed, in which the reaction times and/or the molar amount of the TMP were varied. These new HPBA samples were characterized by MALDI-TOF mass spectrometry, ^{1}H-NMR and SEC to obtain data about their chemical composition and molar mass distributions. The copolymer with the most suitable properties was then selected as plasticizer for PVC samples.

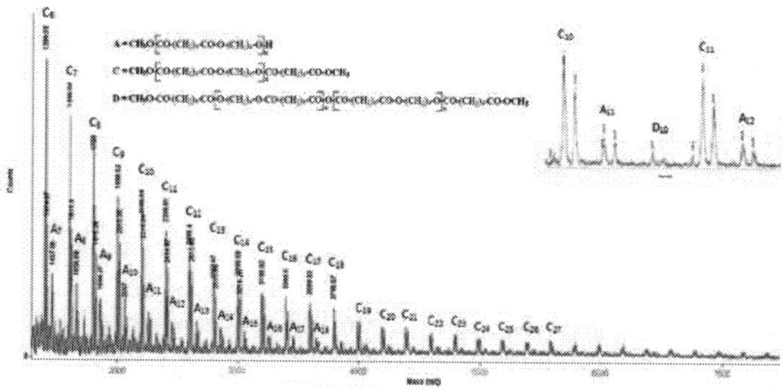

Figure 1. MALDI-TOF mass spectrum of telechelic PBA prepolymer.

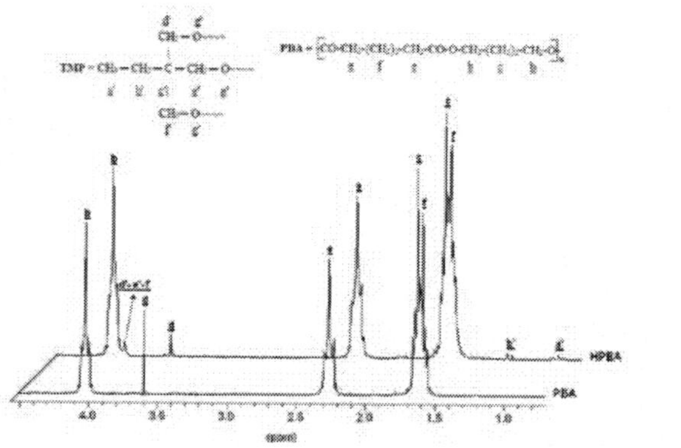

Figure 2. ^{1}H-NMR spectra of PBA and HPBA samples.

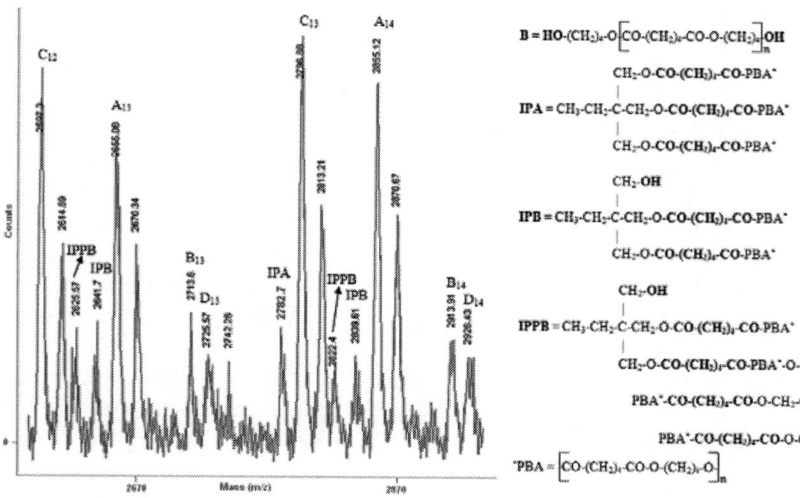

Figure 3. Enlarged section of MALDI-TOF mass spectrum of HPBA sample.

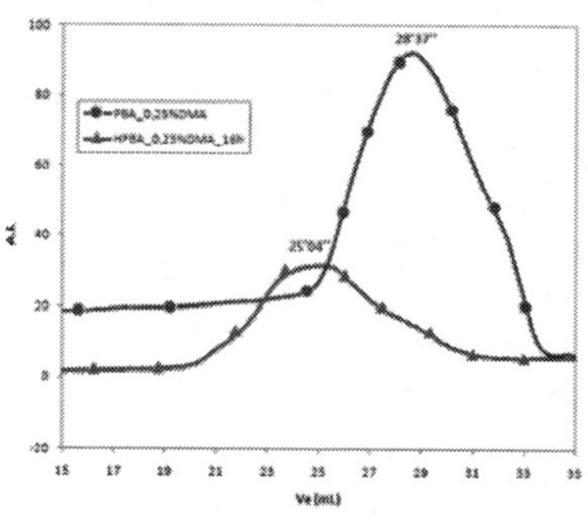

Figure 4. SEC curves of PBA and HPBA samples.

ACKNOWLEDGMENTS

Partial financial support from National Council of Research (CNR) (Rome) and from Italian Ministry for University and Research (MIUR) is gratefully acknowledged. The active collaboration of Dr. Scamporrino Andrea to part of this work is acknowledged.

REFERENCES

1. B. I. Voit and A. Lederer, *Chem. Rev.*, **109**, 5924-5973 (2009).
2. A. Lindström and M. Hakkarainen, *J. Polym. Sci.* **100**, 2180-2188 (2006).
3. A. Lindström and M. Hakkarainen, *J. Polym. Sci.* **104**, 2458-2467 (2007).
4. Y.-K. Han, J. W. Um, S. S. Im, B. C. Kim, *J. Polym. Sci.: Part A: Polym. Chem* **39**, 2143-2150 (2001).

Polylactide Nanocomposites As a New Solution for Eco-compatible Packaging Materials

Loredana Incarnato[a,b], Luciano Di Maio[a,b], Emilia Garofalo[a], Paola Scarfato[a,b]

[a]*Department of Chemical and Food Engineering, University of Salerno, Via Ponte Don Melillo, 84084, Fisciano (SA), Italy*
[b]*Research Centre for NANOMAterials and nanoTEchnology (NANO_MATES) University of Salerno, Via Ponte don Melillo, 84084, Fisciano (SA) Italy*

Abstract. This work focuses on the possibility of improving performance properties of polylactide films for packaging applications by melt mixing silicate nanoparticles within PLA matrix. Nanocomposites at various silicate loadings were produced by cast-film extrusion using two different temperature profiles. Oxygen barrier and mechanical properties of the produced films were investigated and correlated to their nanostructure through analytical techniques sensitive to different aspect of the same morphology, such as TEM, XRD and rheological analyses.

Keywords: polylactide/layered silicate nanocomposites, film-cast extrusion, gas barrier properties.
PACS: Replace this text with PACS numbers; choose from this list: http://www.aip.org/pacs/index.html

INTRODUCTION

Polylactide (PLA) in one of the most important biodegradable polymers, derived from renewable resources and it can often be an alternative for synthetic plastic materials due its interesting physical properties. However, its toughness, thermal stability, and gas barrier properties are not satisfactory.

The field of packaging is one for which the application of polylactide material is highly attractive, and improvements in mechanical and gas barrier properties could allow such polymer to be employed as innovative solution to satisfy the demanding requirements that a modern-day package must fulfill, such as protection, mechanical resistance and biocompatibility. Therefore, it is important to properly modify the PLA in order to improve its properties without losing its biodegradability. In recent studies [1, 3] it was investigated the possibility to modulate the PLA's properties by filling it with selected additives or by functionalizing the polymer with nanosilicates.

Although significant scientific activity has occurred with regards the use of PLA as polymer matrix for layered silicate nanocomposites, surprisingly few publications in the literature focus on film extrusion of polylactide nanocomposites, crucial for their technological exploitation. Moreover, gas barrier properties of PLA hybrid films produced by melt compounding have not been fully investigated and correlated with the degree of dispersion and orientation of silicate platelets inside polymer matrix. The complexity of the final nanostructure makes structure-property relationship not easy to establish and does not allow the observed properties to be explained through conventional composite theories.

In this study PLA based nanocomposites with various organoclay loadings were produced by a twin-screw extruder using two different temperature profiles, in order to point out the effect of both processing conditions and hybrid composition on morphology (silicate dispersion and orientation, matrix crystallinity) of nanocomposites.

All melt-intercalated samples were then processed by cast-film extrusion. Transport properties (oxygen barrier and global migration values) and mechanical performances of the hybrid films were investigated and related to their nanomorphology through structural (TEM and XRD), thermal and rheological analyses.

CP1255, *V[th] International Conference on Times of Polymers (TOP) and Composites,*
edited by A. D'Amore, D. Acierno, and L. Grassia

EXPERIMENTAL

Materials

The silicate used in the preparation of the hybrids was Cloisite30B (supplied by Southern Clay Products, Inc.), an organically modified layered sodium montmorillonite (90 mequiv/100g clay), having an interlayer basal spacing d_{001} = 18.5 Å. Polylactide, under to commercial name of PLA 4032D (~1.5% D-isomer), was used in this study as polymer matrix.

Melt processing

The film production was based on two different extrusion stages. In the first, a twin-screw extruder, fundamental to required degrees of silicate exfoliation, was used to produce PLA nanocomposite pellets. The compounding was conducted usind a Collin ZK25 twin-screw extruder, with co-rotating intermeshing cone-shaped screws (D= 25 mm, L/D=42). A screw speed of 150 rpm and two different temperature profiles (at 180 and 200°C) were used.
The hybrid films at 3 and 6 wt % of silicate were obtained by re-extruding the nanocomposite pellets with a Gimac single screw extruder of L=300 mm, L/D =24 and a rectangular die with dimensions of 0.25x200 mm^2. A screw speed of 50 rpm and a temperature profile identical to the mixing stage were selected.

Characterization

Transmission electron microscopy (TEM) analyses were conducted using a Philips EM 208 at different magnification levels. The images were captured on sections normal to the extrusion direction which had been prepared by microtoming of ultrathin specimens with a Leica Utracut UCT microtome.

Rheological measurements in oscillatory mode were performed with a Rheometric Scientific rotational rheometer in an angular frequency range from 0.1 to 100 rad/s at the temperature of 180 °C using 25 mm diameter parallel plates.

Oxygen permeability of the films was determined with GPC-D Brugger permeability device at 30 °C according to ASTM D1434 procedure.

Thermal behavior of the films was determined using a Mettler differential scanning calorimeter (DSC30), and performing the thermal cycle heating/cooling/heating from 25°C to 200°C at 5°C/min. Crystallinity degrees, X_c, of the different films were determined by the ratio of the sample heat of fusion on heat of fusion of the purely crystalline PLA, i.e., 93.1 J/g [4].

Mechanical tests in tensile mode were performed with a Dynamometer (Instron 4301) imposing a cross speed of 5 mm/min, according to the ASTM D-638 procedure.

RESULTS AND DISCUSSION

Hybrid films were produced using a cast film process and a range of their properties were investigated. Among these the reduction of oxygen permeability is of primary importance for food packaging applications, since oxygen represents one the principal atmospheric contaminants promoting degradation mechanisms. Table 1 reports permeability values (P) and cristallinity degree (X_c) of PLA nanocomposite films at different silicate contents and obtained using two different temperature profiles.

All the hybrid films show a decrease in oxygen permeability with increasing of silicate loading. In particular, for any clay content, PLA nanocomposites films produced at 200°C exhibit the lowest permeability values. The addition of silicate in PLA matrix does not cause significant variations of cristallinity degree, as it can be observed by the X_c values reported in Table 1.

TABLE 1 - Oxygen permeability values (P) and cristallinity degrees (X_c) of PLA nanocomposite films at different silicate loadings.

	Films obtained at 180°C		Films obtained at 200°C	
% wt Cloisite 30B	P [cm^3 mm/m^2d bar]	X_c [%]	P [cm^3 mm/m^2d bar]	X_c [%]
0	17,56 ± 1,03	11	18.75 ± 2,58	11
3	14,61 ± 1,11	12	10,99 ± 2,58	13
6	11,01 ± 0,34	12	7,24 ± 0,43	12

Thus the observed enhancements of oxygen barrier properties for the nanocomposite films can be related to dispersion and orientation of silicate platelets inside the matrix. The presence of the clay layers makes more tortuous the diffusive path for the permeant gas. In particular, PLA hybrids, obtained at higher processing temperature, are characterized by a better exfoliation degree of the silicate inside the polylactide matrix, as verified through rheological and TEM investigations.

The mechanical properties of nanocomposite films were also investigated in order to complete the analysis of film behavior. The hybrid samples processed at 180°C do not exhibit significant variations of the tensile properties compared to the neat matrix, whereas low amount of silicate leads to substantial improvements in stiffness and elongation at break for the nanocomposite films obtained at higher process temperature.

A systematic study about the evolution of physical aging of PLA, in presence of different amount of silicate platelets, was also conducted in this work. Compared to other polymers, the neat polylactide is characterized by fast phenomenon of physical aging. It causes modifications, during time, of some physical properties of PLA such as the ductility.

As reported in Table 2, the hybrid films show a decrease of the relaxation enthalpy values with increasing the clay loading. At longer aging time (about three months since films production), the relaxation enthalpy of pure PLA becomes similar to the corresponding ΔH_{ril} values of the nanocomposite films.

TABLE 2 - Relaxation enthalpy values (ΔH_{ril}) of neat PLA and nanocomposite films, prepared at 200°C with different silicate loadings.

Film sample	Aging time [days]	ΔH_{ril} [J/g]
neat PLA	15	1,12
	90	3,80
PLA + 3% Cloisite 30B	15	0,97
	90	3.70
PLA + 6% Cloisite 30B	15	0,43
	90	3,34

This behavior results from the balance of two opposite effects: the reduced mobility of the polymer amorphous phase, as consequence of the good dispersion of clay platelets inside PLA matrix and the degradation of polylactide macromolecules due to hydrolysis operated by the –OH groups of the organoclay. This last phenomenon becomes more relevant with increasing the aging time, as reported in literature in studies regarding the mechanism of PLA biodegradability in presence of layered silicate [5, 6].

REFERENCES

1. L. Jiang and J. Zhang, *Polymer* **48,** 7632-7644 (2007).
2. S. Sinha Ray and K. Yamada, *Polymer* **44,** 857-866 (2003).
3. S. Sinha Ray and K. Yamada, *Chemistry of Materials* **15,** 1456-1465 (2003).
4. H. Wang and X.S. Sun, *Journal of Polymers and the Environment* **10,** 133-138 (2002).
5. M. A. Paul and C. Delcourt, *Polymer Degradation and Stability* **87**, 535-542 (2005).
6. S. Sinha Ray and P. Maiti, *Macromolecules* **35**, 3104-3110 (2002).

A study on the gate voltage dependence of the activation energy in Meyer-Neldel rule for charge mobility in pentacene OTFTs

Mario Petrosino[a], Riccardo Miscioscia[a,b], Anna de Girolamo del Mauro[b], Romina Rega[b], Valerio Cerri[b], Carla Minarini[b], Alfredo Rubino[a]

[a] *University of Salerno, Department of Information and Electrical Engineering, via Ponte Don Melillo 1, 84084 Fisciano (SA) - Italy*
[b] *Italian National Agency for New Technologies, Energy and Sustainable Economic Development (ENEA), Portici Research Centre, p.le E. Fermi 1, 80055 Portici (NA) - Italy*

Abstract. A temperature analysis of the OTFT having pentacene as channel semiconductor and PMMA as gate dielectric has been performed. The experimental results have been studied by extracting the field-effect mobility of the TFTs and relating it to the sample's temperature. We have found the mobility to follow the Meyer-Neldel rule. This behavior can be considered imputable to the channel carrier hopping. The gate voltage effect on the thermal activation energy for the mobility and the asymptotic parameter has been also taken into account.

Keywords: OTFT, pentacene, carrier mobility.
PACS: 73.61, 85.30

INTRODUCTION

In the last ten years a great attention has been reserved to the analysis and the understanding of the physics of Organic Thin Film Transistors (OTFTs) including the modeling of electrical parameters such as mobility, threshold voltages and non-ideality effects [1][2][3]. Lots of models have been shown in literature about the relation between the carrier mobility, the gate-source voltage and the temperature. The most used model is shown in [4][5][6][7], where a power law is between mobility and temperature has been taken from the a:Si TFT studies relating mobility and temperature as a power law. In this work the Authors show a different analytical law for the above relation, the Meyer-Neldel rule which has been observed in many processes in nature.

EXPERIMENTAL

Substrates have been fabricated on Schott Eagle boron-silicate glass samples 10cm×10cm large. Samples were washed and dried by a standard procedure, soon after this, they were loaded into an evaporation chamber to deposit the gate contacts in order to obtain a bottom-gate top-contacts (BGTC) OTFT structure. A base pressure of 10^{-6} mbars has been kept in the chamber and 50nm of gold have been evaporated on the top of a 10nm chromium layer through a shadow mask without breaking the vacuum. A thin film of polymethylmethacrylate (PMMA M_w=120000) available commercially from Sigma-Aldrich has been spin-coated from a toluene-based solution an then baked on an hot plate at 95°C for two hours to obtain an insulator thickness of 225nm.

On the top of the PMMA gate insulator, patterned pentacene ($C_{22}H_{14}$) islands were formed at a base pressure of 2×10^{-7} mbars at a rate of 0.6Å/s. Gold source and drain contacts were deposited using a third stencil to complete the structure (see figure 1).

CP1255, *V^th^ International Conference on Times of Polymers (TOP) and Composites,*
edited by A. D'Amore, D. Acierno, and L. Grassia

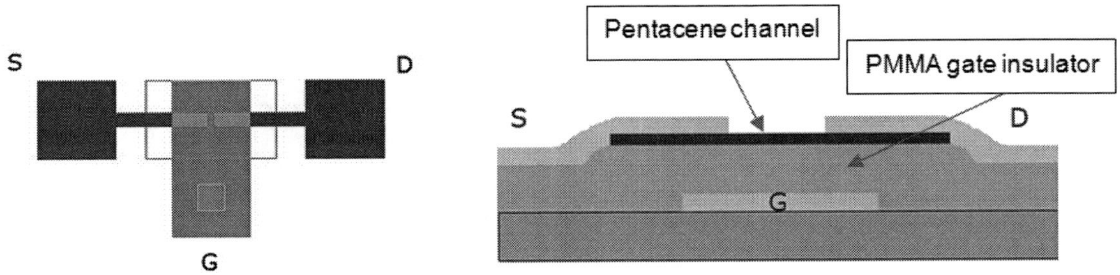

FIGURE 1. Obtained OTFTs layout (seen from the top and laterally).

The fabricated structures have channel lengths of 50 and 70μm and a channel width of 200μm. Output characteristics and trans-characteristics have been measured in vacuum in a Janis cryostat at a pressure of 2×10^{-3} mbars by means of a Keithley SCS4200 instrument. Temperature has been set by a Lakeshore PID controller connected to a thermocouple in contact with the substrate.

DC characteristics have been measured at 300K, 310K, 320K, 330K and 340K and data elaborated to extract thresholds and hole mobilities for this kind of p-type device. By using the standard MOSFET equation and extracting the first derivative of the square root of the drain current we have been also able to evaluate the variation of the mobility versus the gate voltage.

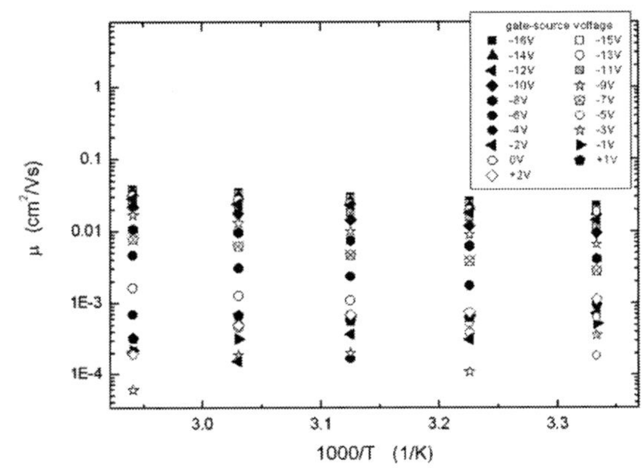

FIGURE 2. Carrier mobility as function of the temperature.

RESULTS AND DISCUSSION

The tested samples have shown threshold voltages about -5 V and a mobility changing with V_{GS} and temperature with a maximum value of 2.25×10^{-2} cm^2/Vs in ambient conditions at V_{GS}=-16V. As shown in Fig. 2, the mobility increases with the temperature for voltages below the threshold, while decreases for upper ones following the Meyer-Neldel rule [8]

$$\mu = \mu_o e^{-\frac{E_a}{kT}}. \tag{1}$$

where μ_o is the asymptotic mobility and E_a the activation energy. This trend has been observed in [8][9] too, while a different relation between carrier mobility and temperature has been found in [4]

$$\mu = \mu_o \frac{N_c}{N_{t0}}\left(\frac{C_i V_g}{qN_{t0}}\right)^{\left(T_c - T\right)/T} \tag{2}$$

noting that in both models the mobility increases with the temperature. In Stallinga's work [8] the $E_a(V_{GS})$ empirical law shows a monotonic trend decreasing with the gate bias. Instead, our samples (Fig. 3) show two linear trends in the operation region of the OTFT and an abrupt change in sign and convexity for $V_{GS}>V_{TH}$.

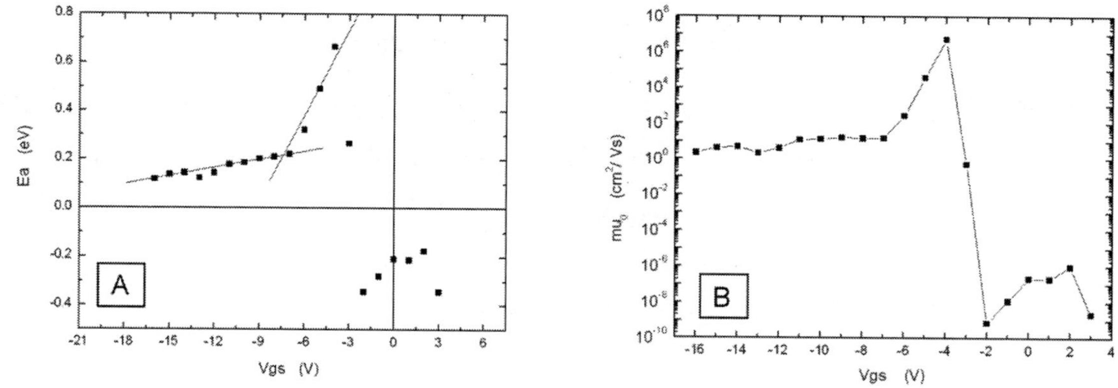

FIGURE 3. Meyer-Neldel parameters of Eq. 2 (activation energy E_a and asymptotical mobility μ_o) as function of the gate-source voltage. In the *A*, two linear trends are observable at high and low biasing, while a change of the sign and convexity is shown for voltages higher than the threshold. In the figure *B*, the asymptotical mobility has a decreasing trend for gate voltages lower than the threshold and decreases to zero after the discontinuity.

The energy is reduced from 0.66 eV at V_{GS}=-4 V to 0.12 eV at V_{GS}=-16 V and decreases with slope 0.12 for $-7<V_{GS}<-4$ and 0.011 for $-16<V_{GS}<-7$. The asymptotical mobility (Fig.2-B) is zero for gate voltages higher than the threshold voltage and decreases with the V_{GS} and a discontinuity is observable for gate biasing near the threshold.

The observed phenomenon may be explained in terms of trapping [8] linearly related to the gate voltage with a change of the regime at V_{GS}=-7V (visible in the lines connection of Fig.2-A), suggesting a variation in the trapping mechanisms and in the response of the trapped/untrapped carrier ratio to the gate bias. Noting that in the observed results the electric field perpendicular to the source-drain current flow changes the mobility in the channel, we think that the hopping in the channel is tuned by the gate-source field by the effect of the microscopic features of the grown films. In fact, because the pentacene grains are deposed maintaining their maximum dimension along the source-drain direction [2], the hops along the vertical direction are more frequent then along the horizontal one, and so the carrier transport in the channel is affected by the gate field.

ACKNOWLEDGMENTS

Support by the TRIPODE project financed by the Ministero dell'Università e della Ricerca (MIUR) is gratefully acknowledged.

REFERENCES

1. G. Horowitz, *J. Mater. Res.*, **7** (2004) 1946-1962.
2. R. Miscioscia, G. Nenna, P. Tassini, A. De Girolamo Del Mauro, P. Vacca, T. Di Luccio, C. Minarini, D. Della Sala, M. Petrosino and A. Rubino, *Proc. of the 5th Int. TFT Conf.,* Paris 2009.
3. M. Petrosino, A. Rubino, R. Miscioscia, A. De Girolamo Del Mauro, C. Minarini, *IEEE Proc. of the Int. Conf. on Sign., Circ. and Syst.*, Djerba 2009.
4. G. Horowitz and M.E. Hajlaoui *Synth. Met.*, **122** (2001) 185-189.
5. M. Estrada, A. Cerdeira, J. Puigdollers, L. Resendiz, J. Pallares, L.F. Marsal, C. Voz and B. Iniguez, *Solid-State Electr.*, **49** (2005) 1009-1016.
6. S. Masouri, M. Mahdouani, A. Oudir, S. Zorai, S.B. Skhil, G. Horowitz and R. Bourguiga, *Eur. Phys. J. Appl. Phys.*, **48** (2009) 30401.
7. Y. Yun, C. Pearson and M.C. Petty, *Journ. of Appl. Phys.*, **105** (2009) 034508.
8. P. Stallinga and H.L. Gomes, Proceedings of the Intern. Symposium "Technologies of Polymer Electronics TPE 04", Rudolstadt/Germany, 28.-30.09.2004, p. 125-129
9. D. Guo, S. Ikeda and K. Saiki, *IEEE Proc. of the Int. Semic. Dev. Res. Symp.*, 2005.

Porous Polyelectrolyte Hydrogels With Enhanced Swelling Properties Prepared Via Thermal Reverse Casting Technique

Aurelio Salerno and Paolo A. Netti

Interdisciplinary Research Centre on Biomaterials (CRIB) and Italian Institute of Technology (IIT), University of Naples Federico II, Piazzale Tecchio 80, 80125 Naples, Italy

Abstract. In this work we investigated the preparation and characterization of porous polyelectrolyte hydrogels via thermal reverse casting technique.

Polyacrylamide hydrogels were synthesized by free-radical crosslinking polymerization into the space of an agarose gel which, after the setting of the chemical gel, was removed to allow the formation of an interconnected porosity pathway. Two different monomer/agarose solution ratios were selected for the reverse casting process and, the resulting hydrogels characterized in terms of morphological, micro-structural and thermal properties, as well as swelling capability in solutions at different ionic strength.

The results of this study demonstrated that proposed technique allowed the design of porous polyacrylamide hydrogels with well controlled pore structures. Furthermore, if compared to non porous polyacrylamide hydrogel, the as obtained hydrogels were characterized by enhanced swelling properties and that, these properties were fine tuned by the appropriate selection of the templating agent concentration.

Keywords: Hydrogels; Polyacrylamide; Porosity; Reverse Casting.
PACS: 82.70.Gg

INTRODUCTION

Polymer hydrogels are materials composed of hydrophilic cross-linked polymer chains, which are either synthetic or natural in origin, able to undergo reversible and discontinuous volume changes in response to variations in environmental conditions, such as solvent composition, temperature, salt concentration and pH [1,2]. This peculiarity has stimulated the interest for these materials in several applications such as smart materials, separation membranes, actuators and sensors as well as for biomedical purposes [1,2]. Therefore, the possibility of fine tuning the swelling properties of hydrogels in response to specific external stimuli may represent a great design advantage for several applications.

It has been recently reported that the creation of an interconnected micro-porous network into polymeric hydrogels may significantly enhance their response to an external stimulus [3-5]. Indeed, the presence of a three-dimensional porous pathway into the polymer network may reduce the solvent transport resistance because of the solvent flow by convection, a faster process if compared to simple diffusion [4]. Furthermore, the presence of an interconnected micro-structure with well controlled porosity and pore size distributions may be strongly required for applications such as tissue engineering scaffolds, for which cells may be able to infiltrate in three dimensions [5].

In this study we prepared novel porous polyacrylamide hydrogels by a reverse casting approach and investigated the effect of the micro-architecture of the pore structure of the hydrogels on their swelling capability.

EXPERIMENTAL

The hydrogels were synthesized by free-radical crosslinking polymerization of sodium methacrylate (SM) and N,N-methylene- bis-acrylamide (MBA). Methacrylamidopropyltrimethil-ammonium chloride (MAPTAC) and ammonium persulfate (APS) were used as cationic comonomer and initiator, respectively. As shown in Figure 1, the

CP1255, *V*th *International Conference on Times of Polymers (TOP) and Composites,*
edited by A. D'Amore, D. Acierno, and L. Grassia

polymerization process was performed into the space of an agarose gel which, after the setting of the polyacrylamide hydrogel, was removed in order to allow the formation of an interconnected porosity pathway.

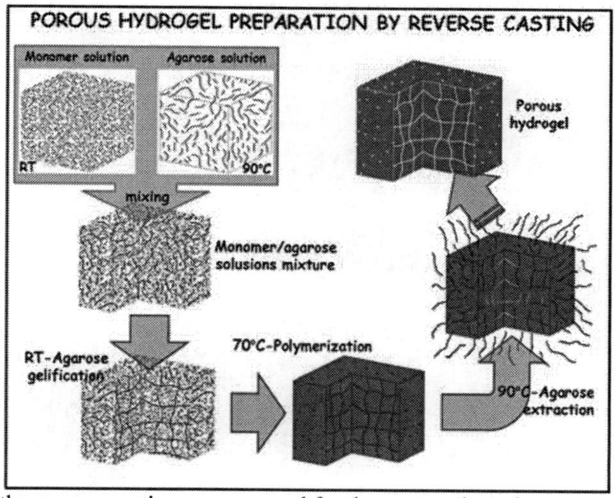

FIGURE 1. Scheme of the reverse casting process used for the preparation of porous polyelectrolyte hydrogels.

Two different porous hydrogels were prepared: the $SMA_{0.5}$ was obtained by using a monomer/agarose solutions ratio equal to 0.5, while the $SMA_{0.25}$ by using a monomer/agarose solutions ratio equal to 0.25. Non porous SM hydrogel was also prepared for comparison.

The obtained hydrogels were characterized by ATR spectroscopy, TGA/DSC tests and SEM analysis to evaluate their chemical composition, thermal properties and morphology, respectively. Furthermore, swelling tests were performed by soaking the samples in water solution at different ionic strength and, the results correlated to the pore structure of the hydrogels in the wet state.

RESULTS AND DISCUSSION

The results of the thermal characterization of the hydrogels are reported in Table 1, clearly indicating differences in the thermal properties of the hydrogels with respect to their formulation.

TABLE (1). Thermal properties of the hydrogels in dry and swollen state. W^f_{H2O} and W^b_{H2O} are the weight of freezable and non-freezable water, respectively. W_G represented the weight of dry hydrogel.

	Dry state		**Swollen state**			
	T_{d1}; T_{d2}; T_{d3} [°C]	**T_g [°C]**	**$^{30-150°C}$Wt-loss [%]**	**T_f [°C]**	**W^f_{H2O}/W_G**	**W^b_{H2O}/W_G**
SM	110.6; 195.3; 426.1	2.05	98.1	2.36	61.2	4.2
$SMA_{0.25}$	108.9; 193.6; 419.4	-3.26	98.9	1.60	99.4	3.8
$SMA_{0.5}$	108.5; 186.3; 415.6	-6.80	99.6	0.28	264.1	1.6

In particular, the TGA/DTGA results evidenced that the dry hydrogels were characterized by three main thermal degradation steps and that the onset of degradation (T_d) decreased after the reverse casting process. Furthermore, the glass transition temperature (T_g) of the hydrogels decreased from 2.05°C for the SM to -6.8°C for the $SMA_{0.5}$. TGA and DSC results of the swollen hydrogels, reported in Table 1, also showed differences in the water absorption properties of the hydrogels. Indeed, the weight loss in the 30 to 150°C temperature range increased with the increase of the agarose concentration in the initial solution, indicating the enhanced water absorbed by the porous hydrogels. These results were corroborated by those of the DSC tests, that also evidenced a significant higher free water content within porous hydrogels, if compared to neat SM.

Swelling tests were carried out on SM, SMA0.25 and SMA0.5 hydrogels in water at RT and by using different ionic strength and, the results reported in Figure 2 along with the morphology of the hydrogels. As shown, hydrogel formulation and ionic strength have a great impact on hydrogel swelling. Indeed, we observed the decrease of the swelling ratio with the increase of the ionic strength of the soaking solution, as a consequence of the reduced water activity of salt solutions and due to the shielding effect exerted by Na^+ and Cl^- ions on the polymeric bulk charges. Furthermore, if compared to neat SM, both $SMA_{0.25}$ and $SMA_{0.5}$ hydrogels evidenced enhanced swelling ratios for all the ionic strength investigated. In particular, the highest swelling ratio, equal to 270, was observed for the $SMA_{0.5}$ at ionic strength = 10^{-7} and, was up to four times higher than that observed for neat SM under the same conditions.

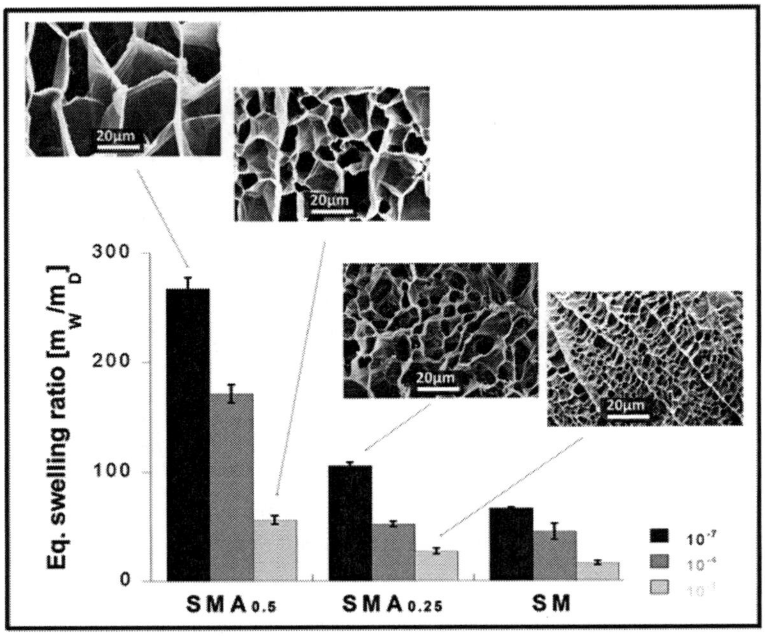

FIGURE 2. Relationship between hydrogel swelling and pore structure at the equilibrium in solutions with different ionic strength.

The SEM micrographs reported in Figure 2 also showed that all the hydrogels were characterized by uniform porosities. Furthermore, in agreement with the results of the swelling tests, lower mean pore size values and narrower pore size distributions were observed at lower agarose concentration and ionic strength.

CONCLUSIONS

In this work we reported a novel approach for the design of porous polyelectrolyte hydrogels by a reverse casting process and we characterized the prepared hydrogels in term of their thermal, swelling and micro-structural properties. The results of this study demonstrated that the proposed technique may allow the inscription of homogeneous and spatially controlled pore structures within the hydrogel and, therefore, .the fine control of their swelling behavior.

ACKNOWLEDGMENTS

The authors gratefully acknowledge Rossella Borzacchiello and Luciano Rosato for their support in the hydrogels preparation.

REFERENCES

1. A. Richter, G. Paschew, S. Klatt, J. Lienig, K. Arndt, H. P. Adler, *Sensors* **8**, 561-581 (2008).
2. B. Kima, N. A. Peppas, *Polymer* **44**, 3701–3707 (2003).
3. D. Ceylan, M. M. Ozmen, O. Okay, *J. Appl. Polym. Sci.* **99**, 319–325 (2006).
4. G. D. Lu, Q. Z. Yan, C. C. Ge, *Polym. Int.* **56**, 1016–1020 (2007).
5. A. Sannino, P. A. Netti, M. Madaghiele, V. Coccoli, A. Luciani, A. Maffezzoli, L. Nicolais, *J. Biomed. Mater. Res.* **79A**, 229–236 (2006).

Weatherability Evaluation of Nanocomposite Polymeric Treatments for Surface Protection of Construction Materials

Paola Scarfato[a,b], Maria Letizia Fariello[a,b], Luciano Di Maio[a,b], Loredana Incarnato[a,b]

[a]*University of Salerno, Department of Chemical and Food Engineering, Via Ponte don Melillo, 84084 Fisciano, Italy*
[b]*Research Centre NANO_MATES, University of Salerno, Via Ponte don Melillo, 84084 Fisciano, Italy*

Abstract. In this work the protective efficacy and stability against UV weathering of polymeric nanocomposites for concrete (CLS) surface protection have been evaluated. In particular, nanocomposite hybrids were prepared dispersing a commercial organomodified montmorillonite (Cloisite 30B) in two different polymeric matrices, one based on fluoroelastomers (Fluoline CP), the other on silane and siloxane (Antipluviol S). The obtained systems were characterized by several techniques (SAXD, DSC, TGA, FT-IR, contact angle measurements, colorimetry), before and after accelerated aging due to UV exposure, in order to evaluate the effect of the nanoscale dispersion of the organoclay on the properties and the UV stability of the treatments.

Keywords: Nanocomposite surface treatments, UV weathering, Concrete protection.
PACS: 81.05.Lg

INTRODUCTION

In last decades many conservators and restorers tested polymeric resins as consolidating and protective agents for building construction materials; in particular synthetic acrylic and vinyl polymers, organosilicone compounds and fluorinated materials were used for preserving new and existing structures [1]. Anyway, up today these classes of products are still under experimentation because often they did not given the expected beneficial effects, mainly due to their low compatibility with the substrate and reduced long-term stability to environmental aggressive agents (water, photo-oxidative weathering, salt attack, mechanical stress, etc.) [2]. Recently nanocomposite systems, based on organic polymers and inorganic layered reinforcement consisting of organomodified silicates whose sheets are few nanometres thick, have emerged as a new class of advanced materials with excellent properties with only a few percent of well-dispersed filler. In particular they exhibit high chemical, thermal and mechanical resistance and reduced gas and vapour permeability at low volume fraction loading of layered silicates (<5%), without spoiling the optical properties of the polymeric matrix [3-4]. In this field, we are conducting a study on the possibility to increase the efficacy of commercial polymeric product used as surface protective agent for concrete and stone restoration by adding a layered organomodified sodium montmorillonite (Cloisite 30B). The experiments were carried out using two commercial polymeric matrices, the fluoroelastomeric Fluoline CP and the silane/siloxanic Antipluviol S. Both resins were loaded with different amounts of organoclay (2, 4 and 6 wt%) and used to prepare film samples by casting procedure. The obtained nanocomposite films were characterized by several techniques (SAXD, DSC, TGA, FT-IR, contact angle measurements, colorimetry), before and after accelerated aging due to UV exposure, in order to evaluate the effect of the organoclay nanoscale dispersion on the properties and the UV stability of the treatments.

EXPERIMENTAL

Materials. Nanocomposite hybrids were prepared using as nanofiller a commercial oragomodified clay, traded as Cloisite 30B (Southern Clay Products, USA), which is a layered sodium montmorillonite organically modified by

CP1255, *V*[th] *International Conference on Times of Polymers (TOP) and Composites,*
edited by A. D'Amore, D. Acierno, and L. Grassia

N,methyl-N,tallow-N,N0,2-hydroxyethyl-ammoniumchloride (90meq/100g clay). The nanofiller was nanodispersed by ultrasonication at three different percentages (2, 4 and 6 wt%) in two types of polymeric resins developed for surface protection of construction materials, commercially named as Fluoline CP (by CTS s.r.l., Italy) and Antipluviol S (by MAPEI S.p.A., Italy). Fluoline CP is a blend of an acrylic polymer and a vinylidenfluoride based polymer in acetone solution. Antipluviol S is a colourless liquid based on silane/siloxane resins in acetone solution. Film samples of neat and nanocomposite resins were produced by solvent casting at room temperature. In the following the polymeric film samples are referred as X-30B Y%, where X is CP or AS for Fluormet CP or Antipluviol S resin, respectively, whereas Y is the weight percentage of Cloisite 30B nanoclay.

Analyses. Small angle X-ray diffraction measurements (SAXD) were performed on nanocomposites films and clay powder with a Rigaku D/MAX -2000 Diffractometer using a Ni-filtered Cu-Kα radiation (40kV, 20mA). Differential scanning calorimetry measurements (DSC) were carried out using a Mettler DSC-30, in the temperature range of -70-300°C, in nitrogen atmosphere. Thermogravimetric analyses (TGA) were performed using a Q500 analyzer in nitrogen atmosphere. ATR FT-IR analyses were carried out using a Nexus Thermo-Nicolet spectrophotometer, in the frequency range of 4000-650 cm^{-1}. Static contact angle measurements were performed with a contact angle system FTA 1000 instrument (First Ten Angstroms, Inc.). Colorimetric analysis was carried out with a CR-410 HEAD colorimeter (Konica Minolta Sensing, Inc.), based on the L*, a* and b* coordinates of the CIELAB space, following the Italian Recommendation NORMAL 43/93 (1993). Artificial UV weathering experiments were performed exposing the samples to UV radiation for 7 days in a climatic chamber at 25°C and R.H.= 50%; at the end of this period the samples were photographed, to compare their chromatic appearance with that of the unaged ones, and resubmitted to ATR FT-IR tests.

RESULTS AND DISCUSSION

In order to verify that the Cloisite 30B achieved a proper nano-dispersion level into the polymeric resins, SAXD measurements were carried out both on the neat silicate powder and on films of Fluoline CP and Antipluviol S at different organoclay content (2, 4 and 6 wt%). As shown in Fig. 1(a), the characteristic basal reflection of the Cloisite 30B centered at about 2θ= 4.9° corresponding to a d-spacing of 1.8 nm, in the SAXD profiles of all the hybrid films appears broadened and shifted at lower 2θ angles. These results suggest that the interlayer distance of the organoclay has increased with both polymeric resins, that is, intercalation/exfoliation of the clay has occurred, as confirmed also by TEM analysis. As an example, in Fig. 1(b) is reported the micrograph of the sample CP-30B 4%.

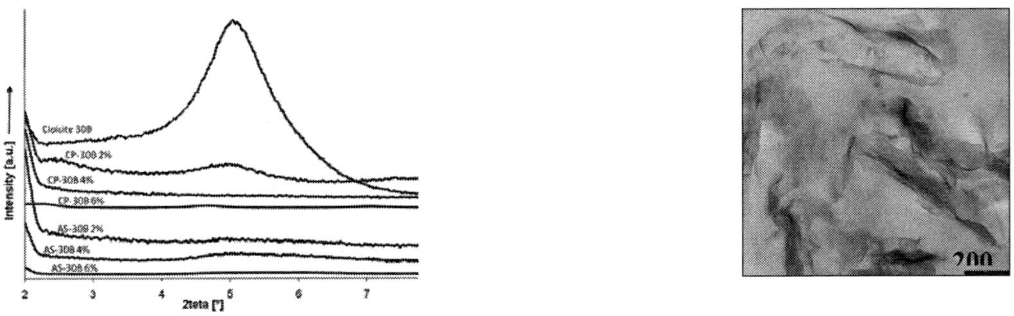

(a) (b)

FIGURE 1. SAXD profiles of neat Cloisite 30B and of Fluoline CP and Antipluviol S cast films at different organoclay content (a); example of TEM micrograph of the sample CP-30B 4% (b).

As a consequence of this nanoscale dispersion of the organoclay inside the polymeric matrices, an improvement of the hydrophobic character of neat resins was obtained, as demonstrated by the increase of about 4-10° of the water contact angle values (data not shown).

The effect of the obtained nanoscale dispersion of the filler on the thermal behavior of the polymeric systems was investigated by means of DSC and TGA analyses, whose results are reported in Tab. 1. As it can be seen, compared to the neat resins, DSC revealed a slight increase in the T_g values of the thermoplastic CP-based hybrids, whereas TGA showed a marked rise in the starting degradation temperature of all nanocomposite films. All these changes demonstrated that the addition of the organoclay, even at very low percentages, significantly improves the thermal stability of the two polymeric matrices.

In order to verify if similar beneficial effects occur towards UV radiation weathering, too, all film samples were exposed for 7 day to UV light in a climatic chamber and their colorimetric variations and FT-IR spectra were measured and compared with that recorded on the unweathered specimens. As for example, in Tab. 2 are reported the colorimetric results obtained on samples unloaded and loaded with 4wt% of organoclay, whereas in Fig. 2 are shown the corresponding FT-IR spectra.

TABLE (1). Glass transition temperature (T_g) and starting degradation temperature (DT_s) of neat and nanocomposite film samples.

organoclay amount	CP-based systems		AS-based systems	
[%]	T_g [°C]	DT_s [°C]	T_g [°C]	DT_s [°C]
0	26	327	*n.d.*	334
2	30	341	*n.d.*	365
4	29	350	*n.d.*	371
6	29	362	*n.d.*	379

TABLE (2). Colorimetric variations of polymeric films after UV exposure.

Film	Unaged films				Films aged for 7 days under UV light			
sample	ΔL^*	Δa^*	Δb^*	$\Delta E_{a,b}$	ΔL^*	Δa^*	Δb^*	$\Delta E_{a,b}$
CP	--	--	--	--	-0.5	0.13	0.2	0.55
CP-30B 4%	-1.5	-0.25	2.0	2.51	-0.7	-0.86	0.3	1.15
AS	--	--	--	--	-1.2	-0.31	0.4	1.30
AS-30B 4%	-0.3	-0.19	0.8	0.88	-0.8	-1.55	0.5	1.81

The data of Tab. 2 evidences that the color changes, measured as a difference in brightness value (L*) and a* and b* coordinates, and the chromatic differences $\Delta E_{a,b}$ before and after UV light exposure are in all cases very small, undetectable to naked eye. Anyway, FT-IR measurements demonstrated the organoclay can effectively reduce the photo-oxidative phenomena due to UV exposure. In particular, the ATR FT-IR spectra of CP-based systems, recorded before and after the exposure of the samples to UV light for 7 days, showed that the increase in the intensity of the carbonyl signal is significantly lower for the hybrid than for the neat resin (Fig. 2).

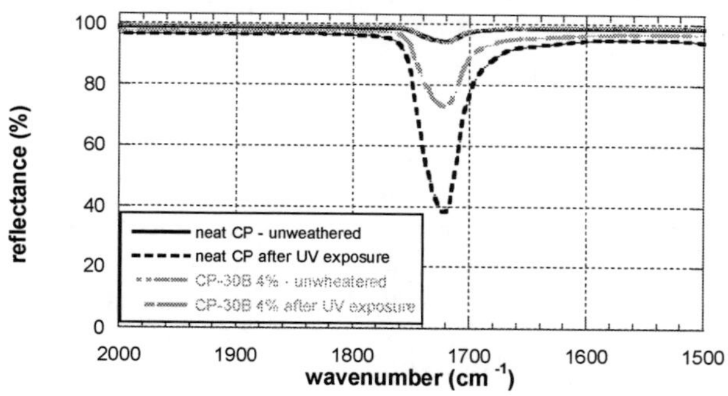

FIGURE 2. ATR FT-IR spectra of CP-based nanocomposite films at different composition, before and after UV exposure.

CONCLUSIONS

The results obtained in the work have shown that the nanoscale dispersion of little amounts of layered silicate particles inside polymeric surface treatments has really promising effects for conservation and protective purposes since, compared to the neat resins, nanocomposite systems can more efficacy delay the weathering phenomena of a substrate. In fact, the nanocomposites were characterized by higher thermal and photo-oxidative stability, improved hydrophobic character and negligible chromatic difference.

REFERENCES

1. M. Robson, Polymers in Conservation, Allen E.S., Edge M. and Horie C.V. Ed., The Royal Society of Chemistry, 1992.
2. M. Favaro, R. Mendichi, F. Ossola, U. Russo, S. Simon, P. Tomasin, P.A. Vigato, *Polym Deg Stab*, 91, 3083-3096 (2006).
3. E.P. Giannelis, *Appl. Organometal. Chem.* 12, 675-680 (1998).
4. C. K.Y. Leung et al., *Journal of Materials in Civil Engineering* 20, 484-492 (2008).

Preparation, Characterization and Release Properties of Nanostructured Polystyrene Microparticles Containing Different Silicates

Paola Scarfato[1,2], Pio Iannelli[3], Pietro Russo[4], Domenico Acierno[4]

[1]*Department of Chemical and Food Engineering, University of Salerno, 84084 Fisciano (SA), Italy*
[2]*Research Centre NANO_MATES, University of Salerno, 84084 Fisciano (SA), Italy*
[3]*Department of Pharmaceutical Sciences, University of Salerno, 84084 Fisciano (SA), Italy*
[4]*Department of Materials and Production Engineering, University of Naples "Federico II", 80125 Napoli, Italy*

Abstract. Nanostructured polystyrene microparticles containing different silicates as nanosized filler were successfully prepared using an emulsification/solvent evaporation method. X-ray diffraction showed that all hybrid systems were in an intercalated morphology. The microparticles display regular geometry, improved thermal resistance and release behavior dependent upon the silicate type in the hybrid, as demonstrated by preliminary investigations performed using benzophenone as model penetrant.

Keywords: polystyrene, nanocomposites, microparticles.
PACS: 81.07.-Pr; 82.35.Np

INTRODUCTION

In recent years many efforts are directed towards the fabrication of polymer nanocomposite hybrids for controlled release (CR) of drugs or other active agents. The great interest in these systems is due to their interesting potential applications in numerous industrial sectors such as pharmaceutical, cosmetic, agrochemical, textiles, etc. [1-2]. In fact, CR formulations can allow either a prolonged duration and stability of action of an active agent and its higher efficacy by tailoring and targeting the release site. A fundamental role in determining the release properties of a CR device is played by the polymer system and by its interaction with the encapsulated active substance. In this regard, host–guest materials made of a polymer and a layered material consisting of thin crystalline inorganic layers with a thickness of a few nanometers can be exploited as matrices for controlled release formulations.

Despite the high prospects offered by polymer-silicate nanocomposite systems for CR applications, up today very few data were reported in literature on the preparation and release behavior of microencapsulated devices based on these matrices [3].

In this work we prepared, by solvent evaporation of oil-in-water (o/w) emulsion, polystyrene-silicate nanocomposites microspheres containing benzophenone (BP) as model active agent. The obtained samples were characterized by several techniques (SAXD, SEM, TGA, UV-Vis, etc.) in order to obtain information about their morphology and release properties.

EXPERIMENTAL PART

Materials. Polystyrene (PS) was purchased by Scientific Polymers Products (SP^2, Inc. USA). Polyvinyl alcohol 80% hydrolyzed (PVA), benzophenone (BP) and chloroform ($CHCl_3$) were analytical reagent grade and purchased by Aldrich. Natural sodium montmorillonite (MMT) and organomodified sodium montmorillonite Cloisite 10A (10A) were supplied by Southern Clay Products (USA), whereas mica (M) was supplied by Ultrafine Mica (UK).

CP1255, *Vth International Conference on Times of Polymers (TOP) and Composites,*
edited by A. D'Amore, D. Acierno, and L. Grassia

Preparation. PS-silicate nanocomposite microparticles, with or without drug embedded, were prepared following a standard procedure involving solvent evaporation of oil-in-water (o/w) emulsion. Typical preparation was performed dissolving 2.0 g of polymer in 20 ml of chloroform and dispersing into the obtained solution different amounts of silicate (MMT, 10A or M at 3, 6 and 9 wt%) and drug (20 wt% of matrix weight, if necessary). The suspension was stirred magnetically and sonicated in water bath for 20 min to promote the polymer intercalation between the silicate layers. Then, this oily phase was emulsified in 150 ml of water containing 0.2 g of PVA using an IKA emulsifier (2000 rpm, 2 min). Agitation was maintained until the chloroform evaporated, so creating solid particles suspended in the water. The suspension was filtered through a standard sieve and washed with water, yielding a free-flowing powder. Following the same procedure, neat PS microparticles were also prepared, for comparison. The PS-silicate nanocomposite microparticles were named using the polymer or silicate abbreviation followed by x/y, where x denotes the silicate weight content and y denotes the drug weight content in the samples.

Characterization techniques. UV spectra of benzophenone were recorded on a Perkin Elmer Lambda 800 Spectrometer, in the range of 200-800 nm. Thermogravimetric analyses (TGA) were performed with a TA Instruments Q500 apparatus, heating microparticle specimens up to 900°C in N_2 atmosphere at 20°C/min. Microparticles size distribution was analyzed using a Beckman Coulter LS13 320 Laser Diffraction Particle Size Analyzer. Scanning electron micrographs (SEM) were obtained using a EVO MA 10 apparatus (ZEISS). Small angle X-ray diffraction (SAXD) spectra were recorded with a Rigaku D-MAX x-ray diffractometer, using a Ni-filtered Cu-Kα radiation (40kV, 200mA).

RESULTS

PS-silicate microparticles, with and without the drug, were preliminary characterized in order to evaluate the microspheres yield, the actual silicate content and the actual drug content in the hybrids. The average values are reported for each sample in table 1. The data show that all systems were prepared with a basically quantitative yield in weight. Moreover, the differences between the nominal and the measured silicate level in the hybrids are less than 7%, indicating that the preparation procedure used to obtain the microspheres has guaranteed a satisfactory control of the samples composition.

TABLE (1). Actual silicate and drug contents, encapsulation yield and particle size distribution for the hybrid microparticles

Sample	Actual silicate content[a] [%]	Actual drug content[b] [%]	Encapsulation yield [%]	PSD [μm]
PS0/0	0	0	-	17 ± 11
MMT3/0	2.9	0	-	23 ± 15
MMT6/0	5.9	0	-	25 ± 17
MMT9/0	8.7	0	-	18 ± 12
M3/0	2.8	0	-	16 ± 11
M6/0	5.7	0	-	12 ± 8
M9/0	8.6	0	-	12 ± 9
10A6/0	6.1	0	-	18 ± 10
PS0/20	0	15.3	92	13 ± 10
MMT6/20	5.8	14.0	84	15 ± 9
M6/20	5.7	16.8	100	11 ± 8
10A6/20	6.0	16.9	100	16 ± 9

[a] from TGA measurements; [b] from UV spectrometry

The morphology of the samples was examined by scanning electron microscopy. Fig. 1 shows the SEM images of PS0/0, 10A 6/0 and 10A 6/20 microspheres, chosen as an example. All samples consist of isolated microparticles with spherical shape and quite smooth, non-porous surface.

In order to verify that the silicates were dispersed on a nanometric scale in the PS matrix, SAXD analyses were carried out on the hybrid systems. The diffractograms show a shift of the basal spacing of the neat silicates towards lower 2θ values, suggesting that a partial intercalation of the polymer chains between the silicate layers has occurred in our experimental conditions, giving an increase of the interstratic gap. The confinement of the PS chains between the silicate platelets markedly improves the thermal resistance of the nanocomposites even at lowest silicate percentages. As demonstrated by TGA measurements, the increase in the values of T_{onset} and DT_{max} for the nanocomposite microspheres is in the range of 10-25°C, depending on the silicate nature.

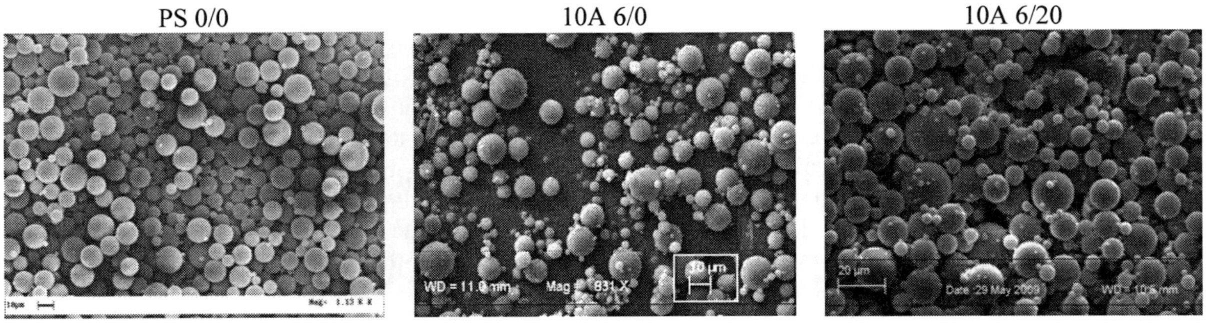

FIGURE 1. SEM images of PS-silicate nanocomposite microspheres at different compositions.

Preliminary in vitro release tests have demonstrated that the incorporation of the silicates into the PS matrix yielded delayed release of the BP from all formulations, as shown in Fig. 2. The kinetic of BP release was linear *vs.* the square root of time, as for pure diffusive release of active molecules from hydrophobic and insoluble matrix.

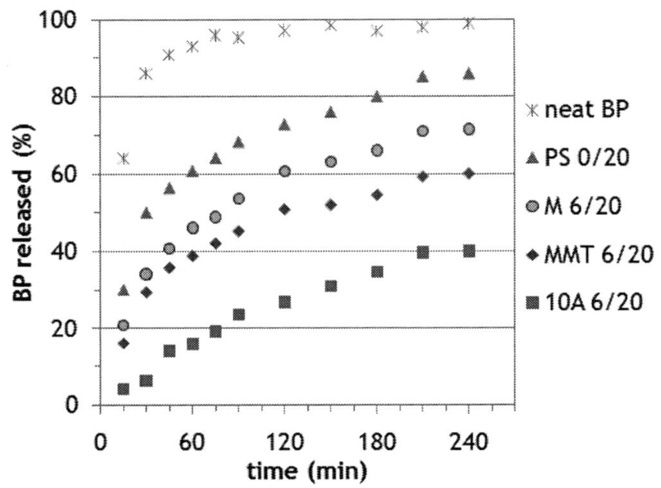

FIGURE 2. BP release profiles from neat PS and PS-silicate nanocomposite microspheres at different compositions.

CONCLUSIONS

In this study PS-silicate nanocomposite microspheres containing BP as model active agent were successfully prepared by solvent evaporation method, with an adequate control of the size and shape of the prepared systems and high entrapment efficiency (> 80%) of the model active agent. All formulations have shown intercalated polymer-silicate morphology and markedly improved thermal resistance. Due to the changes in diffusion characteristics resulting from the addition of the silicates, the BP release profile was significantly modified. The release kinetic is markedly affected by the type of the silicate, so it can be modulated opportunely choosing the filler.

REFERENCES

1. S.H. Cypes, W.M. Saltzman and E.P. Giannelis, *Journal of Controlled Release*, **90(2)**, 163-169 (2003).
2. M. Z. bin Hussein, Z. Z, A.H. Yahaya and D.W. Vui Foo, *Journal of Controlled Release*, **82(2-3)**, 417-427 (2002).
3. M.S. Toprak, B.J. McKenna, J.H. Waite and G.D. Stucky, *Chem. Mater.* **19**, 4263 (2007).

Effect of clay surface silylation and dispersion method on the mechanical properties of epoxy-clay composites

V. Romeo[a], F. Piscitelli[a,b], A.M. Scamardella[a,b], E. Amendola[a], M. Lavorgna[a], G. Mensitieri[b] and D. Acierno[b]

[a] *Institute of Composite and Biomedical Materials, CNR-Italy's National Council of Research, P.le E. Fermi1, 80055 Portici (Na), Italy*

[b] *Department of Materials and Production Engineering, University of Naples "Federico II, P.le Tecchio 80, 80125 Naples, Italy*

Abstract. Epoxy-clay nanocomposites were prepared dispersing both pristine and functionalized sodium montmorillonite powders (1 and 3wt%) in epoxy resin by means of sonication and sonication/ball-milling high energy mixing processes. Silylation reaction of sodium montmorillonite (Na-MMT) was performed by using 3-aminopropyltriethoxysilane (A1100) and N-2-aminoethyl)-3-aminopropyltrimethoxysilane (A1120) as coupling agents. Morphological investigations showed that the MMT stacks are only slightly intercalated. However the surface modification of MMT clays improves the interfacial interaction with epoxy resins and the nanocomposites obtained through sonication exhibit enhanced mechanical properties compared to the nanocomposites prepared from pristine Na-MMT.

Keywords: Na-MMT, Nanocomposites, aminosilanes, epoxy, silylation reaction, sonication, ball-milling.
PACS: 68.35.bm

1 INTRODUCTION

The nanocomposites achieved by dispersion of inorganic nanosized filler (i.e. sodium montmorillonite Na-MMT) in a polymer matrix (i.e. DGEBA-based epoxy resin) often exhibit remarkable properties such as high moduli[1], increased strength and decreased flammability[2]. However a practical problem in the synthesis of nanocomposites is how to disperse the clay in the epoxy. This can be addressed to modify the inorganic clay platelets through ion exchange or chemical silylation and to use a suitable dispersion method. In this work we performed the silylation of MMT clays with different aminosilanes as coupling agents, improving the interfacial interaction between clay and epoxy resin. Two mixing processes, namely sonication (S) and sonication/ high energy ball-milling (SB) were used. The resulting nanocomposites were characterized by SAXS, Dynamic Mechanical Analysis (DMA) and tensile mechanical tests in order to assess the effect of silylation and dispersing method.

2 EXPERIMENTAL

2.1 Materials and methods

Na-MMT, with CEC of 92 meq/100g, was purchased from Southern Clay Products Inc., USA. Both the A1100 and A1120 were obtained from GE Advanced Materials. Glycerol, diglycidyl ether of bisphenol-A (DGEBA) with EEW 187 g/mol eq and meta-xylylenediamine (MXDA) were purchased from Sigma Aldrich.

SAXS spectra were collected by using a SAXSess Instrument manufactured by Anton Paar, with copper Kα radiation, $\lambda = 0.1542$ nm. The scattering data were background subtracted and normalized for primary beam intensity. DMA measurements were performed with a DMA Q800 by TA Instruments using single cantilever geometry operating at a frequency of 1 Hz and 25 μm strain amplitude. The temperature was varied from 0°C to

CP1255, *Vth International Conference on Times of Polymers (TOP) and Composites,*
edited by A. D'Amore, D. Acierno, and L. Grassia

180°C at a heating rate of 3°C/min. Tensile tests were performed according to ASTM-D638-03, using a SANS CMT universal testing machine. An average value of five measures was determined.

2.2 Silylation reaction of Na-MMT and preparation of epoxy-clay nanocomposites

The silylation reaction of Na-MMT was carried out by reacting for 3h at 80°C in glycerol 3g of clay and 3g of aminosilane (i.e. A110 and A1120). The reaction product was separated by centrifugation, washed in distilled water and dried for 3h at 80°C *in vacuum* condition. The functionalized MMTs were denoted as A1100-MMT and A1120-MMT, respectively. The amount of grafted and intercalated aminosilanes was assessed by TGA in nitrogen flow by applying the following relationship[3]:

$$silane\ grafted\ amount\ (mequiv/g) = \frac{10^3 \cdot W_{200-600}}{\left[100 - \left(W_{200-600}\right)\right] \cdot M}$$

where $W_{200-600}$ is the weight loss between 200 and 600°C and M (g/mol) the molecular weight of aminosilanes molecules. The silane amounts resulted 0.182 and 0.365 meq/g for the A1100-MMT and A1120-MMT, respectively.

The DGEBA was mixed with the desired amount of powders (1 and 3 wt%) by using S and SB dispersion methods. A stoichiometric amount of MXDA was added by using a Thinky planetary vacuum mixer. The resulting mixtures were submitted to a complete curing cycle ranging from 30 and 150°C. The nanocomposites samples are denoted as X% A1100-MMT and X% A1120-MMT, where X, equal to 1 or 3, represents the weight percentage of clays added to the resin.

2.3 Experimental results and discussion

SAXS. The SAXS analyses performed on the powders showed that the silylation reaction allows the MMT $d_{spacing}$ to increase with respect to the pristine clay. In detail, the Na-MMT shows an initial $d_{spacing}$ value equal to 1.27 nm that increases to 1.53 nm and 1.89 nm after silylation with A1100 and A1120. Moreover, the S method allows the functionalized-MMT $d_{spacing}$ to slightly increase with respect to the powders (i.e. from 1.53 to 1.71nm related to 1% A1100-MMT nanocomposite), whereas the SB dispersing method makes the $d_{spacing}$ to be reduced to the initial value of the powders. These results confirm that the functionalized-MMTs are only slightly intercalated by epoxy resin. Probably the reactive aminosilanes located on the surface and edges of MMT platelets favour extra galleries polymerization slowing down the diffusion of DGEBA monomers into the galleries and consequently the further intercalation[4].

DMA. The glass transition temperatures (T_g) and the organic network crosslink densities (νe) were assessed by DMA analyses, taking into account the maximum value of tan δ and the storage modulus *E'* value in the rubbery region at 150°C, respectively. In particular, nanocomposites films νe was evaluated according to the following equation[5]: $E' = 3\nu e RT$, where T is 423.15 K, R the gas constant and ν_e the moles number of elastically effective network chains per m^3 of film. The νe and the T_g nanocomposites values, listed in Table 1, show that the S dispersing method makes nanocomposites with enhanced properties compared to the SB samples. Furthermore, probably due to the ammine groups on the platelets surface, the S nanocomposites obtained by using functionalized MMTs show values of T_g and νe higher than those of nanocomposites manufactured with the neat Na-MMT, approaching to those of the neat epoxy. Besides, it is worth noting that in the SB nanocomposites the aminosilanes functionalization is not effective. In fact, their T_g and νe values approach the worse values of nanocomposites obtained by the pristine Na-MMT.

TABLE 1. T_g, E' in the rubbery region at 150°C and ν_e assessed by DMA analyses for the nanocomposites and the neat epoxy resin. (S) indicates the sonication dispersion technique; (SB) the combination of sonication and ball-milling.

Samples	**Tg [°C]** (S)	**E'(MPa) at 150°C** (S)	**Crosslink density [mol/m³]** (×10³) (S)	**T_g [°C]** (SB)	**E' (MPa) [150°C]** (SB)	**Crosslink density [mol/m³]** (×10³) (S)
Neat epoxy resin	124	28.10	2.66	124	28.10	2.66
1% Na-MMT	117	15.30	1.45	110	16.89	1.60
1% A1100-MMT	123	21.36	2.02	108	11.62	1.10
1% A1120-MMT	127	25.99	2.46	111	16.58	1.57
3% Na-MMT	104	11.63	1.10	91	3.01	0.29
3% A1100-MMT	117	23.38	2.22	98	11.01	1.04
3% A1120-MMT	122	23.06	2.18	99	5.671	0.54

Tensile test. The elastic moduli of epoxy resin and nanocomposites with 1%wt of pristine and silylated MMTs are shown in Figure 1. It is worth noting that the nanocomposites moduli increase with increasing of the aminosilanes organic chains length. The enhanced mechanical properties has to be ascribed to better interfacial adhesion due to the reactive inorganic surface between the particles and the epoxy matrix, so that the polymer chains mobility under loading is restricted[6]. Additionally, the S dispersing method provides nanocomposites with higher elastic moduli compared to the nanocomposites obtained by SB dispersing method.

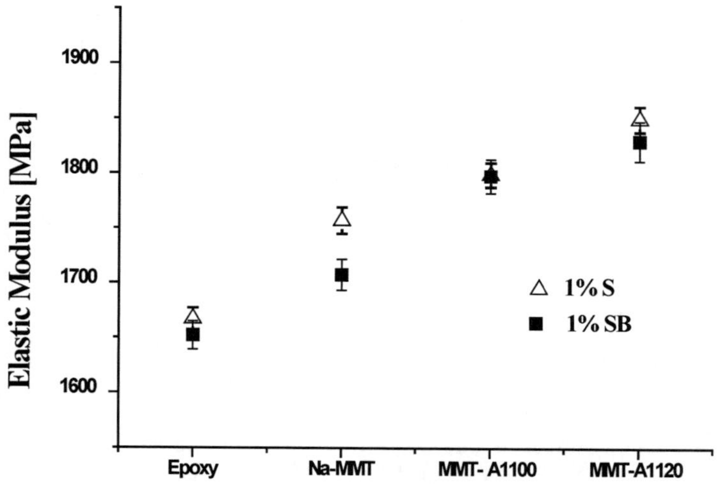

FIGURE 1. Elastic modulus of epoxy and nanocomposites obtained by sonication (S) and sonication/ball-milling (SB) methods.

4 CONCLUSION

The morphological and mechanical characterization of nanocomposites obtained by dispersing trough sonication and sonication/ball milling methods both pristine Na-MMT and silylated-MMT in epoxy resins allows to draw the following conclusions:

1) Even if the MMT clay silylation does not favour the exfoliation of the inorganic platelets, it enhances the interfacial interactions between MMT stacks and epoxy resin;
2) The interfacial interactions improve by using coupling agents such as A1120 characterized by the longer side chains and the higher number of amino reactive groups;
3) The sonication/ball milling method has detrimental effect on the dispersion of MMT particles and consequently on the mechanical properties of the nanocomposites.

ACKNOWLEDGMENTS

The authors thank IMAST and MIUR. This work was supported by the research project MACE coded MIUR n. DM24442. The authors would like to thank Mr. Mario De Angioletti and Mrs. Mariarosaria Marcedula for their technical support.

REFERENCES

1. P. C. Lebaron; Z. Wang and T. J. Pinnavaia, *Appl. Clay Sci.* **15**, 11-29, (1999).
2. J. W. Gilman, *Appl. Clay Sci* **15**, 31-49, (1999).
3. N. Herrera, J. Letoffe, J. Putaux, L. David, E. Bourgeat-Lami, *Langmuir* **20**, 1564-1571, (2004).
4. J. H. Park and S. C. Jana, *Macromolecules* **36**, 2758-2768, (2003).
5. L.W. Hill, *Prog. Org. Coat.***31**, 235-243, (1997).
6. C. L. Wei, M. Q. Zhang, M. Z. Rong, *Comp. Sci Tech.***62**, 1327-1340, (2002).

Probing the Sequestering of Carbon Nanotubes in the PS Domains of SIS Block Copolymer Matrix using Electrostatic Force Microscopy

Laura Peponi[a,*], Luca Valentini[a], Josè M. Kenny[a,b]

[a]*Dipartimento di Ingegneria Civile e Ambientale, Università di Perugia, UdR INSTM, 05100 Terni - Italy.*

[b]*Institute of Polymer Science and Technology, CSIC, Juan de la Cierva 3, 28006 Madrid - Spain*

Abstract. In this work we show the ability of electrostatic force microscopy (EFM) to localize and characterize conducting nanofillers inside the domains of a block copolymer matrix. As case study we report the confinement of carbon nanotubes into polystyrene-b-polyisoprene-b-polystyrene block copolymer (SIS) matrix. In particular, the EFM technique has been used to obtain information with unprecedented imaging of octadecylamine functionalized single-walled carbon nanotubes, sequestered in the self-assembled lamellar polystyrene domains of the SIS block copolymer matrix.

Keywords: Carbon nanotubes, nanocomposites, block copolymer, electrostatic force microscopy.
PACS: 62.23.St

INTRODUCTION

Polymeric matrix nanocomposites can provide useful solutions in advanced sectors such as magnetic-transparent, conductive-transparent, catalytic-magnetic, etc. just combining the matrix properties with the ones own of the nanoreinforcement [1-4]. In particular, carbon nanotubes (CNTs), since their discovery in 1991, became one of the most studied materials due to their unique geometry and exceptional mechanical, electrical and chemical properties which make them attractive items in the field of nanotechnology [5]. However, their integration into a nanostructured polymer matrix remains a critical challenge for their organization into both well-defined straight or curved geometries. In this regards, block copolymers, BC, are an interesting class of nanostructurable polymers able to self-assemble into well-defined, periodic nanoscale morphologies [2-3]. In this case, the chemical incompatibility between the constituent blocks leads to phase separation restricted to the length scale of the polymer molecules (typically 10-100 nm). Due to their intrinsic ability to self-assemble BC can nano-organized in well-defined structures as lamellae, cylinders, spheres, among others, acting as templates to sequester nanofillers. The combination of both nanosize objects and nanostructured templates can lead to unique and complex multifunctional nanocomposite materials. The incorporation of nano-objects in BC provides an opportunity to synthesize polymer nanocomposites that potentially rival the most advanced materials in nature. Strategies for the fabrication of BC-nanocomposites are based on sequestering nano-objects into preferred domains of the block copolymer by manipulating the specific interactions between the nano-objects and polymer segments, followed by inducing microphase separation of the block copolymers. For the block copolymer/particle system, the domain size of the block copolymer and the interaction strength between the copolymer play important roles on the particle dispersion and the equilibrium phase behaviour. Many aspects would be taken into account to obtain a good confinement of

CP1255, *V[th] International Conference on Times of Polymers (TOP) and Composites,*
edited by A. D'Amore, D. Acierno, and L. Grassia

nanofillers in the block copolymer matrix, among them, the geometrical compatibility and the chemical affinity between the nanostructures of the matrix and the shape of the nanoreinforcements. Information about the general compatibility of a particle's geometry with a BC mesostructures can be obtained by considering the symmetry of both the inclusion and the BC host matrix. While, for nanofillers-polymer matrices chemical interaction, due to the typically unfavourable interactions between the particle surface and the host polymer, most particles will need to be pretreated to tailor their surface for compatibilization by using surfactant or "grafting to" technique. So, taking into account that the nanoobject/polymer interface is, together with the nanofiller and the polymer, the third element that defines the final behaviour of these nanostructured materials, in this work, to obtain a well-dispersed hybrid nanocomposites surfactant has been used. In particular in this work, based on our previous works [5-6], we report the ability of dodecanethiol, as surfactant to confine octadecylamine functionalized carbon nanotubes in the polystyrene phase of the block copolymer matrix. The confinement of the tubes is demonstrated by atomic force microscopy (AFM). Moreover, thus taking into account the intrinsic electrical conductivity of the carbon nanotubes electrostatic force microscopy (EFM) has been used to characterize their confinement into the domains of the block copolymer.

EXPERIMENTAL

A linear polystyrene-b-polyisoprene-b-polystyrene block copolymer, Kraton SIS-D1165 with 30 wt.% polystyrene, characterized by a polydispersity index of 1.5 and a number average molecular weight of 58,000 g/mol, measured by means of gel permeation chromatography (GPC) [5], kindly provided by Kraton Polymers LLC, Houston, TX, was used in this work. Octadecylamine functionalized single wall carbon nanotubes, ODA-SWCNTs, (purchased from Aldrich, diameter 2-10 nm, length 0.5-2 μm, purity 80-90 %, were dispersed (1 mg/mL) in toluene (Aldrich). Dodecanethiol (DT, Aldrich) was used as surfactant due to its high solubility with the PS block of the SIS block copolymer, calculated on the basis of Van Krevelen theory [5-6]. DT was added (0.3 mL) to ODA-SWCNTs/toluene mixtures and sonicated for further 15 min. SIS was added both to the ODA-SWCNTs (ODA-SWCNTs/SIS) and DT modified ODA-SWCNT (DT modified ODA-SWCNTs/SIS) suspensions to yield composites with 1 wt% nanotube content. These mixtures have been sonicated for 90 min to obtain the final nanocomposites. AFM was performed in tapping-mode using a NanoScope IV, Veeco, Dimension 3100, equipped with an integrated silicon tip/cantilever having a resonance frequency ~ 300 kHz. Electric force microscopy measurements were performed in the lift-mode (lift height was 150-200 nm, chosen in order to insure a lack of a topography influence) in ambient conditions. EFM was equipped with an integrated Co/Cr coated magnetic force etched silicon probe tip having a resonance frequency around 75 kHz. The imaging has been obtained in tapping mode measuring the electric field gradient distribution above the sample surface by applying a voltage to the cantilever tip.

RESULTS AND DISCUSSION

First, the morphological features of both neat SIS block copolymer and its nanocomposite film were investigated using Tapping Mode AFM (TM-AFM). As report in our previous work [5-6] self-assembled nanostructure for both neat SIS (Fig. 1a) and DT-treated ODASWCNTs/SIS nanocomposites (Fig. 1b) have been obtained.

It is worth to note morphological changes in the self-assembling of the BC matrix, shifting its nanostructure from cylindrical structure oriented parallel to the free surface in the case of the neat matrix, to lamellar nanostructure when 1 wt % DT-treated ODASWCNTs has been added to the matrix. The interlamellar periodicity of about 20 nm is regular along the sample.

This morphological change can be explained taking into account a generic phase diagram of SIS block copolymer. In fact the good affinity between dodecanethiol and the polystyrene (PS) phase of the matrix, indicated by the solubility parameter calculated based on the Van Krevelen theory, induces an enlargement of the PS phase and a consequent shift from cylindrical to lamellar nanostructured morphologies [5-6]. Furthermore is important to note that, the nanotubes can adequate themselves when the lamellae present curved geometry.

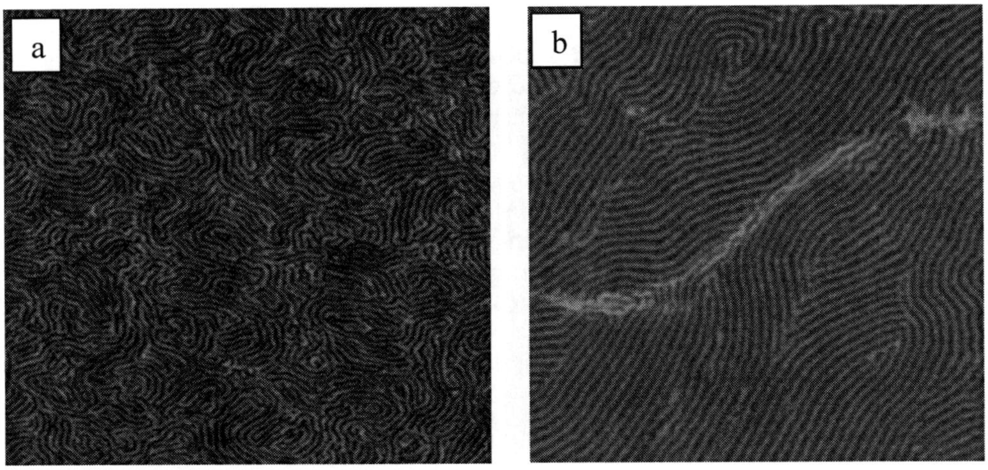

FIGURE 1. 2 x 2 μm TM-AFM phase (right) image of neat SIS (a) and DT-modified ODA-SWCNTs/SIS nanocomposite films.

The confinement of the DT-treatment ODASWCNTs into the PS lamellae is confirmed taking into account the increasing on the glass transition temperatures (Tg) of polystyrene, investigated by DSC and reported on table 1 [5].

TABLE 1. Thermal properties of the analyzed samples.

Sample	PS Tg (°C)	PI Tg (°C)
SIS	78	-58
DT-treated ODASWCNTs/SIS	90	-58

The increment on the Tg value of the PS is due to the interaction between the DT-treated ODA-SWCNTs with the PS phase. Whilst the addition of the ODA-SWCNTs does not affect the Tg of the PI phase. So, the increment of the PS Tg depends on the stiffening of the PS segments caused by the nanotube sequestering

In order to investigate the positioning of the nanotubes into the SIS matrix, the electrical behaviour of the composite films was investigated by EFM (Fig. 2).

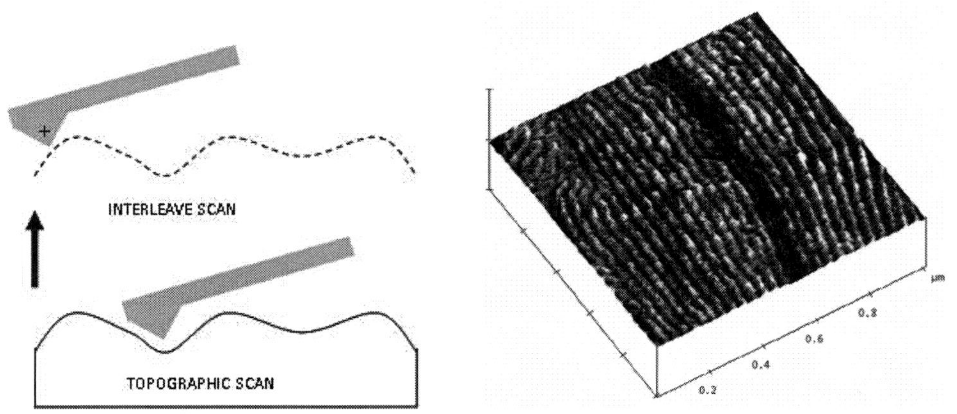

FIGURE 2. EFM lift-mode principle (left); 3D-AFM height image of 1 wt % DT-treated ODASWCNTs/SIS nanocomposite films (right).

This non-invasive technique has been used to qualitatively map the composite surfaces evidencing the contrast between the conductive ODA-SWCNTs and the insulating SIS matrix. As reference we can consider the 3D-AFM height image reporting in Fig. 2b. Figure 3 shows the response of DT-treated ODA-SWCNTs/SIS nanocomposite film when different biases have been applied to the sample surface. The conductive ODA-SWCNTs into the SIS matrix can be clearly distinguished and corresponded with their positioning in the film if the AFM image has been taken into account. This image confirms the confinement of the tubes inside the PS lamellae.

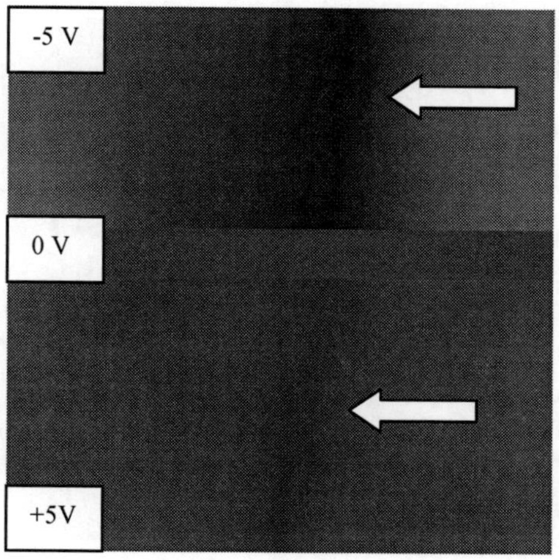

FIGURE 3. EFM phase images (1x1μm) obtained on the DT-modified ODA-SWCNTs/SIS sample varying the bias from negative to positive values passing thorough 0V. The arrows serve to the eyes to indicate the nanotube inside the PS lamella.

This confinement,is in very good agreement with the EFM phase image, where negative voltage applied to the EFM-tip lead to a better visualization of the tube located in the PS-block, as reported by us in previous work [6]. This fact can be explained taken the p-type nature of the carbon nanotubes into account. Since they are confined into the isolated polystyrene lamellae, they need of a negative voltage to construct an high contrast image.

CONCLUSIONS

The use of dodecanethiol as surfactant has promoted the confinement of functionalized carbon nanotubes in a hydrophobic SIS block copolymer matrix. Due to the high affinity of the DT with both the polystyrene block of the SIS matrix and the ODA-SWCNTs, a sequestering of the carbon nanotubes in the PS domains has been achieved. The use of EFM to investigate thin nanocomposite films, brings important new insights on the self-assembling behavior of block copolymer matrices as host for the nanotube positioning, and more specificallym for the imaging of the nanotubes inside the block copolymer domains.

ACKNOWLEDGMENTS

The author would like to acknowledge Prof. I. Mondragon and Dr. A. Tercjak for the AFM facility and for their helpful discussion and support during the AFM measurements.

REFERENCES

1. T. P. Lodge, *Macromol. Chem. Phys.*, **204**, 265-273 (2003).
2. M. Lazzari and M.A. López-Quintela, *Adv. Mater.*, **15**, 1583-1594 (2003).
3. I. W. Hamley, *Nanotechnology*, **14**, R39-R54 (2003).
4. C. Park, J. Yoon, E.L. Thomas, *Polymer,* **44**, 6725-6760 (2003).
5. L. Peponi, L. Valentini, L. Torre, I. Mondragon, J. M. Kenny, *Carbon*, **47**, 2474-2480 (2009).
6. L. Peponi, A. Tercjak, J. Gutierrez, M. Cardinali, I. Mondragon, L. Valentini, J. M. Kenny, *Carbon*, in press (2010).

Preparation and Characterization of Atactic Poly(vinyl alcohol)/Platinum Nanocomposites by Electrospinning

Won Seok Lyoo[a,*], Young Jae Lee[a], Jin Wook Cha[a], Min Jae Kim[a], Sang Woo Joo[b], Yeong Soon Gal[c], Tae Hwan Oh[a], and Sung Soo Han[a]

[a]*Division of Advanced Organic Materials, School of Textiles, Yeungnam University, Gyeongsan 712-749, Korea*
[b]*School of Mechanical Engineering, Yeungnam University, Gyongsan 712-749, Korea*
[c]*Polymer Chemistry Laboratory, College of Engineering, Kyungil University, Gyeongsan 712-701, Korea*

Abstract. Poly(vinyl alcohol) (PVA)/platinum composite nanofibers were successfully prepared by the electrospinning method. Water-based colloidal platinum in a PVA solution was directly mixed without any chemical or structural modifications into PVA polymer fibers to form organic-inorganic composite nanofibers. The PVA/platinum composite nanofibers were characterized by field emission scanning electron microscopy (SEM).

Keywords: PVA, platinum, electrospinning, nanocomposites
PACS: 81

INTRODUCTION

Polymer/inorganic nanoparticle composites have been the focus of extensive research efforts through the past decade. Metals with nano- structures, especially noble metal nanoparticles including platinum, palladium, gold and silver etc, are of considerable interest for their applications in catalysis, batteries, fuel cells, capacitors and sensors. Recently, they have been used vastly in the fields of electroanalytical chemistry and biosensor. The synergistic properties produced by the addition of inorganic nanoparticles include improved modulus and dimensional stability, lower thermal expansion coefficient and gas permeability, higher swelling resistance, and enhanced ionic conductivity, etc. When nanometer-sized particles are applied, these benefits can be achieved at very low loading levels as a result of their unusually large specific surface area.

Ultrafine fibers containing silver nanoparticles will exhibit much stronger antimicrobial activity than conventional microfibers due to their higher surface area-to-volume ratio. Due to the high surface-area-to-volume ratio of the electrospun fibers and the high porosity on the submicrometer length scale of the obtained nonwoven mat, proposed applications for these materials are in areas such as nanofiber-reinforced composites, nanofiber-based supports for enzymes and catalysts, and nanofibrous membranes to be used in many biomedical applications,22-25 including drug delivery, wound healing, and scaffolding for tissue engineering.

EXPERIMENTAL

Materials

Atactic PVA (a-PVA) with number-average degree of polymerization (P_n) of 1700 and degree of saponification (DS) of 99.9% was purchased from DC chemical Co., Ltd., Seoul, Korea. Colloidal nano platinum (Pt) was supplied from MIJI TECH Co., Ltd., Ansan, Korea. Initial concentration of colloidal Pt (APS-WM100®) was 1000 ppm. The diameter of Pt nanoparticles in the colloidal solution were at 50 nm.

CP1255, *V[th] International Conference on Times of Polymers (TOP) and Composites,*
edited by A. D'Amore, D. Acierno, and L. Grassia

Preparation of a-PVA/Pt nanocomposites

In order to prepare the PVA/Pt composite nanofibers, various concentration of a-PVA aqueous solution with small amounts of Pt colloidal solution (0.0, 0.1, 0.2, and 0.4 wt% to amount of PVA) were electrospun. PVA was dissolved in water at 90 °C for 2 h and maintained for 30 min to ensure homogenization. Concentration of PVA aqueous solution was varied from 9 to 13 wt%. To prepare homogeneous polymer solution containing Pt nanoparticles, Pt colloidal solution at 0.1, 0.2, and 0.4 wt% of PVA was mixed to the PVA solution at 60 °C for 30 min. A syringe pump was used to control constant mass flow of the composite solution during the electrospinning. All electrospinning procedures were carried out at room temperature. Applied voltage was ranged from 20 to 30 kV. Tip-to-collector distance (TCD) was varied from 9 to 15 cm.

RESULTS AND DISCUSSION

To prepare fine and even a-PVA/Pt nanocomposites, we performed a series of experiments varying processing parameters, including concentration of a-PVA solution with amount of Pt colloidal solution, applied voltage, TCD. The morphologies of the as-prepared electrospun a-PVA/Pt nanocompsites with different concentration of PVA (9, 11, and 13 wt%) and Pt (0.0, 0.1, 0.2, and 0.4 wt% of the amount of PVA) are shown in Figure 1.

The concentration of a-PVA was 9 wt%, all appeared to show relatively smooth surface morphologies in SEM images, it was also found that the 'bead-on-string' morphology was observed, a direct result of the change of viscosity of spinning solution by addition of Pt colloidal solution. As the amount of Pt colloidal solution increased, the average diameters of a-PVA/Pt nanocomposites were decreased.

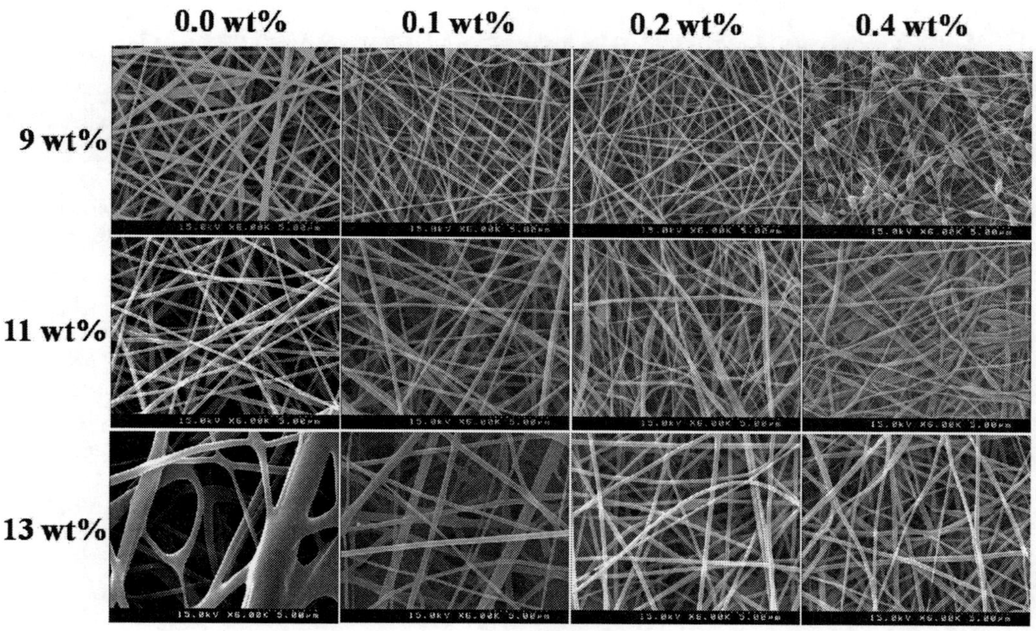

FIGURE 1. SEM photographs of the a-PVA/Pt nanocomposites with different concentration of a-PVA and Pt. The applied voltage and TCD were 25 kV and 12 cm, respectively.

FIGURE 2. SEM photographs of the a-PVA/Pt nanocomposites electrospun from 11 wt% PVA solution with 0.1 wt% of Pt colloidal solution and TCD of 12 cm at different applied voltages: a, 20 kV; b, 25 kV; c, 30 kV.

Figure 2 shows the effects of applied voltage on the morphological changes of electrospun nanocomposites with small amount Pt colloidal solution (0.1 wt% of the amount of a-PVA). With increasing the applied voltage, a-PVA/Pt nanocomposites became thinner. At higher voltages such as 30 kV, the fiber diameter was decreased due to the simultaneous stretching effect of charged fibers. During the travel of the fibers towards the collector, the higher voltage achieved the opportunity to stretch the fiber more and resulted in thin fibers.

FIGURE 3. SEM photographs of the a-PVA/Pt nanocomposites electrospun from 13 wt% PVA solution with 0.2 wt% of Pt colloidal solution and applied voltage of 20 kV at different TCDs: a, 9 cm; b, 12 cm; c, 15 cm.

Figure 3 shows the effects TCD on the morphological changes of electrospun nanocomposites with small amount Pt colloidal solution (0.2 wt% of the amount of a-PVA). The concentration of the base a-PVA solution was 13 wt% carried out a fixed applied voltage 20 kV. As increasing the TCD 9 to 12 cm, the average diameters of a-PVA/Pt nanocomposites were decreased. By the increasing the TCD, the evaporation time of solvent increased with an increase in the elongation time by path length increasing of the jet during its flight to the collector.

ACKNOWLEDGMENTS

This work was supported by grant No. RTI04-01-04 from the Regional Technology Innovation Program of the Ministry of Knowledge Economy (MKE).

REFERENCES

1. D. Okamoto, S. Morita, H. Taguchi, Y. H. Kim, T. Kotaka and H. Tateyama, *Polymer* **41**, 3887-3890 (2000).
2. J. Ramos, A. Millan and F. Palacio, *Polymer* **41**, 8461-8464 (2000).
3. M. Mukherjee, A. Datta and D. Chakravorty, *Appl. Phys. Lett.* **64**, 1159-1161 (1994).
4. X. Zong, K. S. Kim, D. Fang, S. Ran, B. S. Hsiao, B. Chun, *Polymer* **43**, 4403-4412 (2002).

Preparation of Syndiotactic Poly(vinyl alcohol)/Poly(vinyl pivalate/vinyl acetate) Microspheres with Radiopacity Using Suspension Copolymerization and Saponification

Won Seok Lyoo[a]* , Jin Wook Cha[a], Kun Young Kwak[a], Young Jae Lee[a], Han Yong Jeon[b], Yong Sik Chung[c], and Seok Kyun Noh[d].

[a]*Division of Advanced Organic Materials, School of Textiles, Yeungnam University, Gyeongsan, Korea*
[b]*Division of Nano-Systems Engineering, Inha University, Incheon, Korea*
[c]*Department of Textile Engineering, Chunbuk National University, Jeonju, Korea*
[d]*School of Display and Chemical Engineering, Yeungnam University, Gyeongsan, Korea*

Abstract. To prepare Poly(vinyl pivalate/vinyl acetate) [P(VPi/VAc)] microspheres with radiopacity, the suspension copolymerization approach in the presence of aqueous radiopaque nanoparticles was used. After, The P(VPi/VAc) microspheres with radiopacity were saponified in heterogeneous system, and then P(VPi/VAc) microspheres without aggregates were converted to s-PVA/P(VPi/VAc) microspheres of skin/core structure through the heterogeneous surface saponification. Radiopacity of microspheres was confirmed with Computed tomography (CT).

Keywords: P(VPi/VAc), copolymerization, saponification, radiopacity
PACS: 81

INTRODUCTION

Poly(vinyl alcohol) (PVA) obtained by the saponification of poly(vinyl ester) like poly(vinyl acetate) (PVAc) and poly(vinyl pivalate) (PVPi). There have been some reports on the preparation of monodisperse PVAc and PVPi as precursors for PVA through low temperature suspension polymerization of VAc and VPi. According to the study, to obtain polymer beads with a narrow particle size distribution (PSD), a stable steady state monomer droplets is necessary, and excessive coalescence should be excluded at the growth stage, which may be realized by lowering the polymerization temperature.

The general method for preparing s-PVA is the saponification reaction of P(VPi/VAc) in a solution state; a concentrated aqueous solution of sodium hydroxide is dropped into a completely dissolved P(VPi/VAc)/methanol solution. Some problems arise in its application to the embolization because of its irregular surface and broad PSD. It was reported P(VPi/VAc) microspheres were saponified in aqueous alkali solution contaninig sodium hydroxide, sodium sulfate, and methanol and converted to s-PVA from the surface resulting s-PVA/P(VPi/VAc) skin/core structure microspheres.23) This heterogeneous saponification restricted the reaction on the surface of P(VPi/VAc) microspheres dispersed in aqueous alkali solution and was feasible to preserve the spherical shapes of P(VPi/VAc).

EXPERIMENTAL

Materials

VPi and VAc (Shin-Etsu, Tokyo, Japan) were washed with an aqueous solution of NaHSO3 and water and dried over anhydrous CaCl2, followed by distillation in nitrogen atmosphere under reduced pressure. The initiator 2,2'-azobis(2,4-dimethylvaleronitrile) (ADMVN) (Wako Co., Osaka, Japan; 99%) was recrystallized twice from absolute

CP1255, *V[th] International Conference on Times of Polymers (TOP) and Composites,*
edited by A. D'Amore, D. Acierno, and L. Grassia

methanol before use. PVA with a number-average molecular weight of 127,000 and a degree of saponification (DS) of 88% (Aldrich Co., Milwaukee, WI) was used as a suspending agent. TiO_2, Au, Ag and Pt nanoparticles aqueous solutions were purchased from Miji Tech. Co., Ltd., Korean. Ti nanopowder was purchased from Aldrich Co..

Preparation of P(VPi/VAc)/nanoparticles microspheres

To prepare P(VPi/VAc) microspheres with radiopacity, the suspension copolymerization approach in the presence of aqueous radiopaque nanoparticles was used. Suspension agent was dissolved in water under a nitrogen atmosphere and with constant stirring, in a 500 ml three-neck round bottom flask fitted with a condenser. After degassing, VPi, VAc, nanoparticles aqueous solution or powder and ADMVN were added sequentially at a fixed polymerization temperature.

Heterogeneous surface saponification of the P(VPi/VAc)/nanoparticles microspheres

The P(VPi/VAc)/nanoparticles microspheres were slowly added to flask during stirring under a nitrogen atmosphere at a fixed saponification temperature. The flask equipped with a reflux condenser, a thermocouple, a dropping funnel, and a stirring device, alkali solution (sodium hydroxide/sodium sulfate/ethanol/water). After the reaction, the reaction mixture was poured into cold water and kept for 1 day to separate and sink s-PVA/P(VPi/VAc)/nanoparticles microspheres of skin/core structure.

Characterization

The particles size and PSD were measured with video microscopy (VMS) (ICAMSCOPE, SOMETECH, Seoul, Korea). Two hundred microspheres were chosen randomly, and the diameter of each microspheres was measured. Then, the standard deviation (δ) was calculated as follows:

$$\delta=[\sum(d_i-d)^2/(n-1)^{1/2}] \quad (1)$$

where n is the number of microspheres and and d_i are d the diameter of each microsphere and the average diameter, respectively. The dispersion coefficient (ε) was calculated as follows:

$$\varepsilon= \delta /d \quad (2)$$

For ε values less than 0.1, the microspheres can be considered to be monodisperse. The surface morphology of the microspheres was investigated with a scanning electron microscopy (SEM) after gold coating. The microsphere with a skin/core structure were obtained through the control of the saponification time. The skin/core structure of microspheres was confirmed with video microscopy (VMS). Dispersion of nanoparticles in microspheres was observed with transmission electron microscopy (TEM). Radiopacity of microspheres was confirmed with Computed tomography (CT).

RESULTS AND DISCUSSION

The P(VPi/VAc)/nanoparticles microspheres were saponified in heterogeneous system, and then the P(VPi/VAc)/ nanoparticles microspheres without aggregates were converted to s-PVA/P(VPi/VAc) microspheres of skin/core structure through the heterogeneous surface saponification. Figure 1 is VMS photographs of s-PVA/(VPi/VAc) microspheres with skin/core structure by saponification time. According to increasing of saponification time, skin later of s-PVA/(VPi/VAc)/nanoparticles microspheres was gradually thick.

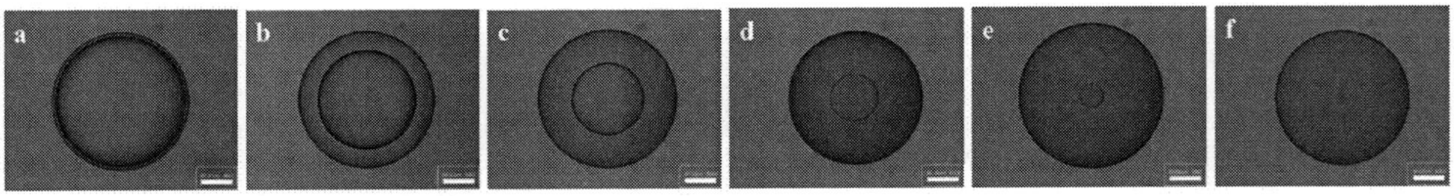

FIGURE 1. VMS photoraphs of s-PVA/(VPi/VAc) microspheres with skin/core structure by saponification time : a, 24 h; b, 48 h; c, 72 h; d, 96 h; e, 120 h; f, 144 h.

Figure 2 is TEM images of P(VPi/VAc)/nanoparticles microspheres with various nanoparticles. The TEM images provides direct evidence that nanoparticles embedded in the P(VPi/VAc)/nanoparticles microspheres. As shown in Figure 2 (a), although cohesion phenomenon happens a little between TiO_2 nanoparticles, TiO_2 nanoparticles of P(VPi/VAc)/TiO_2 microspheres inside were evenly distributed. As shown in Figure 2 (b, c), Au and Ti nanoparticles aggregated in P(VPi/VAc)/Au and P(VPi/VAc)/Ti microspheres inside. Au and Ti nanoparticles were not distributed evenly. Also, Pt nanoparticles aggregated in P(VPi/VAc)/Pt microspheres inside as in Figure 2 (d). But, Ag nanoparticles of P(VPi/VAc)/Ag microspheres inside were not aggregated.

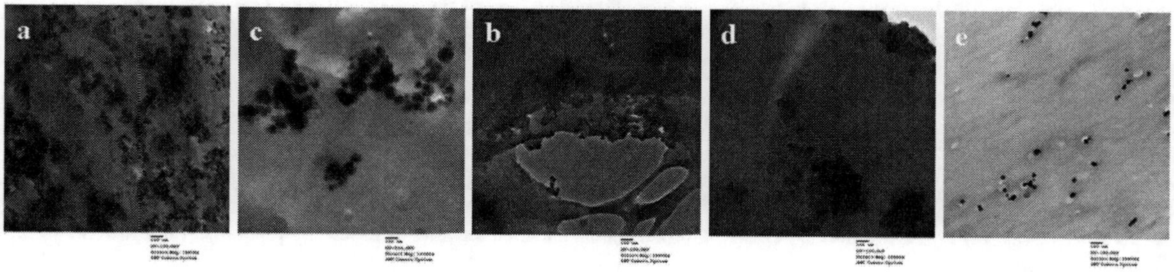

FIGURE 2. TEM images of P(VPi/VAc)/nanoparticles microspheres with various nanoparticles : a, homo; b, TiO_2; c, Au; d, Ti; e, Pt; f, Ag.

Radiopacity of the s-PVA/nanoparticles microspheres was confirmed by CT. Figure 3 is CT photograph of s-PVA/nanoparticles microspheres. The results, The CT photography showed that s-PVA/TiO_2 and s-PVA/Ti microspheres have excellent radiopacity.

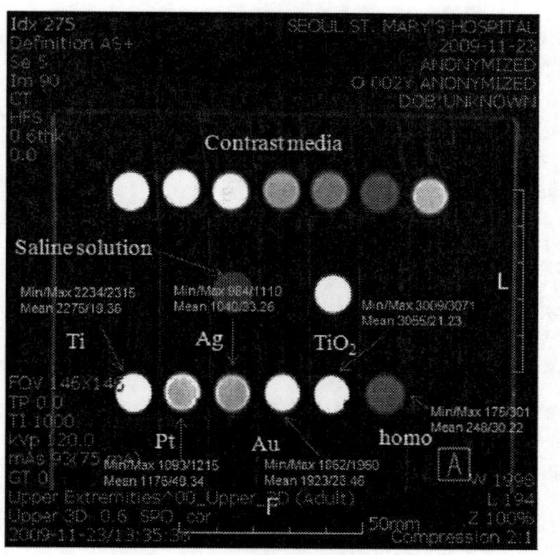

FIGURE 3. CT photograph of s-PVA/nanoparticles microspheres.

ACKNOWLEDGMENTS

This work was supported by grant No. RTI04-01-04 from the Regional Technology Innovation Program of the Ministry of Knowledge Economy (MKE).

REFERENCES

1. H. Masatake and D. Masakazu, *Appl. Catal. A-Gen.* **222**, 427-437 (2001).
2. W. S. Lyoo, J. W. Kwak, J. H. Yeum, B. C. Ji, C. J. Lee and S. K. Noh, *J. Polym. Sci. Pol. Chem.* **43**, 789-800 (2005).
3. Y. B. Kim, Y. A. Kim and K. S. Yoon, *Macromol. Rapid Comm.* **27**, 1247-1253 (2006).

Modeling VOC Sorption and Transport in Glassy Polymeric Membranes

Maria Grazia De Angelis, Luca Olivieri, G.C. Sarti

[a] *Dipartimento di Ingegneria Chimica, Mineraria e delle Tecnologie Ambientali*
Via Terracini 28, Università di Bologna, I-40131 Bologna

Abstract. In this work we evaluated the sorption, diffusion and permeation of a series of volatile organic compounds (VOCs) (acetone, n-butane, n-pentane, n-hexane, ethanol, methanol, chloroform and toluene) into glassy polymers of increasing fractional free volume (FFV): Polycarbonate (PC), Amorphous Teflon AF1600 and AF2400, poly-trimethylsilyl norbornene (PTMSN) and poly[1-(trimethylsilyl)-1-propyne] (PTMSP). Based on some experimental data of sorption and diffusion, and on theoretical and empirical models for the solubility and diffusion coefficients, the permeability for vapor/N_2 mixtures was evaluated. These parameters are useful for the membrane separation processes and for other applications such as chemical sensors. The ideal separation factors of glassy polymeric membranes versus mixtures of VOCs and N_2 were estimated at various pressures and compositions and at 25°C. The selectivity vs. permeability maps for the mixtures considered were plotted, showing that some of these materials show potentially the same selective ability of rubbery polymeric films. In particular it is shown that, the higher the FFV, the better the vapor/gas selectivity.

Keywords: Glassy polymers, membranes, VOC, vapor, transport, modeling.
PACS: 05.70.Ce, 64.70.km, 64.75.Bc, 66.10.cg, 66.30.hk, 66.30.je.

INTRODUCTION AND BACKGROUND

The VOCs emitted from chemical and petrochemical industries represent one of the largest sources of air pollution. Such emissions can be due to losses during the various industrial processes, but can be also present in waste process streams: clearly, the control of VOC emissions is a critical issue for chemical companies.

VOC control methods can be aimed at the reduction of VOC levels in the raw materials, or at the variation of the process operative conditions and equipment; when these measures are not applicable, abatement and recovery processes have to be considered. The recovery of VOCs can be carried on by condensation, absorption, adsorption and membrane processes.[1] In this work we focus on the latter method, and in particular on the predictive methods that can be adopted to estimate, the separation factor of polymeric membrane materials from a small set of experimental data and parameters. It has to be noticed, however, that the same calculations are valuable for the development of membrane-based chemical sensors for the detection of VOCs in the environment. In general, the sensor response is proportional to the sorption level of the vapor in the polymer membrane, and the response time decreases with increasing the diffusion coefficient of the vapor in the film.

In membrane processes the stream entering the membrane splits into a permeate and a retentate stream having different composition. In the removal of VOCs from gas streams one normally has the vapor diluted in an inert gas such as nitrogen: depending on the membrane properties, the permeate stream can be richer in the vapor or in the gaseous component. In the first case, which is normally the preferred one, the low pressure, vapor-enriched permeate stream can be taken to a condenser and the dilute retentate stream to an adsorption column. The current membrane materials for VOC removal are vapor-selective rubbery polymers such as polydimethylsyloxane (PDMS), that exhibit selectivity of the organic vapors over nitrogen varying from 10 to over 100 and are designed to treat gas streams with high loadings of vapors.[1] The rubbery polymers are generally preferred over glassy ones because they are more permeable to vapors than to nitrogen, while the majority of glassy polymers shows the opposite, i.e. size-selective, behavior. In this work we will explore a range of glassy polymers that potentially show vapor

CP1255, *V[th] International Conference on Times of Polymers (TOP) and Composites,*
edited by A. D'Amore, D. Acierno, and L. Grassia

selective behavior, in order to test their applicability in the removal of organic vapors from nitrogen, in a ideal type of binary separation where the vapor molar fraction ranges from 2 to 20%. The permeability P_i of each component i in a polymer, according to the solution-diffusion model, can be estimated as:

$$P_i = D_i \lim_{p_{i2} \to p_{i1}} \frac{C_{i2} - C_{i1}}{p_{i2} - p_{i1}} = D_i \frac{dC_i}{dp_i} \quad (1)$$

where D_i is the diffusion coefficient of the component in the polymer, $p_{i1}(C_{i1})$ and $p_{i2}(C_{i2})$ are the component partial pressures (concentrations) in the downstream and upstream side of the membrane, respectively. The above equation provides the value of permeability as a function of pressure, in the hypothesis of small transmembrane pressure. The selectivity of component i towards component j , $\alpha_{i,j}$, is then obtained for the membrane, in the hypothesis of negligible interactions between the penetrants, i.e. of "ideal" permeation, as:

$$\alpha_{i,j} = \frac{P_i}{P_j} = \alpha_D \cdot \alpha_S = \frac{D_i}{D_j} \cdot \frac{S_i}{S_j} \quad (2)$$

where it can be seen that the selectivity is the product of a diffusivity (α_D) and a solubility (α_S) contribution. In vapor/gas permeation one has normally that α_D <1 and α_S >1.

In this work we evaluate separately the diffusivity and solubility contributions. For the modeling of gas and vapor solubility into the glassy polymers we use the Non Equilibrium Lattice Fluid (NELF) model.[2] In this approach, the gas solubility can be estimated once the pure component parameters for the Lattice Fluid (LF) equation of state are known, which is the case of the present work. Other parameters needed by the model are: *i)* the initial pure polymer density ρ_P^0 at the experimental temperature; *ii)* the adjustable parameter representing the deviation from the regular solution behavior of the gas-polymer mixture, k_{ij}, which can be in first approximation fixed to zero or estimated from the literature; only for swelling penetrants and at moderate pressure, another parameter is required, namely the swelling parameter k_{sw} that is needed to estimate the polymer density, ρ_P, at each penetrant pressure p:

$$\rho_P(p) = \rho_P^0 \left(1 - k_{sw} p\right) \quad (1)$$

the value of k_{sw} can be taken from experimental swelling data or adjusted on experimental solubility data for the penetrant-polymer couple.[2]

The prediction of diffusion data of vapors in polymer is a more challenging problem than the solubility prediction and, in this work, diffusivity values are taken from the literature: in the absence of direct experimental information, the diffusion coefficient is assumed to be constant with concentration and equal to its infinite dilution value. The latter value can be estimated from empirical correlations that relate the infinite dilution diffusivity in a polymer to the critical volume of the diffusing gas or vapor. When experimental data are available, the diffusion coefficient can be obtained as an exponential or polynomial function of concentration.

RESULTS AND DISCUSSION

In Figure 1 we reported the results relative to the modeling of vapors sorption in Teflon AF2400 at 25°C. The experimental data for acetone, chloroform and toluene are taken from ref. 3 while the data of n-C_4 are taken from ref. 4 and those relative to n-C_5 and n-C_6 from ref. 5. The binary parameters used for the prediction are listed in Table 1 while the pure component LF parameters are available in the literature. It can be seen that the solubility of polar vapors decreases with increasing dipolar moment, and that of alkanes decreases with increasing molecular weight, as the model reasonably represents.

In Figure 2 we reported results of the permeability calculation, in the form of a selectivity/permeability plot for the case of acetone/nitrogen mixture, at 10% molar fraction of acetone and total pressure equal to 1 and 2 atm for the various glassy polymers considered. The case of PDMS, taken from the literature is also reported for reference.[6] It can be seen that the higher the fractional free volume of the polymer, the better the performance. Indeed the FFV increases the diffusivity selectivity, which is lower than 1 for vapor/gas separations, as well as the solubility-selectivity.

TABLE 1 Binary parameters for NELF model in AF2400.

Penetrant	Nitrogen	Chloroform	Toluene	Acetone	n-C_4	n-C_5	n-C_6
k_{ij}	0.015	0.11	0.13	0.13	0.14	0.14	0.14
k_{sw} (MPa^{-1})	0	1.2	4.5	1.1	0.4	0.9	4.5

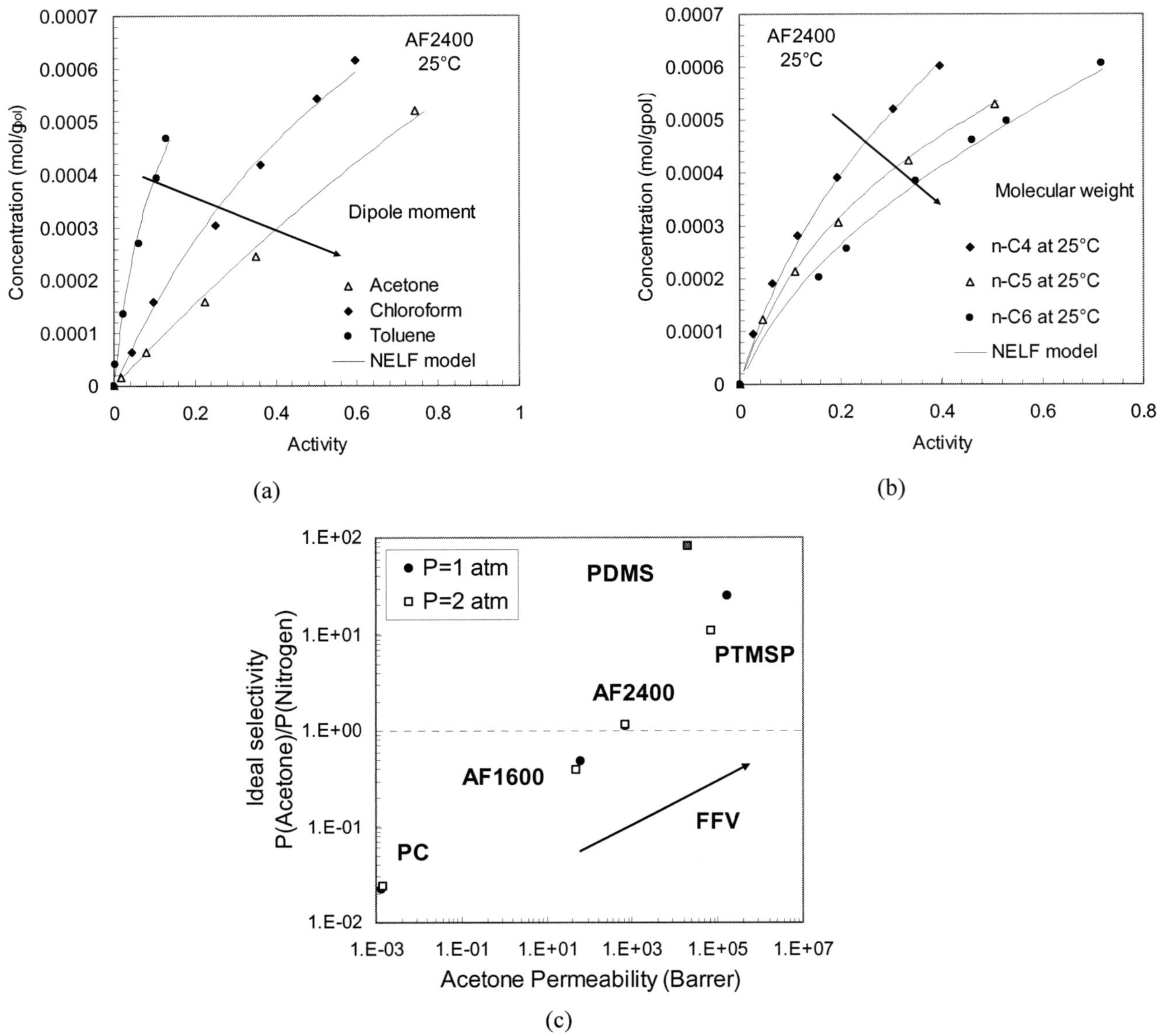

FIGURE 1. solubility of a)polar vapors and b)n-alkanes in AF2400 at 25°C. c) Ideal selectivity acetone/nitrogen at 25°C, versus acetone permeability, for each polymeric membrane, at different total pressure and with molar fraction of acetone equal to 10%.

REFERENCES

1. K. Ohlrogge and K. Stürken, "The separation of organic vapors from gas streams by means of membranes", in *Membrane Technology in the Chemical Industry*, edited by S.P. Nunes and K.-V. Peinemann, Wiley-VCH, Weinheim, 2001.
2. F. Doghieri and G.C. Sarti, *Macromolecules* **29,** 7885-7896 (1996).
3. V.V. Teplyakov et al. *J. Membrane Sci.* **220,** 165–175 (2003).
4. T. C. Merkel et al., *Macromolecules* **36**, 8406-8414 (2003).
5. M. Ghisellini "Gas Solubility in glassy polymers", Ph.D. Thesis, Università di Bologna, 2001.
6. A. Singh et al. *J. Polym. Sci.: Part B: Polym. Phys.* **36**, 289–301 (1998)

Structural relaxation of amorphous PET matrix nanocomposites

A. Greco[a], C. Esposito Corcione[a], A. Cavallo[a] A. Maffezzoli[a],

[a]Department of Engineering for Innovation
University of Salento, Italy

Abstract. The aim of the present paper is the evaluation of the degree of dispersion and intercalation of monodimensional nanofillers in an amorphous thermoplastic matrix. Organically modified montmorillonites (omMMT) were used as monodimensional nanofillers, whereas amorphous PET copolymer (PETg) was used as matrix.

Keywords: **PETg, montmorillonite, nanocomposites**

INTRODUCTION

Polymer nanocomposites are usually studied by means of different techniques, such as wide angle X - ray diffraction (WAXD) and transmission electron microscopy (TEM) the most commonly used. Although WAXD offers a convenient method to determine the average interlayer spacing of the silicate layers in an intercalated nanocomposites, little can be said about the spatial distribution of the silicate layers. On the other hand, TEM is very time-intensive, and only gives information on a very small portion of the sample [1].

In this work, PETg nanocomposites were produced by melt intercalation of omMMT at different temperatures. Nanocomposites were characterized by means of WAXD and TEM analysis, showing that XRD is only able to capture some of the basic features of the nanocomposite morphology. Other macroscopic techniques such as DSC analysis can provide additional data related to the matrix-nanofiller interaction, i.e. the relaxation times of the nanocomposite.

Materials

The materials used were:

a) Amorphous polyethylene terephthalate (PETg), Embrace Shrink Film from Eastman, with an intrinsic viscosity of 0.70 dl/g, and density of 1.30 g/cm^3.

b) Montmorillonite modified with 28 %w/w dimethyl benzohydrogenated tallow (omMMT), tradename Dellite 43 B, supplied by Laviosa, Italy.

The PETg nanocomposites were prepared by melt mixing of PETg with omMMT in a batch mixer (ThermoHaake) Rheomix 600 for 1h, at 150, 180 and 210°C. (samples nano150, nano180 and nano210 respectively) m The content of the nanofiller was held at 8% vol, roughly corresponding to 10 % by weight.

Results and discussion

The results from X-ray analysis are reported in Figure 1 for the nanofiller and the sample nano150 and nano 210. The diffraction peak of the omMMT reveals the presence of lamellar stacks, with an average lamellar spacing, which was determined by the evaluation of the diffraction peak, as d=1.9 nm, as reported in Table 1. After mixing with the PETg, the position of the diffraction peak moves at lower angles as a consequence of an increase of the lamellar spacing, d. As observed in Figure 1 and Table 1, X-ray analysis does not show any relevant difference between the samples processed at different temperatures.

The TEM micrograph of sample nano150, reported in Figure 2 shows the presence of a regular stacking arrangement of the clay layers. Lamellar stacks of MMT are well dispersed in the polymer matrix, being MMT

CP1255, *Vth International Conference on Times of Polymers (TOP) and Composites,*
edited by A. D'Amore, D. Acierno, and L. Grassia

layers still organized in a parallel way. This finding is in very good agreement with WAXD results, evidencing the formation of an intercalated nanocomposite. From a rough estimation of several TEM micrographs, an aspect ratio of lamellar stacks between 6 and 17 was obtained.

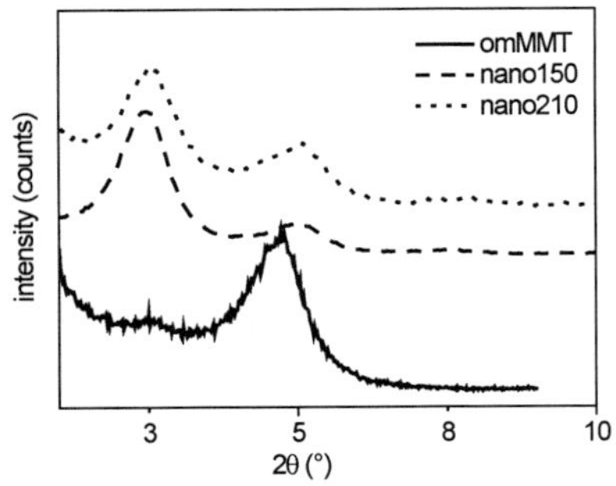

FIGURE 1: X-ray traces of omMMT and nanocomposites with PETg

Sample	2θ	D (nm)	Δh* (J/g)
PETg	4.7	1.9	513±64
Nano150	2.45	3.6	1327±109
Nano180	2.55	3.46	1132±177
Nano210	2.5	3.56	731±90

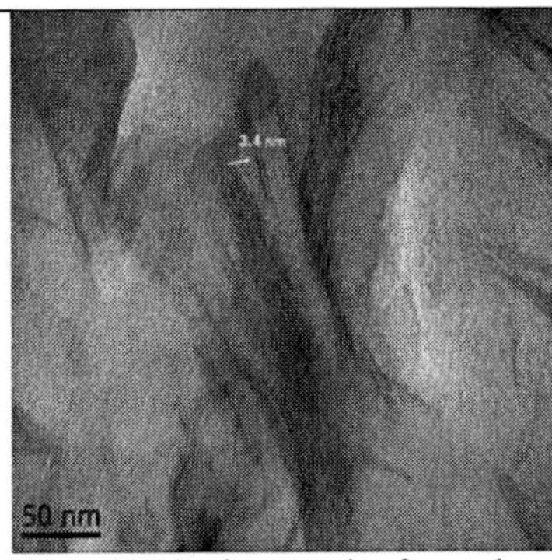

FIGURE 2: TEM micrograph of sample nano150

The DSC trace of PETg during heating are reported in Figure 3, showing a relaxation peak in correspondence of the glass transition temperature T_g [2].

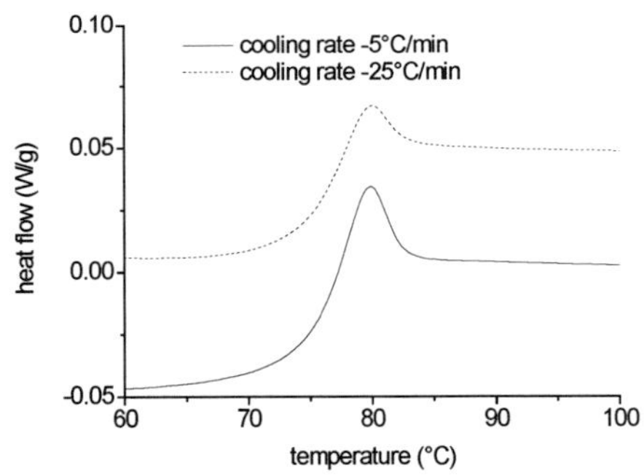

FIGURE 3: rheological curves

The peak is associated to structural relaxation phenomena occurring at temperatures lower than Tg. This phenomen can be characterized using an apparent activation energy, determined as [2]:

$$\frac{d(\ln|q|)}{d(1/T_f)} = -\frac{\Delta h}{R} \tag{1}$$

Where q is the cooling rate and T_f the limiting fictive temperature obtained by the method described by Moynihan et al. [2]. The values determined for the activation energy (Δh^*), reported in Table 1, show that PETg nanocomposites are characterized by a higher activation energy compared to neat matrix. Also, an increase of activation energy with decreasing processing temperature is observed. The activation energy Δh^* is related to the configurational energy barrier and reflects the cooperative nature of the relaxation process [3, 4]. Former studies on enthalpy relaxation of semicrystalline polyimides, indicated that Δh^* increased with decreasing size of the flexible linkages [5]. This is reasonable since a higher chain rigidity is associated with a higher configurational activation threshold. Similarly, in thermosetting resins, the increase in Δh^* with crosslink density is interpreted considering that crosslinks promote restrictive effects on the segmental relaxation. In the case of nanocomposites, as the content omMMT is increased the number of segments anchored/ attached on the surface of silicate layers is also increased. Several flexible chain segments and rigid omMMT stacks can be considered part of a single large branched cooperative unit that needs a higher energy barrier to reach the activated state and to accomplish the corresponding configurational rearrangements [6].Therefore, by decreasing the processing temperature, the increase of the activation energy can be attributed to the higher level of dispersion of the nanofiller, which yields an increase of the interphase area between matrix and nanofiller, which finally causes an increase of the cooperative unit dimensions.
A measure of the relaxation behavior of these nanocomposites can be also obtained by Dynamic Mechanical Analysis (DMA). The G' and G" curves reported in Figure 4 were fitted using the Havriliak-Negami [7] model:

$$G = G_\infty + \frac{\Delta G}{\left[1+(i\omega\tau)^\alpha\right]^\gamma} \qquad (2)$$

By using the Cole-Cole plot, the parameters α, γ, G_∞ and G_0 were determined. The value of the relaxation time, τ, was determined by inversion of eq. 2). The energy of activation for the relaxation time was determined by an Arrhenius fitting of the relaxation times data. The results obtained are in agreement with DSC analysis, showing an increase of the activation energy with decreasing processing temperature.

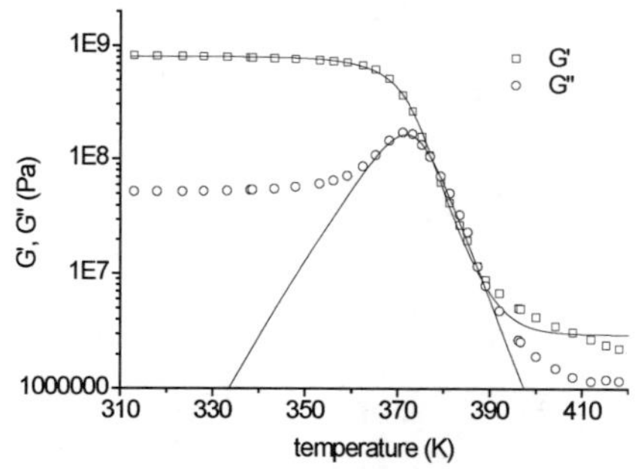

FIGURE 4: Glass transition temperatures of the samples

CONCLUSIONS

In this work, the effect of the addition of the nanofiller on the relaxation times of amorphous PET copolymer was studied by DSC and DMA analysis. Results obtained indicate that addition of the nanofiller involves an increase of the relaxation times, which is in accordance with an increase of the size of cooperative units made of rigid omMMT lamellae and flexible PETg chain segments. The increase of the size of the cooperative units can in turn be attributed to an increase of the interphase area between matrix and nanofiller, which indicates a better level of dispersion of the nanofiller.

References

1 S.S. Ray, M. Okamoto, *Prog Polym Sci* **28**,2003; pp.1539-1641
2 Moynihan CT, Lee SK, Tatsumisago M, Minami T, Thermochimica Acta, 1996; 280-281: 153-162
3 Karasz FE, MacKnight WJ, Macromolecules 1968, 1, 537
4 Solunov CA, Eur Polym J 1999; 35: 1543-1556
5 Cheng SZD, Heberer DP, Janimak JJ, Lien SHS, Harris FW, Polymer 1991; 32: 2053.
6 Lu H, Nutt S, Macromolecules, 2003, 36 (11), 4010-4016
7 Alig I, Tadjbakhsch S, Zosel A, Journal of Polymer Science: Part B: Polymer Physics, 1998; 36: 1703–1711

Torsional Stiffness Verification of an Adhesively Bonded Joint

A. Annicchiarico[1], F. Caputo[2], G. De Angelis[3,4], F. Frascà[3,4], G. Lamanna[2]

[1] *C.R.F. S.C.p.A*
Strada Torino, 50 – 10043 Orbassano (TO), Italy
[2] *Second University of Naples,*
Department of Aerospace and Mechanical Engineering
via Roma, 29 – 81031 Aversa (CE), Italy
[3] *Elasis S.C.p.A.*
via ex-aeroporto sn – 80038 Pomigliano d'Arco (NA), Italy
[4] *IMAST S.c.a.r.l.*
P.le E. Fermi, 1 Loc. Granatello – 80055 Portici (NA) Italy

Abstract. In the present work numerical-experimental analysis for the characterization of a structural adhesive has been performed. The numerical analysis has been carried out through the finite element method by using, for the phases pre / post processing were used commercial programs while for the phase of numerical solution the Abaqus code was used. The experimental analyses were carried out at laboratories of C.R.F. S.C.p.A. by using of a standard quasi static testing machine. Later numerical analysis was performed comparing the torsional stiffness of a vehicle in which the welded connection between the pavilion and the flank has been substituted by bonded one. This comparison has allowed to demonstrate the ability of the bonded joint discussed to provide mechanical performances comparable with those of a welded joint widely used in the automotive industry.

Keywords: Torsional vehicle performances, adhesively bonded joint, epoxy, single lap joint.
PACS: 45.05.+x, 45.10.-b, 45.20.da, 02.70.Dh.

INTRODUCTION

The potential advantages of adhesives over mechanical fasteners have long been recognized; however, industrial acceptance of adhesives for structural applications has been slow. The main reason for this lack of use is that adequate information about the characteristics of adhesives was not available until recently. Development studies by the aerospace industry greatly improved the technical database on structural adhesives and now many industries are convinced that the adhesives can provide bonds that are both cost effective and durable. Interesting methods were proposed by Beevers at al. [1] and Steidler at al. [2].

Structural adhesive bonds replaced both mechanical fasteners and assembly techniques such as soldering and welding in many industrial applications. The advantages of structural adhesive bonding over other joining techniques include: cost savings (including lower labour costs), weight reduction, elimination of stress point concentrations by even distribution of stress over the entire bonded surface, improved load bearing capacity, protective sealing against contamination by liquids or gases, bonding of dissimilar materials (often the adhesive bond line acts as an insulator against galvanic corrosion in metal assemblies), improved fatigue resistance and resistance to vibration and thermal cycling, enhanced structural appearance because protrusions and attachments are eliminated, increased tolerances on machined parts because some structural adhesives can fill gaps left when parts mate poorly.

CP1255, *V^th^ International Conference on Times of Polymers (TOP) and Composites,*
edited by A. D'Amore, D. Acierno, and L. Grassia

ADHESIVE CHARACTERIZATION

Epoxies are by far the most widely used structural adhesives. Adhesives, base resins, curing agents, fillers, and various additives can be formulated to provide high strength bonds to steels, ceramics, glass, wood, thermoplastics, fiber-reinforced composites, rubber and plastic foam.

High tensile strength, up to 70 MPa for reinforced epoxies, makes epoxy bonds especially resistant to impact, vibration and shock. The service temperature range of these adhesives extends from about -70°F to more than 450°F.

The experimental study of the mechanical properties of the adhesive is required for simulating the right behaviour of the component (realized with two steel layers assembled with the adhesive) subjected to thermal cycles.

Depending on the thermal cycle, it is necessary to take into account the deformations of both steel layers (thermal expansion) and adhesive (compression or elongation) in order to forecast, with accuracy, deformations of the front edge of the hood and/or eventual dimensional defects.

Therefore a set of experimental tests has been carried out in order to evaluate the mechanical properties of the adhesive during thermal cycles. A set of linear joints has been realized and tested with traction tests in different conditions of temperature during the cooling phase, within a climatic chamber, in order to simulate the cataphoresis process (figure 1).

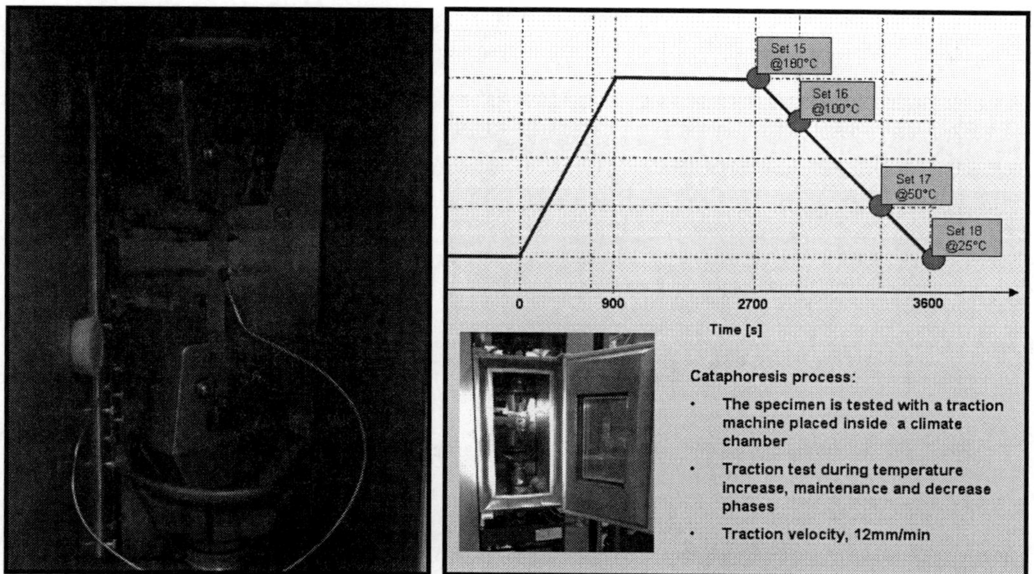

FIGURE 1. Traction test within a climatic chamber (a). Thermal cycle (b).

The specimen size has reported in figure 2. The experimental test provides for an accurate cleaning stage of the two metal sheets; afterwards they are joined by the structural adhesive (0.45 mm tick). A longitudinal quasi static load was applied till reaching the limit load.

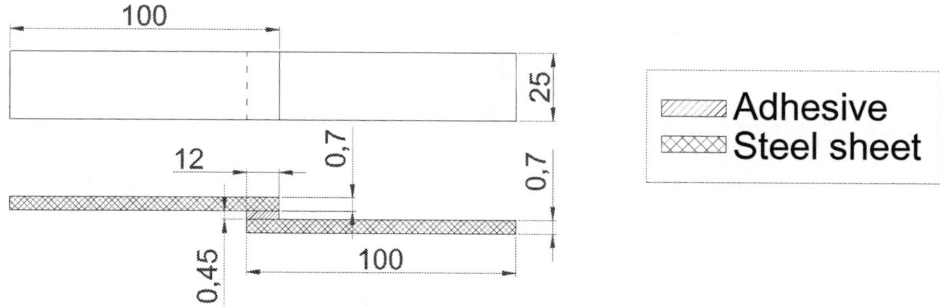

FIGURE 2. Specimen (measures in mm).

A numerical model that represents a single lap joint was built. Two simple specimens with the same attitude were joined by a structural adhesive [3]. The main model's components are: upper specimen, lower specimen and adhesive. In particular: the upper specimen is a planar sheet 25.0 mm wide and 100.0 mm long made of mild steel, it was meshed by shell elements with 0.7 mm thickness; the lower specimen has the same geometric characteristics of the upper one, 1.15 mm from it, an overlap of 12.0 mm exists between specimens (see figure 1); the adhesive is a 12.0 mm x 25.0mm x 0.45 mm prism interposed between the two specimens, 0.35 mm from their reference planes. The sheets were modeled by 3D four nodes shell elements with five degrees of freedom (while the adhesive has been modeled by 3D eight nodes hexahedral elements with four degrees of freedom at each node respectively S4R5 and C3D8T of Abaqus elements library). Two different non linear material models temperature dependent have been used: one for the two steel specimens and the another one for the adhesive component. In order to complete the interactions among the components, a couple of tie contacts was characterized [4]; through this approach there is not initial sliding between adhesive and specimens. The whole numerical model consists of about 16'000 elements and 20'000 nodes and it has a very short time running. The main goal of this section is the characterization of numerical material model associated with adhesive component starting from a standard structural adhesive [5].

The numerical tests have been made at different temperature values: 25°C; 50°C; 85°C; 100°C; 180°C. A calibration process of numerical curves has been carried out for each temperatures by varying the characteristic parameters of the adopted material model. In particular, the value of yield strength and plastic properties of the characteristic curve of material have been differentiated.

The results of such material characterization are shown in figure 3a where the good numerical-experimental correlation for all considered values of temperature must be highlighted. Figure 3b shows the final characteristic curves for the adhesive component at different reference temperatures.

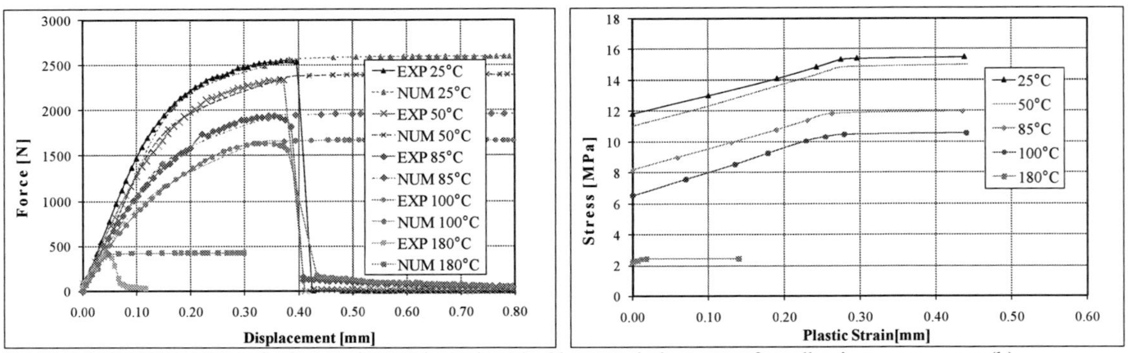

FIGURE 3. Numerical-experimental results (a). Characteristic curves for adhesive component (b).

TORSIONAL STIFFNESS VERIFICATION

The adhesive bonding technique has been applied successfully in many engineering areas. Due to some advantages over conventional methods, the adhesively bonded joints have been studied by many researchers by means of the classical analytical method and the finite element method.

In the automotive field several research projects aim to modify mechanical fasteners in order to replace them with adhesively bonded joints. In this work it has been investigated the coupling between two components of a commercial vehicle as shown on figure 4.

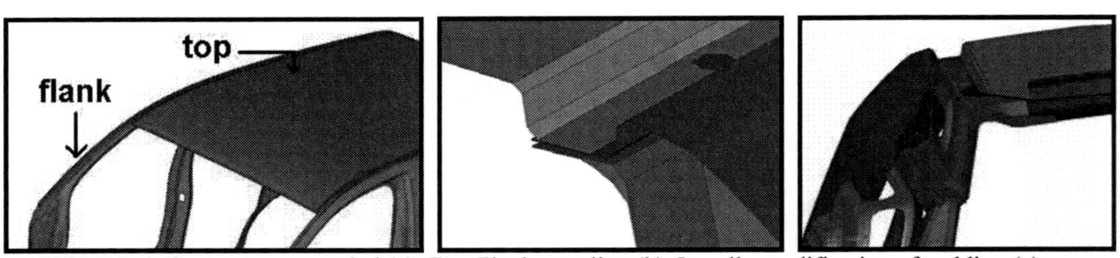

FIGURE 4. Components coupled (a). Top-Flank coupling (b). Locally modification of welding (c).

Actually the welding of the two described parts is one of the last operations during the bodywork assembly. The assembly of the top (figure 4a) is a problem to be considered during the design of many components which are assembled upstream of such top. The pavilion is welded on the flank and on the front and rear crossbars (figure 4b).

There are several technological constrains, connected to the welding process, that restrict the pavilion design. First of all the welding points must be invisible, this is the reason why a small canal must be manufactured to hide them by moulding.

It has to be taken into account the overall dimensions of welding pliers and at last manufacturing experiences says it is convenient to weld at most three sheets but often there are more plies to join, therefore sometimes, some sheets are left out from the welding process (figure 4c).

Replacing this kind of welded joint with an adhesively bonded joint would allow both technological and aesthetics advantages. The design of many components, which usually must be mounted before the pavilion, would be freed by its assemblage and mouldings are no longer necessary.

Numerical analyses on two different top-flank couplings were performed. The torsional stiffness was computed on the following models: whole bodywork of a commercial vehicle with welded top-flank coupling and with adhesively bonded top-flank coupling. The first numerical model is a standard one, widely used in the vehicle's virtual analyses. The second model derives from the first one but without the welded points, which were replaced by an adhesive strip modeled through solid elements (figure 5).

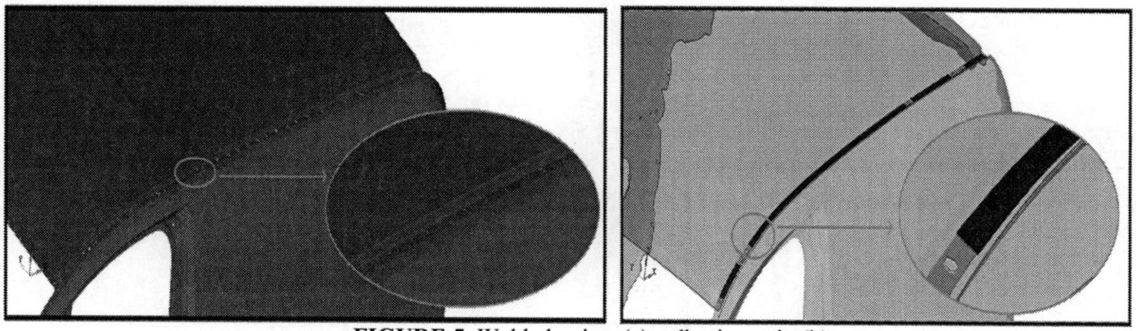

FIGURE 5. Welded points (a); adhesive strip (b).

The numerical models have been subjected to a torsional virtual test in order to get the corresponding torsional stiffness; such parameter has been used to compare the performances of the two joint techniques. In the following table, it has been reported the global torsional stiffness of vehicles.

TABLE a). Torsional stiffness comparison.

Type of joint	Torsional Stiffness k_t	Δk_t
Welded model	726'206 N m/rad	-
Bonded model	728'059 N m/rad	+ 0.25 %

There are no significant differences between the obtained results, therefore there are a lot advantages when adopting an adhesively bonding technique in a top-flank coupling without any loss of performances in terms of global torsional stiffness of the discussed vehicle.

ACKNOWLEDGEMENTS

This work has been performed within the Research Program ASAP (Adesivi compositi Strutturali per Applicazioni nel settore dei trasPorti) founded by the Italian Ministry of University and Research to IMAST (Technological District on composites and Polymeric Materials and Structures).

REFERENCES

1. A. Beevers, S.M. Steidler, J. Duridola, M. Coackley, *J Mater Process Technol* **118**, 96-101 (2001).
2. S.M. Steidler, N. Bonde and H. Ljungquist, *SAE Technical Paper Series* 2003-01-0332.
3. X.R. Xiao, P.H. Foss and J.A. Schroeder, *Int J Adhes Adhes* **24**, 239-246 (2004).
4 ABAQUS Theory Manual, Version 6.6.
5 Product Data Sheet, *4601 Bulk Bulletin*, Dow Chemical, 2001.

Toughening the Macro Defect Free (MDF) Cements

Alberto D'Amore and Luigi Grassia

The Second University of Naples-SUN, Department of Aerospace and Mechanical Engineering, Via Roma 19, 81031 Aversa (CE), ITALY

Abstract. High performance polymer based-cements (i.e. cements having a water to cement ratio, w/a , close to 0.1) almost totally free of defect, have been developed in recent years claiming a great interest for their very high tensile strength compared with traditional cements. These materials can be produced utilizing very simple unit operations and further formed in complex shapes with high surface quality. Indeed these materials have the disadvantage of being too much brittle and for this main reason they are rarely utilized in practice. Toughening the Macro Defect Free Cements (MDF) implies the strict control of the structure. The structure is governed by the process-induced paste modifications. In this paper we show that toughening can be achieved with elastomeric particles having well defined size distribution and physico-chemical characteristics in relation with both the binder polymer and the cement chemistry.

Keywords: Macro Defect Free Cements, Thoughening, Mechanical testing, elastomers.
PACS: 81,82,83

INTRODUCTION

The calcium aluminate cement (CAC) was selected as the best candidate to obtain a MDF (Macro Defect Free) polymer-based mixture exhibiting the highest properties (i.e. the flexural strength or order of 200 MPa) (1,2,3). Polyvinyl alchool/acetate (PVA, 80% hydrolyzed) with a relatively low molecular weight (Mn=25.000 Dalton) was chosen as the hydro-soluble polymer. In all the cement paste included the following components: CAC, PVA, water and glycerol with a percent composition of 84, 6, 9.4 and 0.6 respectively. Starting from the grossly mixed powder a stable and formable paste can be obtained with high shear stress mixing operations that allow a mechano-chemical induced-cement/PVA network formation, as illustrated schematically in Figure 1 (1,2).

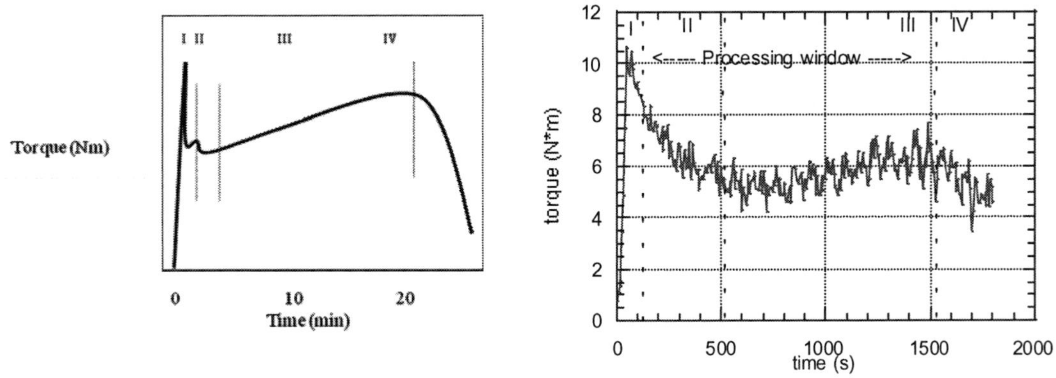

FIGURE 1. Schematic description (left) and real mixing behavior of MDF cements obtained on a Haake Rheocord System 90 torque rheometer (right). I: Transformation of powder to a fluid paste; II: mechano-chemical degradation of PVA;III: particle-polymer network formation; IV network degradation.

Mixing at high shear stress favors a stable paste formation and is followed by shape forming, (for example, hot stamping at 80-100°C and 10 Bar) where the remaining hydration reactions (that occur just at the particle surface,

CP1255, *Vth International Conference on Times of Polymers (TOP) and Composites,*
edited by A. D'Amore, D. Acierno, and L. Grassia

since the low water/cement, w/C, ratio) and the defects minimization allow the consolidation of parts with high strength (in the range 100-200 MPa, depending on the pressure). However the CACs so obtained, having an MDF character, have the disadvantage of being too much brittle and for this main reason they are rarely utilized in practice. Realizing MDF cement with tunable toughness implies the modification of both the binder polymer and its interaction with the cement particles. However, toughening a polymer based high performance cement with elastomers implies the strict control of the structure, including the particle-particle distance, the elastomeric particles being selected having well defined physico-chemical characteristics in relation with both the binder polymer and the cement chemistry. The present paper provides a process for the production of a toughened high performance CAC based material, as mentioned, which exhibits higher toughness compared to previously known standard MDF cements.

TOUGHENING

Among the approaches for toughening brittle (homogeneous) materials, using dispersed toughener phase(s) has been shown to be the most cost effective and can provide substantial toughness improvement. As regards the cement based materials, the toughening process become more complex since the multiphase structure of key material. The whole structure can be regarded as a highly concentrated suspension of un-reacted cement particles embedded in a two-phases (PVA and the elastomeric particles) matrix. The toughening process involves a modification of failure mechanisms of whole material. Rupture of brittle materials consists essentially of sequential crack nucleation and propagation. Toughener phases (i.e. soft particles of a given size and volume fraction) allow the three-dimensional stress state at the crack tip to be released. This condition , however, is fulfilled when a critical particle distance if achieved. The critical particle distance is a function of both the total amount of toughener and the particles size. As a general rule the transition from brittle to tough behavior is a function of both the volume fraction and the size distribution of elastomeric particles. In Figure 2 (left) the interaction of PVA with CAC is illustrated. It was shown (1-8) that the PVA does react with aluminum ions that actually serve as crosslinking agents of the polymer phase. Crosslinking with Ca^{++} ions is less favored thermodynamically (9,10) in respect to aluminum, due to the poly-condensation reactions between PVA e $Al(OH)^{4-}$ that occur in environments at high pH, actually prevailing during the MDF processing (pH= 11.5). In Figure 2 the chemistry of SBR latex (center) and its reaction with CAC (right) is also illustrated.

MATERIALS AND TESTING METHODS

Toughened samples were prepared by using different amount of xSBR latex in a way that the total polymer amount (PVA+SBR) of the entire mixture remained 6% of the total weight. A series of sample were prepared at laboratory scale by using an Hobart mixer for the gross blending of cement powder, water, PVA and glycerol. High shear stress mixing was operated in a double calender were the mixture becomes a stable paste that was then poured in a mould and, with a laboratory press (operating at T=80-100°C and P=10 Bars), plates of consolidated materials of given thickness (1-2 mm depending on the mould) were obtained. Similar unit operations were used at industrial scale that allowed obtaining plates with higher thickness (10-20 mm). Measurements of the thoughening effectiveness were performed on both kind of samples as discussed below. Generally speaking, toughness is a measure of material's resistance to failure. Depending on the application it is usually measured as either the stress or the energy required to fail a specimen under specific loading conditions. More specifically, toughness can be defined as either I) the area under the tensile stress-strain curve, II) the Izod inpact strenght, III the Charpy impact strength, IV) the plain strain critical stress intensity factor (K_{IC}) or V) the plain strain critical energy release rate (G_{IC}). However when the applications are of safety concerns, the worst case scenario toughness definition, that is, either K_{IC} or G_{IC} (or J_{IC}, if the material is extremely ductile), should be considered. Laboratory scale samples were tested by using the compact tension (CT) loading geometry for the sake of evaluating the fracture toughness, K_{IC}. The results in Figure 3 are reported in terms of Weibull cumulative distribution function of K_{IC} (left) as function of the volume fraction of the elastomer with respect to the total polymer amount, together with specimen dimensions and geometry (right). The Charpy impact strength was evaluated on specimens obtained at industrial scale according to ISO 179 by using an instrumented CEAST Impact Tester. The results in terms of energy required to break a test specimen under flexural impact showed a substantial increase of the averaged impact energy, IEav, with increasing the elastomer content passing from IEav=2.64 kJ/m^2 (std.dev=0.46) for standard MDF to EIav=3.93 kJ/m^2 (std.dev.=0.70) for 30% elastomer modified-MDF cement. The results on both kind of specimens are substantially coherent also in respect to the fact that in both cases the reliability decreases with increasing the elastomer content as clearly shown by both the broadness of the Weibull distribution and the amplitude of the standard deviation.

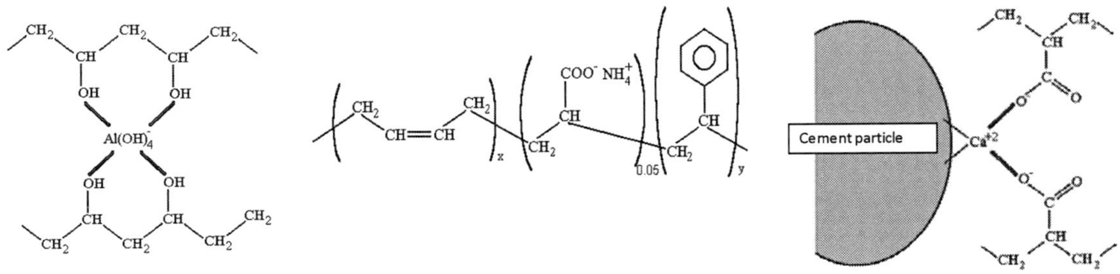

FIGURE 2. Schematic of PVA-CAC reaction (left), chemistry of XSBR (center) and the interaction of the elastomer with the cement particles.

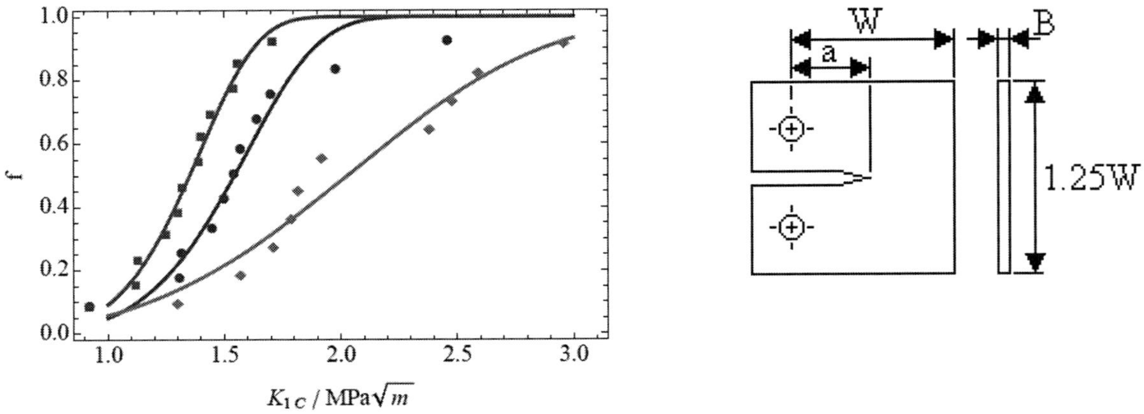

FIGURE 3. The Weibull cumulative distribution function of K_{IC} (left) as function of the volume fraction of the elastomer from standard MDF (squared symbol) to 10% (circle) and 30% elastomer with the specimen geometry (right).

CONCLUSIONS

Substantial toughness improvement of MDF cements has been achieved with elastomeric modification of the binder polymer. The process required a strict control of the size distribution of elastomeric particles in respect to binder polymer domain, the elastomeric inter-particle distance being the key parameter in the toughening mechanisms.

REFERENCES

1. J. D Birchall, *Philos. Trans. R. Soc. London, **A 310**, 31-42 (1983)*
2. J.A. Lewis, M.Boyer, D.P. Bentz, *J.Am. Ceram. Soc.*, 77(3)711-16 (1994)
3. A.J. McHugh, L.S. Tan, , Adv. Cem. Bas. Mat.,1, 2-11 (1993)
4. A. J. McHugh e L. S Tan, Journal of Molecular Science, 31, 3701-3706 (1996).
5. A. J. McHugh e L. S. Tan, Journal mater. Res., 11 (7), 1739-1747 (1996).
6. C. A. Finch, “Polyvinylalcohol: Properties and applications ”, Ed. Wiley & Sons (1973).
7. R. Alfani, P. Colombet, A. D’Amore, N. Rizzo, L. Nicolais J. Materials Science 34 (1999) 5683-5687
8. L. Sartore, M. Penco, F. Bignotti, I. Peroni, M.H. Gil, M.A. Ramos and A. D'Amore, J. Appl. Polym. Sci. 85, 1287-1296 (2002).
9. C.S. Poohr and G.W. Groves, J. Mater. Sci, 23, 657 (1988)
10. S.A. Rodber, S.A.Brooks, W. Sinclair, G.W. Groves and D.U. Double, J. Mater. Sci., 20 , 2853 (1985).

The Anomalies of Glassy Polymers at Nanoscale from the Continuum Mechanics Perspective

Alberto D'Amore and Luigi Grassia

The Second University of Naples-SUN, Department of Aerospace and Mechanical Engineering, Via Roma 19, 81031 Aversa (CE), ITALY

Abstract. Glassy polymers confined at nano-scale do exhibit anomalous thermo-viscoelastic behavior in respect to their bulk response. For example, in ultra-thin polymer films both negative coefficient of thermal expansion (NCTE) and glass transition temperature (Tg) depression are observed, experimentally. Theoretical conjectures have been drawn that qualitatively explain some features of highly confined glassy polymers [1,2,3] yet quantitative prediction of NCTE and Tg depression would allow a reliable design of devices operating at nano-scale. In this paper the thin films anomalies are shortly reviewed and their explanation is tempted with classical existing theories.

Keywords: Ultra-thin polymer films, glassy state, Negative coefficient of thermal expansion, anisotropic viscoelasticity.
PACS: 81,82

INTRODUCTION

Many relevant technological systems are operating at very small dimensional scale and this motivated the need to study the behavior of thin polymer films. In fact, polymeric thin films are more reliable in terms of homogeneity with respect to thin films of "small" inorganic molecules, their "entanglements" providing defect-free structures at nano-scale. However, highly confined macromolecules do exhibit two substantial thermo-viscoelastic anomalies. Supported ultra-thin polymer films show an extraordinarily large negative coefficient of thermal expansion, NCTE, (in the glassy state a temperature increase of about 50°C gives rise to a film "contraction" of about 10%!) [4,5,6]. Again, ones the film is formed (for example by spin coating) spectacular Tg's depressions (50-80°C) are observed [7-13]. Such simultaneous anomalies are rather complex to manage at the design stage of a device operating at nanoscale as the glass transition temperature, Tg, represents the maximum "service" temperature while the negative coefficient of thermal expansion NCTE gives rise to at least unusual thermo-viscoelastic responses. In practice, the thin film's anomalies are observed when the film thickness is smaller than the nominal dimension of the random coil and this circumstance defines the "ultra-thin" polymer films. Concerning the Tg depression, a strict correlation with the molecular weight was widely assessed, experimentally, yet the existence of a NCTE was reported only in a few papers (4,5,6). The phenomenological complexity is further enhanced by the fact that ultra-thin films are required in their glassy state, "the "Glassy State" being one of the open chapters of Materials Science. The response complexity of glassy *isotropic* materials subjected to arbitrary thermal/mechanical histories has been recently substantiated [14,15,16]. Instead, thin films are viscoelastic, *anisotropic structures* as, during the film formation, macromolecules undergo "self-squeezing" and result bi-axially stretched. Furthermore, during the film formation tensile stresses develop that remain frozen-in in way that the glass remains somewhat swollen (roughly 10% with respect to the specific volume of the bulk polymer), giving rise to Tg's depression (1,2,3). In what follow we try to know if looking at existing phenomenological theories of the glassy state and to the mechanics of anisotropic materials, the films anomalies can be somewhat interpreted. To do that we start assuming simple geometrical rules. In particular if we assume that macromolecules with nominal diameter, R_{ee}, are confined to thin films with thickness $h = \frac{1}{2} R_{ee}$

CP1255, *V*th *International Conference on Times of Polymers (TOP) and Composites,*
edited by A. D'Amore, D. Acierno, and L. Grassia

then , at constant volume they pass from a statistical spherical symmetry ($V = V_{ee} \approx R_{ee}^3$) to an ellipsoid-like configuration ($V_{film} \approx ha^2 \approx V_{ee}$). Thus the film undergoes bi-axial stretching ($a = R_{ee}\sqrt{\frac{R_{ee}}{h}}$), a being the equatorial radius of the ellipsoid. The ratio h/R_{ee} may represents one the key modeling parameters as it implies direct correlations with the molecular architecture of the probe polymer (i.e. its molecular weight).

THE DEPRESSION OF GLASS TRANSITION TEMPERATURE

In figure 1 the experimental PVT data of Polycarbonate (LEXAN, GE, Tg=142°C) together with the KAHR (Kovacs, Aklonis, Hutchinson, Ramaos) phenomenological theory predictions (blu curve) [17] are reported. The KAHR theory provide an estimation of the volume variations associated with the Tg change.
From the reported experimental data it follows that

$$\frac{dP}{dT_g} = 2.3\frac{MPA}{°C} \qquad (1)$$

Thus, accounting for Tg depressions of about 60°C (a case that is experimentally reported [4,5,6])

$$P_{T_g=80°C} = P_{T_g=142°C} + 2.3(80-142) \approx -200 MPa \qquad (2)$$

The corresponding volume variation being:

$$(V_{T_g=80°C} - V_{T_g=142}) / V_{T_g=142} \approx 5\%! \qquad (3)$$

The pressure P=-200 MPa seems an unrealistic value. It refers however to a rough estimation, the predictions are based on KAHR theory written for "isotropic" materials. Under these circumstances the order of magnitude of Tg's depression and the spectacular volume variations are coherent with those observed experimentally in thin films.
Thin films are actually swollen (1) and the negative pressures (apart from its calculated value) reveal this circumstance.

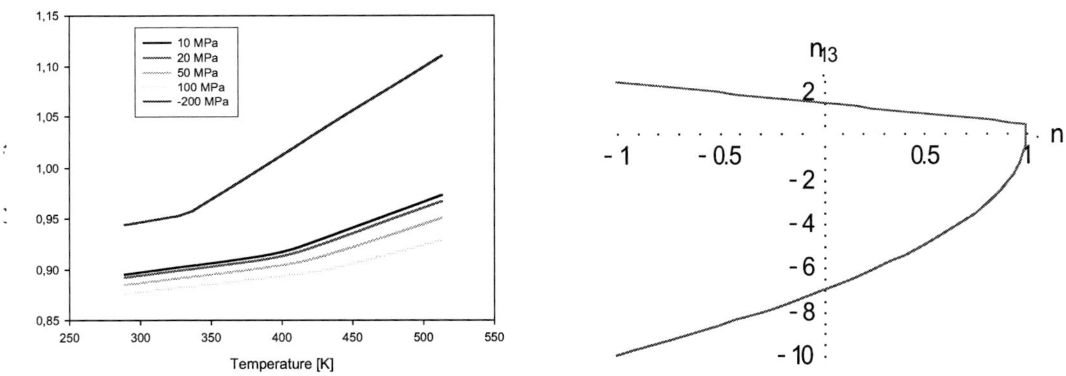

FIGURE 1. Experimental PVT (Pressure-Volume-Temperature) data with predictions based on KAHR phenomenological theory (left) and the "transverse" Poisson's ratio, ν_{13}, as function of the in-plane Poisson's ratio, ν (right).

THE NEGATIVE COEFFICIENT OF THERMAL EXPANSION

Substantial considerations arise when the inherent anisotropy of ultra-thin polymer films is accounted for. In the same Figure 2 (right) are reported the in plane ν and the out of plane, ν_{13}, Poisson's ratio coming out from the fundamental relationships between elastic engineering constants (summarized below) for isotropic orthotropic materials.

$$\nu_{21}\nu_{32}\nu_{13} < \frac{1 - \nu_{21}^2\frac{E_1}{E_2} - \nu_{32}^2\frac{E_2}{E_3} - \nu_{13}^2\frac{E_3}{E_1}}{2} < \frac{1}{2} \qquad |\nu_{13}| < \sqrt{\frac{E_1}{E_2}} \qquad \frac{\nu_{13}}{E_1} = \frac{\nu_{31}}{E_3}$$

The plot of figure 2 refers to transversally isotropic materials ($\nu_{12}=\nu$, $\nu_{13}=\nu_{23}$, $E_2=E_3=E$) where the ratio of the in-plane (E) and out of plane (E_3) moduli, $E/E_3=100$. It can be readily seen that the Poisson's ratio may assume high negative values. Thus, confining the analysis to the elastic case and assuming that the film is perfectly bounded to the support, an increase of temperature, ΔT, induces in-plane compressive deformations, $\varepsilon = -\alpha\Delta T$, α being the in-plane coefficient of thermal expansion. If $\nu_{13} = -\frac{\varepsilon_3}{\varepsilon} \approx -10$, then $\varepsilon_3=10$, $\varepsilon = -10\alpha\Delta T$. But, along the transverse direction no constraints are effective and the transversal deformation, ε_3, is free. Thus, $\varepsilon_3=\alpha_3\Delta T$, and $\alpha_3=-10\alpha$. That is to say that the transversal coefficient of thermal expansion is negative and is one order of magnitude higher than the bulk value, in the glassy region and this conforms to what reported in literature (4,5,6). Of course, thin films are anisotropic structures but neither their viscoelastic nature nor the structural relaxation phenomena can be neglected. In particular, under plane stress conditions, orthotropic materials require four independent viscoelastic functions (namely, E(t), $E_3(t)$ $\nu_{12}(t)$ and $\nu_{13}(t)$) to establish the stress-strain relationships under given loadings. Indeed, one should also recognize that, even in the isotropic case, the different viscoelastic functions do not share the same timescale [15]. It has been shown that the bulk modulus, that is responsible of the volume relaxation phenomena, occurs to within the glassy response of both the shear and tensile moduli. Moreover, in the linear viscoelastic framework the algebraic relationships between the anisotropic viscoelastic functions in the Laplace domain are identical to those holding in the linear elastic case. Thus in thin, anisotropic, viscoelastic films the ratio E/E_3 in the Laplace domain does depend on both the timescales and the relative "*placement*" of viscoelastic functions E(t) and $E_3(t)$.

CONCLUSIONS

The anomalous behavior of ultra-thin polymer films may be qualitatively interpreted with the classical existing theories. The structural complexity arising from the anisotropic and viscoelastic nature of thin films is reflected into the mathematical complexity required to "solve" the problem. The mathematical modeling requires a constitutive link (i.e. a numerical coupling) between four independent viscoelastic functions and the volume relaxation phenomena accounting for both the temperature and the stress-induced relaxation kinetics.

REFERENCES

1. P. G. de Gennes, Glass *Transition in thin polymer films*, The European Physical Journal E, 201-205, (2000)
2. G. Reiter, P.G. DeGennes, Eur. Phys. J. E 6, 25-28 (2001)
3. P.A. O'Connell and G.B. McKenna, Science Vol. 307 (2005)
4. William J. Orts, John H. van Zanten, Wen-li Wu, and Sushil K. Satija Phys. Rev. Lett. **71**, 867 (1993)
5. T. Kanaya, T. Miyazaki, H. Watanabe, K. Nishida, H.Yamano, S. Tasaki, D.B. Bucknall, Polymer, 44, 3769-3773, (2003)
6. M. Mukherjee, M. Bhattacharya, M. K. Sanyal, Th. Geue, J. Grenzer, U. Pietsch, Physical Review E, 66, 061801, (2002)
7. J.L. Keddie, R.A.L. Jones, R.A. Cory, Faraday Discuss. Chem. Soc. 98, 219 (1994)
8. J.S. Sharp, J.A. Forest, Phys. Rev. E 67, 031805 (2003)
9. J.A. Forest, K. Dalnoky-Veress, J.R. Stevens, J.R. Dutcher, Phys. Rew. Letter 77, 2002, (1996))
10. J. A. Forrest, KK. Dalnoki-Veress, J. R. Dutcher, Physical Review E, 56, 5705-5716, (1997)
11. J. S. Sharp, J. A. Forrest, Physical Review Letters, 91, 235701-1, (2003)
12. K. Dalnoki-Veress, J. A. Forrest, C. Murray, C. Gigault, J. R. Dutcher, Physical Review E, 63, 031801, (2001)
13. J. Mattsson, J. A. Forrest, L. Borjesson, Physical Review E, 62, 5187-5200, (2000)
14. Grassia, L. and D'Amore A., Physical Review E **74**, 021504 (2006)
15. Grassia, L. and D'Amore A., Journal of Rheology, March/April, 2009 Volume 53, issue 2, pp. 339-356
16. D'Amore A, Grassia L., Journal of Polymer Science Part B Polymer Physics, Volume 47 Issue 7 , (1 April 2009) p 724-739
17. Kovacs A. J., Aklonis J. J., Hutchinson J. M. Ramos and A. R., J. Polym. Sci., Polym. Phys. Ed. **17**, 1097 (1979)

Modeling the Residual Stresses in Reactive Resins-Based Materials: a Case Study of Photo-Sensitive Composites for Dental Applications

Luigi Grassia and Alberto D'Amore

The Second University of Naples-SUN
Department of Aerospace and Mechanical Engineering
Via Roma 19, 81031 Aversa (CE) Italy

Abstract. Residual stresses in reactive resins-based composites are associated to the net volumetric contraction (shrinkage) arising during the cross-linking reactions. Depending on the restoration geometry (the ratio of the free surface area to the volume of the cavity) the frozen-in stresses can be as high as the strength of the dental composites. This is the main reason why the effectiveness and then the durability of restorations with composites remains quite lower than those realized with metal alloys based materials. In this paper we first explore the possibility to circumvent the mathematical complexity arising from the determination of residual stresses in reactive systems three-dimensionally constrained. Then, the results of our modeling approach are applied to a series of commercially available composites showing that almost all samples develop residual stresses such that the restoration undergoes failure as soon as it is realized.

Keywords: Residual Stress, Dental Composites, photo-reactive resins.
PACS: 81, 82

MODELLING THE RESIDUAL STRESS OF REACTIVE SYSTEMS

The solidification or curing process involves a number of physical, chemical and mechanical phenomena in the composite materials. Generally speaking, the formation of a structure due to the cross linking reactions is the macroscopic effect of heat and pressure inputs: in real terms continuous changes in degree of cure, gelation (the formation of an infinite polymer network as consequence of chemical reaction) and vitrification (the passage from a melt/rubbery to a glassy state) occur. As a consequence of the above phenomena, the evolving chemical and thermal shrinkage give rise to the development of stress and strain within the part [1-3].

In all, thermal shrinkage, chemical shrinkage, time-temperature-degree-of-cure dependent resin properties, viscosity and structural relaxation shrinkage are the matrix properties to control. Furthermore, the evolution of polymer properties is strongly influenced by thermal history that at a given point depends on the transient heat transfer phenomena.

From the micro-mechanical point of view, the process modeling strategies must consider simultaneous interplay of the matrix properties change with the fiber/filler content (including particle/fiber distribution and geometry) and this can be readily achieved by suitable micromechanical model for particle-reinforced composites. Thermosetting system matrix will necessary present a distribution of polymerization reactions at each body point, according to the experienced thermal histories. The temperature distribution can be suitably determined using finite element analysis (FEA); the temperature at given point in the structure depends on the boundary and initial conditions, as well as the materials properties evolution.

The correct analysis then would be an investigation of the temperature distribution with appropriate boundary conditions for the determination of real temperature profile as function of the material point. This analysis would be coupled with a structural analysis, which takes into account the whole set of constraint conditions. Curing raw data are needed to implement the kinetics models can be obtained, by nearly standardized procedures, through isothermal and dynamic DSC (Differential Scanning Calorimetry) tests. The complex interactions of the different phenomena occurring during the curing process could be schematized as follow: a) cross-linking reaction are

CP1255, *Vth International Conference on Times of Polymers (TOP) and Composites,*
edited by A. D'Amore, D. Acierno, and L. Grassia

temperature activated phenomena allowing the transformation of the system from a liquid-like to a solid-like material; b) the structure evolution can be suitable monitored by a degree of conversion parameter, which follows an analytical kinetics mode; c) thermal (specific heat, thermal conductivity) and viscoelastic properties of the material change during the process due to both temperature variations, as for traditional materials, and level of conversion; variation of the specific volume can be originated by three different phenomena, namely: the thermal expansion (or contraction) due to positive (negative) temperature change, the chemical/physical shrinkage due to the densification (polymerization) of the material according to the conversion level, and/or the contraction due to the thermodynamic instability associated with the non-equilibrium kinetics of glass transition (structural relaxation) [1-4]. The experimental evaluation and the modeling of each contribution still represent a difficult task for many researchers, and at the same time it is the main issue for the correct prediction of the residual stress formation in composite systems.

RESULTS AND DISCUSSION

Tensile stress strain experiments have been performed in order to measure the elastic modulus and the tensile strength of commercially available dental restorative materials. The samples are obtained by curing the original material in a bell-shaped stamp with an UV lamp for 40s. Then the samples have been thermally post-cured at 100°C for 24 hours. Both micrometric and nonometric particle reinforced composites have been analyzed and the results of the experimental campaign are reported in figure 1 and 2.

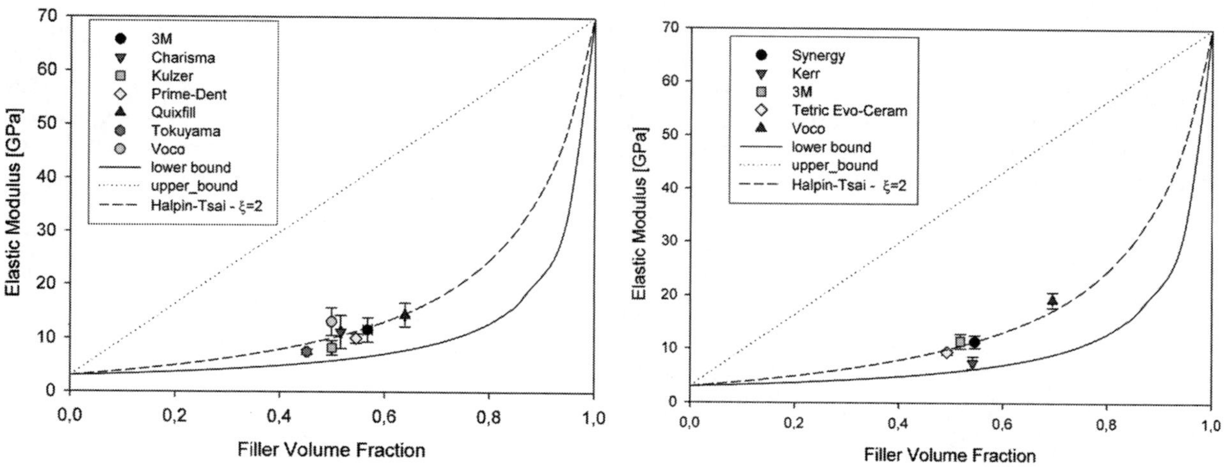

FIGURE 1. The elastic modulus of dental composites filled with micrometric (left) and nanometric (right) particles.

In figure 1 the elastic modulus as function of the proper filler content for several commercially available composites for dental restoration (as indicated) together with the classical Halpin-Tsai model predictions are reported. The parameter ξ indicates the level of efficiency of the filler in terms of both the matrix-filler interaction and the optimal particle size distribution. It should be noted that for many of the studied materials the elastic modulus is well described by using $\xi=2$. Only the samples from Voco and Kerr deviate from this value of ξ. In particular the samples from Voco are characterized by higher value of ξ, whereas the samples from Kerr show a lower value of ξ.

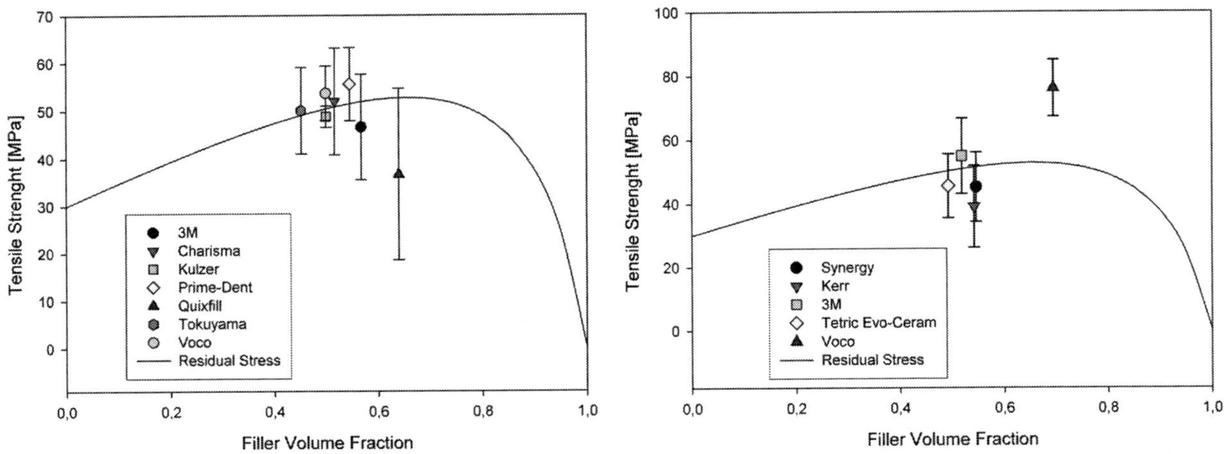

FIGURE 2. The Residual stress (solid line) and the tensile strength of dental composites filled with micro (left) and nano (right) particles.

In figure 2 the tensile strength as function of the filler volume fraction has been reported. Here it is also reported as solid line the modeling prediction of residual stress. We mention that the residual stress level is mainly depends upon the product of the volumetric shrinkage and the stiffness of the material. The volumetric shrinkage is a monotonic decreasing function of the filler volume fraction whereas the stiffness is a monotonic increasing function of the filler volume fraction, consequently the residual stress undergoes a maximum for intermediate value of the filler content. It should be noted that for many of the restorative materials the residual stress level is higher or almost equal to the tensile strength of the corresponding material. Only the noreinforced sample from Voco shows a significant residual strength. Then we can conclude that almost all samples develop residual stresses such that the restoration undergoes failure as soon as it is realized

REFERENCES

1. L. Grassia and A. D'Amore, *Macromol. Symp.* **228**, 1-15 (2005)
2. A. D'Amore, F. Caputo, L. Grassia, M. Zarelli, *Composites: Part A* **37** 556–564 (2006).
3. M. Zarelli, I. Partridge, A. D'Amore *Composites: Part A* **37** 565–570 (2006)..
4. L. Grassia and A. D'Amore Journal of Polymer Science: Part B: Polymer Physics **47** 724–739 (2009)

Tuning by process of the electrical percolation behavior of multiwalled carbon nanotubes/epoxy composites

G. Faiella, M. Zarrelli, V. Antonucci, M. Giordano[a]

[a] *Institute for Composite and Biomedical Materials (IMCB) – National Research Council (CNR)*
P.le E. Fermi 1 80055 Portici (NA) - Italy

Abstract. Electrical properties of carbon nanotubes polymer composites are strictly related not only to the content of the conductive nanoparticles, but also to their network morphology forming within the hosting matrix. The percolation phenomenon is strongly depending on processing conditions and for many composite systems the percolation threshold has been found to be well below the statistical one. In this work, epoxy composites at different multiwalled nanotubes (MWNT) contents have been produced using sonication as dispersion technique. Different electrical properties have been induced in the composites at the same MWNTs content by varying sonication time and curing temperature and the dependence of the percolation phenomenon on the two process parameters has been studied. The electrical conductivity shows an enhancement of even six orders of magnitude for the same MWNTs concentration upon an increase of curing temperature.

Keywords: Carbon nanotubes, composites, percolation.
PACS: 73.63.Fg.Nanotubes

INTRODUCTION

Electrically conductive composites are typically produced my dispersing conductive particles within a polymer insulating matrix. Their electrical conductivity dependence on the filler content can be described by a percolative law, where the transition between the insulator to conductor behavior occurs at a filler concentration named percolation threshold. The percolation threshold is defined as the concentration at which an infinite network of conductive fillers appears in an infinite lattice [1]. In the field of conductive composites, carbon nanotubes (CNTs) have been considered ideal candidates to be used as conductive fillers in polymer matrices, because of their exceptional electrical properties. Their small size as well as the very high aspect ratio provide the possibility to reach high conductivity values at very lower CNTs concentrations than other conductive particles with consequential smaller perturbation of bulk physical properties. However, respect to isotropic particles composites, such as carbon black and metallic particles, the percolation behavior of nanotubes based composites is still not well understood.

Studies on CNTs-polymer composites revealed that a substantial discrepancy between the theoretically calculated and the experimentally evaluated percolation thresholds exists [2, 3]. The theoretical approach, in fact, suffers of the fact that it considers the composite as an equilibrium system, while the occurrence of strong interactions between nanotubes and between nanotubes and matrix provides aggregation phenomena and hence topologies characterized by a kinetic nature and thus non equilibrium features.

In a previous paper [4], authors described how it is possible to obtain a fine tuning of the insulator to conductor transition in epoxy-multiwalled carbon nanotubes (MWNTs) composites at a fixed nanotubes concentration through the development of different tubes network morphologies induced by the variation of process parameters. In line with this argument, here we show how it is possible to induce different percolation behaviors in MWNTs/epoxy composites by varying process parameters.

CP1255, *V[th] International Conference on Times of Polymers (TOP) and Composites,*
edited by A. D'Amore, D. Acierno, and L. Grassia

EXPERIMENTAL

Catalytic carbon vapor deposition (CCVD) MWNT (Nanocyl 3150) with an average diameter of 9.5 nm, an average length less than 1 μm and a purity exceeding 95% have been used. A thermoset bi-component epoxy matrix, based on bisphenol-A/epichlorohydrin (EPON 828) and triethylenetetramine (TETA) as hardener has been used as composites matrix. Nanotubes dispersion has been conducted by sonication in EPON for different sonication times (10, 30, 75 and 120 minutes). Different thermal cycles have been used for the composites curing: 1) 24 hours at 25 °C and 2 hours at 120 °C, 2) 1 hour at 70 °C and 3) 2 hours at 120 °C. In the following, we will refer to the listed curing cycles as curing condition 1, 2 and 3 respectively. A set of composites at various MWNTs concentrations (from 0.01 to 0.1 wt%) has been fabricated for each sonication time and curing temperature, in order to evaluate the influence of these process parameters on the percolation behavior.

Electrical Measurements and Morphological Analysis

Bulk electrical conductivity has been evaluated by means of two probes DC electrical resistance measurements by using a Signatone 1160 probe station connected to a pico-ammeter and voltage source.

In the following, the electrical behaviors of the four groups of composites at different sonication times will be shown firstly as a function of curing temperature, then the dependence of the percolation behavior upon process parameters will be illustrated.

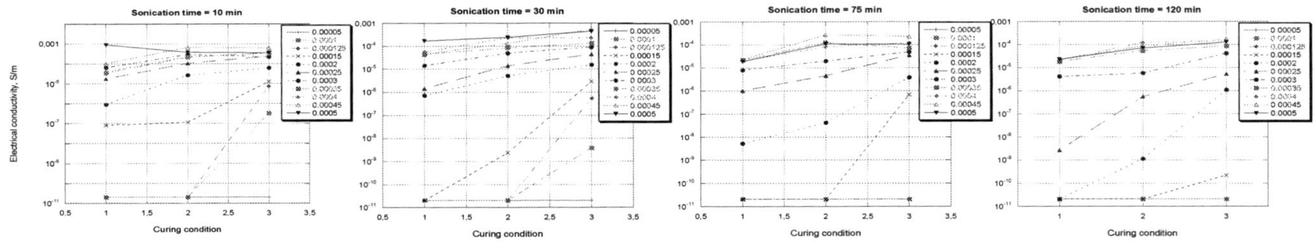

FIGURE 1. Tuning of the electrical conductivity upon variation of curing temperature at fixed MWNTs concentration for the composites produced at four different sonication times: 10 min, 30 min, 75 min, 120 min.

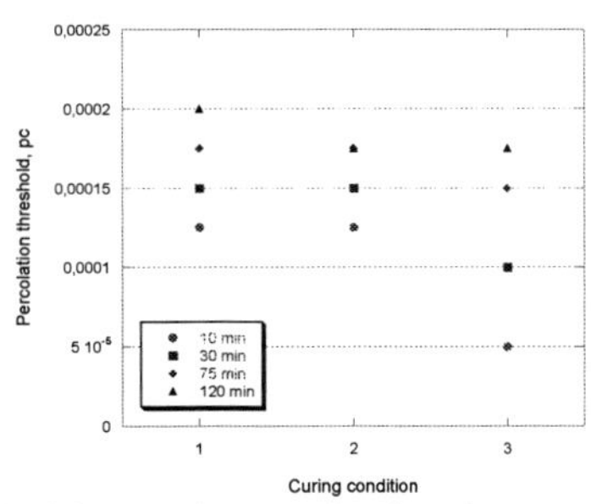

FIGURE 2. Percolation thresholds at various sonication times as a function of curing condition.

In the figures 1, the electrical conductivity of the composites produced by sonication for 10, 30, 75 and 120 minutes at different MWNTs volume fractions are reported as a function of curing condition and it is possible to appreciate the variation of the electrical conductivity with the curing temperature for each MWNTs concentration. In particular, one takes evidence of the fact that only at some concentrations the conductivity appreciably varies with the curing temperature, with increments even of 6 orders of magnitude passing from curing condition 1 to 3.

By the analysis of the percolation behavior, a notable result to be noted is the very low value of percolation threshold for each type of process condition. In fact, the theoretical volume fraction percolation threshold for these MWNTs, evaluated as D/2L, is 0.005, while much lower percolation thresholds have been found respect to the theoretical one, ranging between $5\ 10^{-05}$ and $2\ 10^{-04}$. Moreover, there exists a strong dependence of the percolation threshold on the process parameters: it decreases upon an increase of curing temperature for each sonication time, while it increases with sonication time (figure 2).

Optical microscopy images of the composites showing the temperature induced tuning of the electrical conductivity have been analyzed in order to find a correlation between morphological and electrical properties of the composites. From figure 3 (a) to 3 (d), optical microscopy images of the composites at 10, 30, 75 and 120 minutes of sonication are shown. In particular, the micrographs of composites at the composition characterized by the tunable behavior are shown for each curing temperature. Samples are identified as "*composition_sonication time_curing condition*". An agglomeration process occurs in the composite upon an increase of curing temperature, with the appearance of darker micro-areas where the CNTs organize in micro-structures, the clusters, connecting each other to form percolating networks.

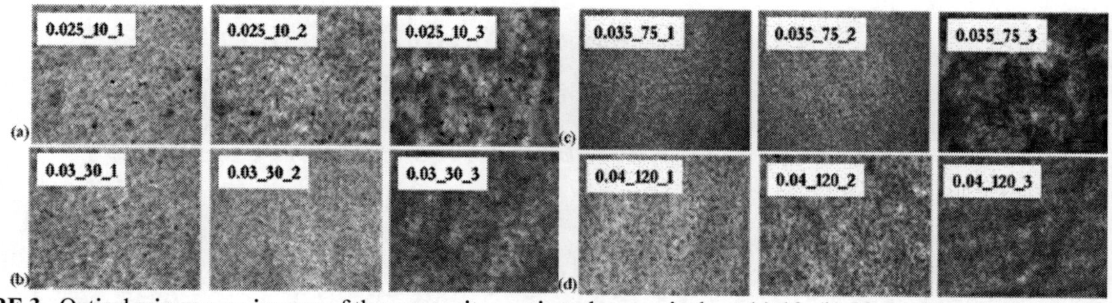

FIGURE 3. Optical microscopy images of the composites sonicated respectively at (a) 10, (b) 30, (c) 75 and (d) 120 minutes

Discussion and Conclusions

Electrical conductivity dependence on curing temperature in our MWNTs-epoxy composites is an important example of "dynamic percolation" and it can be explained by agglomeration phenomena occurring in the composite that lead to the formation of more efficiently conductive networks constituted by nanotubes clusters. Morphological analysis confirms the agglomeration process induced by temperature in those composites showing an increase of conductivity with the curing temperature. On the other hand, from experimental results it appears evident that a prolonged sonication is detrimental for the final electrical properties of the composite. Higher sonication times can induce CNTs shortening, with a consequent reduction of the effective aspect ratio [5]. However, here the most probable reason for the percolation threshold increase has to be found in the finer dispersion due to the higher exfoliation of the tubes within the composite upon an increase of sonication time, that leads to less efficient conductive networks, where the insulating matrix between the tubes and the higher number of contact resistances contribute to decrease the global network conductivity. In order to describe the electrical behavior of the CNTs composites, a multiscale morphological organization of CNTs within the composites has to be considered for the description of the occurring conduction mechanisms, that are responsible for the final global electrical conductivity. At the nanoscale level, CNTs are entangled together forming a structure, the cluster, having a own networking and conductivity. At the microscale, clusters are the percolating elements, that connect each other to form the percolative path, whose electrical behavior can be described by a percolative law. Hence, in these CNTs composites where the percolation threshold was found to be much lower than the statistical one, the percolating element is not the single nanotube, but the CNTs-cluster, produced by dynamic processes, whose intrinsic conductivity depends on the process parameters and on the interactions between tubes and matrix.

The analysis of the different percolation behaviors and the correlation between the derived critical conductivity exponents and the CNTs network morphologies developed within the composites is under study. The used fabrication process gives the possibility to vary process parameters in order to obtain a wide range of electrical conductivity and consequently of different CNTs network morphologies, by starting from the same pristine matrix and filler. To the best of our knowledge, up to now a so huge variation of electrical behavior has been reached only by using different nanotubes and matrices.

REFERENCES

1. D. Stauffer, A. Aharony, *Introduction to percolation theory*, London, Taylor and Francis, 1991.
2. C.A. Martin, J.K.W. Sandler, M.S.P. Shaffer, M.-K. Schwarz, W. Bauhofer, K. Schulte, A.H. Windle, *Comp. Sci. Technol.* **64**, 2309-2316 (2004).
3. J.Z. Kovacs, B.S. Velagala, K. Schulte, W. Bauhofer *Comp. Sci. Technol.* **67**, 922-928 (2007)
4. G. Faiella, F. Piscitelli, M. Lavorgna, V. Antonucci, M. Giordano, *Appl. Phys. Lett.* **95**, 153106-153108 (2009).

Thermal residual stresses in amorphous thermoplastic polymers

Luigi Grassia and Alberto D'Amore

The Second University of Naples-SUN, Department of Aerospace and Mechanical Engineering, Via Roma 19, 81031 Aversa (CE), ITALY

Abstract. An attempt to calculate the internal stresses in a cylindrically shaped polycarbonate (LEXAN-GE) component, subjected to an arbitrary cooling rate, will be described. The differential volume relaxation arising as a result of the different thermal history suffered by each body point was considered as the primary source of stresses build up [1-3]. A numerical routine was developed accounting for the simultaneous stress and structural relaxation processes and implemented within an Ansys® environment. The volume relaxation kinetics was modeled by coupling the KAHR (Kovacs, Aklonis, Hutchinson, Ramos) phenomenological theory [4] with the linear viscoelastic theory [5-7]. The numerical algorithm translates the specific volume theoretical predictions at each body point as applied non-mechanical loads acting on the component. The viscoelastic functions were obtained from two simple experimental data, namely the linear viscoelastic response in shear and the PVT (pressure volume temperature) behavior. The dimensionless bulk compliance was extracted from PVT data since it coincides with the memory function appearing in the KAHR phenomenological theory [7]. It is showed that the residual stress scales linearly with the logarithm of the Biot's number.

Keywordsre, residual stresses glassy polymers, viscoelasticity. structural relaxation
PACS: 81,82

INTRODUCTION

Polymer glasses are rheologically complex materials whose viscoelastic response is a function of the kinetics of structural relaxation. To circumvent the complexity, several experimental loading protocols have been established over the years, but the results have not led to a unifying concept. From a theoretical point of view, we tackled the problem by coupling the structural relaxation kinetics with the constitutive equation for linear viscoelasticity written in the reduced time domain. The constitutive equation for linear viscoelastic behavior is [5-7]:

$$\sigma_{ij} = \delta_{ij}\int_0^{\xi} K(\xi-\xi')\frac{d}{d\xi'}(\varepsilon-\theta)d\xi' + \int_0^{\xi} 2\cdot G(\xi-\xi')\frac{d}{d\xi'}\left(\varepsilon_{ij}-\frac{1}{3}\delta_{ij}\varepsilon\right)d\xi' \qquad (1)$$

where δ_{ij} is the Kronecker delta, G is the shear modulus, K is the bulk modulus, $\varepsilon = \sum_1^3 \varepsilon_{ii}$ is the trace of the strain tensor, and ξ is the reduced time defined in what follows. θ is the strain not directly due to the presence of a stress as it takes into account both the dilatation effects and the strain resulting from structural relaxation phenomena. In the text that follows it will be referred to as *thermal strain*. A constitutive equation is required to evaluate the thermal strain because the linear viscoelasticity equation does not provide information about it. The modeling approach entails the use of reduced time defined as [5-7]:

$$\xi = \int_0^t \frac{\tau_R dt'}{\tau} \qquad (2)$$

CP1255, V^{th} *International Conference on Times of Polymers (TOP) and Composites,*
edited by A. D'Amore, D. Acierno, and L. Grassia

The relaxation time that appears in the expression for the reduced time assumes the simplest possible form: the logarithmic of the relaxation time depends linearly on the temperature, the isotropic part of the stress tensor and the dimensionless volume (i.e. a time-temperature-isotropic stress-dimensionless volume superposition is hypothesized). The explicit expression is [5-7]:

$$\ln\tau = \ln\tau_R - \theta_T(T - T_R) - \theta_\sigma(\sigma - \sigma_R) - (1-x)\left(\frac{\theta_T}{\Delta\alpha} + \frac{\theta_\sigma}{\Delta B}\right)\delta \quad (3)$$

where θ_T and θ_σ are material constants. x is a volume-related nonlinearity parameter.

In order to utilize equation (1) one needs the shear modulus, G, the bulk modulus, K, and an expression for the thermal strain, θ.Different strategies are utilized for the determination of those three functions (G, K, and θ). The shear relaxation modulus, G, is evaluated experimentally by means of classical time temperature superposition procedure. The functional form of the thermal strain, θ, and of the bulk compliance, B, are evaluated with the aid of the KAHR model [6] as:

$$\theta = \int_0^{\xi} \frac{dT}{d\xi'}[\alpha_e - \Delta\alpha \cdot M(\xi - \xi')]d\xi' \quad (4)$$

$$\frac{B_e - B(\xi)}{B_e - B_g} = M(\xi) \quad (5)$$

where B_e and B_g are the relaxed and unrelaxed bulk compliance, respectively. The bulk modulus, K, can be evaluated through its correlation with the bulk compliance in the Laplace domain:

$$K^* = \frac{1}{s^2 B^*} \quad (6)$$

where the K* and B* indicate the Laplace transforms of K and B respectively.

RESULTS AND DISCUSSION

The residual stresses distribution induced in a polycarbonate sample, subjected to a quenching, were evaluated developing a numerical routine coupling a structural and a thermal finite element analysis. In order to start the numerical procedure, a preliminary determination of the temperature distribution vs. time was performed assuming that the stress distribution do not affect the heat transfer problem; the results from this calculation in terms of a three-dimensional temperature distribution have been provided to the numerical routine, based on the optimized parameters (evaluated by fitting PVT data) of the KAHR theory [5-7], and allows the knowledge of time dependence of specific volume at each body point. Then the incremental change in growth (volumetric) strain at each body point can be calculated at a given time according to equation (4).

Finally, imposing the congruent displacement and equilibrium conditions the stresses evolution can be calculated. In the thermal analysis the thermal diffusivity has been assumed constant [1-3, 8]. The ambient temperature T_{amb} and the global convective heat transfer coefficient have been assumed constant during cooling; the initial temperature, T_0, has been assumed uniform and all the external surfaces of the sample have been considered involved in the heat transfer phenomena [1-3, 8].

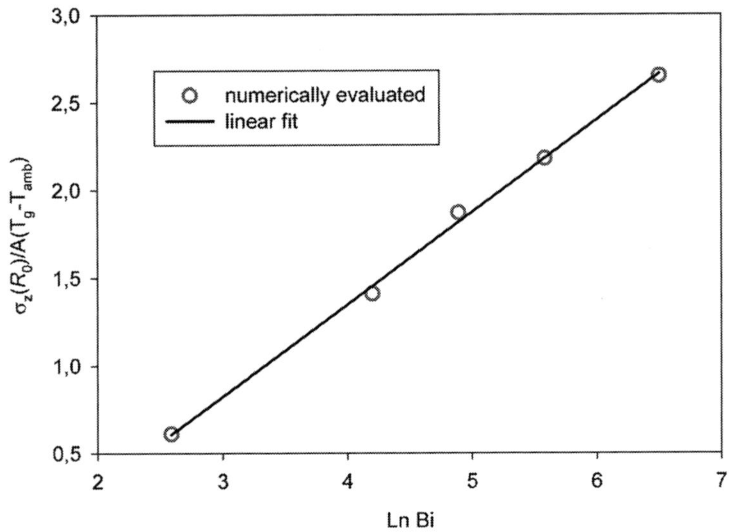

Figure 1. Dimensionless axial residual stress for a polycarbonate cylinder quenched from above Tg to ambient temperature Tamb.

The evaluation of residual stresses will be carried out for cylindrically shaped polycarbonate sample that undergoes a quenching from and T_0=160°C to T_{amb}=25°C. The cylinder length L is 20 times the radius, R_0. The global convective heat transfer coefficient has been assumed equal to 10^3W/m^{2}K (a characteristic value of natural convection in water). Only a quarter of an axial section of the cylinder has been FE modeled due to the axial and transversal symmetries of the structure. A parametric analysis is carried out in order to predict the influence of radial dimensions on the residual stresses. In Figure 1 the axial stress shows a quasi-linear dependence on ln(Bi). This is in agreement with Struik [8] prediction. A linear fitting of dimensionless axial stress vs. ln(Bi) (with E_g = 2800 MPa, a_g=7.3x10^{-5} K^{-1} and ν_g = 0.40 A = $a_gE_g/(1-n_g)$=0.34 MPa/K), shows in our case a slope of 0.52 which is little higher than the Struik predicted value of 0.5.

CONCLUSIONS

In the present work, the contribution of stress and volume relaxation on the residual stress development within a polystyrene cylinder has been studied. A coupling between the KAHR phenomenological model and the viscoelasticity theory was adopted for the determination of specific volume under the thermal histories suffered by each body point as consequence of the cooling treatment. During the cooling process, the mechanical properties of the polycarbonate cylinder change from those characteristic of an entangled melt to those typical of an elastic like-solid material, in the glassy state. The modeling procedure adopted fulfils the physical expectations of the surface in compression and the core in tension, and it predicts that surface residual stress scales linearly with the logarithm of the Biot's number.

REFERENCES

1. L. Grassia and A. D'Amore, *Macromol. Symp.* **228**, 1-15 (2005)
2. A. D'Amore, F. Caputo, L. Grassia, M. Zarelli, *Composites: Part A* **37** 556–564 (2006).
3. M. Zarelli, I. Partridge, A. D'Amore *Composites: Part A* **37** 565–570 (2006)..
4. A. J. Kovacs, J. J. Aklonis, J. M. Hutchinson, A. R. Ramos *J. Polym. Sci., Polym. Phys. Ed.* **17**, 1097 (1979)
5. Grassia, L. and D'Amore A., Physical Review E **74**, 021504 (2006)
6. Grassia L. and D'Amore A,, *J. Polym. Sci., Polym. Phys. Ed*, 724-739 (2009)
7. Grassia, L. and D'Amore A., Journal of Rheology, **53** 339-356 (2009)
8. L.C.E. Struik "Internal Stresses, Dimensional Stabilities and Molecular Orientations in Plastics", John Wiley and Sons Ltd. Chichester, England 1990

Isobaric PVT Behavior of Poly(Carbonate) (PC)

Luigi Grassia and Alberto D'Amore

The Second University of Naples-SUN, Department of Aerospace and Mechanical Engineering, Via Roma 19, 81031 Aversa (CE), ITALY

Abstract. The scaling law for relaxation times, $\tau(V,T)=\Im\left(TV^{\gamma}\right)$, recently proposed by Casalini and Roland, is utilized in the framework of KAHR (Kovacs Aklonis Hutchinson and Ramos) phenomenological theory. With this approach it is shown that the Pressure, Volume, Temperature (PVT) data obtained on Poly(carbonate)(PC) can be reliably predicted, in the region of the alpha relaxation, by using only two fitting parameters, namely: the relaxation time in the reference condition, τ_g, and the fractional exponent, β, that describes the dispersion of the alpha relaxation.

Keywords: scaling law. relaxation times, PC, alpha relaxation
PACS: 81,82

MODELING

Recently, Casalini and Roland [1,2] discussed how the scaling properties can be derived from the temperature, T, and the volume, V, dependences of the entropy, and using the Avramov model [3] derived the following expression for the τ(T, V) dependence [1,2, 4]:

$$\tau(T,V)=\tau_0\exp\left[\left(\frac{A}{TV^{\gamma}}\right)^{\phi}\right] \qquad (1)$$

where τ_0 , A, γ and ϕ are constants. Equation 2 gives a good description of experimental data over a broad dynamic range [1,2,4].
Here, the expression proposed by Casalini and Roland is utilized as the functional form of the relaxation time in the KAHR (Kovacs Aklonis Hutchinson and Ramos) model [10-11] that predicts the volume relaxation kinetic under arbitrary pressure and temperature histories [10-14]. Under isobaric conditions the KAHR equation reduces to [5-9]:

$$V=V_e+V_e\int_0^{\xi}\left[-\left(\alpha_e-\alpha_g\right)\frac{dT}{d\xi'}\right]M(\xi-\xi')d\xi' \qquad (2)$$

where the suffix g and e indicate the glassy and the liquid/equilibrium state, respectively, α is the isobaric thermal expansion coefficient, and M is the memory function defined as follows:

$$M(\xi)=\exp\left[-\left(\xi/\tau_g\right)^{\beta}\right] \qquad (3)$$

where τ_g is the relaxation time in the reference conditions, β is the fractional exponent that describes the dispersion of the alpha relaxation and ξ is the reduced time defined as follows:

CP1255, *V*th *International Conference on Times of Polymers (TOP) and Composites,*
edited by A. D'Amore, D. Acierno, and L. Grassia

$$\xi = \int_0^t \frac{\tau_g dt'}{\tau} \quad (4)$$

where t is the time and τ the relaxation time.

In this paper it is assumed that the relaxation time, τ, has the functional form expressed by equation 2. Indeed equation 2 can be can be rewritten in terms of the isochoric fragility, $m_V = \left.\frac{\partial \log\tau}{\partial(T_g/T)}\right|_{T_g, V=\cos t}$, as follows [2]:

$$\ln\tau = \ln\tau_g + \frac{m_V \ln 10}{\phi}\left[\left(\frac{T_g V_g^{\gamma}}{T V^{\gamma}}\right)^{\phi} - 1\right] \quad (5)$$

Equations (3-6) can be used to predict the PVT response of PC.

RESULTS AND DISCUSSIONS

In what follows we report on a procedure that allowed us to implement equations (2-5). It will be shown that most of the material's functions (namely, α_g , α_e , and the material's properties (γ, M_v and ϕ) appearing in equations (2-5) can be derived directly from opportune PVT data. The parameters τ_g and β remain as fitting parameters that can be evaluated minimizing the sum of squared differences between the model prediction and the experimental data. Under this perspective equations (2-5) differs substantially from the KAHR model that, originally, required the optimization of five parameters [6-9].

A commercially available Poly(Carbonate) (PC) (LEXAN, GE) was chosen as the probe material. The isobaric PVT data were obtained cooling the sample from above to below T_g at 0.5°C/min at four different pressures, namely:10, 20, 50 and 100 MPa.

The Tait's equation[10] is here utilized to fit the data either above or below the T_g in order to obtain both α_e and α_g, in analytical form.

We recall that the relaxation time at different T_gs, obtained cooling the sample at different pressures, should be the same since the cooling rate is the same at each pressure. This allows calculating the parameter γ, since the relaxation time, τ, scales with TV^{γ} . In fact, at the glass transition T_g, the quantity $T_g V_g^{\gamma}$,where V_g is the specific volume at T_g, should be constant. From this assumption the parameter γ is easily calculated and the linear regression analysis gives γ=4.87. This value falls within the range of γ (1.5<γ<6) reported by Casalini and Roland for other polymers [1, 2, 4].

Once γ is known, the isochoric fragility can be obtained as follows:

$$m_V = \frac{a}{1 - \gamma \cdot \alpha_e(T_g, P_0) \cdot T_g(P_0) - b} \quad (6)$$

where $a = 37 \pm 3$ and $b = 0.84 \pm 0.05$. Here $T_g(P_0)$ is extrapolated at atmospheric pressure from the pressure dependence of T_g from isobaric PVT data. From equation (6) we obtain that the isochoric fragility, m_V, is 29.8.

Finally, using the correlation between ϕ and γ:

$$\phi = \frac{20.8 \pm 0.9}{1+\gamma} \quad (7)$$

we found ϕ=3.56. The remaining parameters, τ_g and β are evaluated minimizing the sum of the square differences between the modelling predictions and the experimental results in terms of specific volume obtained cooling the sample from above to below Tg (i.e. by fitting

the

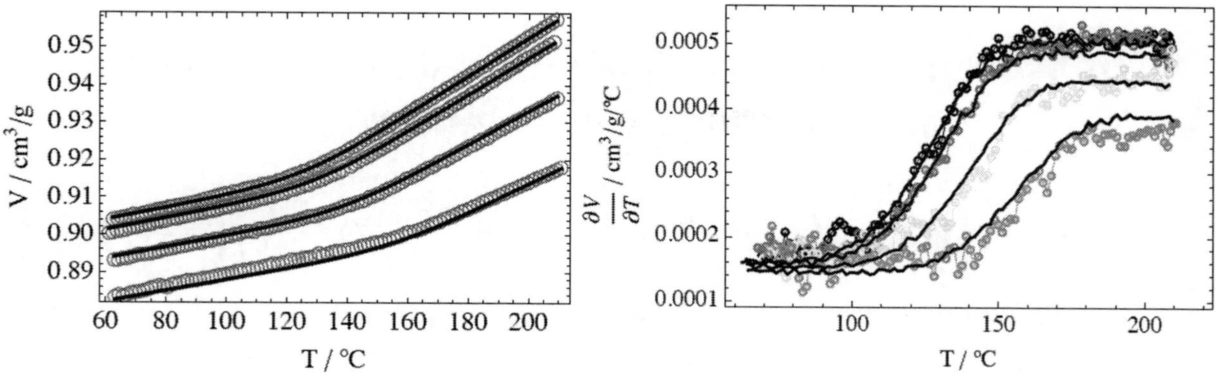

FIGURE 1. The experimental specific volume (left) and its temperature derivative (right) together with the model fitting (solid lines)

experimental PVT data. The values of $\ln(\tau_g)$ and β that minimize the sum of square differences are 8.3 and 0.205, respectively.
The results of the fitting procedure are displayed in figure 1 in terms of the specific volume, V, (left) and its temperature derivative, $\partial V / \partial T$.(right). The proposed modelling approach is able to predict both the asymptotic behaviour, deep in the glassy and in the rubbery state, and the volume relaxation kinetics, by means of a single set of theory's parameters that remains fixed once and for all.

CONCLUSION

The scaling law for the relaxation time, recently proposed by Casalini and Roland, is implemented in the modified KAHR (Kovacs Aklonis Hutchinson and Ramos) model. This approach allows predicting the experimental PVT behaviour of amorphous PC by using only two fitting parameters (instead of the five parameter of the classical KAHR model), namely the relaxation time in the reference condition, τ_g, and the fractional exponent, β, that describes the dispersion of the alpha relaxation. The two model parameters have been calculated by fitting the isobaric PVT data, obtained cooling the sample from above to below T_g. The model is also able to predict the glass transition induced by isothermal compression and the volumetric response to a complex thermal and pressure history.

REFERENCES

[1] R. Casalini, U. Mohanty, C.M. Roland, J. Chem. Phys. 125 (2006) 014505.
[2] R. Casalini, C.M. Roland, Phil. Mag. 87 (2007) 459.
[3] I. Avramov, J. Non-Cryst. Solids 262 (2000) 2585
[4] R. Casalini, C. M. Roland J. Non-Cryst. Solids 353 (2007) 3936-3939
[5] A. J. Kovacs, J. J. Aklonis, J. M. Hutchinson, A. R. Ramos J Polym Sci Part B: Polym Phys 17 (1979) 1097–1162.
[6] A. R. Ramos, A. J. Kovacs, J. M. O'Reilly, J. J. Tribone, J Polym Sci Part B: Polym Phys 26 (1988) 501–513.
[7] L. Grassia, A. D'Amore, Phys. Rev. E 74 (2006) 021504
[8] L. Grassia, A. D'Amore J Polym Sci Part B: Polym Phys 47 (2009) 724–739
[9] L. Grassia, A. D'Amore J. Rheol. 53 (2009) 339-356
[10] Tait, P. G. The Voyage of H. M. S. Challenger Her Majesty's Stationary Office, London, Vol. 2 (1888)

An approach in developing 3D fiber-deposited magnetic scaffolds for tissue engineering

R. De Santis [a], A. Gloria [a], T. Russo [b], U. D'Amora [a], S. Zeppetelli [a], L. Ambrosio [a]

[a] Institute of Composite and Biomedical Materials, National Research Council, P.le Tecchio 80, 80125, Naples, Italy
[b] Department of Materials and Production Engineering, University of Naples "Federico II", P.le Tecchio 80, 80125, Naples, Italy.

Abstract.
Scaffolds should possess suitable properties to play their specific role. In this work, the potential of 3D fiber deposition technique to develop multifunctional and well-defined magnetic poly(ε-caprolactone)/iron oxide scaffolds has been highlighted, and the effect of iron oxide nanoparticles on the biological and mechanical performances has been assessed.

Keywords: Tissue engineering, 3D fiber deposition, scaffolds, poly(ε-caprolactone), iron oxide nanoparticles.

INTRODUCTION

The internal architecture of scaffolds, including pore size, pore shape and interconnectivity, are crucial to their biological and mechanical performances The limitations of conventional scaffold fabrication methods have been frequently reported in literature [1-3], and rapid prototyping techniques have been described as advanced processing methods to develop customised scaffolds with controlled internal microarchitecture. Basically, conventional methods include fiber meshes/fiber bonding, gas foaming, solvent casting/particulate leaching, phase separation, melt molding, freeze drying, solution casting, and emulsion freeze drying [1-3]. However, using conventional scaffold fabrication methods, the control of the internal architecture and interconnectivity may be strongly limited [2, 3]. Conventional scaffold processing techniques lack in precisely controlling pore geometry, pore size, spatial distribution of pores within the scaffold [1-3]. Among the several rapid prototyping methods, 3D fiber deposition technique has emerged as a means to manufacture well-defined and custom-made scaffolds for tissue regeneration, with 100% interconnected pores, also because of its flexibility to process a wide range of materials [2, 3]. The aim of this study was to highlight an approach in developing 3D fiber-deposited magnetic poly(ε-caprolactone)/iron oxide scaffolds, and to investigate the effect of the iron oxide nanoparticles on the biological and mechanical performances.

MATERIALS AND METHODS

Poly(ε-caprolactone) (PCL, M_w=65000–Aldrich) pellets were dissolved in tetrahydrofuran through stirring at room temperature. Polyvinylpyrrolidone (PVP)-coated iron oxide (Fe_3O_4, 99.5%, 25 nm, 0.2 % w PVP-NanoAmor) nanoparticles and ethanol were added to the PCL/THF solution during stirring. A PCL/ Fe_3O_4 weight ratio of 90/10 was used. An ultrasonic bath (Branson 1510 MT) was also used to optimize the Fe_3O_4 nanoparticles dispersion in the polymer solution. Accordingly, a homogenous paste was obtained and then placed under hood. Nanocomposite fibers and scaffolds with a 0°/90° pattern were manufactured by processing PCL/Fe_3O_4 (90/10 w/w) pellets through 3D fiber deposition technique. Using a Bioplotter dispensing machine (Envisiontec GmbH, Germany), nanocomposite scaffolds were built by extruding and alternatively depositing the fibers obtaining a 0°/90° pattern. The stainless steel needle used to extrude PCL/Fe_3O_4 fibers was characterized by an inner diameter of 400 μm. To capture shape and size of the nanocomposite fibers, a micro-computed tomography (MicroCT) was performed using a SkyScan 1072 (Aartselaar, Belgium) system. On the other hand, tensile tests were carried out on PCL and PCL/Fe_3O_4 nanocomposite fibers at speed of 50 mm/min, using a 5566 INSTRON test system. Furthermore,

CP1255, *V^th^ International Conference on Times of Polymers (TOP) and Composites,*
edited by A. D'Amore, D. Acierno, and L. Grassia

confocal analysis was carried out to study human mesenchymal stem cells (hMSCs) adhesion and spreading on PCL/Fe_3O_4 nanocomposite fibers.

RESULTS AND DISCUSSION

Preliminary PCL/Fe_3O_4 scaffolds have been manufactured through 3D fiber deposition technique by suitably optimizing the process parameters.

Cross-sections and 3D model of nanocomposite fibers have been reconstructed using Skyscan's software package, Image J software and Materialise Mimics, that have allowed the image analysis and visualization of the results from MicroCT scanning. The fiber diameter obtained has been firstly assessed, and the presence of Fe_3O_4 nanoparticles and clusters has been shown (Fig. 1).

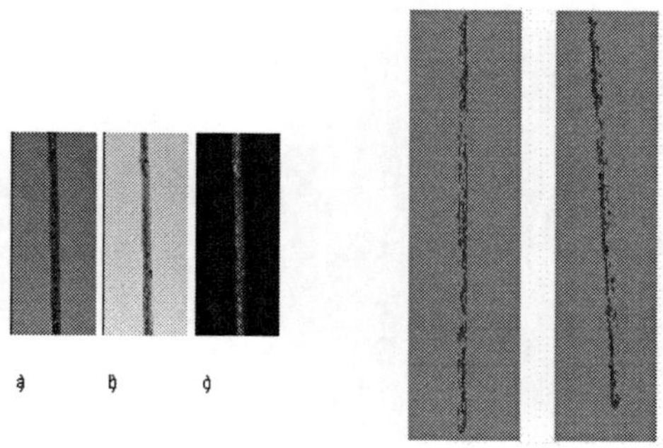

(a) (b)

FIGURE 1. Two different reconstructions of nanocomposite fibers obtained through Image J (a) and Materialise Mimics (b) that highlight the distribution of Fe_3O_4 particles.

Results from tensile test have evidenced a ductile behavior for both PCL and PCL/Fe_3O_4 fibers. Even though the inclusion of Fe_3O_4 nanoparticles strongly reduces the maximum strain, it enhances the modulus (from 571.5 to 640 MPa) and the maximum stress (from 29.3 to 38.1 MPa).

Confocal analysis has shown interesting results in terms of hMSCs adhesion and spreading on PCL/Fe_3O_4 fibers (Fig. 2).

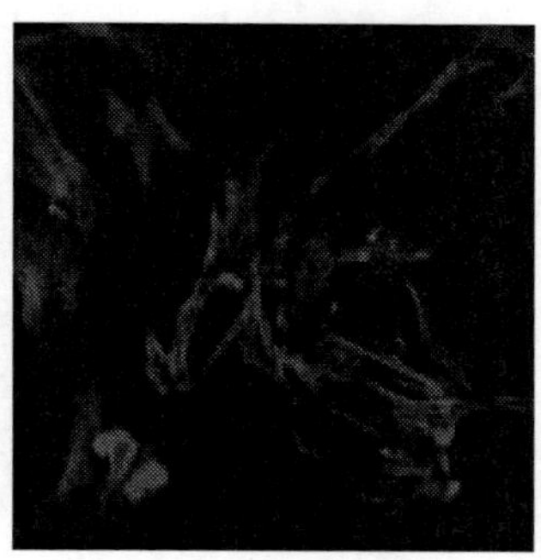

FIGURE 2. An image obtained from confocal analysis on the PCL/Fe_3O_4 fiber, that highlights the phalloidin-labelled actin filaments.

In particular, confocal analysis has highlighted an increase in adhered number and a more evident spreading of hMSCs, if compared to the results usually obtained from the PCL alone fibers.

ACKNOWLEDGMENTS

"Magnetic Scaffolds for in vivo Tissue Engineering" VII PQ project NMP-2007-4.2.3.-1 is gratefully acknowledged for financial support. The authors also thank Mr. Rodolfo Morra and Dr. Gianluca Ametrano for performing mechanical testing and MicroCT scanning, respectively.

REFERENCES

1. E. Sachlos and J.T. Czernuske. *Europ Cells and Mat* **5**, 29–40 (2003).
2. T.M.G. Chu, "Solid freeform fabrication of tissue engineering scaffolds", in *Scaffolding in Tissue engineering* edited by Ma PX, Elisseeff J. Northwest Florida: Taylor and Francis, 2006, 139–153.
3. A. Gloria, T. Russo, R. De Santis, L. Ambrosio. *Journal of Applied Biomaterials & Biomechanics* **7**, 141-152 (2009).

Author Index

A

Abad, M. J., 355
Acierno, D., 168, 187, 190, 206, 376, 379
Afratisa, A., 79
Aglietto, M., 184
Agostiano, A., 89, 92
Ahmetshina, A. I., 252
Albano, C., 278
Alegrìa, A., 172
Almarza, A., 273
Amantia, D., 273
Ambrogi, V., 129, 158, 361
Ambrosio, L., 420
Amendola, E., 206, 379
Anastasiadis, S. H., 79
Andersson, O., 145
Annicchiarico, A., 398
Antonucci, V., 411
Araneda, E., 276
Arbelaiz, A., 290
Ascione, L., 361
Aubouy, L., 273
Audouin, L., 10
Avella, M., 162
Avolio, R., 162
Aydemir, B., 58

B

Balart, J., 313, 323
Balart, R., 307, 313, 319, 326
Barral, L., 355
Battiato, S., 361
Becerra, N., 73
Beiner, M., 107
Benedito, A., 227
Benzarti, K., 101
Bignotti, F., 25
Binetti, E., 89
Bitenieks, J., 333
Bizzarr, V., 117
Bizzarro, V., 342
Boccaccini, A. R., 126
Boccafoschi, F., 16, 352
Bonadies, I., 162
Borbone, F., 300
Boronat, T., 307, 319, 326, 329
Borriello, C., 117, 297
Botta, L., 358
Boucher, V. M., 172
Bouza, R., 355
Bronco, S., 67
Brugger, J., 89
Brument, Y., 155
Buckley, C. P., 4
Buezas, I., 227
Bugatti, V., 178

C

Cadambi, R. M., 138
Cai, M., 95
Calvo, O., 319
Cangialosi, D., 172
Caponi, S., 221
Caprino, G., 165
Capuano, G., 187
Caputo, F., 398
Caramuta, P., 76
Carbone, D., 209
Carella, A., 300
Carfagna, C., 129, 158, 162, 303, 361
Cassulo, G., 120
Castagnet, S., 152
Castiello, S., 67
Cavallo, A., 395
Cerri, V., 367
Cerruti, P., 158
Cha, J. W., 386, 389
Chambon, S., 237
Chang, L., 240
Chodák, I., 287
Chrissopoulou, K., 79
Christodoulides, C., 70
Chung, Y. S., 389
Ciardelli, F., 67, 184
Ciardelli, G., 16, 352
Cicala, G., 148
Coletti, M., 82
Colin, X., 10, 101, 135, 155
Colmenero, J., 172

Colomer, F. J. R., 70
Coltelli, M. B., 67
Coltelli, M.-B., 184
Concilio, S., 175, 178
Conzatti, L., 67
Corcuera, M. A., 290
Corezzi, S., 221
Correc, O., 135
Cosma, P., 89
Covis, M., 278
Cristaldi, G., 148
Csomorová, K., 64
Curri, M. L., 89, 92

D

D'Amato, R., 336
D'Amora, U., 420
D'Amore, A., 402, 405, 408, 414, 417
Davidov, E. Ya., 1
Davletbaev, R. S., 255
Davletbaeva, I. M., 252, 255
Davodi, B., 31, 258
De Angelis, G., 398
De Angelis, M. G., 218
de Girolamo del Mauro, A., 367
De Girolamo Del Mauro, A., 104, 336, 342, 345
De Lavigne, C., 22
De Maria, A., 297
De Micheli, P., 123
De Santis, R., 420
Dettin, M., 52
Di Francia, G., 300, 345
Di Luccio, T., 117
Di Maio, L., 364, 373
Dintcheva, N. T., 213
Djouani, F., 7
Dubnikova, I. L., 1
Durin, A., 123
Dusunceli, N., 55, 58

E

Eceiza, A., 290
Eder, G., 43
Ehsani, M., 339
Eisazade, H., 258
Eisazadeh, H., 31
El Marzy, C., 135
El-Sabbagh, S. H., 249
Elksnite, I., 310
Emanuela Errico, M., 162
Emri, I., 98
España, J. M., 313
Esposito Corcione, C., 395

F

Fages, E., 307
Faiella, G., 411
Favier, A., 193, 196
Fayolle, B., 7
Fenollar, O., 307, 323, 326, 329
Fernández d´Arlas, B., 290
Ferrando, P., 352
Filippone, G., 187, 190
Fioretto, D., 221
Fombuena, V., 313, 323
Fontana, A., 221
Fornes, F., 329
Fotiadou, S., 79
Fox, B. L., 22
Frache, A., 120
Frascà, F., 398
Frick, B., 79
Friedrich, K., 240

G

Gómez Ribelles, J. L., 70
Gal, Y. S., 386
Galdi, M. R., 199
Galindo, B., 227
Galizia, M., 218
García-Sanoguera, D., 323, 326
Garcia-Bernabé, A., 49, 273
Garcia-Sanoguera, D., 329
Gardette, J. L., 213
Gardette, J.-L., 237
Gargallo, L., 73, 276
Gargiulo, M., 231
Garofalo, E., 364
Gaume, J., 237
Gentile, G., 162
Gerard, J. F., 181

Ghamri, N., 123
Ghassemieh, E., 138
Ghezzo, F., 52
Gil-Agustí, M., 49, 273
Giménez, E., 227
Gioffredi, E., 120
Giordano, M., 411
Gloria, A., 420
Glover, A. J., 95
Gomez d'Ayala,, G., 158
González-Gutiérrez, J. P., 49
Gonzalez, G., 278
Gordge, M. P., 126
Grassia, L., 402, 405, 408, 414, 417
Grazia De Angelis, M., 392
Greco, A., 395
Gribkova, O. L., 46
Grimaldi, I. A., 104, 336, 342, 345
Grossbein, H., 37
Guadagno, L., 264, 267
Guenza, M. G., 114
Gumerov, A. M., 252
Guo, Q., 28

H

Häussler, W., 246
Hadjichristidis, N., 276
Han, S. S., 386
Hanemann, T., 13
Hierzenberger, P., 43
Hilton, H. H., 37
Honnef, K., 13
Houdkova, J., 270

I

Iannace, S., 231
Iannelli, P., 175, 178, 376
Iannone, M., 19
Iannuzzo, G., 264, 267
Ilyin, S. O., 132
Ilyina, M. V., 46
Immonen, K., 85
Incarnato, L., 199, 364, 373
Inceoglu, F., 123
Ingrosso, C., 89
Isakova, A. A., 46
Iucci, G., 52
Ivanov, V. F., 46
Ivanova, T., 310
Ivanova, V. T., 46

J

Janigová, I., 287
Jeon, H. Y., 389
Jeun, J.-P., 281, 284
Joo, S. W., 386

K

Kalendova, A., 141, 181
Kalnins, M., 310, 333
Kang, P.-H., 281, 284
Karam, A., 278
Kehagias, N., 92
Kenny, J. M., 348, 382
Keshavarz, T., 126
Kim, H.-B., 281, 284
Kim, J. Y., 89
Kim, M. J., 386
Kindervater, J., 246
Knowles, J. C., 126
Kontou, E., 110
Korneeva, N. V., 61
Kotomin, and V., S. V., 132
Kouzilos, G., 40
Kovarova, L., 141, 181
Kranbuehl, D. E., 95
Kudinov, V. V., 61
Kulichikhin, G., 132
Kulichikhin, V., 202
Kurochkina, Y. E., 46
Kwak, K. Y., 389
Kyritsis, A., 70

L

López, R., 319
La Carrubba, V., 243
La Mantia, F. P., 213, 358
Lainé, E., 152
Lamanna, G., 398
Lamberti, G., 316

Lashkenari, M. S., 31, 258
Lavorgna, M., 379
Lee, J. T., 261
Lee, Y. J., 386, 389
Leiss-Holzinger, E., 43
Leiva, A., 73, 276
Leone, C., 165
Leroux, F., 213
Lesiak, B., 270
Letizia Fariello, M., 373
Llovera, P., 273
Locke, I. C., 126
Loffredo, F., 104, 336, 345
Lomakin, S. M., 1
Longo, P., 264
Lonkar, S., 213
Lopresto, V., 165
Lyoo, W. S., 386, 389

M

Maffettone, P. L., 120
Maffezzoli, A., 395
Magniez, K., 22
Maksimov, R., 310, 333
Malíková, M., 64
Malac, J., 141
Malinconico, M., 158
Malkin, A., 202
Mamalis, A. G., 40
Manceau, M., 237
Mannella, G. A., 243
Manykin, A. A., 46
Mariconda, A., 264
Marieta, C., 290
Masala, S., 117
Massera, E., 300, 345
Mattu, C., 352
McCormick, A., 126
Mecerreyes, D., 92
Meekum, U., 294
Mensitieri, G., 379
Meri, R. M., 333
Messori, M., 218
Michaeli, M., 37
Miglietta, M. L., 300
Milione, S., 352
Minale, M., 224
Minarini, C., 104, 117, 336, 342, 367
Mingmongkol, Y., 294
Miscioscia, R., 367
Mkacher, I., 155
Mošková, D. J., 287
Modesti, M., 52
Mohamed, O. A., 249
Mohammadi, M., 339
Monaco, G., 221
Monchy-Leroy, C., 10
Mondragon, I., 276, 290
Montero, B., 355
Monti, M., 348
Mordon, N., 126

N

Naddeo, C., 264
Natali, M., 348
Nenna, G., 104, 117, 342
Netti, P. A., 370
Nho, Y.-C., 281
Niaounakis, M., 110
Noh, S. K., 389

O

Oh, T. H., 386
Olivieri, L., 392
Ortega, G., 273

P

Pandini, S., 25
Pankaj, S., 107
Pannico, M., 129
Panniello, A., 92
Papa, I., 165
Papaspyrides, C. D., 234
Pappalardo, S., 300
Pascual, J., 307
Pascual, M., 319
Peponi, L., 382
Perier, L., 193, 196
Persico, P., 129, 158, 303, 361
Pesce, E., 117
Petrarca, C., 206
Petrosino, M., 367

Pezzullo, G., 231
Piccarolo, S., 209
Piotto, S. P., 175, 178
Piscitelli, F., 379
Pissis, P., 70
Placido, T., 92
Polichetti, T., 300, 345
Polzonetti, G., 52
Poulose, A. M., 209
Prisacariua, C., 4
Puccini, I., 218
Puglisi, C., 361

Q

Quijano- López, A., 49

R

Radić, D., 276
Radic, D., 73
Rai, R., 126
Raimondo, M., 264, 267
Rakhimkulov, A. D., 1
Re, M., 117
Reboud, V., 92
Recca, G., 148
Rega, R., 367
Renzi, A. I., 303
Richaud, E., 7, 10
Rivaton, A., 237
Romeo, G., 187
Romeo, V., 379
Roso, M., 52
Rossi, F., 221
Roviello, A., 300
Roy, I., 126
Rubino, A., 367
Rueda, L., 290
Russo, G., 316
Russo, P., 168, 376
Russo, S., 264, 267
Russo, T., 420
Rychlý, J., 64

S

Sánchez-Nácher, L., 313, 323, 326, 329
Salerno, A., 370
Salih, V., 126
Samperi, F., 361
Sandoval, C., 73
Santagata, G., 158
Sapurina, I. Yu., 46
Saralegui, A., 290
Sarlin, J., 85
Sarti, G. C., 218, 392
Sartori, S., 16, 352
Savi, S., 184
Scaffaro, R., 358
Scamardella, A. M., 206, 379
Scarfato, P., 364, 373, 376
Schniepp, H. C., 95
Schwarz, M., 297
Scortanua, E., 4
Seo, D.-K., 284
Serafini, P., 16, 352
Shchegolikhin, A.N., 1
Shin, B.-S., 281
Shinyashiki, N., 70
Shtark, A., 37
Silvestri, A., 16, 352
Slouf, M., 181
Song, Y. S., 261
Sorrentino, L., 231
Spena, P., 168
Stathopoulos, A. T., 70
Stifter, D., 43
Striccoli, M., 89, 92
Sun, Y. F., 34

T

Taburoni, E., 67
Terzi, N. U., 55, 58
Thérias, S., 237
Therias, S., 213
Tireau, J., 101
Titomanlio, G., 316
Tonpheng, B., 145
Torre, L., 348
Torres, C. M. S., 92
Toscano, C., 76

V

Vázquez, S., 273
Valentini, L., 382
Valette, R., 123
Van Schoors, L., 101
Verdu, J., 7, 10
Vergnes, B., 123
Villani, F., 104, 336, 345
Ville, J., 123
Vitiello, C., 76
Vittoria, V., 264, 316
Voda, A. S., 28
Vortselas, A. K., 40

W

Wallin, T. J., 95
Wang, X. H., 34
Wool, R. P., 215

Y

Yıldırım, S., 55
Yılmazturk, F., 55
Youn, J. R., 261
Yousefi, A. A., 339
Yu, J., 145

Z

Zaikov, G. E., 1
Zarrelli, M., 411
Zemek, J., 270
Zeng, Y. C., 34
Zeppetelli, S., 420
Zicans, J., 310
Zupancic, B., 98
Zykova, J., 141, 181